Functional Food Carbohydrates

FUNCTIONAL FOODS AND NUTRACEUTICALS SERIES

Series Editor
G. Mazza, Ph.D.
Senior Research Scientist and Head
Food Research Program
Pacific Agri-Food Research Centre
Agriculture and Agri-Food Canada
Summerland, British Columbia

Functional Food Carbohydrates (2007)
Costas G. Biliaderis, Ph.D., and Marta S. Izydorczyk, Ph.D.

Functional Food Ingredients and Nutraceuticals: Processing Technologies (2007)
John Shi, Ph.D.

Dictionary of Nutraceuticals and Functional Foods (2006)
N. A. Michael Eskin, Ph.D. and Snait Tamir, Ph.D.

Handbook of Functional Lipids (2006)
Edited by Casimir C. Akoh, Ph.D.

Handbook of Functional Dairy Products (2004)
Edited by Collete Short and John O'Brien

Handbook of Fermented Functional Foods (2003)
Edited by Edward R. Farnworth, Ph.D.

Herbs, Botanicals, and Teas (2002)
Edited by G. Mazza, Ph.D. and B.D. Oomah, Ph.D.

Functional Foods: Biochemical and Processing Aspects Volume 2 (2002)
Edited by John Shi, Ph.D., G. Mazza, Ph.D., and Marc Le Maguer, Ph.D.

Methods of Analysis for Functional Foods and Nutraceuticals (2002)
Edited by W. Jeffrey Hurst, Ph.D.

Functional Foods: Biochemical and Processing Aspects Volume 1 (1998)
Edited by G. Mazza, Ph.D.

Functional Food Carbohydrates

Edited by
Costas G. Biliaderis
Marta S. Izydorczyk

CRC Press is an imprint of the
Taylor & Francis Group, an **informa** business

CRC Press
Taylor & Francis Group
6000 Broken Sound Parkway NW, Suite 300
Boca Raton, FL 33487-2742

ISBN-13: 978-0-8493-1822-1 (hbk)
ISBN-13: 978-0-367-39016-7 (pbk)

Library of Congress Cataloging-in-Publication Data

Functional food carbohydrates / edited by Costas G. Biliaderis and Marta S. Izydorczyk.
p. cm. -- (Functional foods and nutraceuticals ; 10)
Includes bibliographical references and index.
ISBN 0-8493-1822-X
1. Carbohydrates in human nutrition. 2. Functional foods. I. Biliaderis, Costas G. II. Izydorczyk, Marta S. III. Series: Functional foods & nutraceuticals series ; no. 10.

QP701.F86 2006
612'.01578--dc22 2006044615

Visit the Taylor & Francis Web site at
http://www.taylorandfrancis.com

and the CRC Press Web site at
http://www.crcpress.com

Series Editor's Preface

The Functional Foods and Nutraceuticals Book Series, launched in 1998, was developed to provide a timely and comprehensive treatment of the emerging science and technology of functional foods and nutraceuticals that are shown to play a role in preventing or delaying the onset of diseases, especially chronic diseases. The first nine volumes' titles in the series, *Functional Foods: Biochemical and Processing Aspects*, Volumes 1 and 2, *Herbs Botanicals and Teas*, *Methods of Analysis for Functional Foods and Nutraceuticals*, *Handbook of Fermented Functional Foods*, *Handbook of Functional Dairy Products*, *Handbook of Functional Lipids*, *Dictionary of Functional Foods and Nutraceuticals*, and *Processing Technologies for Functional Foods and Nutraceuticals*, have received broad acceptance by food, nutrition, and health professionals.

The latest volume, *Functional Food Carbohydrates*, edited by Dr. Costas G. Biliaderis (Aristotle University, Thessaloniki, Greece) and Dr. Marta S. Izydorczyk (Grain Research Laboratory, Canadian Grain Commission, Winnipeg, Canada), contains 17 chapters contributed by 31 leading scientists. The first 7 chapters provide in-depth treatments of the peer-reviewed literature on the chemistry, physical properties, technology, safety, and health benefits of cereal beta-glucans, resistant starch, konjac mannan, seed polysaccharide gums, microbial polysaccharides, chitosan, and arabinoxylans. Metabolic and physiological effects of food carbohydrates in relation to cardiovascular disease, obesity, cancer, type 2 diabetes, mineral metabolism, and gastrointestinal tract function, and as modulators of mood and performance are superbly addressed in Chapters 8 to 14. The last part of the book, Chapters 15 to 17, deals with the technological and regulatory aspects of using carbohydrates as functional ingredients in food systems. Chapter 15 examines the major issues concerning the safety and efficacy of pro-, pre-, and synbiotics, and provides the latest data available from clinical studies as well as outlining future areas of utilization. Potential uses of carbohydrates as stabilizers and delivery vehicles of bioactive substances in foods are discussed in Chapter 16, and the current regulatory status for functional food carbohydrates and other nutraceuticals in the U.S., Japan, Canada, and the European Union is summarized in Chapter 17.

Drs. Biliaderis and Izydorczyk have assembled a group of outstanding international contributors in the forefront of carbohydrates science and technology. Together they have produced an excellent reference book that is expected to be a valuable resource for researchers, teachers, students, entrepreneurs, food, nutrition and health practitioners, and all those working in the functional food and nutraceutical industry.

It is hoped that this book will serve to further stimulate the development of functional foods and nutraceuticals and provide consumers worldwide with products and product ideas that prevent diseases and enhance the quality of life of people everywhere.

G. Mazza, Ph.D., FCIFST
Series Editor

Dedication

to

Alexander, Conrad, George,
Lambrini, and Veronica

Preface

Carbohydrates present in food products may vary from small simple sugars to very complex polysaccharides such as those found in plant cell walls. Advances in structural characterization and determination of the physicochemical properties of carbohydrates over the last two decades have unraveled their structural diversity and offered fundamental knowledge, which aids in understanding and predicting functionality as it relates to their end-use applications (e.g., thickening, gelling behavior). As a result, food technologists have concentrated with considerable success on the development and production of a wide range of palatable and marketable food products, using carbohydrates as functional ingredients to modify the appearance, taste, mouth feel, solubility–dispersibility, and stability of such systems. However, until recently, much less attention has been given to the physiological impact and health benefits carbohydrates could have in humans and to the relationship between these nutritional aspects and the chemical constitution and physical properties of the carbohydrate source.

With a rapidly growing international interest and activity in the area of functional foods and nutraceuticals, there is a need to provide a critical overview of the beneficial nonnutritive effects that the various food constituents have on human health. Among them, carbohydrates, which form the bulk of all food consumed by humans, seem to play a central role, showing functional versatility from both a technological and a nutritional viewpoint. For example, research linking the consumption of complex nonstarch polysaccharides with a reduced risk of chronic diseases (including diabetes, cardiovascular diseases, and cancers) is well documented in the literature and has been passed on to consumers with nutritional recommendations to increase their daily dietary fiber intake. Another interesting example to mention is the ability of nonadsorbable carbohydrates, such as fructans (inulin) and fructooligosaccharides, to improve colonic function and stimulate the selective proliferation of beneficial bacteria such as *Bifidobacteria*, which are immunomodulators and can suppress growth of potentially harmful anaerobes in the digestive tract; the potential importance of the changes in colonic function led to the term *prebiotic*.

However, as science begins to uncover some of the physiological roles of carbohydrates, it becomes clear that any dietary recommendations about these constituents should be based on a synthesis of epidemiological, metabolic, animal, and other clinical nutrition data. It also becomes evident that the various physiological effects could be rather specific to certain types of carbohydrates; i.e., biotransformations and functions (physiological and technological) are related to the molecular features–properties of the carbohydrates. In this respect, there is an increased demand to summarize literature information in this rapidly evolving field.

The aim of this book is to address, through clinical and epidemiological evidence, the specific classes of carbohydrates in foods that seem to exert health-enhancing effects, and to discuss the physiological and metabolic roles that different carbohydrates have on disease prevention and management, by focusing attention on certain chronic diseases such as cancer, cardiovascular diseases, diabetes, osteoporosis, various gastrointestinal disorders, etc. The book should be useful to food and nutrition students and scientists, health professionals, and all those who work in the food industry and use carbohydrates as functional ingredients in food formulations and new product development. It aims to cover in a more comprehensive manner a broad range of physiologically active carbohydrate compounds and attempts to provide information on the chemistry, physical properties, processing effects, production, and physiological function of these constituents, by summarizing older literature and the more recent findings, particularly as they have evolved with the development of functional foods and bioactive ingredients as an emerging new science. An attempt was also made to focus on the links between the physicochemical properties and processing of carbohydrates with the health benefits resulting from their regular consumption by humans. Overall, it has been our intention to integrate in this volume the views from authors specializing in the areas of food chemistry and processing, as well as of human nutrition and physiology, in the emerging field of functional foods and nutraceuticals.

Editors

Dr. Costas G. Biliaderis is a professor of the Department of Food Science and Technology, Aristotle University, Thessaloniki, Greece. He received his M.Sc. (1978) and Ph.D. (1980) degrees from the University of Saskatchewan, Canada, and he has held appointments as a research associate with the National Research Council of Canada (1980 to 1981), a project chemist with General Foods, Inc. (1984 to 1985), and an associate professor with the University of Manitoba, Canada (1985 to 1993). An adjunct professor with the University of Guelph, Canada, and a former member of the editorial board of *Carbohydrate Polymers*, Dr. Biliaderis is the author and co-author of numerous journal articles and book chapters in the broad area of physical chemistry of food systems, with a focus on the structure–function relationships of food carbohydrates. His current research interests include the chemistry and physical chemistry of polymeric food carbohydrates (plant and microbial origin), thermal analysis of food constituents, thermophysical properties of polysaccharide blends, and the processing and storage effects on texture and stability of food products and their constituents. Dr. Biliaderis was named a highly cited researcher by ISI-Thomson Scientific (2003) and received several awards for his achievements in research, teaching, and scholarly works.

Dr. Marta S. Izydorczyk is currently a research scientist and program manager of Basic Barley Research at the Grain Research Laboratory, Canadian Grain Commission (CGC), and an adjunct professor at the Department of Food Science, University of Manitoba. She is an active member of the American Association of Cereal Chemists (AACC) International, a member of the AACC Scientific Advisory Panel, and a member of the organizing committee of the 2006 World Grains Summit: Foods and Beverages in San Francisco, U.S. She sits on the editorial board of *Cereal Chemistry.* Dr. Izydorczyk is also a member of the North America Buckwheat Promotional Committee.

Dr. Izydorczyk's areas of expertise are in the molecular structure and physicochemical properties of starch and nonstarch polysaccharides, their chemical and enzymatic modification, and their interactions with proteins, lipids, and other components of cereal grains. Her barley research program is focused on the chemistry and biochemistry of malting barley. Primary objectives are to identify, explain, and clarify the genetic and environmental factors, and the molecular mechanisms that are responsible for the functionality and performance of barley components during malting and brewing. The program also places emphasis on the structure and functional properties of carbohydrate components, e.g., starch, nonstarch polysaccharides, and dietary fiber from barley, and from other economically important prairie grains.

Dr. Izydorczyk holds a bachelor's degree from the Ryerson Polytechnical University in Toronto, a master's degree in food science, and a Ph.D. in food and nutritional sciences from the University of Manitoba.

Contributors

Karin Autio
VTT Biotechnology
Kuopio, Finland

Costas G. Biliaderis
Department of Food Science and Technology
Faculty of Agriculture
Aristotle University of Thessaloniki
Thessaloniki, Greece

Larry Christensen
Psychology Department
University of South Alabama
Mobile, Alabama

Steve W. Cui
Agriculture and Agri-Food Canada
Guelph, Ontario, Canada

Michael N.A. Eskin
University of Manitoba
Winnipeg, Manitoba, Canada

Sharon E. Fleming
Center for Weight and Health
College of Natural Resources
University of California
Berkeley, California

Pirkko Forssell
VTT Biotechnology
Kuopio, Finland

S. Gao
Department of Food and Nutrition
Graduate School of Human Life Science
Osaka City University
Osaka, Japan

Dana E. Gerstein
Center for Weight and Health
College of Natural Resources
University of California
Berkeley, California

Ioannis Giavasis
Department of Food Science and Technology
Faculty of Agriculture
Aristotle University of Thessaloniki
Thessaloniki, Greece

Piia Hakala
VTT Biotechnology
Kuopio, Finland

Shinya Ikeda
Osaka City University
Osaka, Japan

Marta S. Izydorczyk
Grain Research Laboratory
Canadian Grain Commission
Winnipeg, Manitoba, Canada

David D. Kitts
Food Science, Food, Nutrition and Health
University of British Columbia
Vancouver, British Columbia, Canada

David Laaksonen
Department of Medicine
University of Kuopio and Kuopio University Hospital
Kuopio, Finland

Athina Lazaridou
Department of Food Science and Technology
Faculty of Agriculture
Aristotle University of Thessaloniki
Thessaloniki, Greece

Corrado Muzzarelli
Faculty of Medicine
Institute of Biochemistry
Polytechnic University of Marche
Ancona, Italy

Riccardo A.A. Muzzarelli
Faculty of Medicine
Institute of Biochemistry
Polytechnic University of Marche
Ancona, Italy

Hannu Mykkänen
Department of Clinical Nutrition
University of Kuopio
Kuopio, Finland

Päivi Myllärinen
VTT Biotechnology
Kuopio, Finland

K. Nishinari
Department of Food and Nutrition
Graduate School of Human Life Science
Osaka City University
Osaka, Japan

Leo Niskanen
Department of Medicine
University of Kuopio and Kuopio University Hospital
Kuopio, Finland

Gunilla Önning
Biomedical Nutrition
Lund University
Lund, Sweden

Kaisa Poutanen
VTT Biotechnology and University of Kuopio
Food and Health Research Centre
Kuopio, Finland

Lorrene D. Ritchie
Center for Weight and Health
College of Natural Resources
University of California
Berkeley, California

G.C.M. Rouzaud
Food and Microbial Sciences Unit
School of Food Biosciences
University of Reading
Reading, U.K.

Barbara O. Schneeman
Center for Food Safety and Applied Nutrition
Food and Drug Administration
College Park, Maryland

Joanne Slavin
Department of Food Science and Nutrition
University of Minnesota
St. Paul, Minnesota

Donald B. Thompson
Department of Food Science
Penn State University
University Park, Pennsylvania

Gail Woodward-Lopez
Center for Weight and Health
College of Natural Resources
University of California
Berkeley, California

Jerzy Zawistowski
Functional Foods and Nutraceuticals
Forbes Medi-Tech, Inc.
Vancouver, British Columbia, Canada

Contents

1 Cereal β-Glucans: Structures, Physical Properties, and Physiological Functions

Athina Lazaridou, Costas G. Biliaderis, and Marta S. Izydorczyk

CONTENTS

1.1 INTRODUCTION

Mixed-linkage (1→3),(1→4) linear β-D-glucans (β-glucans) are major components of endosperm cell walls of commercially important cereals such as barley, oat, wheat, rye, sorghum, and rice. Cereal β-glucans are linear homopolysaccharides composed of D-glucopyranosyl residues (Glc*p*) linked via a mixture of β-(1→3) and β-(1→4) linkages. The structure features the presence of consecutive (1→4)-linked β-D-glucose in blocks (i.e., oligomeric cellulose segments) that are separated by single (1→3) linkage. Although most of the cellulose segments are trimers and tetramers, longer cellulosic oligosaccharides are also present in the polymeric chains.[1–7] Cereal β-glucans exhibit diversity with respect to their molecular/structural features, such as molecular size, ratio of tri- to tetramers, amount of longer cellulosic oligomers, and ratio of β-(1→4)/-(1→3) linkages. The molecular features of β-glucans seem to be important determinants of their physical properties, such as water solubility, dispersibility, viscosity, and gelation properties, as well as of their physiological function in the gastrointestinal tract.

The physical and physiological properties of β-glucans are of commercial and nutritional importance. Increasing interests in β-glucans during the last two decades are largely due to their acceptance as functional, bioactive ingredients. Cereal β-glucans have been associated with the reduction of plasma cholesterol and a better control of postprandial blood serum glucose and insulin responses in humans and animals.[8–13] The efficacy of oat and barley β-glucans in reducing the risk of coronary heart disease (CHD) has been recognized by the U.S. Food and Drug Administration (FDA), and both oat- and barley-based foods are allowed to carry a claim that they reduce the risk of CHD. The potential application of β-glucans as food hydrocolloids has also been proposed based on their rheological characteristics. In addition to solution viscosity enhancement, β-glucans have been shown to gel under certain conditions.[14–21] β-Glucans can be utilized as thickening agents to modify the texture and appearance in several food formulations or may be used as fat mimetics in the development of calorie-reduced foods. Moreover, lately there have been many efforts to increase the amount of cereal β-glucans in food formulations because of the beneficial physiological function of these polysaccharides. Thus, β-glucan-rich fractions from cereal or purified β-glucans have been successfully incorporated into products, such as breakfast cereals, pasta, noodles, and baked goods (bread, muffins), as well as dairy and meat products.[22–37]

The physical properties of β-glucans, such as solubility and rheological behavior in solution and gel state, are generally controlled by linkage patterns, conformation, and molecular weight or molecular weight distributions.[7,14,15,17–19,21,38–47] On the other hand, the relations between structural features and physiological responses have been only partly investigated.[11]

This chapter surveys most of the recent findings on cereal β-glucans, with regard to their occurrence, structure, and physical and functional properties, including physiological effects. Moreover, procedures of enrichment of cereal grain fractions in β-glucans, as well as methods of extraction, isolation, and purification of β-glucans in laboratory and large scale, are described. An effort is also made to discuss structure–functionality relations of these biopolymers.

1.2 OCCURRENCE

Barley and oat are considered to be major sources of β-glucans. Total β-glucan contents of barley grain range from 2.5 to 11.3% by weight of the kernel, but they usually fall between 4 and 7%.[5,31,44,45,48–73] However, β-glucan levels as high as 13 to 17% have been reported for certain barley genotypes.[57,58,66,67,74–77] β-Glucan contents of oat are also highly variable, ranging from 2.2 to 7.8% in some exceptional lines.[5,48–51,53,54,62,63,78–87] The level of β-glucans in rye, wheat, tricale, sorghum, and rice was found to be much lower: 1.2 to 2.9,[5,48–51,54,88,89] 0.4 to 1.4,[5,48–51,54,90] 0.4 to 1.2,[48–50,54] 0.1 to 1.0,[48,91] and 0.04%,[49] respectively.

The β-glucan content of cereal grains is influenced by both genotypic and environmental factors. However, the genetic background of oat and barley is considered to be far more important than environmental conditions as a determinant of the final β-glucan content of these grains.[12,23,74,76,81,92] It has been reported that feed and six-row barleys may have slightly lower β-glucan levels than malting and other two-row varieties;[74,76] however, Fastnaught et al.[61] have found no difference at β-glucan levels between two-row and six-row head type for any of the traits analyzed. Determination of the β-glucan content in 18 species of *Avena*, including nine diploids, four tetraploids, and five hexaploids grown in one location, showed significant differences between species, as well as between groups based on ploidy.[82] The β-glucan content in the whole grain for Finnish oat cultivars has been found to be higher for the naked oats than for the hulled cultivars.[80,84] The small increase in β-glucan level on going from hulled to hulless barley has been attributed to the fact that there is no β-glucan in the hull, and consequently, it would be expected from a change based solely on the measurement being with or without 11% hull.[93] The increased consumption of barley in human diets as a result of its high dietary fiber content could lead to an increased demand in the future for barleys with high β-glucan contents. An examination of the carbohydrate content of wild barley (*Hordeum spontaneum*) lines revealed β-glucan contents ranging up to 13.2%. Thus, certain wild barley lines high in β-glucan content may represent an important genetic resource for future development of breeding programs.[74,76] Generally, varieties of barley with the anomalous amylose-to-amylopectin ratio have a higher β-glucan content than the varieties with normal starch.[62,63,65,69,70,77,93] Furthermore, it has been demonstrated that waxy barley has significantly higher β-glucan content than nonwaxy; the combination of two recessive genes for waxy and hulless in barley results in a large enhancement of β-glucan levels, by approximately 1.5 to 2 times.[60–63,94] Incorporation of the short-awn gene in waxy hulless barley varieties seems to increase the β-glucan content two to three times.[94] An example is the variety Prowashonupana, a waxy, short-awn, hulless barley containing 14.7 to 17.4% β-glucans.[57,58,66,67,75,77] Beer et al.[62] and Lee et al.,[63] studying different isolines of certain barley cultivars, also found a difference in β-glucan content by the addition of the short-awn gene to a hulless waxy barley, but less significant ranging up to 8.1%. On the other hand, Cervantes-Martinez et al.[95] increased the mean β-glucan content from 5.39 to 5.59% and from 6.35 to 6.60% in two different oat populations, following phenotypic selection.

Differences in β-glucan content within the same genotype have been attributed to environmental effects. The β-glucan content seems to depend not only on the cultivar, but also on the season or growth site (location). The major environmental factor that influences β-glucan levels appears to be the availability of water during grain maturation. Dry conditions before harvest result in high β-glucan levels, and the opposite is true with moist conditions; a functional role for the β-glucan in preventing grain dehydration may be postulated.[40,61,74,76,81,87,92] Moreover, it has been demonstrated that a higher temperature of growing time favors an increase in β-glucan content in barley and oat grain.[61,80,87] It has also been found that β-glucan content had significant positive correlations with grain yield, growing time, and seed size, as well as significant negative correlations with protein content and hull content.[80]

The localization of β-glucans in cereal grains influences the isolation and purification procedures, which aim at fractions/preparations enriched in β-glucans.[96] Compositional analysis of isolated endosperm and aleurone cell walls of barley indicates considerable differences between these cell groups. The endosperm cell walls are built mainly of β-glucans (70 to 75%),[97–99] whereas aleurone cell walls contain smaller amounts of β-glucans (26%).[100] Histochemical evidence has shown that, like barley β-glucan, oat β-glucan is located mainly in the endosperm cell walls, constituting approximately 85% of the wall,[101] with lesser amounts in the aleurone walls.[102] In sorghum, the content of β-glucans in endosperm cell walls was found to vary in the range of 68 to 72%.[103] On the other hand, the main component of wheat and rye aleurone and endosperm cell walls is arabinoxylans. The wheat endosperm cell walls contain about 5% of β-glucans,[104] whereas aleurone cell walls consist of 30% β-glucans.[100] For oat and barley kernels, evidence suggests that differences in β-glucan content reflect differences in cell size and wall thickness in the starchy endosperm.[55,77,105] Microspectrofluorometric imaging showed that the distribution of β-glucan in oat kernel varies in low- and high-β-glucan varieties.[105,106] In a comparison of maps of central cross sections of five oat cultivars of differing β-glucan content (range, 3.7 to 6.4%), there was a trend for a high subaleurone concentration of β-glucan becoming less distinct as the total β-glucan content of the cultivars increased;[105] a relatively even distribution throughout the kernel was observed for the high β-glucan content varieties.[106] The distribution of β-glucan in barley was found to be more uniform, with no high subaleurone concentration of β-glucan in any of the five cultivars examined differing in β-glucan content (range, 2.8 to 11%); the highest concentration was in the central endosperm.[105] Bhatty and MacGregor,[52] using fluorescence microscopy, noticed that β-glucans are uniformly distributed in the endosperm barley cell walls, whereas in a subsequent study Bhatty et al.[55] noticed a concentration of β-glucan in the subaleurone layer, but not for all of the examined hulless cultivars. In wheat, the highest concentration is in the subaleurone layer, with little in the rest of the endosperm.[107] In rye, β-glucan seems to be evenly distributed throughout the grain according to Parkkonen et al.,[88] whereas Harkonen et al.[108] found that over 70% of β-glucans were concentrated in the outer fractions (bran and shorts) of a rye cultivar fractionated by a laboratory-scale roller mill. Moreover, Henry[51] reported that the mean β-glucan content of whole grain was 0.6% for wheat, 4.2% for barley, 3.9% for oats, and 2.5% for rye, while the respective values in the endosperms from these grains were 0.3, 4.1, 1.8, and 1.7%;

these analyses were applied directly to the grain by hand dissection with a scalpel without prior fractionation with a roller mill.

1.3 EXTRACTABILITY

Although high β-glucan content is indicative of high dietary fiber content in cereal grains, it is important to recognize that the soluble component of β-glucan is responsible for a large part of the beneficial activity of β-glucans for health. In various studies it has been found that solubility is influenced by genotypic factors. Generally, solubility of β-glucans from oats seems to be higher than those from barley. Thus, the content of water-soluble β-glucans (%w/w of total β-glucans) extracted from oats and barley at 38°C has been found at the range of 65 to 90%[53,63] and 8 to 71%,[53,56–59,63,66,67,75,77] respectively. The extractability of β-glucan estimated by the content of soluble β-glucans in high-amylose barley, both naked and covered, seems to be relatively low compared to that in normal, zero, and waxy barley genotypes; these results indicate that high-amylose barley cultivars, despite their high content of total β-glucans, might not be the best source of soluble polysaccharides. Similarly, hulless, waxy, short-awn barley genotypes that are often associated with increased total β-glucans were found to have a low percentage of soluble β-glucans.[57,63,66,67] Waxy varieties seem to be a better source for water-soluble β-glucans than the normal starch genotypes in hulless barley.[69–71] On the other hand, Lee et al.,[63] studying different genetic patterns of certain varieties, found that the hulled waxy varieties exhibit a lower ratio of soluble to total β-glucans than the respective normal starch parental types. Moreover, it has been demonstrated by these researchers that the hulless varieties exhibit lower extractability than the parental hulled types.

The extractability of β-glucans is dependent on extraction parameters (type, pH, and ionic strength of solvent, temperature, duration of extraction, and liquid–solids ratio), pretreatments (heating and drying), and presence of enzymes (endogenous or from contaminating microorganisms), as well as the method of milling and particle size; therefore, substantial differences in extractability might be expected with changes of any of these parameters.[109,110] Decreasing particle size and increasing temperature increased extraction efficiency.[110] Defatting did not significantly affect yields of oat gum (β-glucan), whereas enzyme deactivation with hot alcohol treatment decreased the yield of barley and oat β-glucan extracted.[23,75,110–112] Extractability of β-glucans in oat and barley has been investigated by many researchers, but the methodologies used were quite different, and direct comparison of results is difficult. Furthermore, variations in growing conditions may influence both the β-glucan extractability and endogenous β-glucanase activity, which may hydrolyze β-glucans, increasing the extractability and decreasing the molecular size of the polysaccharide fraction.

Graham et al.[113] proposed the use of standardized and physiologically more appropriate extraction conditions. Extraction at high temperature in general gave the highest values for soluble fiber and extraction in acidic buffer the lowest. Bhatty et al.[55] using acidic buffers (pH 1.5) and 1 h extraction at 40°C reported an average of 44.7% extractability of β-glucans from 13 Canadian hulless barley genotypes, whereas Huth et al.,[68] using similar conditions, found that less than 20% of the β-glucan present in hulled barley meal can be extracted. It is worthy to note that in the above two studies particle size of the barley samples seems to differ, a parameter known also to affect the extractability of β-glucans.

Usually, under mild aqueous extraction conditions, a complete extraction of β-glucans from cereal grains cannot be achieved. With increasing extraction temperature[75,83,114,115] and using an alkaline solvent instead of water, the percent of extracted β-glucan increases.[116] The amount of β-glucans extracted from oats and barley by hot water (80 to 100°C) varied from 50 to 70%.[50,62,86,115,117] Upon successive treatments with increasing extraction temperature from room temperature to boiling water, oat and barley β-glucan that could be extracted ranged from 72 to 90%.[75,114,118] Henry[50] reported that 36% of rye grain β-glucan could be extracted by boiling water, while Harkonen et al.[108] found that from the bran, short, and flour fractions of rye obtained after roller milling, 30, 25, and 45%, respectively, of the total β-glucan could be extracted at 30°C.

By contrast, the β-glucans of wheat are unextractable in water at 65°C,[90] but can be recovered together with heteroxylan in alkali extracts.[119] Also, distilled water adjusted to pH 10 with 20% sodium carbonate at 45°C extracted 61 to 64% of the total β-glucan from barley brans and 70% from oat bran.[118] Wood et al.[5] found only ~45% of barley and ~70% of oat β-glucans extracted by carbonate (pH 10) at 60°C, while this buffer could give complete extraction[48] of both oat and barley β-glucans at 80°C if an appropriate liquid–solids ratio was used. Generally, for complete extraction of β-glucans from cereal grains, rather drastic conditions (alkali or dilute perchloric or sulfuric acids) are needed, often inappropriate for potential food uses of the isolates. Extraction with sodium hydroxide (NaOH) increased the percentage of β-glucan extracted to 80 to 100%, but in most cases with decreased molecular size.[54,62,75,99,115,118,120] The extractability of nonstarch polysaccharides, a mixture of arabinoxylans and β-glucans, from wheat bran was found to increase with NaOH concentration (0.1, 0.5, 1.0 *M*) and extraction temperature (25, 60°C); extraction at 60°C resulted in degradation of the polysaccharides at all NaOH concentrations, whereas there was no degradation under extraction at room temperature.[119] The addition of sodium borohydride ($NaBH_4$) to prevent alkaline degradation of polysaccharides reduced the total β-glucan extraction to ~70% at 23°C, whereas at 65°C β-glucan extraction with NaOH–$NaBH_4$ was 100%.[75]

The basis of variation in extractability and solubility of β-glucans is not clear yet. The differences in proportion of water-extractable β-glucan among various barley cultivars had been considered as a heritable trait. Genotypic variation in extractability of β-glucans may be due to variation in the thickness of cell walls;[99] β-glucans in thicker cell walls, such as in the subaleurone endosperm of many oat cultivars, show a greater resistance to extraction.[121] However, wheat β-glucan, despite the thin endosperm cell walls, is extremely resistant to extraction.[90] Water-soluble β-glucans obtained from oat lines with high β-glucan levels (up to 7.8% β-glucan) and a traditional oat line (4.4% β-glucan) exhibited similar extractabilities.[86] Moreover, for two barley cultivars with similar β-glucan content, the proportions of β-glucan extracted under certain conditions (by water at 40 and 90°C and with alkali) differed largely.[115] There are, as yet, no fully satisfactory explanations for the apparent differences in extractability of β-glucans and for the partial insolubility and unextractability of these biopolymers with mild aqueous solvents based on differences in their molecular features or related to associative interactions among cell wall polymeric components.[109,122] The increase in extractability of β-glucans with increasing temperature or ionic strength of the solvent might be explained in terms of differences in proportion of β-(1→3) and β-(1→4) linkages in the polymeric chains, in linkage sequence on the chain, or in the degree of polymerization, which could lead to increasing physical intermolecular associations.[3,7,38,39,41,44]

In many previous studies, it has been demonstrated that with increasing extraction temperature, the molecular size of water-extracted β-glucans from barley and oats increases.[4,38,44,83,99,114,123] On the other hand, other researchers using different temperatures for water extraction or different solvents as extractants found no evidence that ease of extractability is related to molecular weight of oat and barley β-glucans.[3,62,68,98] Woodward et al.[3] found β-glucan extracted from barley flour at 40°C to have relatively more blocks of three or more adjacent (1→4) linkages and fewer (1→3) linkages than those extracted at 65°C and suggested that small differences in the fine structure have an impact on solubility. However, these fractions also differed in protein levels, and one possible explanation for less extractable β-glucans could be their binding into the cell walls with other components. Some evidence for covalent or physical associations between β-glucans and proteins has been reported[4,70,124,125] that might affect the extractability of the polysaccharide. A high frequency of long blocks of adjacent (1→4) linkages could increase the possibility of interactions and junction zone formation with other glucan molecules, or with the heteroxylan and cellulose chains in the cell wall, resulting in a decreased extractability by water. The insoluble fractions are probably held in the wall matrix by entanglement and hydrogen bonding with the other wall components, rather than by covalent bonding.[2,124] Evidence of noncovalent topological associations between β-glucans and arabinoxylans has been recently provided by Izydorczyk and MacGregor[126] for an alkaline-extractable nonstarch polysaccharide fraction from barley. Moreover, it has been found that xylanase, arabinofuranosidase, xyloacetylesterase, and feruloyl esterase all are capable of solubilizing β-glucans from the cell walls of barley endosperm, which indicates that pentosans can restrict the extraction of β-glucans.[127,128] The existence of β-glucans that are not water extractable could be attributed to physical entrapment of β-glucan within a pentosan cross-linked network via phenolic groups in the cell wall. Thus, differences in extractability seem to be consequences of features in microstructure and cell wall organization of the cereal endosperm.

1.4 EXTRACTION, ISOLATION, AND PURIFICATION

A variety of different extraction methods have been used in studies of oat and barley β-glucans in an attempt to maximize yield and viscosity of the extract, and to obtain the most native-like or undegraded material, while avoiding contamination with starch, proteins, and pentosans. A typical isolation and purification protocol for water-extractable oat and barley β-glucans is presented in Figure 1.1, which involves refluxing of whole oat and barley flour, aqueous extraction at low temperatures, a dual-enzyme digestion with heat-stable α-amylase and pancreatin for removal of

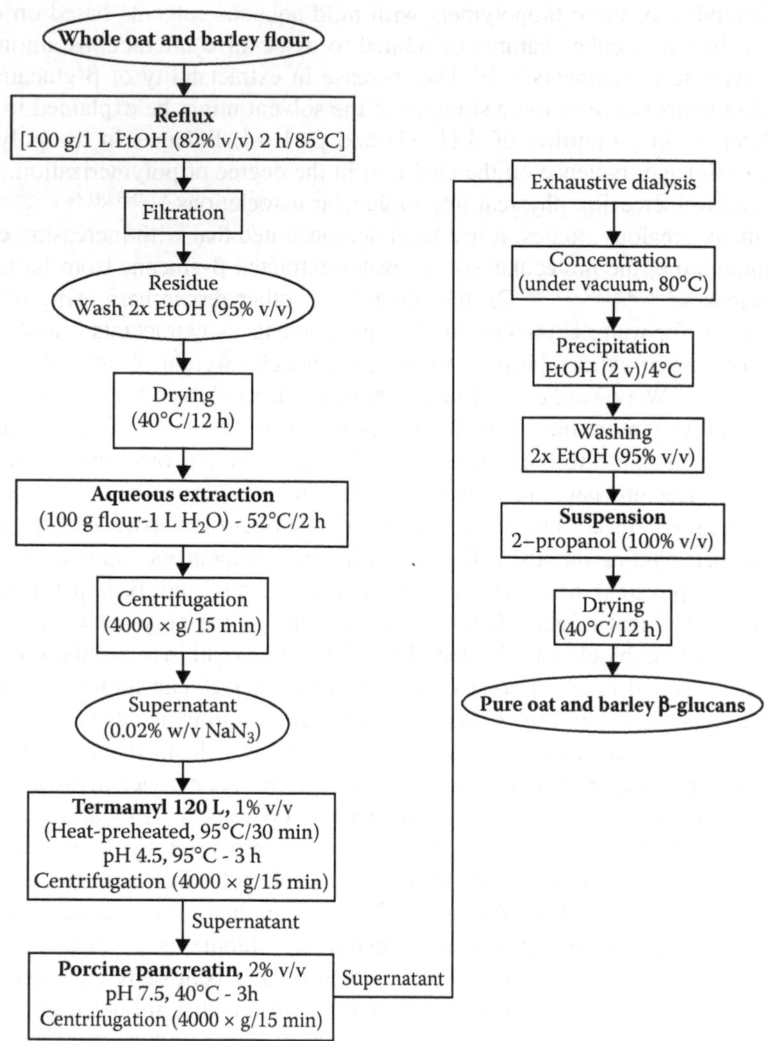

FIGURE 1.1 Extraction–purification laboratory scheme of β-glucans from oat and barley whole flours. (From Lazaridou, A. et al., *Food Hydrocolloids*, 18, 837, 2004. With permission.)

starch and protein contaminants, respectively, dialysis of the extracts for removal of starch and protein hydrolyzates, precipitation of the polysaccharide, and drying by a solvent exchange method.[18]

Endogenous enzymes (β-glucanase) of the seed result in a decreased molecular size of the extracted β-glucan[129] if an enzyme deactivation step is not adopted. A treatment with hot aqueous alcohol (80 to 90% ethanol) is often used as the first step for cereal β-glucan isolation to deactivate any endogenous enzymes present and to remove impurities such as lipids, free sugars, amino acids, small proteins, and some phenolics.[92,107] A similar effect has the defatting treatment with the hot isopropanol and petroleum ether (2:3 v/v) mixture used in the past as the first stage of oat β-glucan isolation.[130] However, there was some evidence that hot alcohol treatment may not destroy the β-glucanase activity; pretreatment of barley and oat with hot ethanol in relatively low concentration (<75%) was found to reduce the β-glucanase activity less than 20%.[75,112,129] Burkus and Temelli[129] found that the refluxing treatment of barley flour with 70% ethanol is not sufficient to stabilize the viscosity of final extracts, whereas boiling of extracts resulted in stable viscosity. Strong alkaline (1 *N* NaOH)[118] as well as acidic extraction conditions (0.1 HCl, 55 m*M* NaCl)[131] have been suggested as alternative pretreatments to inactivate the endogenous β-glucanases.

The β-glucan is usually extracted with hot water (>65°C)[40,123,130,132,133] or sodium hydroxide solutions,[120] or under mild alkaline conditions with a sodium carbonate solution[110–112,134] or acidic conditions.[68] In many early studies, oat and barley β-glucans were extracted with water at temperatures (47 to 52°C) below the gelatinization temperature of starch to minimize starch solubilization.[18,21,43,45] Extraction of β-glucans from barley meal under acidic conditions (KCl/HCl solution, pH 1.5) at 40°C can extract less than 2% of the starch.[68] Wood et al.,[110] using mild alkali (pH 10, 20% Na_2CO_3) for β-glucan isolation, suggested 45°C as the optimum extraction temperature that minimizes starch gelatinization and solubilization (starch content < 0.72% dry basis [db]). However, Bhatty[118] found a starch content of 14% in isolated β-glucan from bran barley under the same conditions (45°C, pH 10, 20% Na_2CO_3), which was reduced to less than 1% after an amylolytic enzyme treatment. Burkus and Temelli[129] reported 8.7% starch content in waxy barley gums extracted at 55°C under the same mild alkaline conditions, which increased to ~25% when the flour was refluxed (70% ethanol) prior to extraction, probably due to gelatinization during refluxing. The starch content in gums from refluxed flour could be reduced to <2% after a purification process with an amylolytic enzyme. In most studies the extraction of β-glucans was carried out simultaneously or followed by incubation of the extracts with amylolytic enzymes, such as salivary or porcine pancreatin α-amylase for removal of starch. Microbial amylases, which often contain β-glucan hydrolases, are avoided unless they are first shown to be free of these enzymes or they are thermostable and can be preheated to eliminate any β-glucanase contaminant activity, such as Termamyl from *Bacillus licheniformis*.[1,38,40,45,123,130,132,133]

Digestion of water extracts by proteolytic enzymes, such as pancreatin, for removal of protein contaminants was used by many researchers, resulting in protein contents between 0.8 and 4% db for their preparations.[18,21,45,123,130,132] An alternative process used by many researchers for minimizing the protein content of isolated cereal β-glucans included the adjustment of pH at 4.5 and removal of precipitated proteins by centrifugation; the protein content of preparations using this treatment ranged from less than 1 to 3.8%.[12,43,79,110–112,134] Moreover, the use of various adsorption materials (celite and vega clay) seems to also be effective for removal of proteins (<1%) from barley water extracts.[38] The addition of a flocculant such as κ-carrageenan into barley extracts with adjusted pH below the isoelectric point of proteins has been used for removing additional amounts of protein.[135]

In addition to increased amounts of solubilized starch, the extraction under alkaline conditions also results in an increased proportion of extracted arabinoxylans.[118,120] The most common treatment used for the purification of β-glucans from arabinoxylans is precipitation by 20 to 40% saturated ammonium sulfate.[3–5,45,123,130] Izydorczyk et al.[38] fractionated by stepwise $(NH_4)_2SO_4$ precipitation barley water extracts at 40 and 65°C, and the obtained subfractions up to 45% saturation contained mostly β-glucans (~97%). The procedure of precipitation by 50% isopropanol (IPA)[23] also afforded considerable purification from pentosans, with a 20:1 ratio of glucose to pentose in the precipitated gum, compared with 9:1 in the original crude solution.

Generally, the purity of β-glucan preparations depends largely on the composition of starting material (β-glucan and contaminant components content), the solvents and conditions employed (pH, temperatures, time and number of extractions), and the treatments applied for isolation and purification. The purity of barley and oat β-glucans that have been isolated in a laboratory scale ranges from 43 to 100%, depending on the above factors.[3,4,17,18,43,45,79,110–112,120,130,132,136] The protocol showed in Figure 1.1, which includes the dual-enzyme digestion with Termamyl and pancreatin of water extracts following dialysis and precipitation with ethanol, provided β-glucan preparations from barley and oat with high β-glucan content (85.3 to 93.2% db). An additional purification step with ammonium sulfate (30 to 37%) precipitation increased the purity of barley water-extractable β-glucans to the range of 91 to 100%.[3,45]

An alternative two-step process, avoiding the use of organic solvents, was developed by Morgan and Ofman[137] for isolating β-glucans of high purity (~90%) from barley flour; the obtained product was named Glucagel. The latter process includes a hot-water extraction (below 55°C), followed by freeze–thaw of the extract, filtration of the formed gelatinous or fibrous precipitate, air drying, redissolving in water at 80°C, and repetition of the freeze–thaw step. However, this process does not involve the deactivation of endogenous enzymes resulting in low molecular weight β-glucans because of the action of these enzymes during the hot-water extraction step.

Isolation and purification of β-glucans from wheat have some additional difficulties because of the very low β-glucan content in wheat and the interactions of this polysaccharide with other cell wall polysaccharides. An isolation–purification procedure of wheat β-glucans extracted from wheat bran has been described by Lazaridou et al.,[18] being essentially the procedure of Cui et al.,[41] with some modifications for further purification of the polysaccharide. This procedure involves digestion of the refluxed wheat bran with the Termamyl, alkaline extraction with 1 *N* NaOH at room temperature for 2 h, and enzyme digestion of the extract with a β-D-xylanase (*Trichoderma viride*, EC 3.2.1.8) preparation (free of β-glucanase activity) for removal of arabinoxylans. The derived oligosaccharides were removed by exhausting dialysis followed by concentration, precipitation with ethanol, and air drying of the precipitated gum after a solvent exchange treatment for the improvement of its water solubility, as proposed by Cui et al.[119] An adjustment of the extract's pH to 4.75 before xylanase treatment was carried out — an optimum pH for the action of this enzyme — but this also added to the removal of some proteins. Despite the xylanase treatment, the purity of β-glucan in the derived wheat gum was relatively low, probably due to intermolecular associations via the unsubstituted regions of xylan chains and the cellulose-like segments from the β-glucan chains that might contribute to enzymic indigestibility of arabinoxylans.

Indeed, it has been found that the alkali-extractable arabinoxylans are characterized by a very low degree of substitution of the xylan backbone, and the alkali-extractable β-glucans by large amounts of contiguously linked β-(1→4) segments.[72,126] Thus, a further step for the wheat β-glucan purification was adopted by Lazaridou et al.,[18] with the preparation of an aqueous concentrated solution of wheat gum and its storage at 25°C until a gel precipitate was formed. The discard of the supernatant resulted in a purified β-glucan gel phase relatively free from other contaminants (arabinoxylans, water-soluble proteins). The gel precipitate was resolubilized in water at 80°C, and the isolation process was followed by precipitation of the polysaccharide with ethanol, solvent exchange treatment, and air drying. The obtained preparation was highly enriched in wheat β-glucans (~82%).

1.5 ENRICHMENT OF CEREAL GRAINS IN β-GLUCAN AND LARGE-SCALE PRODUCTION OF β-GLUCAN CONCENTRATES AND ISOLATES

In the trials of cereal β-glucan extraction described above, only small quantities of β-glucans were isolated, but several β-glucan-enriched grain fractions, concentrates, and isolates have also been prepared in large scale for product development and nutrition studies. These products can be obtained from cereal grains by dry or wet processing procedures or combined dry and wet processing. Thus, β-glucan-enriched fractions have been obtained from cereal grains by dry milling, sieving, and air classification, or wet milling, sieving, and solvent extraction using different solvent systems. These processes aim to partially remove starch from the cereal grain. A large proportion of the starch can be removed from dry cereal flours by sieving and air classification. By sieving, separation of the cell wall material containing the β-glucan mostly occurs in the form of particles, which are larger than the starch granules after milling. Air classification also fosters the separation of the dense starch granules from the lighter cell wall material.

1.5.1 Dry Processing

Dry fractionation of cereals to produce fractions enriched in fibers and β-glucans may provide an economic advantage over methods utilizing solvents to produce β-glucan-rich fractions from cereals. Various dry-milling and sieving processes have been applied to whole cereal grains for production of β-glucan-rich fractions with improved functionality and dietary fiber characteristics. The applications of enrichment of cereal grains in β-glucans in a laboratory and large scale are summarized in Table 1.1 and Table 1.2, respectively. Oat materials are more difficult to fractionate than barley due to high levels of fat; fractionation of oats is successful after removal of fat.[56,134]

The concentration of β-glucans in the oat subaleurone layer, as revealed by microscopic examination,[105,106] has led to milling procedures that produce fractions with concentrated β-glucans, such as oat bran. The thickened cell walls at the aleurone–endosperm junction, and resistance of this region to milling attribution, result in a coarser particle that forms part of the β-glucan-rich bran, although other physical properties of the seed, such as cell size, might influence milling characteristics.[78]

TABLE 1.1
Laboratory-Scale Enrichment of Cereal Grains in β-Glucans by Dry Processing

Starting Material (number of cultivars)	β-Glucan Content (%) of Starting Material	Process		Weight (%) of Coarse Enriched Fractions (yield)	β-Glucan Content (%) of Coarse Enriched Fractions	Enrichment Factor[a]	References
		Milling	Particle Fractionation				
Dehulled oats (2)	5.4–6.7	Falling number mill	Sieving (45 μm)	52–56	8.1–9.5	1.4–1.5	Wood et al.[5]
Hulless oats (2)	4.3–4.5			39–50	7.5	1.7	
Dehulled oats (9)	3.9–6.8	Falling number mill	Sieving (45 μm)	48.3–58.0	5.8–8.9	1.3–1.6	Wood et al.[78]
Hulless oats (2)	4.7			50.8–55.4	6.7–7.2	1.4–1.5	
Oat bran (1)	9.6	Abrasive Udy mill	Sieving (45 μm) and resieving after regrinding (45, 147, 75 μm)	1.7–23.0 (28.5)[b]	26.4–27.2 (22.6)[b]	2.6–2.9 (2.4)[b]	Knuckles et al.[56]
Rolled oat (1)	4.7			5.2–18.3 (36.9)[b]	21.2–23.6 (12.5)[b]	3.7–4.9 (2.7)[b]	
Dehulled barley (2)	5.1–5.3			2.1–20.7 (26.8-27.6)[b]	17.1–22.5 (14.2–14.9)[b]	3.3–4.3 (2.8)[b]	
Hulless barley (2)	6.8–7.2			2.2–30.1 (29.1–48.9)[b]	16.0–21.3 (11.4–19.5)[b]	2.4–3.0 (1.7–2.7)[b]	
Hulless barley (1)	4.6	Cyclotec sample mill (0.5 mm)	Sieving (125 μm)	—	8.5	1.8	Cavallero et al.[31]
Hulled barley (1)	4.5	Hammer mill	Air classification	35.0	8.9	2.0	Knuckles and Chiu[59]
Dehulled barley (1)	5.8	Grinder and pin mill	Sieving (500-43 μm) and air classification	13.2	14.7	2.5	Wu et al.[58]
Hulless barley (1)	8.0			27.3	14.6	1.8	
Defatted hulless waxy barley (1)	19.6			31.0	31.3	1.6	
Hulled feed barley (2)	4.3–4.4	Udy cyclone mill (1.00 mm) and Allis–Chalmers roller mill	Short flow	30.0–30.7	4.9–5.2	1.1–1.2	Bhatty[64]
Hulled malt barley (2)	4.2–4.5			30.5–31.5	5.0–5.4	1.2	
Hulless barley (2)	4.6–5.6			29.8–31.3	6.3–7.8	1.4	
Hulless waxy barley (5)	7.6–11.3			28.4–31.2	10.2–15.4	1.3–1.7	
Hulless normal barley (4)	4.5–5.6			28.9–31.4	6.3–8.1	1.3–1.5	

Rye	1.2	Roller mill	Short flow	Bran	11.0	2.3	1.9	Harkonen et al.[108]
				Shorts	19.0	2.9	2.4	

[a]Ratio of β-glucan concentration in enriched fraction to β-glucan concentration in starting material.

[b]Range of values for all coarse (remained on the screens) enriched fractions obtained after regrinding and resieving through 325-, 100-, and 200-mesh screens; values in paretheses are for the coarse enriched fraction obtained after the first grinding and sieving through a 325-mesh screen.

TABLE 1.2
Large-Scale Enrichment of Cereal Grains in β-Glucans by Dry Processing

Starting Material (number of cultivars)	β-Glucan Content (%) of Starting Material	Process		Weight (%) of Coarse Enriched Fractions (yield)	β-Glucan Content (%) of Coarse Enriched Fractions	Enrichment Factor[a]	References
		Milling	Particle Fractionation				
Dehulled barley (1)	—	Pin mill	Sieving (45 μm)	24.5	18.3	—	Knuckles and Chiu[59]
Hulled barley (1)	4.5	Pin mill	Sieving (45 μm)	19.9	17.7	3.9	
Hulless waxy barley (1)	—	Pin mill	Sieving (45 μm)	25.1	15.75	—	
Defatted oat flakes (1)	5.5	Pin mill	Air classification	38	11.2	2.5	Wood et al.[134]
Dehulled oat (1)	5.6	Pin mill	Air classification and sieving (300 μm)	34	12.8	2.3	
Dehulled oat (1)	5.6	Pin mill	Sieving (355 μm)	43	10.9	1.9	
Hulless normal barley (1)	5.9	Pin mill	Air classification	10.4	13.1	2.2	Vasanthan and Bhatty[60]
Hulless waxy barley (1)	7.2			7.6	23.8	3.3	
Dehulled high-amylose barley (1)	7.8			20.9	21.8	2.8	
Hulled normal barley (1)	4.6	Impact mill	Air classification	21.9	~7	~1.5	Andersson et al.[67]
Hulled high-amylose barley (1)	6.9			18.9	~10	~1.4	
Hulled waxy barley (1)	5.6			17.3	~10	~1.8	
Hulless normal barley (1)	3.8			10.8	~8	~2.1	
Hulless high-amylose barley (1)	7.2			17.7	~12	~1.7	
Hulless waxy barley (1)	5.6			12.4	~11	~2.0	
Hulless waxy barley (1)	17.0			38.9	~23	~1.4	
Dehulled oat (1)	—	Roller mill	Short flow	25	7.9	—	Westerlund et al.,[130] Wikstrom et al.[132]
Rolled oat (1)	—			25	8.4	—	

Hulless barley (1)	5.6	Buhler roller mill	Short flow	28	7.7	1.4	Bhatty[120]
Hulless normal barley (1)	3.4	Pearling to 10% and Buhler roller mill	Short flow	16.3	8.1	2.4	Izydorczyk et al.[71]
Hulless high-amylose barley (1)	6.1			28.9	13.4	2.2	
Hulless waxy barley (1)	5.7			23.3	15.2	2.7	
Rye	1.5	Dehuller and roller mill	Short flow	—	3.3	2.2	Glitso and Bach Knudsen[89]
Wheat	0.4	Debranning commercial system		2.0	1.7	4.3	Dexter and Wood[138]
Wheat	0.5	Debranning commercial system		—	2.6	5.2	Cui et al.[119]

[a]Ratio of β-glucan concentration in enriched fraction to β-glucan concentration in starting material.

According to a definition adopted by the American Association of Cereal Chemists (AACC), oat bran is characterized as the milled fraction that does not exceed 50% of the initial oat groats or rolled oats, has a total β-glucan content of at least 5.5% (dry weight basis), a total dietary fiber of at least 16.0%, and at least one third of the total dietary fiber as soluble fiber.[139]

Fractionation of several oat cultivars into coarse (bran) and fine fractions by a simple dry-milling and sieving procedure gave a 7.4% mean β-glucan content of all the brans, with an average enrichment factor of 1.5 from an average bran yield of 53.3%, close to the maximum suggested in the AACC definition. The value of a particular cultivar as a source of β-glucan in the bran is clearly not solely dependent on the concentration of β-glucan in the groat; it seems that both cell wall thickness and the degree to which this varies throughout the endosperm differ among cultivars.[78] It is, however, possible to increase the β-glucan concentration of the brans by alternative milling procedures designed to improve the fractionation of coarse from fine particles, although this is associated with a decrease in bran yield. Large differences in yields of the rich β-glucan fractions have been demonstrated among the various types of mills used for the dry-milling procedure. Grinding of dehulled barley with the ball mill, roller mill, pin mill, abrasive stone disk mill, and abrasive Udy mill resulted in 21, 65, 51, 70, and 30%, respectively, of the dehulled barley weight remaining on a screen with 45-μm openings.[56] A pilot-scale fractionation of several barley cultivars by pin milling and screening resulted in a coarse fraction containing 16 to 18% β-glucans (Table 1.2) and 40 to 45% total dietary fiber.[59] Application of repeating grinding and sieving techniques yielded fractions with high levels of β-glucans up to 28%.[56,135,140] Furthermore, pearling has been found to increase β-glucan content in coarse fractions, and therefore is used for enrichment of grain fractions in β-glucans alone or in combination with other processes, such as repeated milling/grinding and sieving.[30,56,71,135,140]

Combinations of dry-milling processes or sieving with air classification were also used to improve the β-glucan level in fractions of oat and barley grains in a laboratory and large scale (Table 1.1 and Table 1.2), achieving fractions containing ~7 to 31% total β-glucans, with enrichment factors from 1.4 to 3.3 and yield in the range of ~8 to 39%.[58–60,67,134]

The yields of β-glucan fractions obtained from abrasion milling and subsequent sifting are relatively low, and a prolonged sieving time is required. Roller milling has the capacity for large-scale processing, produces numerous products of highly variable composition, and is used in many cases for production of rich β-glucan fractions in large-scale trials. Quantities of 400 to 500 Kg dehulled oat grains with or without prior steam flaking were processed in a pilot plant roller mill of the type used in commercial wheat milling, yielding three discrete fractions: bran, outer starchy endosperm, and inner starchy endosperm.[130,132] The inner endosperm comprised about 50%, and each of the other fractions about 25%; the content of β-glucans was generally higher in the bran fraction (7.9 to 8.4%) than in the starchy endosperm (2.0 to 2.3%) (Table 1.2).

Bhatty[64] dry milled, in an experimental roller mill, 15 diverse cultivars and genotypes of barley varied in β-glucans from 4.2 to 11.3%. The resulting bran fractions contained β-glucan from 4.9 to 15.4%, with an average yield value and β-glucan enrichment in bran of 30.3% and 1.4-fold, respectively (Table 1.1). Due to poor separation of bran and shorts in barley roller milling, these fractions were combined to obtain a composite bran sample in approximately 30% yield.[64,120] Although for barley bran there is as yet no definition like that of oat bran, Bhatty[141] claims that hulless barley bran of about 30% extraction is a true bran containing the seed coat, germ, aleurone, and subaleurone layers. On the average of bran fractions of 15 different cultivars and genotypes of barley at ~30% extraction, the β-glucan enrichment is about 37%, yielding bran that may contain 7 to 14% soluble fibers — far higher than in oat bran.

Unlike wheat and oats, barley does not have a long tradition of being fractionated by roller milling, and the fractions derived by this process have not yet been standardized in terms of quality, composition, or even terminology. Izydorczyk and coworkers[71,72] roller-milled hulless barley cultivars of variable amylose content and generated mill streams with variable composition. β-Glucans from endosperm cell walls are highly concentrated in the shorts from the reduction system, designated as the fiber-rich fraction (FRF). Generally, for high β-glucan cultivars the FRF yields are greater than 20% (whole barley basis), with β-glucan contents above 15%, having obvious potential as a functional food ingredient (Table 1.2).

Recently, various dry-milling approaches for rye grain have also been investigated to obtain fractions with higher contents of dietary fiber components and to add them to rye breads. Harkonen et al.[108] fractionated a rye cultivar by a laboratory-scale roller mill and obtained bran and short fractions enriched in β-glucans with contents of 2.3 and 2.9%, respectively (Table 1.1). Glitso and Bach Knudsen[89] separated rye by dry milling into three fractions enriched in different rye grain tissue (pericarp/testa, aleurone, and endosperm). The pericarp/testa-enriched fraction, obtained after dehulling of the whole kernels, had a low concentration of β-glucans (0.46%). The dehulled grain was roller milled and, depending on the subsequent sieving procedure, two different fractions were obtained; the highest amount of β-glucans was found in the aleurone-rich fraction (3.3%) (Table 1.2), whereas the β-glucan content was lower in the starchy endosperm (0.75%).

Studies on wheat β-glucans are limited because of their low contents in the grain. However, a newly developed wheat preprocessing technology produces debranned by-products, enriched in β-glucans (up to 1.7%) (Table 1.2), using a commercial system of friction–abrasion technology.[138] Cui and coworkers developed some interest in wheat β-glucan due to this preprocessing technology, which enriches the β-glucan content from 0.5% in whole wheat to 2.6 to 3.0% in one of the bran fractions[107,119] (Table 1.2).

1.5.2 Wet Processing

Methods such as wet milling–sieving also aim to partially remove starch from cereal grain. From 20 to 55% of the barley β-glucan was found to be soluble in water at room temperature,[70,75] whereas only 5% was soluble in ice-cold water.[135] Similarly, little of the cell wall β-glucan is soluble in ethanol or ethanol/water mixtures or aqueous solutions of certain salts. Therefore, for wet milling–sieving, cold water or the latter solvents are used.

Removal of starch from the cell wall material is accomplished by homogenization of barley flour with water saturated with a salt (sodium sulfate) and sieving the slurry;[135] the remaining material on the screen yielded 44% of the starting flour and contained 10.4% β-glucan. From dehulled oats, containing 5.6% β-glucan, Wood et al.,[134] by pin milling, dry sieving, reflux in ethanol, sieving in ethanol (screen, 150 μm), and drying, prepared in a large scale, a fraction with a 16.6% β-glucan level and a 21% yield of the starting material. Vasanthan and Temelli[142] suggested the recovery of β-glucans with wet sieving from ground/milled barley or oat whole grain, or any β-glucan-rich fraction of the grain obtained by dry processes such as milling (i.e., pin, hammer, attrition) and air classification or sieving. This material is mixed with a mixture of ethanol/water (40 to 50% ethanol) at room temperature for 10 to 30 min, and optionally, protease and amylase were used at this step for the removal of starch and proteins. The slurry was separated by a screen (40 to 70 μm), and the retentate portion was washed with the ethanol/water mixture and air dried. This method provided β-glucan concentrates from barley and oat having a β-glucan concentration of 40 to 70%, depending on whether the protease and amylase were employed.

β-Glucan-enriched products from oats and barley were prepared using wet grinding in cold water (8 to 12°C), which may contain an organic solvent (ethanol), about 20% by weight of the water, and following sieving and drying. The β-glucan concentration of the obtained fiber was 18 to 31% and the β-glucan yield, calculated from the β-glucan of the initial materials, was between 75 and 90%.[143,144] In pilot and small industrial scale, oat bran concentrates containing 14.7 to 15.5% β-glucans were produced by wet milling in neutral or acidified cold-water (<14°C) suspensions and shifting. Moreover, two other concentrates have been prepared by wet milling and sieving in ethanol–water suspensions; for wet milling, 70% (v/v) ethanol at 20°C and 90% (v/v) ethanol at 75°C were used, yielding oat bran concentrates with 16.3 and 18.9% β-glucans, respectively.[136]

In general, cereal β-glucan concentrates and isolates resulting from a number of investigations at pilot scale have been obtained by water, acidified water, and aqueous alkali (i.e., NaOH or Na_2CO_3) extraction from whole ground cereal grains or high β-glucan fractions produced by the various aforementioned dry and wet processing of cereal grains. The resultant slurry is then processed by techniques such as filtration, centrifugation, and alcohol precipitation to separate the β-glucan from the slurry (Table1.3). In combination with extraction, treatments with amylases and proteases can be used for removal of starch and proteins.[112,119,120,145,146] Use of α-amylase may yield substantial amounts of glucose, which can provide sweetness to food formulations and promote the formation of colored and bitter products on heating in the presence of amino acids (Maillard reaction products), consuming simultaneously lysine, an essential amino acid. Therefore, the use of β-amylase instead of α-amylase has been suggested in combination with an ultrafiltration step (Table 1.3) for the purification of the product from low molecular weight constituents.[146] Beer et al.[112] examined different recovery methods of β-glucans from oat bran extracts and demonstrated that for production of large amounts of good-quality oat gum rich in β-glucans, an alcoholic precipitation would be the process of choice, but ultrafiltration and dialysis are quite useful alternatives. However, these conventional processes have a number of technical problems, which limit commercial uses

TABLE 1.3
Large-Scale Production of Cereal β-Glucan Concentrates or Isolates by Solvent Extraction

Starting Material	Preprocessing for Enrichment in β-Glucans	Extraction Conditions/Purification Procedures	Recovery Methods	β-Glucan Content of Final Products	References
Barley flour	Dry milling and sieving	Aqueous (4–5°C × 18 h)	Centrifugation and freeze drying	33%	Cavallero et al.,[31] Knuckles et al.[28]
Oat flour	—	Aqueous (90–100°C) with α-amylase digestion	Centrifugation and freeze drying	—	Inglett[145]
Oat flour	Dry milling–sieving and heat treatment	Aqueous (55°C) with β-amylase digestion	Centrifugation or filtration, ultrafiltration, and pasteurization–concentration or drying (spray or freeze drying)	17%	Oste Trantafyllou[146]
Barley and Oat flours	—	Mild alkaline (pH = 8, 50°C × 1 h) and acidification (pH = 4)	Centrifugation, spray drying, and agglomeration	19.4%	Cahill et al.[147]
Oat flakes	Hexane defatting, pin milling–air classification (2000 kg),[a] and refluxing with 70–75% ethanol	Mild alkaline (20% $Na_2(CO_3)$, pH = 10, 45°C × 30 min) and precipitation of proteins at pH = 4.5	Centrifugation, concentration, precipitation with 2-propanol (50%), centrifugation, repeated blending with ethanol or propanol (100%), and centrifugation and air desolventization	78% (18.6 kg)[a]	Wood et al.[134]
			Same as the above process with an additional purification step of the $(NH_4)_2SO_4$ (20% w/v) precipitation	96–98%	Wood et al.[5]
Oat bran concentrates	—	Mild alkaline (20% $Na_2(CO_3)$, pH = 10, 40°C × 30 min) and precipitation of proteins at pH = 4.5	Centrifugation, ultrafiltration, and freeze drying	39.7%	Beer et al.[112]
		Same as the above process with an additional step of pancreatin digestion		62.2%	
Hulless barley (500 kg)[a]	Roller mill (5 kg)[a]	Alkaline (0.25 *N* NaOH, 25°C) with Termamyl digestion	Precipitation with ethanol and freeze drying	50% (0.5 kg)[a]	Bhatty[120]

TABLE 1.3 (continued)
Large-Scale Production of Cereal β-Glucan Concentrates or Isolates by Solvent Extraction

Starting Material	Preprocessing for Enrichment in β-Glucans	Extraction Conditions/Purification Procedures	Recovery Methods	β-Glucan Content of Final Products	References
Wheat flour	Branning with debranning commercial system	Alkaline (0.25 *N* NaOH, 25°C × 1 h) with thermostable -amylase digestion	Precipitation with ethanol and freeze drying	~20%	Cui et al.[119]
Oat flour (7 kg)[a]		Aqueous (<50°C) and heating[b]	Centrifugation, precipitation, and dehydration with 2-propanol (50%), screening and grinding	75% (145 g)[a]	Wang et al.[148]
Barley flour (100 kg)[a]				87% (3.5 kg)[a]	
Waxy barley meal		Aqueous (40–60°C)	Centrifugation, heating (90–95°C × 5 min), centrifugation, ultrafiltration, and drum or spray drying	<50%	Goering and Eslick[149]
Cereal grain or bran or spent brewer's grain	Wet screening	Alkaline (55–65°C × 1–2 h)	Centrifugation, acidification, heating, cooling centrifugation[b]: Concentration, drying milling	45%	Potter et al.[150]
			Evaporation,[c] skimming, filtration, drying milling	50–95%	
			Same as the above process with additional repeated ultrafiltration	75–95%	
Pearled barley	Roller and hammer milling and screening	Aqueous (50°C) with cellulase, xylanase, and amyloglucosidase digestion, centrifugation, concentration (falling film evaporator)	Gelation (0°C × 24 h), washing, centrifugation, and spray drying	~85%	Morgan[135]
			Gelation (–18°C × 48 h and 25°C × 72 h), decanting, and freeze drying	70–90%[d]	Morgan[151]

[a]In parentheses is the amount of starting material or enriched fraction in β-glucan or final concentrate/isolate.
[b]β-Glucanase inactivation or protein coagulation.
[c]For production of a skin (solid film) enriched in β-glucans.
[d]Increasing with repeated freezing–thawing cycles.

by the cost of the isolated material, particularly for food applications. The most common technical problems come about from the viscosity rise in the slurry during extraction of β-glucans, causing clotting of the filter upon filtration and inefficient separation of flour components during centrifugation, as well as from the use of large amounts of organic solvents for precipitation and the high cost for their recycling (distillation).

Van Lengerich et al.[152] suggested digestion with exogenous enzymes (cellulases, hemicellulases, xylanases, and pentosanases) of the aqueous slurry containing the β-glucan grain material to reduce viscosity and optimize separation of insolubles from the extract solution. For the recovery of β-glucan from cereal water extracts without using the precipitation method with organic solvents, Morgan[135,151] suggested gel formation processes (Table 1.3); gelation was induced by various procedures such as resting, shearing, cooling, or freezing the solution for a period. However, the molecular weight of the isolate obtained by this protocol was about 50,000 Da, because no deactivation of endogenous enzymes was carried out before extraction.[135] Later, Morgan[151] proposed a process for controlling the average molecular weight of β-glucan extracted from cereals by managing the extraction time; decreasing the extraction time from 1 h to 30 min resulted in a β-glucan isolate with a molecular weight of more than 100,000 Da.

1.6 STRUCTURAL FEATURES

The structural features of β-glucans are important determinants of their physical properties and functionality, including their physiological responses when they are considered as ingredients in cereal-based foods and other formulated products. These features include ratios of β-(1→4)/β-(1→3) linkages, presence and amount of long cellulose-like fragments, ratios of cellotriosyl/cellotetraosyl units, and molecular size.[96]

Values of molecular weight for cereal β-glucans have been reported in the literature in the range of 31 to 2700 × 10^3, 35 to 3100 × 10^3, 209 to 416 × 10^3, and 21 to 1100 × 10^3 for barley, oat, wheat, and rye, respectively (Table 1.4). Other molecular characteristics of cereal β-glucans obtained by laser light-scattering detectors such as polydispersity index (M_w/M_n) and radius of gyration (R_g) were found ranging from 1.2 to 3.1[154,156,158,159] and 30 to 75,[40,43,45,75,115,123,133,136,157–159] respectively. The apparent discrepancies in the molecular weight estimates of cereal β-glucans might originate from varietal and environmental (growth) factors,[62,71,86] aggregation phenomena (dependent on the structural features and solvent quality),[43,44,133,156,160,161] and the analytical methodology used for the determination of these values (detector, standards).[62,116]

Depolymerization events (endogenous or microbial β-glucanases from contaminating microorganisms) taking place during the extraction step,[110,116,137,140,151,162] as well as differences in extraction and isolation methods (solvent and temperature affect the solubilization) affect the molecular size of the isolated polysaccharide. Increasing temperature of extraction can lead to an increase in molecular size of the extracted cereal β-glucans.[4,38,83,99,114,137] However, an opposite trend has also been observed[3,68,98] that in some cases was attributed to small differences in fine structure.[3]

TABLE 1.4
Molecular Weights of Cereal β-Glucans

Source	Detection Method	Molecular Weight (10^{-3})	References
Barley	Sedimentation velocity	150–290	Woodward et al.[3,153]
	HPSEC with MALLS	80–150	Saulnier et al.[115]
	HPSEC with FD (β-glucan standards)	1300–1500	Beer et al.[62]
	GPC with RI (pullulan standards)	31–560	Morgan and Ofman[137]
	HPSEC with MALLS	100–375	Bohm and Kulicke[40]
	HPSEC with MALLS and RI	~200–600	Gomez et al.[133]
	HPSEC with LALLS, RI, and FD (β-glucan standards)	1700–2700	Wood et al.[116]
	HPSEC with MALLS	570–2340	Knuckles et al.[75]
	HPSEC with FD (β-glucan standards)	708	Cui et al.[41]
	HPSEC with 90° laser LS, DP, and RI	693	Wang et al.[154]
	HPSEC with RI (β-glucan standards)	40–250	Vaikousi et al.[21]
	HPSEC with RI (β-glucan standards)	213	Lazaridou et al.[18]
	HPSEC with MALLS and RI	450–1320	Irakli et al.[45]
Oat	HPSEC with RI (dextran standards)	120–2400	Zhang et al.[83]
	HPSEC with FD (β-glucan standards)	2100–2500	Beer et al.[62]
	HPSEC with FD (β-glucan standards)	1400–1800	Beer et al.[117]
	GPC with RI and FD (β-glucan standards)	1500	Autio et al.[79]
	GPC with MALLS	1100–1500	Malkki et al.[136]
	GPC and FD (β-glucan standards)	600–840	Jaskari et al.[144]
	HPSEC with FD (β-glucan standards)	1200	Beer et al.[117]
	HPSEC with LALLS, RI, and FD (β-glucan standards)	360–3100	Wood et al.[116]
	HPSEC with FD	1100–1600	Johansson et al.[123]
	HPSEC with FD	2060–2300	Aman et al.[155]
	HPSEC FD (β-glucan standards)	1160	Cui et al.[41]
	HPSEC with 90° laser LS, DP, and RI	1700	Wang et al.[156]
	HPSEC with 90° laser LS, DP, and RI	611	Wang et al.[154]
	HPSEC with RI and MALLS	214–257	Roubroeks et al.[157,158]
	HPSEC with RI (β-glucan standards)	35–250	Lazaridou et al.[17,18]
	HPSEC with MALLS and RI	180–850	Skendi et al.[43]
Wheat	HPSEC with FD (β-glucan standards)	267–416	Cui et al.[41]
	HPSEC with RI (β-glucan standards)	209	Lazaridou et al.[18]
Rye	HPSEC with LALLS, RI, and FD (β-glucan standards)	1100	Wood et al.[116]
	HPSEC with LALLS	21	Roubroeks et al.[159]

Note: HPSEC = high-performance size exclusion chromatography; MALLS = multiple-angle laser light-scattering detector; FD = fluorescence detector with calcofluor postcolumn; LALLS = low-angle laser light-scattering detector; RI = refractive index detector; LS = light-scattering detector; DP = differential viscometer detector; GPC = gel permeation chromatography.

Studies of the molecular weight of cereal β-glucans have often dealt with isolated fractions of significantly lower molecular weight than the native cell wall polysaccharides. Several studies also showed that the molecular weight of cereal β-glucans decreases during the isolation and purification procedures.[43,112,116,134,136,144] Usually, harsher extraction conditions, such as high or low pH or prolonged extraction times, especially at high temperatures, can lead to recovery of low molecular size β-glucans.[62,75,116,117,119,137] Moreover, high-speed homogenization, sonication,[109,134] and high shear rates[112,134] were shown to reduce viscosity and molecular weight. Also, the purification step with ammonium sulfate seems to reduce molecular sizes of the extracted cereal β-glucans.[38,116,124]

Many researchers have used several techniques to fractionate β-glucan populations differing in molecular size to study the influence of molecular size on functional properties of β-glucans or to achieve β-glucans with distinct physical and molecular characteristics. Such methods were partial degradation by acid hydrolysis[14,21] or controlled depolymerization by lichenase[157] for different periods of time, as well as ultrasonication.[4,15,40] As aforementioned, Morgan and coworkers[137,151,162] proposed a process for controlling the molecular size of extracted β-glucans by altering the extraction time if no deactivation of endogenous enzymes is applied to the flour. Furthermore, fractionation of oat and barley β-glucans into populations with different molecular sizes has been achieved with ammonium sulfate; with increasing concentration of salt, there was an increasing molecular size of the precipitated β-glucan fractions.[38,39,154] Wang et al.[154] found that the starting β-glucan concentration and temperature also seem to affect the fractionation efficiency. A clear separation of the fractions was possible at values of the overlapping parameter, c[η], lower than ~3.5. Moreover, the higher the temperature, the lower the amount of ammonium sulfate that was necessary to precipitate a fraction of similar M_w.

The fine structure of cereal β-glucans consists predominantly of β-(1→3)-linked cellotriosyl and cellotetraosyl units.[1,3–6,163] Longer cellulosic oligosaccharides in smaller amounts (~5 to 10%), with a degree of polymerization between 5 and 20, have been also identified.[2–3,5–7] Blocks of two or more adjacent (1→3) linkages are absent or present in a very low frequency.[1,2,4–6,41] Moreover, there were some indications for the presence of alternating (1→3) and (1→4) sequences in the cereal β-glucan chain.[157,158] Although the distribution of β-(1→3) linkages in the polysaccharide is not random, statistical analysis of the sequence of cellulose-like oligomers showed a rather random distribution in the polymer chain for cellotriosyl and cellotetraosyl segments.[163,164] From biochemical experiments *in vitro* with active synthases in isolated Golgi membranes, the biochemical features and topology of cereal β-glucan biosynthesis are found to be closely parallel to those of cellulose.[165] According to a current model for biosynthesis of mixed-linkage (1→3),(1→4)-β-glucans proposed by Buckeridge et al.,[165] the (1→3),(1→4)-β-glucan synthase is that of a cellulose core-like synthase that makes cellobiosyl and even-numbered cellodextrin units, and a distinct glycosyl tranferase adds a third glycosyl residue to complete the cellotriosyl and higher odd-numbered units. Further investigation of the synthase activity *in vitro* showed that the cellodextrin unit distribution is altered drastically depending on the uridine diphospate (UDP)–Glc concentration. The suboptimal UDP–Glc concentrations favor the synthesis of longer cellodextrin units in β-glucan, particularly the cellotetraosyl unit, whereas at the highest UDP–Glc concentrations tested, the cellotriose units were predominant of the total polymer synthesized.[165]

Despite the structural similarity of β-glucans from different genera of cereals, as suggested from methylation analysis and their almost identical nuclear magnetic resonance (NMR) spectra, oat, barley, and wheat β-glucans are, in fact, structurally distinct, as shown by quantitative high-performance liquid chromatography (HPLC) analysis of lichenase-released oligosaccharides.[1,5,17,41] The enzyme lichenase, a (1→3),(1→4)-β-D-glucan-4-glucanohydrolase (EC 3.2.1.73), specifically cleaves the (1→4)-glycosidic bond of the 3-substituted glucose residues in β-glucans, yielding oligomers with different degrees of polymerization (DPs). The major products for the cereal β-glucans are 3-O-β-cellobiosyl-D-glucose (DP3) and 3-O-β-cellotriosyl-D-glucose (DP4), but cellodextrin-like oligosaccharides are also released from the polymer regions containing more than three consecutive 4-linked glucose residues. The DP of the long cellulose-like fragments has been found to vary between 5 and 20, with DPs 5, 6, and 9 being the most abundant.[6,7,14,17,18,21,38,43–45,93] The literature data on oligosaccharide distribution of cereal β-glucans from their respective lichenase digests are given in Table 1.5. Generally, the oligosaccharide distribution within the same genera of cereals was found to be similar and showed major differences only between β-glucans of different origins.[5,16,18] The amount of the trisaccharide (DP3) for the β-glucans follows the decreasing order of wheat (67 to 72%), barley (52 to 69%), and oat (53 to 61%), whereas the relative amount of the tetrasaccharide (DP4) follows the increasing order of wheat (21 to 24%), barley (25 to 33%), and oat (34 to 41%). However, the total of tri- and tetrasaccharide of β-glucans is similar among the cereal genera, resulting in a similar total amount (5 to 11%) of cellulose-like oligomers with DP 5 among the cereal β-glucans. The differences in tri- and tetrasaccharide amounts observed among β-glucans from different cereal sources are also reflected in the molar ratio of cellotriose to cellotetraose units (DP3/DP4) following the order of wheat (3.0 to 4.5), barley (1.8 to 3.5), rye (1.9 to 3.0), and oat (1.5 to 2.3); this ratio is considered a fingerprint of the structure of cereal β-glucans.

Literature data also indicate that there are differences within the same genera as well, which could be attributed to genotypic and environmental factors. A narrower range of the DP3/DP4 ratio in domestic cultivars of *Avena sativa* (2.05 to 2.11) than in other cultivars of *Avena* (1.81 to 2.33) has been noticed by Miller et al.[82] Jiang and Vasanthan[69] found the molar ratio of tri- to tetraose units in the β-glucans from waxy barley varieties to be somewhat higher (2.6 to 2.8) than from normal and high-amylose varieties (2.3 to 2.6). Wood et al.[93] also reported a higher DP3/DP4 molar ratio for β-glucans from waxy barleys (3.0) than from nonwaxy cultivars (2.7 to 2.8). Storsley et al.[44] found that the water-extractable (at 45 and 95°C) β-glucans from normal amylose varieties exhibit lower DP3/DP4 molar ratios (2.5 to 2.9) than high-amylose (2.9 to 3.1) and waxy varieties (3.1 to 3.2). In the latter study, a higher amount of long cellulose-like fragments was found in water-extractable β-glucans at 95°C for varieties of barley with the anomalous amylose–amylopectin ratio (12.0 to 17.5%) than for varieties with normal starch (8.2 to 10.7%); however, no significant structural differences were observed among the different barley varieties for water-extractable β-glucans obtained at lower extraction temperatures (45°C).

TABLE 1.5
Structural Features of Cereal β-Glucans

Source	DP3	DP4	DP ≥ 5	Molar Ratio DP3/DP4	(1→4)/(1→3)	References
Oat	—	—	—	2.1–2.3	2.4[c]	Wood et al.[5]
	—	—	—	1.8–2.3	—	Miller et al.[82]
	55.0–58.1[a]	34.2–36.0[a]	7.7–8.9[a]	2.1–2.2	—	Doublier and Wood,[14] Wood et al.[78]
	—	—	—	1.5–1.9	—	Miller and Fulcher[166]
	57.6[b]	34.1[b]	8.2[b]	1.7	—	Izydorczyk et al.[39]
	53.4–53.8[a]	40.4–41.4[a]	—	1.7–1.8	—	Johansson et al.[123]
	58.3[a]	33.5[a]	8.1[a]	2.2	—	Cui et al.[41]
	55.6–55.9[b]	33.6–34.4[b]	7.1–7.5[b]	1.6–1.7	2.4[d,e]	Colleoni-Sirghie et al.[85]
	54.6–56.8[a]	35.3–36.3[a]	7.7–9.2[a]	2.0–2.1	2.3–2.6[e]	Skendi et al.[43]
	56.7[a]	34.6[a]	8.7[a]	2.2	—	Wang et al.[154]
	54.2–60.9[a]	33.8–36.7[a]	3.6–9.7[a]	2.0–2.3	2.4–2.8[e]	Lazaridou et al.[17,18]
	—	—	—	—	2.3–2.6[e]	Dais and Perlin[1]
	—	—	—	—	2.5[d]	Westerlund et al.[130]
	—	—	—	—	2.4[d,e]	Roubroeks et al.[157]
Barley	56–61[a]	28–32[a]	6–13[a]	2.3–2.9	2.2–2.6[c]	Woodward et al.[2,3]
	—	—	—	2.8–3.4	2.4[c]	Wood et al.,[5] Wood[167]
	62.1[a]	29.4[a]	8.4[a]	2.8	—	Wood et al.[78]
	59.2–64.9[a]	25.3–30.4[a]	9.4–10.2[a]	2.6–3.4	2.4[c]	Saulnier et al.[115]
	56.8–61.6[b]	26.1–32.3[b]	10.6–11.2[b]	1.8–2.4	—	Izydorczyk et al.[7,38]
	63.7[a]	28.5[a]	7.8[a]	3.3	—	Cui et al.[41]
	51.8–61.9[a]	28.1–32.1[a]	6.3–12.5[a]	2.3–2.8	—	Jiang and Vasanthan[69]
	61.5–64.3[a]	27.9–30.1[a]	7.8–8.6[a]	2.7–3.0	—	Wood et al.[93]
	59.4–64.3[a]	24.8–31.0[a]	8.2–17.5[a]	2.5–3.2	1.9–2.2[d]	Storsley et al.[44]
	66.0[a]	25.7[a]	8.2[a]	3.4	—	Wang et al.[154]
	62.0–69.3[a]	26.2–29.1[a]	4.5–8.9[a]	2.8–3.5	2.1–2.8[e]	Vaikousi et al.[21]
	62.0–63.3[a]	27.5–29.2[a]	8.8–9.1[a]	2.8–3.0	—	Lazaridou et al.[18]
	57.7–62.4[a]	29.4–32.9[a]	7.7–9.5[a]	2.3–2.8	2.2–2.7[e]	Irakli et al.[45]
	—	—	—	—	1.9–2.3[c]	Balance and Manners[98]
	—	—	—	—	2.3–2.6[e]	Dais and Perlin[1]
	—	—	—	—	2.4[d]	Henriksson et al.[168]
Rye	—	—	—	2.7–3.0	—	Wood et al.[5,78]
	—	—	—	1.9–2.3	2.3[d]	Roubroeks et al.[159]
Wheat	—	—	—	3.0–3.8	—	Wood et al.[5]
	72.3[a]	21.0[a]	6.7[a]	4.5	—	Cui et al.[41]
	67.1[a]	24.2[a]	8.7[a]	3.7	—	Lazaridou et al.[18]

[a]Weight percent from the chromatograms of the lichenase digests.
[b]Mole percent from the chromatograms of the lichenase digests.
[c]Calculated from methylation analysis.
[d]Calculated from ^{1}H-NMR data.
[e]Calculated from ^{13}C-NMR data.

Furthermore, analysis of oat aleurone β-glucan showed a smaller proportion of the cellotetraosyl units than that of the endospermic β-glucan,[6] whereas whole groats and oat bran β-glucans seemed to be having similar trisaccharide/tetrasaccharide ratios.[5] Similarly, Izydorczyk et al.[71] found that among the products obtained from pearling and roller milling of various barley cultivars, the ratio of tri- to tetrasaccharides in β-glucans from pearling by-products was higher than that from flour and a fiber-rich fraction. The β-glucans from the former fraction probably originated from the aleurone tissue, whereas the β-glucans from the latter fraction originated from the endosperm cell walls. Despite the genetic and environmental variations, some discrepancies in the calculated values of the DP3/DP4 ratio for cereal β-glucans might arise from variations in the sensitivity of techniques employed, as well as from the uncertainties in the response factors for the oligosaccharides in the analytical system used because of the lack of appropriate pure oligosaccharide standards.

The calculated ratios of the two types of linkages (1→4) to (1→3) in the native cereal β-glucan structures, based on NMR data and methylation analysis, were found within the range of 1.9 to 2.8 (Table 1.5). Izydorczyk and coworkers[7,38,44] fractionated cereal β-glucan using different aqueous or alkali conditions and different ammonium sulfate concentrations, and obtained fractions exhibiting differences in the (1→4)/(1→3) ratio from 1.9 to 5.3, as well as other distinct molecular/structural characteristics, as described in the following section. Even small differences in the proportions of the two linkages seem to be enough to influence significantly the physical properties of cereal β-glucans, such as the solubility, conformation, and aggregation tendency of the polymeric chains.[3,163]

1.7 PHYSICAL PROPERTIES

1.7.1 Solubility — Solution Behavior

The molecular size and fine structural features of β-glucans play an important role on the solubility and chain conformation or shape, and hence on their rheological properties in solution. The chemical features of cereal β-glucans are reflected in their solubility in water and their extended, flexible chain conformation.[153] The cellulose-like portions of cereal β-glucans might contribute to the stiffness of the molecules in solution.[4] Furthermore, β-glucans containing blocks of adjacent β-(1→4) linkages may exhibit a tendency for interchain aggregation (and hence lower solubility) via strong hydrogen bonds along the cellulose-like segments; the β-(1→3) linkages break up the regularity of the β-(1→4) linkage sequence, making it more soluble and flexible.[163] It has been suggested that this irregular spacing of (1→3)-linked β-glucosyl residues in the β-glucan chain is responsible for the irregular overall conformation of the polysaccharide, and hence the chains are unable to align closely over extended regions, keeping the polysaccharide in solution.[3] On the other hand, it has been reported that helical segments made up of at least three consecutive cellotriosyl residues could constitute conformationally stable motifs (crystalline) in mixed-link (1→3),(1→4)-β-glucan fibers as revealed by x-ray analysis,[169] and that the β-(1→3) linkages could be involved in the ordered conformation of barley β-glucans as revealed by ^{13}C CP/MAS NMR spectroscopy.[162] It is therefore possible that a higher content of cellotriosyl units might impose some conformational regularity in the β-glucan chain, and consequently a higher degree of secondary chain organization of these polymers in solution (i.e., lower solubility).[7,71] Suggestions have been made that differences in the amount of cellotriosyl fragments, long cellulosic oligomers, and the ratio of (1→4)/(1→3) linkages might explain solubility differences among cereal β-glucans in accord with the two previous aggregation mechanisms.[3,7,38,39,41,44]

The reported intrinsic or limiting viscosity ([η]) values for cereal β-glucans vary between 0.28 and 9.6 dl/g, depending largely on the molecular weight of the isolated polymers.[2–4,14,17,18,21,38–40,43,45,129,161,171] Data of the double logarithmic plots of $(\eta sp)_0$ vs. c[η] for several β-glucan preparations differing in fine structure and molecular size superimpose closely, falling into three linear portions, from which the values for the two critical concentrations (c*, c**) have been estimated.[14,17,18,21,43,45] The intermediate zone (semidilute domain), between the dilute ($c < c^*$) and concentrated ($c > c^{**}$) domains, appears to be characteristic of polymers with rigid, rod-like conformation.[170] The observed rheological behavior of β-glucans is not surprising, since the chain of this polymer has been satisfactorily modeled by a partially stiff worm-like cylinder.[2,171] The reported values in the literature for the exponent in the Mark–Houwink viscosity equation ($[\eta] \propto M_w$) varied between 0.6 and 1.1 for oat and barley β-glucans, suggesting an extended random coil conformation.[40,86,154,157,158,172,173] The values of coil overlap parameters c*[η] and c**[η] have been found to vary from 0.12 to 1.25 and 1.41 to 10.1, respectively.[14,17,18,21,43,45] The [η] and c** values for some representative highly pure β-glucan samples isolated from different cereals and their structural features are given in Table 1.6; c** increases with a decrease in molecular size and intrinsic viscosity [η] of the cereal β-glucans.

Rheologically, solutions of cereal β-glucans fall into the category of viscoelastic fluids behaving similarly to the well-characterized random coil-type polysaccharides; i.e., a Newtonian region at low shear rates and a shear-thinning flow at high shear rates are shown. However, the preparation of β-glucan solutions, storage time (i.e., waiting time before analysis), and their thermal history have been proven to affect their rheological behavior.[17,18,21,40,174,175] Time-dependent rheological behavior is revealed by thixotropic loop experiments for cereal β-glucans with certain structural features, implying the formation of intermolecular networks. Freshly prepared cereal β-glucan solutions exhibit typical random coil flow behavior, while with increasing storage time the solutions show an unusual shear-thinning flow behavior at low shear rates. This behavior becomes more pronounced with increasing storage time prior to the rheological testing (Figure 1.2, left). Moreover, the freshly prepared solutions showed no time dependency in the rheological behavior; for experiments with a very short total cycle time (2 min), the curve obtained with increasing shear rate is superimposed on the curve obtained with decreasing shear rate. On the other hand, for thixotropic loop experiments with a longer total cycle time (>50 min), the downward cycle displayed a different profile than the upward cycle (Figure 1.2, right), implying aggregate formation during the test. A strong time-dependent behavior has been observed for mixed-link (1→3),(1→4)-β-glucans with low molecular size and high amounts of cellotriose units or long cellulose-like oligomers in the

TABLE 1.6
Compositional, Structural, and Molecular Features, Gelation, and Melting Characteristics of Some Cereal β-Glucan Isolates[17,18,21]

									Gelation and Melting Parameters (Cure Temperature 25°C)								
									Small Deformation					Large Deformation			DSC
Samples	$M_w \times 10^{-3}$	β-Glucans (% db)	Protein (% db)	DP3 + DP4 (%)	DP5 – DP14 (%)	Molar Ratio DP3/DP4	[η] (dl/g)	c ** (g/dl)	Gelation Time (h)	$I_E(h^{-1})$	G'max (Pa)	tan δ	Melting Temperature (°C)	E-Compression Modulus (kPa)	δ_{TR}-True Stress (kPa)	ε_{TR} (True Strain)	ΔH Values (mJ/mg)
Oat																	
OGL35	35	93.1	3.8	96.4	3.6	2.25	0.67	2.1	0.62[a]	3.36[a]	6970[a]	0.228[a]	64.6[a]	17.24[b]	1.08[b]	0.10[b]	5.30[a]
OGL65	65	95.1	2.1	92.7	7.3	2.30	0.91	1.89	4.91[a]	0.57[a]	2370[a]	0.047[a]	65.1[a]	30.45[b]	4.16[b]	0.22[b]	2.88[a]
OGL110	110	93.8	3.4	91.8	8.2	2.27	1.85	1.18	15.00[a]	0.18[a]	2020[a]	0.069[a]	68.5[a]	19.66[b]	4.45[b]	0.30[b]	1.85[a]
OGL140	140	94.4	3.1	90.9	9.1	2.00	2.41	1.10	42.90[a]	0.03[a]	503[a]	0.258[a]	69.3[a]	10.66[b]	9.64[b]	0.63[b]	1.75[a]
OGL250	250	97.0	0.3	90.8	9.2	1.95	3.83	0.70	n.d.	n.d.	n.d.	n.d.	n.d.	9.37[b]	10.87[b]	0.95[b]	1.68[a]
Barley																	
BGL40	40	96.2	1.47	95.5	4.5	3.49	0.63	n.d.	0.08[b]	50.31[b]	7840[b]	0.07[b]	72.0[b]	44.72[b]	4.17[b]	0.12[b]	4.75[b]
BGL70	70	90.8	1.49	91.6	8.4	2.83	1.11	1.91	1.49[b]	2.16[b]	6120[b]	0.122[b]	64.0[b]	23.67[b]	5.79	0.24[b]	3.38[b]
BGL110	110	89.3	3.4	91.2	8.8	2.80	1.86	1.17	n.d.	n.d.	n.d.	n.d.	n.d.	n.d.	n.d.	n.d.	n.d.
BGL140	140	91.8	0.6	91.1	8.9	2.82	2.19	1.04	4.21[b]	0.45[b]	5159[b]	0.147[b]	67.7[b]	20.98[b]	11.43[b]	0.58[b]	2.61[b]

BGL180	180	93.7	1.0	91.4	8.6	2.8	2.33	0.99	14.25[b]	0.13[b]	4627[b]	0.152[b]	68.3[b]	16.93[b]	17.60[b]	0.68[b]	1.96[b]
BGL210	210	91.5	0.8	91.2	8.8	2.81	2.91	0.97	25.54[b]	0.04[b]	2979[b]	0.176[b]	68.5[b]	16.65[b]	17.71[b]	0.76[b]	1.92[b]
BGL250	250	95.0	1.2	91.2	8.9	2.81	3.01	0.95	47.79[b]	0.02[b]	2938[b]	0.289[b]	68.8[b]	15.96[b]	17.84[b]	0.85[b]	1.89[b]
Oat–Barley–Wheat																	
oat200	203	88.4	4	90.3	9.7	2.13	3.01	0.94	167.40[b]	0.01[b]	505[b]	0.457[b]	67.8[b]	36.62[b]	4.46[b]	0.22[b]	2.72[a]
bar200	213	89.8	0.8	90. 9	9.1	3.04	3.11	1.03	21.20[b]	0.03[b]	1792[b]	0.246[b]	68.7[b]	36.65[b]	6.27[b]	0.38[b]	3.50[a]
whe200	209	82.9	7.1	91.3	8.7	3.66	2.98	1.00	3.30[b]	0.24[b]	6072[b]	0.078[b]	72.4[b]	44.59[b]	18.53[b]	0.51[b]	4.04[a]
oat100	105	93.3	2.9	90.5	9.5	2.12	1.77	1.58	41.80[b]	0.06[b]	794[b]	0.127[b]	66.1[b]	39.71[b]	2.51[b]	0.08[b]	3.10[a]
bar100	107	89.6	3.4	91.2	8.8	2.80	1.86	1.27	10.70[b]	0.19[b]	1905[b]	0.009[b]	68.3[b]	55.74[b]	4.59[b]	0.16[b]	4.1[a]

Note: n.d. = not determined.

[a]β-glucan gels at concentration of 10% (w/v).
[b]β-glucan gels at concentration of 8% (w/v).

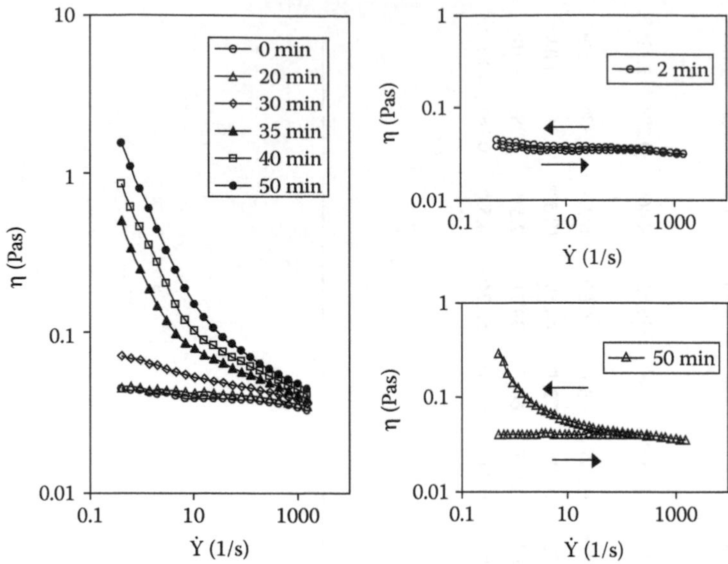

FIGURE 1.2 Effect of storage time on the steady shear viscosity profiles (left) and thixotropic loop experiments with total cycle times of 2 and 50 min (right) at 20°C of 8% (w/v) solution of a low molecular size oat β-glucan (apparent molecular weight calculated as ~35 × 10^3). Arrows indicate the direction of the applied thixotropic loops. (From Lazaridou, A. et al., *Food Hydrocolloids*, 17, 693, 2003. With permission.)

polymeric chain. Also, the departure from the usual flow behavior was noticed in shorter storage periods for solutions with increasing concentration of β-glucans.[17,18,21]

Dynamic and steady shear rheological tests of freshly prepared (or heat treated before measurement) cereal β-glucan solutions usually reveal typical behavior of noninteracting disordered polysaccharides with chain entanglements at the concentrated state.[14,16–18,21,43,45,176,177] For concentrated polymer solutions, the response is liquid — like at lower frequencies (i.e., the loss modulus, G", is larger than the storage modulus, G', and both moduli increase with increasing frequency), while the behavior approaches that of solids — like materials at higher frequencies (i.e., G' is greater than G", and both moduli become largely independent of frequency). For concentrations below the c**, β-glucan solutions exhibit a Newtonian-like behavior, whereas above c** there is a reduction in viscosity (shear thinning) with increasing shear rate. This feature is shifted toward lower shear rates as the concentration and molecular weight of the polysaccharide increase. As expected, with increasing molecular weight, there is an increase in viscosity and the shear-thinning properties of β-glucan dispersions at equivalent polysaccharide concentrations.[17,18,21,43]

The flow curves of β-glucan solutions have often been described[86,132,178,179] by the power law model:

$$\eta = K\ \dot{\gamma}^{\,n-1} \tag{1.1}$$

where η is viscosity, K is consistency index, $\dot{\gamma}$ is shear rate, and n is flow behavior index. As expected, the K values increase and the n values decrease with increasing the concentration and molecular size of the polysaccharide, as well as with decreasing the temperature.[179] The effect of temperature on viscosity of β-glucan solutions has also been modeled by the Arrhenius equation; in this context, estimates of activation energies in the range of 10 to 42 kJ/mol have been obtained.[132,179] Shear-thinning curves of log η vs. log $\dot{\gamma}$ for the disordered polysaccharide solutions all have the same shape, irrespective of primary structure, molecular weight, temperature, solvent environment, and concentration above the onset of coil overlap and chain entanglements (i.e., $c > c^*$). These curves can be superimposed closely if the measured viscosities are expressed as a fraction of η_o, and experimental shear rates are similarly scaled to the shear rate required to reduce η to some fixed fraction of η_0. Reference values commonly used are $\dot{\gamma}_{0.1}$, the shear rate at which $\eta = 0.1\eta_0$, and $\dot{\gamma}_{1/2}$, at which $\eta = \eta_0/2$.[180,181] In this respect, the visocosity data for cereal β-glucan aqueous solutions seem to follow a common master curve using this rationale.[18,21] The general form of shear thinning can be matched with good precision by the following simple relationship:[181]

$$\eta = \eta_0/[1 + (\dot{\gamma} / \dot{\gamma}_{1/2})^{0.76}] \tag{1.2}$$

The experimental data from the generalized shear-thinning profile of cereal β-glucan solutions at certain concentrations (8 and 10% w/v) and temperature (20°C) fit well ($r^2 = 0.99$) to this equation regardless of differences in molecular size and fine structure of the β-glucan isolates.[18,21] The application of the Cox–Merz rule (i.e., the relationship between rheological responses under destructive and nondestructive deformation)[182] for cereal β-glucan preparations, differing in molecular size and polysaccharide concentration in solution, has also been tested. This rule is a further generality observed for disordered polymer chains interacting solely by physical entanglements.[180] Generally, no large departures from the empirical Cox–Merz correlation for cereal β-glucan samples are found, except in the case of high polysaccharide concentration and high molecular size samples.[17,18,40,43,176]

1.7.2 Gelation — Cryogelation

In addition to solution viscosity enhancement on storage, β-glucans have been shown to gel under certain conditions.[15–21,43,45,46,175,179] Cereal β-glucan hydrogels with different molecular characteristics and properties have been obtained under isothermal (5 to 45°C, 4 to 12% w/v polymer concentration) conditions[17,18,21,43,45] as well as after repeated freezing and thawing cycles of relatively dilute polysaccharide solutions (1 to 4% w/v)[19,20]; the later process is called cryogelation and the formed gels are cryogels. The β-glucan hydrogels have been examined by small-strain dynamic rheometry, differential scanning calorimetry (DSC), and large deformation mechanical tests. The gelling ability of β-glucans cured at temperatures above 0°C and the properties of the gels were found to depend on molecular size, fine structure (DP3/DP4 ratio), and concentration of the β-glucans, as well as on gel-curing temperature. Similarly, the cryogelation ability, the phenomenological appearance of the cryogels and their properties, and the yield of cryostructurates were influenced by the initial solution concentration, the number of freeze–thaw cycles, and the molecular features of the β-glucans. The obtained cereal β-glucan gels formed at temperatures above 0°C, and the cryogels belong to the category of physically cross-linked gels whose three-dimensional structure is stabilized mainly by multiple inter- and intrachain hydrogen bonds in the junction zones of the polymeric network. Gels prepared after freeze–thaw cycling at low initial polysaccharide concentrations, such as those usually found in frozen food formulations, are comparable with the gel networks prepared at room temperature and at much higher β-glucan concentrations.[19]

The gelation capacity of aqueous dispersions from cereal β-glucans differing in molecular size and fine structure has been monitored isothermally at different temperatures above 0°C and polymer concentrations; i.e., the time-dependent evolution of G', G", and tanδ was monitored periodically by applying a frequency of 1 Hz and a strain level of 0.1%, conditions in the linear viscoelastic region.[17,18,21,43,45] After an induction period, G' and G" increase with time and the aqueous dispersions begin to adopt gel-like properties (G' > G"). At the end of the gel-curing experiment, the behavior becomes typical of an elastic gel network and the G' attains a pseudo-plateau value, G'_{max} (Figure 1.3). The time where G' crosses G" is considered the gelation time, and the maximum

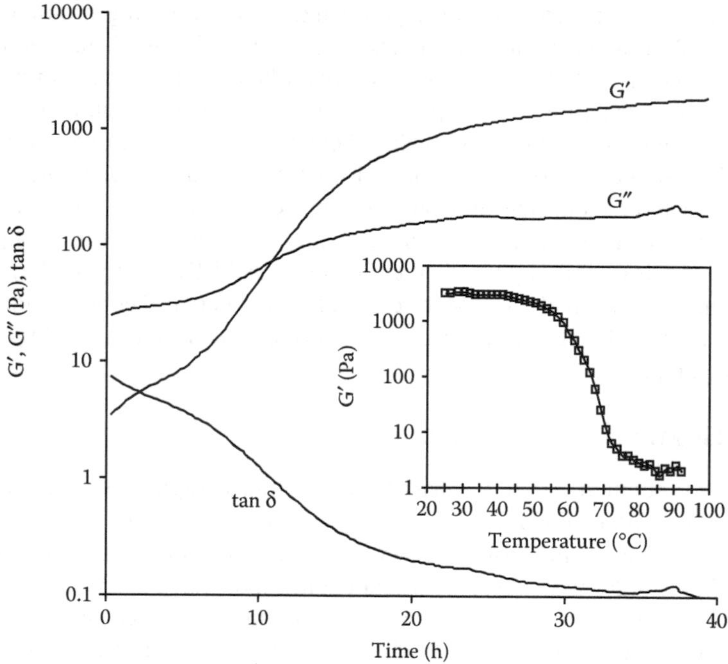

FIGURE 1.3 Gelation kinetics for a representative barley β-glucan (apparent molecular weight of the peak fraction calculated as 100 × 10³) preparation at 8% (w/v) (frequency, 1 Hz; strain, 0.1%; 25°C) and following melting profile (inset) of the formed gel (frequency, 1 Hz; strain, 0.1%; 3°C/min heating rate). (From Lazaridou, A. et al., *Food Hydrocolloids*, 18, 837, 2004.)

slope of the G' curve is reported as an index of the gelation rate; the latter, known as "elasticity increment, I_E", can be calculated as $I_E = (d\log G'/dt)_{max}$.[15] Melting profiles for the β-glucan gels following their formation can be also monitored by dynamic rheometry (Figure 1.3, inset); the melting point (Tm) taken as the temperature where G' becomes equal to G". All of the above parameters from gelation kinetics of different β-glucan samples are summarized in Table 1.6.

Two different gelation models have been proposed in the literature for mixed-linkage (1→3),(1→4)-β-glucans. One involves the side-by-side interactions of cellulose-like segments of more than three contiguous β-(1→4)-linked glucosyl units,[182] and the other the association of chain segments with consecutive cellotriosyl units linked by β-(1→3) bonds.[15] Among cereal β-glucans of equivalent molecular weight, the gelation time decreased and the gelation rate increased in the order of oat, barley, and wheat β-glucans, reflecting the order of the molar ratio DP3/DP4 units (Table 1.6).[15,16,18,46] In addition to the fine structure, the molecular size of the polymer seems to have a strong impact on polysaccharide gelation ability. For samples with a similar distribution of cellulose-like fragments, the gelation time decreases and the gelation rate increases with decreasing molecular weight, possibly due to the higher mobility of the shorter chains, which enhances diffusion and lateral interchain associations.[14,15,17,18,21,43,45,47] The gelation rate also increases with increasing concentration and gel-curing temperatures, reaching a maximum at ~25 to 35°C; at higher temperatures the I_E values decrease.[17,18,21,43] Moreover, incorporation of various sugars, at a concentration of 30% (w/v), to the barley β-glucans gels (6% w/v) led to an increase of the gelation time, following the order of control (without sugar) < glucose < fructose < sucrose < xylose < ribose.[45]

A decreasing molecular weight and an increasing DP3/DP4 ratio in the β-glucan chains yielded increased G'_{max} and decreased tanδ values of the gels (Table 1.6)[18,46]; i.e., with increased amounts of the cellotriosyl units there are longer and more junction zones giving denser cross-linked networks. The dependence of storage modulus on β-glucan concentration (C) followed power law relationships; G' varied as $C^{7.2-7.5}$. When the experimental data of G'_{max} vs. concentration for two representative oat β-glucan samples with apparent molecular weights of 35 and 110 × 10^3 were fitted into the exponential equation, estimates of 3.5 and 4.4% (w/w), respectively, for the critical gelling concentration, C_o, were obtained by extrapolation of this function; C_o is defined as the concentration below which no macroscopic gel is formed.[17]

The Tm values of gels seem to increase with increasing molecular size and amount of DP3 units in the polysaccharide; for cereal β-glucan gels cured at room temperature, the Tm varies within the narrow range of 65 to 72°C (Table 1.6). Although a slower gelation process is noted for the high molecular size β-glucans, the gel network structure consists of structural elements (microaggregates) with better organization, or it involves interchain associations over longer chain segments.[17,18,21] Curing of the gels at higher temperatures (45°C) than room temperature results in increased values for the gelation time and Tm and a decrease in gelation rate and G'_{max}.[17]

All β-glucan isolates from cereals were able to form cryogels at such low initial polysaccharide concentrations (even at 1% w/v), at which these polysaccharides could exist in frozen products. The mechanical spectrum of the solution before freezing was typical of a liquid-like system, but after repeated cryogenic cycles, a gradual transition occurred to the behavior of a weak gel and finally to that of a strong elastic gel (Figure 1.4). These changes were promoted by high initial solution concentration and amount of DP3 units in the polymeric chain, and by low β-glucan molecular size; i.e., it happened at a lower number of cycles with increasing concentration and DP3 segments and decreasing molecular size. Moreover, the G' values for cereal β-glucan cryogels, obtained from their mechanical spectra at 5°C, increase with increasing number of freeze–thaw cycles and the trisaccharide units (DP3), and with decreasing molecular size of the polysaccharide.[19] The concentration at which cryogelation took place was lower than c** of the examined β-glucan isolates[19] and lower than the critical gelling concentrations estimated for oat β-glucan preparations cured at room temperatures.[17] The major reason for the apparent shifting of critical gelling concentration toward lower values, compared to the same process at gel-curing temperatures above zero, is cryoconcentration. This phenomenon happens during freezing of the polymer solutions, which causes an increase of the β-glucan concentration in the still unfrozen regions of the system's bulk phase and results in the promotion of associative interactions among polysaccharide chains.[19] Cryogelation ability is also influenced by food formulation. The presence of sucrose (30% w/v) in a β-glucan dispersion, submitted to repeated cycles of freezing–thawing, induced a significant delay in the transition of β-glucan solutions to a gel state, and the resultant cryogels were weaker.[20]

The DSC thermal scans of cereal β-glucan hydrogels formed after storage at temperature above 0°C and, by cryogelation, exhibited rather broad endothermic gel

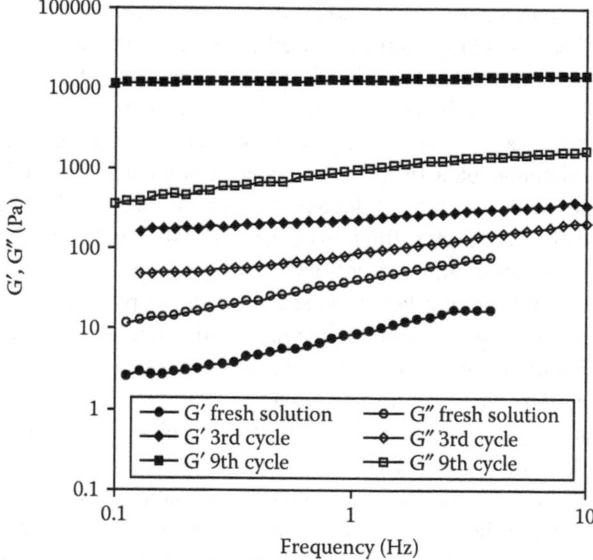

FIGURE 1.4 Mechanical spectra (0.1% strain, 5°C) of unfrozen fresh solution, and cryogels obtained after specified number of freeze–thaw cycles for wheat β-glucan (apparent molecular weight calculated as 200 × 10^3) at 3% w/v initial concentration. (From Lazaridou, A. and Biliaderis, C.G., *Food Hydrocolloids*, 18, 933, 2004. With permission.)

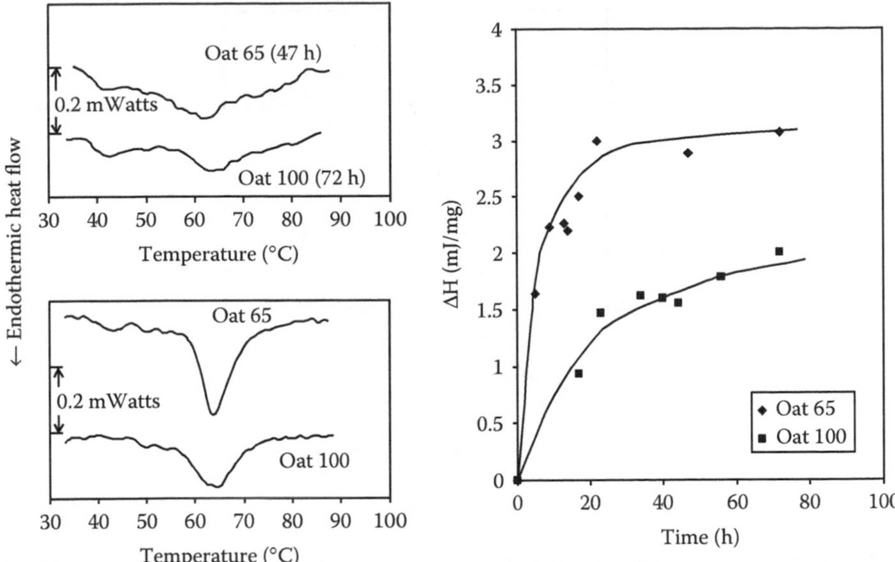

FIGURE 1.5 DSC thermal curves of oat β-glucan gels obtained from solutions of 10% w/v concentration cured at 25°C for specified time periods (left, top) and of oat β-glucan cryogels obtained from solutions of 3% w/v concentration submitted to nine freeze–thaw cycles (left, bottom), and time dependence of apparent melting enthalpy of the β-glucan gels cured at 25°C (concentration, 10% w/v) (right); heating rate = 5°C/min. Samples named as oat 65 and oat 100 had apparent molecular weights of 65 and 100 × 10^3, respectively. (From Lazaridou A. et al., *Food Hydrocolloids*, 17, 693, 2003; Lazaridou, A. and Biliaderis, C.G., *Food Hydrocolloids*, 18, 933, 2004.)

sol transitions at 55 to 80°C (Figure 1.5, left).[17–19,21,43] Similar to the rheological responses, DSC kinetic data showed that the rate of endotherm development during gel curing at room temperature (Figure 1.5, right),[17,18] as well as the apparent melting enthalpy values (plateau ΔH) of the gels, increases with decreasing molecular size[17,21] and with increasing the DP3/DP4 ratio (Table 1.6).[17,18] The ΔH values of the cryogels also increase with increasing the DP3/DP4 ratio and with decreasing molecular size of the cereal β-glucan.[20] Moreover, the melting temperature of all gel networks studied, as obtained by DSC, seem to increase with the molecular size and the amount of cellotriosyl units of the β-glucans.[17–19,21]

Variations in mechanical properties of cereal β-glucan gels have also been revealed by large deformation compression tests. For samples with similar molecular size, an increase in the DP3/DP4 ratio resulted in higher values of firmness and gel strength and a decrease in brittleness; compression modulus (E), true stress (σ_{TR}), and strain (ε_{TR}) obtained from the stress–Hencky strain curves (Figure 1.6) define the firmness, strength, and brittleness of the gel, respectively. An increase in strength and a decrease in brittleness of the β-glucan gels cured at room temperature were found with increasing concentration, molecular size, and DP3/DP4 ratio of the polysaccharide (Table 1.6).[17,18,21] Similar to the gels cured at room temperature,

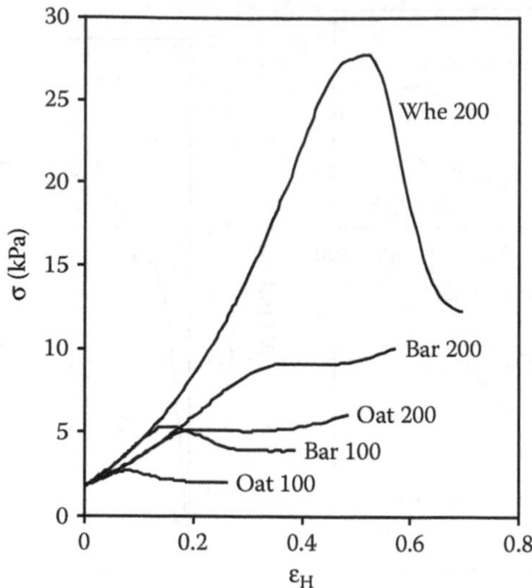

FIGURE 1.6 Compression stress (σ)–Hencky strain (ε_H) curves for gels of cereal β-glucans (concentration, 8% w/v; gel-curing temperature, 25°C). The samples are named oat (oat β-glucan), bar (barley β-glucan), and whe (wheat β-glucan), and their calculated apparent molecular weights were 100 and 200 × 10^3. (From Lazaridou, A. et al., *Food Hydrocolloids*, 18, 837, 2004. With permission.)

cryogels from high molecular size β-glucans seem to exhibit better organized and stronger structures than their low molecular size counterparts, implying that networks obtained from high molecular size β-glucans involve interchain associations over longer chain segments. Therefore, high molecular size β-glucans can form better organized and stronger gels than low molecular size samples, from the point of view of macrostructure. In contrast with the results obtained for gels formed at 25°C, the strength of cryogels submitted to large deformations was the highest for oat, lowest for wheat, and intermediate for barley β-glucans — a fact that was attributed to differences in the nature of the network microstructure among samples.[20] Microscopic images of cereal β-glucan gels aged 7 days at 5°C revealed that the microstructure was not homogenous and there was a coarsening of gel structure as the β-(1→3)-linked cellotriosyl unit content increased.[47]

Other factors having an impact on textural properties of cereal β-glucan gels are polysaccharide concentration, gel-curing temperature, and formulation. Generally, gel strength and firmness increase and brittleness decreases with concentration.[15,17,21,173] Moreover, gel structure formation at a higher temperature (45°C) than room temperature results in less firm, weaker, and less brittle gels.[17] Compression tests of mixed gels from barley β-glucan (6% w/v) and low molecular weight sugars (30% w/v) revealed that the mechanical properties of gels are affected by the type of added sugar. The strength of the gels, as expressed by the true stress, increased with the addition of glucose, fructose, and sucrose, but decreased with ribose. Moreover, the Young's modulus of most mixed gels increased with the addition of hexose sugars, such as glucose, fructose, and sucrose, but declined significantly with the pentose sugars, such as xylose and ribose.[45]

1.7.3 Processing Effects on Physical Properties

The physiological benefits and better technological performance of soluble, highly viscous fibers over their insoluble counterparts increased the interest in investigating processing factors that have an impact on solubility, extractability, and viscosity and molecular weight of β-glucan in isolates, concentrates, or cereal products and grains.[183]

Although some β-glucan preparations obtained from cereal grains by dry- or wet-milling procedures are products with high content of total β-glucans, they might not be the best source of soluble or high-viscosity polysaccharides. Jaskari et al.[144] found the amount of soluble β-glucans (at 37°C for 2 h) to be greater (34%) in a dry-milling preparation containing lower β-glucans than in a wet-milling product (24%) containing significantly more β-glucans. Heating increased the amount of soluble β-glucan, on average, from 29 to 84%. The molecular weight of β-glucans was 8.4×10^5 from oat bran obtained by dry milling and 6×10^5 from fiber-concentrated oat bran obtained by wet milling, and remained unchanged during hydrothermal treatment; however, the later preparations developed much higher viscosity than the former as a result of the hydrothermal treatment. Knuckles et al.[56] reported 59% β-glucan solubility for low-fiber oat brans and 55% for higher-fiber oat brans. Similarly, Malkki et al.[136] found the highest β-glucan extractability at 40°C in an untreated oat bran (90.8%), whereas for the preparations from this bran enriched in β-glucans by wet milling with neutral cold water, acidic cold water, hot-mixture ethanol–water, and cold-mixture ethanol–water, the respective values were 48.0, 76.7, 40.4, and 44.0%. The viscosity and molecular weight of β-glucans were reduced at acidic conditions under concentration and trypsin treatment. β-Glucan from the cold-water, wet-milled concentrate had the highest molecular weight and a higher hydrodynamic volume than β-glucan isolated from the other concentrates.[136]

Since it is postulated that solubility and viscosity of β-glucans are responsible for the biological activity of these polymers, several thermal, enzymic, and physical treatments of cereal grains were assessed as potential strategies for increasing the level and viscosity of β-glucans in cereal grains or extracts. Processing and storage, which involve structural breakdown and restructuring, may therefore increase their availability and solubility. Prolonged storage of baked corn bread at 20, 4, and –20°C appeared to increase extractability and measurability of β-glucans in corn bread; β-glucan content increased from ~0.35 to ~0.5 after 7 days of storage.[184] Freezing and frozen storage for 3 weeks of soup preparations containing β-glucans did not affect the molecular weight or concentration values of these polysaccharides. The effects of freezing on high and low molecular weight oat β-glucan preparations have been compared. Freezing decreased the viscosity of very high molecular weight β-glucan ($M_w = 1.7$ to 2×10^6) slightly, had no effect on the viscosity of intermediate molecular weight β-glucans ($M_w = 0.9$ to 1.4×10^6), and increased the viscosity of the low molecular weight products ($M_w = 60$ to 160×10^3). Molecular size and pH have a notable impact on viscosity of β-glucans during freezing and frozen storage.

Noticeable shear-thinning effects were also observed when β-glucan dispersions were first subjected to freeze–thaw cycling, implying formation of network structures, a process known as cryogelation,[19,20] as discussed in detail in the previous section. Overall, the thermal history and structural features of β-glucan preparations seem to significantly affect the rheological properties of these polysaccharides.

Physical treatments such as stirring and sonication were found to increase extractability of β-glucans from barley, but a reduction of the molecular size of the polysaccharide was observed. In alkaline solutions, both treatments also decreased the molecular weight of β-glucans. The decrease in molecular weight is related to the energy applied.[70] On the other hand, a study of flow behavior of barley β-glucans under acidic solutions (pH 2.0) after a high-shear treatment (24,000 rpm, 4 to 30 min at 25°C, at 3 and 1% w/v concentrations) showed no pronounced changes in viscosity compared with the control solutions, where no high-shear treatment was applied, implying a stable viscosity of β-glucans under high-shear homogenization.[185]

Response surface methodology has been used[20] for the expression of viscosity variations occurring during thermal processing of acidic media containing barley β-glucans of two different molecular weights (apparent peak molecular weight = 140 and 250 × 10^3). The pH range (~2.7 to 4.3) used in this study was typical of the pH usually met in acidic food products like fruit and tomato juices, while the temperature (~70 to 95°C) and time (~4 to 75 min) ranges reflected well the conditions under which the pasteurization process of such products occurs in most of the food industries. The viscosity reductions, as a consequence of acid hydrolysis, were dependent on pH, temperature, and time, factors that were well fitted in second-order polynomial equations. Acid hydrolysis effects were more pronounced for the high molecular weight sample, indicating potential differences in the flow behavior of liquid products containing β-glucans of different molecular size, during thermal processing of acidic products.[20] A similar experimental design was adopted to examine the effects of β-glucan concentration (1 to 6% w/w), sucrose (10 to 40% w/w), and salt (3 to 12% w/w) on the viscosity of the aqueous phase of a mayonnaise-type dispersion, in which β-glucans can be used as a stabilizer of the oil-in-water dispersion, containing a fixed vinegar concentration 2% (v/w). All the factors exerted strong influence on the steady shear viscosity (significant positive linear effects), while their interaction effects were negative, indicating that the viscosity increases were smaller when the factors were combined at high levels.[20]

Moisture and heat processing were found to influence β-glucan availability in sorghum grain. The β-glucan content from 0.12% in unprocessed sorghum increased up to 0.55, 1.14, or 3.05% by processing the grains for various periods with soaking at 37°C or autoclaving at either 130 or 120°C, respectively.[91] Zhang et al.[83] found that steaming of oat groats reduced the amount of β-glucans that could be extracted, compared with raw or roasted grain, but the extracts from steamed grain had much greater viscosity, despite the fact that the average relative molecular mass values of β-glucans among raw, roasted, and steamed oat samples were equivalent. These authors assumed that dry heat treatment (roasting) might increase intramolecular associations in the polymeric chain, whereas moist heat (steaming) might disrupt intramolecular cross-linking in native β-glucan, replacing intramolecular bonding with water binding, and therefore allowing a linear chain configuration to generate enhanced rheological properties. An increase of solubility and β-glucan content when barley was extruded has also been reported.[27,186] The physical properties of extruded preparations from barley meal and their extracts have been investigated and compared with an autoclaved product.[68] The amount of extracted β-glucan at 40°C was more influenced by the feed moisture than by the mass temperature during processing. Using a KCl/HCl solution (pH 1.5) for acidic extraction, less than 20% of the β-glucans present in barley meal and approximately 60% of the β-glucans from extruded and autoclaved products could be extracted. Extrusion was found to be connected with a partial depolymerization of β-glucans. Despite the decrease of molecular size of β-glucans during hydrothermal treatments, the macromolecular state of β-glucans was preserved during extrusion; the molecular weight of β-glucans extracted from barley meal was 160×10^3, whereas from the autoclaved product it was less than 100×10^3 and from extrudates, depending on the extrusion temperature, it was between 80 (170°C) and 125×10^3 (150°C). The rheological properties of the examined products were essentially affected by the amount and molecular weight of β-glucans. The highest acid extract viscosity (AEV) was found for the untreated barley meal. On the other hand, Izydorczyk et al.[70] reported that hydrothermal treatments, such as autoclaving and steaming of barley, had no significant effect on extractability of β-glucans, but prevented enzymic hydrolysis of β-glucans, and thereby substantially improved their molecular weight and the viscosity development of barley slurries. Therefore, hydrothermal treatments of barley grains were suggested as a potential pretreatment before incorporation of the grain into food systems. In contrast, roasting barley grains at 100°C did not result in any increase of viscosity.[70]

Enzymic rather than physical treatments proved to be more important in accounting for the increased extractability of β-glucans.[70,125,144] Enzymic treatments with a phytase preparation on oat bran showed an increase in β-glucan extractability, and simultaneously a partial degradation of the polysaccharide ($M_w = 600 \times 10^3$) to a product of $M_w = 40 \times 10^3$, which resulted in reduction of the viscosity of the oat bran slurry.[144] On the other hand, Izydorczyk et al.[70] found that the addition of enzymes (protease and esterase) in combination by hydrothermal treatments (autoclaving and steaming) on barley grain resulted not only in increased β-glucan solubilization, but also in improved molecular weight of β-glucans. Such treatments, therefore, might have a potential to positively affect the physiological responses to barley β-glucans in human diets.

Robertson et al.[125] found that barley β-glucan extractability increased to around 50% during cooking at 100 to 190°C, from about 30% in the raw barley flour. Distinct differences were observed between two cooking methods of oatmeal, gradually from room temperature or rapidly by addition to boiling water of rolled oats. The gradually cooked oatmeal had more solubilized β-glucan and gave a more viscous supernatant; the solubilized β-glucan was the major contributor to the viscosity. Microscopic examination revealed differences in the appearance of the two differently cooked samples; the disruption of the endospermic cell walls of the gradually cooked samples was greater than for the rapidly cooked samples.[187] In a latter study,[188] rolled-oat porridge was prepared by conventional cooking and microwave cooking to determine and compare the effects of these cooking methods on the texture, viscosity, and amount of solubilized β-glucans. The porridge prepared by microwave cooking was more grainy, less viscous, and with less solubilized β-glucans than conventional cooking. The difference was slight with short cooking duration (1 min) but increased drastically with prolonged cooking (20 min); on average, β-glucan content doubled when conventional cooking was extended to 20 min. Thicker oat flakes (regular-cooking rolled oats) released less solubilized β-glucan than did thinner oat flakes (quick-cooking rolled oats) upon cooking. Moreover, microscopic examination revealed that the cell walls of the rolled oats prepared by microwave cooking were less disrupted than those of conventional cooking. These differences were attributed to the effect of stirring; the microwave-heated rolled oats were stirred relatively less often than the conventionally cooked samples.[188]

The extractability and molecular weight of β-glucan in oat bran, oat bran muffins, and oat porridge, and the changes taking place during processing and storage, have been studied.[117,189] The original brans used, prior to mixing with other ingredients and cooking, showed about 25 to 30% β-glucan extractability and a M_w of 1 to 2 × 10^6. Cooking of oat bran muffins resulted in reduction of the M_w to 600 to 950,000, but increased the percentage of β-glucan solubilized about threefold, to 55 to 85%. During frozen storage, extractable β-glucans decreased by >50% in all muffins, but no change in the peak M_w of β-glucans was detected; the decline in solubility of the β-glucan possibly reflects changes in molecular organization during frozen storage.[117]

β-Glucans in ready-to-eat breakfast cereals (oat bran, multigrain flakes) were found to range in molecular weight from 600 to 2930 × 10^3.[116] Åman et al.[155] also reported large differences in the average molecular weight and the molecular weight distributions of various oat products, which are very likely to be of nutritional importance; oats, different oat fractions, and experimental and commercial oat-based foods were examined in this study. Oats, rolled oats, oat bran, and oat bran concentrates all had high average molecular weights (2060 to 2300 × 10^3 g/mol) and essential monomodal distributions. The M_w for the oat-containing experimental foods ranged from 450 to 1920 × 10^3, with extruded flakes, macaroni, and muffins having high molecular weights and pasteurized apple juice, fresh pasta, and tea cakes containing degraded β-glucans. Also, the average molecular weight was found to vary from 240 to 1670 × 10^3 in different types of oat bran-based breads baked with almost the same ingredients. The large particle size of the bran and short fermentation time limited the β-glucan degradation during baking and in yeast-leavened bread; the polymodal distribution of β-glucan in these breads implied the enzymatic nature of this degradation. Of the commercial oat foods studied, porridge made of rolled oats, a breakfast cereal product, and an extruded oat product showed high M_w (1930 to 2010 × 10^3); the crisp bread (950 × 10^3) and the yogurt-like product (830 × 10^3) contained medium degraded β-glucans; and bread loaf, fried pancakes, pancake batter, and fermented oat soup had highly degraded β-glucans (190 to 630 × 10^3). In general, processing such as baking, including a fermentation step, fresh pasta preparation, and production of fermented soup and pancake batter all seem to result in extensive degradation of oat β-glucan. It is evident from this study that heat treatment is relatively lenient for the β-glucan, while prolonged treatment at lower temperatures may result in extensive enzymatic degradation.[155] Another study[190] also reported a slight decrease in molecular weight of oat β-glucans (from 200×10^3 to 80 to 100×10^3), during production of cereal β-glucan beverages containing oat β-glucan concentrates, whereas production of beverages with a low molecular weight (40×10^3) barley β-glucan concentrate did not cause any further decrease in the molecular weight of β-glucans. On the other hand, the viscosity of the barley concentrate was higher than the viscosity of the oat concentrate as a result of the higher gelling capacity of the barley β-glucan.[190]

1.8 APPLICATIONS IN FOOD SYSTEMS

The acceptance of cereal β-glucans as a functional, bioactive ingredient over the last two decades has increased the interest in incorporation of cereal β-glucans in food formulations. The addition of flour, bran, or β-glucan-rich fractions from oat or barley into cereal-based foods, such as breakfast cereals, pasta, and baked goods (bread, muffins), which increased the amount of β-glucans in these products, has been investigated. Moreover, the use of cereal β-glucan concentrates or isolates as thickening agents, stabilizers, and fat mimetics to modify the texture and appearance in calorie-reduced, low-fat foods has been studied.

1.8.1 Cereal-Based Foods

Oatmeal, oat flakes (rolled oats), and oat bran have found a substantial number of applications in breakfast hot cereals, and oat flour is a frequently used ingredient in cold cereals and oatmeal in granola cereals and bars. The high fat of oat flour tends to restrict its utilization in highly expanded extruded cereal products because of oxidative rancidity problems.[191] On the other hand, barley is an excellent material for extrusion into ready-to-eat (RTE) cereal products.[27,186] Barley flour from four cultivars and a hulless waxy cultivar (Wanubet) was incorporated into extruded rice products.[27] Consumer sensory panelists rated cereals extruded from the four 50:50 barley–rice blends and the 65:35 Wanubet barley–rice blend higher than the 100% rice cereal for crispness and color, and rated them similar to the rice cereal for flavor and overall acceptability; these products contained >2% (w/w) β-glucans. Instead, cereals produced by 100% barley flour had limited expansion and high bulk densities.

In the past, oat and barley flours have not been extensively used in yeast-leavened bread, since they are less desirable than wheat flour due to the lack of gluten and their higher soluble fiber content leading to greater water retention.[186,191] In addition to undesirable effects on crumb texture and loaf volume, incorporation of whole oat or barley flour or fiber-rich fractions from cereals into bread makes the color of the end product darker. Bhatty,[22] studying the bread-making properties of hulless barley, suggested that 5% or possibly 10% barley flour could be added without seriously affecting loaf volume and bread appearance. However, the renewed interest in high-fiber foods has expanded the utilization of cereal fractions enriched in β-glucans. Several bakery products have been prepared using oat and barley high β-glucan fractions such as bran, and shorts or even water-extracted fractions. Delcour et al.[192] reported that addition of oat β-glucan increases the volume of gluten–starch loaves, most likely by increasing the viscosity of dough, while Wang et al.[193] claimed that β-glucan might play a role in improvement of bread crumb grain by stabilizing air cells in the dough and preventing their coalescence. Recent investigations on the contribution of β-glucans to the pasting characteristics of oat flours indicate the influence these polysaccharides have on such properties.[194,195] Significant correlations have been established between β-glucan concentration and the pasting parameters, such as pasting peak viscosity after amylolysis, which might be explained by an increase of water-binding capacity of oat flours with high β-glucan concentration.[194] However, research on experimental foods in which wheat flour was replaced by enriched β-glucan fractions revealed some difficulties in bread manufacturing. Krishnan et al.[196] substituted white wheat flour at 10 and 15% levels with commercial oat bran and found that the dough properties and bread quality were dependent on substitution levels and particle size reduction of bran; the water absorption requirements of oat bran doughs increased and the loaf volume decreased with increasing levels of bran and reduction of bran particle size. The 10% blends (large- and intermediate-size bran) produced doughs of good stability, and the 10% bran breads had better loaf volume, grain, and texture than 15% breads. A taste panel preferred the 15% bran breads made with large-size bran and 10% bran breads made with intermediate-size bran. Newman et al.[29] observed that incorporation at 28% of a fiber-rich fraction from waxy hulless barley cultivar in wheat flour depressed the loaf volume of yeast pan bread dramatically, although sensory analysis rated the fiber-enriched bread equal to a bread made from the same base wheat flour enriched with 50% whole wheat flour. The potential of fiber-rich fractions (β-glucan content, 8.1 to 15.2% total, 2.8 to 7.3% soluble) from hulless barley cultivars produced by a simple roller-milling procedure[72] as functional bread ingredients has recently been examined.[197,198] The added fiber-rich fraction was adjusted from 11 to 12%, depending on the barley genotype (waxy, high amylose, normal starch), to result in a constant addition of 2.5 g of β-glucans and arabinoxylans per 100 g of flour. A sponge-and-dough process using a high-protein-content wheat flour supplemented with xylanase resulted in breads with reasonable quality; bread made with the fiber-rich fraction from waxy barley exhibited the best volume and crumb structure, whereas bread with added fiber-rich fraction from the high-amylose genotype was the worst. Knuckles et al.[28] reported an increase in water absorption in bread dough, time to peak at farinograph, and mixing time by addition of β-glucan-enriched barley fractions. However, bread in which a dry-milled/sieved barley fraction containing 19% β-glucans replaced 20% of the standard flour was judged acceptable in laboratory acceptance tests, although the loaf volume was reduced and the color was slightly darker than the control. Moreover, for breads in which the wheat flour was substituted by 5% with a water-extracted barley fraction, containing 33% β-glucans, the color scores and acceptability improved over those breads made with the dry-milled/sieved barley fraction; the former and the latter breads contained 3.6 w/w (1.1% soluble) and 1.9% w/w (0.9% soluble) β-glucans, respectively. Similarly, Cavallero et al.[31] produced bread with a high amount of β-glucans by replacing the wheat flour with 50% whole-grain flour (4.6% β-glucan content), a 50% dried sieved fraction (8.5% β-glucan), or a 20% water-extracted fraction (33.2% β-glucan) from barley. The bread with the 20% water-extracted barley fraction, despite its highest β-glucan content (6.3% total, 5.7% soluble) among all the experimental breads, showed the best scores for sensory attributes (color, flavor/aroma, texture), even comparable to the control.

Muffins prepared using 100% barley flour exhibited equal or higher overall acceptability scores than the control wheat muffins, although volume and density were influenced by barley cultivar; the average muffin contained about 0.8 to 1 total and 0.4 to 0.5 g soluble β-glucans.[25] Hudson et al.[26] replaced the wheat flour in a traditional muffin recipe by oat bran (-glucan content = 9.3%) and a dry-milled/sieved high β-glucan barley fraction (-glucan content = 18.9%) at levels of 100 and 40%, respectively, and the resulting products were acceptable. These higher-fiber muffins contained more β-glucans (3.8 to 4.5%), moisture, protein, and minerals, and had 20 to 30% fewer calories than the commercial muffins. In a later study,[29] barley shorts obtained from roller milling of a waxy hulless cultivar and containing 18% dietary fiber were incorporated into baked products, such as muffins, biscuits, and sugar cookies, at 23 to 30% of the wheat flour, to double the soluble dietary fiber content without sacrificing product quality; the overall acceptability scores of the baked products prepared with the barley milling fraction were similar to those for the standard products.

Barley fractions, enriched in β-glucan, have been used in mixtures with semolina to produce pasta with acceptable sensory properties. Pastas in which a dry-milled/sieved barley fraction containing 19% β-glucans replaced the wheat semolina by 20% provided increased amounts of β-glucans (4.1 total and 2.1% w/w soluble); the fortified products had acceptable sensory quality scores, although they were darker in color than the control. Moreover, substitution of semolina by 20% with a water-extracted barley fraction containing 33% β-glucans yielded β-glucan-enriched pastas (7.1 total and 6.3% w/w soluble) with quality characteristics comparable to the control.[28] Functional pastas were also prepared by replacing 50% of durum wheat semolina with dry-milled/sieved barley pearling by-products enriched in β-glucans (9.1 to 10.5% db), adding 5% vital wheat gluten and adopting a high-temperature drying processing. Although the products were darker and less yellow than those of durum wheat pasta, the modified barley pastas had good cooking qualities with regard to stickiness, bulkiness, firmness, and total organic matter released in rinsing water, as well as a much higher β-glucan content (4.3 to 5.0% db).[30] Recently, roller-milling fractions prepared by Izydorczyk and coworkers[72] from hulless barley genotypes with variable amylose content were added to pasta[34] and noodles,[35,36] resulting in products with satisfactory textural attributes. When hulless barley flour replaced wheat semolina at 40%, color of pasta was adversely affected. However, when barley was pearled 15% before milling, the color of barley flour–semolina blends was reasonably acceptable. Replacing 40% of semolina with high-amylose or normal starch hulless barley flours made spaghetti slightly less firm, whereas replacement by zero amylose hulless barley flour made spaghetti substantially firm. Addition of the fiber-rich fraction derived from roller milling (10.3 to 18.3% β-glucans) did not impact spaghetti firmness. Color was unfavorably affected by incorporation of the fiber-rich barley fractions into pastas, but when a sufficient fraction from the high or zero amylose cultivar (11%) was added to increase the β-glucan content by 2%, the color was reasonably acceptable.[34] Yellow alkaline noodles containing a blend of wheat flour and the hulless barley flour (20 and 40%)

were prepared with minimal processing difficulties. The addition of any barley flour changed noodle appearance due to significant decreases in noodle brightness and yellowness, combined with increased redness and "speckiness." The addition of flour to a 40% level increased water absorption, cooking time, and cooking losses, as well as firmness. Moreover, the starch characteristics (amylose/amylopectin ratio) seemed to play an important role in altering the texture of noodles.[35] Replacement of 25% wheat flour to white salted and yellow alkaline noodles with the fiber-rich fraction obtained from roller milling posed no problem in noodle processing, although water absorption had to be substantially increased; however, the products were significantly darker and contained more brown specks than the wheat flour control noodles. The addition of the fiber-rich fraction improved the cooked yellow alkaline noodles' texture, as evidenced by increased firmness and chewiness. Overall, fresh yellow alkaline and white salted noodles enriched by fiber-rich barley fraction offer convenience due to short cooking time, improved nutritional quality, and acceptable cooking quality, particularly when intended for health-conscious consumers.[36]

1.8.2 Applications as Fat Mimetics, Stabilizers, and Thickening Agents

Many cereal fractions enriched in β-glucans or β-glucan concentrates with fat replacement potential have been developed and described in numerous studies and have also been patented. The development of a weak gel network structure by cereal β-glucans with certain structural characteristics and under certain conditions[17,18,21] is a desirable attribute in water-continuous, low-fat spreads. Moreover, cereal β-glucans show potential as fat replacers due to their highly viscous nature and water-binding, foaming, and emulsion-stabilizing capabilities.[199,200] Kontogiorgos et al.[200] investigated the effects of pure barley and oat β-glucans on rheological and creaming behavior of concentrated egg yolk-stabilized model emulsions (salad dressing model). The high molecular weight β-glucans (apparent $M_w \sim 110 \times 10^3$) stabilized the oil-in-water (o/w) emulsions mainly by increasing the viscosity of the continuous phase, while the low molecular weight β-glucans (apparent peak $M_w \sim 40 \times 10^3$) influenced emulsion stability through network formation in the continuous phase. The comparison of egg yolk with the Tween 20-stabilized emulsions demonstrated that the egg yolk acts synergistically with β-glucans, offering protection from creaming even at low oil volume fractions.

Oatrim, a product from enzymatically hydrolyzed oat flour or bran, containing oat β-glucans (1 to 10% db) and amylodextrins, has been proposed by Inglett[24,201–203] as a fat mimetic in a gel form substituting for shortening in oatmeal–raisin cookies. This product is being used experimentally in various reduced-fat, low-calorie formulations and soluble fiber-enriched foods, such as meats, muffins, cakes, oat fiber milk, frozen desserts, yogurt, sour cream, cheese spreads, salad dressings, sauces, gravies, soups, mayonnaise, margarine, snacks, breakfast cereals, and candy products. Jenking and Wild[204] proposed a cereal or grain enzymic hydrolysate, which could be used in combination with a hydrocolloid (carrageenan or blend xanthan/locust bean gum) as a fat mimetic in low-fat meat products (frankfurters, beef and pork sausage patties). The fortified products were characterized by organoleptic features comparable to the full fat product, good cooking yields, and the benefit of β-glucans; thermoirreversible gels useful for providing fat mimetic attributes can also be prepared according to this method. Inglett also described the production of dietary fiber zero-calorie gels (Z-Trim gels) from corn, wheat bran, oat hulls, and a variety of other insoluble fibers (soybean, rice, peas) by a multistage alkaline–high-shear process.[205] The combination of Z-Trim with other hydrocolloids, including Oatrim, provided different options of generating food textures in low-calorie snacks, hamburger and other meat products, cheese, and some baked foods. In a later study, by a heat-shearing treatment of oat and barley substrates, Inglett[206] produced hydrocolloidal formulations, which are rich in soluble dietary fiber, principally β-glucan (1 to 15%). Dispersions of these materials are smooth in texture and display the properties of a dairy cream, coconut cream, fat imitation, or substitute for shortening. Moreover, the incorporation of steamed or wet cooked kibbles or flakes with elevated levels of β-glucan obtained from pearled barley has been proposed in meat loaf as a fat mimetic and in sauces (gravies, tomato sauces) and soups as a thickener;[207] the addition of barley kibbles in burgers at concentrations from 5 to 20% gave a juicy and tasty patty with reduced animal fat content. Reduced-fat (12% w/w) breakfast sausages formulated with barley β-glucan gum (76.2% purity) were successfully made at a level of the gum that provides 0.3 to 0.7% β-glucan in the product.[32] It appeared that the main advantage of β-glucan gum is the improvement of the water binding within a meat system, without having any significant effect on product texture or flavor if added at a 0.3% level to the breakfast sausage formulations. Low-fat white-brined cheese has also been manufactured from bovine milk (70% fat reduction) containing two levels, 0.7 and 1.4% (w/w), of a commercial oat β-glucan concentrate with 22.2 and 20.7% β-glucan and amylodextrin content, respectively.[37] For the cheese made with the β-glucan concentrate the yield, the extent of proteolysis, and the levels of short-chain fatty acids (lactic, acetic, and butyric) increased compared to the low-fat control products. Large deformation mechanical testing and the sensory ratings showed an improvement in texture of the low-fat cheeses containing β-glucan concentrates, with no significant differences between a full fat control and the low-fat cheese with a 0.7% level of β-glucan concentrate. However, the color, flavor, and overall impression scores were significantly inferior to those of a typical white-brined cheese product, particularly for the product made with a high level of β-glucan concentrate (1.4%).

Efficient, selective, and economical methods have been patented[208] for producing lactose-free, nondairy, ready-to-use milk substitute (oat milk beverage, yogurt, ice cream, oat-based cream, whipped cream, and buttermilk) cereal dispersions having intact β-glucans, proteins, and natural sugars, as well as playing the role of stabilizer, while retaining the aroma and flavor of natural cereals. The dispersions prepared from synergistic enzymatic hydrolysis can be homogenized, subjected to UHT, and aseptically packaged or pasteurized and kept refrigerated until used or evaporated and subsequently spray dried to yield a stable powder.

Salovaara and Kurka[209] manufactured a snack food, yogurt-like, fermented product, based on oat bran (~5% v/v), containing living microorganisms (*Streptococcus thermophilus*, *Lactobacillus bulgaricus*, *Lactobacillus acidophilus*, *Lactobacillus casei*); this product was tasty and combined two health benefits, dietary fibers and probiotic microorganisms. Recently, it has been introduced into the market as a functional snack food from oat bran having a texture like yogurt, but totally free from milk or other animal products; this product can be associated with beneficial health effects because it is a good source of β-glucan (0.75 g per serving), low in fat (0.7 g/100 g), lactose-free, and cholesterol-free, as well as containing the probiotic bacteria *Bifidobacterium lactis* Bb12 and *L. acidophilus* LA5.[210]

In recent studies, the possibility of incorporation of cereal β-glucan preparations into beverage (fruit juice) and soup (shrimps and dill, mushrooms, lentil with ham) food prototypes, as well as the consumer acceptability of such products, has been examined.[33] The molecular weight of β-glucan had a significant effect on the sensory thickness of the beverage and soup samples containing β-glucan. From a technological point of view, the more processed β-glucan preparations are easier to be incorporated into a beverage and soup system in amounts sufficient for achieving a physiologically functional amount of β-glucan in a product.[33] However, the relationship between physiological functionality and molecular weight also has to be kept in mind. Incorporation of cereal fractions corresponding to 4 g of β-glucan in one serving of soup (400 g) or 5 g of β-glucan in 500 ml of beverage is feasible from a technological — as well as from a product acceptance — point of view.[190] Generally, food products with added β-glucans must be carefully designed, because for consumer acceptability, the health benefit perception of such products might not be enough to counterbalance a poor taste of the formulation.

Morgan[151] proposed the use of a low molecular size barley β-glucan isolate, obtained by water extraction without deactivation of the endogenous enzymes and a process of freezing–thawing, as a food additive in several applications. In addition to the incorporation of this β-glucan isolate into food formulations such as cakes, an Italian-style dressing (2% w/v) as a fat mimetic, and ice cream (5% w/v) for inhibition of ice crystal formation and keeping the smooth mouth feel, Morgan[151] has suggested its potential use as film-forming agent and a matrix for slow release (encapsulation). Skendi et al.[43] confirmed the film-forming ability of β-glucans and found the observed range of tensile strength values (20 to 80 MPa) for oat β-glucan films to be comparable to many medium-strength commercial synthetic films, making them potentially useful as a biodegradable edible food packaging material. These researchers found that the mechanical properties of the casted β-glucan films are affected from the amount of plasticizer (water, polyol) and the molecular size of the polysaccharide. Water as well as sorbitol, added as a co-plasticizer at a 15% (w/w) level, improved the extensibility, but decreased the mechanical strength of the β-glucan film. Moreover, all the mechanical parameters (tensile modulus and strength, percentage elongation) exhibited higher values for a high molecular weight sample than for a low molecular weight sample at certain moisture and sorbitol levels.

1.9 NUTRITIONAL IMPACT — HEALTH EFFECTS

In addition to the health benefits from the replacement of fat in various food formulations by preparations enriched in β-glucan, associated with lower fat and energy intakes,[211] cereal β-glucans display all the physiological properties that have been attributed to dietary fiber (DF). Recent epidemiological data indicate that diets rich in high DF from whole cereal grains are associated with lower risks of type 2 diabetes and obesity, as well as mortality and incident of cardiovascular diseases and ischemic stroke (atherosclerosis risk), through multiple metabolic pathways, such as decrease of plasma cholesterol, reduction in blood lipids and blood pressure, enhancement of insulin sensitivity, and improvement in blood glucose control.[67,212–215] Since a 1% reduction in serum cholesterol could reduce heart disease mortality by 2%, even modest reductions could have a dramatic effect when viewed in the context of the total hypercholesterolemic population.[216,217] Soluble fibers can lower serum cholesterol by 10 to 20%, theoretically reducing risk for coronary heart disease (CHD) by 20 to 40%.[218] A regular intake of whole-grain foods was found to be associated with a 26% reduction in risk for CHD.[67] Cereal whole grains enriched in soluble DF by improving glucose tolerance and reducing insulin exert additional health benefits for type 2 diabetic people, as well as delay or reduce the risk of developing this disorder.[212,215,219,220]

Various physiological functions of DF are attributed to several physicochemical properties of these nondigestible polymeric carbohydrates: water-holding capacity, swelling, diffusion-suppressing ability (viscosity, gel formation), binding properties, and ease or resistance to bacterial degradation and fermentation. Thereby, DF is regarded as acting on the modifications of nutrient absorption, enterohepatic sterol metabolism, regulation of intestinal microflora, and decomposition and fermentation in the intestinal tract, and further on the motility modulation of the digestive tract, gastrointestinal transit time, and bulk and weight of stool.[221] Soluble fiber increases intestinal transit time, delays gastric emptying, and slows glucose absorption. These and other actions lower postprandial blood glucose concentrations, decrease serum cholesterol, and produce other metabolic effects. Insoluble fiber decreases intestinal transit time, increases fecal bulk, delays glucose absorption, and slows starch hydrolysis; these effects improve the function of the large intestine and generally alter gastrointestinal function, but do not lower serum glucose or cholesterol.[218] Therefore, grains with high levels of soluble β-glucans, such as oats, rye, and barley, are generally more effective in improving serum cholesterol levels, blood glucose, and insulin responses than wheat, which contains predominantly insoluble dietary fiber.

1.9.1 Hypocholesterolemic Effect

Oat products show prominent hypercholesterolemic effects in humans and in experimental animals, since they selectively lower the atherogenic serum or low-density lipoprotein (LDL) cholesterol concentration, while raising the protective or antiatherogenic or high-density lipoprotein (HDL) cholesterol concentration, or at least raising the ratio of HDL:LDL cholesterol concentrations.[8,136,222,223] A serum cholesterol-reducing effect was achieved in rats receiving variously prepared oat bran concentrates at about 20% concentration of the feed, an amount corresponding to 3.3% β-glucan in the feed; the serum total cholesterol levels were lowered 10 to 30% relative to the initial cholesterol levels and the levels of the control group receiving cellulose in the diet.[136] Moreover, several animal studies showed that barley grain can also offer a high-fiber, cholesterol-lowering alternative to oats.[8,57,222–225] After a 40-day feeding trial in rats by barley flour containing 6.2%

total and 5.4% soluble β-glucans, significant reductions were observed in the levels of total cholesterol (39%), LDL cholesterol (61%), and triglycerides (21%), and a significant elevation in the level of HDL cholesterol (34%) in serum, compared with a control casein diet.[224]

The cholesterol-lowering activities of oats and barley are commonly attributed to the β-glucan fractions. Studies using purified oat β-glucan preparations strongly suggest that this is the most active component in oats,[226,227] although oat lipids and proteins,[227] as well as phytosterols,[228] tocotrienols,[223] and certain phenolics,[109] may also make some contribution to the cholesterol-lowering effects. Two different factors, soluble fibers and tocotrienols (components of barley oil), have been implicated in the cholesterol metabolism of different animals and humans fed barley-based diets with an additive effect on serum lipid parameters.[12,186,223,229–232] Kalra and Jood[224] demonstrated that the content of total and soluble β-glucans in barley flour used as feed appeared to be strong predictors of cholesterol lowering in the serum and liver of rats.

In 1997, the U.S. Food and Drug Administration (FDA) approved a health claim for the use of oat-based foods for lowering the risk of heart disease and passed a unique ruling that allowed oat bran to be registered as the first cholesterol-reducing food at a dosage of 3 g of β-glucan per day from 0.75 g of β-glucan per serving.[233] A similar claim for Oatrim has recently been allowed.[234] Most clinical studies on cereal β-glucan have been with intact cell wall material, as in flours and bran. In their meta-analysis, Ripsin et al.[235] identified 12 studies that had investigated the relationship between consumption of oats and blood total cholesterol in 1503 free-living subjects with an age range from 23 to 73 years, and mean cholesterol levels between 4.6 and 7.1 mmol/l. A statistically significant decrease of –0.13 mm/l, with an average daily dose of oat soluble fiber of 3.2 g, was calculated for the 12 trials. In a subsequent meta-analysis report, Brown et al.[236] identified 25 trials for a total of 1600 subjects, including normal healthy individuals, hyperlipidemics, and diabetics, within an age range of 26 to 61 years. The mean initial total, LDL, and HDL cholesterol values were 6.31, 4.40, and 1.28 mmol/l, respectively, and the amount of soluble fiber consumed daily ranged from 1.5 to 13.0 g. In the analysis of the practical dose range from 2 to 10 g/day of oat soluble fiber, there were significant reductions in blood total (–0.040 mmol/l per 1 g of fiber) and LDL (–0.037 mmol/l per 1 g of fiber) cholesterol. The triacylglycerols and HDL cholesterol were not significantly influenced. Bell et al.[217] summarized several clinical studies on cholesterol-lowering effects of various oat products, mostly oat bran, and reported that diets providing from 3.4 to 7.5 g oat β-glucans lower total serum cholesterol and LDL cholesterol in the ranges of 2 to 19% and 9 to 23%, respectively. Jenkins et al.[237] investigated the reduction of serum lipid risk factors for cardiovascular disease in hyperlipidemic adults consuming a test (high-fiber) or control diet for 1 month. The high-fiber diet included four servings/day of foods containing β-glucan that delivered 8 g/day more soluble fiber than did similar, unsupplemented foods in the control diet; test and control foods included breakfast cereals, breads, pasta meals, bakery products, crisps, and smoothies. The high-fiber diet compared with the control diet reduced total cholesterol by 2.1%, the total-to-HDL cholesterol ratio by 2.9%, and the LDL-to-HDL cholesterol ratio by 2.4%, whereas small reductions in blood pressure were also found after the high-fiber diet. These researchers applied the Framingham cardiovascular disease risk equation to the data and confirmed a reduction in risk of 4.2%, which is likely to be significant on a population basis.

The relatively large amount of oat cereal that must be consumed to achieve an intake of 3 g of β-glucan led to the development of oat and barley β-glucan concentrates. Incorporation of an oat gum containing 66% β-glucan into rat diets at a 5% level showed reduction of serum and liver cholesterol and an increase in HDL cholesterol.[238] An extract of β-glucan from waxy hulless barley, containing 56% β-glucans, was incorporated into flour tortillas to provide 2 g of soluble fiber as β-glucan per serving. These flour tortillas were incorporated into rat diets and resulted in a reduction of plasma LDL cholesterol, although total cholesterol and triglycerides did not differ.[239] Delaney et al.[13] formulated the experimental diet of hamsters to include β-glucan (2, 4, or 8 g/100 g) by addition of β-glucan concentrates prepared from oats and barley by water extraction and several purification steps and containing 65 and 78% β-glucans, respectively. For diets with a higher concentration of β-glucan than 4 g/100 g, dose-dependent decreases in plasma total and LDL cholesterol concentrations were observed in hamsters fed with the β-glucan compared with control hamsters; in this study, the cholesterol-lowering potency of β-glucans in hypercholesterolemic hamsters was approximately identical whether its origin was oat or barley. However, decreased HDL cholesterol concentrations were also observed in hamsters consuming high concentrations of β-glucan (8 g/100 g) from oats and barley.[13,228]

Debate still exists as to the mechanisms by which a soluble fiber such as β-glucan exerts its hypolipidemic effect. Several authors reviewing this subject[12,216–218,226,227,240] reported the following possible mechanisms: modification of bile acid (BA) absorption and metabolism, interference with lipid absorption and metabolism, production of short-chain fatty acids (SCFAs) from fiber fermentation in the colon, alterations in concentration of or sensitivity to insulin and other hormones, and indirect effect of replacement of dietary saturated fat and cholesterol by soluble fibers. The lowering cholesterol effect is believed to occur through a combination of these mechanisms rather than by a single mechanism.

It has been suggested that cereal β-glucans decrease absorption and readsorption of cholesterol, bile acids, and their metabolites by inducing high viscosity of the gastrointestinal tract contents, which reduces the diffusion rate of nutrients. The increase of the intestinal viscosity causes the digesta to hold on to extra water, which slows its movement. It has been demonstrated that β-glucanase supplementation in animal-fed diets increased lipids compared to diets without enzyme supplementation.[225] Moreover, higher fecal excretion of neutral sterols and cholesterol[13,239,241] or decreases in liver cholesterol[222,224] have been demonstrated in trials with animals fed enriched β-glucan products, suggesting that β-glucans interfere with fat absorption. In rats fed with diets containing β-glucans from various oat bran concentrates, a dependence of the hypocholesterolemic action on extractability, viscosity, pseudoplastic flow behavior, hydrodynamic properties, and molecular weight of β-glucans was found, supporting the hypothesis that the mechanism of action is largely based on the high viscosity induced by β-glucans.[136]

Moreover, fibers bind or trap bile acids in the digestive tract, which results in prevention of their absorption or reabsorption and thereby a reduction of the bile acid pool circulating back to the liver. These actions stimulate production of more bile acids derived from cholesterol that is either made endogenously or captured from the circulation. It has been postulated that consumption of β-glucans increases the bile acid synthesis and excretion in the feces, as well as excretion in ileostomy subjects.[240–245] Bowles et al.[131] showed by solid-state ^{13}C-NMR spectroscopy that there is no direct binding between bile acid salt molecules and specific sites on the β-glucan polymer, supporting the proposition that the ability of β-glucan to inhibit readsorption of bile acids is a function of its high viscosity in aqueous solutions (possibly gel network formation in the intestinal environment), rather than any specific binding or complexation. However, these researchers noted that it is possible that binding might occur between β-glucan and micelles from bile and fatty acids, rather than the isolated bile acid salts alone. On the other hand, Huth et al.[68] found that barley meal, as well as autoclaved and extruded products from barley, interacted with bile acids at pH values of 5.0 to 6.5, similar to those present in the small intestine, but there was no direct effect of the molecular weight of the β-glucan present in these products with the strength of the interactions, suggesting that viscosity in media might not play the dominant role in the binding mechanism of bile acids.

The fermentation of soluble fibers in the large intestine by colonic bacteria leads to the production of short-chain fatty acids (SCFAs), acetate, propionate, and butyrate, which inhibit the hepatic cholesterol synthesis by limiting the action of 3-hydroxy-3-methylglutaryl (HMG)–coenzyme A (CoA) reductase, the rate-limiting enzyme required for cholesterol biosynthesis. In rats, diets enriched in cereal β-glucan increased the SCFAs in the colon and feces.[241] During *in vitro* fermentation of pancreatin-digested barley meal or barley extruded and autoclaved products with human feces flora, higher amounts of SCFA were found for the hydrothermally treated substrates; in this study, the solubility of β-glucan was increased by the hydrothermal treatments.[68] A decrease of HMG-CoA reductase activity and an increase in cholesterol 7 -hydroxylase activity, an enzyme that breaks cholesterol in the synthesis of bile acids, were correlated with the presence of β-glucan in the chicken diet.[223]

Hormone secretion from the gut and pancreas is altered by fiber intake, resulting in improved insulin sensitivity and glucose tolerance. After fiber-supplemented meals, insulin secretion is significantly lower than after low-fiber meals providing the same quantities of nutrients. Soluble fibers delay gastric emptying and intestinal absorption of nutrients by developing high viscosity in the intestinal contents, thereby reducing postprandial hypoglycemia and insulin secretion. Carbohydrate and lipid metabolism are closely interrelated. Insulin has been reported to increase hepatic cholesterol synthesis; therefore, if fibers decrease carbohydrate absorption and insulin secretion, they may indirectly contribute to the hypocholesterolemic effects.[216,217,227,246,247]

1.9.2 Hypoglycemic Effect

The hypoglycemic effect is measured as a glycemic index (GI), which is the area under the curve of the glucose responses to a carbohydrate-containing food compared to either a specific glucose dose or a specific intake of white bread.[248] Clinical studies with diets containing foods enriched in oat and barley β-glucans revealed a reduction of GI and GII (insulinemic response); these studies included nondiabetic and type 2 diabetic subjects, as well as moderately hypercholesterolemic and overweight individuals.[31,109,190,237,246,249–252] Human studies in healthy subjects fed 1.8 to 14.5 g of purified oat gum (80% β-glucan) in 500 ml of water revealed a significant reduction in blood glucose and insulin rise relative to the gum-free control trial.[9,11,167,253] Purified oat gum added to a meal to give a β-glucan dose similar to that of an oat bran meal gave postprandial glucose and insulin responses similar to those from oat bran in both nondiabetic and type 2 diabetic subjects.[109] Cavallero et al.[31] fed nondiabetic individuals four types of breads: one was from 100% wheat flour and the rest were enriched in barley β-glucans by replacing the wheat flour with 50% whole-grain flour or 50% dried sieved fraction or 20% water-extracted fraction from barley. All breads contained the same amount (50 g) of available carbohydrates. The barley fraction-fortified breads exhibited reduced glycemic indices compared to the plain wheat flour bread; the 20% water-extracted fraction bread had a 28% lower GI than the control. This study confirms the effectiveness of cereal β-glucans in reducing blood glucose levels, even in foods with a high GI, such as bread.

It is likely that the mechanism by which β-glucans decrease the postprandial glucose response is the result of not only high viscosity in the gastrointestinal track, but also the reduction of starch digestion by α-amylase. Izydorczyk et al.[36] measured the *in vitro* digestibility of starch in noodles and showed that replacement of 25% wheat flour to noodles with a fiber-rich fraction from barley decreases significantly the glucose release compared to the plain wheat noodles. These researchers speculated that the nonstarch polysaccharides of the fiber-rich fraction, including β-glucans, decrease the accessibility of starch-degrading enzymes to their starch substrate by encapsulation of the starch granules or by interaction between the fiber components and starch granules or by changing the water distribution within the dough matrix, and thus affecting the structure of the protein–starch matrix. Micrographs showed that the noodle matrix surrounding the fiber inclusions in fiber-rich fraction-supplemented noodles, both before and after cooking, was more compact than the relatively porous matrix of the control noodles.

1.9.3 Determinants of Hypocholesterolemic and Hypoglycemic Effects

The variable or conflicting data occurring in various studies on β-glucan physiological effects may be related to the subjects, the design of the trials, the daily dose of ingested fiber and number of servings, the vector used, and, moreover, the variability in the physical and structural properties of the polysaccharide preparations (isolates, concentrates, etc.).

The property of soluble fibers to lower the glucose and insulin response is most likely to be found in subjects for whom lowering glucose and insulin is an improvement, that is, hypercholesterolemic, older, less slim, and noninsulin-dependent diabetic subjects. Effects are less significant if subjects are young, fit, and have normal glucose and insulin responses.[219,220,254] It seems that higher initial blood cholesterol levels in subjects would likely produce more significant results in lowering the serum cholesterol concentrations.[217] Inclusion of oat bran in hypercholesterolemic men's diets seems to reduce the serum cholesterol and LDL cholesterol concentrations by 5.4 to 19% and 8.5 to 26%, respectively; the magnitude of the reduction also depends on the dose.[241,244,255,256]

Furthermore, oat bran may have an additional beneficial effect in hypercholesterolemic individuals by reducing plasma triglycerides.[256] In a clinical study with normal (cholesterol < 200 mg/dl) and hypercholesterolemic men, the subjects were fed an amount of cookies (100 g) equivalent to 2.6 g/day of soluble fiber from oat bran, and a 28% reduction of plasma triglycerides was observed after 8 weeks, but only for the hypercholesterolemic individuals. Furthermore, incorporation of 5 g of β-glucan into different foods (such as breakfast cereals, bread, tea cakes, muffins, pasta, macaroni, apple beverages) as an oat bran concentrate can reduce serum total and LDL cholesterol levels by 6 and 9%, respectively, in hypercholesterolemic individuals when consumed daily for 3 weeks.[257] Using a purified oat gum preparation (80% β-glucan) in a diet with a daily dose of 5.8 g of oat β-glucans, split into two servings of 2.9 g and given to hypercholesterolemic subjects either in soft drinks or dissolved in water, resulted in a significant reduction of serum total and low-density lipoprotein (LDL cholesterol) by 9% compared to the initial values, without affecting HDL cholesterol, while the triglyceride levels remained relatively unchanged.[10]

Furthermore, mildly hypercholesterolemic individuals can reduce total and LDL cholesterol levels with incorporation of oat bran in their diet;[258,259] consumption of orange juice containing β-glucans from oat bran (daily intake, 5 g of β-glucan) can decrease LDL cholesterol by 6.7% and the ratio of total to HDL cholesterol by 5.4% compared with the control, without changing HDL cholesterol and triacylglycerol concentrations.[259] A recent study[260] on children and adolescents with mild to moderate hypercholesterolemia showed that consumption of a ready-to-eat cereal products in two servings that provided 3 g/day β-glucan, in combination with a low-saturated fat and -cholesterol diet, produced a modest, but clinically relevant, reduction of 5.3% in LDL cholesterol. The response appeared to be most pronounced (9.2% reduction) among subjects with a body mass index below the median value of 25.7 kg/m^2. Oat bran added to a low-fat diet of moderately hypercholesterolemic subjects (5.92 to 8.02 mmol/l) significantly decreased serum cholesterol and LDL cholesterol by 13 and 17.1%, respectively, and improved the lipid ratio (HDL:LDL) by 78%, without having any affect on triglycerides and HDL cholesterol.[261] However, in a clinical study on mild to moderate hyperlipidemic individuals, the proposed FDA dosage of β-glucan, provided by 20 g of oat bran concentrate, did not significantly reduce total or LDL cholesterol.[262] Romero et al.[256] indicated that oat bran is efficacious in lowering plasma LDL cholesterol and improving the LDL/HDL ratio in normal subjects and in those with lipid disorders when an effort is made to reduce the amount of dietary fat. This can also explain the fact that a diet enriched with 10 g of total dietary fiber from oat bran was found to reduce postprandial lipemia in healthy (normolipidemic) adults,[263] although incorporation of β-glucans in the diet can usually do very little to further lower a cholesterol level already in the normal range.[254]

The dose dependence of postprandial insulinemia reduction was revealed in an early study, in which oat gum was incorporated into the diet of rats;[238] however, in this study there was no reduction in glycemia, a fact that was attributed to the small meal size (0.75 g/100 g of body weight). In a later study,[11] the dose–response of oat gum on plasma glucose and insulin levels of healthy humans consuming oat gum in the range 1.8 to 14.5 g after an oral glucose load of 50 g was investigated. By increasing the dose of oat gum, the plasma glucose and insulin responses were successively reduced relative to a control without gum reaching a plateau dose–response at an intake of ~ 6 g of β-glucan (maximum effective dose). It is also worthy to note that the dose–response for a solid meal might differ from that for a liquid.[11] In a recent study by Cavallero et al.,[31] a linear decrease of glycemic index of nondiabetic humans was found with increasing barley β-glucan content in bread consumed by these subjects; the GI was related to the total β-glucan content (TBG) by the equation

$$GI = 91.27 - 3.68\ TBG \quad (r^2 = 0.96) \tag{1.3}$$

In the case of lowering blood serum cholesterol levels by β-glucans, the dose–response relations are not well established. However, in hypercholesterolemic rats after treatment diets containing 0 to 10%, by weight, dietary fiber from a high-fiber oat flour, linear inverse relationships were found between the amount of the oat flour intake and the fasting serum and liver cholesterol and triglyceride concentrations.[227]

In some cases, the lack of physiological action of β-glucan preparations was attributed to the reduction of molecular size of the polysaccharide occuring during the isolation stage.[190,264] Diets with 4.2 to 4.3% β-glucan from Oatrim, having a M_w of 0.39 to 0.477 × 10^6, that was fed to hamsters reduced serum LDL cholesterol up to 67% compared to a control.[228] However, in a human study on mildly hypercholesterolemic men, Keogh et al.[264] found no significant effects of a barley-enriched β-glucan fraction on factors used as markers for cardiovascular and diabetes risks (total, LDL, and HDL cholesterol, triacylglycerol, fasting and postprandial glucose), although the β-glucan supplements were incorporated into several cooked and processed foods (bread, waffles, muffins, spaghetti Bolognese, chicken curry, cakes, cookies) in a daily intake of 8.1 to 11.9 g of β-glucan, much higher than the efficacious dose of 3 g/day recommended by the FDA for reduction of cardiovascular disease risk. The authors attributed such findings to the possible reduction of molecular size or extractability of β-glucans of the β-glucan preparation (Glucagel) as a result of the isolation method used (extraction without inactivation of endogenous β-glucanases) or from the freezing storage or baking of the product during the intervention period.

In a recent human study,[190] diets with beverages enriched with oat β-glucan preparation (daily β-glucan intake, 5 g) showed reduction of serum concentrations of total and LDL cholesterol in a healthy population by 4.8 and 7.7%, respectively, as well as a significant decrease of postprandial glucose and insulin levels. However, a beverage enriched with 5 g of β-glucans from a barley preparation did not affect the blood lipid levels or postprandial glucose and insulin levels significantly compared with a control beverage, probably because the latter preparation contained β-glucan with a lower molecular size (40×10^3) than that from the oat preparation (80 to 100×10^3).

The viscosity development of intestinal contents after consuming oat or barley products depends on both the amount and molecular size of the solubilized β-glucans. In human studies, after glucose solutions containing oat gum are consumed, an inverse linear relationship between log (viscosity) and postprandial glucose and insulin response was found in the dose range examined (1.3 to 10.5 g of pure β-glucan).[11] This relationship showed that 79 to 96% of the changes in plasma glucose and insulin are attributable to viscosity, and that changes occur at relatively low doses and viscosities; reduction of the meal viscosity by partial acid hydrolysis of oat gum reduced or eliminated the capacity to decrease postprandial glucose and insulin levels. These results suggest that reduction in postprandial glucose and insulin levels is dependent on both the molecular weight of the β-glucan and the dose given. Similarly, Malkki et al.[136] found no convincing evidence that the molecular size of β-glucan alone can predict the cholesterol-lowering potency of β-glucan. The hypocholesterolemic property of barley seems to be related to factors such as barley and starch type, total content and degree of solubility of the β-glucans, extract viscosity, and environmental growing conditions.[229] In several animal studies on diets supplemented with cereal fiber or cereal fractions, it was demonstrated that the content of β-glucans, especially soluble β-glucans, molecular weight of β-glucans, and endogenous β-glucanase activity, which are factors affecting extract viscosity, appeared to be strong predictors of the serum cholesterol response.[57,186,222] Wood et al.[11] reported that the apparent viscosity range of aqueous oat gum solutions consumed by healthy humans that improved the postprandial glucose and insulin responses was within the range of 20 to 8000 mPa·sec (at 30 sec^{-1}), with little additional effect achieved above 5000 mPa·sec. The regression analysis of their experimental data suggested that even with viscosity as low as 10 mPa·sec, an average 12 to 13% reduction in peak plasma glucose could be observed. Additionally, Wood[189] found a relationship between the change in peak blood glucose levels (G) and polysaccharide concentration (c), and the weight-average molecular weight (M_w) of the β-glucan:

$$G = 7.93 - 0.68 \log_{10}(c) - 1.01 \log_{10}(M_w) \quad (r^2 = 0.88) \tag{1.4}$$

As already discussed in detail above, food processing can change the solubility and, therefore, the availability of β-glucan, as well as its molecular size. Overall, in most of the studies, it seems that food processing improves the physiological activity of β-glucans by increasing availability (cooking, extrusion) and that polymer molecular size, although it might be reduced (milling, stirring, pumping), is still effective in plasma cholesterol, glucose, and insulin response improvements.[225,227,228,251] Animal studies demonstrated that β-glucans that have been reduced in molecular mass by an order of magnitude are still able to reduce plasma cholesterol levels.[265] Wood[189] has also noted that the net result of food processing could improve the bioavailability or bioactivity of the β-glucan that is related to the product of (C × M_w).

However, the viscosity in the gastrointestinal tract is the most critical for the physiological effects of β-glucans. It is important to take into account the modifications of the fiber matrix during the transit small intestine when assessing the dietary response to a fiber source. There is also some evidence for changes in solubility and molecular size of β-glucan during digestion. Beer et al.[117] examined oat bran, oat bran muffins, and oat porridge to determine changes in the β-glucans that might affect the development of viscosity in the gut. The β-glucan was extracted using hot water and a thermostable α-amylase and by an *in vitro* system including a sequence of amylase and pepsin to crudely mimic the human digestion process. Hot-water extraction yielded 50 to 70% of the total β-glucan present in oat bran and rolled oats, with an apparent M_w in the range of 1.4 to 1.8 × 10^6; the *in vitro* enzyme digestion system resulted in solubilization of 12 to 33% of the total β-glucan with similar M_w. In muffins, 30 to 85% of total β-glucan was solubilized by *in vitro* digestion, with a major difference in extractability among muffins from different recipes; this increase was thought to be caused by degradation of β-glucans during cooking, as revealed by the peak M_w, which was lower in all muffins than in the original bran. Robertson et al.[125] showed that endogenous proteases from the small intestine were capable of enhancing the extractability of the barley β-glucan measured *in vitro* to a level similar to that found in ilea effluent from patients fed an acute barley-based diet. The extractability measured *in vivo* was significantly higher than that measured in the original barley, but without protease treatment. Increased extractability can be accompanied by a reduction in M_w of the β-glucan in the upper gut, 10% of the total β-glucan recovered being of $M_w < 12{,}000$. Similarly, Sundberg et al.[266] observed some depolymerization of β-glucans in ileostomates of consumed breads of different cereals. Thus, unextractable β-glucans have the potential to behave as a source of soluble fiber because of the modification during transit in the upper gut.[125] The molecular size of β-glucan from the stomach and intestines of hamsters was 100,000 daltons, even lower than the reduced size of the fiber from processing treatments.[228,265] Wood and coworkers have also demonstrated a significant depolymerization of oat β-glucan during digestion in the small intestine of rats and chicks.[5,167] The maximum β-glucan concentration was found in the ileal section, where most of the starch was absorbed. The molecular weight of oat β-glucan extracted from the intestinal content was lower relative to the diet, whereas there was a further decrease in cecum and feces.[116] Dongowski et al.[241] demonstrated that β-glucan was highly fermentable in the rat cecum, while β-glucan was not found in feces. In a recent study,[190] the concentration of β-glucans in the ileostomy contents from ileostomic patients after consuming meals containing β-glucans first increased and then decreased as the β-glucan passed forward in the gut. No decrease (four subjects) or a slight decrease (two subjects) in molecular weight of β-glucan was observed during the transit in the gut.

It is also noteworthy that a real food matrix is very different from a mixture of fully hydrated gums; i.e., the cellular structure and interaction with other components will influence the physiological response of the polysaccharides. In a clinical study,[259] the enriched β-glucan preparation from oat bran consumed with orange juice was more effective in lowering total and LDL cholesterol concentrations and the ratio of total to HDL cholesterol than was the same preparation administered in bread and cookies; these differences in efficacy were attributed to the food matrix or food processing. In a recent study,[190] β-glucan-enriched food prototypes gave different results on the lipid and glucose metabolism in humans. Beverages enriched with β-glucans had significant cholesterol-lowering effects in hyperlipidemic subjects, but a β-glucan-enriched ready meal (soup), providing 4 g of β-glucan daily, given to hyperlipidemic or diabetic subjects did not influence the lipid or glucose metabolism. The oat β-glucans used in the beverage or the ready meal had similar molecular weights. Explanations for the lack of effect of the β-glucan ready meal were attributed to the nature of the matrix, which may not be appropriate, and the single dose, which may be insufficient to induce a permanent effect. The amount of β-glucans used might also have been too low to exert a significant effect due to the difficulty to raise the β-glucan content in a single meal and maintain its palatability.

In general, the cereal β-glucans incorporated into various foods seem to be an acceptable and palatable ingredient at doses that exhibit physiological responses.[211,237,250] However, consumption of diets high in fiber is perceived to result in undesirable gastrointestinal symptoms, such as constipation, diarrhea, and flatulence. The amounts of β-glucan that must be consumed to achieve statistically and clinically significant physiological benefits are large, and its consumption at one time might produce gastrointestinal side effects in many individuals. However, a few subjects in clinical studies using β-glucan-enriched products have been reported to have mild to moderate gastrointestinal symptoms.[211,217,237] Moreover, in a recent study,[268] the toxicity of β-glucan-enriched soluble fiber from barley in rats was evaluated on dietary administration at concentrations of 0.7, 3.5, and 7.0% β-glucan for 28 days. The highest used concentration of β-glucan corresponds with an overall intake of 5.6 g of β-glucan/kg of body weight/day, which for a 60-kg adult corresponds to 336 g of β-glucan/day. The results of this study demonstrated that consumption of concentrated barley β-glucan was not associated with any obvious signs of toxicity in rats, even following consumption of those large quantities.

The potency of soluble fibers to slow intestinal transit and increase transit time has also been attributed to the gel formation properties of these materials. The formed gel network could act like a molecular sieve in the small intestine. Large molecules could pass rapidly through the system, but smaller molecules would be trapped in the various pores for variable lengths of time. This fact may slow down the digestion and absorption of various nutrients by decreasing the interaction between nutrients and digestive enzymes.[226] However, the possible relations between the gelling properties of cereal β-glucans and their physiological functions have not been examined yet. As already discussed in previous sections, in addition to molecular size, other structural features, such as the ratios of β-(1→4)/β-(1→3) linkages, the presence and amount of long cellulose-like fragments, and the ratios of cellotriosyl/cellotetraosyl units, can influence interchain aggregation phenomena, and thereby the gelling ability of cereal β-glucans, as well as the physical properties of the obtained gels. There are no reports in the literature investigating the possible relations between those structural features, and therefore the gelling behavior of β-glucans and their physiological effects.

1.9.4 Other Physiological Effects

In many animal studies,[109,223–225,239] diets enriched in β-glucan from oats or barley showed a retardation of growth and a reduction in feed intake and weight gain, whereas in clinical studies,[254] there was no evidence for changes in weight body after a diet enriched in cereal β-glucan. However, some human studies[226,242] indicated that oat bran reduced intestinal transit time and increased fecal weights, which could improve laxation.

It has been suggested that diets high in dietary fiber, such as cereal β-glucans, may have a protective effect against colon cancer.[8,12,226,240,241,267] High levels of SCFAs derived from fermentation of undigested β-glucan in the lower part of the intestinal track by intestinal microflora, particularly of butyrate, are important for a healthy development of mucosa.[8] In a recent study it was shown that diets in humans containing oats, barley, and their extracts increase breath hydrogen and methane production, parameters that are considered markers of colonic fermentation.[267] It has been postulated *in vivo* that SCFA levels in the cecal, colon, and feces of rats increase after consumption of enriched barley β-glucan diets. Moreover, the proportions of secondary BA were lower in feces of rats fed the barley-containing diets than in the controls. Secondary BAs, particularly lithocholic acid (LCA) and deoxycholic acid (DCA), are considered to be promoting factors in the pathogenesis of colon cancer. Furthermore, unfermented, swollen dietary fibers dilute the BA in the colon. Butyrate and DCA appear to interact in a complex and antagonistic manner to selectively modulate crypt base and surface proliferation in the rat colon.[241] The undigested insoluble dietary fibers from cereals increase fecal weight and speed intestinal transit, which decrease the opportunity for both nutrients and fecal mutagens to interact with the intestinal epithelium.[240] A recent study[269] on mice fed barley β-glucan (100 mg/kg of body weight) showed that β-glucans play a role in reducing chromosomal damage induced by antineoplasmic drugs and may provide a decrease of secondary tumor risk.

Furthermore, there is evidence that cereal β-glucans have a prebiotic effect showing selectivity to *Lactobacillus* species.[241,270,271] Dongowski et al.[241] found that the numbers of coliforms and *Bacteroides* were lower in rats fed barley β-glucan-enriched diets than in those fed control diets, and the numbers of *Lactobacillus* were greater; however, these effects were small. In a clinical study,[270] *Lactobacillus plantarum* strain 299 has been shown to have the highest capacity among 19 strains of *Lactobacillus* to ferment oatmeal and to survive the acidity of the stomach and the bile acid content of the small intestine; only 5 of the 19 tested strains could be identified in the stools. On the other hand, Crittenden et al.[272] examining *in vitro* a range of prebiotic and other intestinal bacteria for their ability to ferment barley β-glucan and found that β-glucan was fermented by *Bacteroides* spp. and

Clostridium beijerinckii, but was not by lactobacilli, bifidobacteria, enterococci, or *Escherichia coli*, indicating that these bacteria do not possess glycosidases capable of hydrolyzing this polysaccharide. However, lactobacilli and bifidobacteria may be able to cross-feed on oligomers resulting from the hydrolysis of β-glucans by other intestinal bacteria, such as bacteroides. Jaskari et al.[271] prepared oligomers from oat β-glucan that resist digestion in the upper gut by enzymatic hydrolysis with lichenase of *Bacillus subtilis* and compared them with fructooligomers and raffinose in their ability to act as growth substrates for prebiotic (*Lactobacillus* and *Bifidobacterium*) and intestinal (*Bacteroides*, *Clostridium*, and *E. coli*) strains *in vitro*. The β-glucan oligomers enhanced the growth of health-promoting prebiotic strains compared with intestinal bacterial growth, but not to a significant level; however, oat β-glucooligomers were almost as effective as raffinose and better than fructooligomers. *Lactobacillus rhamnosus* GG differed significantly from all other bacteria in its ability to degrade β-glucan oligomers, consuming over 70% of the total oligosaccharides in the growth media after 24 h of incubation. Whereas the other starter strains poorly utilized these oligomers, *L. acidophilus* strains digested them by 17% and *Bifidobacterium* strains by 4% after 24 and 72 h incubation, respectively. The 13 different intestinal bacterial strains tested were also bad at degrading and utilizing β-glucan oligosaccharides during the incubation time, which was 72 h for all other gut bacteria except *E. coli*, which was incubated for 24 h. The highest degree of consumption of oligomers was 22%, achieved by *Clostridium difficile*, whereas for the rest of the intestinal bacteria it was <7%.[271]

Moreover, Yun et al.[273] found that β-glucan treatment increased the resistance to *Eimeria vermiformis* infection in immunosuppressed mice fed a oat β-glucan concentrate (68.2% purity); protozoa of the genus *Eimeria* cause coccidiosis, a clinically important intestinal disease of domestic animals. Recently, Delaney et al.[274] showed that consumption of concentrated barley β-glucan (64% purity) did not cause treatment-related inflammatory or other adverse effects in mice when blended into mouse feed at concentrations of 0.7, 3.5, and 7.0% β-glucan. However, another recent study *in vitro*[190] showed that oat β-glucan can increase the inflammatory responses of intestinal cells. An enhanced immune system may protect against infections, but high-inflammation markers in the circulation increase the risk of cardiovascular disease. Further research is needed to see if these results can be extrapolated to the *in vivo* situation and if these findings should be interpreted as beneficial.

REFERENCES

1. Dais, P. and Perlin, A.S., High-field, C-N.M.R. spectroscopy of β-D-glucans, amylopectin, and glycogen, *Carbohydr. Res.*, 100, 103, 1982.
2. Woodward, J.R., Fincher, G.B., and Stone, B.A., Water-soluble (1→3),(1→4)-β-D-glucans from barley (*Hordeum vulgare*) endosperm. II. Fine structure, *Carbohydr. Polym.*, 3, 207, 1983.

3. Woodward, J.R., Phillips, D.R., and Fincher, G.B., Water-soluble (1→3),(1→4)-β-D-glucans from barley (*Hordeum vulgare*) endosperm. IV. Comparison of 40°C and 65°C soluble fractions, *Carbohydr. Polym.*, 8, 85, 1988.
4. Varum, K.M. and Smidsrod, O., Partial chemical and physical characterization of (1→3),(1→4)-β-D-glucans from oat (*Avena sativa* L.) aleurone, *Carbohydr. Polym.*, 9, 103, 1988.
5. Wood, P.J., Weisz, J., and Blackwell, B.A., Molecular characterization of cereal β-D-glucans. Structural analysis of oat β-D-glucan and rapid structural evaluation of β-D-glucans from different sources by high-performance liquid chromatography of oligosaccharides released by lichenase, *Cereal Chem.*, 68, 31, 1991.
6. Wood, P.J., Weisz, J., and Blackwell, B.A., Structural studies of (1→3)(1→4)-β-D-glucans by ^{13}C-nuclear magnetic resonance spectroscopy and by rapid analysis of cellulose-like regions using high-performance anion-exchange chromatography of oligosaccharides released by lichenase, *Cereal Chem.*, 71, 301, 1994.
7. Izydorczyk, M.S., Macri, L.J., and MacGregor, A.W., Structure and physicochemical properties of barley non-starch polysaccharides. II. Alkali-extractable β-glucans and arabinoxylans, *Carbohydr. Polym.*, 35, 259, 1998.
8. Klopfenstein, C.F., The role of cereal beta-glucans in nutrition and health, *Cereal Foods World*, 33, 865, 1988.
9. Braaten, J.T. et al., Oat gum lowers glucose and insulin after an oral glucose dose, *Am. J. Clin. Nutr.*, 53, 1425, 1991.
10. Braaten, J.T. et al., Oat β-glucan reduces blood cholesterol concentration in hypercholesterolemic subjects, *Eur. J. Clin. Nutr.*, 48, 465, 1994.
11. Wood, P.J. et al., Effect of dose and modification of viscous properties of oat gum on plasma glucose and insulin following an oral glucose load, *Br. J. Nutr.*, 72, 731, 1994.
12. Bhatty, R.S., The potential of hull-less barley, *Cereal Chem.*, 76, 589, 1999.
13. Delaney, B. et al., β-Glucan fractions from barley and oats are similarly antiatherogenic in hypercholesterolemic Syrian golden hamsters, *J. Nutr.*, 133, 468, 2003.
14. Doublier, J.L. and Wood, P.J., Rheological properties of aqueous solutions (1→3)(1→4)-β-D-glucan from oats (*Avena sativa* L.), *Cereal Chem.*, 72, 335, 1995.
15. Bohm, N. and Kulicke, W.M., Rheological studies of barley (1→3)(1→4)-β-D-glucan in concentrated solution: mechanistic and kinetic investigation of the gel formation, *Carbohydr. Res.*, 315, 302, 1999.
16. Cui, W. and Wood, P.J., Relationships between structural features, molecular weight and rheological properties of cereal β-D-glucan, in *Hydrocolloids*, Part 1, *Physical Chemistry and Industrial Applications of Gels, Polysaccharides and Proteins*, Nishinari, K., Ed., Elsevier Science B.V., Amsterdam, 2000, p. 159.
17. Lazaridou, A., Biliaderis, C.G., and Izydorczyk, M.S., Molecular size effects on rheological properties of oat β-glucans in solutions and gels, *Food Hydrocolloids*, 17, 693, 2003.
18. Lazaridou, A., Biliaderis, C.G., Micha-Screttas, M., and Steele, B.R., A comparative study on structure-function relations of mixed linkage (1→3), (1→4) linear β-D-glucans, *Food Hydrocolloids*, 18, 837, 2004.
19. Lazaridou, A. and Biliaderis, C.G., Cryogelation of cereal β-glucans: structure and molecular size effects, *Food Hydrocolloids*, 18, 933, 2004.
20. Vaikousi, H. and Biliaderis, C.G., Processing and formulation effects on rheological behavior of barley β-glucan aqueous dispersions, *Food Chem.*, 91, 505, 2005.

21. Vaikousi, H., Biliaderis, C.G., and Izydorczyk, M.S., Solution flow behavior and gelling properties of water-soluble barley (1→3), (1→4)-β-glucans varying in molecular size, *J. Cereal Sci.*, 39, 119, 2004.
22. Bhatty, R.S., Physicochemical and functional (breadmaking) properties of hull-less barley fractions, *Cereal Chem.*, 63, 31, 1986.
23. Wood, P.J., Oat β-glucan: structure, location, and properties, in *Oats: Chemistry and Technology*, Webster, F.H., Ed., American Association of Cereal Chemists, St. Paul, MN, 1986, p. 121.
24. Inglett, G.E., USDA's Oatrim replaces fat in many food products, *Food Technol.*, 44, 100, 1990.
25. Newman, R.K., McGuire, C.F., and Newman, C.W., Composition and muffin-baking characteristics of flour from four barley cultivars, *Cereal Foods World*, 35, 563, 1990.
26. Hudson, C.A., Chiu, M.M., and Knuckles, B.E., Development and characteristics of high-fiber muffins with oat bran, rice bran, or barley fiber fractions, *Cereal Foods World*, 37, 373, 1992.
27. Berglund, P.T., Fastnaught, C.E., and Holm, E.T., Physicochemical and sensory evaluation of extruded high-fiber barley cereals, *Cereal Chem.*, 71, 91, 1994.
28. Knuckles, B.E., Hudson, C.A., Chiu, M.M., and Sayre, R.N., Effect of β-glucan barley fractions in high-fiber bread and pasta, *Cereal Foods World*, 42, 94, 1997.
29. Newman, R.K., Ore, K.C., Abbott, J., and Newman, C.W., Fiber enrichment of baked products with a barley milling fraction, *Cereal Foods World*, 43, 23, 1998.
30. Marconi, E., Graziano, M., and Cubadda, R., Composition and utilization of barley pearling by-products for making functional pastas rich in dietary fiber and β-glucans, *Cereal Chem.*, 77, 133, 2000.
31. Cavallero, A., Empilli, S., Brighenti, F., and Stanca, A.M., High (1→3, 1→4)-β-glucan barley fractions in bread making and their effects on human glycemic response, *J. Cereal Sci.*, 36, 59, 2002.
32. Morin, L.A., Temelli, F., and McMullen, L., Physical and sensory characteristics of reduced-fat breakfast sausages formulated with barley β-glucan, *J. Food Sci.*, 67, 2391, 2002.
33. Lyly, M. et al., Influence of oat β-glucan preparations on the perception of mouthfeel and on rheological properties in beverages prototypes, *Cereal Chem.*, 80, 536, 2003.
34. Dexter, J.E., Izydorczyk, M.S., Marchylo, B.A., and Schlichting, L.M., Texture and color of pasta containing mill fractions from hull-less barley genotypes with variable content of amylose and fibre, in *Proceedings of the 12th International ICC Cereal and Bread Congress*, Cauvain, S.P., Salmon, S.S., and Young, L., Eds., Woodhead Publishing Ltd., Cambridge, U.K., 2004, p. 488.
35. Hatcher, D.W. et al., Quality characteristics of yellow alkaline noodles enriched with hull-less barley flour, *Cereal Chem.*, 82, 60, 2005.
36. Izydorczyk, M.S. et al., The enrichment of Asian noodles with fiber rich fractions derived from roller milling of hull-less barley, *J. Sci. Food Agric.* 85, 2094, 2005.
37. Volikakis, P., Biliaderis, C.G., Vamvakas, C., and Zerfiridis, G.K., Effects of a commercial oat β-glucan concentrate on the chemical, physico-chemical and sensory attributes of a low-fat white-brined cheese product, *Food Res. Int.* 37, 83, 2004.
38. Izydorczyk, M.S., Macri, L.J., and MacGregor, A.W., Structure and physicochemical properties of barley non-starch polysaccharides. I. Water-extractable β-glucans and arabinoxylans, *Carbohydr. Polym.*, 35, 249, 1998.
39. Izydorczyk, M.S., Biliaderis, C.G., Macri, L.J., and MacGregor, A.W., Fractionation of oat (1→3), (1→4)-β-D-glucans and characterization of the fractions, *J. Cereal Sci.*, 27, 321, 1998.

40. Bohm, N. and Kulicke, W.M., Rheological studies of barley (1→3)(1→4)-β-D-glucan in concentrated solution: investigation of the viscoelastic flow behaviour in the sol state, *Carbohydr. Res.*, 315, 293, 1999.
41. Cui, W., Wood, P.J., Blackwell, B., and Nikiforuk, J., Physicochemical properties and structural characterization by two-dimensional NMR spectroscopy of wheat β-D-glucan: comparison with other cereal β-D-glucans, *Carbohydr. Polym.*, 41, 249, 2000.
42. Lazaridou, A., Vaikousi, H., and Biliaderis, C.G., Molecular size effects on gelation of water-soluble barley and oat β-glucans, in *Gums and Stabilizers for the Food Industry*, Vol. 12, IRL Press, Washington, DC, 2003, p. 108.
43. Skendi, A., Biliaderis, C.G., Lazaridou, A., and Izydorczyk, M.S., Structure and rheological properties of water soluble β-glucans from oat cultivars of *Avena sativa* and *Avena bysantina*, *J. Cereal Sci.*, 38, 15, 2003.
44. Storsley, J.M. et al., Structure and physicochemical properties of β-glucans and arabinoxylans isolated from hull-less barley, *Food Hydrocolloids*, 17, 831, 2003.
45. Irakli, M., Biliaderis, C.G., Izydorczyk, M.S., and Papadoyannis, I.N., Isolation, structural features and rheological properties of water-extractable β-glucans from different Greek barley cultivars, *J. Sci. Food Agric.*, 84, 1170, 2004.
46. Tosh, S.M., Wood, P.J., Wang, Q., and Weisz, J., Structural characteristics and rheological properties of partially hydrolyzed oat β-glucan: the effects of molecular weight and hydrolysis method, *Carbohydr. Polym.*, 55, 425, 2004.
47. Tosh, S.M. et al., Evaluation of structure in the formation of gels by structurally diverse (1→3)(1→4)-β-D-glucans from four cereal and one lichen species, *Carbohydr. Polym.*, 57, 249, 2004.
48. Prentice, N., Babler, S., and Faber, S., Enzymatic analysis of beta-D-glucans in cereal grains, *Cereal Chem.*, 57, 198, 1980.
49. McCleary, B.V. and Glennie-Holmes, M., Enzymic quantification of (1→3)(1→4)-β-D-glucan in barley and malt, *J. Inst. Brew.*, 91, 285, 1985.
50. Henry, R.J., A comparison of the non-starch carbohydrates in cereal grains, *J. Sci. Food Agric.*, 36, 1243, 1985.
51. Henry, R.J., Pentosan and (1→3),(1→4)-β-glucan concentration in endosperm and wholegrain of wheat, barley, oats and rye, *J. Cereal Sci.*, 6, 253, 1987.
52. Bhatty, R.S. and MacGregor, A.W., Gamma irradiation of hulless barley: effect on grain composition, β-glucans and starch, *Cereal Chem.*, 65, 463, 1988.
53. Åman, P. and Graham, H., Analysis of total and insoluble mixed-linked (1→3),(1→4)-β-D-glucans in barley and oats, *J. Agric. Food Chem.*, 35, 704, 1987.
54. Carr, J.M., Glatter, S., Jeraci, J.L., and Lewis B.A., Enzymic determination of β-glucan in cereal-based food products, *Cereal Chem.*, 67, 226, 1990.
55. Bhatty, R.S., MacGregor, A.W., and Rossnagel, B.G., Total and acid-soluble β-glucan content of hulless barley and its relationship to acid-extract viscosity, *Cereal Chem.*, 68, 221, 1991.
56. Knuckles, B.E., Chiu, M.M., and Betschart, A.A., β-Glucan-enriched fractions from laboratory-scale dry milling and sieving of barley and oats, *Cereal Chem.*, 69, 198, 1992.
57. Newman, R.K. et al., Comparison of the cholesterol-lowering properties of whole barley, oat bran, and wheat red dog in chicks and rats, *Cereal Chem.*, 69, 240, 1992.
58. Wu, Y.V., Stringfellow, A.C., and Inglett, G.E., Protein- and β-glucan enriched fractions from high-protein, high-β-glucan barleys by sieving and air classification, *Cereal Chem.*, 71, 220, 1994.
59. Knuckles, B.E. and Chiu, M.M., β-Glucan-enrichment of barley fractions by air classification and sieving, *J. Food Sci.*, 60, 1070, 1995.

60. Vasanthan, T. and Bhatty, R.S., Starch purification from waxy, normal and high amylose barleys by pin milling and air classification, *Cereal Chem.*, 72, 379, 1995.
61. Fastnaught, C.E., Berglund, P.T., Holm, E.T., and Fox, G.T., Genetic and environmental variation in beta-glucan content and quality parameters of barley for food, *Crop Sci.*, 36, 941 1996.
62. Beer, M.U., Wood, P.J., and Weisz, J., Molecular weight distribution and (1→3)(1→4)-β-D-glucan content of consecutive extracts of various oat and barley cultivars, *Cereal Chem.*, 74, 476, 1997.
63. Lee, C.J., Horsley, R.D., Manthey, F.A., and Schwarz, P.B., Comparisons of β-glucan content of barley and oat, *Cereal Chem.*, 74, 571, 1997.
64. Bhatty, R.S., β-Glucan content and viscosities of barleys and their roller-milled flour and bran products, *Cereal Chem.*, 69, 469, 1992.
65. Bhatty, R.S., β-Glucans and flour yield of hull-less barley, *Cereal Chem.*, 76, 314, 1999.
66. Andersson, A.A.M. et al., Chemical and physical characteristics of different barley samples, *J. Sci. Food Agric.*, 79, 979, 1999.
67. Andersson, A.A.M., Andersson, R., and Åman, P., Air classification of barley flours, *Cereal Chem.*, 77, 463, 2000.
68. Huth, M., Dongowski, G., Gebhard, E., and Flamme, W., Functional properties of dietary fibre enriched extrudates from barley, *J. Cereal Sci.*, 32, 115, 2000.
69. Jiang, G. and Vasanthan, T., MALDI-MS and HPLC quantification of oligosaccharides of lichenase-hydrolyzed water-soluble β-glucan from ten barley varieties, *J. Agric. Food Chem.*, 48, 3305, 2000.
70. Izydorczyk, M.S. et al., Variation in total and soluble β-glucan content in hulless barley: effects of thermal, physical, and enzymic treatments, *J. Agric. Food Chem.*, 48, 982, 2000.
71. Izydorczyk, M.S., Jacobs, M., and Dexter, J.E., Distribution and structural variation of nonstarch polysaccharides in milling fractions of hull-less barley with variable amylose content, *Cereal Chem.*, 80, 645, 2003.
72. Izydorczyk, M.S. et al., Roller milling of Canadian hull-less barley: optimization of roller milling conditions of mill streams, *Cereal Chem.*, 80, 637, 2003.
73. Nielsen, J.P. and Munk, L., Evaluation of malting barley quality using exploratory data analysis. I. Extraction of information from micro-malting data of spring and winter barley, *J. Cereal Sci.*, 38, 173, 2003.
74. MacGregor, A.W. and Fincher, G.B., Carbohydrates of the barley grain, in *Barley: Chemistry and Technology*, MacGregor, A.W. and Bhatty, R.S., Eds., American Association of Cereal Chemists, St. Paul, MN, 1993, p. 73.
75. Knuckles, B.E., Yokoyama, W.H., and Chiu, M.M, Molecular characterization of barley β-glucans by size-exclusion chromatography with multiple-angle laser light scattering and other detectors, *Cereal Chem.*, 74, 599, 1997.
76. Jadhav, S.J., Lutz, S.E., Ghorpade, V.M., and Salunkhe, D.K., Barley: chemistry and value-added processing, *Crit. Rev. Food Sci.*, 38, 123, 1998.
77. Andersson, A.A.M., Andersson, R., Autio, K., and Åman, P., Chemical composition and microstructure of two naked waxy barleys, *J. Cereal Sci.*, 30, 183, 1999.
78. Wood, P.J., Weisz, J., and Fedec, P. Potential for β-glucan enrichment in brans derived from oat (*Avena sativa* L.) cultivars of different (1→3),(1→4)-β-D-glucan concentrations, *Cereal Chem.*, 68, 48, 1991.
79. Autio, K. et al., Physical properties of (1→3),(1→4)-β-D-glucan preparates isolated from Finnish oat varieties, *Food Hydrocolloids*, 5, 513, 1992.

80. Saastamoinen, M., Plaami, S., and Kumpulainen, J., Genetic and environmental variation in β-glucan content of oats cultivated or tested in Finland, *J. Cereal Sci.*, 16, 279, 1992.
81. Cho, K.C. and White, P.J., Enzymatic analysis of β-glucan content in different oat genotypes, *Cereal Chem.*, 70, 539, 1993.
82. Miller, S.S., Wood, P.J., Pietrzak, L.N., and Fulcher, R.G., Mixed linkage beta glucans, protein content, and kernel weight in *Avena* species, *Cereal Chem.*, 70, 231, 1993.
83. Zhang, D., Doehlert, D.C., and Moore, W.R., Rheological properties of (1→3),(1→4)-β-D-glucans from raw, roasted, and steamed oat groats, *Cereal Chem.*, 75, 433, 1998.
84. Wilhelmson, A. et al., Development of a germination process for producing high β-glucan, whole grain food ingredients from oat, *Cereal Chem.*, 78, 715, 2001.
85. Colleoni-Sirghie, M., Fulton, D.B., and White, P.J., Structural features of water soluble (1,3) (1,4)-β-glucan and traditional oat lines, *Carbohydr. Polym.*, 54, 237, 2003.
86. Colleoni-Sirghie, M. et al., Rheological and molecular properties of water soluble (1,3) (1,4)-β-D-glucans from high β-glucan and traditional oat lines, *Carbohydr. Polym.*, 52, 439, 2003.
87. Colleoni-Sirghie, M. et al., Prediction of β-glucan concentration based on viscosity evaluations of raw oat flours from high β-glucan and traditional oat lines, *Cereal Chem.*, 81, 434, 2004.
88. Parkkonen, T., Harkonen, H., and Autio, K., Effect of baking on the microstructure of rye cell walls and protein, *Cereal Chem.*, 71, 58, 1994.
89. Glitsø , L.V. and Bach Knudsen, K.E., Milling of whole grain rye to obtain fractions with different dietary fibre characteristics, *J. Cereal Sci.*, 29, 89, 1999.
90. Beresford, G. and Stone, B.A., (1→3)(1→4)-β-D-glucan content of *Triticum* grains, *J. Cereal Sci.*, 1, 111, 1983.
91. Niba, L.L. and Hoffman, J., Resistant starch and β-glucan levels in grain sorghum (*Sorghum bicolor* M.) are influenced by soaking and autoclaving, *Food Chem.*, 81, 113, 2003.
92. Stone, B.A. and Clarke, A.E., *Biology of (*1→3*)-β-Glucans*, La Trobe University Press, Melbourne, 1992.
93. Wood, P.J. et al., Structure of (1→3)(1→4)-β-D-glucan in waxy and nonwaxy barley, *Cereal Chem.*, 80, 329, 2003.
94. Fox, G.J., Food Ingredients Derived from Viscous Barley Grain and the Process of Making, U.S. Patent 5,614,242, 1997.
95. Cervantes-Martinez, C.T. et al., Selection for greater β-glucans content in oat grain, *Crop Sci.*, 41, 1085, 2001.
96. Fulcher, R.G. and Rooney Duke, T.K. Whole-grain structure and organization: Implications for nutritionists and processors, in *Whole-Grain Foods in Health and Disease*, Marquart, L., Slavin, J.L., and Fulcher, R.G., Eds., American Association of Cereal Chemists, St. Paul., MN, 2002, p. 9.
97. Fincher, G.B., Morphology and chemical composition of barley endosperm cell walls, *J. Inst. Brew.*, 81, 116, 1975.
98. Balance, G.M. and Manners, D.J., Structural analysis: an enzymatic solubilization of barley endosperm cell walls, *Carbohydr. Res.*, 61, 107, 1978.
99. Ahluwalia, B. and Ellis, E., Studies of β-glucan in barley, malt and endosperm cell walls, in *New Approaches to Research on Cereal Carbohydrates*, Hill, R.D. and Munck, L., Eds., Elsevier Science Publishers B.V., Amsterdam, 1985, p. 285.
100. Bacic, A. and Stone, B.A., Chemistry and organization of aleurone cell wall components from wheat and barley, *Austr. J. Plant Physiol.*, 8, 475, 1981.

101. Miller, S.S., Fulcher, R.G., Sen, A., and Arnason, J.T., Oat endosperm cell walls. I. Isolation, composition, and comparison with other tissues, *Cereal Chem.*, 72, 421, 1995.
102. Fulcher, R.G., Morphological and chemical organization of the oat kernel, in *Oats: Chemistry and Technology*, Webster, F.H., Ed., American Association of Cereal Chemists, St. Paul, MN, 1986, p. 47.
103. EtokAkpan, O.K., Enzymic degradation and nature of the endosperm cell walls of germinating sorghums and barley, *J. Sci. Food Agric.*, 61, 389, 1993.
104. Mares, D.J. and Stone, B.A., Studies on wheat endosperm. I. Chemical composition and ultrastructure of the cell walls, *Austr. J. Biol. Sci.*, 26, 793, 1973.
105. Miller, S.S. and Fulcher, R.G., Distribution of (1→3),(1→4)-β-D-glucan in kernels of oats and barley using microspectrofluorometry, *Cereal Chem.*, 71, 64, 1994.
106. Fulcher, R.G. and Miller, S.S., Structure of oat bran and distribution of dietary fiber components, in *Oat Bran*, Wood, P.J., Ed., American Association of Cereal Chemists, St. Paul, MN, 1993, p. 1.
107. Cui, S.W., *Polysaccharide Gums from Agricultural Products: Processing, Structure and Functionality*, Technomic Publishing Company, Lancaster, PA, 2001.
108. Härkönen, H., Pessa, E., Suortti, T., and Poutanen, K., Distribution and some properties of cell wall polysaccharides in rye milling fractions, *J. Cereal Sci.*, 26, 95, 1997.
109. Wood, P.J., Physicochemical characteristics and physiological properties of oat (1→3),(1→4)-β-D-glucan, in *Oat Bran*, Wood, P.J., Ed., American Association of Cereal Chemists, St. Paul, MN, 1993, p. 83.
110. Wood, P.J., Siddiqui, I.R., and Paton, D., Extraction of high viscosity gums from oats, *Cereal Chem.*, 55, 1038, 1978.
111. Wood, P.J., Paton, D., and Siddiqui, I.R., Determination of β-glucan in oats and barley, *Cereal Chem.*, 54, 524, 1977.
112. Beer, M.U., Arrigoni, E., and Amado, R., Extraction of oat gum from oat bran: effects of process on yield, molecular weight distribution, viscosity and (1→3)(1→4)-β-D-glucan content of the gum, *Cereal Chem.*, 73, 58, 1996.
113. Graham, H., Groen Rydberg, M.B., and Åman, P., Extraction of soluble dietary fiber, *J. Agric. Food Chem.*, 36, 494, 1988.
114. McCleary, B.V., Purification of (1→3)(1→4)-beta-D-glucan from barley flour, in *Methods in Enzymology*, Vol. 160, Wood, W.A. and Kellogg, S.T., Eds., Academic Press, San Diego, 1988, p. 511.
115. Saulnier, L., Gevaudan, S., and Thibault, J.F., Extraction and partial characterization of β-glucan from the endosperms of two barley cultivars, *J. Cereal Sci.*, 19, 171, 1994.
116. Wood, P.J., Weisz, J., and Mahn, W., Molecular characterization of cereal β-glucans. II. Size-exclusion chromatography for comparison of molecular weight, *Cereal Chem.*, 68, 530, 1991.
117. Beer, M.U., Wood, P.J., Weisz, J., and Fillion, N., Effect of cooking and storage on the amount and molecular weight of (1→3)(1→4)-β-D-glucan extracted from oat products by an *in vitro* digestion system, *Cereal Chem.*, 76, 705, 1997.
118. Bhatty, R.S., Extraction and enrichment of (1→3),(1→4)-β-D-glucan from barley and oat brans, *Cereal Chem.*, 70, 73, 1993.
119. Cui, W., Wood, P.J., Weisz, J., and Beer, M.U., Nonstarch polysaccharides from preprocesssed wheat bran: carbohydrate analysis and novel rheological properties, *Cereal Chem.*, 76, 129, 1999.
120. Bhatty, R.S., Laboratory and pilot plant extraction and purification of β-glucans from hull-less barley and oat brans, *J. Cereal Sci.*, 22, 163, 1995.

121. Wood, P.J. and Fulcher, R.G., Interactions of some dyes with cereal β-glucans, *Cereal Chem.*, 55, 952, 1978.
122. Izydorczyk, M.S. and Biliaderis, C.G., Structural and functional aspects of cereal arabinoxylans and β-glucans, in *Novel Macromolecules in Food Systems*, Doxastakis, G. and Kiosseoglou, V., Eds., Elsevier Science B.V., Amsterdam, 2000, p. 361.
123. Johansson, L. et al., Structural characterization of water soluble β-glucan of oat bran, *Carbohydr. Polym.*, 42, 143, 2000.
124. Forrest, I.S. and Wainwright, T., The mode of binding of β-glucans and pentosans in barley endosperm cell walls, *J. Inst. Brew.*, 83, 279, 1977.
125. Robertson, J.A., Majsak-Newman, G., Ring, S.G., and Selvendran, R.R., Solubilization of mixed linkage (1→3),(1→4)-β-D-glucans from barley: effects of cooking and digestion, *J. Cereal Sci.*, 25, 275, 1997.
126. Izydorczyk, M.S. and MacGregor, A.W., Evidence of intermolecular interactions of β-glucans and arabinoxylans, *Carbohydr. Polym.*, 41, 417, 2000.
127. Bamforth, C.W. and Kanauchi, M.K., A simple model for the cell wall of the starchy endosperm in barley, *J. Inst. Brew.*, 107, 235, 2001.
128. Kanauchi, M. and Bamforth, C.W., Release of β-glucan from cell walls of starchy endosperm of barley, *Cereal Chem.*, 78, 121, 2001.
129. Burkus, Z. and Temelli, F., Effect of extraction conditions on yield, composition, and viscosity stability of barley β-glucan gum, *Cereal Chem.*, 75, 805, 1998.
130. Westerlund, E., Andersson, R., and Åman, P., Isolation and chemical characterization of water-soluble mixed-linked β-glucans and arabinoxylans in oat milling fractions, *Carbohydr. Polym.*, 20, 115, 1993.
131. Bowles, R.K., Morgan, K.R., Furneaux, R.H., and Coles, G.D., ^{13}CP/MAS NMR study of the interaction of bile acids with barley β-D-glucan, *Carbohydr. Polym.*, 29, 7, 1996.
132. Wikstrom, K., Lindahl, L., Andersson, R., and Westerlund, E., Rheological studies of water-soluble (1→3),(1→4)-β-D-glucans from milling fractions of oat, *J. Food Sci.*, 59, 1077, 1994.
133. Gomez, C. et al., Physical and structural properties of barley (1→3),(1→4)-β-D-glucan. I. Determination of molecular weight and macromolecular radius by light scattering, *Carbohydr. Polym.*, 32, 7, 1997.
134. Wood, P.J., Weisz, J., Fedec, P., and Burrows, V.D., Large-scale preparation and properties of oat fractions enriched in (1→3),(1→4)-β-D-glucan, *Cereal Chem.*, 66, 97, 1989.
135. Morgan, K.R., Process for Extraction of β-Glucan from Cereals and Products Obtained Therefrom, International Patent Application PCT/NZ01/00014, WO 01/57092, 2001.
136. Malkki, Y. et al., Oat concentrates: physical properties of β-glucan and hypocholesterolemic effects in rats, *Cereal Chem.*, 69, 647, 1992.
137. Morgan, K.R. and Ofman, D.J., Glucagel, a gelling β-glucan from barley, *Cereal Chem.*, 75, 879, 1998.
138. Dexter, J.E. and Wood, P.J., Recent applications of debranning of wheat before milling, *Trends Food Sci. Technol.*, 7, 35, 1996.
139. Anon., AACC committee adopts oat bran definition, *Cereal Foods World*, 34, 1033, 1989.
140. Morgan, K.R., Extraction of Beta-Glucan from Cereals, International Patent Application PCT/NZ02/00129, WO 03/014165, 2003.
141. Bhatty, R.S., Hull-less barley bran: a potential new product from an old grain, *Cereal Foods World*, 40, 819, 1995.

142. Vasanthan, T. and Temelli, F., Grain Fractionation Methods and Products, International Patent Application PCT/CA01/01358, WO 02/27011, 2002.
143. Lehtomaki, I., Karinen, P., Bergelin, R., and Myllymaki, O., Beta-Glucan Enriched Alimentary Fiber, U.S. Patent 5,183,677, 1993.
144. Jaskari, J. et al., Effect of hydrothermal and enzymic treatments on the viscous behavior of dry- and wet-milled oat brans, *Cereal Chem.*, 72, 625, 1995.
145. Inglett, G.E., Method for Making a Soluble Dietary Fiber Composition from Cereals, U.S. Patent 5,082,673, 1992.
146. Oste Trantafyllou, A., Method for the Isolation of a β-Glucan Composition from Oats and Products Made Therefrom, International Patent Application PCT/SE99/01913, WO 00/24270, 2000.
147. Cahill, A.P., Jr., Fenske, D.J., Freeland, M., and Hartwig, G.W., Beta-Glucan Process, Additive and Food Product, U. S. Patent 2002/0106430, 2002.
148. Wang, L., Lynch, I.E., and Goering, K., Production of β-Glucan and β-Glucan Product, U.S. Patent 5,512,287, 1996.
149. Goering, K.J. and Eslick, R.F., Process for Recovery of Products from Waxy Barley, U.S. Patent 5,013,561, 1991.
150. Potter, R.C., Fisher, P.A., Hash, K.R., Sr., and Neidt, J.D., Method for Concentrating β-Glucan, U.S. Patent 6,323,338, 2001.
151. Morgan, K.R., Beta-Glucan Products and Extraction Processes from Cereals, U.S. Patent 6,426,201, 2002.
152. Van Lengerich, B.H., Gruess, O., and Meuser, F.P., Beta-Glucan Compositions and Process Therefore, U.S. Patent 2003/0153746, 2003.
153. Woodward, J.R., Phillips, D.R., and Fincher, G.B., Water-soluble (1→3),(1→4)-β-D-glucans from barley (*Hordeum vulgare*) endosperm. I. Physicochemical properties, *Carbohydr. Polym.*, 3, 143, 1983.
154. Wang, Q., Wood, P.J., Huang, X., and Cui, W., Preparation and characterization of molecular weight standards of low polydispersity from oat and barley (1→3)(1→4)-β-D-glucan, *Food Hydrocolloids*, 17, 845, 2003.
155. Åman, P., Rimsten, L., and Andersson, R., Molecular-weight distribution β-glucan in oat-based foods, *Cereal Chem.*, 81, 356, 2004.
156. Wang, Q., Wood, P.J., and Cui, W., Microwave assisted dissolution of β-glucan in water: implications for the characterization of this polymer, *Carbohydr. Polym.*, 47, 35, 2002.
157. Roubroeks, J.R., Mastromauro, D.I., Andersson, R., and Åman, P., Molecular weight, structure, and shape of oat (1→3),(1→4)-β-D-glucan fractions obtained by enzymatic degradation with lichenase, *Biomacromolecules*, 1, 584, 2000.
158. Roubroeks, J.R. et al., Molecular weight, structure, and shape of oat (1→3),(1→4)-β-D-glucan fractions obtained by enzymatic degradation with (14)-β-D-glucan 4-glucanohydrolase from *Trichoderma reesei*, *Carbohydr. Polym.*, 46, 275, 2001.
159. Roubroeks, J.R., Andersson, R., and Åman, P., Structural features of (1→3),(1→4)-β-D-glucan and arabinoxylan fractions isolated from rye bran, *Carbohydr. Polym.*, 42, 3, 2000.
160. Varum, K.M., Smidsrod, O., and Brant, D.A., Light-scattering reveals micelle-like aggregation in the (1→3),(1→4)-β-D-glucans from oat aleurone, *Food Hydrocolloids*, 5, 497, 1992.
161. Grimm, A., Kruger, E., and Burchard, W., Solution properties of β-D-(1,3)(1,4)-glucan isolated from beer, *Carbohydr. Polym.*, 27, 205, 1995.

162. Morgan, K.R. et al., A ^{13}C CP/MAS NMR spectroscopy and AFM study of the structure of Glucagel™, a gelling β-glucan from barley, *Carbohydr. Res.*, 315, 169, 1999.
163. Buliga, G.S., Brant, D.A., and Fincher, G.B., The sequence statistics and solution conformation of a barley (1→3, 1→4)-β-D-glucan, *Carbohydr. Res.*, 157, 139, 1986.
164. Staudte, R.G., Woodward, J.R., Fincher, G.B., and Stone, B.A., Water-soluble (1→3)(1→4)- β-D-glucans from barley (*Hordeum vulgare*) endosperm. III. Distribution of cellotriosyl and cellotetraosyl residues, *Carbohydr. Polym.*, 3, 299, 1983.
165. Buckeridge, M.S. et al., Mixed linkange (1→3),(1→4)-β-D-glucans of grasses, *Cereal Chem.*, 81, 115, 2004.
166. Miller, S.S. and Fulcher, R.G., Oat endosperm cell walls. II. Hot-water solubilization and enzymatic digestion of the wall, *Cereal Chem.*, 72, 428, 1995.
167. Wood, P.J., Evaluation of oat bran as a soluble fibre source. Characterization of oat β-glucan and its effects on glycaemic response, *Carbohydr. Polym.*, 25, 331, 1994.
168. Henriksson, K. et al., Hydrolysis of barley (1→3), (1→4)-β-D-glucan by a cellohydrolase II preparation from *Trichoderma reesei*, *Carbohydr. Polym.*, 26, 109, 1995.
169. Tvaroska, I., Ogawa, K., Deslandes, Y., and Marchessault, R.H., Crystalline conformation and structure of lichenan and barley β-glucan, *Can. J. Chem.*, 61, 1608, 1983.
170. Cuvelier, G. and Launay, B., Concentration regimes in xanthan gum solutions deduced from flow and viscosity properties, *Carbohydr. Polym.*, 6, 321, 1986.
171. Gomez, C. et al., Physical and structural properties of barley (1→3),(1→4)-β-D-glucan. II. Viscosity, chain stiffness and macromolecular dimensions, *Carbohydr. Polym.*, 32, 17, 1997.
172. Christensen, B.E. et al., Macromolecular characterization of three barley β-glucan standards by size-exclusion chromatography combined with light scattering and viscometry: an inter-laboratory study, *Carbohydr. Polym.*, 45, 11, 2001.
173. Burkus, Z. and Temelli, F., Determination of the molecular weight of barley β-glucan using intrinsic viscosity measurements, *Carbohydr. Polym.*, 54, 51, 2003.
174. Gomez, C. et al., Physical and structural properties of barley (1→3),(1→4)-β-D-glucan. III. Formation of aggregates analysed through its viscoelastic and flow behavior, *Carbohydr. Polym.*, 34, 141, 1997.
175. Tosh, S.M., Wood, P.J., and Wang, Q., Gelation characteristics of acid-hydrolyzed oat beta-glucan solutions solubilized at a range of temperatures, *Food Hydrocolloids*, 17, 523, 2003.
176. Autio, K., Rheological properties of solutions of oat β-glucans, in *Gums and Stabilizers for the Food Industry*, Vol. 4, Phillips, G.O., Wedlock, D.J., and Williams, P.A., Eds., IRL Press, Oxford, 1988, p. 488.
177. Ren, Y. et al., Dilute and semi-dilute solution properties of (1→3),(1→4)-β-D-glucan, the endosperm cell wall polysaccharide of oats (*Avena sativa* L.), *Carbohydr. Polym.*, 53, 401, 2003.
178. Autio, K., Myllymaki, O., and Malkki, Y., Flow properties of solutions of oat β-glucans, *J. Food Sci.*, 52, 1364, 1988.
179. Papageorgiou, M. et al., Water extractable (1→3,1→4)-β-D-glucans from barley and oats: an intervarietal study on their structural features and rheological behaviour, *J. Cereal Sci.*, 42, 213, 2005.
180. Morris, E.R. et al., Concentration and shear rate dependence of viscosity in random coil polysaccharide solutions, *Carbohydr. Polym.*, 1, 5, 1981.
181. Morris, E.R., Shear-thinning of 'random coil' polysaccharides: characterization by two parameters from a simple linear plot, *Carbohydr. Polym.*, 13, 85, 1990.

182. Cox, W.P. and Merz, E.H., Correlation of dynamic and steady shear flow viscosities, *J. Polym. Sci.*, 28, 619, 1958.
183. Fincher, G.B. and Stone, B.A., Cell walls and their components in cereal grain technology, in *Advances in Cereal Science and Technology*, Pomeraz, Y., Ed., American Association of Cereal Chemists, St. Paul, MN, 1986, p. 207.
184. Niba, L.L., Effects of storage period an temperature on resistant starch and β-glucan content in cornbread, *Food Chem.*, 83, 493, 2003.
185. Vaikousi, H., Isolation, Molecular Characterization and Rheological Properties of Water-Soluble Barley β-Glucans, M.Sc. thesis, Aristotle University of Thessaloniki, Thessaloniki, Greece, 2003.
186. Newman, R.K. and Newman, C.W., Barley as a food grain, *Cereal Foods World*, 36, 800, 1991.
187. Yiu, S.H., Wood, P.J., and Weisz, J., Effects of cooking on starch and β-glucan of rolled oats, *Cereal Chem.*, 64, 373, 1987.
188. Yiu, S.H., Weisz, J., and Wood, P.J., Comparison of the effects of microwave and conventional cooking on starch and β-glucans in rolled oats, *Cereal Chem.*, 68, 372, 1991.
189. Wood, P.J., Relationships between solution properties of cereal β-glucans and physiological effects: a review, *Trends Food Sci. Technol.*, 13, 313, 2002.
190. Quality of Life and Management of Living Resources Programme, Key Action 1: Food, Nutrition and Health, Design of Foods with Improved Functionality and Superior Health Effects Using Cereal Beta-Glucans, final report, QLK1-2000-00535, 2004.
191. Webster, F.H., Oat utilization: past, present, and future, in *Oats: Chemistry and Technology*, Webster, F.H., Ed., American Association of Cereal Chemists, St. Paul, MN, 1986, p. 426.
192. Delcour, J.A., Vanhamel, S., and Hoseney, R.C., Physicochemical and functional properties of rye nonstarch polysaccharides. II. Impact of a fraction containing water-soluble pentosans and proteins on gluten-starch loaf volumes, *Cereal Chem.*, 68, 72, 1991.
193. Wang, L., Miller, R.A., and Hoseney, R.C., Effects of (1→3)(1→4)-β-D-glucans of wheat flour on breadmaking, *Cereal Chem.*, 75, 629, 1998.
194. Colleoni-Sirghie, M., Jannink, J.L., and White, P.J., Pasting and thermal properties of flours from oat lines with high and typical amounts of β-glucan, *Cereal Chem.*, 81, 686, 2004.
195. Zhou, M., Robards, K., Glennie-Holmes, M., and Helliwell, S., Effects of enzyme treatment and processing on pasting and thermal properties of oats, *J. Sci. Food Agric.*, 80, 1486, 2000.
196. Krishnan, P.G., Chang, K.C., and Brown, G., Effect of commercial oat bran on the characteristics and composition of bread, *Cereal Chem.*, 64, 55, 1987.
197. Dexter, J.E., Izydorczyk, M.S., Preston, K.R., and Jacobs, M., The enrichment of bread with a fibre-rich fraction derived from roller milling of hull-less barley, in *Proceedings of the ICC-SA Bread and Cereals Symposium, Advances in Cereal Science and Technology, World Perspectives*, Fowler, A., Ed., 2005 (CD-ROM).
198. Izydorczyk, M.S. and Dexter, J.E., Barley/milling and processing, in *Encyclopedia of Grain Sciences*, Elsevier, Oxford, 2004, p. 57.
199. Burkus, Z. and Temelli, F., Stabilization of emulsions and foams using barley β-glucan, *Food Res. Int.*, 33, 27, 2000.
200. Kontogiorgos, V., Biliaderis, C.G., Kiosseoglou, V., and Doxastakis, G., Stability and rheology of egg-yolk-stabilized concentrated emulsions containing cereal β-glucans of varying molecular size, *Food Hydrocolloids*, 18, 987, 2004.

201. Inglett, G.E., Method for Making a Soluble Dietary Fiber Composition from Oats, U.S. Patent 4,996,063, 1991.
202. Inglett, G.E. and Grisamore, S.B., Maltodextrin fat substitude lowers cholesterol, *Food Technol.*, 45, 104, 1991.
203. Inglett, G.E. and Warner, K.A., Amylodextrin containing β-glucan from oats as a fat substitute in some cookies and candies, *Cereal Foods World*, 37, 589, 1992.
204. Jenking, R.K. and Wild, J.L., Dietary Fiber Compositions for Use in Foods, U.S. Patent 5,585,131, 1996.
205. Inglett, G.E., Development of a dietary fiber gel for calorie-reduced foods, *Cereal Foods World*, 42, 382, 1997.
206. Inglett, G.E., Soluble Hydrocolloid Food Additives and Their Preparation, International Patent Application PCT/US99/17897, WO 00/07715, 2000.
207. Fox, G.J., High Viscosity Cereal and Food Ingredient from Viscous Barley Grain, U.S. Patent 6,238,719, 2001.
208. Oste Trantafyllou, A., Non-Dairy, Ready-to-Use Milk Substitute, and Products Made Therewith, U.S. Patent 6,451,369, 2002.
209. Salovaara, H. and Kurka, A.M. Food Product Containing Dietary Fiber and Method of Making Said Product, International Patent Application PCT/FI91/00157, WO 91/17672, 1991.
210. Anon., Oat specialist shows how to build a new category, *New Nutr. Business*, 8, 29, 2003.
211. Hallfrisch, J. and Behall, K.M., Evaluation of foods and physiological responses to menus in which fat content was lowered by replacement with Oatrim, *Cereal Foods World*, 42, 100, 1997.
212. Liu, S., Intake of refined carbohydrates and whole grain foods in relation to risk of type 2 diabetes mellitus and coronary heart disease, *J. Am. Coll. Nutr.*, 21, 298, 2002.
213. McKeown, N.M. et al., Whole-grain intake is favorably associated with metabolic risk factors for type 2 diabetes and cardiovascular disease in the Framingham offspring study, *Am. J. Clin. Nutr.*, 76, 390, 2002.
214. Steffen, L.M. et al., Associations of whole-grain, refined grain, and fruit and vegetable consumption with risks of all-cause mortality and incident coronary artery disease and ischemic stroke: the Atherosclerosis Risk on Communities (ARIC) study, *Am. J. Clin. Nutr.*, 78, 383, 2003.
215. Steffen, L.M. et al., Whole grain intake is associated with lower body mass and greater insulin sensitivity among adolescents, *Am. J. Epidemiol.*, 158, 243, 2003.
216. Anderson, J.W. and Bridges, S.R., Hypocholesterolemic effects of oat bran in humans, in *Oat Bran*, Wood, P.J., Ed., American Association of Cereal Chemists, St. Paul, MN, 1993, p. 139.
217. Bell, S. et al., Effect of β-glucan from oats and yeast on serum lipids, *Crit. Rev. Food Sci. Nutr.*, 39, 189, 1999.
218. Anderson, J.W. et al., Dietary fiber and coronary heart disease, *Crit. Rev. Food Sci. Nutr.*, 29, 95, 1990.
219. Hallfrisch, J. and Behall, K.M., Mechanisms of the effects of grains on insulin and glucose responses, *J. Am. Coll. Nutr.*, 19, 320S, 2000.
220. Davy, B.M. et al., High-fiber oat cereal compared with wheat cereal consumption favorably alters LDL-cholesterol subclass and particle numbers in middle-aged and older men, *Am. J. Clin. Nutr.*, 76, 351, 2002.
221. Schneeman, B., Dietary fibre and gastrointestinal function, in *Advanced Dietary Fibre Technology*, McCleary, B.V. and Prosky, L., Eds., Blackwell Science, Oxford, 2001, p. 168.

222. Kahlon, T.S., Chow, F.I., Knuckles, B.E., and Chiu, M.M., Cholesterol-lowering effects in hamsters of β-glucan-enriched barley fraction, dehulled whole barley, rice bran, and oat bran and their combinations, *Cereal Chem.*, 70, 435, 1993.
223. Peterson, D.M. and Qureshi, A.A., Effects of tocols and β-glucan on serum lipid parameters in chickens, *J. Sci. Food Agric.*, 73, 417, 1997.
224. Kalra, S. and Jood, S., Effect of dietary barley β-glucan on cholesterol and lipoprotein fractions in rats, *J. Cereal Sci.*, 31, 141, 2000.
225. Sundberg, B., Pettersson, D., and Åman, P., Nutritional properties of fibre-rich barley products fed to broiled chickens, *J. Sci. Food Agric.*, 67, 469, 1995.
226. Anderson, J.W. and Chen, W.-J.L., Cholesterol-lowering properties of oat products, in *Oats: Chemistry and Technology*, Webster, F.H., Ed., American Association of Cereal Chemists, St. Paul, MN, 1986, p. 309.
227. Shinnick, F.L. and Marlett, J.A., Physiological responses to dietary oats in animal models, in *Oat Bran*, Wood, P.J., Ed., American Association of Cereal Chemists, St. Paul, MN, 1993, p. 113.
228. Yokoyama, W.H., Knuckles, B.E., Stafford, A., and Inglett, G., Raw and processed oat ingredients lower plasma cholesterol in the hamster, *J. Food Sci.*, 63, 713, 1998.
229. Newman, R.K., Newman, C.W., and Graham, H., The hypocholesterolemic function of barley β-glucans, *Cereal Foods World*, 34, 883, 1989.
230. Wang, L., Newman, R.K., Newman, C.W., and Hofer, P.J., Effect of barley oil on serum cholesterol in chicks, *Cereal Foods World*, 35, 819, 1990.
231. Bhatty, R.S., Nonmalting uses of barley, in *Barley: Chemistry and Technology*, MacGregor, A.W. and Bhatty, R.S., Eds., American Association of Cereal Chemists, St. Paul, MN, 1993, p. 55.
232. Wrick, K.L., Functional foods: cereal products at the food-drug interface, *Cereal Foods World*, 38, 205, 1993.
233. Anon., Food labeling: soluble dietary fibre from certain foods and coronary heart disease, *Federal Register*, 67, 61773, 1997.
234. Anon., Food labeling: health claims; oats and coronary heart disease, *Federal Register*, 62, 3584, 2002.
235. Ripsin, C.M. et al., Oat products and lipid lowering. A meta-analysis, *JAMA*, 267, 3317, 1992.
236. Brown, L., Rosner, B., Willet, W.W., and Sacks, F.M., Cholesterol-lowering effects of dietary fiber: a meta-analysis, *Am. J. Clin. Nutr.*, 69, 30, 1999.
237. Jenkins, D.J.A. et al., Soluble fiber intake at a dose approved by the US Food and Drug Administration for a claim of health benefits: serum lipid risk factors for cardiovascular disease assessed in a randomized controlled crossover trial, *Am. J. Clin. Nutr.*, 75, 834, 2002.
238. Wood, P.J. et al., Physiological effects of β-D-glucan rich fractions from oats, *Cereal Foods World*, 34, 878, 1989.
239. Hecker, K.D., Meier, M.L., Newman, R.K., and Newman, C.W., Barley β-glucan is effective as a hypocholesterolaemic ingredient in foods, *J. Sci. Food Agric.*, 77, 179, 1998.
240. Jones, P.J., Clinical nutrition. 7. Functional foods: more than just nutrition, *CMAJ*, 166, 1555, 2002.
241. Dongowski, G., Huth, M., Gebhardt, E., and Flamme, W., Dietary fiber-rich barley products beneficially affected the intestinal tract of rats, *J. Nutr.*, 132, 3704, 2002.
242. Anderson, J.W. et al., Hypercholesterolemic effects of oat-bran or bean intake for hypercholesterolemic men, *Am. J. Clin. Nutr.*, 40, 1146, 1984.

243. Andersson, M., Ellegard, L., and Andersson, H., Oat bran stimulates bile acid synthesis within 8 h as measured by 7-hydroxy-4-cholesten-3-one, *Am. J. Clin. Nutr.*, 76, 1111, 2002.
244. Kirby, R.W. et al., Oat-bran intake selectively lowers serum low-density lipoprotein cholesterol concentrations of hypercholesterolemic men, *Am. J. Clin. Nutr.*, 34, 824, 1981.
245. Lia, A. et al., Oat beta-glucan increases bile acid excretion and a fiber-rich barley fraction increases cholesterol excretion in ileostomy subjects, *Am. J. Clin. Nutr.*, 62, 1245, 1995.
246. Wood, P.J., Oat β-glucan-physicochemical properties and physiological effects, *Trends Food Sci. Technol.*, 2, 311, 1991.
247. Bourdon, I. et al., Postprandial lipid, glucose, insulin, and cholecystokinin responses in men fed barley pasta enriched with β-glucan, *Am. J. Clin. Nutr.*, 69, 55, 1999.
248. Wolever, T.M.S., The glycemic index: methodology and clinical implications, *Am. J. Clin. Nutr.*, 54, 846, 1991.
249. Hallfrisch, J., Scholfield, D.J., and Behall, K.M., Diets containing soluble oat extracts improve glucose and insulin responses of moderately hypercholesterolemic men and women, *Am. J. Clin. Nutr.*, 61, 379, 1995.
250. Yokoyama, W.H. et al., Effect of barley β-glucan in durum wheat pasta on human glycemic response, *Cereal Chem.*, 74, 293, 1997.
251. Van Der Sluijs, A.M.C. et al., Effect of cooking on the beneficial soluble β-glucans in Oatrim, *Cereal Foods World*, 44, 194, 1999.
252. Hallfrisch, J., Scholfield, D.J., and Behall, K.M., Physiological responses of men and women to barley and oat extracts (Nu-trimX). II. Comparison of glucose and insulin responses, *Cereal Chem.*, 80, 80, 2003.
253. Wood, P.J. et al., Comparison of viscous properties of oat and guar gum and the effects of these and oat bran on glycemic index, *J. Agric. Food Chem.*, 38, 753, 1990.
254. Beer, M.U., Arrigoni, E., and Åmado, R., Effects of oat gum on blood cholesterol levels in healthy young men, *Eur. J. Clin.*, 49, 517, 1995.
255. Anderson, J.W. et al., Oat-bran cereal lowers serum total and LDL cholesterol in hypercholesterolemic men, *Am. J. Clin. Nutr.*, 52, 495, 1990.
256. Romero, A.L., Romero, J.E., Galaviz, S., and Fernandez, M.L., Cookies enriched with psyllium or oat bran lower plasma LDL cholesterol in normal and hypercholesterolemic men from northern Mexico, *J. Am. Coll. Nutr.*, 17, 601, 1998.
257. Lia-Amundsen, A., Haugum, B., and Andersson, H., Changes in serum cholesterol and sterol metabolites after intake of products enriched with an oat bran concentrate within a controlled diet, *Scan. J. Nutr.*, 47, 68, 2003.
258. Kestin, M., Moss, R., Clifton, P.M., and Nestel, P.J., Comparative effects of three cereal brans on plasma lipids, blood pressure, and glucose metabolism in mildly hypercholesterolemic men, *Am. J. Clin. Nutr.*, 52, 661, 1990.
259. Kerckhoffs, D.A.J.M., Hornstra, G., and Mensink, R.P., Cholesterol-lowering effect of β-glucan from oat bran in mildly hypercholesterolemic subjects may decrease when β-glucan is incorporated into breads and cookies, *Am. J. Clin. Nutr.*, 78, 221, 2003.
260. Maki, K.C. et al., Lipid responses to consumption of a beta-glucan containing ready-to-eat cereal in children and adolescents with mild-to-moderate primary hypercholesterolemia, *Nutr. Res.*, 23, 1527, 2003.
261. Gerhardt, A.L. and Gallo, N.B., Full-fat rice bran and oat bran similarly reduce hypercholesterolemia, *J. Nutr.*, 128, 865, 1998.

262. Lovegrove, J.A., Clohessy, A., Milon, H., and Williams, C.M., Modest doses of β-glucan do not reduce concentrations of potentially atherogenic lipoproteins, *Am. J. Clin. Nutr.*, 72, 49, 2000.
263. Cara, L. et al., Effects of oat bran, rice bran, wheat fiber, and wheat germ on postprandial lipemia in healthy adults, *Am. J. Clin. Nutr.*, 55, 81, 1992.
264. Keogh, G.F. et al., Randomized controlled crossover study of the effect of a highly β-glucan-enriched barley on cardiovascular disease risk factors in mildly hypercholesterolemic men, *Am. J. Clin. Nutr.*, 78, 711, 2003.
265. Yokoyama, W.H., Knuckles, B.E., Wood, D., and Inglett, G.E., Food processing reduces size of soluble beta-glucan polymers without loss of cholesterol-reducing properties, *Bioactive Compounds Foods ACS Symp. Ser.*, 816, 105, 2002.
266. Sundberg, B. et al., Mixed-linked β-glucan from breads of different cereals is partly degraded in the human ileostomy model, *Am. J. Clin. Nutr.*, 64, 878, 1996.
267. Hallfrisch, J. and Behall, K.M., Physiological responses of men and women to barley and oat extracts (Nu-trimX). I. Breath hydrogen, methane, and gastrointestinal symptoms, *Cereal Chem.*, 80, 76, 2003.
268. Delaney, B. et al., Evaluation of the toxicity of concentrated barley β-glucan in a 28-day feeding study in Wistar rats, *Food Chem. Toxicity*, 41, 477, 2003.
269. Tohamy, A.A., El-Ghor, A.A., El-Nahas, S.M., and Noshy, M.M., β-glucan inhibits the genotoxicity of cyclophosphamide, adriamycin and cisplatin, *Mutat. Res.*, 541, 45, 2003.
270. Bengmark, S., Use of some pre-, pro- and synbiotics in critically ill patients, *Best Pract. Res. Clin. Gastroenterol.*, 17, 833, 2003.
271. Jaskari, J. et al., Oat β-glucan and xylan hydrolysates as selective substrates for *Bifidobacterium* and *Lactobacillus* strains, *Appl. Microbiol. Biotechnol.*, 49, 175, 1998.
272. Crittenden, R. et al., *In vitro* fermentation of cereal dietary fibre carbohydrates by prebiotic and intestinal bacteria, *J. Sci. Food Agric.*, 82, 781, 2002.
273. Yun, C.-H. et al., β-(1→3, 1→4) Oat glucan enhances resistance to *Eimeria vermiformis* infection in immunosuppressed mice, *Int. J. Parasitol.*, 27, 329, 1997.
274. Delaney, B. et al., Repeated dose oral toxicological evaluation of concentrated barley β-glucan in CD-1 mice including a recovery phase, *Food Chem. Toxicity*, 41, 1089, 2003.

2 Resistant Starch

Donald B. Thompson

CONTENTS

2.1 INTRODUCTION: WHAT IS RESISTANT STARCH?

This simple question belies the underlying complexity of the topic. The concept of resistant starch (RS) first arose due to inconsistencies in the analysis of total dietary fiber (TDF).[1] From the beginning, RS has been troubling to the food analytical chemist. Although dietary fiber is a physiological concept, it has been understood as a physiological concept that would be amenable to chemical analysis because the carbohydrate polymer composition of dietary fiber was understood to be due only to nonstarch polysaccharide (NSP).[2] The idea that only NSPs contributed to fiber was reasonable because the only enzyme available in the small intestine to hydrolyze polysaccharides is α-amylase; thus, NSP cannot be hydrolyzed in the small intestine. Nevertheless, in early *in vitro* analyses of TDF, the value for a particular sample varied according to the way the sample was treated prior to analysis. As it turned out, the assumption that starch was fully digested in the analytical procedure was faulty, as even a boiling treatment in the presence of heat-stable α-amylase was not sufficient to digest all of the starch in some samples, and the proportion of the undigested starch was affected by prior treatment. Thus, the concept of enzyme-resistant starch was born from a problem in the attempted chemical analysis of TDF. Ensuing research showed the RS to be of physiological significance.

Recognition of the concept of RS led to two analytical problems: how to take RS into account in the determination of TDF and how to determine RS itself. For the food analytical chemist, the former problem in chemical analysis is difficult, and the latter is intractable because *RS is not a chemical entity.* Confusion ensues whenever RS is mistakenly considered to be a chemical entity.

RS is properly defined in physiological terms as the sum of starch and starch digestion products not absorbed in the small intestine.[3,4] Champ and colleagues have described various *in vivo* techniques that have been used to quantify RS.[5,6] Recently, ileostomy data and intubation data were compared, leading the researchers to conclude that the ileostomy model for determination of RS was the more reliable one.[7] An alternative *in vivo* approach is to monitor RS indirectly, based on recovery of selected products of its colonic fermentation, primarily breath hydrogen,[8–11] but also breath methane.[12,13] This approach has been described as semiquantitative.[5] Although each method has advantages and disadvantages, there is general agreement that if enough RS is eaten that it can be of physiological significance for some individuals.[14–16]

Direct *in vivo* RS determination techniques have been used to show that the amount of RS in a particular starch-containing material varies greatly among normal subjects, and also for the same subject on different days.[6,17] This observation clearly shows that RS cannot be considered to be a chemical entity, but is instead operationally defined for each individual in a particular replicate of a test. If one wants to evaluate the RS content of a sample using human subjects, one must be satisfied with a mean value for a group.

2.2 *IN VITRO* ANALYSIS OF RS

Due to the difficulty in working with human subjects, there has been strong motivation to develop *in vitro* analytical methodology for RS. It must be stressed that any *in vitro* method would only be meaningful upon validation of the results by comparison with results using human subjects.[6,17]

The outcome of an *in vitro* analytical procedure for RS would be a numerical value for the RS in a material. This value must agree with the mean value obtained with human subjects. Moreover, ideally one would want to compare values for a range of starch-containing materials differing in RS levels and get reasonably good agreement for all types of samples. Validation of this sort has been accomplished for several analytical strategies.[17–19] Assuming that the agreement were acceptable, one would feel justified in using the *in vitro* procedure for new samples. One very large advantage of an *in vitro* procedure for RS is that the variability among replicates and the overall precision in the analysis would be far lower than for RS determination in a human population. Thus, one might use an *in vitro* method to determine whether minor changes in RS had occurred due to a processing treatment. Nevertheless, caution is still required in applying the *in vitro* method to a material for which validation has not been accomplished, as Danjo et al.[9] showed for a heat–moisture-treated starch.

The literature of RS in foods and ingredients is exceptionally difficult to read due to the absence of a single method of *in vitro* analysis. Among the many methods in the literature, a major distinction is whether the amylolytic digestion occurs at 37°C or at 100°C.[5,20] The former makes sense in that body temperature is insufficient to accomplish much in the way of a physical change in the starch present; it is the basis for the Englyst method,[3] the Goni method,[21] and the recently approved RS method.[22] The high temperature is used in the current TDF method[23] (AOAC Method 985.29), and the result has practical significance on current U.S. food labels, on which RS cannot be declared but TDF can be. It is important to note that the TDF method produces a value that includes some, but not all, of the RS,[24,25] and the proportion of the RS included in the TDF value can vary with processing treatment.[26–28] Thus, it is problematic to use the TDF method to determine RS levels for design of nutrition experiments. This distinction has not always been apparent in the nutrition literature; confusion about the use of the TDF method is apparent in some recent reports.[29,30]

In vitro methods of RS analysis may be either direct or indirect. With respect to obtaining a value for RS that agrees with human studies, for accuracy it makes little difference which is chosen. However, McCleary[31] has pointed out that a direct method should give greater precision. For those methods that determine RS directly by collecting some material and analyzing it,[18,22] one might be tempted to say that the material collected is in fact RS. However, that assumption is problematic unless it could be somehow validated in the human.[32] One must bear in mind that analytical validation of the RS method concerns the RS value and not the nature of any material obtained.

Recently, a method for RS has been developed[18,22] and approved (AOAC Method 2002.02, American Association of Cereal Chemists (AACC) Method 32-40). This method is a direct one, based on the analysis of α-glucan materials precipitated from a 50% ethanol suspension after extended α-amylolysis.

2.3 *IN VITRO* ISOLATION OF RS

For a given meal fed to a particular person at a particular time, one can imagine that this nonabsorbed starch could be recovered entirely and its composition analyzed. However, the task of recovery is difficult to accomplish with a normal, intact person. Patients who have no large intestine (ileostomists) may be used as research subjects to this end because the effluent of the small intestine may be collected in a bag, but the resulting nonabsorbed starch may be affected by the unusual status of these patients. The intubation procedure to remove material at a specific intestinal location has also been criticized as interfering with the passage of the digesta. Recovery of RS by this means is problematic as well. From among a group of human subjects, one can recover variable amounts of material that is, by operational definition, RS in each unique individual. However, there is no reason to expect that the recovered RS from patient A will be identical in chemical composition to the recovered RS from patient B. Moreover, if the amount of RS from patient B is less than the amount of RS from patient A, one cannot even say with confidence that the RS from patient B is the more resistant fraction of the RS from patient A. Thus, while it is possible to estimate a mean value for the RS content of a material using human subjects, one must exercise great caution about the inferences made regarding the nature of any actual material isolated.

Isolation of RS is especially problematic when one considers that the physiological definition includes both starch and starch digestion products. Many times, what is actually recovered is the solid-phase material, for example, residual partially digested granules. One should assume that recovery is only partial, since some starch digestion products are likely lost. Additionally, if the recovered material is the basis for an analytical RS value that agrees with the RS value determined in humans, one might be inclined to think that the RS in the human was limited to solid-phase material as well, an idea almost certainly wrong and certainly not consistent with the RS definition above. Moreover, if the *in vitro* RS value agreed with the human RS value, then the putative RS isolated by an *in vitro* RS procedure would include some solid-phase material in excess of that which would contribute to RS in the human.

2.4 THE NATURE OF RS

In some of the early *in vitro* work about the nature of RS, the TDF method was employed to monitor changes in putative RS. The reasonable assumption (although untested) was that material resistant to digestion by heat-stable α-amylase at 100°C would certainly be RS. However, the serious flaw in this logic was the idea that the amount of this material would reflect the RS content of the starting material, or even that the material would be representative of the RS in the starting material. Even the idea that the amount of this material resistant to heat treatment and enzyme-digestion would reflect a constant proportion of RS in a sample is problematic. This early thinking led to some confusion and to the idea that the RS was what one collected when one applied the TDF method to a starch sample. Since this material tended to give a differential scanning calorimetry (DSC) endotherm with a peak well above 100°C,[33] the erroneous inference that this DSC endotherm was a means of monitoring RS levels was problematic as well. We now know that one could more properly say that when one observed this DSC endotherm, one also observed RS; nevertheless, one would need to guard against the inference that the absence of this peak indicated the absence of RS, or that the enthalpy of this peak should be correlated with the actual RS value. Most properly, one might infer that some material behaving as RS is apparently well enough organized molecularly that the DSC endotherm is associated with it. The chemical nature of material resistant to α-amylolysis will be considered below.

2.5 SOURCES OF RS

Having established that RS is not a chemical entity, one must be cautious with the very idea of something being a source of RS. Most properly, one should describe materials that lead to significant mean values of RS when tested in human subjects. In the discussion below, the term *RS* will be used for convenience. At times, the wording might seem to imply that RS is a chemical entity. One should keep in mind that RS is not a chemical entity, avoiding this conceptual trap.

Further confusion about the nature of RS ensues from the way the term is commonly used in commerce. Commercial RS-containing starch ingredients are generally sold as "resistant starch." These materials should be considered RS-containing, since a large and variable fraction is fully digestible. Untreated high-amylose maize starch granules are not RS per se, but they do give high RS values on analysis. When one considers the physical properties of ingredients, it is even more important to be aware of the difference between the resistant component of an ingredient and the RS-containing ingredient, as the water-holding behavior for the portion contributing to the RS value is likely less than that for the non-RS portion.

Englyst originally described three types of RS: type 1, due to physical inaccessibility of the starch, for example, in whole grains or large particulates; type 2, due to the poor digestibility of untreated starch granules from certain sources; and type 3, due to the physical reassociation, or retrogradation, of molecules subsequent to cooking.[3] Subsequently, type 4 was added to account for resistance due to chemical modification.[34,35] The classification according to type is justified to the extent that these types represent distinct reasons for the poor starch digestibility. In many cases, the classification is helpful, but at times it can lead to confusion, as it is possible to imagine how a particular starch sample might include all four types.

Type 1 RS is conceptually trivial, although at times quantitatively important.[26] It is evident that the extent that a physical barrier of some sort precludes access by the enzyme to the starch substrate, digestion will be severely compromised. Type 2 RS occurs in a limited number of plant starches, most notably potato, banana, and high-amylose maize.[3] Because most potato starch is eaten cooked, RS from raw potato is not an important dietary consideration. Although bananas are most commonly eaten raw, the starch is quantitatively important only in the unripe banana, and so banana starch is also not of dietary importance (with the exception of plantain bananas[36]). High-amylose maize starch (HAMS) is commercially marketed as a useful source of RS. It has been known for some time that high-amylose maize genotypes containing the *ae* gene are poorly digested.[37,38] What is useful about many of these HAMSs is that they are also difficult to cook at 100°C to the point of disrupting molecular structure. Thus, it appears that if these high-RS starches are included into a formulation and the mixture cooked at atmospheric pressure, the processed product will still show appreciable RS by analysis.[39]

It is an open question why some untreated starch granules are poorly digested and the reasons may differ for different starches.[40] For HAMS it appears that external portions of the granule are particularly resistant to digestion.[41] Whether existence of this intractable layer might allow one to consider the nature of enzyme resistance in the granule interior to be type 1 is a matter of definition regarding scale. We and others have studied enzyme resistance as a means of exploring the nature of granule structure.[40,41]

Type 3 RS can form from several types of starches, the common feature being that some sort of molecular reassociation apparently precludes the enzyme's ability to address the substrate.[42] In commerce, RS-containing ingredients have commonly been manufactured from HAMS, as will be described further below. The features responsible for poor digestion are thought to involve ordered structures, such as double helices, double helices packed into crystalline array, amylose–lipid complexes, and amylose–lipid complexes packed into crystalline array. Gidley and associates have studied putative RS isolated from putative type 3 RS, and they suggested that a three-dimensional matrix of indigestible material might include ordered but noncrystalline regions that are inaccessible to the enzyme.[43] That molecular ordering is involved in formation of type 3 RS is generally accepted and, for this reason, classical polymer science principles have been brought to bear on the problem of manufacturing ingredients with high levels of RS.

Type 3 RS may form during common processes applied to normal starches in traditional food products when the processing treatment allows gelatinization and subsequent retrogradation. Thus, wheat bread,[44,45] corn bread,[46] or cooked potatoes[47] can contain low but dietarily significant amounts of RS. Hoover and Zhou reviewed the recent literature on hydrolysis of legume starches. They point out that processed legumes tend to have higher RS than cereals or potatoes, noting that retrogradation after cooking can lead to a three- to fivefold increase in RS.[48]

Type 4 RS may be understood to result from a chemical modification that precludes complete enzyme hydrolysis. Cross-linking of starches makes the granule not only difficult to cook, but also difficult to digest.[49,50] Extensive substitution reactions have recently been shown to inhibit enzyme digestion in the small intestine,[51] even though the granules likely became easier to gelatinize as a result of the modification. One assumes that the substitution interferes with the fruitful binding of monomer units to the subsites of the enzyme active site.

2.6 RS AND DIETARY FIBER

Should RS be considered a component of TDF? Currently, the AOAC method for TDF leads to inclusion of part of the RS value in the TDF value that may be claimed on a food label. A committee of the AACC recently deliberated concerning the question of whether RS should be included as dietary fiber. One key question was: How would RS vary as a result of processing treatments? The outcome of the debate was the AACC position that it would be acceptable to continue to include the RS that was determined in the TDF analysis because this RS would likely remain in the product after further treatment.[52,53] As yet, this debate and the current position of AACC are of no legal consequence for food labels. Moreover, the debate took place prior to the acceptance of a method of analysis for RS. McCleary and Monaghan have suggested that the TDF procedure be modified so that the RS would reflect the full RS value.[18] It is clear that the overlapping of values from the current TDF procedure and the newly approved RS procedure makes it very difficult to calculate an accurate total of all the nonstarch TDF plus all the RS.

2.7 WHAT IS THE FATE OF RS IN THE GUT?

One may consider starch digestion to be a function of the nature of the starch and the nature of the amylolytic enzyme. The enzyme for the human is α-amylase, chiefly that from the pancreas. Colonna et al.[42] reviewed the molecular effects of physical modification of starch and then reviewed the factors that limit α-amylolysis in heterogeneous phases: diffusion of the enzyme, porosity of the starch matrix, adsorption of enzyme to substrates, and catalysis. Based on monitoring the time course of digestion with porcine pancreatic α-amylase, it appears that the rate of digestion is commonly initially rapid and becomes slower and slower as the starch substrate moves through the small intestine.[25] Although to an extent the rate of transit through the small intestine would be expected to influence RS levels, variation in transit time would affect the period when rate is already slow because most starch that can be digested already has been. Once the undigested starch reaches the large intestine, it encounters an extensive array of microflora with a much enhanced complement of amylolytic enzymes.[16] For most RS-containing sources, little or no RS is recovered in the feces. One must assume that the starch resistant to pancreatic α-amylase is hydrolyzed by the gut enzymes and then fermented. It is on the basis of this assumption that one may indirectly estimate RS in the human by monitoring selected fermentation products (see above).

A major reason for the interest in RS is that this undigested material apparently has been for some time an important contribution to the microbial substrate reaching the large intestine. The interaction among (1) the indigenous gut microflora, (2) the residual food and other material that serve as substrate for the microflora, and (3) the colonic tissues of the host organism appears to be an important determinant of the health of the host.[54] When psyllium[55] or wheat bran[56] is present in the diet with RS, the colonic region of fermentation is shifted distally. One might assume that judicious selection of the RS-containing ingredient might be done with a similar effect.[57]

2.8 NUTRITIONAL IMPORTANCE OF RS

Any starch material will be heterogenous with respect to susceptibility to amylolytic enzymes. Consequently, for any RS-containing sample we expect a distribution of enzyme susceptibility, with the least susceptible portion contributing to the RS value.[58] Although Englyst et al.[3] have shown that raw potato starch is poorly hydrolyzed, the *in vitro* data indicate that there is no clear biochemical distinction between the resistant and nonresistant portions, as hydrolysis continues after the 120-min digestion period in their *in vitro* method.[3,25] The same behavior is evident for HAMSs[25] and for a combination of corn and tapioca starch.[59]

The functional division of the human digestive tract into the small intestine and colon is at the essence of the concept and definition of RS. Digestion and absorption of glucose from starch occur in the small intestine. The rate and extent of hydrolysis in the small intestine are both important nutritionally.[60] An altered rate of digestion may lead to an altered glycemic index (glycemic index is due to the properties of the starch, as well as to the influence of nonstarch materials on the metabolism of absorbed glucose[61]), and an altered extent of digestion means an altered level of RS. If RS were to make up part of the fixed amount of carbohydrate in the meal employed to determine the glycemic index, a lower blood glucose response would be anticipated on that basis alone. Although determination of the glycemic index should not include the RS portion of the carbohydrate when determining the amount of carbohydrate tested, as this portion is by definition not glycemic,[62] the nonglycemic nature of RS is not always taken into acount.[63,64] As explained above, it is difficult to know with accuracy or precision the actual RS value; therefore, it is difficult to know the value that should be employed to correct for the RS portion of the carbohydrate in the test meal. Given the large intersubject variation for RS determined *in vivo*, the difficulty in making an appropriate correction is of concern, as it confounds the glycemic index value obtained. In a recent development related to RS and the glycemic index, it appears that previous consumption of a type 2 RS enhances insulin sensitivity in a subsequent, non-RS-containing meal (other types of RS were not evaluated).[65] Thus, RS can complicate glycemic index determination in at least two ways. Vonk et al. actually used the glycemic response as a means of estimating the proportion of an RS-containing ingredient that is glycemic. They showed that the glucose that was released from HAMS tended to appear in the blood more slowly than the glucose released from common cornstarch.[66]

RS has been considered to have beneficial effects similar to those of some forms of dietary fiber;[11,12,28,36,67–70] some have considered RS to be a nutraceutical. Many of the beneficial effects of RS and dietary fiber are indirect, through their role as an energy substrate for maintenance of the large and diverse colon microflora and through the influence of microbial fermentation by-products.[71] Surprisingly, recent estimates of the quantitative importance of dietary fiber vs. RS as a fermentation substrate for maintenance of the colon microflora suggest that RS is likely as important as, and possibly more so than, dietary fiber in a Westernized diet.[16]

Depending on its form, RS may or may not be further hydrolyzed in the colon, but when it is, the glucose released will be rapidly fermented. The fermentation products include short-chain fatty acids (SCFAs): acetate, propionate, and butyrate. Uptake of these products in the colon accounts for the 1 to 2 kcal/g attributed to the RS value.[72] An altered rate of hydrolysis in the colon may lead to the desired RS fermentation occurring in a different portion of the colon, with any physiological effects of the fermentation products accruing to the specific colonic region.[57] Evidence is accumulating that the precise nature of an RS material may lead to differing nutritional effects.[16,29,57,70,73–76] Consequently, the various means of manufacturing RS-containing ingredients (see below) may produce materials with different nutritional effects. This prospect is only beginning to be explored.

Butyrate is a preferred energy source for the absorbing colonic epithelial cells, having a beneficial effect on the host tissue.[70,72] The possible complexity of the microbiology of butyrate formation has been recently reviewed.[77] The acetate and propionate may reach the liver and affect metabolism there.[78] Fermentation is associated with a variety of health benefits,[79–81] including the potential beneficial influence on immune function as mediated by metabolic products of the microflora.[70,82] Rapid production of butyrate may exceed the capacity for utilization by the colonic tissues, leading to butyrate reaching the portal blood.[57,83] While the butyrate in portal blood would not itself influence the colon, it may lead to benefits for cells in other locations.[83] Slower fermentation might lead to later, and thus more distal, butyrate production, thus having a beneficial effect on those regions of the colon most susceptible to cancerous growth.[57]

In addition to being a fermentation substrate, RS also may act as a prebiotic,[84–86] favorably influencing the ecology of the microbial flora in the large intestine. Topping et al.[16] argue that RS can be beneficial with respect to probiotics as well. They explain that continued ingestion of RS serves to extend the viability of some probiotic organisms that exist in the colon from a prior meal. They make the creative argument that when RS and probiotics are consumed together (the combination has been termed a *synbiotic*), a prebiotic role of RS may be due to its ability to protect some of the ingested organisms on their hazardous path to the colon, effectively increasing the beginning levels of the desirable species once the colon is reached. Once both the probiotic organisms and the RS are present in the colon, the RS would then confer a selective advantage through its role as substrate for a portion of the probiotic organisms.

2.9 MANUFACTURING OF RS-CONTAINING INGREDIENTS

The literature describing manufacture of RS-containing ingredients can be difficult to follow due to the different RS analytical methods employed in the various reports. Most of the work described below was done prior to the availability of an approved analytical method for RS. In reading the literature, one must be particularly careful to distinguish methods in which digestion occurs at 100°C from those in which digestion occurs at 37°C. In general, the latter temperature tends to give higher RS values, but no consistent proportion can be assumed. All four RS types can be manipulated by processing treatments, and combinations of RS types are possible as well.

2.9.1 Type 1 RS

The literature shows that size-reduction treatments can lead to decreased RS, but contains little about how to increase type 1 RS by processing treatments. The enzyme resistance of type 1 RS is based on the inaccessibility of the enzyme to the starch, and it does not depend on the intrinsic enzyme resistance of the starch itself. In theory, any source of starch could be trapped within a particle in such a way to make it inaccessible to enzymes. Both particle size and particle integrity during digestion would be pertinent factors. If this type of RS were part of an ingredient to be used in a manufactured food, the ultimate RS level in the food would depend on the stability of the entrapping material to the conditions during manufacturing of the formulated food.

2.9.2 Type 2 RS

2.9.2.1 RS Levels in Untreated Granular Starch

Although it is well known that native starch granules from potato or banana are highly resistant to digestion,[3] there is little information about how specific cultivars of these species might be selected for high or low RS levels.

HAMSs from a variety of different mutant maize genotypes are not well digested.[37,38] HAMS is the result of one or more endosperm mutations that alter the nature and proportion of the amylose and amylopectin fractions.[87] While the adjective "high-amylose" implies that the amylose fraction is enriched, the longer chains of the constituent amylopectin also contribute to a higher apparent amylose content.[87] Thus, reports that the amylose/amylopectin ratio influences RS values are not entirely accurate. Most HAMSs are from endosperm containing the *amylose-extender* (*ae*) gene, and it has been known for some time that for *ae*-type maize starches the level of amylose varies considerably with the genetic background.[88] Even for a particular genotype, selection of a starch cultivar for an altered amylose content (even if the amylose value is not completely meaningful structurally, differences observed may guide the selection) may also be a means of modulating the RS level. One company has marketed a maize starch with an exceptionally high amylose content as a high-RS ingredient.[35] It is clear that judicious choice of genetic makeup of the plant is a strategy for manipulating type 2 RS.

Berry first showed that the amylose level in maize starches generally correlates with RS levels.[89] Because it was reported that high-amylose starch granules with decreasing granule size have higher amylose content and lower digestibility,[90] fractions of high-amylose starches were prepared in an attempt to modulate the RS value. As determined by the TDF method, McNaught et al.[91] found that there is an optimal granule size to obtain maximal RS values; it is interesting that this granule size fraction is not associated with the the highest amylose content.

Unlike potato or banana, a good portion of the initial RS value for HAMS survives boiling in excess water, and thus a fraction appears to be stable to common processing conditions.[25] This starch is of practical interest as a type 2 RS source. Moreover, this difference in physical behavior suggests that the nature of the type 2 RS in HAMS is likely to be different than the nature of the type 2 RS of potato or banana. Gelatinization of normal starches is well understood to involve loss of granule order, swelling, and selective leaching of amylose from the granule, among other behaviors. When HAMS is cooked in excess water, it is difficult to document appreciable swelling or selective amylose leaching.[92] After initial heating of any native starch above the gelatinization temperature range, when the sample is observed by DSC on immediate reheating, a portion of the initially observed enthalpy is no longer observed. For normal starches, this portion of the initial enthalpy is understood to be associated with loss of granule order from amylopectin crystallites immediately prior to granule swelling. However, for HAMS this portion of the initial enthalpy is largely observed below 90°C, far below the temperature necessary to disperse the granule.[93] Thus, it would seem that for HAMS the loss of original granule structure during dispersion is not closely associated with disordering of the crystallites of branched-chain components, as is the case for normal starch. HAMS granules do not gelatinize in the same sense as do normal starch granules, but rather, they lose their structure gradually over a broad temperature range, thus leaving only the increasingly thermostable elements as the heating temperature increases.

2.9.2.2 Enhancement of RS Levels in Granular Starch

Various combinations of temperature, moisture, and time may be employed to enhance RS levels while maintaining a granule nature. The granule may be subjected to conditions that alter RS levels without gelatinization or melting. The physical treatment of granular starch to modify structure without gelatinization or melting has been termed *hydrothermal treatment*.[94,95] The native structure of the granule is semicrystalline. Hydrothermal treatments may be considered to improve the order of the crystalline fraction, to enhance the proportion of this fraction, or to have both effects. Since mobility of the amorphous regions is thought to be a prerequisite for either of these changes, hydrothermal treatments will be effective only above the glass transition temperature (T_g) of the amorphous components. Thus, hydrothermal modifications only occur when the component polymers of the amorphous regions within the granules are in a rubbery and mobile state.[95] Since T_g varies inversely with moisture, the moisture content dictates what minimum temperature is required. Although the original component polymers of the granules may be reasonably well ordered and packed as semicrystalline material in the native granule, they exist in a metastable state; consequently, there is the prospect of enhancing the stability of this physical state.

Jacobs and Delcour[95] have divided hydrothermal treatments into heat–moisture treatment (HMT), at moisture levels below 35%, and annealing (ANN), at moisture levels greater than or equal to 40%. By either type of hydrothermal treatment, it is possible to enhance the RS value without destroying granular structure. Because both approaches generate more highly ordered structures based on the initial structural order, they would both be considered annealing as the term is used in the polymer literature.[96] Starch annealing has been shown to influence enzyme susceptibility.[97] HMT has also been shown to influence amylase adsorption onto starch.[98]

At moisture contents above about 60%, loss of the granule structure occurs at or near the gelatinization temperature; therefore, for these moisture levels, ANN processing temperatures must remain below this temperature to maintain or improve the level of type 2 RS. As moisture content decreases from about 35%, the initial loss of the granule structure occurs at higher temperatures, as a result of what has loosely been considered a melting transition, as observed by DSC. Thus, HMT can be accomplished at these low moisture contents even at temperatures above the gelatinization temperature that is observed with excess water. At intermediate moisture levels (between 40 and 60%), the granule structure is lost by a combination of gelatinization and melting.[99–101] In this moisture range the process of heating to just below the gelatinization temperature has also been considered ANN.

If the resistance of starch to α-amylase action is in some way related to the ordered regions in the granule, it is reasonable to suppose that an alteration of the granule structure to produce a more ordered, and thus more stable, physical state can be associated with a more enzyme-resistant starch granule. As a consequence of an enhanced granule stability produced by hydrothermal treatments, an enhanced RS content might be expected.

Würsch described two procedures for generating RS from HAMS by annealing.[102] One procedure was for annealing of HAMS below 100°C, generating a product with 42% RS by digestion at 37°C. The second procedure was for annealing of HAMS followed by debranching, which led to a product with 30% RS by the TDF method. Shi and Trzasko[103] described treatment of HAMS at temperatures from 60 to 160°C, with temperature varying inversely according to the moisture content so that birefringence was not lost. On heating HAMS at 100°C at 37% water for 1 to 4 h, about 40% RS was determined by the TDF analysis, compared to 12% RS in the initial HAMS. In a subsequent improvement,[104] inorganic salts were added to inhibit swelling, thus helping retain granular structure at a higher temperature for a particular moisture content.

Gunaratne and Hoover have recently compared the effects of HMT on a variety of tuber and root starches, pointing out that these starches tend to be more susceptible to HMT than legume or cereal starches. They attributed changes caused by HMT to disruption of crystallites and to disruption of helices in the amorphous regions. They confirmed earlier work in which HMT accomplished at least a partial B-to-A polymorphic transition.[105]

Haralampu and Gross[106] heated HAMS to partially swell but not rupture the granules, and then debranched the starch and allowed it to retrograde. The product was then annealed at 90°C, leading to about 30% RS by TDF analysis.

We have shown that partial acid hydrolysis of HAMS prior to ANN or HMT can increase RS as determined by the TDF method, compared to ANN or HMT without partial acid hydrolysis.[25] We believe that limited acid hydrolysis enhances the mobility of the molecules to allow more efficient rearrangement. For 120 or 140°C HMT temperatures, RS by the TDF method increased (from about 18% for the HAMS to as high as about 60%) even as RS by the Englyst method decreased. We ascribed this behavior to creating a higher proportion of thermally stable RS, but at the expense of the total RS.[25] We have highlighted the advantages of producing boiling-stable granular RS, as this form of granular RS should be stable to subsequent thermal treatments.[107]

2.9.3 Type 3 RS

The production of type 3 RS involves retrogradation of starch molecules subsequent to gelatinization or dispersion of the native starch granules. Some recently proposed strategies involve partial depolymerization either before or after native granule structure is lost. Selective hydrolysis before thermal treatment is done to allow increased polymer mobility for molecular rearrangement, and hydrolysis after thermal treatment is done to enhance the proportion of RS in the ingredient. Shi and Jeffcoat[108] described details of the chemical nature of highly resistant portions from several RS-containing products (although they termed the material "highly pure" RS, it might alternatively be described as highly resistant starch, as it represented the most resistant portion of the starch ingredient). They showed that the molecular nature of the resistant material from the type 3 RS-containing sample was similar to what had been observed previously for retrograded amylose:[109–111] a narrow distribution of linear materials. The molecular weight of the linear resistant material from the granular (type 2) RS-containing samples gave a much broader distribution, leading Shi and Jeffcoat to suggest that granule structure can be an important factor in enzyme resistance for HAMSs.[108] Our recent work supports this suggestion. We utilized the Englyst method[3] to collect the solid-phase material at the conclusion of the *in vitro* digestion of native HAMS samples.[41] After debranching, the chain length profiles from native HAMSs and the putative RS from HAMSs were not different, suggesting that regional differences in granule organization, not differences in molecular structure, accounted for the difference in enzyme susceptibility.

Although type 3 RS is often attributed to amylose retrogradation,[33,111] retrograded amylopectin can also contribute to type 3 RS.[112,113] RS from retrograded amylopectin is completely lost on heating to 100°C. Klucinec and Thompson[114] have shown that dispersions of amylose and amylopectin apparently can interact on cooling. Although it may be that RS would result from this interaction, this RS would not be stable at 100°C. Nevertheless, the presence of amylopectin has been shown to detract from ordering of amylose as observed by DSC.[114,115] Debranching of amylopectin was first shown by Berry to be effective in generating RS from amylopectin.[89] This treatment has the additional advantage that the adverse effect of amylopectin on amylose ordering is apparently minimized as well.

Even starches with normal levels of amylose (<30%) can be processed by heat and moisture to produce some type 3 RS. For example, wheat starch will produce low levels of RS as a result of gelatinization and cooling in bread baking.[44,45] However, the proportion of RS from processed normal starches is currently too low for these starches to be useful in manufacturing RS-containing ingredients.

HAMS is the preferred starting material for producing most high-RS ingredients. Because HAMS is not readily gelatinized, the question may remain of whether RS values for RS-containing ingredients based on treatment of HAMS represent type 2, type 3, or a combination of both. For untreated HAMS the DSC endotherm on initial heating in excess water has two readily identifiable components: one below about 90°C, lost after initial heating, and attributed to the branched molecules in the raw starch; and one between 90 and 110°C, observed on an immediate reheating, and attributed to melting of amylose–monoacyl lipid complexes. Sievert and Pomeranz showed that an endotherm around 140°C was associated with HAMS after variable autoclaving/cooling cycles.[33] They also noted that although native HAMS gives an appreciable RS value, no such high-temperature endotherm is evident for it.[33] Moreover, in a HAMS-derived RS-containing material, the first of the two endotherms for native HAMS may be absent without the presence of an endotherm at or above 120°C.[25] Since the DSC endotherm marking the transition that might be considered gelatinization can be lost without creation of a DSC endotherm associated with one sort of type 3 RS, DSC cannot be used to monitor RS, and this high-temperature endotherm cannot be used to distinguish between types 2 and 3 RS for HAMS. One should certainly not assume that thermal processing of HAMS at 100°C will result in type 3 RS. This complex situation is an example that illustrates that in some circumstances, attempts to distinguish between types of RS may be more confusing than helpful.

The DSC endotherms above 120°C that have been observed for some RS preparations from HAMS have been attributed to amylose retrogradation.[33] When this high-temperature endotherm is present in a treated HAMS, it clearly indicates a form of type 3 RS. However, it is important to recall that type 3 RS may exist without this endotherm. Monoacyl lipid has been shown to provide the central stabilizing ligand for helical amylose inclusion complexes. Monoacyl lipid interferes with amylose double-helical association, and consequently with RS formation.[116–118] We have shown that amylose association on cooling, as observed by DSC, is enhanced when native monoacyl lipid levels are reduced prior to heating.[93]

Use of the TDF method to quantitate type 3 RS after processing treatments might lead one to conclude inappropriately that type 3 RS is entirely stable to boiling. The nature and amount of type 3 RS isolated by a lower-temperature digestion have been compared to those of type 3 RS isolated using a boiling/digestion treatment.[119] Putative type 3 RS was prepared from maize using the lower-temperature digestion; it was then subjected to a boiling/digestion treatment. The authors observed that about one third of the putative RS from the initial digestion was further hydrolyzed by the boiling/digestion treatment. Both the activity of the heat-stable bacterial enzyme and the high temperature contributed to the additional hydrolysis.[119] Others have compared the amount of RS from HAMS that could be determined by the two types of analysis. One group found about 29% RS by the 37°C treatment, but only 14% RS using a boiling treatment.[20] In a separate study of RS from autoclaved HAMS, the 37°C procedure produced an RS value of 39%, whereas the boiling treatment gave a value of only 21% RS.[120] Consequently, for these various samples the boiling/digestion procedure allows more complete amylolytic digestion.

Several reports describe the manufacture of material assumed to be type 3 RS from HAMS. Sievert and Pomeranz[33] described the use of autoclave/cooling cycles to produce RS. They investigated autoclaving at temperatures of 121, 134, and 148°C, and explored up to 20 autoclave/cooling cycles. By this approach they were able to produce up to 40% RS, as determined by the TDF method. They studied the autoclaved starches and the portion of the RS isolated by the TDF method, and showed that the high-temperature DSC endotherm of the intact starches correlated with a similar endotherm from the RS. The authors noted that although they showed a positive association between the amount of recovered RS and the high-temperature endotherm, RS is not a well-defined material. As noted above, one must bear in mind that the TDF procedure would select for that portion of the RS that would be most likely to contribute to this endotherm. Subsequent work by this group employed refined autoclave/cooling strategies for preparing RS.[120–123]

Based on the polymer crystallization theory, Eerlingen et al.[124] hypothesized that the formation of type 3 RS can be considered a crystallization process of amylose in a partially crystalline system. To test that hypothesis, they autoclaved wheat starch in excess water, and then they controlled the crystallization conditions by cooling the gelatinized starch to 0, 68, or 100°C. They followed the RS value over time using the TDF method. They found that initially (~15 min) an increased RS value was favored by the lowest crystallization temperature (0°C), while for long storage times (>10 h) an increased RS value was favored by the highest crystallization temperature (100°C). The authors interpreted these results based on the theory of polymer nucleation and growth.[124]

Eerlingen et al.[111] also studied the influence of amylose chain length on type 3 RS formation by analyzing the starch isolated by the TDF method. They found that the chain length of the crystalline regions was independent of the amylose chain lengths originally used to form the RS. As degree of polymerization (DP) of the initial amylose increased to 260, the RS values increased with the amylose chain length, reaching a maximum value of ~28% RS. Longer chain lengths than DP 260 yielded less than the 28% RS.

Cairns et al.[125] prepared gels from dispersed amylose. After storage for 24 h at 37°C, they studied the enzyme-resistant portion. They suggested that resistance was due to discontinuous crystals, with amorphous regions contained within the crystals and protected by them. Gidley et al.[43] treated HAMS by autoclaving at 134°C, subsequently isolating the putative type 3 RS using a 42°C amylolytic digestion, according to Berry.[89] They showed that this residual material was 60 to 70% double helical, but only 25 to 30% crystalline. The authors suggested that double helices were aggregated imperfectly into B-type crystallites, with single-chain material present as imperfections. Thus, crystallinity was not the major feature in the putative RS they studied. Crystallinity is also clearly not the primary explanation for enzyme resistance of type 2 RS, as the crystallinity of raw potato starch is high, while that of raw HAMS is much lower, and yet both are high in RS. Shamani and colleagues have recently demonstrated that by manipulation of temperature and moisture conditions, RS with either the A or B crystal polymorph can be formed on retrogradation, the A type occurring on crystallization at higher temperature.[126]

We have shown that when HAMS is heated in excess water to 140°C, an endotherm above 140°C is observed on reheating, but when HAMS is heated to 160°C, no high-temperature endotherm is observed on reheating,[93] and we interpret this behavior as an indication that the 160°C temperature allowed melting of the most stable structures. Others have shown that heating amylose to or above 160°C is required to remove all initial structure.[127] We have suggested that these residual stable structures may be important in the formation of RS subsequent to heating. Heating of RS to 200°C in water has been shown to eliminate the characteristic high-temperature endotherm on reheating, whereas treatment at 100°C does not eliminate this endotherm.[123]

Iyengar et al.[128] developed a method for manufacturing type 3 RS that involved the dispersion of HAMS by cooking an aqueous suspension, incubating at 60 to 120°C, and then holding at 4°C. Because intact amylopectin otherwise reduced the retrogradation rate, an enzymatic debranching step was suggested prior to the 60 to 120°C incubation in order to accelerate crystallization. The RS value could be increased by subjecting the retrograded material to enzyme or acid hydrolysis, resulting in a predominately crystalline material that was about 92% resistant to digestion at 37°C. A related method for manufacturing type 3 RS was described by Chiu et al.[129] and Henley and Chiu.[130] This method followed the same general debranching strategy of Iyengar et al.; the innovation of Chiu et al. was that cooling by extrusion could replace subjecting the dispersion to cooling and heating cycles to produce RS, thus dramatically reducing the process time.

Vasanthan and Bhatty[131] described a type 3 RS-containing product made from HAMS and other starches using an initial heating at 100°C and retrogradation at

4°C. Samples were then subjected to partial acid hydrolysis prior to heating to 100 to 140°C in 30% water. By the TDF method, an RS value of over 25% was achieved for HAMS. Haynes et al.[132] described the manufacture of a particularly thermostable type 3 RS, with a DSC endotherm peak above 140°C. The process was designed to avoid lower-melting crystals and amylose–lipid complexes. The initial heating step is above the amylose–lipid complex melting temperature, but said to be below the melting point of the type 3 RS. One or more subsequent cycles of nucleation (at about 60°C) and propagation (at above 120°C) are called for, first at a temperature that precludes amylopectin crystallization, and then at a temperature above melting of the amylose–lipid complex, to destroy any amylose–lipid complexes that might have formed during nucleation. Haynes et al. produced a material of up to 35% RS, as determined by the TDF method.[132] One might consider this temperature cycling protocol to be an improvement over the original temperature cycling of Sievert and Pomeranz,[33] as the earlier treatment would have had a similar effect of removing amylose–lipid complexes and amylopectin crystals while retaining the more stable amylose–amylose interactions with each heating.

Kettlitz et al.[133] produced an RS-containing ingredient of largely DP 10 to 35 by beginning with a potato or tapioca maltodextrin, and simultaneously debranching and allowing retrogradation. The product was about 55% RS as determined by 37°C incubation and had a DSC peak at about 105°C. Although the RS value was not thermally stable, the advantage of this product is in its enhanced *in vitro* production of butyrate. The potential health-promoting contributions of this material have recently been reviewed.[70] Others have produced a more heat-stable butyrogenic RS-containing material from synthetic short-chain amylose.[134]

2.9.4 Type 4 RS

The digestibility of modified starches has been studied from the perspective of caloric availability and safety,[135] but until recently not from the point of view of using these techniques to intentionally modulate RS levels. The use of a cross-linking chemical modification specifically for manufacture of RS from normal starches of several different plants has been described,[49,50] wherein it is shown that generation of distarch phosphodiester cross-linkages can be modulated to produce products of 40 to 98% RS by the TDF method. More recently, Annison et al. showed that a substitution chemical modification could be used to generate high levels of RS from normal maize starch. They employed the particularly innovative strategy of substituting with SCFAs that are generated by fermentation of RS, thereby accomplishing higher levels of SCFAs when the RS was degraded in the colon.[51] Dextrinization would be another way to produce RS, as enzyme resistance of pyrodextrins has been reported.[136,137]

2.10 CONCLUSIONS

RS is not a chemical entity, but rather, it is defined with reference to the human digestive tract for a population of healthy individuals. The newly accepted *in vitro* method for RS determination may help generate increased clarity with respect to research on RS manufacture. Nevertheless, there is a risk that the availability of an approved RS method might obscure our ability to perceive the diverse chemical nature of materials contributing to the RS values.

RS is generally found as a component of an RS-containing ingredient or food. The goal of including an ingredient high in RS should be to combine physical functionality, processing stability, and nutritional functionality. The physical functionality of the RS-containing ingredient is required for appropriate physical characteristics of the food, such as texture, water-holding capacity, etc. The processing stability of RS is important to preserve the nutritional functionality of the RS-containing ingredient. The nutritional functionality of the RS-containing ingredient involves both resistance to digestion in the small intestine and fermentation in the colon.

Although categorization of RS into types 1, 2, 3, and 4 can provide a conceptual basis for differentiating among RS materials, these categories may be a constraint to better understanding of RS because they may contribute to the idea that there is homogeneity within a type, and to the idea that the differentiation among the types is straightforward. Digestion of starch-containing materials can be manipulated by both chemical and physical means, but much research remains to be done to understand the process and nature of the products. Eventually we should be able to design and produce starch materials having a desired rate and extent of digestion, and for the undigested starch, desirable characteristics of fermentation in the colon.

REFERENCES

1. Englyst, H.N., Wiggins, H.S., and Cummings, J.H., Determination of the non-starch polysaccharides in plant foods by gas-liquid chromatography of constituent sugars as alditol acetates, *Analyst*, 107, 307, 1982.
2. Englyst, H.N. et al., Dietary fiber and resistant starch, *Am. J. Clin. Nutr.*, 46, 873, 1987.
3. Englyst, H.N., Kingman, S.M., and Cummings, J.H., Classification and measurement of nutritionally important starch fractions, *Eur. J. Clin. Nutr.*, 46, S33, 1992.
4. Asp, N.-G. and Bjorck, I., Resistant starch, *Trends Food Sci. Technol.*, 3, 111, 1992.
5. Champ, M. et al., *In vivo* techniques to quantify resistant starch, in *Complex Carbohydrates in Foods*, Cho, S.S., Prosky, L., and Dreher, M., Eds., Marcel Dekker, New York, 1999, p. 157.
6. Champ, M., Kozlowski, F., and Lecannu, G., *In-vivo* and *in-vitro* methods for resistant starch measurement, in *Advanced Dietary Fibre Technology*, McCleary, B.V. and Prosky, L., Eds., Blackwell Science Ltd., Oxford, 2001, p. 106.
7. Langkilde, A.M., Champ, M., and Andersson, H., Effects of high-resistant-starch banana flour (RS2) on *in vitro* fermentation and the small-bowel excretion of energy, nutrients, and sterols: an ileostomy study, *Am. J. Clin. Nutr.*, 75, 104, 2002.
8. Rumessen, J.J., Hydrogen and methane breath tests for evaluation of resistant carbohydrates, *Eur. J. Clin. Nutr.*, 46, S77, 1992.
9. Danjo, K. et al., The resistant starch level of heat moisture: treated high amylose cornstarch is much lower when measured in the human terminal ileum than when estimated *in vitro*, *J. Nutr.*, 133, 2218, 2003.
10. Olesen, M., Rumessen, J.J., and Gudmand-Høyer, E., The hydrogen breath test in resistant starch research, *Eur. J. Clin. Nutr.*, 46, S133, 1992.

11. Muir, J.G. et al., Resistant starch in the diet increases breath hydrogen and serum acetate in human subjects, *Am. J. Clin. Nutr.*, 61, 792, 1995.
12. van Munster, I.P., Tangerman, A., and Nagengast, F.M., Effect of resistant starch on colonic fermentation, bile acid metabolism, and mucosal proliferation, *Dig. Dis. Sci.*, 39, 834, 1994.
13. van Munster, I.P. et al., Effect of resistant starch on breath-hydrogen and methane excretion in healthy volunteers, *Am. J. Clin. Nutr.*, 59, 626, 1994.
14. Elmstahl, H.L., Resistant starch content in a selection of starchy foods in the Swedish market, *Eur. J. Clin. Nutr.*, 56, 500, 2002.
15. Baghurst, K.I., Baghurst, P.A., and Record, S.J., Dietary fibre, non-starch polysaccharide, and resistant starch intakes in Australia, in *CRC Handbook of Dietary Fiber in Human Nutrition*, Spiller, G.A., Ed., CRC Press, Boca Raton, FL, 2001, p. 583.
16. Topping, D.L., Fukushima, M., and Bird, A.R., Resistant starch as a prebiotic and synbiotic: state of the art, *Proc. Nutr. Soc.*, 62, 171, 2003.
17. Englyst, H.N. et al., Measurement of resistant starch *in vitro* and *in vivo*, *Br. J. Nutr.*, 75, 749, 1996.
18. McCleary, B.V. and Monaghan, D.A., Measurement of resistant starch, *J. AOAC Int.*, 85, 665, 2002.
19. Muir, J.G. and O'Dea, K., Validation of an *in vitro* assay for predicting the amount of starch that escapes digestion in the small intestine of humans, *Am. J. Clin. Nutr.*, 56, 540, 1993.
20. Champ, M., Determination of resistant starch in foods and food products: interlaboratory study, *Eur. J. Clin. Nutr.*, 46, S51, 1992.
21. Goni, I. et al., Analysis of resistant starch: a method for foods and food products, *Food Chem.*, 56, 445, 1996.
22. McCleary, B.V., McNally, M., and Rossiter, P., Measurement of resistant starch by enzymatic digestion in starch and selected plant materials: collaborative study, *J. AOAC Int.*, 85, 1103, 2002.
23. Prosky, L. et al., Determination of insoluble, soluble, and total dietary fiber in foods and food products: interlaboratory study, *J. Assoc. Off. Anal. Chem.*, 71, 1017, 1988.
24. Haralampu, S.G., *In-vivo* and *in-vitro* digestion of resistant starch, in *Advanced Dietary Fibre Technology*, McCleary, B.V. and Prosky, L., Eds., Blackwell Science Ltd., Oxford, 2001, p. 413.
25. Brumovsky, J.O. and Thompson, D.B., Production of boiling-stable granular resistant starch by partial acid hydrolysis and hydrothermal treatments of high-amylose maize starch, *Cereal Chem.*, 78, 680, 2001.
26. Muir, J.G. et al., Food processing and maize variety affects amounts of starch escaping digestion in the small intestine, *Am. J. Clin. Nutr.*, 61, 82, 1995.
27. Muir, J.G. and O'Dea, K., Measurement of resistant starch: factors affecting the amount of starch escaping digestion *in vitro*, *Am. J. Clin. Nutr.*, 56, 123, 1992.
28. Rabe, E., Effect of processing on dietary fiber in foods, in *Complex Carbohydrates in Foods*, Cho, S.S., Prosky, L., and Dreher, M., Eds., Marcel Dekker, New York, 1999, p. 395.
29. Saito, K. et al., Effect of raw and heat-moisture treated high-amylose corn starch on fermentation by the rat cecal bacteria, *Starch*, 53, 424, 2001.
30. Kishida, T. et al., Heat moisture treatment of high amylose cornstarch increases its resistant starch content but not its physiologic effects in rats, *J. Nutr.*, 131, 2716, 2001.
31. McCleary, B.V., Measurement of dietary fibre components: the importance of enzyme purity activity and specificity, in *Advanced Dietary Fibre Technology*, McCleary, B.V. and Prosky, L., Eds., Blackwell Science Ltd., Oxford, 2001, p. 89.

32. Faisant, N. et al., Structural discrepancies in resistant starch obtained *in vivo* in humans and *in vitro*, *Carbohydr. Polym.*, 21, 205, 1993.
33. Sievert, D. and Pomeranz, Y., Enzyme-resistant starch. I. Characterization and evaluation by enzymatic, thermoanalytical, and microscopic method, *Cereal Chem.*, 66, 342, 1989.
34. Eerlingen, R.C. and Delcour, J.A., Formation, analysis, structure and properties of type III enzyme resistant starch, *J. Cereal Sci.*, 22, 129, 1995.
35. Brown, I., Complex carbohydrates and resistant starch, *Nutr. Rev.*, 54, S115, 1996.
36. Björck, I., Starch: nutritional aspects, in *Carbohydrates in Food*, Vol. 74, Eliasson, A.-C., Ed., Marcel Dekker, New York, 1996, p. 505.
37. Wolf, M.J., Khoo, U., and Inglett, G.E., Partial digestibility of cooked amylomaize starch in humans and mice, *Starch*, 29, 401, 1977.
38. Sandstedt, R.M., Strahan, D., Ueda, S., and Abbot, R.C., The digestibility of high-amylose corn starches compared to that of other starches. The apparent effect of *ae* gene on susceptibility on amylase action, *Cereal Chem.*, 39, 123, 1962.
39. Brown, I.L. et al., Resistant starch: plant breeding, applications, development and commercial use, in *Advanced Dietary Fibre Technology*, McCleary, B.V. and Prosky, L., Eds., Blackwell Science Ltd., Oxford, 2001, p. 401.
40. Gerard, C. et al., Amylolysis of maize mutant starches, *J. Sci. Food Agric.*, 81, 1281, 2001.
41. Evans, A. and Thompson, D.B., Resistance to alpha-amylase digestion in four native high-amylose maize starches, *Cereal Chem.*, 81, (1): 31–37, 2003
42. Colonna, P., Leloup, V., and Buléon, A., Limiting factors of starch hydrolysis, *Eur. J. Clin. Nutr.*, 46, S17, 1992.
43. Gidley, M.J. et al., Molecular order and structure in enzyme-resistant retrograded starch, *Carbohydr. Polym.*, 28, 23, 1995.
44. Björck, I. et al., On the digestibility of starch in wheat bread: studies *in vitro* and *in vivo*, *J. Cereal Sci.*, 4, 1, 1986.
45. Rabe, E. and Sievert, D., Effects of baking, pasta production, and extrusion cooking on formation of resistant starch, *Eur. J. Clin. Nutr.*, 46, S105, 1992.
46. Niba, L., Effect of storage period and temperature on resistant starch and beta-glucan content in cornbread, *Food Chem.*, 83, 493, 2003.
47. Englyst, H.N. and Cummings, J.H., Digestion of polysaccharides of potato in the small intestine of man, *Am. J. Clin. Nutr.*, 45, 423, 1987.
48. Hoover, R. and Zhou, Y., *In vitro* and *in vivo* hydrolysis of legume starches by alpha-amylase and resistant starch formation in legumes: a review, *Carbohydr. Polym.*, 54, 401, 2003.
49. Seib, P.A. and Woo, K., Food Grade Starch Resistant to Alpha-Amylase and Method of Preparing the Same, U.S. Patent 5,855,946, 1999.
50. Woo, K.S. and Seib, P.A., Cross-linked resistant starch: preparation and properties, *Cereal Chem.*, 79, 819, 2002.
51. Annison, G., Illman, R.J., and Topping, D.L., Acetylated, propionylated or butyrylated starches raise large bowel short-chain fatty acids preferentially when fed to rats, *J. Nutr.*, 133, 3523, 2003.
52. Anon., The definition of dietary fiber, *Cereal Foods World*, 46, 112, 2001.
53. DeVries, J., On defining dietary fibre, *Proc. Nutr. Soc.*, 62, 37, 2003.
54. Bourlioux, P. et al., The intestine and its microflora are partners for the protection of the host: report on the Danone Symposium "The Intelligent Intestine," held in Paris, June 14, 2002, *Am. J. Clin. Nutr.*, 78, 675, 2003.

55. Morita, T. et al., Psyllium shifts the fermentation site of high-amylose cornstarch toward the distal colon and increases fecal butyrate concentration in rats, *J. Nutr.*, 129, 2081, 1999.
56. Govers, M.J. et al., Wheat bran affects the site of fermentation of resistant starch and luminal indexes related to colon cancer risk: a study in pigs, *Gut*, 45, 840, 1999.
57. Martin, L.J.M. et al., Potato and high-amylose maize starches are not equivalent producers of butyrate for the colonic mucosa, *Br. J. Nutr.*, 84, 689, 2000.
58. Thompson, D.B., Strategies for the manufacture of resistant starch, *Trends Food Sci. Technol.*, 11, 245, 2000.
59. Seal, C.J. et al., Postprandial carbohydrate metabolism in healthy subjects and those with type 2 diabetes fed starches with slow and rapid hydrolysis rates determined *in vitro*, *Br. J. Nutr.*, 90, 853, 2003.
60. Björck, I. and Asp, N.-G., Controlling the nutritional properties of starch in foods: a challenge to the food industry, *Trends Food Sci. Technol.*, 5, 213, 1994.
61. Johnston, K.L., Clifford, M.N., and Morgan, L.M., Coffee acutely modifies gastrointestinal hormone secretion and glucose tolerance in humans: glycemic effects of chlorogenic acid and caffeine, *Am. J. Clin. Nutr.*, 78, 728, 2003.
62. Wolever, T.M.S. and Jenkins, D.J.A., Effect of dietary fiber and foods on carbohydrate metabolism, in *CRC Handbook of Dietary Fiber in Human Nutrition*, Spiller, G.A., Ed., CRC Press, Boca Raton, FL, 2001, p. 321.
63. Raben, A. et al., Resistant starch: the effect on postprandial glycemia, hormonal response, and satiety, *Am. J. Clin. Nutr.*, 60, 544, 1994.
64. Behall, K.M. and Howe, J.C., Effect of long-term consumption of amylose vs. amylopectin starch on metabolic variables in human subjects, *Am. J. Clin. Nutr.*, 61, 334, 1995.
65. Robertson, M.D. et al., Prior short-term consumption of resistant starch enhances postprandial insulin sensitivity in healthy subjects, *Diabetologica*, 46, 659, 2003.
66. Vonk, R.J. et al., Digestion of so-called resistant starch sources in the human small intestine, *Am. J. Clin. Nutr.*, 72, 432, 2000.
67. Annison, G. and Topping, D.L., Nutritional role of resistant starch: chemical structure vs. physiological function, *Ann. Rev. Nutr.*, 14, 297, 1994.
68. Ranhotra, G., Resistant starch: health aspects and food uses, in *Advanced Dietary Fibre Technology*, McCleary, B.V. and Prosky, L., Eds., Blackwell Science Ltd., Oxford, 2001, p. 424.
69. de Deckere, E.A.M., Kloots, W.J., and vanAmelsvoort, J.M.M., Resistant starch decreases serum total cholesterol and triacylglycerol concentrations and fat accretion in the rat, *J. Nutr.*, 123, 2142, 1993.
70. Brouns, F., Kettlitz, B., and Arrington, E., Resistant starch and the butyrate revolution, *Trends Food Sci. Technol.*, 13, 251, 2002.
71. Phillips, J. et al., Effect of resistant starch on faecal bulk and fermentation-dependent events in humans, *Am. J. Clin. Nutr.*, 62, 121, 1995.
72. Smith, J.G. and German, J.B., Molecular and genetic effect of dietary derived butyric acid, *Food Technol.*, 87, 1995.
73. Heijnen, M.-L. A., Berg, G.J.v.d., and Beynen, A.C., Dietary raw versus retrograded resistant starch enhances apparent but not true magnesium absorption in rats, *J. Nutr.*, 126, 2253, 1996.
74. Schulz, A.G.M., van Alemsvoort, J.M.M., and Beynen, A.C., Dietary native resistant starch but not retrograded resistant starch raises magnesium and calcium absorption in rats, *J. Nutr.*, 123, 1724, 1993.

75. Birkett, A. et al., Resistant starch lowers fecal concentrations of ammonia and phenols in humans, *Am. J. Clin. Nutr.*, 63, 766, 1996.
76. Dongowski, G., Schmiedl, D., and Jacobasch, G., Beneficial effects of resistant starch in colon depend on the structure of RS type 3, *FASEB J.*, 17, A306, 2003.
77. Pryde, S.E. et al., The microbiology of butyrate formation in the human colon, *FEMS Microbiol. Lett.*, 217, 133, 2002.
78. Rumessen, J.J., Franck, Y.S., and Gudmand-Høyer, E., Acetate in venous blood for determination of carbohydrate malabsorption, *Eur. J. Clin. Nutr.*, 46, S135, 1992.
79. Cassidy, A., Bingham, S.A., and Cummings, J.H., Starch intake and colorectal cancer risk: an international comparison, *Br. J. Cancer*, 69, 937, 1994.
80. Heijnen, M.-L.A. et al., Limited effect of consumption of uncooked (RS2) or retrograded (RS3) resistant starch on putative risk factors for colon cancer in healthy men, *Am. J. Clin. Nutr.*, 67, 322, 1998.
81. Hylla, S. et al., Effects of resistant starch on the colon in healthy volunteers: possible implications for cancer prevention, *Am. J. Clin. Nutr.*, 67, 136, 1998.
82. Saemann, M.D., Bohmig, G.A., and Zlabinger, G.J., Short-chain fatty acids: bacterial mediators of a balanced host-microbial relationship in the human gut, *Wien. Klin. Wochenschr.*, 114, 289, 2002.
83. Knudsen, K.E.B. et al., New insight into butyrate metabolism, *Proc. Nutr. Soc.*, 62, 81, 2003.
84. Brown, I. et al., Fecal numbers of bifidobacteria are higher in pigs fed *bifidobacterium longum* with a high amylose cornstarch than with a low amylose cornstarch, *J. Nutr.*, 127, 1822, 1997.
85. Gibson, G.R. and Roberfroid, M.B., Dietary modulation of the human colonic microbiota: introducing the concept of prebiotics, *J. Nutr.*, 125, 1401, 1995.
86. Wang, X. et al., Manipulation of colonic bacteria and volatile fatty acid production by dietary high amylose maize (amylomaize) starch granules, *J. Appl. Microbiol.*, 93, 390, 2002.
87. Klucinec, J. and Thompson, D.B., Fractionation of high-amylose maize starches by differential alcohol precipitation and chromatography of the fractions, *Cereal Chem.*, 75, 887, 1998.
88. Bear, R.P. et al., Development of "Amylomaize": corn hybrids with high amylose starch, *Agron. J.*, 50, 598, 1958.
89. Berry, C.S., Resistant starch: formation and measurement of starch that survives exhaustive digestion with amylolytic enzymes during the determination of dietary fibre, *J. Cereal Sci.*, 4, 301, 1986.
90. Knutson, C.A. et al., Variation in enzyme digestibility and gelatinization behavior of corn starch granule fractions, *Cereal Chem.*, 59, 512, 1982.
91. McNaught, K.J. et al., High Amylose Starch and Resistant Starch Fractions, U.S. Patent 5,977,454, 1999.
92. Daniels, H.E. and Thompson, D.B., Granule swelling and thermal behavior of several waxy and high amylose maize starches as influenced by native lipid, in *American Association of Cereal Chemists Annual Meeting Program*, 1995, p. 314.
93. Boltz, K.W. and Thompson, D.B., Initial heating temperature and native lipid affects ordering of amylose during cooling of high-amylose starches, *Cereal Chem.*, 76, 204, 1999.
94. Stute, R., Hydrothermal modification of starches: the difference between annealing and heat/moisture-treatment, *Starch*, 44, 205, 1992.
95. Jacobs, H. and Delcour, J., Hydrothermal modifications of granular starch, with retention of the granular structure: a review, *J. Agric. Food Chem.*, 46, 2895, 1998.

96. Fisher, D.K. and Thompson, D.B., Retrogradation of maize starch after thermal treatment within and above the gelatinization temperature range, *Cereal Chem.*, 74, 344, 1997.
97. Jacobs, H. et al., Impact of annealing on the susceptibility of wheat, potato and pea starches to hydrolysis with pancreatin, *Carbohydr. Res.*, 305, 193, 1998.
98. Kurakake, M. et al., Adsorption of alpha-amylase on heat-moisture treated starch, *J. Cereal Sci.*, 23, 163, 1996.
99. Colonna, P. and Mercier, C., Gelatinization and melting of maize starches with normal and high amylose phenotypes, *Phytochemistry*, 24, 1667, 1985.
100. Whittam, M.A., Noel, T.R., and Ring, S.G., Melting behaviour of A- and B-type crystalline starch, *Int. J. Biol. Macromol.*, 12, 359, 1990.
101. Jang, J.K. and Pyun, Y.R., Effect of moisture content on the melting of wheat starch, *Starch*, 48, 48, 1996.
102. Würsch, P., Production of resistant starch, in *Complex Carbohydrates in Foods*, Cho, S.S., Prosky, L., and Dreher, M., Eds., Marcel Dekker, New York, 1999, p. 385.
103. Shi, Y.-C. and Trzasko, P.T., Process for producing amylase resistant granular starch, U.S. Patent 5,593,503, 1997.
104. Chiu, C.-W., Shi, Y.-C., and Sedam, M., Process for Producing Amylase-Resistant Granular Starch, U.S. Patent 5,902,410, 1999.
105. Gunaratne, A. and Hoover, R., Effect of heat-moisture treatment on the structure and physicochemical properties of tuber and root starches, *Carbohydr. Polym.*, 49, 425, 2002.
106. Haralampu, S.G. and Gross, A., Granular Resistant Starch and Method of Making, U.S. Patent 5,849,090, 1998.
107. Thompson, D.B. and Brumovsky, J.O., Manufacture of Boiling-Stable Granular Resistant Starch by Acid Hydrolysis and Hydrothermal Treatment, U.S. Patent 6,468,355, 2002.
108. Shi, Y.-C. and Jeffcoat, R., Structural features of resistant starch, in *Advanced Dietary Fibre Technology*, McCleary, B.V. and Prosky, L., Eds., Blackwell Science Ltd., Oxford, 2001, p. 430.
109. Jane, J.-L. and Robyt, J.R., Structure studies of amylose-V complexes and retrograded amylose by action of alpha amylases, and a new method for preparing amylodextrins, *Carbohydr. Res.*, 132, 105, 1984.
110. Leloup, V.M., Colonna, P., and Ring, S.G., Physicochemical aspects of resistant starch, *J. Cereal Sci.*, 16, 253, 1992.
111. Eerlingen, R.C., Deceuninck, M., and Delcour, J.A., Enzyme-resistant starch. II. Influence of amylose chain length on resistant starch formation, *Cereal Chem.*, 70, 345, 1993.
112. Eerlingen, R.C., Jacobs, H., and Delcour, J.A., Enzyme resistant starch. V. Effect of retrogradation of waxy maize starch on enzyme susceptibility, *Cereal Chem.*, 71, 351, 1994.
113. Russell, P.L., Berry, C.S., and Greenwell, P., Characterization of resistant starch from wheat and maize, *J. Cereal Sci.*, 9, 1, 1989.
114. Klucinec, J. and Thompson, D.B., Amylose and amylopectin interact in retrogradation of dispersed high-amylose starches, *Cereal Chem.*, 76, 282, 1999.
115. Sievert, D. and Würsch, P., Amylose chain association based on differential scanning calorimetry, *J. Food Sci.*, 58, 1332, 1993.
116. Sievert, D. and Würsch, P., Thermal behavior of potato amylose and enzyme-resistant starch from maize, *Cereal Chem.*, 70, 333, 1993.

117. Eerlingen, R.C., Cillen, G., and Delcour, J.A., Enzyme-resistant starch. IV. Effect of endogenous lipids and added sodium dodecyl sulfate on formation of resistant starch, *Cereal Chem.*, 71, 170, 1994.
118. Szczodrak, J. and Pomeranz, Y., Starch-lipid interactions and formation of resistant starch in high-amylose barley, *Cereal Chem.*, 69, 626, 1992.
119. Würsch, P. and Koellreutter, B., Susceptibility of resistant starch to alpha-amylase, *Eur. J. Clin. Nutr.*, 46, S113, 1992.
120. Sievert, D., Czuchajowska, Z., and Pomeranz, Y., Enzyme-resistant starch. III. X-ray diffraction of autoclaved amylomaize VII starch and enzyme-resistant starch residues, *Cereal Chem.*, 68, 86, 1991.
121. Sievert, D. and Pomeranz, Y., Differential scanning calorimetry studies on heat-treated starches and enzyme-resistant starch residues, *Cereal Chem.*, 67, 217, 1990.
122. Czuchajowska, Z., Sievert, D., and Pomeranz, Y., Enzyme-resistant starch. IV. Effects of complexing lipids, *Cereal Chem.*, 68, 537, 1991.
123. Gruchala, L. and Pomeranz, Y., Enzyme-resistant starch: studies using differential scanning calorimetry, *Cereal Chem.*, 70, 163, 1993.
124. Eerlingen, R.C., Crombez, M., and Delcour, J.A., Enzyme-resistant starch. I. Quantitative and qualitative influence of incubation time and temperature of autoclaved starch on resistant starch formation, *Cereal Chem.*, 70, 339, 1993.
125. Cairns, P. et al., Physicochemical studies using amylose as an *in vitro* model for resistant starch, *J. Cereal Sci.*, 21, 37, 1995.
126. Shamani, K., Bianco-Peled, H., and Shimoni, E., Polymorphism of resistant starch type III, *Carbohydr. Polym.*, 54, 363, 2003.
127. Doublier, J.L. et al., Effect of thermal history on amylose gelation, *Progr. Colloid Polym. Sci.*, 90, 61, 1992.
128. Iyengar, R., Zaks, A., and Gross, A., Starch-Derived, Food Grade, Insoluble Bulking Agent, U.S. Patent 5,051,271, 1991.
129. Chiu, C.-W., Henley, M., and Altieri, P., Process for Making Amylase Resistant Starch from High Amylose Starch, U.S. Patent 5,281,276, 1994.
130. Henley, M. and Chiu, C.-W., Amylase Resistant Starch Product from Debranched High Amylose Starch, U.S. Patent 5,409,542, 1995.
131. Vasanthan, T. and Bhatty, R.S., Enhancement of resistant starch (RS3) in amylomaize, barley, field pea and lentil starches, *Starch*, 50, 286, 1998.
132. Haynes, L. et al., Process for Making Enzyme-Resistant Starch for Reduced-Calorie Flour Replacer, U.S. Patent 6,013,299, 2000.
133. Kettlitz, B.W. et al., Highly Fermentable Resistant Starch, U.S. Patent 6,043,229, 2000.
134. Schmiedl, D. et al., Production of heat-stable, butyrogenic resistant starch, *Carbohydr. Polym.*, 43, 183, 2000.
135. Wurzburg, O.B., Nutritional aspects and safety of modified food starches, *Nutr. Rev.*, 44, 74, 1986.
136. Wang, Y.-J., Kozlowski, R., and Delgado, G.A., Enzyme resistant dextrins from high amylose corn mutant starches, *Starch*, 53, 21, 2001.
137. Laurentin, A. et al., Preparation of indigestible pyrodextrins from different starch sources, *J. Agric. Food Chem.*, 51, 5510, 2003.

3 Konjac Glucomannan

K. Nishinari and S. Gao

CONTENTS

3.1 INTRODUCTION

Konjac glucomannan (KGM) is a polysaccharide that is extracted from the tuber of *Amorphophallus konjac* K. Koch. It forms a thermally stable gel in the presence of an alkaline coagulant and has been used in Japanese traditional dishes for a long time.[1] Konjac gels are boiled with vegetables, mushrooms, meat, surimi, etc. KGM is a dietary fiber that cannot be hydrolyzed by digestive enzymes in humans and is considered to be good for health.

It has recently attracted much attention because it forms a gel by mixing with xanthan or κ-carrageenan in the absence of alkaline coagulant.[2] Since this gel can be made at low pH, many kinds of dessert jellies containing various fruit juices have appeared in the market in Japan. Finally, konjac gum has also been used as binding material in pet foods in Europe. With the advent of an aged society, the number of aged persons with difficulty in mastication and deglutition increases and becomes a serious problem. KGM is expected to act as a good texture modifier to control the rheological properties of various foods. Contamination causes another problem, and KGM as a biodegradable material has attracted even more attention than before.

In the present work, the physicochemical properties and functionality of konjac glucomannan are reviewed.

3.2 STRUCTURE AND MOLECULAR WEIGHT OF KONJAC GLUCOMANNAN

The constitution of konjac glucomannan was the object of many previous investigations, and the results show that konjac glucomannan consists of glucose and mannose, and the ratio is 1.0 to 1.6.[3] It has been believed that there are some branching points at the C-3 of the mannose unit.[4] However, a recent study shows that the branching point is C-6 of glucosyl units and the degree of branching is about 8%. The ratio of terminal glucosyl units to mannosyl units is ca. 2.[5]

KGM contains some acetyl groups (about 1 acetyl group per 19 sugar residues)[6] that confer the water solubility to the konjac polymer. It forms a gel upon heating in the presence of alkali, and the role of alkali is believed to remove the acetyl groups. The peak at 1730 cm^{-1} in the Fourier-transformed infrared (FTIR) absorbance spectra of KGM film, which was attributed to the absorption by acetyl groups, disappeared by alkali treatment,[7,8,20] as shown in Figure 3.1.

3.2.1 The Man/Glc Ratio

The ratio of mannose (Man) to glucose (Glc) for KGM samples with different molecular weights, prepared by enzymatic degradation, was determined by high-performance liquid chromatography (HPLC).[8] The Man/Glc ratio was calculated from the peak areas of mannose and glucose detected with a refractive index (RI) detector, and was approximately 2.0 for all the fractions, higher than the previously reported[3] 1.6. It was close to the ratio 2.1 reported for glucomannan extracted from Scotch pine.[9] The experimental fact that the Man/Glc ratio was not much different for a nondegraded specimen and for enzymatically degraded specimens suggests that there is no block structure in konjac glucomannan because glucose and mannose residues do not exhibit a difference in enzyme reactivity.

3.2.2 Fractionation

To obtain fractions with different molecular weights, konjac glucomannan powder was dissolved in water. Methanol as a precipitant was added, and the solution was

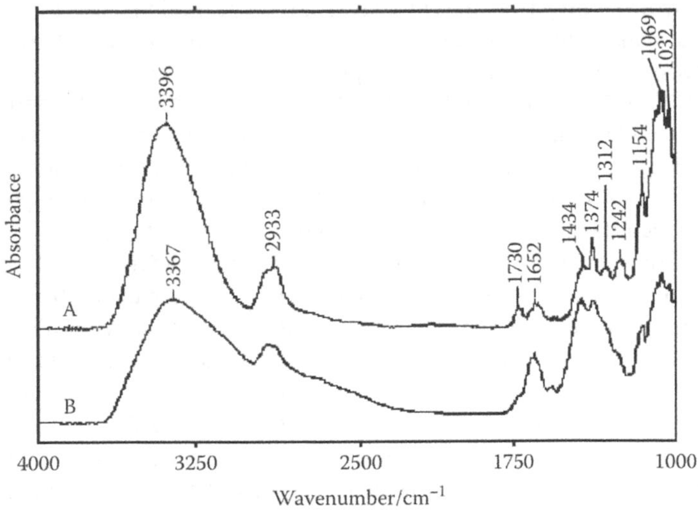

FIGURE 3.1 Absorbance FTIR spectra of LM4 with the assignment of the main bands. A: Untreated. B: Treated with Na_2CO_3. (From Zhang, H. et al., *Biopolymers*, 59, 38, 2001. With permission.)

kept at 30°C. Although a clear phase separation could not be achieved as has been done for many solutions of synthetic polymers in organic solvents, cloud-like flakes appeared in the solution after 1 day. The flakes were removed and methanol was added again. By repeating this procedure, four fractions (F1 to F4) were obtained.[10]

Fractions with different molecular weights were also prepared by Shimizu Chemical Co. (Hiroshima, Japan) using an enzymatic degradation method. The native and nondegraded konjac glucomannan (ND) were treated with an enzyme (SP-249, Novo Nordisk A/S, Copenhagen, Denmark) for different reaction times at ambient temperature, and four fractions with low molecular weights, LM1 to LM4, were obtained.[11]

3.2.3 Molecular Weight

The molecular weight of each KGM fraction was determined by gel permeation chromatography (GPC) at room temperature. KGM was dissolved in cadoxen to obtain a dilute solution because of its very low solubility in aqueous solutions.[1] Cadoxen is known to dissolve cellulose and to give a clear, colorless, and stable solution.[12] A colorless solution allows the use of an RI detector,[13] and the much lower viscosity of the cadoxen solution compared to that of aqueous solution is a great advantage for GPC analysis. The peak corresponding to the highest molecular weight fraction (8.53×10^5) of pullulan appeared at 11.73 min, and that corresponding to the lowest one (5.80×10^3) appeared at 15.07 min. The standard pullulan samples[14] gave a good linear calibration curve (data not shown). The weight-average molecular weight (Mw) for five KGM samples decreased with increasing the reaction time with the enzyme. The molecular weight distribution was not narrow judging from the chromatograms. The elution curve for ND showed the second peak in the lower molecular weight region. The sample did not flow homogeneously through the columns, probably due to the high viscosity. This behavior was typically observed in a highly viscous solution, and the true average molecular weight could be higher. Even in cadoxen, which forms a KGM solution with far lower viscosity than in water, it was difficult to prepare a suitable solution for GPC analysis, especially in the case of the highest molecular weight fraction. Native (nondegraded) KGM is known as a gum with very high viscosity, and it is difficult to determine the molecular weight of native KGM by GPC. The ND sample contains molecules with molecular weights higher than 1.0×10^6, but no lower than 1.0×10^4 molecular weight fraction. A similar wide distribution of molecular weights was also reported for aqueous konjac mannan.[15] It ranged from 4.0×10^4 to more than 1.0×10^6 and the polymer eluted over almost the entire range of elution volumes for the GPC gels (4.0×10^4 to 2.0×10^7). The enzyme treatment increased the ratio of lower molecular weight polymers; however, the fraction of higher than 1.0×10^6 molecular weight still remained, even in the lower molecular weight fraction LM1. The exclusion limit for the GPC column was assumed to be 5.0×10^7 because there are no standard polymers with higher than 1.0×10^6 molecular weight. It is difficult to assign a molecular weight for the earlier eluted fraction. Therefore, high accuracy for the molecular weight obtained from GPC should not be expected. It is certain, however, that the KGM fractions have different molecular weights.

3.3 SOLUTION PROPERTIES OF KONJAC GLUCOMANNAN

3.3.1 INTRINSIC VISCOSITY

The intrinsic viscosity was determined in water and in 4 *M* urea for the four fractions, F1 to F4.[10] The intrinsic viscosity observed in 4 *M* urea was higher than that observed in water for all the fractions. The conformation of konjac glucomannan molecules may be more expanded in urea because hydrogen bonds between hydroxyl groups in konjac glucomannan molecules are broken.

A fraction with a high molecular weight began to deviate from the straight line in the plot of the reduced viscosity vs. konjac glucomannan concentration at a lower concentration than a fraction with a lower molecular weight. This has also been observed for pullulan[14] and many other polymers.

3.3.2 ZERO-SHEAR SPECIFIC VISCOSITY

The relationship between the logarithm of the zero-shear specific viscosity (η_{sp0}) and the logarithm of the KGM concentration (c) is shown in Figure 3.2. Log η_{sp0} increased linearly with increasing log c when the value of log η_{sp0} was lower than about 1. The double logarithmic plots of the zero-shear specific viscosity η_{sp0} against the coil overlap parameter ($c[\eta]$) are shown in Figure 3.3. For dilute solutions, slopes of the plots were close to 1.4 for all the fractions, as observed for many polysaccharide solutions. It was not possible to obtain a clear inflection

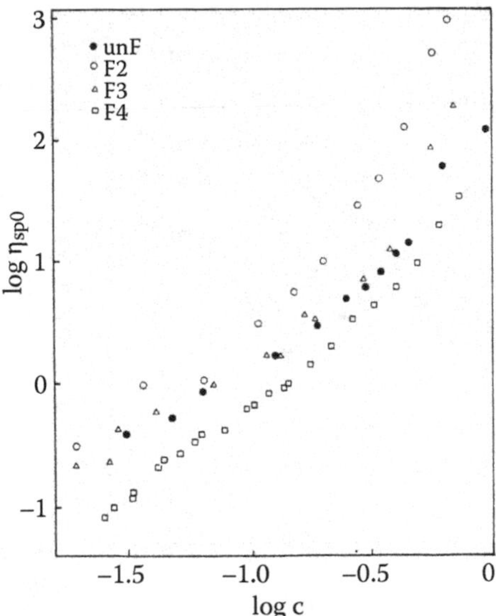

FIGURE 3.2 Concentration dependence of the zero-shear specific viscosity (η_{sp0}) of konjac glucomannan solution. UnF (unfractionated material, ●), F2 (○), F3 (Δ), F4 (□). (From Kohyama, K. and Nishinari, K., *Jpn. Agric. Res. Q.*, 31, 301, 1997. With permission.)

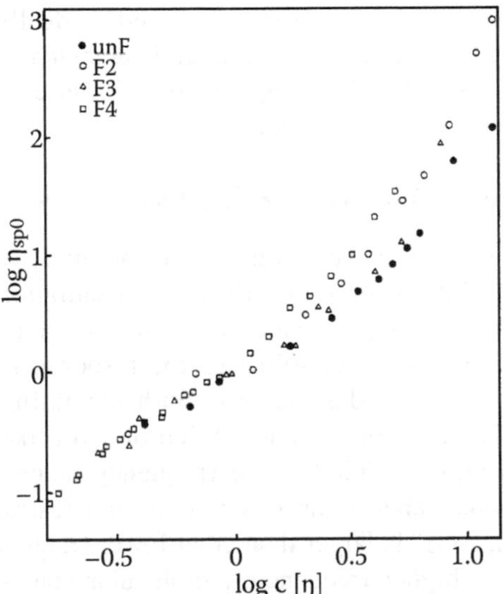

FIGURE 3.3 Dependence of the zero-shear specific viscosity (η_{sp0}) of konjac glucomannan solution on the coil overlap parameter, c[η]. Symbols are the same as in Figure 3.2. (From Kohyama, K. and Nishinari, K., *Jpn. Agric. Res. Q.*, 31, 301, 1997. With permission.)

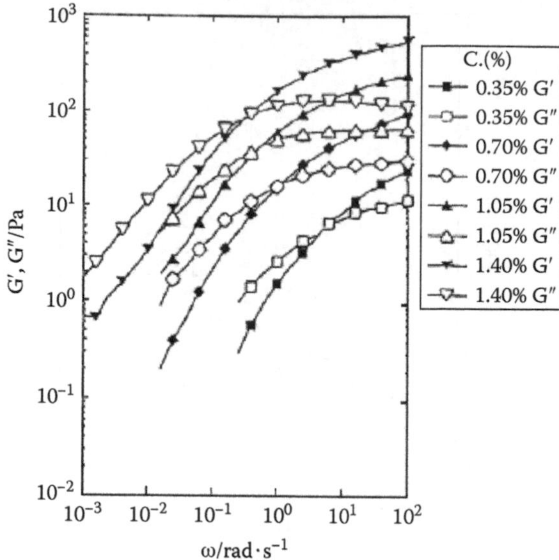

FIGURE 3.4 Frequency dependence of G' and G" of KGM aqueous dispersions in various concentrations. C: concentration of KGM (wt %) (From Yoshimura, M. et al., *Carbohydr. Polym.*, 35, 71, 1998. With permission.)

point of the curves in Figure 3.3, because the low solubility of konjac glucomannan made it difficult to prepare solutions that exhibited large c[η] values. However, the inclination of the slope of the curves increased gradually with increasing log c[η], suggesting that significant coil overlap and entanglement had already started when c[η] > 1. The onset of coil overlap occurs at lower concentrations for KGM molecules than for other polysaccharides.

3.3.3 Dynamic Viscoelasticity of KGM Dispersions

The frequency dependence of the storage shear modulus, G', and the loss shear modulus, G", for KGM dispersions of different concentrations is shown in Figure 3.4. The behavior of KGM plus water is typical of a concentrated polymer solution. In concentrated polymer solutions the response is liquid-like, i.e., G" is larger than G' and both moduli increase with increasing frequency at lower frequencies, while the behavior approaches that of solid-like materials, i.e., G' is larger than G", and both moduli become frequency independent at higher frequencies. The molecular chains can disentangle and rearrange during the long period of oscillation, so G" is larger than G' at lower frequencies in concentrated polymer solutions. At higher frequencies, molecular chains cannot disentangle during the short period of oscillation, so G' is larger than G", because the entanglement points play the role of a temporary cross-linking junction zone. Moreover, the crossover frequency of G' and G" shifted as expected to lower frequencies with increasing KGM concentration.

3.3.4 Other Solution Properties

The partial specific volume of konjac glucomannan was determined as a function of pH by density measurements[16] and showed a steep rise at around pH = 11.5 and 3, as shown in Figure 3.5. This suggests that the conformational change is necessary for gel formation of konjac glucomannan. The typical results of light-scattering measurements, carried out for acetylated konjac glucomannan Ac21 prepared by using acetic anhydride in the presence of zinc chloride as a catalyst, are shown in Figure 3.6.[17] From the intercept with the ordinate and the slopes

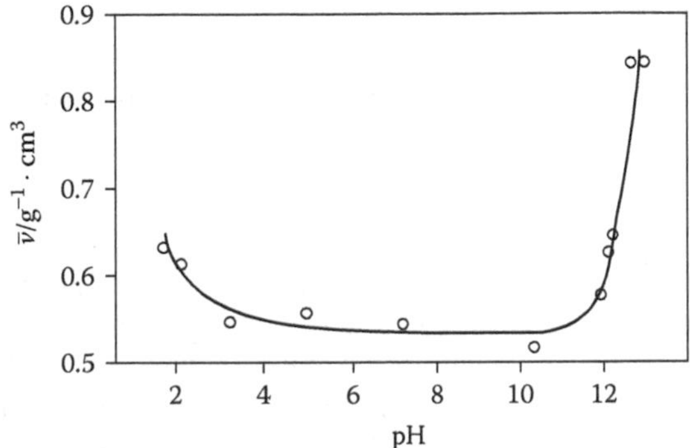

FIGURE 3.5 Apparent partial specific volume of konjac glucomannan as a function of pH at 25.0°C. The pH was adjusted by addition of HCl or NaOH. Concentration of konjac glucomannan: 0.22 w/w%. (From Kohyama, K. and Nishinari, K., in *Gums and Stabilisers for the Food Industry 5*, Phillips, G.O. et al., Eds., IRL Press, Oxford, 1990, p. 459. With permission.)

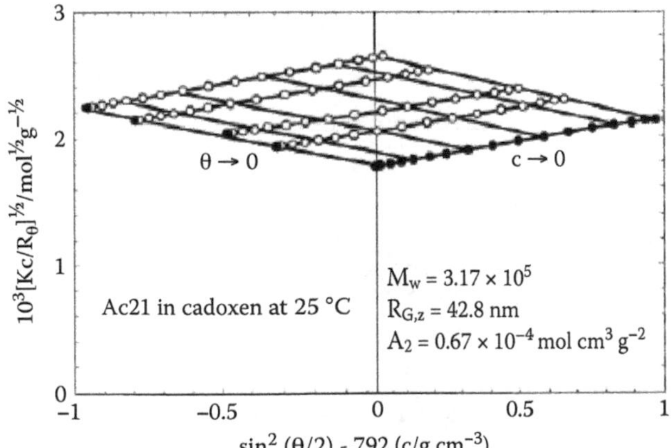

FIGURE 3.6 Zimm plot of the acetylated KGM sample Ac21 in cadoxen at 25°C.[17]

against the concentration and scattering angle, weight-average molecular weight, Mw, second virial coefficient, A_2, and z-average radius of gyration, $R_{G,z}$, were evaluated, respectively.

KGM forms a biphasic liquid crystal (LC) phase in water at 7 wt% concentration and becomes completely anisotropic above 10 wt% concentrations, as observed by polarized optical microscopy.[18] No order–disorder transition was observed when the LC solutions were heated on the hot stage. Circular dichroism spectra show positive bands at 210 and 290 nm for the LC phase and the shear-induced birefringence, respectively. Increases in the intensity of wide-angle x-ray diffraction patterns of the films cast from LC solutions provided further evidence of the existence of mesophase in the KGM solutions.

3.4 GELATION OF KGM IN THE PRESENCE OF ALKALI

3.4.1 GELATION KINETICS OF KGM WITH DIFFERENT MOLECULAR WEIGHTS

The time dependence of G' and G" or 2% aqueous dispersions of KGM with molecular weights from 2.56 to 5.96 × 10^5 in the presence of 0.02 mol/l sodium carbonate at 60°C is shown in Figure 3.7.[19] Time t = 0 was defined as the time when the alkali was added. The observed curve for G' as a function of time was fitted well by an equation for the reaction of the first order. The gelation time t_0, at which G'

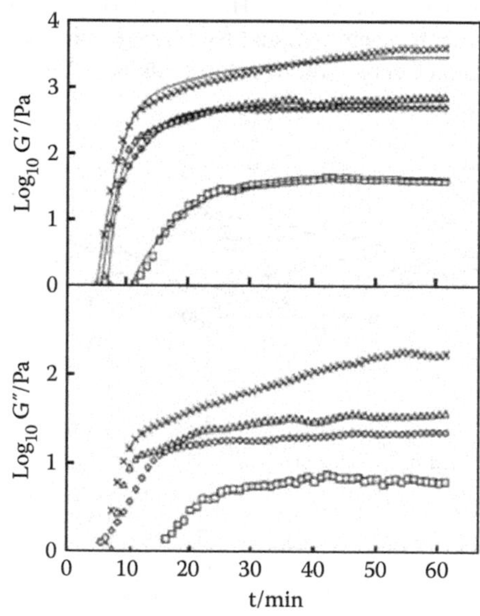

FIGURE 3.7 Time dependence of G' and G" for 2% aqueous KGM dispersions with different molecular weights. □, LM1; ◇, LM2; △, LM3; ×, LM4. Measurement temperature: 60°C. Symbols represent the experimental values and the solid lines represent the calculated curves. (From Yoshimura, M. and Nishinari, K., *Food Hydrocoll.*, 13, 227, 1999. With permission.)

began to deviate from the baseline, became shorter and the saturated G' increased with increasing molecular weight. The time dependence of G' and G" for aqueous dispersions of KGM with the molecular weight of 2.56×10^5 and different concentrations from 1.0 to 3.0% in the presence of 0.02 mol/l sodium carbonate at 65°C was also fitted well by an equation for the reaction of the first order (data not shown). The gelation time t_0 became shorter and the saturated G' increased with increasing concentration of KGM.

In the time dependence of G' and G" for 2 and 3% aqueous dispersions of KGM with the molecular weight of 4.38×10^5 at temperatures from 40 to 90°C in the presence of 0.02 mol/l sodium carbonate, the maximum of G' was observed at temperatures above 75°C.[19,20] To examine whether the maximum of G' is a real phenomenon or an artifact caused by the slippage, the penetration force was measured using a spherical Teflon plunger of radius 5 mm.[20] This method has the advantage that it is free from slippage. The normalized penetration force observed at different temperatures from 50 to 90°C did not show any maximum as a function of time, indicating that the maximum observed in G' by the oscillatory shear mode mentioned above was induced by the slippage. The apparent maximum in G' in the study of gelation is a notorious problem and has sometimes been reported erroneously.[20]

The time dependence of G' and G" for aqueous dispersions of KGM with the molecular weight of 1.17×10^6, which was not degraded by enzyme, with different concentrations from 0.5 to 2.0% in the presence of 0.02 mol/l sodium carbonate at 80°C did not show any maximum. The gelation time t_0 became shorter and the saturated G' increased with increasing concentration, as in enzymatically treated samples. The slippage may be due to disentanglement of the adsorbed chains of KGM during gelation. When the surface of the fixture is predominantly occupied by the free water molecules, catastrophic slippage takes place.[20]

A molecular level description of the time course of the gelation of KGM was presented by Williams et al.,[21] and the role of alkali addition was considered in detail. Nuclear magnetic resonance (NMR) relaxometry was utilized as a complementary methodology to mechanical spectroscopy to probe events occurring as a prelude to network formation, and high-resolution NMR was used to follow the deacetylation process. It was shown that the addition of alkali plays an important solubilizing role as well as facilitating the deacetylation of the chain. Deacetylation is important both in reducing the inherent aqueous solubility of the polymer and in progressively negating the alkali-induced polyelectrolytic nature of the polysaccharide chain via reaction-induced pH changes. Figure 3.8a shows spin–spin relaxation time T_2 obtained from 1% KGM samples after the addition of alkali, measured during a series of temperature ramps (at 1°C min^{-1}) from 20°C to 50, 60, 70, and 80°C and subsequent holding periods. The initial linear-like response merely reflects the temperature dependence of the chain mobility, although it should be remembered that at this concentration there is an initial solubilizing effect of alkali. After this initial period (some 25 to 30 min of ramping at 1°C min^{-1}), it is clear that the observed T_2 begins to drop. The simplest interpretation here is that the chain mobility now begins to decrease owing to the aggregation of material that has undergone alkali-induced deacetylation. It is clear from the rheological data shown in Figure 3.8b that with a continued temperature increase, the modulus increases with the rate dependent on the holding temperature. There also seems to be some evidence of a delay period at the lower temperature (50°C) after the final temperature is reached before the elastic modulus is detected to rise.

It is proposed that observed induction periods following alkali addition (during which the elastic modulus does not rise) are not simply deacetylation delays but are related to the aggregation kinetics of the deacetylated material, although the two processes may appear indistinguishable. A summary of the proposed mechanism is given in Figure 3.9.

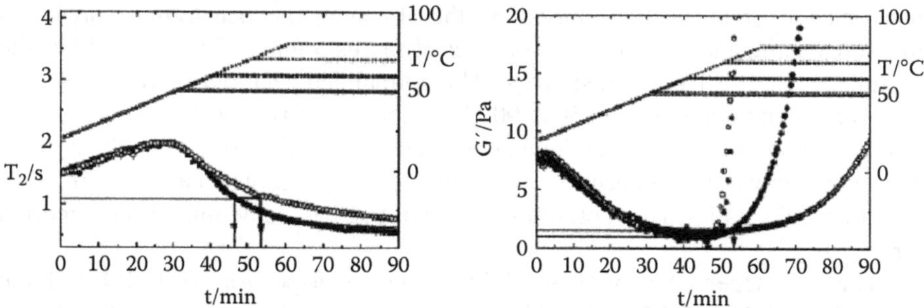

FIGURE 3.8 (a) T_2 values obtained from 1% KGM samples after the addition of alkali, measured during a series of temperature ramps (at 1°C min^{-1}) from 20°C to 50, 60, 70, and 80°C and subsequent holding periods. (b) Corresponding elastic modulus (measured at 1 Hz and 0.5% strain) for a 1% KGM solution under the same thermal regime as the NMR experiments (a) with identical alkali addition. In both cases, the data are given in the same symbols as the corresponding temperature ramp (also shown) under which they were obtained. (From Williams, M.A. et al., *Biomacromolecules*, 1, 440, 2000. With permission.)

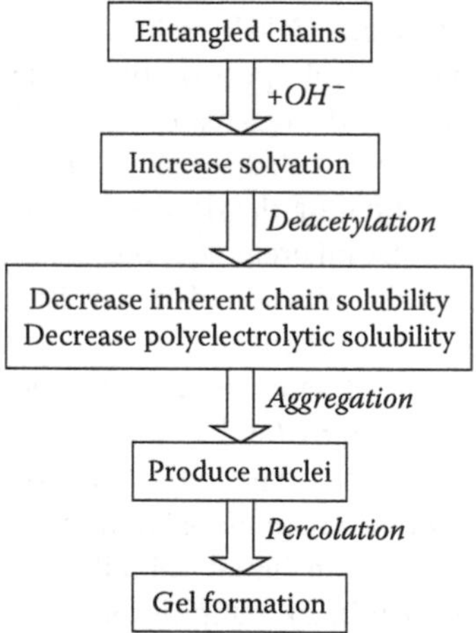

FIGURE 3.9 Schematic diagram of the proposed gelation mechanism for KGM. (From Williams, M.A. et al., *Biomacromolecules*, 1, 440, 2000.)

To understand the role of acetyl groups in the gelation of KGM solutions, several fractions of KGM with different degrees of acetylation were prepared.[17] Five fractions with different acetylation levels were obtained. The degree of acetylation (DA) (the weight percent of acetyl-substituted residues in KGM backbone) ranged from 1.6 (not treated) to 5.3%.

The molecular weight determined by light scattering in cadoxen was decreased to about the half value by the deacetylation.[17] However, the various acetylation reaction conditions, such as different acetylation temperature and the amount of catalyst, did not lead to a remarkable difference in molecular weight. Therefore, a slight difference in molecular weight between the acetylated samples was not taken into account, and the attention was paid mainly to the influence of DA. The mechanism of the main chain scission was explored later by Gao and Nishinari[23] to get better or ideal samples with different DAs without changing the molecular weight.

Figure 3.10a shows the time dependence of G' of 2.0 wt% aqueous dispersions of KGM with different DAs in the gelation process at 45°C in the presence of Na_2CO_3. The concentration of Na_2CO_3 was fixed as 0.2 wt%. The value of G' at $t = 0$ for a native fraction Rs was far larger than that for acetylated KGM samples. However, G' of Rs was finally overtaken by that of acetylated KGM samples with time elapsing. The gelation time, t_{gel}, defined as the time of the crossover of G' and G", as shown in Figure 3.10b, became longer with increasing DA. Although the gelation time determined in this way slightly changed with the frequency, this method was adopted for the simplicity.[22] It is expected that the deacetylation reaction and further aggregation process for KGM samples with higher DAs need a longer time than those for KGM with lower DAs. The G' of all the samples increased rapidly in the beginning of gelation and finally attained the plateau values. It took a longer time for KGM with higher DAs to reach the saturated value of G'. The G' of the native KGM, Rs, attained maximum values in ca. 245 min (ca. 4 h), while the G' of a KGM fraction with the highest DA, Ac32, still continued to increase even after 2300 min (ca. 38 h). The reason why Ac32 showed a different behavior is interpreted

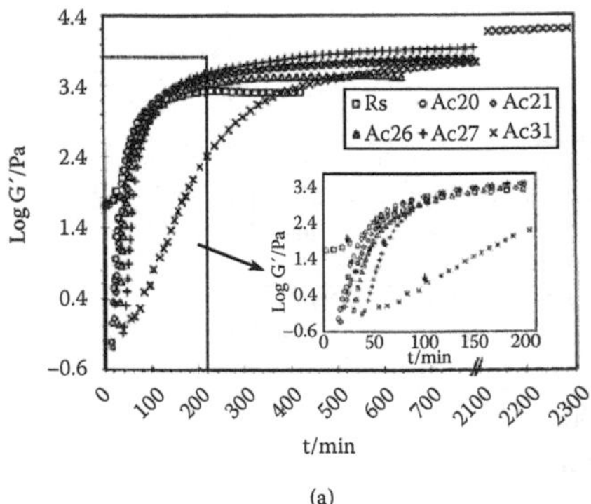

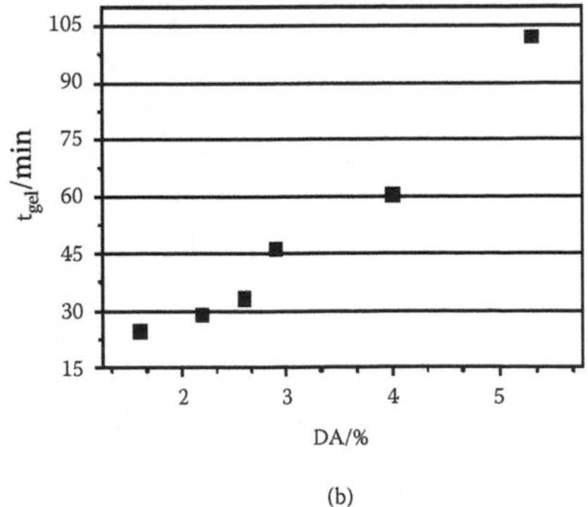

FIGURE 3.10 (a) Time dependence of G' of 2.0 wt% aqueous dispersions of KGM with different DAs in the gelation process at 45°C in the presence of Na_2CO_3. The concentration of Na_2CO_3 was fixed as 0.2 wt%. (b) Gelation time t_{gel} of the KGM aqueous dispersions against DA. (From Huang, L. et al., *Biomacromolecules*, 3, 1296, 2002. With permission.)

as follows: the DA of Ac32 is the highest, and the largest amount of alkali is required. The present concentration of Na_2CO_3 solution was insufficient for 2 wt% Ac32 dispersion to form a gel in a short time. It was difficult to observe the exact plateau values of G' for some acetylated KGM samples in this experimental condition due to the long measuring time.

After a certain amount of time (data not shown), both storage and loss moduli, G' and G", of Rs aqueous dispersions, in the presence of Na_2CO_3 at a fixed concentration ratio of Na_2CO_3 to the degree of acetylation of KGM (0.1), were found to increase monotonically and attained plateau values, the saturated storage modulus (G'_{sat}) and loss modulus (G''_{sat}) respectively, as was observed previously. t_{gel} increased sharply with decreasing concentration, and G'_{sat} increased with increasing concentration. Since molecular chains are close to each other at higher concentrations, the probability of the formation of a junction zone is higher than that at lower concentrations. Gelation would begin even before the complete loss of acetyl groups at higher concentrations; therefore, t_{gel} became shorter with increasing concentration of KGM, as expected.[17]

If the alkaline concentration (C_{Al}) to DA was fixed to a constant (0.1), a similar time course of G' for all the samples, except the fraction with the highest DA, would be observed as shown in Figure 3.11. It indicates that C_{Al}/DA plays a crucial role in the gelation process. The gelation rate is determined mainly by the ratio of alkaline concentration to the degree of acetylation. However, the fact that Ac32 shows an exceptional behavior should be explored in the near future.

Recently, KGM samples with different degrees of acetylation were prepared without changing the molecular weight.[23] The acetylation of KGM was carried out using acetic anhydride in the presence of pyridine as the catalyst. By changing the reaction temperature or the amount of pyridine, a series of acetylated samples with

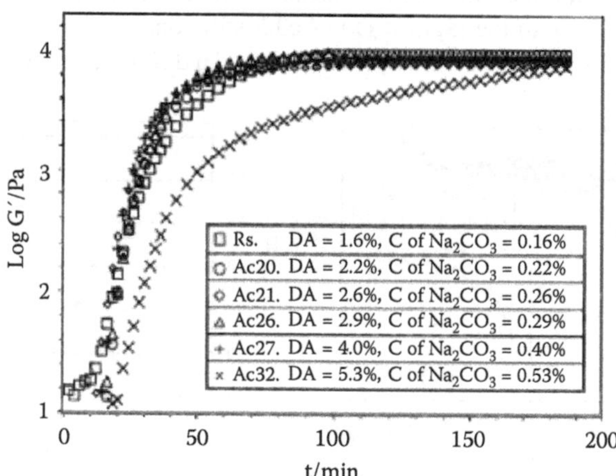

FIGURE 3.11 Time dependence of G' of 2.0 wt% KGM aqueous dispersions in the presence of Na_2CO_3 at 60°C. The ratio of alkaline concentration to the degree of acetylation was fixed to a constant (0.1). (From Huang, L. et al., *Biomacromolecules*, 3, 1296, 2002. With permission.)

a DA range from 1.38 to 10.1 wt% was obtained. Table 3.1 lists the reaction conditions and some parameters of the acetylated KGM samples. The intrinsic viscosity of native KGM (Rs) decreased by about 9% from 557 $cm^3\ g^{-1}$ of Rs to 500 ± 20 $cm^3\ g^{-1}$ of each acetylated sample. Comparably, the intrinsic viscosity of native KGM decreased by 28% after acetylation by using zinc chloride as the catalyst in a previous work.[17] A slight decrease in the viscosity-average molecular weight from 12.0 × 10^5 of native KGM to 9.70 to 11.0 × 10^5 of acetylated KGM products was observed. This indicates that pyridine is a milder catalyst than zinc chloride in the acetylation of KGM.

The effect of the degree of acetylation (DA) on the gelation behaviors upon addition of sodium carbonate to native and acetylated KGM samples was studied by dynamic viscoelastic measurements.[23] At a fixed alkaline concentration (C_{Na}), both the critical gelation times (t_{cr}) and the plateau values of storage moduli (G'_{sat}) of the KGM gels increased with increasing DA (shown in Figure 3.12). However, at a fixed ratio of alkaline concentrations to values of DA (C_{Na}/DA), similar t_{cr} and G'_{sat} values independent of DA were observed (shown in Figure 3.12). On the whole, increasing KGM concentration or temperature shortened the gelation time and enhanced the elastic modulus for KGM gel.

In conclusion, it was suggested that the deacetylation leads to the aggregation of stiffened molecular chains. In the presence of excessive alkali, the gelation proceeds too fast, resulting in a gel with a smaller elastic modulus. The final elastic modulus of gels depends strongly on the gelation rate. The KGM gel is thermoirreversible and the rearrangement of network chains does not seem to occur as in cold-set gels, such as gellan gels and κ-carrageenan gels. It was reported recently that the helix–coil transition, which occurs in self-supporting gels of gellan induced by temperature change,[24] also occurs in κ-carrageenan gels by immersion in salt solutions.[25] In the gelation of KGM, the gelation rate was shown to be governed by the concentration and molecular weight of the polymer, temperature, and DA and

TABLE 3.1
Effect of the Amount of Pyridine and Temperature on the Extent of Acetylation and the Results of Viscosity Measurements of the Products

Sample	Rs	Ac1	Ac2	Ac3	Ac4	Ac5	Ac6	Ac-D
Pyridine (ml)	—	0.5	1	1.5	2	2.5	2.5	10
Temperature (°C)	—	40	40	40	50	50	80	40
Time (h)	—	2	2	2	2	2	2	3
DA (%)	1.38	4.13	4.47	4.82	5.85	7.40	7.57	10.15
DS	0.05	0.16	0.18	0.19	0.23	0.30	0.31	0.42
$[\eta]^a$ ($cm^3\ g^{-1}$)	557	493	520	486	500	524	487	480
$M_v^a\ 10^{-5}$	12.0	10.1	10.9	9.88	10.3	11.0	9.91	9.70

[a]The viscosity measurements of KGM cadoxen solutions were carried out at 25 ± 0.02°C by using an Ubbelohde-type viscometer. The viscosity-average molecular weights (M_v) of the KGM samples were calculated according to $[\eta] = 3.55 \times 10^{-2} M^{0.69}$.[8]

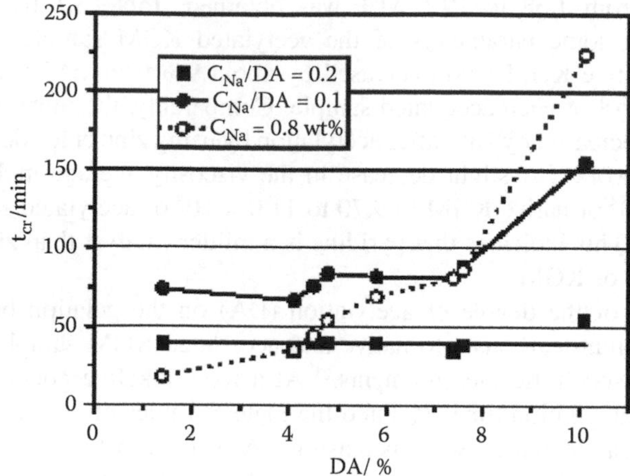

FIGURE 3.12 Plots of t_{cr} of the KGM samples as a function of DA in the case of C_{Na}/DA = 0.2 (solid square) and C_{Na}/DA = 0.1 (solid circle) and in the case of the concentration of sodium carbonate (C_{Na}) as 0.8 wt% (hollow circle). (From Gao, S. and Nishinari, K., *Biomacromolecules*, 5, 175, 2004.)

alkaline concentration, and it was also shown that slower gelation leads to the stronger gels. Although molecular forces responsible for gel formation in both cold-set gels and KGM gels are believed to be hydrogen bonds, tighter stacking occurs in KGM gels than in other thermoreversible gels, such as gellan, carrageenan, agarose, etc. Stacking might be tighter in solutions of polymers with longer persistent length. Unfortunately, the persistent length of KGM is not known because of its poor solubility in water. The atomic force microscopy (AFM) observation will be useful to gain insight on this point.

3.5 MIXTURE OF KONJAC WITH OTHER POLYSACCHARIDES

3.5.1 Konjac–Xanthan Mixtures

Since the discovery of the synergistic effect of xanthan and locust bean gum, there have been many investigations on the interaction between two different polysaccharides.

Individually, neither konjac nor xanthan dilute aqueous solution forms a gel at a neutral pH and room temperature; however, on mixing, a gel can be produced. Annable and her coworkers[26] studied the interaction of these polymers and found that the mixture on heating above 55°C formed a gel. This temperature is lower than the conformational transition temperature of xanthan molecules (Figure 3.13). It was concluded, therefore, that the junction zones in mixed gels are formed by the association of konjac and the helical form of xanthan.[26] The effects of salts were also examined, and it was found that their addition shifted the gelation temperature of the mixture to lower temperatures. This was attributed to the self-association of

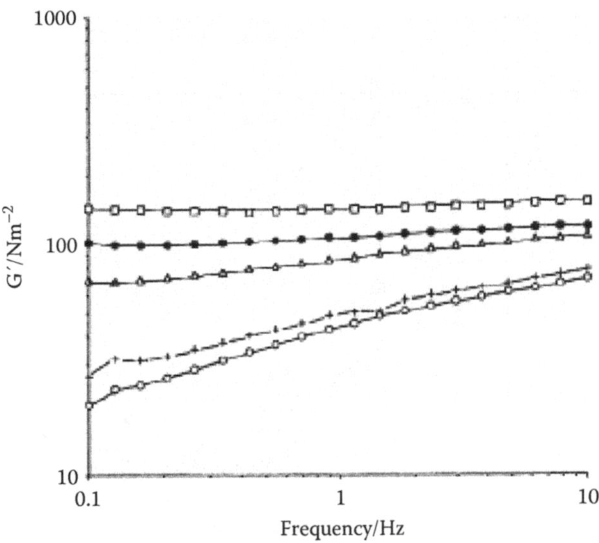

FIGURE 3.13 G' as a function of frequency for xanthan in 0.04 mol/dm^3 NaCl in the presence of glucomannan measured at 25°C after heating to 25°C (o), 35°C (+), 45°C (△), 55°C (●), and 65°C (□). Cooling rate, approximately 1°C/min. (From Annable, P. et al., *Macromolecules*, 27, 4204, 1994.)

xanthan molecules rather than the association of konjac and xanthan (Figure 3.14).[26] The reason why gelation temperature shifted to lower temperature by the addition of salt is that electrolyte promotes xanthan self-association at the expense of xanthan–KGM interaction.

3.5.2 Konjac–κ-Carrageenan Mixtures

Although κ-carrageenan can form a gel, in Japan the mixture of konjac and κ-carrageenan has been used to produce dessert jellies containing fruit juices. The texture of the mixed gel is more rubber-like than that of κ-carrageenan gels. Williams et al. observed two exothermic peaks in cooling differential scanning calorimetry (DSC) curves for mixtures of konjac and κ-carrageenan, with the mixing ratio KGM/CAR from 0.1/0.5 to 0.2/0.4 (total polysaccharide concentration, 0.6 wt%), as shown in Figure 3.15.[27] Only one exothermic peak was observed for mixtures with CAR content below 0.3/0.3, and this was attributed to the formation of an ordered structure by the interaction between konjac glucomannan and κ-carrageenan. When the CAR content became higher than 0.45/0.15, the lower temperature peak began to appear in addition to the exothermic peak originating from the formation of the ordered structure by the interaction between konjac glucomannan and κ-carrageenan. Exothermic enthalpy per unit mass of κ-carrageenan accompanying gelation of the mixtures was significantly smaller than that for κ-carrageenan alone, indicating that the gel formation of κ-carrageenan is strongly affected by the presence of konjac glucomannan. Electron spin resonance (ESR) spectra for mixtures

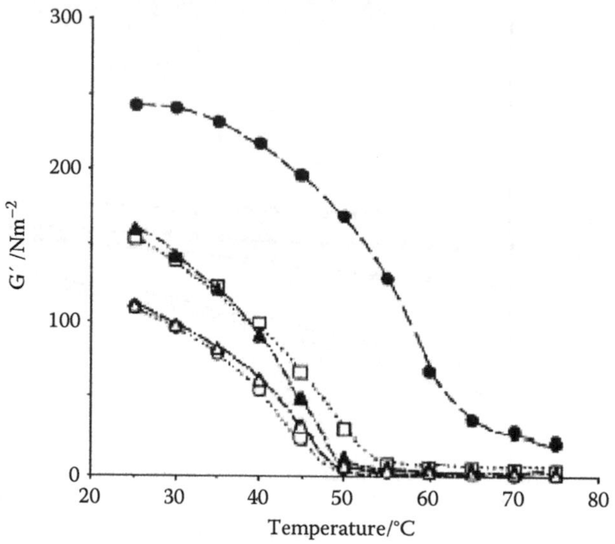

FIGURE 3.14 G' as a function of temperature for xanthan–KGM mixtures (1:1, 0.6 wt%) in the presence and absence of monovalent cations (0.04 mol/dm^3), measured at 3 Hz. H_2O (●), NaCl (o), KCl (□), CsCl (▲), NH_4Cl (△). (From Annable, P. et al., *Macromolecules*, 27, 4204, 1994.)

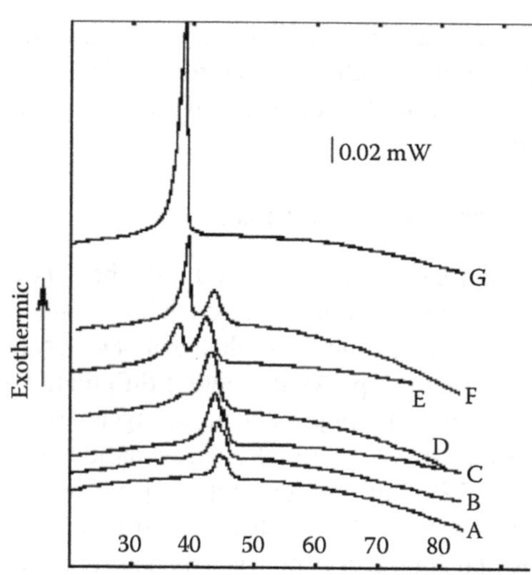

FIGURE 3.15 DSC cooling curves for various ratios of κ-carrageenan–konjac mannan mixtures in the presence of 50 m*M* KCl (0.6% total polymer concentration). Scanning rate was 0.1°C/min. % κ-carrageenan/% konjac mannan: (A) 0.1/0.5, (B) 0.2/0.4, (C) 0.3/0.3, (D) 0.4/0.2, (E) 0.45/0.15, (F) 0.5/0.1, and (G) 0.6/0. (From Williams, P.A. et al., *Macromolecules*, 26, 5441, 1993.)

and konjac glucomannan alone showed that the segmental motion of konjac glucomannan is reduced by the interaction with κ-carrageenan molecules.

Kohyama et al.[8] carried out a large deformation extension measurement on mixed gels of konjac glucomannan with different molecular weights and κ-carrageenan molded into a ring shape, and found that the breaking stress increased with increasing molecular weight of KGM, indicating that KGM chains contribute to the network structure. Since the gel-to-sol transition temperature of mixed gels did not depend so much on the molecular weight of KGM, it was suggested that the KGM chains interact weakly with CAR; this can contribute to strengthening the network mechanically, but does not affect its thermal stability to a great extent.

3.5.3 Konjac–Gellan Mixtures

The rheological properties of mixtures of konjac glucomannan with different molecular weights and gellan (GELL) were also examined.[28] The elastic modulus of mixtures of gellan with a lower and a medium molecular weight KGM as a function of mixing ratio exhibited a maximum at a certain KGM content, while that with higher molecular weight KGM increased with increasing KGM content. The relaxational strength of mixtures, defined as the difference between the loss shear moduli as a function of temperature at a lower temperature and at a higher temperature, was smaller than that of gellan gum alone, indicating that the gel formation of gellan is strongly affected by the presence of KGM. The decrease in the relaxational strength was more significant in mixtures with higher molecular weight KGM, indicating that the higher molecular weight KGM inhibits the gel formation of gellan. The storage modulus of mixtures of gellan and KGM with a medium molecular weight showed a maximum at the mixing ratio gellan/KGM = 0.3/0.5. The effect of sodium chloride and calcium chloride on this mixture was examined by rheology and DSC. The midpoint transition temperature shifted to higher temperatures with increasing concentration of sodium chloride or calcium chloride, while the relaxational strength showed a maximum at a certain concentration of calcium chloride, as shown in Figure 3.16.[29] The sodium or calcium ions shield the electrostatic charge of the carboxyl groups of gellan and promote the self-aggregation of gellan and network formation by the attachment of KGM molecules on the surface of gellan aggregates. Excessive salt addition promotes the self-aggregation of gellan and hinders the attachment of KGM, thus leading to phase separation.[29]

The values of T_s, T_m, or ΔH_m determined from DSC as a function of KGM concentration indicated that KGM little influenced the thermal stability of gellan solutions in comparison with sugars; however, KGM with relatively lower molecular weight significantly increased the G' observed from the rheological measurement. Thus, in the presence of KGM with relatively lower molecular weight, the effective concentration of gellan may increase by immobilizing water molecules in gellan solutions, so that KGM with relatively lower molecular weight could indirectly promote the helix–coil transition of gellan molecules. Therefore, the effect of the KGM with relatively lower molecular weight on the rheological and thermal properties of gellan solution seems to be essentially different from that of sugar, which stabilizes junction zones of gellan molecules by forming hydrogen bonds. KGM

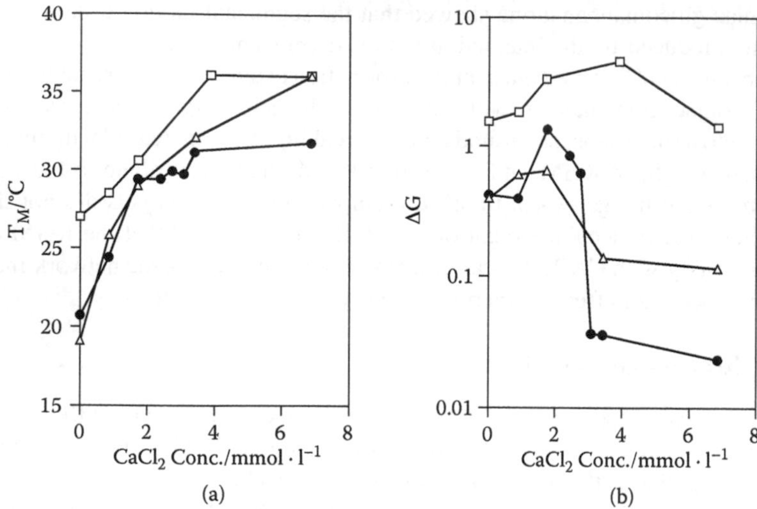

FIGURE 3.16 Dependence of the midpoint temperature of transition T_M(a) and the relaxation strength ΔG(b) for mixtures with GELL/KGM = 0.3/0.5 (total polysaccharide concentration, 0.8%), or 0.3 or 0.8% GELL solutions on the $CaCl_2$ concentration: (●) mixture, (△) 0.3% GELL, and (□) 0.8% GELL. (From Miyoshi, E. et al., *J. Agric. Food Chem.*, 44, 2486, 1996.)

with relatively higher molecular weight may inhibit helix formation in gellan molecules; moreover, it may hinder further aggregation of gellan helices, because KGM with relatively higher molecular weight decreased both ΔG observed by rheological measurement and ΔH_m observed by DSC.[30]

3.5.4 Konjac–Acetan Mixtures

Diffraction patterns of deacetylated acetan and KGM blend fibers were recorded on a microfocus x-ray generator.[31] One of the best patterns is shown in Figure 3.17. Background scattering extends to the outer edge, yet few sharp Bragg reflections and some diffuse spots stand out in the interior. The distribution of intensity is consistent with good axial orientation and short-range lateral organization of the helices in the fiber. The meridional reflection on the sixth layer line suggests that the binary complex is a sixfold helix of pitch 55.4 Å. A molecular modeling study incorporating this information reveals that a double helix in which one strand is acetan and the other glucomannan is stereochemically feasible. While the backbone and side groups are sufficiently flexible to allow the chains to associate with the same or opposite polarity, the parallel model is superior in terms of unit cell packing. The results are compatible with the observed synergy, namely, the weak gelation behavior of the complex. The molecular model can be generalized for the binary system when acetan is replaced by xanthan, or glucomannan by galactomannan.

3.5.5 Konjac–Starch Mixtures

When the konjac was added to a starch dispersion, both the elastic modulus and the breaking stress of the resulting gel increased. However, following storage for a long

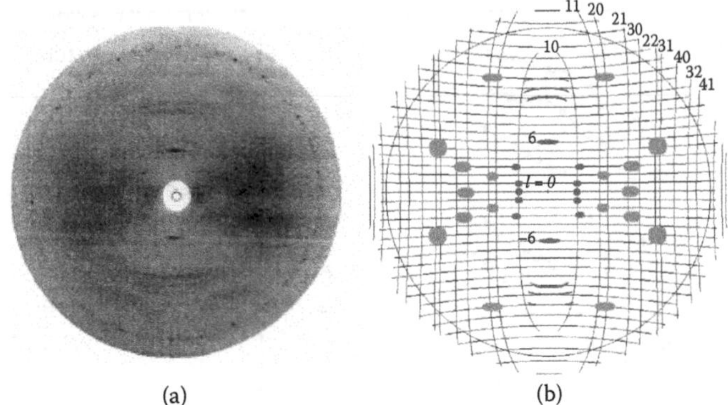

FIGURE 3.17 (a) Diffraction pattern from a fiber of the 1:1 acetan:glucomannan complex characterized by a sharp meridional reflection on the sixth layer line and a few diffuse spots. The ring is from calcite (d-spacing, 3.035 Å) for internal calibration. (b) A schematic of the pattern is shown superposed on the Bernal chart for the hexagonal unit cell a = 17.3 and c = 55.4 Å. The (hk) indices of the row lines, and layer lines 0, 6, and –6 are marked. (From Chandrasekaran, R. et al., *Carbohydr. Res.*, 338, 2889, 2003. With permission.)

time (2 weeks), the starch gel containing konjac showed smaller values for these parameters. The retrogradation ratio, $\Delta H_2/\Delta H_1$, the regelatinization enthalpy observed in the second-run DSC heating curve divided by the gelatinization enthalpy observed in the first-run DSC heating curve, showed the same tendency to rheological observation. Therefore, the addition of a small amount of konjac to starch promotes the retrogradation for short storage. However, it prevents the retrogradation after long storage.[32] The addition of a small amount of konjac to the gelatinized dispersion of starch was found to be effective in preventing the syneresis and the consequent phase separation, as shown in Figure 3.18.[33]

3.6 SOLID-STATE PROPERTIES OF KONJAC GLUCOMANNAN

3.6.1 Dielectric, Viscoelastic, and Broad-Line NMR Studies of Konjac Glucomannan Films

KGM films were prepared by a conventional casting method and the dielectric, viscoelastic, and broad-line NMR measurements were carried out to elucidate the relationship between the chemical structure of KGM and its physicochemical properties.[34] The molecular motion of solid KGM is discussed in comparison with amylose,[35] pullulan,[36] dextran,[37] and cellulose derivatives.[38] The experimental finding showed that the dielectric relaxation strength, $\Delta\varepsilon$ (ca. 2.2), of KGM is smaller than that of amylose (3.0)[35] but larger than those of pullulan (1.5)[37] and dextran (0.83).[37] It is considered that the magnitude of $\Delta\varepsilon$ is mainly determined by the rotational motion of hydroxymethyl groups because the sample has no acetyl groups (no signal of the carbonyl carbon was perceived at 180 ppm in the ^{13}C cross-polarization (CP) and magic angle spinning (MAS) CP/MAS NMR spectrum). The number of hydroxymethyl groups in KGM and in amylose is considered the same, while the

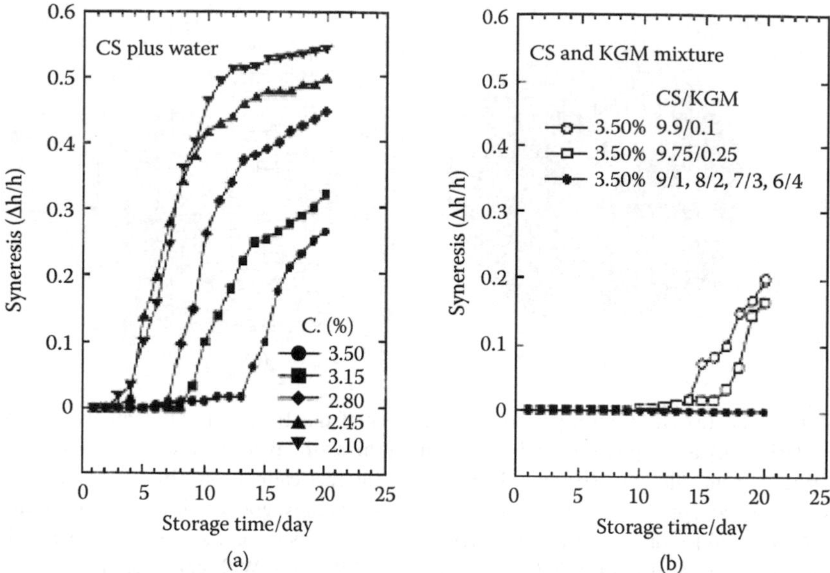

FIGURE 3.18 Syneresis of a cornstarch (CS) dispersion (a) and a CS–KGM mixture (b). C, concentration of cornstarch: ●, 3.50 wt%; ■, 3.15 t%; ◆, 2.80 wt%; ▲, 2.45 wt%; and ▼, 2.10 wt%. CS/KGM mixing ratio: -○-, 3.50 wt% 9.9/0.1; -□-, 3.50 wt% 9.75/0.25; and -●-, 3.50 wt% 9/1, 8/2, 7/3, 6/4. (From Yoshimura, M. et al., *Carbohydr. Polym.*, 35, 71, 1998.)

number in pullulan is about two thirds of that. Dextran has very few hydroxymethyl groups. Therefore, the order of magnitude of the dielectric relaxation strength for these polysaccharides can be understood in terms of the content of hydroxymethyl groups, as discussed above.[38] According to these experimental results, the rotational motion of the hydroxymethyl groups in KGM is slightly hindered in comparison with the motion of the hydroxymethyl groups in amylose.

The complex viscoelastic constant $c^* = c' + ic''$ at 10 Hz was determined by detecting the sinusoidal strain and stress at both ends of the film. The results from temperature dependence of the viscoelastic coefficients c' and c'' showed that the real part c' decreased monotonically with increasing temperature. The value of c'' for the nondried sample showed a large peak at about –50°C, and this peak decreased in height when the sample was dried, just as in the case of the dielectric coefficient results. For dry KGM, a peak appeared at about –100°C, which is attributed to the rotation of hydroxymethyl groups attached to the C-5 atom in mannose and glucose residues. As the moisture content decreased, the value of c' became larger at temperatures below about –20°C, but it became smaller at higher temperatures. The water plays the role of a plasticizer at temperatures higher than the ambient temperature.

Results from temperature dependence of the second moment of KGM, amylose, pullulan, and dextran showed that the second moment of KGM at lower temperatures is smaller than those of the other three polysaccharides.[39] The second moment decreased rapidly at about 0 and 60°C. The temperatures at which the second moment decreases rapidly are higher than those in the case of the other polysaccharides. This suggested that the movements of hydroxymethyl groups in KGM are slightly hindered in comparison with the other three polysaccharides.

3.6.2 Biodegradable Material

The widespread use of synthetic polymer film materials has caused serious pollution problems and, as a result, biodegradable films from renewable resources have attracted much attention. Blend films were prepared by casting the mixtures of 7 wt% KGM aqueous solution and 2 wt% chitosan (CH) in acetic acid aqueous solution.[40] Crystallinities of the blend films measured by wide-angle X-ray diffraction decreased with the increase of KGM. The thermostability, tensile strength, and breaking elongation of the blend films in dry state were obviously higher than those of both pure KGM and chitosan films. The tensile strength of the dry blend film achieved the maximum of 73.0 MPa when the weight ratio of chitosan to KGM was 7:3, which is in line with the observation from scanning electron microscopy of a homogenous morphology of the blend film with the KGM content of 30 wt%. The structure analysis indicated that there is a strong interaction between KGM and chitosan that results from intermolecular hydrogen bonds. The water solubility of the blend films was improved by blending with KGM, in contrast to pure chitosan film. Therefore, KGM/CH blend films have promising application as coatings of pills, because they have good mechanical properties in the dry state and can be dissolved in aqueous medium.

To enhance the mechanical properties of konjac glucomannan film in the dry state and to find an application of konjac glucomannan in the food preservation domain, blend transparent film was prepared by blending 3 wt% sodium alginate aqueous solution with 4.5 wt% konjac glucomannan aqueous solution and coagulating in a mixture of water–ethanol–sodium hydroxide (9:10:1 by weight).[41] Crystallinities of blend films increased with the increase of sodium alginate. The tensile strength and breaking elongation of the blend films in the dry state were 3 ~ 4 times of those for both pure sodium alginate and konjac glucomannan films. The tensile strength of the dry blend film achieved 77.8 MPa when the retention of sodium alginate in the film was 27.9 wt%, as shown in Figure 3.19, indicating the high miscibility of the two components at this composite ratio, as confirmed by the observation from scanning electron microscopy. The structure analysis indicated that there was a strong interaction between konjac glucomannan and sodium alginate, resulting from the intermolecular hydrogen bonds. The moisture content and degree of water swelling of the blend films were increased due to the introduction of sodium alginate. Results from the film-coating preservation experiment on litchi and honey peach, two kinds of traditional fruits abundant in China, both having a high content of syrup, showed that this blend film had a certain degree of water-holding ability. Both the fruit weight loss rate and the rot rate of the experimental fruit groups decreased by various degrees in comparison with those of the control groups.

Abundant hydroxyl groups in konjac glucomannan facilitate the chemical modification in konjac glucomannan by replacing the hydrogen atoms of the hydroxyl groups with others. Nitro-konjac glucomannan (NKGM) was synthesized by inho-

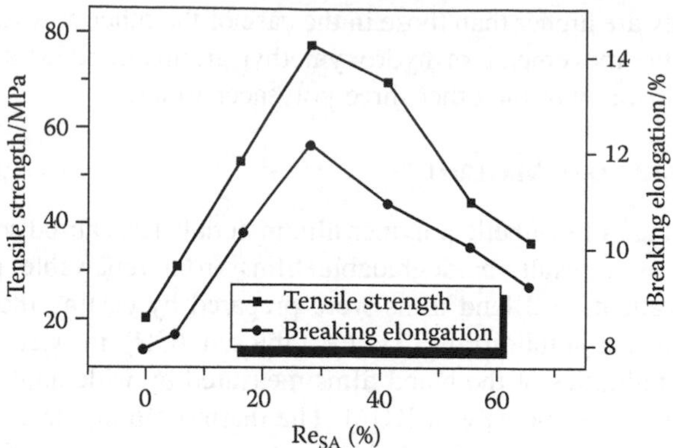

FIGURE 3.19 Tensile strength and breaking elongation of the pure and blend films against the content of sodium alginate (Re_{SA}). (From Xiao, C. et al., *J. Appl. Polym. Sci.*, 77, 617, 2000.)

mogeneous reaction in the presence of KGM as the raw material and the mixture of fuming acid–sulfuric acid–phosphorous pentoxide (10:1:4, by weight) as the nitration agent. Results from elemental analysis and light-scattering measurements showed that its degree of substitution and weight-average molecular weight are 2.4 and 4.75×10^4, respectively.[42] Interpenetrating polymer network (IPN) materials, a kind of mutual polymer blend held together by permanent entanglement between two or more distinctly cross-linked polymers, have drawn much attention due to the special properties brought about by interlocking of polymer chains. Normally, the IPN technique is applied to synthetic polymers to obtain composite materials meeting various demands. Semi-IPN materials were successfully synthesized from castor oil-based polyurethane (PU) prepolymer and 10 ~ 40 wt% NKGM. Structural analysis showed that the intermolecular interaction of hydrogen bonding between NKGM and PU exists in the semi-IPN sheets, resulting in the miscibility. When NKGM content in an IPN sheet was 20 wt%, the mechanical properties, thermostability, and light transmittance were significantly higher than in the PU sheet. The NKGM in the IPN sheets plays an important role in the enhancement of the tensile strength and in accelerating the cure. This new material has not only performance similar to that of PU but also biodegradability.[42]

To investigate the effect of the molecular weight of NKGM on the properties of this IPN material, eight NKGM products with different weight-average molecular weights, from 2.86×10^4 to 14.1×10^4, determined by light-scattering measurements were synthesized by controlling the nitration reaction time.[43] Semi-IPN sheets with the content of NKGM of 20 wt% were prepared from castor oil-based PU and the NKGM products. A relatively broad mechanical loss peak appeared on the thermograms of dynamic mechanical analysis for the IPN sheets, and a single glass transition (Tg) appeared on the DSC curves for the IPN sheets. The optical transmittances of the IPN sheets in the wavelength of 400 ~ 800 nm were all higher than that of the PU sheet. These results indicated the good miscibility between the two polymers. It is worth noting that the tensile strengths of the IPN composite sheets were

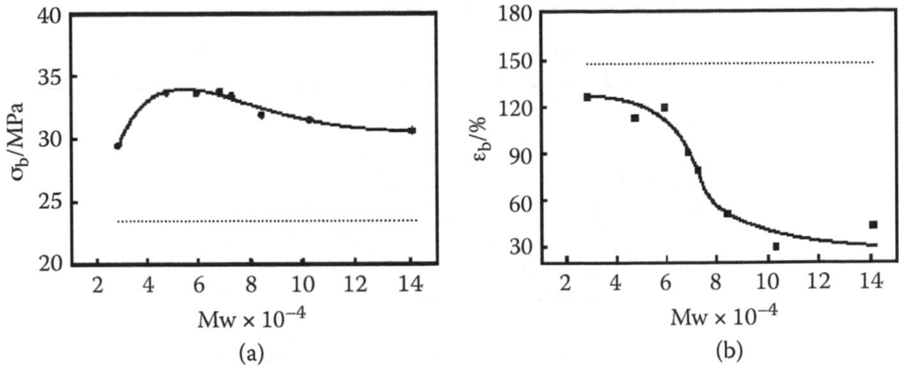

FIGURE 3.20 Effect of molecular weights (Mw) of NKGM on (a) tensile strength (σ_b) and (b) breaking elongation (ε_b) of films PU and UNK. (…) represents σ_b or ε_b of PU. (From Gao, S. and Zhang, L., *Macromolecules*, 34, 2202, 2001.)

obviously higher than that of pure PU, and the tensile strength and breaking elongation of the IPN sheets increased with decreasing NKGM molecular weight from 8.44×10^4 to 4.75×10^4, as shown in Figure 3.20. NKGM with relatively low molecular weight plays a role in plasticization, acceleration of curing, and enhancement of interaction between PU and NKGM.[43]

By changing the synthesis route, namely, first blending NKGM with castor oil in butanone and thereafter adding toluene diisocyanate to promote the polymerization reaction, novel semi-IPN materials are synthesized.[44] Compared to pure PU, the α-transition peak on DMA thermograms for each IPN sheet becomes sharper and shifts to a higher temperature position, and Tg on the DSC curves of the IPN sheets increases with increasing the content of NKGM. These experimental results are different from those in the case mentioned above, where PU prepolymer was first synthesized and then NKGM was added in the presence of a chain extender and catalyst. This is because blending NKGM with castor oil before polymerization has hindered the formation of a PU network. As a result, the amount of networks decreased and the microphase separation occurred in the composite sheets, resulting in the decrease of optical transmittance. In this case, NKGM predominantly plays a role in filling and enhancing PU, resulting in the increase of tensile strength. The stress–strain curves of the composite sheets reflected the transition from elastomer to plastics with an increase of NKGM.[44]

Castor oil can be considered a polyether-type polyol, and the polyurethane synthesized from castor oil and isocyanate is called poly(ether urethane). To investigate the effect of the type of PU on the structure and properties of the composite materials, semi-IPN sheets were prepared by mixing poly(ester urethane) synthesized by the polymerization reaction between polyethylene glycol adipate and toluene diisocyanate in the presence of 2,2-bis(hydroxyl methyl) propionic acid as an extender with NKGM in tetrahydrofuran.[45] These poly(ester urethane)/NKGM semi-IPN sheets showed good miscibility in a NKGM content range of 10 to 90%. It is mainly attributed to the penetration of linear NKGM into poly(ester urethane) and

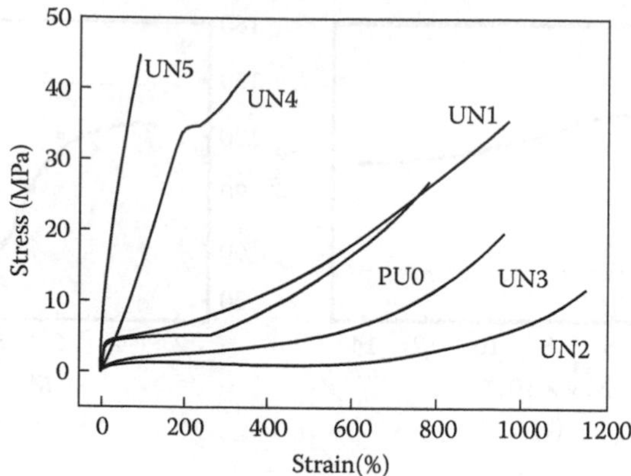

FIGURE 3.21 Tensile stress–strain curves of the semi-IPN sheets. NKGM contents in the sheets of UN1, UN2, UN3, UN4, UN5, and PU0 are 10, 20, 30, 40, 50 and 0 wt% respectively. (From Gao, S. and Zhang, L., *J. Appl. Polym. Sci.*, 90, 2224, 2003. With permission.)

the intimate entanglement with PU chains. However, this relatively strong interaction restricted the motion of PU soft segment, resulting in the elevation of Tg. Figure 3.21 shows the tensile stress–strain curves of the semi-IPN sheets, where the NKGM contents in the sheets of UN1, UN2, UN3, UN4, UN5, and PU0 are 10, 20, 30, 40, 50 and 0 wt%. The semi-IPN composite sheet showed the best mechanical properties (35 MPa in tensile strength and 960% in breaking elongation) when the NKGM content was 10 wt%. When the NKGM content was more than 40%, the semi-IPN composite films showed the tensile stress–strain behavior of stiff and elastic materials and had an obviously higher σ_b than pure PU film, which can be used as plastics.[45]

On the basis of the above results, a semi-IPN system is an effective method to obtain novel biodegradable materials by interpenetrating linear natural polymer into a polyurethane network, and it provides a novel way for modification and exploitation of natural polymers such as polysaccharides.

3.7 PHYSIOLOGICAL FUNCTIONS OF KGM

Dietary fiber has recently received recognition for reducing the risk of developing diabetes and heart disease, because it has a therapeutic benefit in prediabetic metabolic conditions. KGM has been chosen as a dietary fiber because it represents a polysaccharide with one of the highest viscosities.

Vuksan et al.[46] studied the effect of KGM fiber on type 2 diabetic patients, as an adjunct to conventional treatment, on a cluster of coronary heart disease (CHD) risk factors: hyperglycemia, hyperlipidemia, and hypertension. The experimental observation showed that the application of KGM supplementation in a high-risk diabetic study group demonstrated simultaneous improvement in all three diet-modifiable risk factors, indicating a reduction in overall CHD risk. Furthermore, KGM-supplemented therapy may require lower drug dosages and improve overall cost-effectiveness and acceptability of treatment.

Vuksan et al.[47] screened 278 free-living subjects between the ages of 45 and 65 years from the Canadian-Maltese Diabetes Study. A total of 11 (age, 55 ± 4 years; body mass index (BMI), 28 ± 1.5 kg/m^2) were recruited who satisfied the inclusion criteria: impaired glucose tolerance, reduced high-density lipoprotein (HDL) cholesterol, elevated serum triglycerides, and moderate hypertension. After an 8-week baseline, they were randomly assigned to take either KGM fiber-enriched test biscuits (0.5 g of glucomannan per 100 kcal of dietary intake, or 8 to 13 g/day) or wheat bran fiber (WB) control biscuits for two 3-week treatment periods separated by a 2-week washout. The diets were isoenergetic, metabolically controlled, and conformed to National Cholesterol Education Program Step 2 guidelines. Serum lipids, glycemic control, and blood pressure were the outcome measures. Decreases in serum cholesterol, apolipoprotein, and serum fructosamine were observed during KGM treatment compared with the WB control. Blood glucose, insulin, triglycerides, HDL cholesterol, and body weight remained unchanged. These indicated that a diet rich in high-viscosity KGM improves glycemic control and lipid profile, suggesting a therapeutic potential of KGM fiber in the treatment of the insulin resistance syndrome.

Vuksan and his collaborators[48] have selected two such promising and functionally complementary therapies for further investigation as potential treatment alternatives for type 2 diabetes: KGM and American ginseng (AG). They have generated a mounting body of evidence to support the claim that rheologically selected, highly viscous KGM, and AG with a specific composition may be useful in improving diabetes control, reducing associated risk factors such as hyperlipidemia and hypertension, and ameliorating insulin resistance. KGM has a demonstrated ability to modulate the rate of absorption of nutrients from the small bowel, whereas AG has postabsorptive effects. Consequently, it appears that KGM and AG are acting through different, yet complementary, mechanisms: KGM by increasing insulin sensitivity and AG likely by enhancing insulin secretion.

The glycemic and insulinemic increments with respect to type 2 diabetic subjects after a standard breakfast with glucomannan-enriched biscuits and common slices of toast containing the same amounts of carbohydrates and calories have been evaluated.[49] The basal serum values of glucose and C-peptide were similar in the 2 days of the test. The mean increments of glucose and C-peptide were significantly higher ($p < 0.001$) after slices of toast than after glucomannan-enriched biscuits. In conclusion, the results show a reduction in glycemic increments after breakfast with glucomannan-enriched biscuits. The decreased insulin secretion and the reduction of insulin need can preserve the functional reserve of beta-cells.

Seventy-two type 2 diabetic subjects were given konjac food for 65 days. The data analyzed by multiple F test indicate that the fasting blood glucose (FBG) and the 2-h postprandial blood glucose (PBG) on the 30th and 65th days after the food was ingested were significantly reduced ($p = 0.001$, $p < 0.001$, respectively), as was the glycosylated hemoglobin level at the end of the trial ($p < 0.05$). The final FBG and PBG of the subjects with an initial FBG (FBG-O) greater than 200 mg% decreased on the average by 51.8 and 84.6 mg%, respectively; those with an FBG-O of 150 to 200 mg% had decreases of 24.1 and 68.7 mg%; and those with an FBG-O less than 150 mg% had decreases of 4.8 and 21.4 mg%. No significant changes in blood lipid indexes were observed, except that the triglyceride values of subjects with hypertriglyceridemia (>200 mg%) significantly decreased by 118.7 mg%. It was concluded that konjac food is very useful in the prevention and treatment of hyperglycemia.[50]

The effects of the soluble fiber KGM on serum cholesterol concentrations were investigated in 63 healthy men in a double-blind crossover, placebo-controlled study.[51] After a 2-week baseline period, the subjects were given 3.9 g of KGM or placebo daily for 4 weeks. After a washout period of 2 weeks, crossover took place, followed by another 4 weeks of treatment. The subjects were encouraged not to change their ordinary diets or general lifestyle during the investigation. KGM fibers reduced total cholesterol (TC) concentrations by 10% ($p < 0.0001$), low-density lipoprotein cholesterol (LDL-C) concentrations by 7.2% ($p < 0.007$), triglycerides by 23% ($p < 0.03$), and systolic blood pressure by 2.5% ($p < 0.02$). High-density lipoprotein cholesterol (HDL-C) and the ratio of LDL-C to HDL-C did not change significantly. No change in diastolic blood pressure or body weight was observed. No adverse effects were observed. The results of this study show that glucomannan is an effective cholesterol-lowering dietary adjunct.

Cairella et al.[52] studied the behavior of body weight, blood glucose, total serum cholesterol, and hunger and satiety sensation in 30 patients treated for 60 days with a 1200-kcal diet plus either placebo or glucomannan. All the variables considered show that the low-calorie diet plus glucomannan is more effective than the low-calorie diet alone. Extensive clinical studies[53–55] have been carried out to clarify the effect of KGM fiber on obesity for children or adult patients. The results showed a significant mean weight loss using highly purified glucomannan after a period of treatment. Importantly, no significant side effects were observed in treated patients. On the basis of the marked ability to satiate patients and the positive metabolic effects, glucomannan diet supplements have been found to be particularly efficacious and well tolerated even in the long-term treatment of severe obesity.

Chronic constipation is a very frequent disease in Western countries. Recently, the efficacy and acceptability of glucomannans in 93 patients affected with chronic constipation was tested.[56] The multicentric, open, and noncontrolled study was divided into an initial phase (treatment with 1 g of glucomannans t.i.d. for 1 month) and a maintenance phase (1 g b.i.d. for 1 month). Both objective parameters (number of days per week with bowel movements and number of enemas) and abdominal symptoms were evaluated. After one month, all assessed parameters showed a statistically significant improvement lasting through the second month. Glucomannans were well accepted and devoid of relevant side effects. In conclusion, considering their efficacy and tolerability, they can be proposed as an ideal therapeutic tool in the management of chronic constipation symptoms.

Inadequate dietary fiber intake is a widely accepted explanation for chronic constipation in children with severe brain damage. To evaluate the efficacy of glucomannan as a treatment for chronic constipation, 20 children with severe brain damage and chronic constipation were randomly assigned to double-blind treatment with either glucomannan or placebo for 12 weeks. Stool habits, total and segmental gastrointestinal transit times, and anorectal motility were evaluated in all children before and after the treatment period. It was shown that glucomannan significantly increased stool frequency, whereas the effect of the placebo was not significant. Laxative or suppository use was significantly reduced by glucomannan but was not affected by the placebo. Clinical scores of stool consistency were significantly improved, and episodes of painful defecation per week were significantly reduced by glucomannan but not by the placebo. However, neither glucomannan nor the placebo had a measurable effect on total and segmental transit times. It seems that in neurologically impaired children, glucomannan improves stool frequency but has no effect on colonic motility.[57]

REFERENCES

1. Nishinari, K., Konjac Glucomannan in *Novel Food Macromolecules*, Doxastakis, G. and Kiosseoglou, V.D., Eds., IRL Press, Oxford, 2000, p. 373.
2. Nishinari, K., Williams, P.A., and Phillips, G.O., Review of the physico-chemical characteristics and properties of konjac mannan, *Food Hydrocoll.*, 6, 199, 1992.
3. Kato, K. and Matsuda, K., Studies on the chemical structure of konjac mannan, *Agric. Biol. Chem.*, 33, 1446, 1969.
4. Maeda, M., Shimahara, H., and Sugiyama, N., Detailed examination of the branched structure of konjac glucomannan, *Agric. Biol. Chem.*, 44, 245, 1980.
5. Katsuraya, K. et al., Constitution of konjac glucomannan: chemical analysis and ^{13}C NMR spectroscopy, *Carbohydr. Polym.*, 53, 183, 2003.
6. Maekaji, K., The mechanism of gelation of konjac mannan, *Agric. Biol. Chem.*, 38, 315, 1974.
7. Maekaji, K., A method for measurement and kinetic analysis of the gelation process of konjac mannan, *Nippon Nogeikagakukaishi*, 52, 251, 1978.
8. Kohyama, K., Iida, H., and Nishinari, K., A mixed system composed of different molecular weights konjac glucomannan and kappa carrageenan, *Food Hydrocoll.*, 7, 213, 1993.
9. Chanzy, H. et al., Crystallization behavior of glucomannan, *Biopolymers*, 21, 301, 1982.
10. Kohyama, K. and Nishinari, K., New application of konjac glucomannan as a texture modifier, *Jpn. Agric. Res. Q.*, 31, 301, 1997.
11. Kohyama, K., Sano, Y., and Nishinari, K., A mixed system composed of different molecular weights konjac glucomannan and kappa-carrageenan: II, *Food Hydrocoll.*, 10, 229, 1996.
12. Henley, D., A macromolecular study of cellulose in the solvent cadoxen, *Arkiv. Kemi.*, 18, 327, 1961.
13. Schwald, W. and Bobleter, O., Characterization of nonderivatized cellulose by gel permeation chromatography, *J. Appl. Polym. Sci.*, 35, 1937, 1988.
14. Nishinari, K. et al., Solution properties of pullulan, *Macromolecules*, 24, 5590, 1991.
15. Clegg, S.M., Phillips, G.O., and Williams, P.A., Determination of the relative molecular mass of konjac mannan, in *Gums and Stabilisers for the Food Industry 5*, Phillips, G.O., Wedlock, D.J., and Williams, P.A., Eds., IRL Press, Oxford, 1990, p. 463.
16. Kohyama, K. and Nishinari, K., Dependence of the specific volume of konjac glucomannan on pH, in *Gums and Stabilisers for the Food Industry 5*, Phillips, G.O., Wedlock, D.J., and Williams, P.A., Eds., IRL Press, Oxford, 1990, p. 459.
17. Huang, L. et al., Gelation behavior of native and acetylated konjac glucomannan, *Biomacromolecules*, 3, 1296, 2002.

18. Dave, V. et al., Liquid crystalline, rheological and thermal properties of konjac glucomannan, *Polymer*, 39, 1139, 1998.
19. Yoshimura, M. and Nishinari, K., Dynamic viscoelastic study on the gelation of konjac glucomannan with different molecular weights, *Food Hydrocoll.*, 13, 227, 1999.
20. Zhang, H. et al., Gelation behaviour of konjac glucomannan with different molecular weights, *Biopolymers*, 59, 38, 2001.
21. Williams, M.A. et al., A molecular description of the gelation mechanism of konjac mannan, *Biomacromolecules*, 1, 440, 2000.
22. Huang, L., Kobayashi, S., and Nishinari, K., Dynamic viscoelastic study on the gelation of acetylated konjac glucomannan, *Trans. Mater. Res. Soc. Jpn.*, 26, 597, 2001.
23. Gao, S. and Nishinari, K., Effect of degree of acetylation on gelation of konjac glucomannan, *Biomacromolecules*, 5, 175, 2004.
24. Nitta, Y. et al., Helix-coil transition in gellan gum gels, *Trans. Mater. Res. Soc. Jpn.*, 26, 621, 2001.
25. Watase, M. and Nishinari, K., The rheological study of the interaction between alkali metal ions and kappa-carrageenan gels, *Colloid Polym. Sci.*, 260, 971, 1982.
26. Annable, P., Williams, P.A., and Nishinari, K., Interaction in xanthan-glucomannan mixtures and the influence of electrolyte, *Macromolecules*, 27, 4204, 1994.
27. Williams, P.A. et al., Investigation of the gelation mechanism in kappa-carrageenan/konjac mannan mixtures using differential scanning calorimetry and electron spin resonance spectroscopy, *Macromolecules*, 26, 5441, 1993.
28. Nishinari, K. et al., Rheological and DSC studies on the interaction between gellan gum and konjac glucomannan, *Carbohydr. Polym.*, 30, 193, 1996.
29. Miyoshi, E., Takaya, T., and Nishinari, K., Effects of sodium chloride and calcium chloride on the interaction between gellan gum and konjac, *J. Agric. Food Chem.*, 44, 2486, 1996.
30. Miyoshi, E., Takaya, T., and Nishinari, K., Effects of glucose, mannose and konjac glucomannan on the gel-sol transition in gellan gum aqueous solutions by rheology and DSC, *Polymer Gels Netw.*, 6, 273, 1998.
31. Chandrasekaran, R., Janaswamy, S., and Morris, V.J., Acetan: glucomannan interactions: a molecular modeling study, *Carbohydr. Res.*, 338, 2889, 2003.
32. Yoshimura, M., Takaya, T., and Nishinari, K., Effects of konjac-glucomannan on the gelatinization and retrogradation of corn starch as determined by rheology and differential scanning calorimetry, *J. Agric. Food Chem.*, 44, 2970, 1996.
33. Yoshimura, M., Takaya, T., and Nishinari, K., Rheological studies on mixtures of corn starch and konjac-glucomannan, *Carbohydr. Polym.*, 35, 71, 1998.
34. Kohyama, K. et al., Dielectric, viscoelastic and broad-line NMR study of konjac glucomannan films, *Carbohydr. Polym.*, 17, 59, 1992.
35. Nishinari, K. and Fukada, E., Viscoelastic, dielectric, and piezoelectric behavior of solid amylose, *J. Polym. Sci. Polym. Phys. Ed.*, 18, 1609, 1980.
36. Nishinari, K., Horiuchi, H., and Fukada, E., Viscoelastic and dielectric properties of solid pullulan, *Rep. Prog. Polym. Phys. Jpn.*, 23, 759, 1980.
37. Nishinari, K., Shibuya, N., and Kainuma, K., Dielectric relaxation in solid dextran and pullulan, *Makromol. Chem.*, 186, 433, 1985.
38. Nishinari, K. et al., Molecular motions in cellulose derivatives, in *Viscoelasticity of Biomaterials*, ACS Symposium Series 489, Glasser, W. and Hatakeyama, H., Eds., Oxford University Press, Inc., New York, 1992, p. 357.

39. Nishinari, K. and Tsutsumi, A., Studies on molecular motion of polysaccharides in the solid state by broad-line nuclear magnetic resonance, *J. Polym. Sci. Polym. Phys. Ed.*, 22, 95, 1984.
40. Xiao C. et al., Blend films from chitosan and konjac glucomannan solutions, *J. Appl. Polym. Sci.*, 76, 509, 2000.
41. Xiao, C., Gao, S., and Zhang, L., Blend films from konjac glucomannan and sodium alginate solutions and their preservative effect, *J. Appl. Polym. Sci.*, 77, 617, 2000.
42. Gao, S. and Zhang, L., Semi-interpenetrating polymer networks from castor oil-based polyurethane and nitrokonjac glucomannan, *J. Appl. Polym. Sci.*, 81, 2076, 2001.
43. Gao, S. and Zhang, L., Molecular weight effects on properties of polyurethane/nitrokonjac glucomannan semi-interpenetrating polymer networks, *Macromolecules*, 34, 2202, 2001.
44. Gao, S. and Zhang, L., Effect of the synthesis route on the structure and properties of polyurethane/nitrokonjac glucomannan semi-interpenetrating polymer networks, *J. Appl. Polym. Sci.*, 90, 1948, 2003.
45. Gao, S. and Zhang, L., Synthesis and characterization of poly (ester urethane)/nitrokonjac glucomannan semi-interpenetrating polymer networks, *J. Appl. Polym. Sci.*, 90, 2224, 2003.
46. Vuksan, V. et al., Konjac-mannan (glucomannan) improves glycemia and other associated risk factors for coronary heart disease in type 2 diabetes. A randomized controlled metabolic trial, *Diabetes Care*, 22, 913, 1999.
47. Vuksan, V. et al., Beneficial effects of viscous dietary fiber from Konjac-mannan in subjects with the insulin resistance syndrome: results of a controlled metabolic trial, *Diabetes Care*, 23, 9, 2000.
48. Vuksan,V. et al., Konjac-mannan and American ginsing: emerging alternative therapies for type 2 diabetes mellitus, *J. Am. Coll. Nutr.*, 20, 370S, 2001.
49. Melga, P. et al., Dietary fiber in the dietetic therapy of diabetes mellitus. Experimental data with purified glucomannans, *Riv. Eur. Sci. Med. Farmacol.*, 14, 367, 1992.
50. Huang, C.Y. et al., Effect of konjac food on blood glucose level in patients with diabetes, *Biomed. Environ. Sci.*, 3, 123, 1990.
51. Arvill, A. and Bodin, L., Effect of short-term ingestion of konjac glucomannan on serum cholesterol in healthy men, *Am. J. Clin. Nutr.*, 61, 585, 1995.
52. Cairella, M. and Marchini, G., Evaluation of the action of glucomannan on metabolic parameters and on the sensation of satiation in overweight and obese patients, *Clin. Ther.*, 146, 269, 1995.
53. Walsh, D.E., Yaghoubian, V., and Behforooz, A., Effect of glucomannan on obese patients: a clinical study, *Int. J. Obes.*, 8, 289, 1984.
54. Vita, P.M. et al., Chronic use of glucomannan in the dietary treatment of severe obesity, *Minerva Med.*, 83, 135, 1992.
55. Vido, L. et al., Childhood obesity treatment: double blinded trial on dietary fibres (glucomannan) versus placebo, *Padiatr. Padol.*, 28, 133, 1993.
56. Passaretti, S. et al., Action of glucomannans on complaints in patients affected with chronic constipation: a multicentric clinical evaluation, *Ital. J. Gastroenterol.*, 23, 421, 1991.
57. Staiano, A. et al., Effect of the dietary fiber glucomannan on chronic constipation in neurologically impaired children, *J. Pediatr.*, 136, 41, 2001.

4 Seed Polysaccharide Gums

Steve W. Cui, Shinya Ikeda, and Michael N.A. Eskin

CONTENTS

4.1 INTRODUCTION

Seed polysaccharides are one of the most important categories of plant-originated gums used in the food industry, as they play important roles in both food processing and improving the mouth feel and texture of food products. The occurrence of polysaccharides in plant seeds is mainly in three forms: as nonstarch polysaccharide food reserve material (e.g., guar, locust bean, etc.), as mucilages in the seed coats (e.g., psyllium seed, flaxseed, yellow mustard seed, etc.), and as cell wall materials of seed cotyledons and endosperms (e.g., tamarind and soybean seeds). The chemical compositions, fine structures, and physical and functional properties of these polysaccharides vary significantly with plant sources, growing environments, and method of production. Similar to other polysaccharides described in this book, seed polysaccharide gums found broad applications in the areas of foods, cosmetics, pharmaceuticals, and medicines due to their ability to interact with water and manipulate the flow behavior of water-based systems. Seed polysaccharides (nonstarch) are also an important source of dietary fiber, which may exhibit bioactivities such as reducing calorie intakes, controlling blood glucose and insulin levels, and reducing the risks of heart diseases and colon cancer. This chapter intends to review the most recent advances on seed polysaccharide gums with an effort to cover their sources and basic structures, processing technologies, molecular characteristics, physical and functional properties, and applications. Information on physiological effects, health benefits, and regulatory status will also be provided when available.

4.2 SEED GUMS AS FOOD RESERVE: GALACTOMANNANS

Galactomannans are a group of storage polysaccharides from various plant seeds that reserve energy for germination in the endosperm. There are four major sources of seed galactomannans: locust bean (*Ceratonia siliqua*), guar (*Cyamopsis tetragonoloba*), tara (*Caesalpinia spinosa* Kuntze), and fenugreek (*Trigonella foenum-graecum* L.). Among these, only locust bean and guar gums are of considerable industrial importance. The use of tara and fenugreek gums is limited due to availability and price. Other sources of galactomannans have also been explored in the literature, but no commercial potential can be expected in the near future.

Seed galactomannans consist essentially of a linear (1→4)-β-D-mannopyranose backbone with varied amounts of side groups of single 1→6-linked α-D-galactopyranosyl units. The molar ratio of galactose to mannose varies with plant origin but is typically in the range of 1.0:1.0 ~ 1.1, 1.0:1.6 ~ 1.8, 1.0:3.0, and 1.0:3.9 ~ 4.0 for fenugreek, guar, tara, and locust bean gums, respectively. The conformation of the 1→4-linked β-D-mannan backbone is similar to that of cellulose, so that it does not dissolve in water. The galactose side groups are considered to sterically disturb the interchain association and crystallization, thereby imparting certain water solubility to the galactomannans. As a result, the solubility of the galactomannans increases with the degree of galactosyl substitution: fenugreek and guar gums are readily dissolved in cold water, but heating is needed to reasonably solubilize locust bean gum in water.

Galactomannans are widely utilized in the industry due to their suitable functional properties, such as thickening, binding, and stabilizing abilities. These functional properties are led by rheological behavior of galactomannans in an aqueous phase and also by intermolecular binding in certain conditions. Hydrated galactomannan molecules occupy a large hydrodynamic volume in aqueous solution and control the rheological behavior of the entire solution. Galactomannans themselves are nongelling agents, while some galactomannans show synergistic interactions with other polysaccharides such as agar, xanthan, carrageenan, and yellow mustard gum to form a three-dimensional gel network in appropriate conditions.[1–4] Galactomannans thus find a wide range of applications as texture modifiers or stabilizers in food and pet food industries.

4.2.1 Locust Bean Gum

4.2.1.1 Source

Locust bean gum is also called carob bean gum since it is separated and refined from the endosperm of the seed of the carob tree. Carob is a long-lived evergreen tree that grows to about 10 m in height in 10 to 15 years after germination and yields large brown fruits called carob pods. The sickle-shaped carob pods are 10 to 20 cm long and 2 to 4 cm wide and contain 10 to 15 oval-shaped seeds or kernels. Locust bean has been used as a food ingredient in the eastern Mediterranean region since ancient times and is currently produced mostly in Spain, Italy, Cyprus, and other Mediterranean countries.

4.2.1.2 Method of Production

The production of commercial locust bean, guar, and tara gums is similar, involving separation of endosperms from the seed hull and germ, followed by grinding and sifting of the endosperm into fine particle-sized flour. Further purification is made by repeated alcohol washings. Most commercial gums contain more than 80% galactomannan. The quality or purity of the final gum product depends on the extent of endosperm separation and de-hulling. Remaining fragments of the hull may appear as dark specks and deteriorate product quality.

The carob pod contains approximately 10% of seeds by weight. The endosperm is commonly called splits because two spherical halves constitute the endosperm surrounding the germ. The splits are milled to obtain flours that give a cloudy solution when dissolved in water. The flour is then dispersed into hot water and insoluble particles are removed by diatomaceous earth filtration. The clarified solution is then precipitated using isopropyl alcohol, washed with alcohol, pressed, dried, ground, and sieved. The final product is a white to cream-colored powder that should give a clear solution when dissolved in water.

4.2.1.3 Chemistry and Structural Features

The molar ratio of galactose to mannose of locust bean gum is approximately 1:4. The distribution of D-galactosyl residues along the backbone chain can be random, blockwise, and ordered, where there are long runs of unsubstituted mannosyl units and block condensation of galactosyl units.[5–8]

Like many polysaccharides found in nature, galactomannans are highly polydispersed. The average molecular weight varies significantly, typically ranging from 0.3 to 2 million, depending on the source of seed, growing and harvesting conditions, and manufacturing processes. The galactomannan molecule is considered to adopt an extended ribbon-like structure at the solid state and a semiflexible coil-like conformation in solution.[9] The flexibility of the mannan backbone in solution seems to be limited to some extent since galactomannan oligosaccharides tend to precipitate in an aqueous solution, which is not the case with an ideally flexible polysaccharide chain such as pullulan.[10] Nevertheless, the persistence length, a measure of the flexibility of a polymer chain, has been evaluated to be ca. 3–5 nm for a molecularly solubilized locust bean gum using the pressure cell solubilization method.[11] This value corresponds to the end-to-end length of only several linearly connected glucopyranoses and is much smaller than the value normally found for a food polysaccharide; in the latter case, aggregation was frequently not considered.

4.2.1.4 Functional Properties and Applications

Locust bean gum is an effective thickener or stabilizer since it provides a relatively high viscosity at low concentrations. Locust bean gum is only partially soluble in cold water. Care should be taken, however, if a galactomannan sol is heated above 80°C to achieve complete solubilization, since heating to such an extent may cause oxidative–reductive depolymerization of the main chain and reduction of viscosity of the final solution.[12,13] In the pH range from 4 to 9, the viscosity of a locust bean gum solution is fairly stable, while the viscosity decreases with pH increasing above 9 or decreasing below 4.[14,15] In an acidic condition, acid-catalyzed hydrolysis may occur especially on heating. Locust bean gum is relatively stable against mechanical distortion. It is therefore recommended to moderately apply heat with stirring to prepare a homogeneous gum solution.

The thickening ability of locust bean gum depends on various factors, such as molecular weight distribution, polymer concentration, shear rate, solubilization methods, etc.; it is often regarded as a less viscous galactomannan than guar and tara gums. Locust bean gum solutions ordinarily exhibit pseudoplastic steady-flow behavior or shear thinning at high shear rates, but Newtonian flow behavior may be observed at low shear rates.[16] Viscosity values in the Newtonian plateau can be used to evaluate molecular characteristics. The intrinsic viscosity is determined based on the concentration dependence of Newtonian viscosities of dilute solutions and is related to the molecular weight through a power-law equation called the Mark–Houwink–Sakurada equation. The value of the power-law exponent α is known to represent the stiffness of the polymer chain and the nature of polymer–solvent interactions. In the case of locust bean gum, the α value has been reported to be 0.77, suggesting that water is a relatively good solvent for the gum and that the molecule behaves essentially as a flexible coil.[11] A higher α value of 0.98 was reported for highly polydispersed samples.[16]

Locust bean gum does not normally form a gel, while a weak gel can be obtained upon freeze–thaw treatment[17] or in the presence of a large amount of sucrose.[18] There is also a useful synergistic increase in viscosity or gel strength by blending locust bean gum with certain helix-forming or rigid polysaccharides, including xanthan, κ-carrageenan, and yellow mustard gum.[1–4] X-ray fiber diffraction studies on xanthan–locust bean gum mixtures have revealed new diffraction patterns that are absent in individual polysaccharides, suggesting intermolecular binding between these two types of polysaccharides.[19,20] The formation of a three-dimensional polymer network is then possible if the intermolecularly bound sections play a role as junction zones that are connected by unbound sections of the molecular chains. However, no such preferential intermolecular alignment has been evidenced between κ-carrageenan and locust bean gum in the x-ray fiber diffraction studies, indicating random aggregation between galactomannan and the surface of κ-carrageenan crystallites or only a small degree of intermolecular binding.[19,20]

Locust bean gum is one of the most extensively utilized gums in the world. In the food and pet food industries, the gum is widely used as texturizing, thickening, and stabilizing agents, usually in the amount of <1% of the product weight. One of the important applications of locust bean gum is in ice cream products. It improves the smoothness of the body and handling properties, and gives uniformity of the product and desired resistance to melting.[21] Locust bean gum forms a structured gel network after freezing and temperature cycling between −18 and −10°C; this would explain why locust bean gum-stabilized products are more stable against temperature fluctuations. The mechanism of preventing crystallization has been attributed to its ability to limit the rate of growth of the ice crystals during recrystallization without affecting the initial ice crystal formation processes.[21,22] It is recommended to add a small amount of κ-carrageenan together with locust bean gum in ice cream and other dairy products since syneresis or whey separation caused by the incompatibility between locust bean gum and milk proteins can be prevented.[23] The synergy between locust bean gum and κ-carrageenan produces a cohesive gel and is used in meat and dessert products.

Other food applications include bakery products, pie fillings, sauces, dressings, and creams. Locust bean gum has been used for improving the texture and stability of starch-based products. The viscosity of a mixture of gelatinized starch and locust bean gum is normally much higher than that of starch or galactomannan alone, demonstrating a strong synergy.[24] Gelatinized starch paste can be regarded as a suspension of swollen starch granules in a continuous aqueous phase containing amylose leached out from the granules during gelatinization. Since locust bean gum is likely to be confined in the aqueous phase, its local concentration or effective concentration is anticipated to be higher than the bulk concentration. Thus, it is not surprising that the overall viscosity is higher than what is expected based on the bulk concentration. Other factors influencing properties of starch–galactomannan mixtures are possible reduction in the amount of amylose leaching, suppression of granule swelling, and acceleration of retrogradation since galactomannans reduces the amount of water available for starch. The textile industry uses locust bean gum, sometimes in combination with starch, as a sizing agent and a thickener for paints. Other users of locust bean gum are the pharmaceutical, cosmetics, mining, oil drilling, and construction industries.

4.2.1.5 Physiological Properties and Health Benefits

Locust bean gum is regarded as a dietary fiber and a potential dietary supplement in weight control and treatment of diabetes and hyperlipidemia. Generally observed physiological effects of dietary fibers are often considered as a result of absorption or binding of nutrient compounds by fibers at a high level of viscosity that slows down the mechanical disruption of foods and flow of nutrients in the gut, eventually leading to a lower degree of absorption of nutrients.[25] Locust bean gum has been shown to decrease postprandial plasma glucose and insulin levels in diabetic humans.[26] Hypocholesterolemic effects have also been confirmed for LBG, but underlying mechanisms remain unclear; influence of the gum on the secretion of the gastrointestinal hormones has been indicated.[27]

4.2.2 Guar Gum

4.2.2.1 Source

The development of guar gum resulted from the shortage of locust bean gum in the 1940s. The guar plant is an annual summer legume that is cultivated mainly in western India and eastern Pakistan, and to a lesser extent in tropical areas, such as South and Central America, Africa, Brazil, Australia, and the semiarid regions of the U.S. Southwest. The plant grows to about 1 m in approximately 5 months and yields pods slightly smaller than the carob pod. The guar pod contains no more than 10 seeds, in which the endosperm corresponds to 35% by weight.

4.2.2.2 Method of Production

The endosperm of guar is also called splits because of its structure. The manufacturing process of guar gum is similar to that of locust beam gum: milling of splits, dissolution in hot water, removal of insoluble materials, alcohol precipitation, washing, pressing, drying, grinding, and sieving. Final gum properties depend on the extent of separation. Contamination by rotten black seeds in the raw material causes specks in the gum and a darker color of the powder. Residual germ may cause enzymatic degradation of the gum.

4.2.2.3 Chemistry and Structural Features

The molar ratio of galactose to mannose of guar gum is approximately 1:2. The side group substitution occurs irregularly: side groups are arranged mainly in pairs and triplets.[6] The average molecular weight varies, typically up to a few million, depending on growth and manufacturing factors. The persistence length of molecularly solubilized guar gum has been reported to be ca. 4 nm, similar to that of molecularly solubilized locust bean gum.[11] It thus seems that the intrinsic flexibility of the mannan backbone itself is little influenced by galactose substitution. Differences in functional properties between locust bean and guar gums may rather reflect differences in their solubility and tendency toward intra- and intermolecular aggregation that are relevant to the extent of galactose substitution.

4.2.2.4 Functional Properties and Applications

Due to the high degree of substitution, guar gum is cold water soluble and serves as an effective thickener or stabilizer in the food and other industries. The thickening ability of commercially available guar gum is often higher than that of locust bean gum if no special care is taken for solubilization. Solution properties are fairly stable over the wide pH range of 4 to 10, but the viscosity significantly decreases at pH over 10.[15,28] The hydration rate reduces in the presence of salts and other water-binding agents, such as sucrose.[29–31]

Rheological properties of guar gum solutions share many common features with those of locust bean gum. Guar gum solutions show pseudoplastic steady-flow behavior at high shear rates, and Newtonian domains can be observed at sufficiently low shear rates.[32,33] In the Mark–Houwink–Sakurada equation, which describes the relationships between the intrinsic viscosity and molecular weight, the value of the exponent α of guar gum is reported to be 0.70 to 0.75.[16,32] Additionally, recent ultra-small-angle light-scattering studies have probed the presence of large aggregates in the order of 10 to 100 μm in guar gum aqueous solutions,[34] suggesting that the cold-water solubility does not necessarily guarantee molecular solubilization of guar gum. Dynamic rheological properties of guar gum have features typical of an ordinary random coil polymer. Mechanical spectra, the frequency dependence of the storage (*G*155) and loss (*G*") moduli, show transitions from dilute to semidilute behavior with increasing gum concentration. The relaxation frequency, i.e., *G*' – *G*" crossover frequency, shifts toward lower frequencies with increasing concentration. All mechanical spectra can be superposed onto a single master curve by dividing the moduli by the concentration and shifting the spectra laterally along the frequency axis (Figure 4.1).[32]

Similar to locust bean gum, guar is regarded as a nongelling gum, while synergistic interactions are also observed between guar and other gums, such as xanthan,[3] κ-carrageenan,[1,2] and yellow mustard gum.[4] The magnitude of a synergistic increase in gel strength is much smaller than that of locust bean gum. The fact that the synergistic interactions are more pronounced with lowering galactose contents of galactomannans seems to be in line with the hypothesis that unsubstituted smooth regions of the galactomannan chain bind to xanthan or κ-carrageenan helices, and that a coupled gel network is eventually formed.[35,36] Another plausible explanation is that the galactomannans are incompatible with xanthan or κ-carrageenan, and that the effective galactomannan concentration becomes higher than the bulk concentration in the presence of these helix-forming polysaccharides.[2]

Borate is known to cause gelation of guar gum.[37] The borate ion reacts with the *cis*-hydroxyl groups of guar gum, such as the hydroxyl groups at the 2- and 3-positions of the backbone mannose and those at the 3- and 4-positions of the galactose side

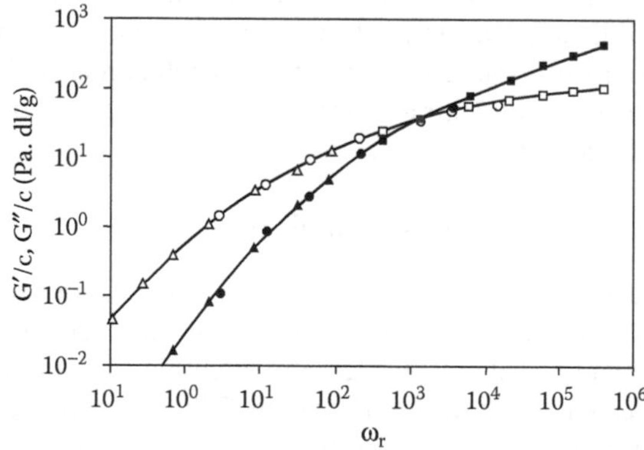

FIGURE 4.1 Master curve derived from frequency–concentration superposition of guar galactomannans. (Replotted from Robinson, G. et al., *Carbohydr. Res.*, 107, 17, 1982. With permission.)

residues. A gel can be formed at room temperature in the presence of borax in alkaline solution, while the gel melts on heating. The pH and temperature dependences of gelling properties can be explained by the pH and Arrhenius type temperature dependences of the dissociation and binding equilibrium constants of the borate ion.

Guar gum is also an effective starch modifier. Gelatinized starch paste containing guar gum can be regarded as a suspension of swollen starch granules dispersed in a guar gum solution.[24] Since the local concentration of guar gum is expected to be higher than the bulk concentration, the viscosity of a starch–guar gum mixture is normally much higher than that of starch or galactomannan alone. Amylose that leaches out of granules during gelatinization should reside in a liquid phase as well. Incompatibility between amylose and guar gum may cause a further increase in the effective concentration of guar gum.[38]

The supply of guar gum is stable not only because the plant origin can be harvested every year, but also because sufficient stock is constantly available. Just like locust bean gum, guar gum has a wide range of applications in food products as a thickener and stabilizer. For example, guar gum is used for obtaining mouth feeling in soup products, improving extensibility of cheese spread and low-sugar jam, and preventing syneresis in meat products. In contrast to locust bean gum, which is susceptible to freeze–thaw processes, guar gum is tolerant to repeated "freeze–thaw cycles" and prevents the formation of large ice crystals during freezing. Guar gum is used to achieve consistency of wheat flour dough elasticity and the water retention property, which vary significantly depending on the gluten content of the flour. A commercial guar gum ingredient may generate yellow color in an alkaline condition, which can be used for adjusting the appearance of products.

4.2.2.5 Physiological Properties and Health Benefits

Physiological activities of guar gum have been relatively well established. By the end of the 1970s, the efficacy of guar gum in reducing plasma cholesterol levels and postprandial hyperglycemia was well known.[39,40] More recent studies have ensured that guar gum can be an effective tool for the dietary management of elevated plasma total and low-density lipoprotein cholesterol if administered into normal daily diets.[41] Physiological effects of guar gum are considered due to not only an increased viscosity of digesta, but also to fermentation of the gum by the colonic microflora.[42] Guar gum is, however, a potential deterrent to palatability since organoleptic characteristics of guar-containing food products tend to be poor because of high levels of viscosity. Efforts have been made to suppress excessive thickening and enhance the amount of intake by partially hydrolyzing the gum.[43] Enzymatic hydrolysis has been shown to be effective in controlled depolymerization of guar gum.[44]

4.2.3 Tara Gum

4.2.3.1 Source

The tara tree is a spiny evergreen shrub that grows to 3 to 10 m. The tree is found in South American countries, including Bolivia, Peru, Ecuador, Chile, and Colombia, but the major cultivator is Peru. The flat pod of tara is about 10 cm long and 1 to 3 cm wide and contains several seeds.

4.2.3.2 Method of Production

Tara gum is extracted from the endosperm with warm or hot water. The industrial manufacturing process is essentially the same as that of locust beam gum and guar gum.

4.2.3.3 Chemistry and Structural Features

The molar ratio of galactose to mannose of tara gum is approximately 1:3. The side group substitution is considered to have random and blockwise distributions.[45] The average molecular weight typically ranges from 0.2 to 2 million. The persistence length and the Mark–Houwink–Sakurada exponent of molecularly solubilized tara gum have been reported to be 3 nm and 0.79, respectively,[11] similar to those of locust bean and guar gums. The intrinsic flexibility of galactomannan backbone thus seems to be insensitive to the degree of galactose substitution.

4.2.3.4 Functional Properties and Applications

The galactose substitution ratio of tara gum is in the middle of highly substituted cold-water-soluble guar and the less substituted locust bean gum, which is only a little soluble in cold water. Tara gum is partially soluble in cold water up to about 80% of the dry weight. The thickening ability of tara gum is higher than that of locust bean gum and comparable to that of guar gum. The freeze–thaw stability is superior to that of locust bean gum.

Tara gum also exhibits synergistic interactions with xanthan[46] and κ-carrageenan.[1] The magnitude of synergistic increments in gel strength is larger than the highly substituted guar gum but smaller than the less substituted locust bean gum. Therefore, there is a trend in interactions between galactomannans and helix-forming polysaccharides that synergistic effects are progressively enhanced with decreasing galactose content of the galactomannans. Similar to the case of locust bean gum–κ-carrageenan mixed gels, x-ray fiber diffraction patterns of tara gum–κ-carrageenan mixed gels are essentially identical to those of pure κ-carrageenan, suggesting the absence of discrete intermolecular binding between the two polysaccharides.[47]

Tara gum is still a little exploited functional food ingredient despite its moderate functional efficiency. Organoleptic characteristics of tara gum are said to be much better than those of guar gum.

4.2.3.5 Physiological Properties and Health Benefits

There is little documented information available regarding physiological properties and health benefits of tara gum. An earlier animal model study indicated that diets containing high doses of tara gum (5%) demonstrated depressions in body weight gain greater than 10% for dosed groups relative to their respective controlled groups.[48]

4.2.4 Fenugreek Gum

4.2.4.1 Source

Fenugreek, *Trigonella foenum-graecum*, is a ca. 60-cm-tall annual leguminous plant native to southern Europe and western Asia and has a long history as a culinary and medicinal herb since ancient times. The plant is grown in northern Africa, the Mediterranean, western Asia, northern India, and currently in Canada.[49] The fenugreek pod is about 3 to 10 cm long and 1 cm wide and contains 10 to 20 seeds with a size of about 2 to 3 mm. The seed has a strong aroma and is somewhat bitter in taste.

4.2.4.2 Method of Production

Fenugreek gum is extracted from the endosperm or ground whole seed with water or dilute alkali, and the yield varies from 13.6 to 38%, depending on the variety/cultivar and extraction methods.[50] Commercial fenugreek gum products, such as Fenupure and Fenu-life, contain over 80% galactomannans with about 5% proteins. Laboratory preparation involving pronase (a nonspecific protease isolated from *Streptomyces griseus*) treatments produces gum products of much higher purity with less than 0.6% protein contaminants.[49]

4.2.4.3 Chemistry and Structural Features

The molar ratio of galactose to mannose of fenugreek gum is approximately 1:1. A reported value of the molecular weight is 1.4 million, compared to 1.3 and 1.2 million for commercial guar and locust bean gum, respectively.[49] The radius of gyration of fenugreek gum was 75 nm, which is in agreement with the range obtained by experiments and modeling for a gum with equal galactose and mannose contents and a molecular weight of 1.4 million.[9,51]

4.2.4.4 Functional Properties and Applications

Solution properties of fenugreek gum are typical of a random coil polymer.[49] In steady-flow rheological tests, fenugreek gum solutions exhibit pseudoplastic behavior at high shear rates. Mechanical spectra of locust bean, guar, and fenugreek gums with comparable molecular weights are similar in shape, but with slightly smaller moduli values for fenugreek gum. Fenugreek gum is a nongelling galactomannan and shows resistance against freeze–thaw treatments. In addition, little synergistic interaction was observed between fenugreek gum and other gums.[52]

Polysaccharides are generally considered to be non-surface active, although in practice some surface activity of a gum specimen may be observed in experiments. Such apparent surface activity has been attributed to the presence of small amounts of protein impurities.[53] Fenugreek gum was reported to possess substantial surface activity and be able to produce stable oil-in-water emulsions with moderately small droplet sizes (2 to 3 μm).[54–56] In addition, physical separation of protein residues from the crude gum sample did not reduce the surface activity.[54] It is therefore tempting to assume that the surface activity is an intrinsic property of the polysaccharide. However, a protease-treated fenugreek gum sample containing less than 0.6% protein (calculated from nitrogen content) exhibited a reduced surface activity.[49] These conflicting results may solely indicate that the protease treatment is more effective in eliminating residual proteins. Protein components in fenugreek gum ingredients may be tightly associated with polysaccharide chains and capable of dominating some surface activity-related functional properties.

4.2.4.5 Physiological Properties and Health Benefits

The seed of fenugreek is a popular food ingredient in northern Africa and southern and western Asia, where it is consumed in various ways in ordinary daily diets. One of the most popular uses is as an appetite-stimulating spice in curries, while the testa of the fenugreek seed contains steroidal saponins that are amphiphilic, highly surface active agents and may find pharmaceutical applications. Fenugreek seed powders have been reported to have antidiabetic and hypocholesterolemic properties in humans.[57] The galactomannan fiber appears to delay gastric emptying and contributes to antihyperglycemic effects. Hypocholesterolemic effects have been attributed to binding of cholesterol with both the saponins and galactomannan fibers in the digestive tract and subsequent excretion in the feces. The conversion of hepatic cholesterol to bile salts may also increase.[57] Inhibitory effects of fenugreek seed powder on colon carcinogenesis have been reported and attributed to suppressed activities of a carcinogen-liberating enzyme, β-glucuronidase, and mucinase, which hydrolyzes the protective mucin and exposes the underlying intestine cells to carcinogenic toxins.[58]

4.3 SEED GUMS AS CELL WALL MATERIALS

4.3.1 Xyloglucan from Tamarind Seed

4.3.1.1 Source

Xyloglucan is a member of a group of so-called hemicelluloses that are plant cell wall polysaccharides composed of a cellulosic backbone and some branches. An important biological function of xyloglucan is that it binds to cellulose microfibrils and controls the rigidity of the cell wall, thereby controlling cell growth.[59–62] Xyloglucan polymers incorporated into growing plant cells have been found to increase the elastic modulus of the cell wall and suppress cell elongation.[62] Similarity in the backbone structure between xyloglucan and cellulose should facilitate noncovalent association of xyloglucan and cellulose.[63–65] In contrast, xyloglucan oligosaccharides activate xyloglucan endotransglycosylase that degrades the polymer in the presence of the oligosaccharides, loosen the cell wall, and promote cell elongation.[62] Additionally, xyloglucan oligosaccharides induce phytoalexin that protects a plant body from microbial infection, indicating their potential application as an environment-friendly pesticide.[66]

Xyloglucan occurs widely in the primary cell wall of higher plants, while the major source of commercially available food-grade xyloglucan is storage xyloglucan in the seed of tamarind tree (*Tamarindus indica*), indigenous to India and Southeast Asia.[66] Recent studies have revealed potentially therapeutic benefits of xyloglucan extracted from the seed of *Detarium senegalense* Gmelin, the flour of which has been used as a thickening agent in traditional Nigerian foods.[67,68]

4.3.1.2 Methods of Production

Xyloglucan may account for 20 to 30% of the dry weight of the primary cell wall of dicotyledons and nongraminaceous monocotyledons. Tamarind seed xyloglucan is the only seed xyloglucan currently produced on an industrial scale.[66] Tamarind seeds are washed with water, heated, de-hulled, and ground to provide tamarind kernel powder (TKP). Tamarind seed xyloglucan (TSX) is a water-soluble polysaccharide fraction prepared from TKP. TSX is a permitted food additive in Japan, Korea, and Taiwan. The purity of commercially available TSX can be improved by repeating solubilization in water and precipitation using alcohol.

4.3.1.3 Chemistry and Structural Features

Xyloglucan has a backbone of 1→4-linked β-D-glucopyranosyl residues, three quarters of which is substituted with α-D-xylopyranose at the 6-position. Some of the xylopyranosyl residues are substituted at the 2-position with β-D-galactopyranose. In the living cell wall, a part of the galactosyl residues are further substituted at the 2-position by an α-L-fucosylpyranose. Four types of structural units, or monomers, have been identified in extracted xyloglucan: Glc_4Xyl_3 heptasaccharide, two types of Glc_4Xyl_3Gal octasaccharides, and $Glc_4Xyl_3Gal_2$ nonasaccharide (Figure 4.2). Tamarind and detarium seed xyloglucans slightly differ in the content of the galactosyl residue (Table 4.1).[69] Synchrotron-radiated small-angle x-ray scattering (SAXS) and molecular dynamics simulation studies on tamarind xyloglucan monomers have suggested that they can be regarded as flat ellipsoids in shape.[70] The evaluated length of the shortest semiaxis is constantly 0.22 nm for all types of monomers, while the cross-sectional width increases from 0.62 nm for the heptasaccharide to 0.71 nm for the octasaccharide and 0.75 nm for the nonasaccharide with increasing number of side-chain residues. The longest axis, corresponding to the backbone length, decreases from 1.49 nm for the heptasaccharide to 1.43 nm for the octasaccharide and 1.41 nm for the nonasaccharide, indicating that the β-glucan backbone with a larger number of side-chain residues is more arched or twisted.

Reported values of the molecular weight evaluated based on light scattering vary from 880,000 to 1,160,000 for tamarind xyloglucan to 2,690,000 for detarium xyloglucan.[69,71] X-ray fiber diffraction analyses have confirmed that xyloglucan adopts an extended twofold helix conformation similar to cellulose in solid state.[70] The conformation in an aqueous solution of xyloglucan polymer has been investigated using light scattering and synchrotron-radiated SAXS.[71] The evaluated cross-sectional radius of gyration is 0.58 nm for tamarind xyloglucan and 0.49 nm for detarium xyloglucan. Earlier SAXS studies have reported a cross-sectional radius of gyration value of tamarind xyloglucan to be 0.29 nm.[72] The light-scattering profiles of tamarind xyloglucan suggest an extended and stiff molecular chain with the Kuhn

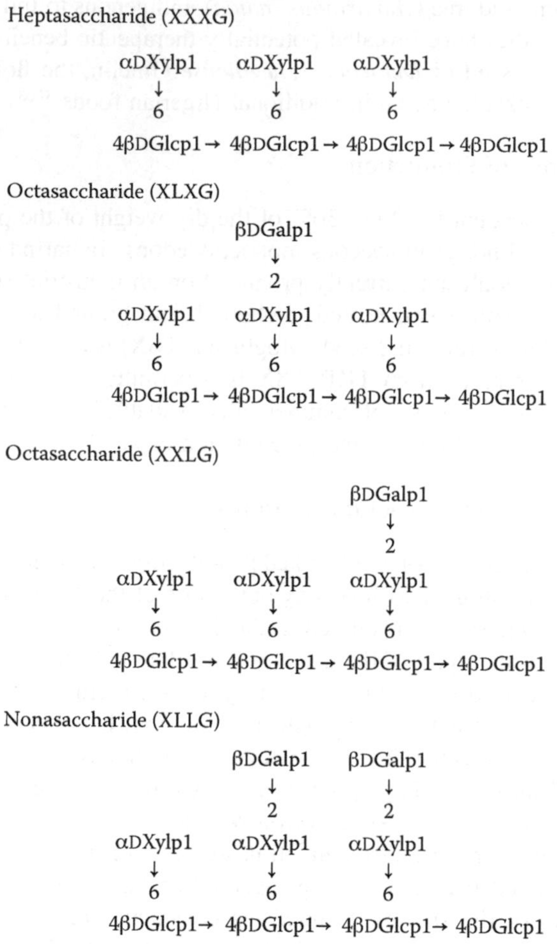

Heptasaccharide (XXXG)

αDXylp1 αDXylp1 αDXylp1
↓ ↓ ↓
6 6 6
4βDGlcp1→ 4βDGlcp1→ 4βDGlcp1→ 4βDGlcp1

Octasaccharide (XLXG)

βDGalp1
↓
2
αDXylp1 αDXylp1 αDXylp1
↓ ↓ ↓
6 6 6
4βDGlcp1→ 4βDGlcp1→ 4βDGlcp1→ 4βDGlcp1

Octasaccharide (XXLG)

βDGalp1
↓
2
αDXylp1 αDXylp1 αDXylp1
↓ ↓ ↓
6 6 6
4βDGlcp1→ 4βDGlcp1→ 4βDGlcp1→ 4βDGlcp1

Nonasaccharide (XLLG)

βDGalp1 βDGalp1
↓ ↓
2 2
αDXylp1 αDXylp1 αDXylp1
↓ ↓ ↓
6 6 6
4βDGlcp1→ 4βDGlcp1→ 4βDGlcp1→ 4βDGlcp1

FIGURE 4.2 Structural features of tamarind and detarium seed xyloglucans. (Adapted from Wang, Q. et al., *Carbohydr. Res.*, 283, 229, 1996.)

TABLE 4.1
Comparison of Composition of Monosaccharide and Oligosaccharide Tamarind and Detarium Xyloglucans[75]

Source	Oligosaccharides				Monosaccharides		
	XXXG	XLXG	XXLG	XLLG	Xylose	Galactose	Glucose
Tamarind	1	0.42	2.07	6.2	1	0.51	1.34
Detarium	1	0.3	5.6	6.2	1	0.46	1.33

Note: X = xylose-substituted glucose residue; L = galactosylxylose-substituted glucose residue; G = unsubstituted glucose residue.

segment length, a measure of chain rigidity, ranging from 108 to 184 nm.[70] The backbone is supposed to be twisted like cellobiose in a solution.[63] Atomic force microscopy has been utilized to directly visualize tamarind xyloglucan polymers.[73] Xyloglucan molecules spread onto the molecularly flat surface of mica appear as largely linear chains with some branches. The widely ranging contour length of the main chain, approximately 0.1 to 1.5 μm, suggests a highly polydisperse nature of the molecular weight. The heights of chains, measures of the cross-sectional chain diameter, are fairly uniformly about 0.6 nm. On the other hand, reported light-scattering profiles of detarium xyloglucan are inconsistent with those of a linear polymer, but more similar to those of a branched polymer with long side-arm chains.[74] SAXS profiles of enzymatically carboxylated detarium xyloglucan have also supported the hypothesis that the main chain of detarium xyloglucan is highly branched, and thus the overall conformation is more compact than an extended tamarind xyloglucan molecule.

4.3.1.4 Functional Properties and Applications

The intrinsic viscosity has been determined to be 6.0 dl/g for tamarind xyloglucan[72] and 8.9 dl/g for detarium xyloglucan,[69] consistent with the higher molecular weight of detarium xyloglucan. Steady-flow characteristics of dilute solutions of xyloglucan are described essentially as Newtonian. Solutions at higher xyloglucan concentrations (ca. >0.5% w/w) exhibit a constant viscosity at relatively low shear rates, and at higher shear rates, shear thinning is observed.[74] The onset of shear thinning shifts to a lower shear rate with increasing xyloglucan concentration. Dynamic rheological properties of xyloglucan solutions are similar to those of ordinary polymer solutions. Mechanical spectra exhibit a transition from a spectrum typical of dilute polymer solutions ($G' < G''$ at all frequencies) to that of semidilute polymer solutions ($G' < G''$ at low frequencies and $G' > G''$ at higher frequencies) with increasing concentration.[74,75] Rheological properties are stable against heat (e.g., 100°C for 2 h) and pH (e.g., for 45 days in the presence of 2.25% acetic acid and 1.0% salt), making this polysaccharide a promising candidate for a physically functional food ingredient.[66] The absence of ionic groups in the molecule may indicate insensitivity of solution properties to the presence of salts. Tamarind xyloglucan can be used as a starch replacer in food products since its solution properties are similar to gelatinized starch but more stable against heat, pH, and mechanical distortion.

Tamarind xyloglucan can form a gel in the presence of alcohol[76] or a large amount (40 to 70% w/w) of sugar.[77] Gels made with alcohol are hard and melt at a lower temperature than gels with sugar. SAXS studies have confirmed the absence of ordered structures in alcohol-induced gels at the nanometer scale, suggesting that cross-linking domains are composed of randomly aggregated polysaccharide chains, and the side groups prevent substantial aggregation or precipitation.[76] Sugar-induced gels are elastic and have good water-holding properties. Freeze–thaw processes can make the gels harder and more elastic.

Xyloglucan is categorized as an amyloid that exhibits a characteristic blue color when an iodine–potassium iodide solution is added. In the case of iodine–amylose reaction, amylose molecules transform into single helices and form inclusion compounds with iodine. The architecture of an iodine–xyloglucan complex may be different: an iodine molecule is held between two laterally associated xyloglucan chains. Supporting evidence of this structural model is the fact that a thermoreversible gel is formed in the presence of iodine at a sufficiently high concentration.[66] It is likely that an iodine–xyloglucan complex plays a role as a cross-link in such a gel network. Tamarind xyloglucan is also known to form a gel with polyphenols such as catechin.[66]

Not much information is available regarding interactions of xyloglucan with polysaccharides other than cellulose. Effects of xyloglucan on gelatinization and retrogradation of corn starch have been investigated, but no significant interaction has been recognized.[75] Synergistic effects on dynamic viscoelasticity have been found between xyloglucan and gellan, a microbial-produced gelling polysaccharide, and the synergistic effect is attributed to the exclusion effects of highly hydrophilic xyloglucan molecules that effectively increase the local concentration of the gellan polysaccharide.[73]

The galactoxylose branch of xyloglucan is considered to generate steric hindrance and prevent intermolecular association. Thus, ordinary xyloglucan is soluble in cold water and forms a gel only in the presence of alcohol or a substantial amount of sugar. It has been revealed that enzymatic elimination of more than 35% of original galactose residues with β-galactosidase imparts gelling ability to the modified xyloglucan.[78] Additionally, sol–gel transition behavior of this galactose-cleaved xyloglucan (up to ca. 60% of the total galactose residues) presents a quite unique feature: a sol of modified xyloglucan turns into a gel at a certain temperature on heating, indicating involvements of intermolecular association driven by hydrophobic interactions, and the gel melts at a higher temperature on further heating.[78] Cross-linking domains in such a gel are considered to be composed of randomly aggregated polysaccharide chains, similar to the case of alcohol- or sugar-induced gelation of unmodified xyloglucan. The unique temperature sensitivity of enzyme-modified xyloglucan may indicate its potential use as a controlled drug delivery material.

4.3.1.5 Physiological Properties and Health Benefits

Xyloglucan is regarded as a dietary fiber that can increase the viscosity of digesta in the stomach and small intestine to reduce the rate and extent of absorption of nutrients. A dietary fiber would also have an impact on fermentable intestinal bacteria as prebiotics. Tamarind xyloglucan has been shown to reduce plasma and liver cholesterol levels in rats on high-cholesterol diets.[79] However, a high viscosity can be regarded as a drawback since texture or mouth feel of food products is also influenced. A strategy to reduce the viscosity is to reduce the molecular weight of the polymer. Partially hydrolyzed xyloglucans have been reported to be less viscous but maintain hypocholesterolemic effects, albeit to a lesser extent than the intact polymer.[79] Other studies have reported that tamarind xyloglucan prevents suppression of immune responses in mice exposed to ultraviolet irradiation.[80] Detarium xyloglucan has been found to have promise in the treatment of diabetes and hyperlipidemia.[67,68,81] Postprandial plasma glucose and insulin levels of healthy human subjects significantly decreased after consumption of meals supplemented with detarium seed flour.[68] Detarium seed flour also significantly reduced plasma cholesterol levels in rats.[81]

4.3.2 Soluble Soybean Polysaccharides (Soya Fiber)

4.3.2.1 Source

Water-soluble soybean polysaccharide (SSPS) or soya fiber is the cell wall material of the cotyledons in soybeans. Soya fiber is also referred to as soybean hemicellulose. Commercial SSPS or soya fiber is extracted from okara, a by-product produced during production of soybean protein isolates and tofu.[82]

4.3.2.2 Method of Production

Soybean fiber or SSPS is produced on an industrial scale in Japan by Fuji Oil Co. Ltd. It is extracted from the by-product okara in the production of soy protein isolates or tofu. The industrial production of soya fiber is similar to the procedure described below: Okara in dilute acid (pH 5, adjusted by hydrochloric acid) is autoclaved at 120°C for 1.5 h.[82,83] After cooling to room temperature, the suspension is centrifuged at 10,000 *g* for 30 min to remove the residue. The remaining SSPS in the residue is washed with distilled water and centrifuged again. The supernatants are combined and then concentrated and dried. This process gives a yield of about 45% based on the starting okara material.

4.3.2.3 Chemistry and Structural Features

Soybean fiber contains 64 to 80% soluble fiber, 4.8 to 14% proteins, and 6 to 8.6% ash. Highly purified SSPS contains 83.6% carbohydrates, 4.7% proteins, and 5.3% ash. SSPS is composed of galactose (42 to 46%), arabinose (18 to 27%), rhamnose (4.8 to 6.5%), xylose (2.3 to 5.1%), and small amounts of fucose and glucose (1 to 3%), in addition to the presence of galacturonic acid (18 to 24%)[82,83]

Soy fiber or SSPS is a highly branched pectic polysaccharide, and the backbone consists of two types of structures: a galacturonan (GN) and a rhamnogalacturonan (RG).[83] The RG structure is comprised of a diglycosyl repeating unit: (1→4)-α-D-galacturonic acid-(1→2)-α-L-rhamnopyranose units with β-D-galactan side chains. The side chains are substituted with L-fucosyl and L-arabinosyl residues, which are linked to the C-4 of the rhamnosyl residues. The degree of polymerization of the side chains is estimated to be 43 to 47, which is longer than those of fruit pectins. SSPS is highly polydispersed in molecular weight. Three to four fractions of molecular weight ranging from 4700 to 542,000 Da, based on pullulan standards, were identified by gel permeation chromatography.[83,84] A structure model of SSPS was proposed by Nakamura and coworkers[83] (Figure 4.3).

4.3.2.4 Functional Properties and Applications

SSPS is a low-viscosity gum and its flow behavior is similar to that of gum arabic (Figure 4.4). It is soluble in cold water and exhibits Newtonian flow behavior at a

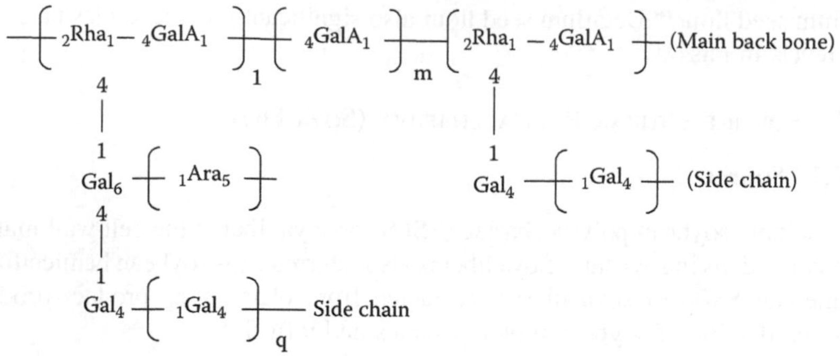

$$-\left[{}_2\mathrm{Rha}_1 - {}_4\mathrm{GalA}_1\right]_1\left[{}_4\mathrm{GalA}_1\right]_m\left[{}_2\mathrm{Rha}_1 - {}_4\mathrm{GalA}_1\right]- \text{(Main back bone)}$$

4 | 1 Gal$_6$ $-\left[{}_1\mathrm{Ara}_5\right]-$ 4 | 1 Gal$_4$ $-\left[{}_1\mathrm{Gal}_4\right]_q-$ Side chain

4 | 1 Gal$_4$ $-\left[{}_1\mathrm{Gal}_4\right]-$ (Side chain)

FIGURE 4.3 Structural features of soluble soybean polysaccharides. (Adapted from Maeda, H., in *Handbook of Hydrocolloids*, Phillips, G.O. and Williams, P.A., Eds., CRC Press, Boca Raton, FL, 2000, p. 309.)

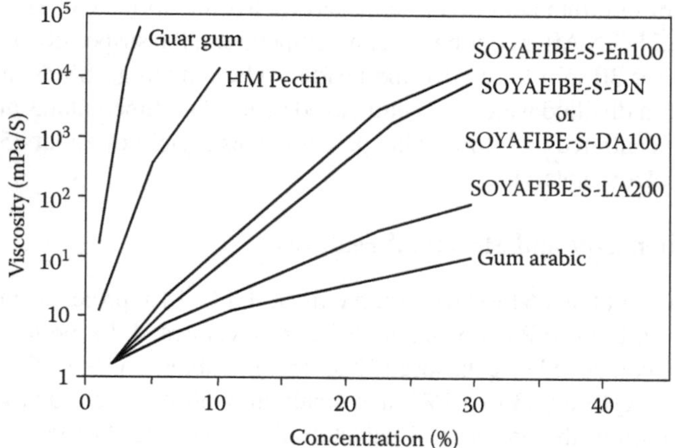

FIGURE 4.4 Comparison of viscosity–concentration relations of soluble soybean polysaccharides (SSPSs) and some commercial gums. (Reproduced from Maeda, H., in *Handbook of Hydrocolloids*, Phillips, G.O. and Williams, P.A., Eds., CRC Press, Boca Raton, FL, 2000, p. 309.)

concentration as high as 10% (w/w). However, when the concentration is increased to 20% (w/w), it exhibits shear-thinning flow behavior. Unlike pectins, the viscosity of SSPS solution is not sensitive to salts, including NaCl, $CaCl_2$, and KCl. However, the viscosity increases with the increase of sugar concentration (e.g., sucrose). The viscosity of SSPS solution is also pH and temperature sensitive: the viscosity tends to decrease with the decrease of pH and increase of temperature, and the process is reversible.[82]

SSPS could be used in the food industry as a stabilizing and thickening agent. For example, SSPS exhibits superior properties regarding volume, color, and crumb grain when added to a Japanese sponge cake at 4%.[82] Addition of 10% SSPS has less effect than wheat bran on color and surface smoothness of Chinese steamed bread, but has a stronger detrimental effect on volume and texture.[82] Recent applications of SSPS include stabilizing milk proteins under acidic conditions.[85] It requires lower SSPS concentrations than pectin to stabilize and disperse protein particles, and the fine milk protein structures (about 0.6 μm) and prevent agglomeration. The mechanism for stabilization of milk proteins was attributed to the steric effect of the thick layer of the neutral sugar side chains; this is different from pectin, which stabilizes milk proteins mainly by electrostatic repulsion.[85] A more detailed application of SSPS in food systems is described by Maeda.[82]

4.3.2.5 Physiological Properties and Health Benefits

SSPS is classified as a food ingredient and food additive with no limitation of application in Japan.[82] Recently, the European Commission Health and Consumer Protection Directorate-General evaluated the safety of soybean hemicellulose (soy fiber) as a food additive and concluded that the use of soybean hemicellulose in food products at levels ranging from 0.5 to 30 g/kg is acceptable.[86] In a study for supporting a GRAS (generally recognized as safe) notice submission to the U.S. Food and Drug Administration (FDA), SSPS was fed to approximately 6-week-old Sprague Dawley rats at three dosages for 3 months. The animal toxicity test revealed no observed adverse effect at intake levels of 2.43 g/kg of body weight for males and 2.91 g/kg of body weight for females.[87] Tests on animal models also revealed that SSPS appeared to have favorable effects on the intestinal functions, as it is partially metabolized into organic acid by enteric bacteria and it effectively shortens the gastrointestinal transit time in rats.[87] Ingestion of SSPS also could improve postgastrectomy osteopenia and partially prevent calcium malabsorption in rats. A patented method describes the use of SSPS as a cholesterol oxide-adsorbing agent.[88] All these data demonstrate that SSPS is a safe and effective dietary fiber in addition to its functional properties.

4.4 GUMS FROM SEED COAT: MUCILAGE

4.4.1 Psyllium Mucilage

4.4.1.1 Source

Psyllium is the common name of the plant genus *Plantago*. The seed coat (husk or hull) of the plant contains high levels of mucilage (dietary fiber) and has a long history for medicinal use. Psyllium husks from *Plantago ovato* and *Plantago psyllium* are commercially produced in several European countries, Pakistan, and India, while the U.S. is the world's largest importer of psyllium husk, mainly for pharmaceutical firms to make fiber-based laxative products. Recently, there has been a significantly increased interest in psyllium due to the FDA's approval that producers of certain foods containing soluble fiber from psyllium seed husk (PSH), such as some breakfast cereals, may claim that these foods, as part of a diet low in saturated fat and cholesterol, may reduce the risk of coronary heart disease (CHD).[89]

4.4.1.2 Method of Production

Psyllium mucilage is produced by mechanical milling/grinding of the seed coat of psyllium seed, and the yield is about 25% of the seed weight. The milled seed coat is a white fibrous material that absorbs water quickly to give clear, colorless mucilaginous gel.

4.4.1.3 Chemistry and Structural Features

Psyllium mucilage is a heteroxylan consisting of (1→3)- and (1→4)-mixed-linked β-D-xylopyranosyl backbone chains, with side chains attached at the 2- and 3-positions of the 1,4-linked β-D-xylopyranosyl residues. The side chains are composed of β-D-xylopyranosyl and α-L-arabinofuranosyl residues. In addition, a pectic fraction may also exist containing 1→4-linked α-D-GalpA residues and small amounts of 1,2,4-linked Rha*p* and 1,3- 1,6-, and 1,3,6-linked Galp.[90]

4.4.1.4 Functional Properties and Applications

Psyllium gum does not dissolve completely in water but swells to a mucilagenous dispersion with the general appearance of wallpaper paste. A 2.0% psyllium gum dispersion exhibits gel-like structure, similar to that of xanthan gum, which generates a weak gel network by entanglement or association of rigid and ordered molecular structures. Increasing the concentration of psyllium gum from 1 to 2% gives a significant increase in gel strength. However, freshly prepared solutions/dispersions of psyllium gum (1%) show flow properties similar to those of disordered coils, with a Newtonian plateau at low shear rate. Upon aging, psyllium gum solutions form cohesive gels and show obvious syneresis. The gels continue to contract on storage over long periods (to about 30% of their original volume after 3 months). This contraction process can be accelerated by freezing and thawing cycles. However, psyllium gum is stable in high-salt solutions (e.g., 2.5 *M* NaCl) that are formed by neutralization of the alkaline extract over a prolonged storage, with no evidence of gelation or precipitation.

Psyllium mucilage is used in ice creams and frozen deserts as a thickener and stabilizer. It is also used with other gums in bakeries to replace wheat gluten. For example, in preparation of gluten-free bread, psyllium gum and hydroxypropylmethylcellulose (HPMC) at 2 and 1%, respectively, are added to rice flour to give a loaf volume close to that of hard wheat control.[91] The effectiveness of the psyllium–HPMC in bread making is attributed to the network stabilizing properties of psyllium that stabilize the gas cells formed during proving, and preventing them from collapsing during the initial stages of heating in the baking oven. Further stabilization of the gas cells is due to the gelation of HPMC. Psyllium mucilage is also used in preparing highly soluble fiber foods, such as ready-to-eat breakfast cereals and nutritional bars for their cholesterol-reducing effect.[92] In the U.S., the challenge is to incorporate the required amount of psyllium into one serving of a food product in order for the cholesterol-lowering claim to be allowed on the label.[92]

Reducing gelling properties or water-holding capacity will help to include more psyllium in a food product. To achieve this, several techniques have been used to modify psyllium mucilage, such as physical, mechanical, and enzymatic means.[93] Lower-grade psyllium husk is used in landscaping as a water-binding agent for improving water retention of newly seeded grass areas or improving the transplanting success of woody plants. Psyllium husk is also used as an environmental friendly binder for fixing joints of interlocks in the landscape industry.

4.4.1.5 Physiological Effects and Health Benefits

Psyllium mucilage exhibited significant bioactivities in both animal models and human subjects. For example, psyllium has been shown to reduce total and LDL cholesterol in animals and humans. A meta-analysis of 12 studies on 404 adults with mild to moderate hypercholesterolemia concluded that psyllium-enhanced cereal product reduced total and LDL cholesterols in 5 and 9%, respectively.[94] Other studies demonstrated that psyllium mucilage lowers plasma LDL cholesterol levels by 6 to 20% in mildly hypercholesterolemic individuals. Animal studies have been carried out to elucidate the mechanisms by which psyllium mucilage brings about a reduction of cholesterol.[95] Psyllium mucilage increased the activity of cholesterol 7 alpha-hydrolylase, which is the rate-limiting enzyme in bile acid synthesis in guinea pigs.[96] Diet intake containing 10 to 20% psyllium mucilage significantly increased the gastrointestinal mucin level in rats, which correlated with the reduction of cholesterol. The increase of mucin level caused by psyllium mucilage may protect these organs as well as alter nutrient absorption.[97] The mechanism of the LDL cholesterol-lowering action of psyllium was examined in the hamster, and the major effect is exerted at the level of LDL cholesterol production.[95] In a human study, 125 patients with type 2 diabetes were treated with 5 g of psyllium mucilage for 6 weeks. It was found that the levels of fasting plasma glucose, total plasma cholesterol, LDL cholesterol, and triglyceride were significantly reduced ($p < 0.05$), while the HDL cholesterol level increased significantly ($p < 0.01$).[98] It was also demonstrated that intake of psyllium reversed the hypercholesterolemic effect of trans fatty acids in rats.[99]

Many epidemiological and experimental studies suggest that a low-fat, high-fiber, high-calcium diet has a strong protective effect against colon cancer. The combination of wheat bran and psyllium mucilage exhibited a synergistic effect in inhibiting earlier phases of carcinogenesis.[100]

Psyllium gum has been used as a demulcent in dysentery, erosion of intestines, dry coughs, burns, excoriations, and inflammations of the eyes.[101,102] It is extensively used as a bulk laxative. The water-holding capacity and gelling property of psyllium gum can be used to delay and reduce allergic reactions by holding toxins and allergens in the gel structure. Although there were no adverse effects for consumption of psyllium seed or husk, the recommended daily intake is between 10 and 30 g/day in divided doses. It is also recommended that psyllium mucilage be kept well hydrated before taking. However, some individuals may be allergic to psyllium mucilage; thus, caution must be taken in this regard.[103,104]

4.4.2 Flaxseed Gum

4.4.2.1 Source

Flaxseed is the seed of flax plant (*Linum usitatissimum*), a member of the *Linaceae* family. Flaxseed has a flat and oval shape with average dimensions of 5 mm in length, 2.5 mm in width, and 1.5 mm in thickness.[105] The seed coat (testa) contains a thick mucilage (epidermis) layer that can be easily identified by microscopy. This mucilaginous material is a secondary cell wall polysaccharide that is soluble in cold water. The high content and easy extraction of mucilage make flaxseed gum a potential commercially viable gum.

4.4.2.2 Method of Production

4.4.2.2.1 Extraction of Gum from Flaxseed

Flaxseed gum's characteristic of easy dissolution in cold water allows the use of mild conditions to extract the gum from raw materials. A typical extraction procedure is to soak the seeds in water at various temperatures with stirring for 3 to 16 h.[52] Temperature and time have a significant effect on the yield and composition of flaxseed gum.[106] For example, higher temperature could increase the gum yield from 5 to 9%, but it also increases the content of proteins in the gum.[106] The protein contaminants can be partially removed by treatment with Vega clay.[106,107] There is an optimum extraction condition that gives a relatively high yield of gum (ca. 8%) with low levels of protein contaminants (<8%).[108] The optimum condition was identified as a temperature of 85 to 90°C, a pH between 6.5 and 7.0, a water:seed ratio of 13:1, and an extraction time of 2.5 to 3 h.[108]

The aqueous extract is then filtered or centrifuged to remove solid particles, and the supernatant is precipitated in organic solvent, followed by drying and grinding to obtain powder products. Spray drying is an alternate way of drying flaxseed gum. However, the final products often have lower viscosity due to the high outlet temperature.[109]

4.4.2.2.2 Extraction of Flaxseed Gum from Flaxseed Meal

Since linseed oil is the primary product of flaxseed, a large amount of flaxseed meal is produced as a by-product from the oil crushing industry. Solvent-extracted meal could be separated by air or screened to obtain kernel and hull fractions.[110] The hull fraction is then extracted with water (water:solid ratio, 30:1; pH 4.5; temperature, 60 to 80°C). The extracted liquid is centrifuged or filtered, adjusted to pH 7, concentrated by rotary evaporation, and spray dried. The meal cake can also be extracted with 5% sodium chloride, followed by centrifugation with activated carbon, and alcohol precipitation. The flaxseed gum produced from commercial meals can be useful as an emulsifying agent for chocolate milk and other food products because it frequently contains high levels of proteins. Iron salt can be added to the extraction solvent to prevent extraction of tannin pigment.[111]

4.4.2.2.3 Extraction from Hull

Because the mucilage is deposited in the outer layer of the seed coat, it will be more efficient to extract the gum from the hull if it can be separated from the kernel. A recently patented technology enables the separation of flaxseed into a hull fraction and a kernel fraction on an industrial scale.[112,113] Flaxseed hull produced by this technology is rich in fiber, lignans, and other nutrients. The hull is first extracted with organic solvents to give a high-lignan product, and the residue is further extracted with water to produce flaxseed gum. The extracted gum can be recovered by spray drying or precipitation in organic solvents.

4.4.2.3 Chemistry and Structural Features

Flaxseed gum contains 50 to 80% carbohydrates, 4 to 20% proteins, and 3 to 9% ash. The large variation in chemical composition is mostly due to the raw materials, such as varieties and growing conditions, and, more importantly, to the form or part of the material used for extraction (e.g., whole seed, hulls or meal, etc.). The extraction solvent, pH, temperature, and other processing conditions also have a significant influence on the chemical composition of flaxseed gum.[106,107,114–119] Table 4.2 displays the variations of chemical and monosaccharide composition of flaxseed gums extracted from different flaxseed cultivars and breeding lines.[116]

Flaxseed gum has two major fractions: a neutral polysaccharide mainly composed of xylose, arabinose, and galactose, and an acidic polysaccharide consisting of D-galactose, L-rhamnose, and D-galacturonic acid. The neutral arabinoxylan has a (1→4)-β-D-xylosyl backbone to which arabinose and galactose side chains are attached at positions 2 and 3.[115] The acidic polysaccharide has a backbone consisting of (1→2)-linked α-L-rhamnopyranosyl and (1→4)-linked D-galactopyranosyluronic

TABLE 4.2
Variation of Monosaccharide Compositions of Flaxseed Gum from Different Cultivars[116]

	Flaxseed Gums[a]			
	Norman	**Omega**	**Foster**	**84495**
Uronic acid (%)[b]	21.0	25.1	23.9	15.7
Sugar composition (%)[c]				
Rhamnose	21.2	27.2	25.6	12.8
Fucose	5.0	7.1	5.8	3.0
Arabinose	13.5	9.2	11.0	18.1
Xylose	37.4	28.2	21.1	42.5
Galactose	20.0	24.4	28.4	18.4
Glucose	2.1	3.6	8.2	3.7

[a]Flaxseed gums were extracted from four cultivars/breeding lines.
[b]On dry base.
[c]Relative composition.

TABLE 4.3
Intrinsic Viscosity of Flaxseed Gum and Its Fractions[114]

Flaxseed Gum[a]	Intrinsic Viscosity (ml g^{-1}, in 1 *M* NaCl)
Norman	
CFG[b]	483.0
NFG	530.4
AFG	248.4
84495	657.8
Foster	434.0
Omega	536.6

Note: CFG = crude flaxseed gum; NFG = neutral flaxseed gum; AFG = acidic flaxseed gum.

acid residues with side chains of fucose and galactose. The ratio of L-rhamnose, L-fucose, L-galactose, and D-galacturonic acid is about 2.6:1:1.4:1.7.[119,120] The molecular weight of the neutral polysaccharide fraction is much higher than the acidic fraction, as revealed by size-exclusion chromatography, which is in agreement with their hydrodynamic volumes (as indicated by intrinsic viscosity), as shown in Table 4.3.[114]

4.4.2.4 Functional Properties and Applications

Flaxseed gum exhibits Newtonian flow behavior at low concentrations and shear-thinning flow behavior at high concentrations. However, the broad variation in chemical composition allows some flaxseed gum to exhibit stronger rheological properties, such as formation of gel, while others may behave like a viscoelastic fluid.[113] There is an apparent correlation between the structural features of flaxseed gum and its rheological properties: gums that contain high levels of neutral polysaccharides (arabinoxylan) are more viscous and exhibit shear-thinning flow behavior; in contrast, gums that contain higher levels of acidic polysaccharide have a much lower viscosity and exhibit Newtonian flow behavior.[116] Evidence has shown that the neutral arabinoxylan is responsible for the high viscosity and shear-thinning flow behavior due to its higher molecular weight. In contrast, the acidic polysaccharide fraction has a much smaller molecular weight, and therefore exhibits lower viscosity and Newtonian flow behavior. The roles of neutral and acidic polysaccharides were verified by examining their molecular weight and rheological properties separately.[115,118]

Dynamic oscillatory rheological tests showed that flaxseed gums extracted from different varieties exhibited wide variations in their viscoelastic properties. It can be a typical viscoelastic fluid or a real gel when examined at the same concentration (1 to 3%).[118] The viscoelastic properties of flaxseed gum are also correlated to the amount and molecular weight of the neutral fraction (arabinoxylan).[114,118]

The pH of the gum solution has a significant effect on the flow behavior and viscosity of flaxseed gum. The lowest viscosity for flaxseed gum is observed at pH 2. As the pH increases, the viscosity increases steadily until pH 8, at which the viscosity is three times its value at pH 2. Further increase of pH results in decrease of viscosity.[106] Depending on the source and chemical composition of the gum, the effect of pH on solution viscosity varies slightly, but generally follows the trends of high viscosity at neutral pHs and lower viscosities in both high and low pH regions.

Similar to other gums, flaxseed gum can be used as a thickener and stabilizer in food products. Flaxseed gum affects bread-making properties including pasting, dough rheology, and baking. Adding flaxseed gum to bread formulations improved the grain texture of the bread loaves.[121] The cells were more elongated, and the texture was silky and softer after storage. The arabinoxylan component may have played a role in its ability to delay the firming. The gum may have a mellowing effect on the gluten, which allows for greater expansion during the fermentation and baking stages.[121]

The addition of flaxseed gum could improve muffin height and volume, without changing the texture significantly; 0.5% flaxseed gum (flour basis) could replace 0.1% xanthan gum or guar gum to produce muffins with good volume and texture.[122] In the evaluation of the effect of flaxseed gum on the stability of a model salad dressing, pH affected the stability of the emulsion the most. The most stable emulsion occurred at pH 6, while the least stable occurred at pH 2; this observation correlates with the viscosity dependence of pH: flaxseed gum exhibits the highest viscosity at pH 6 and the lowest viscosity at pH 2, as described previously. Flaxseed gum appears to act as a steric stabilizer, which means sufficient adsorption of flaxseed gum is required to cover the particle surface completely to prevent two particles from approaching one another.[122]

Flaxseed gum is also considered to be a better water-in-oil emulsifier than Tween 80, gum arabic, or gum tragacanth. Concentrations of 0.5 to 1.5% flaxseed gum are suitable to stabilize oil/water emulsions. In food applications, flaxseed gum has been used as egg white substitutes in bakery products and ice creams. The strong buffering action of flaxseed gum also makes it useful in the manufacture of fruit drinks. Other applications of flaxseed gum include use in the printing, textile, and cigar industries. A paper product using flaxseed gum as a deflocculant has good tensile and flexural (tear) strength.

4.4.2.5 Physiological Properties and Health Benefits

A diet supplement with partially defatted flaxseed reduced the level of LDL cholesterol in serum, and flaxseed gum was identified as the likely responsible active ingredient.[123] It has been suggested that flaxseed gum could also be used in medicinal preparations — ointments. Pastes containing flaxseed gum are effective in the treatment of furunculosis, carbunculosis, impetigo, and ecthyma.[124] Flaxseed gum has been used as a bulk laxative, a cough emollient agent, and a stabilizer in barium sulfate suspensions for x-ray diagnostic preparations.[125,126] Tablets prepared with flaxseed gum have improved disintegration and a slower rate of drug release. The stringy and fast drying properties of flaxseed gum make it suitable in hairdressing preparations, hand cream formulations, and denture adhesives. At 2.5% concentration, flaxseed gum is a good base for an eye ointment.

Flaxseed gum solution has been used as a saliva substitute due to its lubricating and moisture-retaining characteristics, resembling those of natural saliva.[127] It is effective for patients suffering from dryness of the mouth (xerestomia), and particularly suitable for reducing the dryness of the mouth at night.[127]

4.4.3 Yellow Mustard Gum

4.4.3.1 Source

Yellow mustard gum (YMG) is a mucilage deposited in the epidermal layer of yellow mustard seed (*Sinapis alba).* It is soluble in water and can be extracted from whole seed or seed coat (bran).

4.4.3.2 Method of Production

Extraction from Seed

Yellow mustard gum is relatively easy to extract with water because it is in the epidermal layer of the testa and it is highly soluble in water. The whole seed can be extracted with water with a seed-to-water ratio of 1:10. The viscous extract is filtered or centrifuged to remove any solid particles. The supernatant can be recovered by precipitation in alcohol. The fibrous gelatinous material from alcohol precipitation is redissolved and precipitated again in alcohol, to give a snow-white fibrous product with a yield of ~5%.

Extraction from Bran

Using whole seed as a source of YMG is not economically feasible since there is no practical use for the seeds after the water extraction. In North America, substantial amounts of mustard meats are used as food ingredients, while the bran is a by-product. Since mucilage is deposited in the epidermal layer of the seed coat, extraction of mucilage from the bran could be commercially viable (1) to reduce the processing cost and (2) to allow utilization of the by-product. The yield of YMG from bran varied from 15 to 25%, depending on the extraction conditions.[128,129] The extraction process generally includes (1) defatting of the bran with a mixture of hexane, ethanol, and water; (2) water extraction (1:20 ratio) and separation by centrifugation; and (3) precipitation in ethanol and drying.

4.4.3.3 Chemistry and Structural Features

YMG contains 80.4% carbohydrates, 4.4% protein, and 15% ash. Dialysis against distilled water could reduce the ash content to 4.8% with a corresponding increase in carbohydrates from 80.4 to 91.1%. YMG is a mixture of complex polysaccharides containing six neutral sugars and two uronic acids. Of the monosaccharides, glucose (23.5%) is the predominant neutral sugar, followed by galactose (13.8%), mannose (6.1%), rhamnose (3.2%), arabinose (3.0%), and xylose (1.8%); the uronic acids are about 14.7%, including galacturonic acid and glucuronic acid.[129]

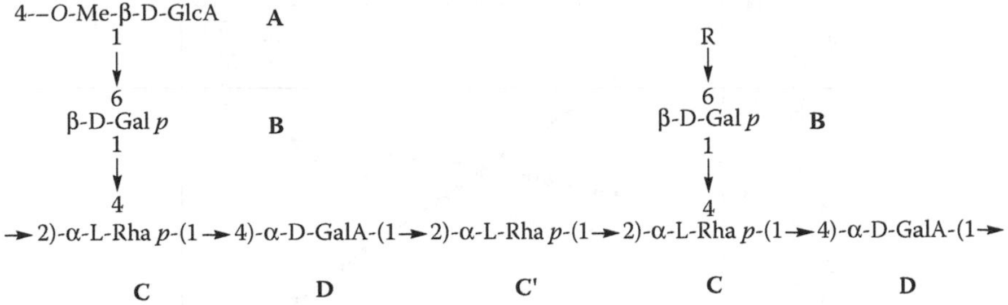

FIGURE 4.5 Structural features of yellow mustard rhamnogalacturonan. (Reproduced from Cui, W. et al., *Carbohydr. Res.*, 292, 173, 1996. With permission.)

The crude YMG can be separated into two fractions according to solubility in water: a water-soluble (WS) and a water-insoluble (WI) fraction. The WS is the major fraction and exhibits shear-thinning flow behaviors observed for YMG.[129] Subsequent fractionation and structural analysis revealed that WS is composed of a neutral polysaccharide and an acidic polysaccharide. The neutral polysaccharide is essentially a cellulose-like polymer that contains 1,4-linked β-D-glucose as the backbone chain with ether groups (i.e., ethyl and propyl) randomly distributed at the C2-, 3-, and 6-positions. Ether groups present along the cellulosyl backbone chain behave as "kinks" that prevent the formation of crystalline structures due to interchain interactions, therefore rendering the polysaccharide water soluble.[52]

The structure of the acidic polysaccharide is a rhamnogalacturonan, which has a rhamnogalacturonan backbone with branches consisting of galactose and glucuronic acid. An average repeating structure unit was elucidated from methylation analysis, one- and two-dimensional nuclear magnetic resonance (NMR) spectroscopy, and characterization of oligosaccharides released from partial hydrolysis. The rhamnogalacturonan consists of →2)-α-L-Rhap-(1→4)-α-D-GalA-(1→ backbone chain, with side chains composed of glucuronic acid and galactosyl residues at the 4-position of 50% of the 2-linked α-L-rhamnosyl residues, as shown in Figure 4.5.[130]

4.4.3.4 Functional Properties and Applications

YMG is the only naturally produced gum that resembles xanthan gum by (1) exhibiting shear-thinning flow behavior at low concentrations (0.3% and above), (2) forming a weak gel structure, and (3) interacting synergistically with galactomannans. The shear-thinning flow behavior and weak gel structure of YMG are shown in Figure 4.6.

When yellow mustard gum is mixed with galactomannans, the viscosity of the mixed systems increases significantly, accompanied by the changes of viscoelastic characteristics.[4] YMG and locust bean gum (LBG) blends are prepared by mixing the two gums in the same concentration according to designed proportions and heating at 80°C for at least 30 min before rheological measurements. The 1:9 (LBG:YMG) blend exhibited gel-like properties where G' is much greater than G'' and the two moduli are less independent of frequency (Figure 4.7). The synergism

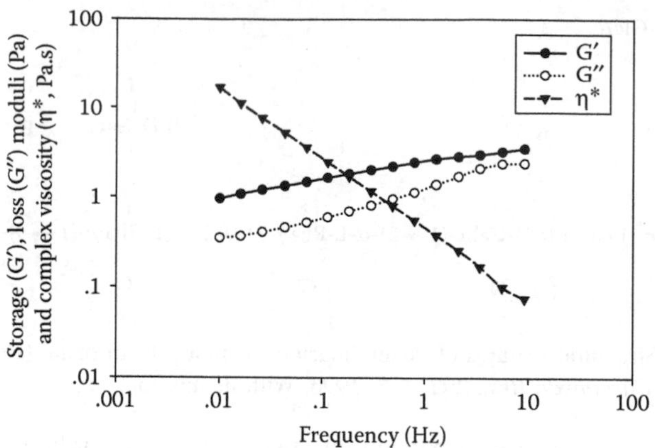

FIGURE 4.6 Mechanical spectrum of yellow mustard gum at 1.0% (w/w) and 23°C.

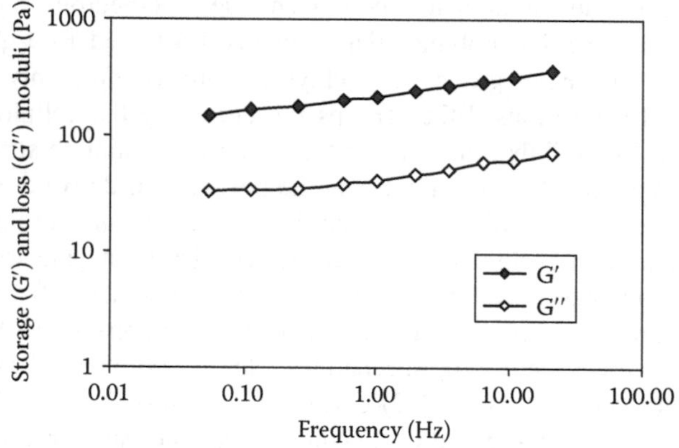

FIGURE 4.7 Mechanical spectra of yellow mustard gum and locust bean gum at the ratio of 1:9 at 2% total polymer concentration, 23°C.

of yellow mustard gum with galactomammans follows the order of LBG > guar gum > fenugreek gum (Figure 4.8), which is consistent with the order of manose:galactose ratios (4:1, 2:1, and 1:1, respectively). YMG–LBG mixtures can form a gel even at 0.1% (w/w) total polymer concentration. Both polymers behave as viscoelastic fluids at 0.1% polymer concentration.[4]

The observed improvement in gelling behavior of YMG by addition of small amounts of LBG at low polymer concentrations (e.g., 0.1% w/w) suggests that the synergistic interactions that occurred between LBG and YMG are through a cooperative association of long stretches of the two polymers into mixed junction zones.[131] That heating is necessary for the interaction suggests that the associative synergistic interactions between galactomannans and YMG require melting the ordered structures in YMG and galactomannans. The proposed mixed-junction model for YMG and LBG is analogous to that of mixtures of xanthan and LBG.[131] Evidence has also

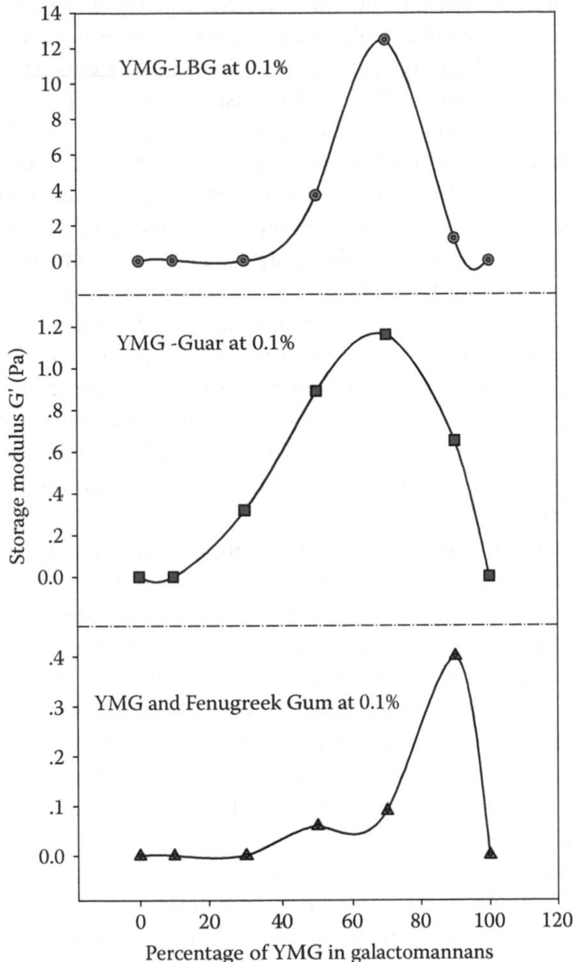

FIGURE 4.8 Storage modulus G' (Pa) of yellow mustard gum and three galactomannans at different ratios.

shown that the water-soluble (1→4)-linked β-D-glucan is the active component that can synergistically interact with galactomannans.

Because of charged groups on the polysaccharide chains, the effects of pH and salt on viscosity are significant. Viscosity at a fixed shear rate (93.32 sec^{-1}) increases in both the lower and higher pH regions. An increase in temperature results in a continuous reduction in viscosity.[129]

The surface active property of YMG makes it one of the best stabilizers for oil/water emulsion-based products.[129] Increasing YMG concentration up to 0.05% can substantially reduce the surface tension. The surface tension and interfacial activities of some plant hydrocolloids (guar and locust bean gums, etc.) are ascribed to the presence of proteins. However, the surface activity of YMG and its fractions appears independent of the level of protein contaminant. Emulsifying capacity and emulsion stability of YMG and its fractions are much higher than those of most of the commercial gums.[129]

The applications of gums/hydrocolloids in food systems are based on their functional properties. Since YMG has demonstrated rheological properties similar to those of xanthan gum, similar applications can be expected for YMG, including synergistic interactions with galactomannans. For example, a salad cream product using a YMG:LBG (9:1) blend (at 0.3% w/w gum concentration) exhibited favorable stability and rheological properties comparable to those of commercial products containing xanthan gum with or without alginates.[132] A commercial salad dressing produced in New Zealand was stabilized with yellow mustard mucilage.

There is a substantial amount of commercial ground yellow mustard products that contain ~5% YMG and are used in processed meats as condiments or bulking agents. The addition of a small amount of LBG to commercial yellow mustard flour will significantly increase the gelling strength and improve the rheological behavior of the products. For example, LBG was added to yellow mustard flour before being blended with meat systems, enhancing the rheological characteristics of the meat products.

Yellow mustard gum was also found to improve the texture and stability of a number of different plant starches. A strong synergy was observed between mixtures of yellow mustard gum and gelatinized wheat, rice, and pea starches.[133,134] A marked increase in viscosity was observed when yellow mustard gum was added to wheat and rice starches, which was attributed to entanglement with amylose released from the starch granule as well as with swollen amylopectin molecules.[134] In addition, syneresis of both of these starches was substantially decreased in the presence of the gum. Similar results were also observed for buckwheat, pea, and corn starches.[135] The ability of yellow mustard gum to modify the properties of plant starches should find applications in food and nonfood products.

Yellow mustard gum has recently been commercially produced in Canada for food and cosmetic applications. A skin care moisturizing lotion was prepared using YMG according to a formula including 15% YMG preparation.[136] The prepared moisturizing lotion has a yield stress and a shear-thinning flow behavior with a favorable hand feeling compared with lotions prepared with other commercial gums.[136] The solution of yellow mustard gum has a high viscosity in a wide range of pHs. It is an ideal stabilizer for salad dressings and fruit juice concentrates.

4.4.3.5 Physiological Properties and Health Benefits

Yellow mustard gum is a dietary fiber that is expected to exert some physiological effects typical for dietary fibers, including regularizing colonic function, normalizing serum lipid levels, and attenuating the postprandial glucose response, and perhaps suppressing appetite. A study by Begin and coworkers[137] examined the effect of YMG and other soluble fibers on glycemia, insulinemia, and gastrointestinal function in the rat. They found that YMG, guar gum, oat β-glucan, and carboxymethylcellulose all significantly decreased postprandial insulin levels at 45 min, indicating a slowdown in glucose absorption. YMG decreased insulinemia primarily by delaying gastric emptying, while the other fibers increased intestinal contents and consequently decreased absorption.[137] Viscosity was considered the major contributing factor to the improved insulin status at peak time. Since viscosity of YMG increases at acid pH, it exerts a stronger gastric effect than the other fibers examined. Jenkins and coworkers demonstrated that incorporating mustard fiber into white bread at levels not affecting palatability had a modest but significant effect of reducing the glycemic index of the bread in both normal and diabetic human volunteers.[138] A significant reduction in percent peak rise in postprandial blood glucose was also observed.

The anticancer potential of yellow mustard gum was recently observed by Eskin et al. using male Sprague Dawley and female Zucker obese rats as models of sporadic and obesity-associated colon cancer, respectively.[139] At 8 weeks of age, rats were injected with azoxymethane, a specific colon carcinogen, and maintained on a basal diet with or without 5% yellow mustard gum for 6 weeks. A significant ($p < 0.05$) decrease in the number of advanced aberrant crypt foci (ACF > 7 crypts) was found in the case of Sprague Dawley rats, which accounted for a 50% reduction compared to the control diet. In the case of the Zucker obese rats, a 60% inhibition in the development of large colonic ACF was observed. These studies pointed to the potential of yellow mustard mucilage as a nutraceutical in the treatment of colon cancer.

4.5 CONCLUSIONS

The latest available information has been presented on a number of important and potentially important seed polysaccharide gums. Of the four galactomannans discussed, locust bean and guar gum are by far the most widely used and best understood. Both are starch modifiers and are used as thickeners and stabilizers. Fenugreek gum is becoming more prominent because of its considerable surface activity and its ability to stabilize oil-in-water emulsions, although its mechanism is far from understood. Fenugreek gum-based dietary fiber products are commercially available. The least known galactomannan, tara gum, has comparable thickening properties to guar gum but better organoleptic characteristics. Commercial tara gum produced in South America is now available.

Xyloglucan, a hemicellulose in the cell wall of tamarind seeds, has properties similar to those of gelatinized starch but is more stable. It is used as a starch replacer, such as a thickening agent in Nigerian foods. Soya fiber, a water-soluble polysaccharide in the cell wall of soybean cotyledons, is a low-viscosity gum with functional properties similar to those of gum arabic. It has considerable potential as a stabilizing agent in the beverage industry.

Plant mucilages reviewed were from psyllium, flaxseed, and yellow mustard. Of these, psyllium is the mostly widely used as a thickener and stabilizer in such products as ice cream and frozen desserts. Flaxseed gum, similar to other gums, can be used as a stabilizer. The variability in properties between some flaxseed gums due to the different amounts of acidic and neutral polysaccharides present may offer flaxseed gums a broad range of functionality. A relative newcomer among these mucilages is yellow mustard, which is the only naturally produced gum to resemble xanthan gum. Examination of this gum identified a unique β-glucan that appears responsible for some of its starch-thickening properties.

In addition to the physical properties of the seed polysaccharides, many contribute important health benefits for preventing or slowing the onset of chronic diseases. For example, locust bean and guar gums both act as dietary fibers reducing cholesterol levels and postprandial hyperglycemia. Fenugreek seed powders were also reported to have antidiabetic and hypocholesterolemic properties. Seed gum polysaccharides such as xyloglucan also reduce postprandial plasma glucose and insulin levels. Of the seed coat gums, psyllium is used extensively in the manufacture of laxative products. Flaxseed gum has also been used as a laxative, as well as in tablet form to slow down the rate of drug release. Yellow mustard gum was shown to also reduce postprandial plasma glucose and, more recently, to have anticancer properties. It is evident from the previous discussions that plant seed polysaccharides have considerable potential as functional food and nutraceutical ingredients.

REFERENCES

1. Fernandes, P., Goncalves, M., and Doublier, J.-L., A rheological characterization of kappa-carrageenan/galactomannan mixed gels: a comparison of locust bean gum samples, *Carbohydrate Polymers*, 16, 253, 1991.
2. Fernandes, P., Goncalves, M., and Doublier, J.-L., Rheological behaviour of kappa-carrageenan/galatomannan mixtures at a very low level of kappa-carrageenan, *Journal of Texture Studies*, 25, 267, 1994.
3. Schorsch, C., Garnier, C., and Doublier, J.-L., Viscoelastic properties of xanthan/galactomannan mixtures: comparison of guar gum with locust bean gum, *Carbodrate Polymers*, 34, 165, 1997.
4. Cui, W. et al., Synergistic interactions between yellow mustard polysaccharides and galactomannans, *Carbohydrate Polymers*, 27, 123, 1995.
5. Hoffman, J. and Svensson, S., Studies of the distribution of the D-galactosyl side chain in guaran, *Carbohydrate Research*, 65, 65, 1978.
6. McCleary, B. et al., The fine structures of carob and guar galactomannans, *Carbohydrate Research*, 139, 237, 1985.
7. Daas, P., Schols, H., and de Jongh, H., On the galactosyl distribution of commercial galactomannans, *Carbohydrate Research*, 329, 609, 2000.
8. Daas, P. et al., Toward the recognition of structure-function relationships in galactomannans, *Journal of Agriculture and Food Chemistry*, 50, 4282, 2002.
9. Petkowicz, C.L.O., Reicher, F., and Mazeau, K., Conformational analysis of galactomannans: from oligomeric segments to polymeric chains, *Carbohydrate Polymers*, 37, 25, 1998.
10. Nishinari, K. et al., Solution properties of pullulan, *Macromolecules*, 24, 5590, 1991.
11. Picout, D. et al., Pressure cell assisted solution characterization of polysaccharides. 2. Locust bean gum and tara gum, *Biomacromolecules*, 3, 761, 2002.
12. Kök, M.S., Hill, S.E., and Mitchell, J.R., A comparison of the rheological behaviour of crude and refined locust bean gum preparations during thermal processing, *Carbohydrate Polymers*, 38, 261, 1999.

13. Kök, M.S., Hill, S.E., and Mitchell, J.R., Viscosity of galactomannans during high temperature processing: influence of degradation and solubilisation, *Food Hydrocolloids*, 13, 535, 1999.
14. Goycoolea, F.M., Morris, E.R., and Gidley, M.J., Viscosity of galactomannans at alkaline and neutral pH: evidence of 'hyperentanglement' in solution, *Carbohydrate Polymers*, 27, 69, 1995.
15. Kunisaki, N. and Sano, Y., *Food Polysaccharides*, Saiwaishobo, Tokyo, 2001 (in Japanese).
16. Doublier, J. and Launnay, B., Rheology of glactomannan solutions: comparative study of guar gum and locust bean gum, *Journal of Texture Studies*, 12, 151, 1981.
17. Tanaka, R., Hatakeyama, T., and Hatakeyama, H., Formation of locust bean gum hydrogel by freezing-thawing, *Polymer International*, 45, 118, 1998.
18. Richardson, P. and Norton, I.T., Gelation behavior of concentrated locust bean gum solution, *Macromolecules*, 31, 1575, 1998.
19. Cairns, P. et al., X-ray fibre-diffraction studies of synergistic, binary polysaccharide gels, *Carbohydrate Research*, 160, 411, 1987.
20. Morris, V.J., Gelation of polysaccharides, in *Functional Properties of Food Macromolecules*, Hill, S.E., Ledware, D.A., and Mitchell, J.R., Eds., Aspen Publishers, Gaithersburg, MD, 1998, p. 143.
21. Regand, A. and Goff, H.D., Structure and ice recrystallization in frozen stabilized ice cream model systems, *Food Hydrocolloids*, 17, 95, 2003.
22. Goff, H.D., Ferdinando, D., and Schorsch, C., Fluorescence microscopy to study galactomannan structure in frozen sucrose and milk protein solutions, *Food Hydrocolloids*, 13, 353, 1999.
23. Thaiudom, S. and Goff, H.D., Effect of κ-carrageenan on milk protein polysaccharide mixtures, *International Dairy Journal*, 13, 763, 2003.
24. Alloncle, M. et al., A rheological characterization of cereal starch-galactomannan mixtures, *Cereal Chemistry*, 66, 90, 1989.
25. Ellis, P., Rayment, P., and Wang, Q., A physico-chemical perspective of plant polysaccharides in relation to glucose absorption, insulin secretion and the entero-insular axis, *Proceedings of the Nutrition Society*, 55, 881, 1996.
26. Feldman, N. et al., Enrichment of an Israeli ethnic food with fibres and their effects on the glycaemic and insulinaemic responses in subjects with non-insulin-dependent diabetes mellitus, *British Journal of Nutrition*, 74, 681, 1995.
27. Evans, A. et al., Relationship between structure and function of dietary fibre: a comparative study of the effects of three galactomannans on cholesterol metabolism in the rat, *British Journal of Nutrition*, 68, 217, 1992.
28. Wang, Q., Ellis, P.R., and Ross-Murphy, S.B., The stability of guar gum in an aqeous system under acidic conditions, *Food Hydrocolloids*, 14, 129, 2000.
29. Whistler, R.L. and Hymowitz, R., *Guar: Agronomy, Production, Industrial Use and Nutrition*, Purdue University Press, West Lafayette, IN, 1979.
30. Richardson, P.H., Willmer, J., and Foster, T.J., Dilute solution properties of guar and locust bean gum in sucrose solutions, *Food Hydrocolloids*, 12, 339, 1998.
31. Wang, Q., Ellis, P.R., and Ross-Murphy, S.B., Dissolution kinetics of guar powders. II. Effects of concentration and molecular weight, *Carbohydrate Polymers*, 53, 75, 2003.
32. Robinson, G., Ross-Murphy, S.B., and Morris, E.R., Viscosity-molecular weight relationships, intrinsic chain flexibility, and dynamic solution properties of guar galactomannan, *Carbohydrate Research*, 107, 17, 1982.

33. Cheng, Y., Brown, K., and Prud'homme, R.K., Preparation and characterization of molecular weight fractions of guar galactomannans using acid and enzymatic hydrolysis, *International Journal of Biological Macromolecules*, 31, 29, 2002.
34. Gittings, M. et al., Structure of guar in solutions of H_2O and D_2O: an ultra-small-angle light-scattering study, *Journal of Physical Chemistry B*, 104, 4381, 2000.
35. Dea, I.C.M., and Clark, A.H., and Mcleary, B.V., Effect of galactose-substitution patterns on the interaction properties of galactomannans, *Carbohydrate Research*, 147, 275, 1986.
36. Mcleary, B.V. et al., Interaction properties of D-galactose-dependent guar galactomannan samples, *Carbohydrate Polymers*, 4, 253, 1984.
37. Kesavan, S. and Prud'homme, R.K., Rheology of guar and HPG cross-linked by borate, *Macromolecules*, 25, 2026, 1992.
38. Closs, C. et al., Phase separation and rheology of aqueous starch/galactomannan systems, *Carbohydrate Polymers*, 39, 67, 1999.
39. Jenkins, D. et al., Dietary fiber and blood lipids: reduction of serum cholesterol in type II hyperlipidemia by guar gum, *American Journal of Clinical Nutrition*, 32, 16, 1979.
40. Wolever, T. et al., Guar gum and reduction of post-prandial glycaemia: effect of incorporation into solid food, liquid food, and both, *British Journal of Nutrition*, 41, 505, 1979.
41. Haskell, W. et al., Role of water-soluble dietary fiber in the management of elevated plasma cholesterol in healthy subjects, *American Journal of Cardiology*, 69, 433, 1992.
42. Velázquez, M. et al., Effect of oligosaccharides and fibre substitutes on short-chain fatty acid production by human faecal microflora, *Anerobe*, 6, 87, 2000.
43. Slavin, J.L. and Greenberg, N.A., Partially hydrolysed guar gum: clinical nutrition uses, *Nutrition*, 19, 549, 2003.
44. Cheng, Y. and Prud'homme, R.K., Enzymatic degradation of guar and substituted guar galactomannans, *Biomacromolecules*, 1, 782, 2000.
45. McCleary, B.V., Purification and properties of a β-D-mannoside mannohydrolase from guar, *Carbohydrate Research*, 101, 75, 1982.
46. Tako, M., Synergistic interaction between xanthan and tara-bean gum, *Carbohydrate Polymers*, 16, 239, 1991.
47. Cairns, P., Miles, M., and Morris, V.J., X-ray diffraction studies of kappa-carrageenan-tara gum mixed gels, *International Journal of Biological Macromolecules*, 8, 124, 1986.
48. Melnick, R.L. et al., Chronic effects of agar, guar gum, gum arabic, locust-bean gum, or tara gum in F344 rats and $B6C3F_1$ mice, *Food and Chemical Toxicology*, 21, 305, 1983.
49. Brummer, Y., Cui, W., and Wang, Q., Extraction, purification and physicochemical characterization of fenugreek gum, *Food Hydrocolloids*, 17, 229, 2003.
50. Andrews, P., Hough, L., and Hones, K.N., Mannose-containing polysaccharides. II. The galactomannan of fenugreek seed, *Journal of American Chemistry*, 74, 2744, 1952.
51. Brummer, Y., Physicochemical and Structural Characterization of Fenugreek Gum, M.Sc. thesis, University of Guelph, Ontario, 2001.
52. Cui, W., *Polysaccharide Gums from Agriculture Products: Processing, Structure and Functional Properties*, Technomic Publishing Company, Lancaster, PA, 2001.
53. Dickinson, E., Hydrocolloids at interfaces and the influence on the properties of dispersed systems, *Food Hydrocolloids*, 17, 25, 2003.

54. Garti, N. et al., Fenugreek galactomannans as food emulsifiers, *Food Science and Technology*, 30, 305, 1997.
55. Huang, X., Evaluation of Hydrocolloid Gum as Stabilizers in Oil/Water Emulsion: Emulsion Stability, Interfacial Activity and Rheological Properties, M.Sc. thesis, University of Guelph, Ontario, 2000.
56. Huang, X., Kakuda, Y., and Cui, W., Hydrocolloids in emulsions: particle size distribution and interfacial activity, *Food Hydrocolloids*, 15, 533, 2001.
57. Al-Habori, M. and Raman, A., Antidiabetic and hypocholesterolaemic effects of fenugreek, *Phytotherapy Research*, 12, 233, 1998.
58. Devasena, T. and Menon, V.P., Fenugreek affects the activity of beta-glucuronidase and mucinase in the colon, *Phytotherapy Research*, 17, 1088, 2003.
59. Hayashi, M. et al., Structure and biosynthesis of the xylose-containing carbohydrate moiety of rice alpha-amylase, *European Journal of Biochemistry*, 191, 287, 1990.
60. Hayashi, T., Xyloglucans in the primary cell wall, *Annual Review of Plant Physiology and Plant Molecular Biology*, 40, 139, 1989.
61. de Lima, D.U. and Buckeridge, M.S., Interaction between cellulose and storage xyloglucans: the influence of the degree of galactosylation, *Carbohydrate Polymers*, 46, 157, 2001.
62. Takeda, T. et al., Suppression and acceleration of cell elongation by integration of xyloglucans in pea stem segments, *Proceedings of the National Academy of Sciences of the United States of America*, 99, 9055, 2002.
63. Levy, S. et al., Simulations of the static and dynamic molecular conformations of xyloglucan: the role of the fucosylated sidechain in surface-specific sidechain holding, *Plant Journal*, 1, 195, 1991.
64. Levy, S., Maclachlan, G., and Staehelin, L.A., Xyloglucan sidechains modulate binding to cellulose during *in vitro* binding assays as predicted by conformational dynamics simulations, *Plant Journal*, 11, 373, 1997.
65. Finkenstadt, V.L., Hendrixaon, T.L., and Millane, R.P., Models of xyloglucan binding to cellulose microfibrils, *Journal of Carbohydrate Chemistry*, 14, 601, 1995.
66. Nishinari, K., Yamatoya, K., and Shirakawa, M., Xyloglucan, in *The Handbook of Hydrocolloids*, Phillips, G.O. and Williams, P.A., Eds., Woodhead Publishing Ltd., Cambridge, U.K., 2000, p. 247.
67. Ellis, P.R., Polysaccharide gums: their modulation of carbohydrate and lipid metabolism and role in the treatment of diabetes mellitus, in *Gums and Stabilisers for the Food Industry 7*, Phillips, G.O., Williams, P.A., and Wedlock, D.J., Eds., Oxford University Press, Oxford, 1994, p. 277.
68. Onyechi, U., Judd, P.A., and Ellis, P.R., African plant foods rich in non-starch polysaccharides reduce postprandial blood glucose and insulin concentrations in healthy human subjects, *British Journal of Nutrition*, 80, 419, 1998.
69. Wang, Q. et al., A new polysaccharide from a traditional Nigerian plant food: *Detarium senegalense* Gmelin, *Carbohydrate Research*, 283, 229, 1996.
70. Taylor, I.E.P. and Atkins, E.D.T., X-ray diffraction studies on the xyloglucan from tamarind (*Tamarindus indica*) seed, *FEBS Letter*, 181, 300, 1985.
71. Urakawa, H., Mimura, M., and Kajiwara, K., Diversity and versatility of plant seed xyloglucan, *Glycoscience and Glycotechnology*, 14, 355, 2002.
72. Gidley, M.J. et al., Structure and solution properties of tamarind-seed polysaccharide, *Carbohydrate Research*, 214, 299, 1991.
73. Ikeda, S. et al., Single-phase mixed gels of xyloglucan and gellan, *Food Hydrocolloids*, 18, 669, 2004.

74. Wang, Q. et al., Solution characteristics of the xyloglucan extracted from *Detarium senegalense* Gmelin, *Carbohydrate Polymers*, 33, 115, 1997.
75. Yoshimura, M., Takaya, T., and Nishinari, K., Effects of xyloglucan on the gelatinization and retrogradation of corn starch as studied by rheology and differential scanning calorimetry, *Food Hydrocolloids*, 13, 101, 1999.
76. Yamanaka, S. et al., Gelation of tamarind seed polysaccharide xyloglucan in the presence of ethanol, *Food Hydrocolloids*, 14, 125, 2000.
77. Salazar-Montoya, J.A., Ramos-Ramirez, E.G., and Delgado-Reyes, V.A., Changes of the dynamic properties of tamarind (*Tamarindus indica*) gel with different saccharose and polysaccharide concentrations, *Carbohydrate Polymers*, 49, 387, 2002.
78. Shirakawa, M., Yamatoya, K., and Nishinari, K., Tailoring of xyloglucan properties using an enzyme, *Food Hydrocolloids*, 12, 25, 1998.
79. Yamatoya, K., Shirakawa, M., and Baba, O., Effects of xyloglucan on lipid metabolism, in *Hydrocolloids*, Part 2, *Fundamentals and Applications in Food, Biology, and Medicine*, Nishinari, K., Ed., Elsevier, Amsterdam, 2000, p. 405.
80. Strickland, F.M. et al., Inhibition of UV-induced immune suppression and interleukin-10 production by plant oligosaccharides and polysaccharides, *Photochemistry and Photobiology*, 69, 141, 1999.
81. Bell, S. et al., An investigation of the effects of two indigenous African foods, *Detarium microcarpum* and *Cissus rotundifolia*, on rat plasma cholesterol levels, Proceedings of *Nutrition Society*, 52, 372A, 1993.
82. Maeda, H., Soluble soybean polysaccharide, in *Handbook of Hydrocolloids*, Phillips, G.O. and Williams, P.A., Eds., CRC Press, Boca Raton, FL, 2000, p. 309.
83. Nakamura, A. et al., Structural studies by stepwise enzymatic degradation of the main backbone of soybean soluble polysaccharides consisting of galacturonan and rhamnogalacturonan, *Bioscience, Biotechnology, and Biochemistry*, 66, 1301, 2002.
84. Furuta, H. and Maeda, H., Rheological properties of water-soluble soybean polysaccharides extracted under weak acidic conditions, *Food Hydrocolloids*, 13, 267, 1999.
85. Nakamura, A. et al., Effect of soybean soluble polysaccharides on the stability of milk protein under acidic conditions, *Food Hydrocolloids*, 17, 333, 2003.
86. European Commission of Health and Consumer Protection Directorate-General, *Opinion of the Scientific Committee on Food on Soybean Hemicellulose*, Scientific Committee on Food SCF/CS/ADD/EMU/185 Final, 2003.
87. Takahashi, T. et al., A soluble soybean fiber: a 3-month dietary toxicity study in rats, *Food and Chemical Toxicology*, 41, 1111, 2003.
88. Takahashi, T. et al., Physiological effects of water-soluble soybean fiber in rats, *Bioscience, Biotechnology, and Biochemistry*, 63, 1340, 1999.
89. Food and Drug Administration, *Soluble Fiber from Certain Foods and Coronary Heart Disease*, FDA, HHS, Final Rule (UIUC), The director of the office of the Federal Register approves of the incorporation by reference in accordance with 5 U.S.C.552 (a) and 1 CFR Part 51 of certain publications in 21 CFR 101.81 (c) (2) (ii) (B), 1998.
90. Samuelsen, A.B. et al., Structural studies of a heteroxylan from *Plantago major* L. seeds by partial hydrolysis, JPAEC-PAD, methylation and GC-MS, ESES and ESMS/MS, *Carbohydrate Research*, 315, 312, 1999.
91. Haque, A., Morris, E.R., and Richardson, R.K., Polysaccharide substitutes for gluten in non-wheat bread, *Carbohydrate Polymers*, 25, 337, 1994.
92. Childs, N.M., Marketing functional foods: what have we learned? An examination of the metamucil, benefit, and heartwise introductions as cholesterol-reducing ready-to-eat cereals, *Journal of Medicinal Food*, 2, 11, 1999.

93. Yu, L. and Perret, J., Effects of solid-state enzyme treatments on the water-absorbing and gelling properties of psyllium, *Swiss Society of Food Science and Technology*, 36, 203, 2003.
94. Olson, B.H. et al., Psyllium-enriched cereals lower blood total cholesterol and LDL cholesterol, but not HDL cholesterol, in hypercholesterolemic adults: results of a metaanalysis, *Journal of Nutrition*, 127, 1973, 1997.
95. Turley, S.D. and Dietschy, J.M., Mechanisms of LDL-cholesterol lowering action of psyllium hydrophillic mucilloid in the hamster, *Biochimica et Biophysica Acta.*, 1255, 177, 1995.
96. Vergara-Jimenez, M., Furr, H., and Fernandez, M.L., Pectin and psyllium decrease the susceptibility of LDL to oxidation in guinea pigs, *Journal of Nutritional Biochemistry*, 10, 118, 1999.
97. Satchithanandam, S. et al., Effects of dietary fibers on gastrointestinal mucin in rats, *Nutrition Research*, 16, 1163, 1996.
98. Rodriguez-Moran, M., Guerrero-Romero, F., and Lazcano-Burciaga, G., Lipid- and glucose-lowering efficacy of plantago psyllium in type II diabetes, *Journal of Diabetes and Its Complications*, 5, 273, 1998.
99. Fang, C., Deitary psyllium reverses hypercholesterolemic effect of trans fatty acids in rats, *Nutrition Research*, 20, 695, 2000.
100. Alabaster, O., Tang, Z., and Shivapurkar, N., Dietary fiber and the chemopreventive modelation of colon carcinogenesis, *Mutation Research*, 350, 185, 1996.
101. Wisker, E. et al., Digestibilities of energy, protein, fat and nonstarch polysaccharides in a low-fiber diet and diets containing coarse or fine whole meal rye are comparable in rats and humans, *Journal of Nutrition*, 126, 481, 1996.
102. Montague, J.E., *Psyllium Seed: The Latest Laxative*, Montague Hospital for Intestinal Ailments, New York, 1932.
103. Suhonen, R., Bjorksten, F., and Kantola, I., Anaphylactic shock due to ingestion of psyllium laxative, *Allergy*, 38, 363, 1983.
104. James, J.M. et al., Anaphylactic reactions to a psyllium-containing cereal, *Journal of Allergy and Clinical Immunology*, 88, 402, 1991.
105. Freeman, T.P., Structure of flaxseed, in *Flaxseed in Human Nutrition*, Cunnane, S.C., and Thompson, L.U., Eds., AOCS Press, Champaign, IL, 1995, p. 11.
106. Mazza, G. and Biliaderis, C.G., Functional properties of flax seed mucilage, *Journal of Food Science*, 54, 1302, 1989.
107. Fedeniuk, R.W. and Biliaderis, C.G., Composition and physicochemical properties of linseed (*Linum usitatissimum* L.) mucilage, *Journal of Agricultural and Food Chemistry*, 42, 240, 1994.
108. Cui, W. et al., Optimization of an aqueous extraction process for flaxseed gum by response-surface methodology, *Lebensmittel-Wissenschaft-und-Technologie*, 27, 363, 1994.
109. Oomah, B.D. and Mazza, G., Optimization of a spray drying process for flaxseed gum, *International Journal of Food Science and Technology*, 36, 135, 2001.
110. BeMiller, J.N., Quince seed, psyllium seed, flaxseed and okra gums, in *Industrial Gums*, Whistler, R.L. and BeMiller, J.N., Eds., Academic Press, New York, 1973, p. 331.
111. Tomoda, G. and Asami, Y., Mucilage from linseed or linseed meal, *Chemistry Abstracts*, 46, 10654, 1950.
112. Cui, W. and Han, N.F., Process and apparatus for flaxseed component separation. U.S. Patent 7,022,363, 2006.

113. Cui, W. and Mazza, G., Methods for Dehulling of Flaxseed, Producing Flaxseed Kernels and Extracting Lignans and Water-Soluble Fibre from the Hulls, Canadian Patent CA 216,7951, 2002.
114. Cui, W. and Mazza, G., Physicochemical characteristics of flaxseed gum, *Food Research International*, 29, 397, 1996.
115. Cui, W., Mazza, G., and Biliaderis, C.G., Chemical structure, molecular size distributions, and rheological properties of flaxseed gum, *Journal of Agricultural and Food Chemistry*, 42, 1891, 1994.
116. Cui, W., Kenaschuk, E., and Mazza, G., Influence of genotype on chemical composition and rheological properties of flaxseed gums, *Food Hydrocolloids*, 10, 221, 1996.
117. Susheelamma, N.S., Isolation and properties of flaxseed mucilage, *Journal of Food Science*, 24, 103, 1987.
118. Wannerberger, K., Nylander, T., and Nyman, M., Rheological and chemical properties of mucilage in different varieties from linseed (*Linum usitatissimum*), *ACTA Agriculturae Scandinavica*, 41, 311, 1991.
119. Muralikrishna, G., Salimath, P.V., and Tharanathan, R.N., Structural features of an arabinoxylan and a rhamnogalacturonan derived from linseed mucilage, *Carbohydrate Research*, 161, 265, 1987.
120. Erskine, A.J. and Jones, J.K.N., The structure of linseed mucilage. Part I, *Canadian Journal of Chemistry*, 35, 1174, 1957.
121. Garden-Robinson, J., Flaxseed gum: extraction, composition, and selected applications, in *Proceedings of the 55th Flax Institute of the United States*, 1994, p. 154.
122. Stewart, S., and Mazza, G., Effect of flaxseed gum on quality and stability of a model salad dressing, *Journal of Food Quality*, 23, 373, 2000.
123. Jenkins, D.J. et al., Health aspects of partially defatted flaxseed, including effects on serum lipids, oxidative measures, and *ex vivo* androgen and progestin activity: a controlled crossover trial, *American Society for Clinical Nutrition*, 69, 395, 1999.
124. Aliev, R.K., New galenical preparations from flax seeds, *American Journal of Pharmacy*, 118, 439, 1944.
125. Boichinov, A., Akhtardzhiev, K., and Kolev, D., A mucous substance (polyuronide) for medicinal and technical purposes obtained from linseed groats, *Chemistry Abstracts*, 66, 68861z, 1967.
126. Tufegdzic, N., Tufegdzic, E., and Georgijevic, A., Floculation and stabilization of barium sulfate suspensions by hydrophilic colloids, *Chemistry Abstracts*, 62, 8944f, 1965.
127. Attstrom, R. et al., Saliva Substitute, U.S. Patent 5,260,282, 1993.
128. Vose, J.R., Chemical and physical studies of mustard and rapeseed coats, *Cereal Chemistry*, 51, 658, 1974.
129. Cui, W., Eskin, N.A.M., and Biliaderis, C.G., Chemical and physical properties of yellow mustard (*Sinapis alba* L.) mucilage, *Food Chemistry*, 46, 169, 1993.
130. Cui, W. et al., NMR characterization of a 4-O-methyl-beta-D-glucuronic acid-containing rhamnogalacturonan from yellow mustard (*Sinapis alba* L.) mucilage, *Carbohydrate Research*, 292, 173, 1996.
131. Morris, E.R., in Mixed polymer gels in *Food Gels*, Harris, P., Ed., Elsevier Applied Science, London, 1990, p. 291.
132. Cui, W. and Eskin, N.A.M., Interaction between yellow mustard gum and locust bean gum: impact on a salad cream product, in *Gums and Stabilisers for the Food Industry 8*, IRL Press, Oxford, 1996, p. 161.

133. Liu, H. and Eskin, N.A.M., Interactions of native and acetylated pea starch with yellow mustard mucilage, locust bean gum and gelatine, *Food Hydrocolloids*, 12, 37, 1998.
134. Liu, H., Eskin, N.A.M., and Cui, S.W., Interaction of wheat and rice starches with yellow mustard mucilage, *Food Hydrocolloids*, 17, 863, 2003.
135. Liu, H., Eskin, N.A.M., and Cui, S.W., Effect of yellow mustard mucilage on functional and rheological properties of buckwheat and pea starches, *Food Chemistry*, 95, 83, 2006.
136. Cui, W. et al., Extraction process and use of yellow mustard gum, U.S. Patent 6,194,016, 2001.
137. Begin, F. et al., Effect of dietary fibers on glycemia and insulinemia and on gastrointestinal function in rats, *Canadian Journal of Physiology and Pharmacology*, 67, 1265, 1988.
138. Jenkins, A.L. et al., Effect of mustard seed fiber on carbohydrate tolerance, *Journal of Clinical Nutrition Gastroenterology*, 2, 81, 1987.
139. Eskin, N.A.M., Raju, J., and Bird, R., Novel mucilage fractions from *Sinapis alba* L. (Mustard) reduces azoxymethane-induced colonic aberrant colonic crypt foci in F344 and Zucker obese rats, *Phytomedicine* (in press), 2006.

5 Microbial Polysaccharides

Ioannis Giavasis and Costas G. Biliaderis

CONTENTS

5.1 INTRODUCTION

Functional microbial polysaccharides have been used for many years as part of the human diet or as a medical resource. In the Far East, for instance, hot-water-soluble extracts of mushroom (mainly polysaccharides) have been used for hundreds of years for medicinal purposes.[1] These practices of traditional Eastern medicine paved the way for modern scientific studies on medicinal microbial polysaccharides. It is known today that many fungi, including common mushrooms, as well as yeasts, lactic acid bacteria, and other bacteria, excrete extracellular polysaccharides (EPSs) as part of their metabolism, or contain polysaccharides in their cell wall, which exhibit antitumor, immunostimulatory, hypocholesterolemic, hypoglycemic, or other effects beneficial to human health. When an immune response is involved, their protective action is usually mediated by the stimulation of the immune system rather than direct interaction with the agent of pathogenicity; therefore, they are also referred to as biological response modifiers (BRMs).[2] The above render these compounds useful to the design of functional foods, or the production of bioactive ingredients, known as nutraceuticals. Any food products that exert a positive impact on health, in addition to their nutritional contribution, can be considered functional foods.[3] The world market for this type of food is currently growing, and the scientific interest in this field is high, as a result of the realization by consumers of the importance of food to the quality of life.[4] Functional microbial polysaccharides, in particular, constitute a very interesting group of bioactive compounds for the food industry, since many of them derivate from edible fungi (mushrooms) or other microorganisms that are safe for use in food — generally recognized as safe (GRAS).[5,6] Moreover, in the context of food applications, some of these biomolecules may also contribute to improved food texture and physical stability of food dispersions, in parallel with their health benefits.

In comparison to the polysaccharides and other functional carbohydrates obtained from plant sources, the functional polysaccharides isolated from microorganisms or microbial culture media offer the advantages of a well-controlled production process, chemical characteristics, and availability.[7] However, the economic efficiency, purity, and polysaccharide yields in some of these microbial processes are still a matter of concern, and appropriate strategies need to be developed for process optimization.

In the present chapter, the physiological effects of the main functional polysaccharides of microbial origin are described, with an emphasis on the relation between chemical structure and functional properties. Also, the production processes are discussed, as well as the potential effects of food processing on the properties and effectiveness of these biopolymers.

5.2 TYPES AND SOURCES OF FUNCTIONAL POLYSACCHARIDES

5.2.1 GLUCANS

Some of the most important and effective biological response modifiers (BRMs) are β-D-glucans. The majority of them have a main chain of (1→3)-β-D-glucopyranosyl units, usually with side glucopyranosyl groups attached by (1→6) linkages.

Most of the β-D-glucans with BRM activity come from the mycelia, fruit bodies, or liquid culture media of *Basidiomycetes*, while a few are produced by *Ascomycetes* and *Oomycetes*.[2,5] Table 5.1 demonstrates some of the most important β-D-glucans with antitumor or immonustimmulatory activity and the microorganisms that produce them.

Two of the most well studied immunostimulating (1→3)-β-D-glucans are lentinan, a cell wall polysaccharide, and schizophyllan, an extracellular polysaccharide, both synthesized by higher *Basidiomycetes* mushrooms.[2,5] Notably, *Lentinus elodes*,

TABLE 5.1
Some Antitumor and Immunostimulating β-D-Glucans and Their Main Producer Microorganisms

Glucan Name	Source	References
AM-ASN	*Amanita muscaria*	Kiho et al., 1992[8]
AS-I	*Cochliobolus miyabeanus*	Nanba and Kuroda, 1987[9]
CI-6P	*Cordyceps cicadae*	Kiho et al., 1989[10]
	Pythium aphanidermatum	Blaschek et al., 1992[11]
	Ganoderna applanatum	Usui et al., 1981[12]
CO-1	*Cordyceps ophioglossoiodes*	Yamada et al., 1984[13]
Curdlan	*Agrobacterium* sp.	Lee et al., 1999[14]
CW II, HW II	*Ganoderma lucidum*	Sone et al., 1985[15]
Glomerellan	*Glomerella cingulata*	Gomaa et al., 1991[16]
Glucan component of Zymozan	*Saccharomyces cerevisiae*	Manners et al., 1973[17]
Glucan I	*Auricularia auricula-judae*	Sone et al., 1978[18]
		Misaki et al., 1981[19]
Grifolan	*Grifola fondosa*	Ohno et al., 1985a[20]
		Nanba et al., 1987[21]
GU	*Grifola umbellata*	Miyazaki et al., 1978[22]
HA	*Pleurotus ostreatus*	Yoshioka et al., 1985[23]
H-3-B	*Cryptoporous volvatus*	Kitamura et al., 1994[24]
Lentinan	*Lentinus elodes*	Sasaki and Takasaka, 1976[25]
Pachyman	*Poria cocos*	Saito et al., 1968[26]
Pestalotan	*Pestalotia* sp. 815	Misaki et al., 1984[27]
PGG	*Saccharomyces cerevisiae*	Jamas et al., 1991[28]
	Fomes pinicola	Misaki, 1984[29]
PHYT-G, A1, Glucan P	*Phytophthora parasitica*	Fabre et al., 1984[30]
		Blatchek et al., 1987[31]
		Bruneteau et al., 1988[32]
P-I	*Laetisaria arvalis*	Aouadi et al., 1991[33]
PVG	*Peziza vesiculosa*	Ohno et al., 1985b[34]
		Ohno et al. 1985c[35]
Schizophyllan	*Schizophyllum commune*	Tabata et al., 1981[36]
Scleroglucan	*Sclerotium glucanicum*	Rinaudo and Vincendon, 1982[37]
SSG	*Sclerotinia sclerotiorum*	Ohno et al., 1986[38]
T-4-N	*Dictyophora indusiata*	Hara et al., 1991[39]
Tylopilan	*Tylopilus felleous*	Defaye et al., 1988[40]
VVG	*Volvariella volvacea*	Misaki et al., 1986[41]
		Kishida et al., 1989[42]
N/A	*Agaricus blazei*	Mizuno et al., 1990[43]
N/A	*Alcaligenes faecalis* var. *myxogenes*	Sasaki et al., 1979[44]
N/A	*Candida albicans*	Ishibashi et al., 2002[45]
N/A	*Phellinus linteus*	Hwan et al., 1996[46]
N/A	*Pleurotus tuber-regium*	Zhang et al., 2001[47]

the producer of lentinan, is the most common edible mushroom in Japan.[48] These two polysaccharides are characterized by a main chain of (1→3)-β-D-glucose units to which (1→6)-β-D-glucose side groups are linked. These branches occur at every third main chain unit; thus, the degree of branching (DB), the ratio of number of branch units per number of main chain units, for both of these glucans is 0.33. Their average molecular weight (MW) has been reported at 500,000 and 450,000 Da, respectively.[49] Scleroglucan is another functional extracellular polysaccharide excreted by phytopathogenic fungi (serving as a means of attachment to a plant surface), which is very similar in chemical structure to schizophyllan,[50] but has a higher MW, which is in the region of 2.2 to 5.7 × 10^6 Da.[51,52] The scleroglucan repeating unit is shown in Figure 5.1. In their soluble form, the above glucans, like many others, have a single- or triple-helix conformation, the latter being stabilized via internal hydrogen bonds.[52,53] These two structural conformations of β-D-glucans are often (at least partly) interchangeable upon polysaccharide denaturation and renaturation,[45,52–54] and along with DB and MW are factors that influence the functionality of BRMs (discussed in detail in Section 5.5). The triple helix of many β-D-glucans may exist not only in hydrated but also in anhydrous form, depending on the relative humidity (RH). Changes in RH (above or below a certain boundary) can reversibly transform the triple helix from the one form to the other.[55]

Other important bioactive glucans include grifolan, glomerellan, and curdlan. Grifolan, produced by the fungus *Grifola fondosa*, is a gel-forming (1→3)-β-D-glucan branched with (1→6)-β-D-glucose residues at every third glucopyranosyl unit of the main chain (DB of 0.33).[56] Glomerellan, a (1→3)-β-D-glucose with branches of mono-, di-, and trisaccharides in decreasing order of abundance and a reported MW of 6.7 × 10^5 Da, is isolated from the filtrate of liquid cultures of *Glomerella cingulata*.[16] Curdlan, a linear (1→3)-β-D-glucan of bacterial origin (from *Agrobacterium* sp.),[14] with no side chains, has little or no immunostimulatory activity in its native form, but when modified with sulfate, sulfoalkyl, or glycosyl branch units, it has shown significant therapeutic effects.[57,58] Another interesting β-D-glucan is obtained by oxidation of the cell wall of the pathogenic fungus *Candida albicans*. The original particulate form of this glucan is insoluble to H_2O or NaOH, but it can be solubilized in aprotic solvents (e.g., dimethyl sulfoxide (DMSO)).[45,59] As with other cell wall glucans, the insoluble particles consist of linear, long 1,3-β-glucosyl

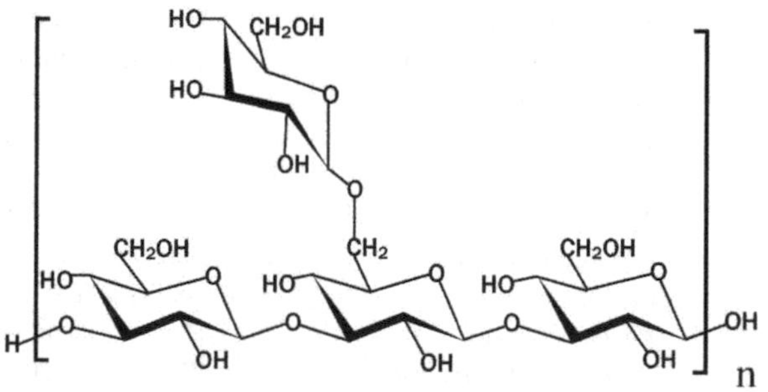

FIGURE 5.1 General structure of scleroglucan.

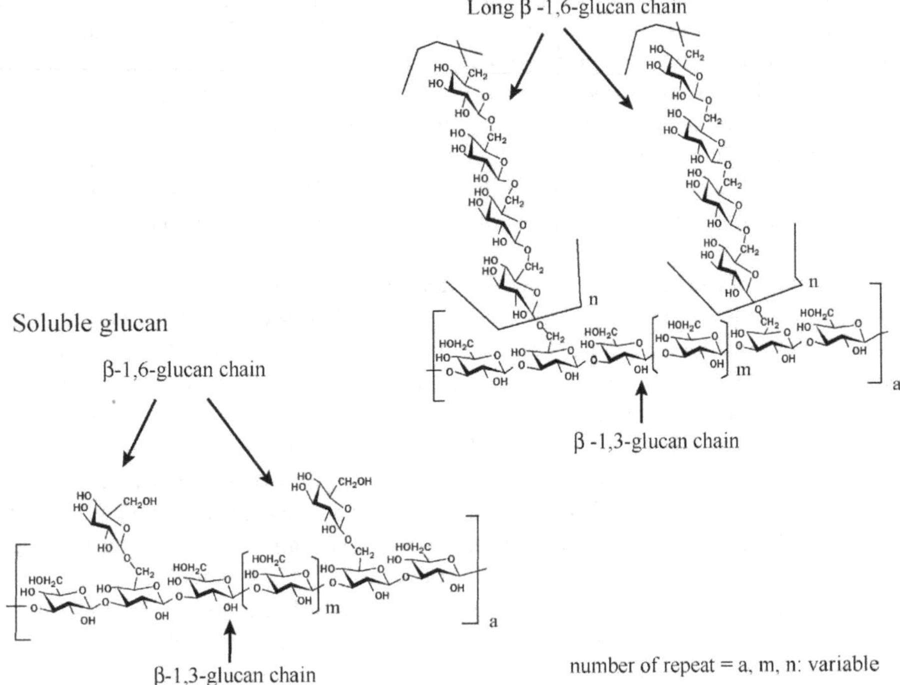

FIGURE 5.2 General primary structure of particulate (insoluble) and soluble β-D-glucans from fungal cell wall. (From Ishibashi, K. et al., *Int. Immunopharm.*, 2, 1109, 2002. With permission.)

and 1,6-β-glucosyl segments, while the soluble derivatives are made up of 1,3-β-glucosyl units branched with short 1,6-β-glucosyl moieties (Figure 5.2), with both types having significant and distinct immunostimulatory activity.[45]

Several microorganisms produce or contain more than one type of biologically active glucan. For instance, *Saccharomyces cerevisiae*, one of the most common food-grade microbes, synthesizes an extracellular (1→3)-β-D-glucan with a DB of 0.2 in the wild-type strain, while a genetically engineered strain yields PGG (Poly-Glucotriosyl-Glucopyranose, also known as Betafectin), a commercial therapeutic (1→3)-β-glucan with a DB of 0.5.[28] Substituted (e.g., sulfated) derivatives of this glucan with increased solubility have also been prepared.[60] The same yeast is used as the basis for the production of zymosan, a complex substance with widespread use in immunopharmacological studies. Its major component is a functional cell wall β-glucan of *S. cerevisiae*, but it also contains yeast mannans, proteins, and nucleic acids, which seem to contribute to the immunoactive properties of zymosan.[61] The particulate β-glucan contains insoluble long (1→3)- and (1→6)-glucosyl segments, which can be sulubilized after chemical oxidation.[61,62] The structure of particulate and solubilized glucans of *S. cerevisiae* can be depicted by the generalized model shown in Figure 5.2.[45]

Other microorganisms may produce heteroglycans (i.e., polysaccharides with a glucan backbone and side groups of mannose, xylose, arabinose, ribose, galactose, or glucuronic acid) or proteoglycans, with or without concomitant production of pure β-glucans.[5] For instance, cultured mycelia of the basidiomycetal fungi *Coriolous versicolor* CM-101 yield a proteoglucan, known as krestin with significant immune-enhancing effects.[63,64] Another strain, *C. versicolor* Cov-1, synthesizes a similar polysaccharide–peptide molecule (PSP) with different peptide content. The polysaccharide component of both of these proteoglycans is a β-glucan.[63,64]

Agaricus blazei, a well-known edible and medicinal mushroom originating from Brazil, contains at least seven antitumor polysaccharides in its fruit body.[43] The water-soluble fraction of these biopolymers includes a β-(1→6)-, β-(1→3)-glucan, an acidic β-(1→6)-, α-(1→4)-glucan, and an acidic β-(1→6)-, α-(1→3)-glucan.[43] Interestingly, these bioactive glucans have a main chain of β-(1→6)-glycopyranose, unlike the characteristic β-(1→3)-linked backbone of most known glucans. *A. blazei* also contains a soluble proteoglucan with a α-(1→4)-glucan main chain and β-(1→6) branches at a ratio of 4:1. It is composed of more than 90% glucose and has an MW of 380,000 Da.[65] In addition, several water-insoluble, immunostimulating polysaccharides have been extracted from fruit bodies of *A. blazei*, namely, three heteroglucans (AG-2, -3, and -6),[66] a proteoglucan, and a xyloglucan.[67] AG-2 and AG-3 are made up of glucose, galactose, and mannose in molar ratios of 74.0:15.3:10.7 and 63.6:17.6:12.7, respectively, while AG-6 contains glucose and ribose in the molar ratio 81.4:12.6.[66] The proteoglucan contains 50.2% polysaccharide (β-(1→6)-glucan) and 43.3% protein, while the xyloglucan consists of glucose and xylose at a molar ratio of 10:2.[67] Moreover, submerged cultures of *A. blazei* produce other medicinal polysaccharides under controlled process conditions, different from those contained in the fruit body of the microorganism, namely, a protein–heteroglucan containing mainly glucose and mannose,[68] as well as a different β-(1→2)-, β-(1→3)-glucomannan.[69] Recently, another extracellular protein–polysaccharide complex with significant antitumor activity was obtained from the filtrate of liquid culture of the same fungus. The polysaccharide component of this biomolecule contained mainly mannose, as well as glucose, galactose, and ribose, and had an MW in the range of 100,000 to 10,000,000 Da.[67]

Ganoderma lucidum is another fungus (of the *Basiomycetes* family) that yields a variety of medicinal polysaccharides. It has been used for decades in East Asia in therapies of several illnesses, in the form of dry fungi powder. A typical β-(1→3)-glucan with β-(1→6)-glucose branches at C-6 has been solubilized from fruit bodies of the fungi, although there has been some controversy with respect to the degree of branching of the final isolate.[49,70] Misaki et al.[49] reported a glucan with a DB of 0.06, while Bao et al.[70] separated a highly branched glucan. Antitumor β-(1→3)-D-glucan with (1→6) side chains was also produced in submerged cultures of mycelia of *G. lucidum*. It was a water-soluble exopolysaccharide with an MW of 1.2×10^6.[71] Besides these β-D-glucans, two complex heteroglucans from fruit bodies of *G. lucidum* have been identified.[70] The one contained (1→4)-α-D-glucopyranosyl residues with glucose branches at C-6 and (1→6)-β-D-galactopyranosyl residues with galactose branches at C-2. The other consisted of β-glucopyranosyl residues with

(1→3), (1→4), (1→6) linkages and (1→6)-linked β-D-mannopyranosyl residues. An antiviral, protein-bound, and water-soluble polysaccharide was also isolated from the fruit bodies of this fungus.[72] Additionally, distinct polysaccharides were found in spores of *G. lucidum*: a linear (1→3)-α-D-glucan that can be further modified by addition of sulfate and carboxymethyl units,[73] as well as a heteroglucan with a β-D-glucan backbone and side groups of mono-, di-, and oligosaccharides.[74]

This notable diversity in structure and composition of polysachharides derived from the same microorganism is partly due to the composition of the nutrient medium and the cultivation/fermentation process conditions adopted.[5] Also, in fungi the type of polysaccharides may differ significantly among fruit bodies, mycelia, and spores, and the total amount of polysaccharides in fruit bodies is usually higher than that found in cultured mycelia. In addition, the fractionation and purification methods can have an impact on the structure and amount of the extracted biomolecules (e.g., polysaccharide degradation and changes in DB can occur during the harsh extraction steps from fungal fruit bodies).[5,71]

5.2.2 EPS from Lactic Acid Bacteria

Lactic acid bacteria (LAB) are some of the most thoroughly studied microorganisms and are historically recognized as safe (GRAS) for use in food. Apart from their key role in the production of fermented dairy, meat, and other food products, and their contribution to the sensorial attributes of food, they have recently received increasing attention due to the potential therapeutic properties of some extracellular polysaccharides (EPSs) that several LAB produce.[75] These biopolymers are usually heteropolysaccharides in the form of slime or a capsule surrounding the cell membrane.[75] The main producers of such bioactive compounds identified so far are strains of *Lactobacillus delbrueckii* ssp. *bulgaricus* (*L. bulgaricus*),[76] *Lactobacillus helveticus*,[77] *Lactobacillus casei*,[78] *Lactococcus lactis* ssp. *cremoris* (*L. cremoris*),[79] *Bifidobacterium adolescentis*,[80] and *Bifidobacterium longum*.[81]

L. bulgaricus strains excrete different polysaccharides with variable chemical composition, not all of which exhibit immunopotentiating properties. Two EPS fractions, a neutral polysaccharide (NPS) and an acidic polysaccharide (APS), have been isolated from culture supernatants of *L. bulgaricus* OLL 1073R-1,[76] while another strain, NCFB 2483, also synthesizes a NPS. These EPSs contain glucose and galactose in the molar ratio (1.0):(1.3 to 1.6), whereas the acidic fraction contains an additional 0.1% of phosporous, which differentiates the APS from the NPS.[82] Indeed, this phosphopolysaccharide with an estimated MW of 1.2×10^6 exhibited immunostimulating activity, unlike the NPS. Additionally, another neutral EPS from the *L. bulgaricus* strain NCFB 2483 provoked no immune responses.[76] The phosphate group in the EPSs of strain OLL 1073R-1 was reported to be a key component, acting as trigger of macrophage functions.[76,82]

Lactococcus cremoris SBT 0495 also produces another functional phosphopolysaccharide with both interesting rheological and bioactive properties, viilian, which has been isolated from a Finnish, fermented milk product (viili).[74] Viilian is composed of glucose, galactose, rhamnose, and phosphate with a molar ratio of 2:2:1:1. Its structure is shown in Figure 5.3.[83] Viilian is separated from an initial

α-L-**Rha***p*
1
↓
2
→4)-β-D-**Glc***p*-(1→4)-β-D-**Gal***p*-(1→4)-β-D-**Glc***p*-(1→
3
|
α-D-**Gal***p*-1-**PO**$_4^-$

FIGURE 5.3 Structure of the repeating unit of viilian produced by *Lactococcus lactis* ssp. *cremoris* SBT 0495. (From Oba, T. et al., *Arch. Microbiol.*, 171, 343, 1999. With permission.)

protein–polysaccharide mixture excreted by the microorgranism, after elimination of the protein moiety, the concentration of which varies in analogy with the protein content of the culture medium.[84]

In contrast to the above extracellular LAB polysaccharides, *Bifidobacterium* species contain antitumor polysaccharides in their cell walls. A water-soluble polysaccharide extracted from *Bifidobacterim adolescentis* strain M101-4 was examined by Hosono et al.[80] It consisted of major residues of -4-galactopyranose-1, -4-glucopyranose-1, and -6-glucopyranose-1, and minor residues of galactofuranose-1 and -6-galactofuranose-1. The characteristic galactofuranosyl segments were unique among all previously studied soluble biopolymers of Gram-positive bacteria.[80] Another crude polysaccharide isolated from *Bifidobacterium longum* strains was found to be responsible for the antimutagenic effects of skim milk fermented with *B. longum*, although its chemical composition was not fully elucidated.[81]

These examples of LAB polysaccharides with therapeutic properties show the potential for introduction of new functional food products in the market. Since the addition of LAB (or their bioproducts), in particular, in food is already practiced for technological reasons, and their acceptance by consumers is ensured, the production of food or nutraceuticals containing LAB polysaccharides as active compounds (such as fermented milk, cheese, and yogurt) is probably a realistic prospect for food scientists and the food market. However, careful selection of the proper strain is necessary, as the desired physiological functions may be specific to a certain strain.

5.2.3 Other Microbial Polysaccharides

A significant group of biopolymers with biological activities are glycans. Glycans are polysaccharides with molecules other than glucose in their main chain.[5] They are classified as mannans, galactans, xylans, fructans, and fucans, according to the sugar units of the chain backbone. Heteroglycans contain combinations of the above sugars and side chains of mannose, galactose, fucose, xylose, arabinose, glucose, and glucuronic acid.[5]

Levans are representative immunomodulatory glycans. They are extracellular polysaccharides composed solely of fructose (fructans). They are produced by the bacteria *Zymomonas mobilis*[85] and *Aerobacter levanicum*.[86] Catalazans et al. purified several levan fractions, with weight-average MW ranging from 3.5 to 10.7 $\times 10^5$ Da.[87] The same authors suggested that a levan fraction with an MW of ~5 × 10^5 Da had the highest antitumor activity. Since *Z. mobilis* is also a large ethanol producer in the industry, levans could be utilized as valuable by-products of ethanol distilleries that are already in place. Alternatively, new processes designed for maximal levan productivity from *Z. mobilis* should aim at minimizing ethanol biosynthesis.[88]

Several fungi also produce glycans with immune-enhancing effects. A bioactive galactomannan has been isolated from fruit body extracts of a common, edible mushroom, *Morchella esculenta.*[89] It is composed of major residues of mannose (62.9% molar content) and galactose (20.0%), and minor residues of N-acetyl glucosamine (7.9%), glucose (6.5%), and rhamnose (2.7%). The MW of the biopolymer was estimated to be ~10^6 Da.[89] This high MW glycan was clearly different from three heteropolysaccharides obtained from liquid fermentation media of *M. esculenta*; the latter isolates were of low MW (11.5 to 44 × 10^3 Da) and have not been reported to exert any immunostimulating effects.[89]

Fruit bodies of the mushroom *Sarcodon aspratus* also contain a highly branched immunomodulating fucogalacatan with unusual structure. It is made up of a main chain of α-(1→6)-linked galactopyranose with side groups of β-(1→2)-galactopyranosyl, as well as side residues of α-(1→2)-L-fucosyl-α-(1→4)-D-galactopyranose.[90] The postulated structure of this biomolecule is shown in Figure 5.4.[90] Other therapeutic-immunopotentiating mushroom glycans include an arabinogalactan from *Pleurotus citrinopileatus*,[91] a mannofucogalactan from *Fomitella fraxinea*,[92] a mannan,[93] a fucomannogalactan from *Dictyophora indusiata*,[94] a mannogalactan from *Pleurotus pulmonarius*,[95] a mannogalactofucan from *Grifola fondosa*,[96] and a xylan from *Hericium erinaceus*.[97]

Alginates are a family of microbial polysaccharides with multiple dietetic (regulation of lipid and glucose metabolism) and curative properties. They are produced either by brown (mainly) and red algae or by the bacteria *Azotobacter vinelandii* and *Pseudomonas aeruginosa*.[98,99] In the present chapter, the focus will be on the bacterial alginate, and in particular that from *A. vinelandii*, since *P. aeruginosa* is a pathogenic microorganism and its use in food applications is unlikely. However, much of the information registered herein on the chemical and physiological properties of bacterial alginate is also valid for alginates from seaweed sources. Alginic acid and the sodium, calcium, potassium, and other salt forms are safe for use in

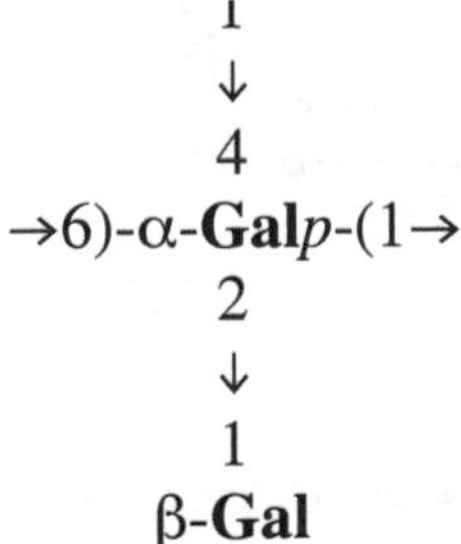

FIGURE 5.4 Postulated chemical structure of fucogalactan from *Sarcodon aspratus*. (From Mizuno, M. et al., *Immunopharmacology*, 46, 113, 2000. With permission.)

food (GRAS) and are used in foodstuffs as thickeners, stabilizers, or gelling agents. Typical examples of use of alginate in food are in the production of jams, sweets, juices and soft drinks, ice cream, soups, sauces, margarine, milk shakes, liquors, structured meat, milk, and fish products.[99] Recent research has revealed a number of biological effects of alginates, which, in combination with the multiple existing food applications of alginate, render these biopolymers potential key ingredients in the manufacture of functional foods or neutraceuticals.

In terms of chemical composition and structure, alginates are unbranched, binary co-polymers of (1→4)-linked β-D-mannuronic acid (M) and α-L-guluronic acid (G) with varying composition (content of the M and G groups) and chain length.[98,99] Also, microbial alginates are acetylated on some mannuronic acid residues.[99] Homopolymeric M and G regions are normally interspaced with alternating residues of both acids (MG groups) (Figure 5.5). Both the sequence of the two acids in the MG region and the length and frequency of the G block are characteristic for *Azotobacter* alginate.[99] Alginate biosynthesis is initiated with the accumulation of poly-D-manuronic acid polymer extracellularly, which is then partly converted into poly-L-guluronic acid through the action of an extracellular C-5-epimerase. The activity of this enzyme depends on the presence of calcium ions. Calcium is also vital for the formation of the characteristic structure of alginate gels. The metal ions bind to the carboxyl and hydroxyl groups of adjacent guluronic acid residues, resulting in the so-called egg-box conformation (Figure 5.5).[98–101] Alginate gels

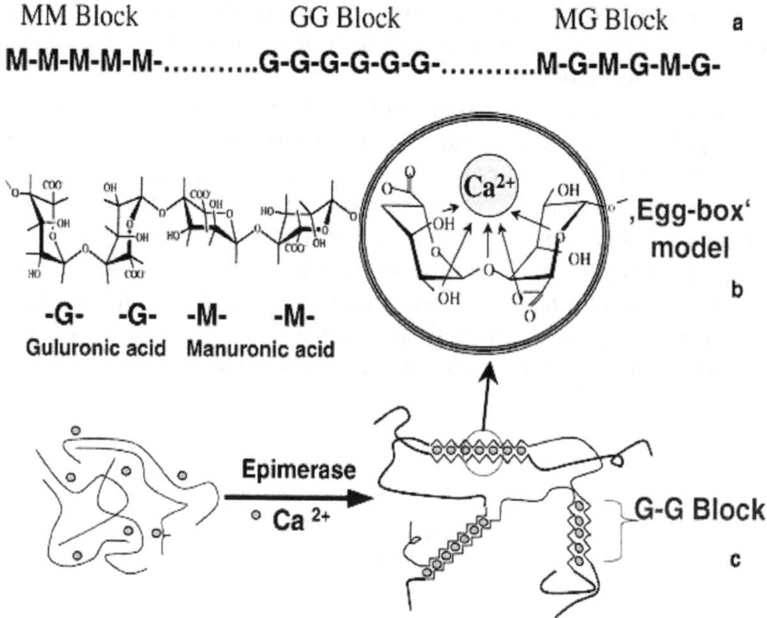

FIGURE 5.5 Structure of *Azotobacter vinelandii* alginate. (a) Block structure. (b, c) The calcium ion-dependent epimerization process and the formation of gel, according to the so-called egg-box model. (From Sabra, W. et al., *Appl. Microbiol. Biotechnol.*, 56, 315, 2001. With permission.)

cannot be formed unless the guluronic acid content of alginate is at least 20%, indicating the important role of guluronic acid in the physicochemical properties of the biopolymer.[98]

Last but not least, xanthan, a well-known gum in the food industry, is another microbial polysaccharide with reported biological functions. Xanthan is produced by the phytopathogenic bacterium *Xanthomonas campestris* and consists of glucose, mannose, and glucuronic acid as major components, and pyruvate and acetate as minor components.[102] Structurally, it is characterized by a cellulosic (D-glucosyl) backbone, with a trisaccharide side chain of internal D-mannose, D-glucuronic acid, and external (terminal) D-mannose linked on every second glucose molecule of -(1→4) cellulosic backbone. Pyruvic acid diketal groups (pyruvate) are fixed to the 4,6-position of the terminal mannose residues of the trisaccharide side chain, while O-acetyl groups are located at the 6-position of the internal mannose.[102] Although there is some variance in the reports on the molecular weight of xanthan, this is usually in the range of 4 to 12×10^6 Da.[103] Xanthan possesses Food and Drug Administration (FDA) approval for use in food and currently dominates the food market in the field of thickeners, stabilizers, and gelling agents, due to the high yields and low production cost of the biopolymer.[104] Despite the potential functional properties of xanthan, as displayed by some researchers in experiments with rats,[105] Castro et al. questioned any practical physiological effects of xanthan in humans, arguing that high doses of xanthan, incompatible with the food applications, are probably needed to bring about the desired biological effects.[106] Thus, in the absence of clinical experiments on humans, it is still ambiguous whether or not xanthan can be established as a functional ingredient. More research on this matter is needed, since the possible biological effects of such a widely used food additive would have a great impact on the dietetic value of many foodstuffs.

5.3 PRODUCTION

Knowledge of the biosynthesis of functional microbial polysaccharides and of the production and purification conditions is crucial for the commercialization of these biopolymers. The improvement of polysaccharide formation and the molecular modification of these biopolymers via the control of microbial physiology can open new prospects for the manufacture of these biopolymers and their use in food. Some of these issues will be covered in the present section.

5.3.1 Biosynthesis

Microbial polysaccharides can appear in the form of an extracellular capsule attached to the cell membrane, or be excreted in the environment or a bioprocess fluid as an extracellular slime, or alternatively be a part of the cell wall material.[107–111] Polysaccharides, including those with biological-therapeutic functions, serve several purposes in cell development and maintenance, depending on their origin. For many prokaryotic and eukaryotic microorganisms, including mushrooms, polysaccharides are constitutive components of the cell structure (lipopolysaccharides, glycolipids, or glycoproteins of the cell membrane and cell envelope), offering rigidity and mechanical stability to the cell.[111] Another significant aspect of polysaccharides is that they have a high capacity (higher than other macromolecules, such as proteins and nucleic acids) for carrying biological information, due to their structural diversity (monosaccharide composition, linkage distribution). This is particularly important for higher organisms, such as mushrooms, where polysaccharides function as biological markers in cell recognition and cell–cell interactions.[5,112] In addition, microbial polysaccharides often act as a protective barrier against unfavorable environmental conditions (such as high or low pH, ionic strength, oxygen stress, desiccation, presence of antibiotics), controlling the diffusion of other molecules into the cell, as well as the export of molecules out of the cell.[50,113–115] Also, microbial polysaccharides may serve as a means of attachment to other cells.[50,116] In other cases, microorganisms may synthesize polysaccharides to use them as an energy reserve; that is, the cells are able to decompose the polysaccharides with hydrolytic enzymes and utilize those as a carbon source.[99,117]

Although it is difficult to summarize the biosynthetic route of all microbial polysaccharides due to the diversity of the available metabolic pathways, there are some major steps in cell metabolism that are encountered in many of these biopolymers. Biosynthesis begins with the assimilation of a carbon source by the cells. The carbon source may be a monosaccharide such as glucose, fructose, or lactose, a disaccharide like saccharose or lactose, or a more complex substance, which is degraded to mono-/disaccharides before being imported into the cell.[108,118,119] The transfer of the mono-/disaccharide in the cell is usually performed by a hexokinase or a permease or a transferase, which translocates the sugar through the cell membrane, or by a transport system involving the coupling of sugar translocation with Adenosine Triphosphate (ATP) hydrolysis (ATPase) and H^+ release, or alternatively the coupling of sugar import with the simultaneous import (symport) or export (antiport) of ions or solutes through the cell membrane.[108,118,119] Sugar phosphorylation takes place before, during, or after the mono-/disaccharide entrance into the cell, and eventually phosphated monomers form sugar nucleotides, which are the precursors (building blocks) of the final polymer.[50,99,108,116,118,119] Sugar nucleotides are activated sugars that can take part in interconversion reactions (e.g., epimerization, decarboxylation, dehydrogenation) that yield different sugar monomers. For polymerization to take place, the sugar moieties of the sugar nucleotides are sequentially transferred via the action of glycosyl transferases onto a lipid (e.g., isoprenoid) carrier, located in the internal or external part of the cell (cytoplasmic) membrane.[50,108,116,119] As the repeating unit is built on the lipid carrier via glycosyl transferases and polymerized via polymerases, molecules other than sugars (such as acetyl groups from sugar catabolism) may also be anchored onto it. The activity and availability of glycosyl transferases, polymerases, and lipid carriers have a significant impact on the degree of polymerization.[99,108,119] After the final polymer is assembled, either it becomes part of the structural components of the cell membrane, or it is excreted extracellularly (in the form of a capsule or a slime). The mechanisms that halt polymerization after some point and lead to the release of the polysaccharide are not well elucidated so far, but it is believed that glycosyl transferases and polymerases are again involved in this process.[99,119]

Two examples of microbial polysaccharide biosynthesis are shown in Figure 5.6

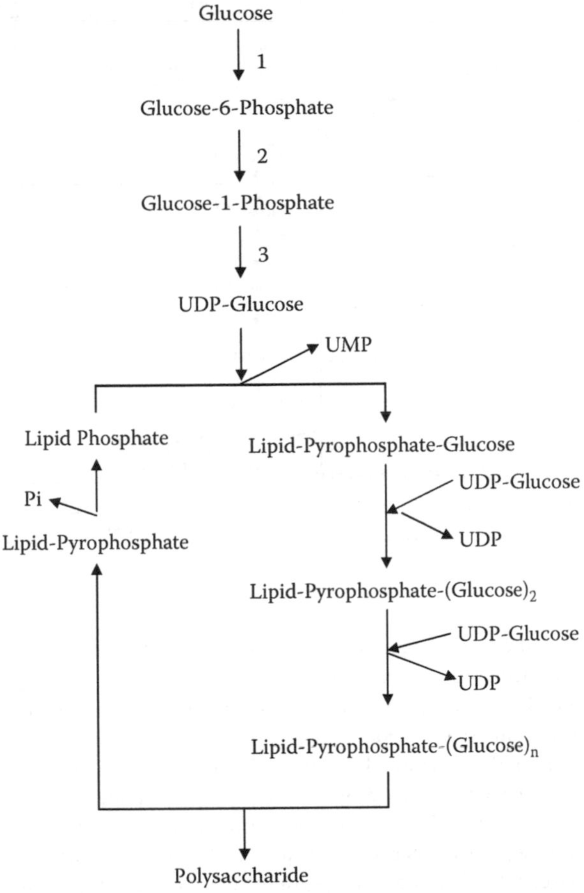

FIGURE 5.6 Possible pathway for scleroglucan synthesis: (1) hexokinase, (2) phosphoglucomutase, (3) phosphoglucose isomerase. Abbreviations: UDP, uridine diphosphate; UMP, uridine monophosphate. (From Giavasis, I. et al., in *Biopolymers*, Vol. 8, Steinbuchel, A., Ed., Wiley-VCH Verlag GmbH, Weinheim, 2002, chap. 2. With permission.)

and Figure 5.7. The first depicts the biosynthetic route of a fungal glucan (scleroglucan) (Figure 5.6),[50] while the second shows the pathway for heteropolysaccharide (HePS) synthesis in some LAB (Figure 5.7).[119] Uridine diphospate (UDP) and thimidine diphosphate (TDP) are the two main nucleotides to which monosaccharides are attached to form sugar nucleotides. Figure 5.7 also describes the catabolism of the carbon source (lactose and galactose), which produces energy (ATP) and metabolites that are essential not only for cell growth and maintenance, but also for polysaccharide biosynthesis.

Another significant aspect of biosynthesis for several polysaccharides is their degradation or depolymerization by hydrolytic enzymes (lyases) produced by the

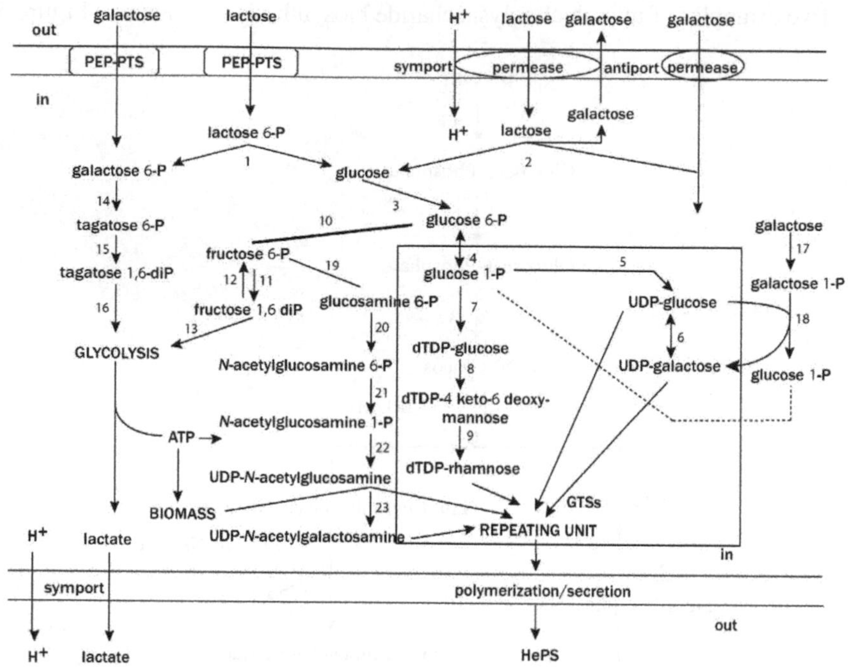

FIGURE 5.7 Schematic representation of metabolic pathways for lactose/galactose catabolism and heteropolysaccharide synthesis in lactic acid bacteria. Lactose/galactose transport may take place via a lactose/galactose-specific phosphotransferase primary transport system, a specific permease primary transport system, or a secondary transport system involving lactose–galactose antiport or lactose-H+ symport. Numbers represent enzymes: (1) phospho-β-galactosidase, (2) β-galactosidase, (3) glucokinase, (4) phosphoglucomutase, (5) UDP–glucose pyrophosphorylase, (6) UDP–galactose-4-epimerase, (7) dTDP–glucose pyrophosphorylase, (8) dehydratase, (9) epimerase reductase, (10) phosphoglucose isomerase, (11) 6-phosphofructokinase, (12) fructose-1,6-biphosphatase, (13) fructose-1,6-diphosphate aldolase, (14) galactose-6-phosphate isomerase, (15) tagatose-6-phosphate kinase, (16) tagatose-1,6-diphosphate aldolase, (17) galactokinase, (18) galactose-1-phosphate–uridylyl transferase, (19) glutamine–fructose-6-phosphate transaminase, (20) glucosamine–phosphate acetyltransferase, (21) acetylglucosamine–phosphate mutase, (22) UDP–glucosamine pyrophosphorylase, (23) UDP–N-acetylglucosamine-4-epimerase. (From DeVuyst, L. et al., *Int. Dairy J.*, 11, 687, 2001.)

same microorganism that synthesizes the biopolymer; degradation can be either partial, yielding polymers of lower MW or oligomers, or complete, leading to sugar monomers. These degrading enzymes are often involved in polysaccharide biosynthesis, cleaving parts of the biopolymer before it is released, thus affecting the degree of polymerization and the molecular weight. Additionally, they may act when the cells enter the death phase, as part of the general cell lysis mechanism, or under conditions of carbon limitation. Furthermore, they may become activated when polysaccharide accumulation exceeds an acceptable limit for the cells (in bioreactors where high productivity is achieved), causing oxygen or mass transfer limitations, or in other cases, they play a role in the pathogenicity of some microbes.[99,116,117,120–123]

For instance, scleroglucan can be hydrolyzed to glucose molecules via the action of β-1,3-glucanase (endo-acting) and β-glucosidase. These enzymes are activated toward the end of the fermentation process if the carbon source is depleted.[117] Similarly, polysaccharide hydrolysis by glycohydrolases of EPS-producing lactic acid bacteria often occurs upon prolonged cultivation.[124] Other examples of functional polysaccharides enzymically degraded by the producer microorganism include alginates, which can be enzymaticaly decomposed by hydrolases of *Azotobacter vinelandii*;[120] the glucan of *Candida albicans*, which can be degraded by a group of *Candida* β-1,3-glucanases;[125] and lentinan, which is susceptible to glucanases of the fruit bodies *Lentinus elodes* during storage of the mushroom at relatively high temperatures (20°C). *In vitro* digestion of lentinan by these enzymes showed that the glucan can be totally decomposed to glucose.[126] *S. cerevisiae* strains also possess glucanases and α- and β-glucosidases,[127–129] which may depolymerize the glucans synthesized by the yeast.

The control of such enzymic degradation is crucial in the production of bioactive microbial polysaccharides. Polysaccharides vulnerable to lyases may have low yields, and their desired structure, MW, and DB may be altered, affecting their functionality. Therefore, the action of lyases needs to be limited, e.g., by choosing process parameters (temperature, pH, oxygen tension, carbon source concentration, time of harvest) that do not favor such enzymic hydrolysis, or by means of genetic engineering, aiming at deleting genes responsible for lyase activity. However, the deletion of genes encoding lyase synthesis or activation may be detrimental to the synthesis of the desired polysaccharide, if these enzymes are also involved in the biosynthetic and polymerization steps.[122,123]

5.3.2 Bioprocess Conditions

The bioprocess conditions during the production of microbial functional polysaccharides influence significantly microbial physiology, and thus the concentration and structure of the final product. In several bioprocesses the composition, molecular weight, and chemical structure of polysaccharides can be altered by different process conditions. Apart from the case of fruit bodies of (medicinal) mushrooms, which are cultivated on solid media (soil enriched with nutrients, sawdust, etc.) over a long period (up to months) under partly controlled conditions, all other processes for polysaccharide production take place in bioreactors (or fermentors) using a liquid medium, which offers easy and automated control of the process. Such processes can be batch (only one addition of carbon source at the beginning of the process), fed-batch (additional supply of carbon source at some point of the process), or continuous (continuous addition of process medium and simultaneous withdrawal of equal amount of medium and product), although the industrial fermentations for microbial polysaccharide production are usually either batch or fed-batch. Of course, the environmental parameters that influence the numerous polysaccharide processes do not have the same effects in all processes, since each individual microorganism has different responses to process conditions, such as medium composition, temperature, pH, oxygen, agitation, and aeration. The impact of process parameters depends on the relation between cell growth and product synthesis. Microbial polysaccharide synthesis may follow either growth-associated or nongrowth-associated kinetics; in the former, the polysaccharide is synthesized mainly during the exponential growth phase, whereas in the latter the polysaccharides are formed mainly toward or after the end of the growth phase (secondary metabolites). Polysaccharide syntheses with mixed kinetic response also exist.[108,109,124]

For several bioactive polysaccharides the physiology of the producer microorganism has not been studied yet, and reports on their production process are scarce. Based on the available data, some general principles regarding the effect of process parameters will be cited below, including examples from some processes.

5.3.2.1 Composition of the Process Medium

The composition of the process medium or fermentation medium (or fluid) is a key factor that influences the progress of the process, and careful design of the medium is essential in process optimization. In well-defined media the carbon source is usually a monosaccharide (e.g., glucose or fructose) or a disaccharide (e.g., sucrose or lactose), but in some processes complex carbon sources such as molasses, hydrolyzed starch, or dairy whey have been used, due to their low cost.[50,108,109,130–132] Also, a nitrogen source is needed to enhance cell growth. Nitrogen may be supplied in organic (e.g., peptone, yeast extract, corn steep liquor, soy hydrolyzate) or inorganic (e.g., ammonium or nitrate salts) form. In addition, phosphorous and sulfur ions are often included in the process medium, contributing to aminoacid synthesis, ATP synthesis (phosphorous), the uptake of other nutrients, and sometimes acting as pH buffers. Apart from the above macronutrients (used in comparatively high concentrations), micronutrients (in low concentrations) such as Mg, K, Na, Ca, or other metal ions are often added to the process medium, serving as enzyme co-factors or as components of the polysaccharide structure. The fermentation medium may also contain vitamins, aminoacids, organic acids, or other substances, which assist cell growth or stimulate biopolymer synthesis.

In scleroglucan production, a typical fermentation medium contains (per liter of water) glucose or sucrose (30.0 to 35.0 g), $NaNO_3$ (3.0 g), KH_2PO_4 (1.3 g), KCl (0.5 g), $MgSO_4 \times 7H_2O$ (0.5 g), $FeSO_4 \times 7H_2O$ (0.05 g), $ZnSO_4 \times 7H_2O$ (3.3 mg), and yeast extract (YE) (1.0 g). Citric acid (0.7 g/l) and thiamine × HCl (3.3 g/l) may also be added.[50,133] Under these conditions, scleroglucan concentrations of 8.5 to 10.0 g/l can be obtained. However, these values are considerably lower than the maximal concentrations reported for other polysaccharides, such as xanthan (27.0 g/l).[134] The increase in initial sugra concentration might make more carbon available to glucan biosynthesis, but, in some cases, growth inhibition might occur when the sugra content exceeds a certain level, having an adverse effect on polysaccharide synthesis.[135] Schizophyllan can normally reach concentrations similar to that of scleroglucan, but when citric acid (up to 5.0 g/l) was added to the fermentation medium, cell growth rate decreased and glucan production rate (g glucan/l/h) and glucan product yield (g glucan/g glucose) showed a 6-fold increase,[136] making the process more economic.

The type and amount of nitrogen source may also affect biopolymer formation. Generally, high levels of nitrogen favor cell growth at the expense of scleroglucan synthesis, and more glucan is produced when ammonium is replaced by nitrate ions.[137] The inhibitory effect of ammonium on polysaccharide synthesis has also been demonstrated in alginate-producing cultures of *Azotobacter vinelandii*.[99] A high initial carbon/nitrogen (C/N) ratio and N depletion (toward the end of the growth phase) stimulate the production of (nongrowth-associated) polysaccharides, such as xanthan[138] and curdlan.[139] Furthermore, the number of side pyruvate groups of xanthan was found to be higher when the total N content of the process medium increased, which in turn affected the rheological properties of the biopolymer;[140] viscosity of xanthan solutions rose when pyruvilation degree increased.

Medium formulation for glucan production by *S. cerevisiae* should aim at maximizing biomass, since the glucan is a component of the cell wall. The yeast forms ethanol (which can restrict growth) in high quantities, especially if a critical concentration of carbon source is exceeded (at least 100 g/l of glucose or sucrose is usually used for ethanol production). A high C/N ratio or N limitation also stimulates ethanol synthesis and reduces not only cell growth but also the amount of glucans in the cell wall.[141,142] Thus, high initial N concentration or relatively low carbon concentration should be preferred for glucan production. Furthermore, significant variance in polysaccharide composition of the cell wall of *S. cerevisiae* with the nature of carbon source has been demonstrated.[142] Similarly, the production yield of polysaccharides by submerged cultures of *Ganoderma lucidum* was found to be influenced by medium composition.[143,144]

Lactic acid bacteria (LAB) have different requirements in medium composition, depending on the individual species and the relation between growth and polysaccharide synthesis. It has been established that, in general, functional polysaccharides from mesophilic species (*Lactococcus cremoris*, *Lactobacillus casei*) are not growth associated, while thermophilic species (*Lactobacillus bulgaricus*, *Lactobacillus helveticus*) produce growth-associated EPSs.[119,145] For example, the absence of N limitation (optimal balance between carbon and nitrogen) and the inclusion of vitamins in the process medium favor EPS synthesis from *L. bulgaricus*.[146] Lactose and glucose are often used as carbon sources for polysaccharide production from LAB, while fructose and mannose usually lead to low EPS yields. Milk, whey, or whey permeate (deproteinized) can also be used.[131] The choice of carbon source in LAB cultures may influence not only the quantity, but also the composition of the EPSs (e.g., ratio of sugar monomers in heteropolysaccharides).[147] Therefore, special care must be taken when functional polysaccharides are produced, since changes in chemical composition may affect their functionality. For this reason, the use of well-defined media is probably preferable. On the other hand, new bioactive LAB polysaccharides with novel composition might be designed in this way, i.e., by careful selection of the type of carbon source.

5.3.2.2 Temperature

Temperature influences the formation of bioactive polysaccharides in two ways: first, it affects cell growth, which in turn has an impact on biopolymer synthesis;

second, it may affect the activity of key enzymes involved in polysaccharide biosynthesis.[108,109,116,124] For example, the control of temperature is crucial in scleroglucan synthesis. Glucan formation occurs in the range of 20 to 37°C; the optimal temperature for synthesis is 28°C, while the optimal temperature for cell growth is above 28°C. Additionally, below 28°C by-product (oxalic acid) formation takes place, reducing glucan production.[50,148] For most polysaccharide-producing fungi, yeast, and bacteria, the process temperature is around 30°C, which often represents a compromise between the optimal temperature for growth and that for biopolymer synthesis. Levan from *Zymomonas mobilis*[88] and exopolysaccharides from *Ganoderma lucidum*[144] are also produced at 30°C. Although a temperature of 30°C is often used for xanthan production, it has been shown that in the range of 25 to 34°C, the xanthan MW is highest at 25°C.[149] Shu and Yang also noticed that process temperature can affect the link between xanthan and cell growth.[150] They found that at 24°C or lower, xanthan formation began toward the end of the exponential growth phase, whereas at 27°C or higher, it followed growth-associated kinetics.

When batch cultures of *S. cerevisiae* were grown at 22, 30, and 37°C, the highest growth rate and cell wall content of the yeast occurred at 30°C.[142] However, more total β-glucans were contained in the cell wall of yeasts cultivated at 37°C.[142]

The optimal temperature for functional EPS formation by LAB varies considerably between mesophiles and thermophiles. Low temperatures between 18 and 20°C enhance polysaccharide production by *Lactococcus cremoris* and *Lactobacillus casei* strains. These temperatures are suboptimal for growth, which is preferable for nongrowth-related kinetics.[131,145] Moreover, it has been noted that since there is antagonism for the isoprenoid lipid carriers between cell wall synthesis and EPS formation, the slow growth under low temperatures allows more lipid carriers to be used for polysaccharide production.[108] Conversely, high temperatures (37 to 42°C) are applied in EPS production by thermophiles (*Lactobacillus bulgaricus* and *Lactobacillus helveticus*).[131,151]

5.3.2.3 pH

The control of pH via automatic addition of alkali or acid in a bioreactor is essential for most polysaccharide processes (uncontrolled pH usually leads to low yields or even cell death). According to Kang and Cotrell,[152] polysaccharide production by fungi is generally optimal in the pH range of 4.0 to 5.5. However, many fungi are acidophilic and grow readily at lower pH values. Thus, Wang and McNeil[153] designed a bi-staged process for scleroglucan production where growth was stimulated in the first stage at pH 3.5, and after a sufficient amount of cells was produced, pH was shifted to 4.5 in the second stage to promote glucan formation. Under these conditions oxalic acid synthesis was also reduced (by 10%).

In a similar mode, a bi-staged process for *Ganoderma lucidum* exopolysaccharides with pH shift from 3.0 to 6.0 resulted in a remarkable increase in EPS concentration, namely, 20.1 g/l EPS was obtained, compared to 14.0 g/l when the pH was controlled at 6.0, 6.5 g/l when the pH was adjusted at 3.0, and 4.1 g/l when the pH was uncontrolled.[143]

Low environmental pH also promotes cell wall synthesis in *S. cereviciae* and results in increased resistance to (1→3)-β-glucanase action.[154] Decreased sensitivity of *S. cerevisiae* cell wall to zymolyase at low pH was also noted by Aguilar-Uscanga and François,[142] who found that the yeast grown at pH 5.0 had the highest growth rate and cell wall content (with high mannan concentration in the cell wall), whereas the proportion of total β-glucans and β-1,6-glucans in the cell wall was higher at a pH of 4.0.[142]

For many bacterial polysaccharides, a neutral pH is preferable.[107] Thus, xanthan is produced at pH 7.0,[107,149] while for alginate synthesis a pH of 7.0 to 7.2 is normally used.[155] In most cases, the optimal pH for EPS synthesis by lactic acid bacteria is close to 6.0.[131] Nevertheless, it has been suggested that higher pH values may be beneficial to polysaccharide production by *Lactobacillus bulgaricus*, because they cause prolongation of the exponential growth phase, during which much of the EPS is formed (growth-associated kinetics).[153] Also, the control of pH results in greater EPS yields, compared to uncontrolled pH conditions where the medium is progressively acidified. In fact, the impact of pH adjustment seems to be greater than that of supplementation with nutrients.[131,146,156] This could pose a limitation in the industrial exploitation of polysaccharide-producing LAB and their addition in functional fermented dairy products, since relatively low EPSs must be expected under uncontrolled pH conditions.

5.3.2.4 Agitation, Aeration, and Dissolved Oxygen

These additional fermentation parameters determine the efficiency of mass and oxygen transfer and uptake, as well as the rate of metabolism and metabolites release. They are interconnected to some extent with one another, and are particularly important for all polysaccharide processes, because the viscous fluid that is usually developed after some point in the process makes mixing of the fermentation fluid problematic. Usually, a high agitation (stirring) rate (above 250 rpm) is essential in microbial polysaccharide production. In xanthan production, biopolymer synthesis was a function of stirring rate in an agitation range of 200 to 800 rpm, as a result of improved nutrient uptake by the microorganism.[157] In the scleroglucan process, despite the importance of agitation in effective mixing, it was revealed that the glucan is sensitive to shear-induced depolymerization under vigorous agitation (reduction in MW).[158]

The effect of aeration and oxygen on polysaccharide formation is sometimes controversial as well. Several studies on polysaccharide synthesis support the idea that oxygen enhances biopolymer formation and production rate. For instance, this has been observed for xanthan[157] and curdlan.[159] On the other hand, it has been proposed for scleroglucan and schizophyllan that low dissolved oxygen (DO) in the bioreactor, expressed as a percentage of air saturation with oxygen, and especially DO limitation could enhance their biosynthesis, in contrast to cell growth.[158] This might happen either via the induction of oxygen-sensitive biosynthetic enzymes or via the direction of carbon flux primarily toward glucan synthesis, under growth-limiting conditions. Additionally, high DO is believed to be responsible for radical-induced degradation of xanthan[160] and scleroglucan[54] under oxidative stress conditions.

During exopolysaccharide production by *Ganoderma lucidum*, the biopolymer concentration and production rate rose as the agitation rate increased in the range of 100 to 400 rpm.[161] Also, comparing a DO of 10% (of air saturation) with a DO of 25% showed that at low DO cell growth was restricted, due to oxygen limitation occurring in mycelia aggregates in the fermentor, but EPS production by *G. lucidum* was elevated.[162] Restriction (but not cessation) of growth of *S. cerevisiae* occurred at a DO of 0% (absence of aeration), compared to a process with the DO above 50%. In the second process, cell growth rate and cell wall content were higher, whereas mannan and glucan concentrations of the cell walls did not differ significantly between the two processes.[142] The alginate process exhibits an increase in biopolymer formation with increased agitation rate (from 300 to 700 rpm) and under relatively high aeration.[155] Lastly, polysaccharide production by lactic acid bacteria usually takes place under low agitation (e.g., 100 rpm), since EPS content is rarely above 1 g/l and the low viscosity allows efficient mixing. Minimal or no aeration is applied, because most strains are microaerophilic or can grow anaerobically.[131]

5.3.3 Isolation and Purification

Two main procedures for polysaccharide isolation from microorganisms exist, depending on whether the biopolymers are part of the cell wall, or excreted by the cells as a capsule or slime (exopolysaccharides). In the former case, isolation is based on extraction from cell mass, while in the latter, centrifugation or filtration is applied to separate product from cells.

For instance, the isolation of cell wall β-D-glucans from *S. cerevisiae* follows the subsequent steps.[163]

After harvesting, the fermentation medium is centrifuged and yeast cells are separated in the sediment, which is then extracted with 6% (w/v) NaOH at 60°C. Polysaccharides, oligosaccharides, and sugars are extracted and subsequently dialyzed in water (30 min of stirring). Thus, soluble and low MW substances are removed with water and the remaining fraction (polysaccharides) is centrifuged and the sediment suspended in 3% (w/v) NaOH at 90°C for 2 h. A second round of centrifugation and sediment dialysis against water follows to further purify the glucan. Finally, the purified glucan is neutralized and extracted with 4% (v/v) H_3PO_4 (18°C, 2 h). The wet purified glucan is then lyophilized, air dried, or spray dried to form the final product (powder).

A similar but simpler method for isolation of cell wall mannans from *S. cerevisiae*, *Candida albicans*, and *Candida utilis* is described elsewhere[164] and includes extraction of cell debris with 2% (w/v) KOH, removal of impurities from the extract with Fehling's reagent, and purification of mannans. Isolation of mushroom polysaccharides follows similar procedures, but the cells (fruit bodies) are homogenized or milled before extraction. Specifically, lentinan was isolated after extraction of the homogenized and lyophilized sample with hot (boiling) water for 10 h, subsequent filtration of the suspension to remove insoluble matter, and ethanol precipitation of the water-soluble polysaccharide, which was then centrifuged and lyophilized.[126] A second extraction cycle after filtration could increase the degree of purity of lentinan. Extraction protocols of other bioactives from commercial medicinal mushrooms (e.g., antioxidants from *Ganoderma lucidum*, *Ganoderma tsugae*, and *Coriolus versicolor*) have also been described.[165]

For the isolation of several polysaccharides from the same microorganism, a stepwise extraction of the different fractions is adopted, and chromatographic techniques can be employed for improved purification. For example, hot-water-soluble polysaccharide extracts from fruit bodies of *Agaricus blazei* were fractionated and purified after ethanol precipitation, ion exchange chromatography, gel filtration, and affinity chromatography. In this way, four different antitumor biopolymers were separated.[43] The water-insoluble heteroglycans were obtained by successive extraction with 1% (w/v) ammonium oxalate, 5% (w/v) sodium hydroxide, 20% (w/v) sodium hydroxide, and 5% (w/v) lithium chloride–dimethylacetamide, yielding four polysaccharide fractions. These were precipitated with ethanol and purified chromatographically.[166]

Isolation of extracellular polysaccharides is usually easier and faster, since the main step of filtration or centrifugation (to remove cells) is shorter than extraction procedures. The speed of centrifugation, filter pore size, and duration of centrifugation or filtration depend on the nature and viscosity of the polysaccharide, and on whether EPS is in the form of slime or capsule (the latter is more difficult to separate from cells). Especially for capsular EPSs, which are closely attached to the cell membrane, a sample pretreatment is usually applied before centrifugation or filtration, to facilitate dissociation of EPSs from cells. This may include alkali addition (e.g., NaOH) or heat treatment of the sample (e.g., at boiling temperature). Sometimes the sample is preheated in saline solution or a mixture of water and phenol. Sonication of cell suspension or sample autoclaving can also be used as a treatment to release capsular EPSs from cells.[167]

For isolation of scleroglucan, samples of the fermentation medium were initially neutralized (process pH was 4.5) with addition of alkali and diluted fourfold with distilled water. After a heating pretreatment (80°C, 30 min), samples were homogenized in a blender and centrifuged (27,5000 *g*, 20 min, 10 to 15°C). The polysaccharide was collected from the supernatant and precipitated by the addition of an equal amount of 96% (v/v) ethanol. After precipitation was completed, scleroglucan was recovered by filtration and dried (105°C).[137] Similar techniques are used for isolation of other water-soluble EPSs, but naturally the pretreatment stages (if any), the intensity of centrifugation, and the volume and type of alcohol vary.

In LAB cultures, isolation and purification of EPS are also based on centrifugation and alcohol precipitation of dilute samples, but often include several steps for removal of proteins and impurities, which are especially essential if the cells are grown on dairy whey or milk. Deproteinization of the samples, after removal of cells, is achieved by treatment with trichloroacetic acid, proteinases, or heating. Ion exchange chromatography and high-performance liquid chromatography (HPLC) can be applied to lyophilized polysaccharides for fractionation and further purification.[67,119]

Water-insoluble exopolysaccharides are separated in a somewhat different way. For example, curdlan from *Agrobacterium* ATCC 31750 (former *Alcaligenes faecalis* ssp. *myxogenes*) was isolated as follows: Samples of fermentation broth were diluted with water and centrifuged (8000 *g*, 30 min, 4°C). The pellet containing cells and curdlan was washed with 0.01 *N* HCl and precipitated again by centrifugation. Subsequently, curdlan was solubilized by addition of 0.5 *N* NaOH for 1 h. Cells were removed by centrifugation (8000 *g*, 30 min, 4°C), and curdlan present in the supernatant was recovered (as a pellet) after addition of 2.0 *N* HCl, and finally dried.[168]

It has to be noted that after isolation and purification of the polysaccharide, an additional treatment (derivatization) may be applied for the formulation of the final product, which may aim at increasing its solubility or immunomodulating–prophylactic activity. For instance, particulate microbial glucans are often solubilized by carboxy(m)ethylation, sulfation, sulfoethylation, phosphation, or chemical oxidation.[5,169–171] This is done to increase the activity of therapeutic polysaccharides and to avoid possible side effects of insoluble or hardly soluble glucans after intraveneous or intraperitoneal administration (inflammation, pain, granuloma formation, microembolism, hepatosplenomegaly, circulatory collapse).[5,169,170] In some cases, the derivatives are conjugated with other active compounds or drugs to further improve their performance.[171] This solubilization step could also be useful in a food context, where the functional polysaccharides may need to be dissolved in the water phase of the nutraceutical preparation, so that aggregation of the active compound is prevented, or gel formation and thickening effects of the polysaccharide are facilitated. In addition, ultrasonic treatment can be used to aid the derivatization of the functional polysaccharides. Also, ultrasounds have been applied for targeted depolymerization of polysaccharides to produce lower MW derivatives, which can be more readily dissolved in water, and in some cases are more effective than the original biopolymer.[169] The latter procedure can also be performed by chemical and enzymatic degradation, but ultrasonication is advantageous in that it does not change the chemical nature of the biopolymer, apart from reducing its molecular weight.[169]

5.4 PHYSIOLOGICAL FUNCTIONS

There are numerous studies on the physiological functions of therapeutic or dietary microbial polysaccharides in the literature. Many of these polysaccharides exhibit antitumor, immunomodulatory, or antiviral–antimicrobial effects. Others have hypocholecterolemic–hypolipidemic, hypoglycemic, antioxidant, or anticoagulatory properties. A description of most of these physiological functions and the mechanism of action of the bioactive polysaccharides will be attempted here.

5.4.1 Antitumor–Immunomodulatory Effects

Most of the research on bioactive microbial polysaccharides so far is centered on potential antitumor properties and has resulted in the commercialization of several immunoactive biopolymers. There are several direct and indirect ways of anticancer activity of microbial polysaccharides, the most important of which are (1) prevention of oncogenesis by oral consumption, (2) direct antitumor action against existing tumors, (3) immunopotentiation against tumors in conjunction with chemotherapy or radiotherapy, and (4) prevention of metastasis. Selected examples of this activity will be given below.

Schizophyllan and lentinan are two of the most common immunotherapeutic glucans and have been used clinically for cancer treatment since 1986, usually in conjunction with chemotherapy or radiotherapy. In general, they have low toxicity even at high doses (lethal dose of lentinan for mice, rats, dogs, and monkeys was above 100 mg/kg).[2] Clinical studies showed that when schizophyllan was administered along with antineoplastic drugs, it prolonged the lifespan of patients suffering from lung or gastric cancer. It has also been used for the therapy of stage II or III cervical cancer.[5,172] Moreover, schizophyllan was shown to possess antiradiation properties, restoring mitosis of bone marrow cells previously suppressed by antitumor drugs.[173] Lentinan is also used successfully in the treatment of patients with gastric, colorectal, or breast cancer, and the prevention of metastasis, without having any toxic side effects in humans.[174,175] For both schizophyllan and lentinan, early administration, i.e., before or simultaneously with initiation of chemotherapy or irradiation, improves their effectiveness.

The antitumor effects of lentinan and other bioactive glucans are due to their mitogenic activity, which causes several immune responses, such as natural killer (NK) cell activation and T-cell-mediated cytotoxicity, stimulation of monocytes, increased production of cytokines — e.g., interferons (IFNs), interleukines (ILs), and tumor necrosis factor-α (TNF-α) — activation of peripheral mononuclear cells (PMNCs), and stimulation of phagocytosis by neutrophils.[5,50,52,176,177] In other words, the cells, proteins, or hormones of the natural immune system are stimulated. For example, proliferation of NK of peripheral mononuclear cells above normal levels or enhancement of immunoglobulin production by PMNCs after intake of lentinan in cancer patients with declined numbers of PMNCs demonstrated the immunopotentiating activity of the glucan.[176–178] Schizophyllan and lentinan were also found to stimulate secretion of TNF-α by human monocytes.[52] In a study with gastric carcinoma patients, injection of a single dose of 2 mg of lentinan led to significantly increased production of cytokines (interleukins 1-α and 1-β and TNF-α) by monocytes of peripheral blood mononuclear cells (PBMCs) 3, 5, and 7 days after the injection.[179] Also, lentinan caused complete regression of Sarcoma 180 in mice at a dose of 1 mg/kg for 10 days.[176] In addition, lentinan can restore the suppressed activity of helper T-cells in tumor-bearing subjects, and thus aid the restoration of immune responses or activate the normal or alternative pathways of the complement system.[176,180]

In many cases, the presence of immunocompetent cytotoxic T-cells of the host immune system is necessary for the development of the antitumor activity of lentinan, schizophyllan, and *Phytophthora* glucan.[181,182] It has been suggested that this may happen because these and other therapeutic polysaccharides are T-cell adjuvants. However, immune responses (to β-D-glucans) not involving T-cells also exist, such as increased numbers of macrophage or activation of neutrophils and NK cells, which can destroy malignant cells.

Furthermore, the immunoenhancing ability of lentinan seems to be related to its modulation of hormonal factors involved in tumor growth. It was reported that intake of thyroxin or hydrocortisone inhibited the antitumor functions of lentinan.[176] Also, restriction of anticancer activity of lentinan and other polysaccharides occurred after pretreatment of the glucan with antimacrophage agents, such as carrageenans.[182] This is of particular interest for food scientists and the formulation of nutraceuticals, since it indicates that some food additives or compounds should be avoided or removed before the introduction of a bioactive polysaccharide.

Sinofilan, the immunopharmacological form of scleroglucan, is also used clinically for the treatment of cancer. The antitumor activity of scleroglucan is believed to be mediated by the increase in macrophages in the presence of soluble glucan.[50] It has been reported that scleroglucan has a high affinity for human monocytes and possesses two main biological properties: the stimulation of phagocytic cells, and monocyte, neutrophil, and platelet hemopoietic activity.[50,183] Sclerotinia sclerotiorum glucan (SSG), the pharmacological form of scleroglucan, exhibits antitumor–immunomodulating activities when administered either parenterally or orally.[2,184] Oral administration, uncommon among other glucans, is important in eliminating side effects and pain. Also, effective oral administration of a polysacharide may be indicative of the potential effectiveness of consumption of foods containing this biopolymer.

The glucans deriving from *S. cerevisiae* have various immunostimulating effects. PGG enhances the production of cytokines and monocyte and neutrophil phagocytosis.[28] Zymosan containing mainly insoluble β-glucan (70% (w/w)) from *S. cerevisiae* activates neutrophils (PMNCs) through a trypsine-sensitive recognition mechanism and exhibits mitogenic activity.[2,170] Interestingly, the β-glucan of the wild type of yeast (named cerevan), exhibited higher mitogenic activity in rats than zymosan.[170] Also, soluble derivatives of *S. cerevisiae* glucan maintain or increase their antitumor properties. Glucan sulfate, sulfoethyl glucan, carboxymethyl glucan, and oxidized glucan have shown significant antitumor, antimutagenic, and radioprotective activity, e.g., via stimulation of T-cells of lymphocytes, B-cells, and macrophages, and enhancement of hemopoiesis.[2,60,169,170]

Another commercialized antitumor glucan, Krestin (or Polysaccharide-K, PSK), has broad immunostimulating and antineoplastic scope. It has been administered orally or intravenously as a clinical medicine in many types of cancer, along with other drugs, and has been found to be especially effective in the prevention of metastasis when injected on a tumor site. It acts either directly on the tumor cells or indirectly by boosting the immune system, e.g., by enhancing phagocytosis by macrophages.[1,173] The various glucans of *Agaricus blazei* fruit bodies or cultures are also known for their antitumor functions. A soluble acid-treated fraction of proteoglucan extracted from fruit bodies of the fungi activated natural killer cells with marked tumoricidal activity, which could infiltrate mice tumors. This bioactive proteoglucan inhibited tumor growth and led to the induction of tumor apoptosis (death).[65] The proteoglucan was also found to be inactive toward normal mouse splenic mononuclear cells, indicating that it is only cytotoxic upon tumor cells.[65] β-Glucans, α-glucans, and a xyloglucan from *A. blazei* fruit bodies, as well as glucomannans and a proteomannan from cultured mycelia, also possess antitumor properties.

In a study on the immunopharmacological activities of a solubilized (oxidized) form of *Candida* soluble β-D-glucan (CSBG), Tokunaka et al.[59] concluded that the biopolymer was characterized by the following properties: stimulation of interleukin-6 synthesis of macrophages *in vitro*, antagonistic effect for zymosan-mediated TNF-α synthesis, induction of lipopolysaccharide-mediated TNF-α synthesis, enhancement of hematopoietic activity, restoration of leukocytes after induced leukopenia, antitumor activity against the ascites form of tumor, and high vascular permeability, which is important to the production of cytokines and other immune responses.

Although linear, native curdlan does not provoke immune responses; substituted derivatives of curdlan (etherified, sulfated, sulfoethylated, sulfoalkylated) were proved to have significant antitumor activity against Sarcoma 180. They were responsible for high mitogenic activity, stimulation of macrophages, and induction of phagocytosis.[2] Other antitumor glucans include *Ganoderma lucidum* glucan,[15] glomerellan,[16] and glucans from *Auricularia auricula-judae*,[19] *Cordyceps cicadae*, *Dictyophora indusiata*,[93] *Pleurotus ostreatus*,[23] *Pleurotus tuber-regium*,[47] and others. Furthermore, the fucogalactan of *Sarcodom aspratus* was found to exhibit mitogenic functions. The biopolymer elicited the release of TNF-α and nitric oxide (free radical with tumoricidal and microbicidal activities) in microphages of mice. The *in vitro* stimulation of TNF-α synthesis after a 50 μg/ml fucogalactan dose was higher than that obtained after a 500 μg/ml lentinan dose.[90]

Levans of *Zymomonas mobilis* also exert antitumor and radioprotective activity. This was evident by the generation of mononuclear cells, the increase in peripheral leukocytes and spleen cell antibodies, the stimulation of macrophages, and the changes in weight and survival of tumor-bearing mice.[87] Alginate is also capable of stimulating the immune system, by inducing the secretion of cytokines, e.g., tumor necrosis factor, TNF-α, interleukin-1, and interleukin-6.[185] The antitumor properties of alginate are strongly connected to macrophage stimulation.[98]

Moreover, functional EPSs from lactic acid bacteria are characterized by antitumor or immunostimulatory activity *in vitro* and *in vivo*. An extracellular phosphopolysaccharide from *Lactococcus lactis* ssp. *cremoris* can enhance lymphocyte mitogenicity, macrophage cytostaticity, and cytokine production in macrophages.[79,186] In addition, antigen-specific antibody production was stimulated by the same EPSs at a dose of 100 to 500 μg/mouse.[84] Similarly, when the extracellular phosphopolysaccharide of *Lactobacillus delbrueckii* spp. *bulgaricus* was injected intraperitoneally at a 100 mg/kg dose in mice, the number of intraperitoneal macrophages increased and macrophage phagocytosis was augmented *in vivo* and *in vitro*, while the cytostatic activity of thioglycolate-induced macrophages upon tumor cell lines (Sarcoma 180 and P388) was markedly improved *in vitro*, after a 6-h treatment with phosphopolysaccharide solution (10 to 100 μg/ml).[82] Mitogenic activity was also exhibited by *Bifodobacterium adolescentis* M101-4 exopolysaccharide *in vitro*.[80] Notably, Sreekumar and Hosono[81] reported that skimmed milk fermented with *Bifodobacterium longum* had antimutagenic properties against heterocyclic amines, mediated via the binding of the EPSs to the mutagens. Even though these results were not verified *in vivo* or in clinical studies, they suggest potential prophylactic benefits from the consumption of dairy products fermented with bifidobacteria.

5.4.2 Antimicrobial–Antiviral Effects

Apart from their antitumor effects, schizophyllan and lentinan are also useful as immunostimulatory agents acting against microbial infections, such as tuberculosis (by *Mycobacterium tuberculosis*) and *Listeria monocytogenes* infection. The antimicrobial activity of lentinan is mediated by augmentation of phagocytosis of microbial cells by neutrophils and macrophages.[2,172,174] In a study on the microbiocidal activity of lentinan, it was found that lentinan induced activation of peritoneal macrophages against *Salmonella enteritis* and especially bronchoalveolar macrophages against *Staphylococcus aureus*, possibly due to adaptive transformation of the macrophages via interactions with lentinan.[187] Also, lentinan showed high antiviral activity against influenza virus, prevented proliferation of polio virus, increased host resistance to HIV virus, and limited the toxicity of AZT, a common drug used against the HIV virus. The mechanism of antiviral activity of this glucan is believed to be linked to induction of interferon.[188] Inhibition of HIV replication was also achieved by curdlan sulfate *in vitro*. Possibly curdlan sulfate impedes the attachment of the virus onto the host cell, although encapsulation of the virus by the glucan, after being internalized into the cell, might also occur.[189] Scleroglucan is another β-glucan with antiviral functions. It has been proposed that the mode of action of scleroglucan against herpes simplex virus (HSV) and rubella virus is based on the binding of the bioactive glucan to glycoproteins of the host cell membrane, which hinders interactions between the virus and the host cell plasma. However, this inhibition mechanism is only effective at an early stage of the infection, i.e., before the host cell is infected with various RNA.[190,191] In a similar study, the antiherpetic activity of an acidic protein-bound polysaccharide from *Ganoderma lucidum* was attributed to the binding of the biomolecule with HSV-specific glycoproteins of the cell membrane, which are responsible for virus attachment and penetration.[72]

Glucans from *S. cerevisiae* also exert significant antimicrobial effects. Particulate (but not soluble) glucan from baker's yeast, as well as SSG-glucan from *Sclerotinia sclerotiorum*, inhibited growth of *Mycobacterium tuberculosis in vitro*, especially when administered at the time of infection initiation.[192] Additionally, administration of 0.5 to 4.0 mg/kg dose of PGG-glucan from *S. cerevisiae* reduced significantly the counts of antibiotic-resistant *S. aureus* in the blood of contaminated mice.[193] The enhanced clearance of *S. aureus* by PGG is accompanied by an increase in monocytes and neutrophils, as well as a stimulation of oxidative microbicidal activity by neutrophils. Also, when used in combination with traditional antibiotics, PGG enhanced drug effectiveness against *S. aureus*.[193]

Schizophyllan has also shown significant antiviral activity against Sendai virus (otherwise lethal to mice) by both oral and intraperitoneal administration.[194] Also, sizofiran, the commercial type of pharmacological schizophyllan, could stimulate the immune responses of patients against hepatitis B virus, by increasing interferon-gamma levels, and the proliferative response of peripheral blood mononuclear cells (PBMCs).[195]

Furthermore, levan (fructan) from *A. levanicum* was used successfully for oral immunization against pneumonia caused by *P. aeruginosa*.[86] Oral co-administration of 1000 μg of levan and 10 μg of cholera toxin from *P. aeruginosa* increased resistance of humans to pulmonary infection, by increasing levan-specific titers of serum immunoglobulin A (sIgA) in the lungs. This preventive treatment was more effective when initiated in the first hours of the hemorrhage rather than 4 days later.[86]

Finally, alginate has displayed significant functions as a prebiotic and the ability to regulate partly the intestinal microflora. In studies on the dietary effect of alginate on human fecal microflora, it was shown that consumption of 10 g of sodium alginate per day raised the concentration of bifidobacteria and reduced the population of *Enterobacteriacae* and the occurrence of lecithinase-negative clostridia.[98] Also, when used alone or in combination with drugs in the treatment of ulcer or gastritis, alginate suppressed the population of *Helicobacter pylori* and promoted intestinal biocenosis, which reduced the number of staphylococci and bacteria of the genus *Proteus*.[98]

5.4.3 Hypocholesterolemic–Hypoglycemic and Other Effects

Alginates are polysaccharides with strong hypocholesterolemic effects. These biopolymers are resistant to gastrointestinal hydrolases in humans, and thus hardly digestible, unless they are depolymerized and fermented by microbial enzymes in the intestine. Research showed a reduction of plasma cholesterol in mice, following an alginate-rich diet (50 g/kg). This is likely to result from the interruption of enterohepatic circulation of bile acids, which leads to higher liver sterol output and bile acid excretion in the feces.[196]

The effectiveness of alginate in reducing blood glucose has also been studied, and alginate products have been used to assist the treatment of patients suffering from diabetes. This probably occurs through the decreased diffusion of glucose in the gastrointestinal tract, due to the high viscosity and gelling properties of alginate.[98] Other important biological functions of alginate include the prevention of gastroesophageal influx and epigastric burning, management of dyspepsia, and treatment of gastritis and gastroduodenal ulcer.[98,197] Drugs and patented products are already in use for these purposes (Gaviscon, Algitec, Gastralgin). Apart from the antimicrobial activity mentioned in Section 5.4.2, alginates mediate the normalization of esophageal and gastric pH, reduce acidic refluxes, facilitate the regeneration of mucus membrane in the stomach, and suppress gastrointestinal inflammation by forming a protective gel layer with the gastric juices, covering the mucus membrane (gastric dressing).[98,197] Additionally, sodium alginate is known for its antitoxic properties, especially the ability to suppress bioabsorption of harmful heavy or radioactive metals, such as strondium and cadmium.[98,198] Also, a calcium alginate-containing bioactive food additive (Detoxal) is currently used for its antitoxic effects (e.g., against chemically induced hepatitis, where it helps to control lipid peroxidation and reduce lipid and glycogen content in the liver).[98]

Xanthan was also shown to significantly reduce cholesterol levels in rats when supplemented at 1% (w/w) in a daily diet with a cholesterol content of 2% (w/w).[105] Nevertheless, as mentioned in Section 5.2.3, the effectiveness of xanthan in hypercholesterolemic humans is still uncertain,[106] and its characterization as a bioactive compound is still tentative.

Also, hypoglycemic effects of polysaccharides from *Ganoderma lucidum* fruit bodies have been reported. Although the mechanism of action is not well studied, it is possible that attachment of the polysaccharides to intestinal surfaces reduces and decelerates glucose absorption.[199]

Another aspect of bioactive microbial polysaccharides is their antioxidative properties. Some mushrooms are known for antioxidative functions. Although these are primarily due to the phenol content of mushrooms, polysaccharides also contribute to antioxidant activity.[165,200] For instance, polysaccharide isolates (protein-bound or not) from liquid cultures of *Antrodia camphorata* were partly responsible for inhibition of lipid peroxidation.[200] Also, an antioxidative effect of yeast cell wall mannan isolates, namely, glucomannan from *Candida utilis*, mannan from *S. cerevisiae*, and mannan from *Candida albicans*, has been reported.[164] In addition, antioxidant properties against chemically induced DNA oxidative damage have been demonstrated by polysaccharides of *Lentinus elodes*, *G. lucidum*, and *Coriolous versicolor*.[201] Antioxidative action against hydroxy radicals was also shown by a sulfoethyl glucan derivative of *S. cerevisiae*.[169] These effects are mainly due to the scavenging ability of these biopolymers toward reactive oxygen species and other free radicals.[164,169,200,201]

Lastly, other biological functions of microbial polysaccharides have been reported, such as antithrombotic effects (e.g., of curdlan sulfate or alginate)[98,202] or wound-healing properties (e.g., of PGG-glucans and alginate),[2,28,98] and probably merit further investigation.

5.5 STRUCTURE–ACTIVITY RELATIONSHIPS

The determination of the relationship between chemical structure and physical properties and the functionality of bioactive microbial polysaccharides is of great importance for understanding the mechanism of action and the targeted improvement of these biopolymers (by regulation of bioprocess conditions, chemical modification–derivatization, or genetic engineering methods), as well as for the preservation of the beneficial activities in a food environment and through food processing. Nevertheless, great controversy exists over this relationship, and variable results concerning structure–activity links have been published for several polysaccharide isolates.

The most important attributes that influence the activity of microbial polysaccharides are the molecular weight, degree and type of branching, existence of conformational order (random coil vs. single, double, or triple helix), chemical substitution of polysaccharide derivatives, and solubility.

Many researchers suggest that the β-(1,3) linkage of the main chain is vital for the expression of antitumor activity by glucans. Glucans containing mainly (1,6) linkages generally exhibited lower antitumor activity than glucans with a (1,3)-linked backbone.[5] This might be explained to some extent by the presence of a β-(1,3)-glucan receptor in macrophages, which enables macrophage activation by β-(1,3)-glucans. However, immunoactive glucans with a β-(1,6)-linked backbone have also been isolated, which shows that the β-(1,3)-linked main chain is not always a prerequisitive for antitumor activity of glucans.[2,5,57]

The degree of branching (DB) also affects the biological functionality of polysaccharides. Table 5.1 demonstrates that glucans with low, medium, or high DB may possess immunostimulatory properties. For antitumor activity of glucans, perhaps a moderate branching (DB between 0.20 and 0.33) is preferable.[2] The highly branched glucans of *Auricularia auricula-judae* (DB = 0.75) and *Pestalotia* sp. 815 (DB = 0.67), as well as the almost linear glucan of *Ganoderma lucidum* (DB = 0.06), are weakly active against tumors.[2,19,29,49] However, native PGG had low immunopotentiating activity when DB was 0.2, but genetically engineered PGG with a DB of 0.5 was more effective and showed a 35-fold higher affinity for the β-glucan receptor of human monocytes and neutrophiles.[28] Increased branching leads to weaker interchain associations, which favors the formation of single helices. These structural differences resulting from different DBs may relate to the variance in biological activity of polysaccharides of different origin and with different DBs.[28] Cleary et al.[203] found that between two modified β-D-glucans prepared from *S. cerevisiae*, those with more β-(1,6) side chains (20% (w/w) of the total molecule) led to higher stimulation of macrophages than their less branched counterparts (5% branches). This was ascribed to the higher ability of (highly) branched glucans to cross-link glucan receptors in macrophages, thus increasing the immunostimulatory effect.[203]

The absence of branching often impedes immunomodulatory activity (e.g., of curdlan), because long, unbranched polysaccharides tend to be insoluble, and solubility seems to play an important role in polysaccharide–cell interactions. Interestingly, branched (modified) curdlans exhibited significant antitumor activity.[2] Nevertheless, branched curdlans did not have the ability to stimulate macrophages and induce phagocytosis,[181] which indicates that the structural requirements (of polysaccharides) for stimulation of immune cells may be different than those for antitumor activity or other bioactivities.

The effect of molecular weight on the biological functionality of microbial polysaccharides is also surrounded by some controversy. High MW usually promotes antitumor and immunomodulating activity of glucans. It has been suggested that size (MW) is more important than structure in microbial glucans, because they usually act via a nonspecific immune mechanism.[2] Furthermore, molecular weight can affect the ordered conformation of glucans: high MW schizophyllan (above 100,000 Da) exists in a triple helical structure, whereas low MW schizophyllan (below 50,000 Da) forms a single helix, but not a triple helix.[204] Schizophyllan[204] and the glucans isolated from *Phytophthora parasitica*[30] exhibit high antitumor activity when their MW is high (100,000 to 200,000 Da), while their lower MW fractions (5000 to 50,000 Da) are reported to be biologically inactive. On the contrary, the low MW (10,000 to 20,000 Da) β-glucans of *Pythium aphanidermaticum* have significant antitumor effects.[11] A positive link between high MW and immunomodulatory activity of *S. cerevisiae* particulate β-D-glucans was observed when glucans with an MW of 500,000 to 4,000,000 Da were compared.[203] Conversely, Sandula et al.[169] reported that solubilized (carboxymethylated and sulfoethylated) glucans of the same yeast had high mitogenic, radioprotective, and antimutagenic activity when their MW was reduced ultrasonically (to 90,000 to 100,000 Da, from an initial 300,000 to 600,000 Da). The contradiction in the last two examples suggests that the way MW affects the biological functionality of glucans depends partly on the solubility of the molecule.

In variance with β-(1,3)-glucans, the immunopharmacological functionality of heteroglycans is usually not (strongly) influenced by differences in MW.[43,67,205] The ability of α-(1,3)-glucuronoxylomannans of *Tremella fuciformis* to induce interleukin-6 production by human monocytes resulted mainly from the existence of an α-(1,3) backbone and was barely influenced by the molecular weight of the biopolymer.[205] Tumor inhibition by levans (fructans) of *Zymomonas mobilis* was dependent on MW, with the molecular fraction of 211,000 Da being the most active.[87] Levan fractions with higher or lower MW were less active, which suggests that the isolation and purification procedures have an impact on obtaining the most preferable molecular fractions.

The bioactivity of alginates is also affected by the molecular size of the polysaccharide. High MW is desirable for traditional food applications where it is used as a thickener or gelling agent, but low MW alginates are more soluble, and thus preferable when solutions of low viscosity are prepared.[206] Kimura et al.[206] studied the functional properties of sodium alginates of 10,000 (AG-1), 50,000 (AG-5), and 100,000 (AG-10) Da produced from a natural, high MW sodium alginate (2,700,000 Da, AG-270) after heat treatment at 130°C (for 250, 120, and 100 min, respectively). They found that AG-5 and AG-10 increased cholesterol excretion into feces and lowered blood glucose and insulin levels in a manner similar to that of natural alginate (AG-270), while AG-5 intake reduced blood glucose but did not affect insulin levels. AG-1 increased neither cholesterol excretion nor glucose tolerance. This implies that intense and prolonged heat treatment of alginate in a food environment should be avoided for its functional properties to be preserved.

The composition of alginate is of even greater importance for its bioactivity and functionality. High content of mannuronic acid is connected to higher induction of cytokinesis and antitumor activity, while guluronic acid appears to be nonstimulating for the immune system.[98,99] In fact, it has been claimed that in pharmaceutical preparations (pure solutions), guluronate should be eliminated because it may cause unwanted side effects, such as antibody generation.[101] On the other hand, guluronate is much more effective in binding radioactive molecules and heavy metals than mannuronate, and alginate with a high (G)/(M) ratio is an effective antitoxic, radioprotective agent. For example, a single administration of sodium alginate enriched with guluronic acid caused a fourfold decrease in the retention of strondium in humans.[98] Furthermore, an animal diet with alginate rich in guluronic acid resulted in reduced consumption of food and slower increase of body weight in comparison with a diet with mannuronate-rich alginate. However, liver adiposity could be better controlled with consumption of alginate saturated with mannuronic acid.[98]

The structural conformation and the presence of an ordered or nonordered structural domain play a substantial role in the functionality of many bioactive microbial polysaccharides, although different biopolymers are affected in different ways. The ordered structure in medicinal mushroom β-(1,3)-glucans is often considered essential for their immunostimulating functions.[5] When the triple helix of lentinan was denatured with dimethyl sulfoxide (DMSO), urea, or NaOH, the tertiary structure was lost and the antitumor activity was lowered, according to the degree of denaturation.[207] The loss of triple helical structure was also detrimental for the immunological effects of schizophyllan.[208] In a complete reversal of these results, Saito et al.[209] concluded that lentinan and schizophyllan were better immunostimulants when they acquired a single-helix conformation. These differences imply that a single helix and triple helix may both have immunostimulatory effects, but they have different modes of action and affect the immune system in different ways. For instance, macrophage nitric oxide synthesis and limulus factor G activation are dependent on this tertiary structure. However, other immunological effects, such as interferon synthesis and colony stimulating factor, do not depend on the tertiary structure.[5]

In other studies, immunostimulation by antitumor polysaccharide from *Pythium aphanidermatum* was higher when the biopolymer had a single helical conformation,[11] while the antitumor effects of glomerellan were independent of ordered structure.[210] Kulicke et al. examined scleroglucan from *Sclerotium rolfsii* and *Sclerotium glucanicum* and the fungal glucans synthesized by *Monilia fructigena* and *Monilia fructicola* and supported that immunological activity did not require or was not favored by helically ordered structures (the coil-like scleroglucan from *S. rolfsii* was the best activator of human blood monocytes).[211] Suzuki et al.[212] attempted to elucidate the impact of ordered structure on the immunological characteristics of microbial glucans such as curdlan, grifolan, schizophyllan, and SSG-glucan. They found that the triple helix activated the Alternative Pathways of Complement (APC) in the immune system more strongly than did the single helix, while single helical conformation was more efficient in stimulating the Classical Pathways of Complement (CPC). Activation of CPC by the single helix was dependent on the degree of branching and occurred via the binding of the glucan to immunoglobulin in serum.[212] Of course, both systems are crucial for the expression of immune responses, and structural conformation seems to influence the way glucans are recognized by each of these systems.

Another factor that influences the performance of bioactive microbial polysaccharides is their solubility and the chemical substitution (with sulfate, phosphate, carboxymethyl, or hydroxyl groups) of the original biopolymer. The degree of branching (DB) or the degree of substitution (DS) enhances water solubility, but only to a certain limit. The antitumor activity of many glucans is promoted by high solubility, which seems to depend primarily on the number and type of hydrophilic (anionic) groups positioned on the outer surface of the helix.[2,169] Particulate β-glucans of *S. cerevisiae* were solubilized after ultrasonication and chemical derivatization (with carboxyethyl and sulfoethyl compounds).[169] Carboxymethylated glucans with DS ranging from 0.4 to 1.15 were water soluble, while samples with DS below 0.4 were insoluble or only partly soluble. Optimal biological activity of carboxymethylated glucans was in the DS range of 0.6 to 0.8, and the immunomodulating activity of carboxymethyl glucans with DS above 1.0 gradually diminished. A comparatively lower DS was needed to achieve solubility of sulfoethyl *S. cerevisiae* glucans; glucans with DS = 0.3 were totally soluble when sulfoethyl groups were introduced, indicating the importance of the ionic substituents.[169]

The role of chemical substitution with ionic groups in enhancing biological activity has been demonstrated in many studies. Curdlan sulfate[2,189] and sulfated schizophyllans[213] inhibited the growth of HIV virus, depending on their sulfur content. In fact, the latter was the most determinant in HIV treatment by SPG, compared to the molecular weight or the chemical nature of the polysaccharide component.[213] Other examples of enhanced immunostimulation after solubilization by chemical substitution include the carboxymethylated β-glucan from *Pleurotus ostreatus* (pleuran)[214] and the carboxylated linear α-(1,3)-glucans from the fruit bodies of *Amanita muscaria* and *Agrocybe aegerita.*[215] Carboxymethylated SSG displayed antitumor properties with DSs up to 0.14, above which its activity gradually decreased. It was postulated that this might be due to changes in the electric charge that modulates the electrostatic binding of the polysaccharide to a receptor or ligant.[216]

Another form of derivatized biopolymers with increased solubility and immunological activity are polyols. Glucan polyols or glucan polyaldehydes of grifolan, pestalotan, and *Auricularia* sp. glucan, formed after oxidation and subsequent reduction of the original polymeric structure, were characterized by enhanced antitumor potency, which was a function of the polyol content.[5,19,27]

Despite the above literature on the positive link between solubility and polysaccharide functionality, some microbial polysaccharides are more effective or only active in insoluble (particulate) form. For instance, the particulate β-(1,3)-glucan from *Candida albicans* cell wall was more active in stimulating human leukocytes, specifically TNF-α, interleukin-α, and hydrogen peroxide production,[45] than dimethyl sulfoxide-soluble samples. Also, lentinan was effective against mouse sarcoma (80 to 98% inhibition), but hardly soluble in water.[174] In addition to this, particulate (and modified) glucans of *S. cerevisiae* that were insoluble even for low MW preparations proved to be efficient stimulators of macrophages.[203]

The presence of ionic groups not only favors the solubility of microbial polysaccharides, but also changes their charge and perhaps their ability to bind to other molecules or cells. In LAB functional polysaccharides it appears that the presence of phosphate groups determines the extent of biological activity. Mitogenic activity was lost after dephosphorylation of *Lactobacillus bulgaricus* phosphopolysaccharide, while a similar but neutral EPS by the same microorganism had no mitogenic activity. Although the exact role of phosphate groups in LAB EPS is not clear, it may be connected to direct stimulation of B-cell mitogenicity, as was shown for inorganic sulfate, or the ability to adhere strongly to intestinal epithelian cells, as was evident for *Lactobacillus cremoris* phosphopolysaccharide.[76,82,84,186] It has to be noted, however, that it is still not known whether and to what extent these phosphate groups retain their functionality after interactions with other polysaccharides, or with proteins, or through pH changes, as might occur in a real food environment.

Although there is a fair amount of literature on the structure–functionality relationship of many microbial polysaccharides, as shown above, most of these studies examine these molecules from a pharmacological perspective, where the pure bioactive compound is tested *in vitro* or administered in soluble or particulate preparations directly into the body of an animal or patient. To date there has been only limited research on how these biopolymers act in a real food environment (incorporated as food ingredients) and how food processing practices (pasteurization, freezing, etc.) and the interaction with other food compounds may affect their structure, MW, physicochemical characteristics, and eventually their functional properties. These are parameters that have to be taken into account in designing and marketing functional food products with certain bioactive function.

Hromadkova et al.[163] compared the effect of different drying methods, namely, solvent exchange (GE), lyophilization (GL), and spray drying (GS), on the biological activity of the particulate form of glucans from *S. cerevisiae*. They observed that immunomodulatory activity was twice as high with GS as with GL and GE. Also, GS dispersion had the lowest apparent viscosity, which perhaps denotes a relationship between biopolymer functionality and rheological behavior of aqueous dispersions of the polysaccharide isolates. Thus, it was concluded that spray drying is the most suitable method for preparing particulate β-glucan suspensions.

The effect of mechanical shear on polysaccharide MW and hydration properties is another important aspect. Although reports on microbial polysaccharides are scarce, it has been demonstrated for hypocholesterolemic cereal β-glucans that mechanical shear can reduce their molecular mass without having an adverse effect on their ability to reduce plasma cholesterol.[217] Other processing parameters such as temperature (heating), pressure, or combined thermomechanical treatments (e.g., extrusion) may change the structure, porosity, solubility, and water retention properties of microbial polysaccharides added in food, as has been shown for polysaccharides of plant origin.[218,219]

5.6 FUTURE PERSPECTIVES

Certainly, more research regarding the effectiveness and fate of microbial polysaccharides in a food matrix is needed, to test their biological activity after food consumption, to identify the optimal process conditions for production and stability of functional food products, and to choose the most suitable biopolymers in this context. Along with the bioactive properties of these microbial biopolymers, their thermal, rheological, and mechanical properties must be evaluated under conditions encountered during food processing operations. Existing examples of plant functional polysaccharides and their functionality as food components under various food processing conditions may be used as a guide in this effort.[218,219] The testing of applicability and effectiveness of microbial polysaccharides as food ingredients should take into consideration not only the health benefits of microbial biopolymers as these are exhibited *in vitro*, but also the molecular interactions with other major food components and the physicochemical and sensory attributes of the final (functional) food product. This is an area needing further research. Also, more clinical tests in humans with a number of potential bioactive polysaccharides are expected to boost the production of functional foods. Only a few of the bioactive polysaccharides described in this review have been tested in humans (patients), and their effectiveness and potential side effects in realistic conditions have to be evaluated before their acceptance by the public. Taking into account the lengthy procedures of clinical trials usually needed before new health-related products are granted approval by the FDA or other regulatory organisms, the potential application of polysaccharides, especially from safe (GRAS) microorganisms already used in food or as food (such as polysaccharides from edible mushrooms, *S. cerevisiae*, or lactic acid bacteria), gains particular interest, because the products of such safe microorganisms would probably need less time to make their way to the market.

The optimization of bioprocess conditions and the study of microbial physiology are necessary for rendering microbial polysaccharides more economical and readily available to the food market. Many of the producer microorganisms are not well studied by the biotechnologists, and there is great scope for metabolic engineering and enhancement of polysaccharide synthesis, which will make microbial bioactive polysaccharides very competitive (high yield and purity) against those derived from plants. The close control of process (fermentation) conditions by on-line or at-line techniques would contribute to this purpose.[220–222] The utilization of unconventional, nondefined growth media, such as molasses, corn syrups, or dairy whey, can also offer alternative routes for their production at reduced production costs.[130,168] Moreover, advances in molecular biology and the manipulation of microbial genetics may allow the design of tailor-made and less expensive microbial polysaccharides for broad use as nutraceuticals or functional food ingredients in the future. Research in this field, although relatively limited for microbial polysaccharides, includes studies on EPSs from lactic acid bacteria and bacterial alginates.[99,119,223,224] Genes homologous to proteins involved in polysaccharide biosynthesis and genes encoding the key enzymes, glycosyl transferases, have been identified and studied by heterologous expression and functional studies. This offers the possibility for control and overexpression of these genes in the host microorganism or other microbial models, which can enhance polysaccharide production and also lead to structurally engineered biopolymers. Generating designer polysaccharides may prove to be vital for the utilization of these biopolymers in functional foods, since their biologic activity is closely linked to their structure.

All of the above will contribute to a better understanding of the production and practical application and use of functional microbial polysaccharides and, finally, to their wider acceptance by the market and consumers. The latter, after all, will determine the success of the new generation of functional foods and nutraceuticals formulated with these biopolymers.

REFERENCES

1. Hobbs, C., *Medicinal Mushrooms: An Exploration of Tradition, Healing and Culture*, Botanica Press, Santa Cruz, CA, 1995.
2. Bohn, J.A. and BeMiller, J.N., (1→3)-β-D-Glucans as biological response modifiers: a review of structure-functional activity relationships, *Carbohydr. Polym.*, 28, 3, 1995.
3. Kwak, N.S. and Jukes, D.J., Functional foods. Part 1. The development of a regulatory concept, *Food Contrib.*, 12, 99, 2001.
4. Hardy, G., Nutraceuticals and functional foods: introduction and meaning, *Nutrition*, 16, 688, 2000.
5. Wasser, S.P., Medicinal mushrooms as a source of antitumor and immunomodulating polysaccharides, *Appl. Microbiol. Biotechnol.*, 60, 258, 2002.

6. Oda, M., Hasegawa, H., Komatsu, N., Kambe, M., and Tsuchiya, F., Antitumor polysaccharide from *Lactobacillus* sp., *Agric. Biol. Chem.*, 47, 1623, 1983.
7. Reshetnikov, S.V., Wasser, S.P., and Tan, K.K., Higher Basidiomycota as a source of antitumor and immunostimulating polysaccharides, *Int. J. Med. Mushrooms*, 3, 361, 2001.
8. Kiho, T., Katsuragawa, M., Nagai, K., Ukai, S., and Haga, M., Structure and antitumor activity of a branched (1→3)-β-D-glucan from the alkaline extract of *Amanatia muscaria*, *Carbohydr. Res.*, 224, 237, 1992.
9. Nanba, H. and Kuroda, H., Potentiating effect of β-glucan from *Cochliobolus miyabenus* on host-mediated antitumor activity in mice, *Chem. Pharm. Bull.*, 35, 1289, 1987.
10. Kiho, T., Ito, M., Yoshida, I., Nakai, K., Hara, C., and Ukai, S., Polysaccharides in fungi. XXIV. A (1→3)-β-D-glucan from the alkaline extract of the insect-body portion of Chan hua (fungus *Cordyceps cicadae*), *Chem. Pharm. Bull.*, 37, 2770, 1989.
11. Blaschek, W., Kasbauer, J., Kraus, J., and Franz, G., *Pythium aphanidermatum*: culture, cell wall composition, and isolation and structure of antitumor storage and solubilised cell wall (1→3), (1→6)-β-D-glucans, *Carbohydr. Res.*, 231, 293, 1992.
12. Usui, T., Iwasaki, Y., Hayashi, K., Mizuno, T., Tanaka, M., Shinkai, K., and Arawaka, M., Antitumor activity of water-soluble β-D-glucan elaborated by *Ganoderma applanatum*, *Agric. Biol. Chem.*, 45, 323, 1981.
13. Yamada, H., Kawaguchi, N., Ohmori, T., Takeshita, Y., Taneya, S.-I., and Miyazaki, T., Structure and antitumor activity of an alkali-soluble polysaccharide from *Cordyceps ophioglossoides*, *Carbohydr. Res.*, 125, 107, 1984.
14. Lee, J., Lee, I., Kim, M., and Park, Y., Optimal control of batch processes for production of curdlan from *Agrobacterium* species, *J. Ind. Microbiol. Biotechnol.*, 23, 143, 1999.
15. Sone, Y., Okuda, R., Wada, T., Kishida, E., and Misaki, A., Structure and antitumor activities of the polysaccharides isolated from fruiting body and the growing culture of mycelium of *Ganoderma lucidum*, *Agric. Biol. Chem.*, 49, 2641, 1985.
16. Gomaa, K., Kraus, J., Franz, G., and Roper, H., Structural investigations of glucans from cultures of *Glomerella cingulata*, *Carbohydr. Res.*, 217, 153, 1991.
17. Manners, D.J., Mason, A.J., and Patterson, J.C., The structure of a β-(1→3)-D-glucan from yeast cell walls, *Biochem. J.*, 135, 19, 1973.
18. Sone, Y., Kakuta, M., and Mizaki, A., Isolation and characterization of polysaccharides of "Kikurage" fruit body of *Auricularia auricula-judae*, *Agric. Biol. Chem.*, 42, 417, 1978.
19. Misaki, A., Kakuta, M., Sasaki, T., Tanaka, M., and Miyaji, H., Studies of interrelation of structure and antitumor effects of polysaccharides: antitumor action of periodate-modified, branched (1→3)-β-D-glucan of *Auricularia auricula-judae* and other polysaccharides containing (1→3)-glycosidic linkages, *Carbohydr. Res.*, 92, 115, 1981.
20. Ohno, N., Adachi, Y., Suzuki, I., Sato, K., Oikawa, S., and Yadomae, T., Structural characterization and antitumor activity of the extract from matted mycelium of cultured *Grifola fondosa*, *Chem. Pharm. Bull.*, 33, 3395, 1985.
21. Nanba, H., Hamaguchi, A., and Kuroda, H., The chemical structure of an antitumor polysaccharide in fruit bodies of *Grifola fondosa* (Maitake), *Chem. Pharm. Bull.*, 35, 1289, 1987.
22. Miyazaki, T., Oikawa, N., Yamada, H., and Yadomae, T., Structural examination of antitumor, water-soluble glucans from *Grifola umbellata* by use of four types of glucanase, *Carbohydr. Res.* 65, 235, 1978.

23. Yoshioka, Y., Tabeta, R., Saito, H., Uehara, N., and Fukuoka, F., Antitumor polysaccharides from *P. ostreatus*: isolation and structure of a β-glucan, *Carbohydr. Res*, 140, 93, 1985.
24. Kitamura, S., Hori, T., Kurita, K., Takeo, K., Hara, C., Itoh, W., Tabata, K., Elgsaeter, A., and Stokke, B.T., An antitumor, branched (1→3)-β-D-glucan from a water extract of fruiting bodies of *Cryptoporus volvatus*, *Carbohydr. Res*, 263, 111, 1994.
25. Sasaki, T. and Takasaka N., Further study of the structure of lentinan, an anti-tumor polysaccharide from *Lentinus elodes*, *Carbohydr. Res.*, 47, 99, 1976.
26. Saito, H., Misaki, A., and Harada, T., A comparison of the structure of curdlan and pachyman, *Agric. Biol. Chem.*, 32, 1261, 1968.
27. Misaki, A., Kawaguchi, K., Miyaji, H., Nagae, H., Hokkoku, S., Kakuta, M., and Sasaki, T., Structure of pestalotan, a highly branched (1→3)-β-D-glucan elaborated by *Pestalotia* sp. 815, and the enhancement of its antitumor activity by polyol modification of the side chains, *Carbohydr. Res.*, 129, 209, 1984.
28. Jamas, S., Easson, D.D., Ostroff, G.R., and Onderdonk, A.B., PGG-glucans. A novel class of macrophage-activating immunomodulators, *ACS Symp. Ser.*, 469, 44, 1991.
29. Misaki, A., Antitumor polysaccharides. Interrelation between their structures and antitumor effects, *Chem. Abstr.*, 102, 55469a, 1984.
30. Fabre, I., Bruneteau, M., Ricci, P., and Michel, G., Isolation and structural study of *Phytophthora parasitica* glucans, *Eur. J. Biochem.*, 142, 99, 1984.
31. Blaschek, W., Schutz, M., Kraus, J., and Franz, G., *In vitro* production of specific polysaccharides: isolation and structure of an antitumor active β-glucan from *Phytophthora parasitica*, *Food Hydrocoll.*, 1, 371, 1987.
32. Bruneteau, M., Fabre, I., Perret, J., Michel, G., Ricci, P., Joseleau, J.P., Kraus, J., Schneider, M., Blaschek, W., and Franz, G., Antitumor active β-D-glucans from *Phytophthora parasitica*, *Carbohydr. Res.*, 175, 137, 1988.
33. Aouadi, S., Heyraud, A., Seigle-Murandi, F., Steiman, R., Kraus, J., and Franz, G., Structure and properties of an extracellular polysaccharide from *Laetisaria arvalis*. Evaluation of its antitumor activity, *Carbohydr. Polym.*, 16, 155, 1991.
34. Ohno, N., Miura, T., Suzuki, I., and Yadomae, T., Antitumor activity and structural characterization of polysaccharide fractions extracted with cold alkali from a fungus, *Peziza vesiculosa*, *Chem. Pharm. Bull.*, 33, 2564, 1985.
35. Ohno, N., Miura, T., Suzuki, I., and Yadomae, T., Purification, antitumor activity and structural characterization of β-(1→3)-glucan from *Peziza vesiculosa*, *Chem. Pharm. Bull.*, 33, 5096, 1985.
36. Tabata, K., Ito, W., Kojima, T., Kawabata, S., and Misaki, A., Ultrasonic degradation of schizophyllan, an antitumor polysaccharide produced by *Schizophyllum commune* Fries, *Carbohydr. Res.*, 89, 121, 1981.
37. Rinaudo, M. and Vincendon, M., ^{13}C NMR structural investigation of scleroglucan, *Carbohydr. Polym.*, 2, 135, 1982.
38. Ohno, N., Suzuki, I., and Yadomae, T., Structure and antitumor activity of a β-(1→3)-glucan isolated from the culture filtrate of *Sclerotinia sclerotiorum* IFO 9395, *Chem. Pharm. Bull.*, 34, 1362, 1986.
39. Hara, C., Kumazawa, Y., Inagaki, K., Kareko, M., Kiho, T., and Ukai, S., Mitogenic and colony-stimulating factor-inducing activities of polysaccharide fractions from the fruit bodies of *Dictyophora indusiata* Fisch, *Chem. Pharm. Bull.*, 39, 1615, 1991.
40. Defaye, J., Kohlmunzer, S., Sodzawiczny, K., and Wong, E., Structure of an antitumor, water-soluble glucan from the carpophores of *Tylpilus felleus*, *Carbohydr. Res.*, 173, 316, 1988.

41. Misaki, A., Nasu, M., Sone, Y., Kishida, E., and Kinoshita, C., Comparison of structure and antitumor activity of polysaccharides isolated from Fukorotake, the fruiting body of *Volvariella volvacea*, *Agric. Biol. Chem.*, 50, 2171, 1986.
42. Kishida, E., Sone, Y., and Misaki, A., Purification of an antitumor-active, branched (1→3)-β-D-glucan from *Volvariella volvacea*, and elucidation of its fine structure, *Carbohydr. Res.*, 193, 227, 1989.
43. Mizuno, T., Hagiwara, T., Nakamura, T., Ito, H., Shimura, K., Sumiya, T., and Asakura, A., Antitumor activity and some properties of water-soluble polysaccharides from "Himematsutake", the fruiting body of *Agaricus blazei* Murrill, *Agric. Biol. Chem.*, 54, 2889, 1990.
44. Sasaki, T., Abiko, N., Nitta, K., Takasuka, N., and Sugino, Y., Antitumor activity of carboxymethylglucans obtained by carboxymethylation of (1→3)-β-D-glucan from *Alcaligenes faecalis* var. *myxogenes* IFO 13140, *Eur. J. Cancer*, 15, 211, 1979.
45. Ishibashi, K., Miura, N.N., Adachi, Y., Ogura, N., Tamura, H., Tanaka, S., and Ohno, N., Relationship between the physical properties of *Candida albicans* cell wall β-glucan and activation of leucocytes *in vitro*, *Intern. Immunopharm.*, 2, 1109, 2002.
46. Hwan, M.K., Sang, B.H., Goo, T.O., Young, H.K., Dong, H.H., Nam, D.H., and Ick, D.Y., Stimulation of humoral and cell mediated immunity by polysaccharide from mushroom *Phellinus linteus*, *Int. J. Immunopharm.*, 18, 295, 1996.
47. Zhang, M., Cheung, P.C.K., and Zhang, L., Evaluation of mushroom dietary fiber (nonstarch polysaccharides) from sclerotia of *Pleurotus tuber-regium* (Fries) Singer as a potential antitumor agent, *J. Agric. Food Chem.*, 49, 5059, 2001.
48. Maeda, Y.Y., Takahama, S., and Yonekawa, H., Four dominant loci for the vascular responses by the antitumor polysaccharide, lentinan, *Immunogenetics*, 47, 159, 1998.
49. Misaki, A., Kishida, E., Kakuta, M., and Tabata, K., Antitumor fungal β-(1→3)-D-glucans: structural diversity and effects of chemical modification, in *Carbohydrates and Carbohydrate Polymers*, Yalpani, M., Ed., ATL Press, Mount Prospect, IL, 1993, p. 116.
50. Giavasis, I., Harvey, L.M., and McNeil, B., Scleroglucan, in *Biopolymers*, Vol. 8, Steinbuchel, A., Ed., Wiley-VCH Verlag GmbH, Weinheim, 2002, chap. 2.
51. Lecacheux, D., Mustiere, Y., and Panaras, R., Molecular weight of scleroglucan and other extracellular microbial polysaccharides by size-exclusion chromatography and low angle laser scattering, *Carbohydr. Polym.*, 6, 477, 1986.
52. Falch, B.H., Espevik, T., Ryan, L., and Stokke, B.T., The cytokine stimulating activity of (1→3)-β-D-glucans is dependent on the triple helix conformation, *Carbohydr. Res.*, 329, 587, 2000.
53. Bluhm, C., Deslandes, Y., Marchessault, R., Perz, S., and Rinaudo, M., Solid-state and solution conformations of scleroglucan, *Carbohydr. Res.*, 100, 117, 1982.
54. Hjerde, T., Stokke, B.T., Smidsrod, O., and Christensen, B.E., Free-radical degradation of triple-stranded scleroglucan by hydrogen peroxide and ferrous ions, *Carbohydr. Polym.*, 37, 41, 1998.
55. Ogawa, K. and Toshifumi, Y., X-ray diffraction of polysaccharides, in *Polysaccharides: Structural Diversity and Functional Versatility*, Dumitriu, S., Ed., Marcel Dekker, New York, 1998, p. 101.
56. Adachi, Y., Suzuki, Y., Ohno, N., and Yadomae, T., Adjuvant effect of grifolan on antibody production in mice, *Biol. Pharm. Bull.*, 21, 974, 1998.
57. Osawa, Z., Morota, T., Hatanaka, K., Akaike, T., Matsusaki, K., Nakashima, H., Yamamoto, N., Suzuki, E., Miyano, H., Mimura, T., and Kaneko, Y., Synthesis of sulphated derivatives of curdlan and their anti-HIV activity, *Carbohydr. Polym.*, 21, 283, 1993.

58. Demleiter, S., Kraus, J., and Franz, G., Synthesis and antitumor activity of derivates of curdlan and lichenan branched at C-6, *Carbohydr. Res.*, 226, 239, 1992.
59. Tokunaka, K., Ohno, N., Adachi, Y., Tanaka, S., Tamura, H., and Yadomae, T., Immunopharmacological and immunotoxicological activities of a water-soluble β-(1→3)-D-glucan, CSBG, from *Candida* spp., *Int. J. Immunopharm.*, 22, 383, 2000.
60. Williams, D.L., Pretus, H.A., McNamee, R.B., Jones, E.L., Ensley, H.E., and Browder, I.W., Development of a water-soluble, sulfated (1→3)-β-D-glucan biological response modifier derived from *Saccharomyces cerevisiae*, *Carbohydr. Res.*, 235, 247, 1992.
61. Ohno, N., Miura, T., Miura, N.N., Adachi, Y., and Yadomae, T., Structure and biological activities of hypochlorite oxidized zymosan, *Carbohydr. Polym.*, 44, 339, 2001.
62. Miura, N.N., Ohno, N., Adachi, Y., and Yadomae, T., Characterization of sodium hypochloride degradation of β-glucan in relation to its metabolism *in vivo*, *Chem. Pharm. Bull.*, 44, 2137, 1996.
63. Hiroshi, S. and Takeda, M., Diverse biological activity of PSK (Krestin), a protein-bound polysaccharide from *Coriolous versicolor* (Fr.) Quel, in *Mushroom Biology and Mushroom Products*, Chang, S.T., Buswell, J.A., and Chiu, S.W., Eds., Chinese University Press, Hong Kong, 1993, p. 237.
64. Ooi, V.E.C. and Liu, F., Immunomodulation and anti-cancer activity of polysaccharide-protein complexes, *Curr. Med. Chem.*, 7, 715, 2000.
65. Fujimiya, Y., Suzuki, Y., Oshiman, K., Kobori, H., Moriguchi, K., Nakashima, H., Matumoto, Y., Takahara, S., Ebina, T., and Katakura, R., Selective tumoricidal effect of soluble proteoglucan extracted from the basidiomycete *Agaricus blazei* Murill, mediated via the natural killer cell activation and apoptosis, *Cancer Immunol. Immunother.*, 46, 147, 1998.
66. Cho, S.M., Park, J.S., Kim, K.P., Cha, D.Y., Kim, H.M., and Yoo, I.D., Chemical features and purification of immunostimulating polysaccharides from the fruit bodies of *Agaricus blazei*, *Kor. J. Mycol.*, 27, 170, 1999.
67. Mizuno, T., Medicinal properties and clinical effects of culimary-medical mushroom Agaricus blazei Murrill (Agaricomycetidae), *Int. J. Med. Mushrooms*, 4, 32, 2002.
68. Hikichi, M., Hiroe, E., and Okubo, S., European Patent 0939082, 1999.
69. Tsuchida, H., Mizuno, M., Taniguchi, Y., Ito, H., Kawade, M., and Akasaka, K., Japanese Patent 11-080206, 2001.
70. Bao, X.F., Wang, X.S., Dong, Q., Fang, J.N., and Li, X.Y., Structural features of immunologically active polysaccharides from *Ganoderma lucidum*, *Phytochemistry*, 59, 175, 2002.
71. Lee, K.H., Kang, T.S., Moon, S.O., Lew, I.D., and Lee, M.Y., Fractionation and antitumor activity of the water soluble exo-polysaccharide by submerged cultivation of *Ganoderma lucidum* mycelium, *Kor. J. Appl. Microbiol. Biotechnol.*, 24, 459, 1996.
72. Eo, S.K., Kim, Y.S., Lee, C.K., and Han, S.S., Possible mode of antiviral acivity of acidic protein bound polysaccharide isolated from *Ganoderma lucidum* on herpes simplex viruses, *J. Ethnopharm.*, 72, 475, 2000.
73. Bao, X., Duan, J., Fang, X., and Fang, J., Chemical modification of the (1→3)-α-D-glucan from spores of *Ganoderma lucidum* and investigation of their physicochemical properties and immunological activity, *Carbohydr. Res.*, 336, 127, 2001.
74. Bao, X., Liu, C., Fang, J., and Li, X., Structural and immunological studies of a major polysaccharide from spores of *Ganoderma lucidum* (Fr.) Karst, *Carbohydr. Res.*, 332, 67, 2001.

75. Adachi, S., Lactic acid bacteria and the control of tumours, in *The Lactic Acid Bacteria in Health and Disease*, Wood, J.B., Ed., Elsevier Applied Science, London, 1992, p. 233.
76. Kitazawa, H., Harata, T., Uemura, J., Saito, T., Kaneko, T., and Itoh, T., Phosphate group requirement for mitogenic activation of lymphocytes by an extracellular phosphopolysaccharide from *Lactobacillus dulbreeckii* sp. *bulgaricus*, *Int. J. Food Microbiol.*, 31, 99, 1998.
77. Oda, M., Hasegawa, H., Komatsu, S., and Tsuchiya, F., Anti-tumor polysaccharide from *Lactobacillus* sp., *Agric. Biol. Chem.*, 47, 1623, 1983.
78. Matsuzaki, T., Nagaoka, M., Nomoto, K., and Yokokura, T., Antitumor effect of polysaccharide-peptidoglycan complex (PS-PG) of *Lactobacillus casei* YIT 9018 on Meth A, *Pharmacol. Ther.*, 18, 51, 1990.
79. Kitazawa, H., Toba, T., Itoh, T., Kumano, N., Adachi, S., and Yamaguchi, T., Antitumoral activity of slime-forming encapsulated *Lactococcus lactis* sp. *cremoris* isolated Scandinavian ropy sour milk "viili," *Anim. Sci. Technol.*, 62, 277, 1991.
80. Hosono, A., Lee, J., Ametani, A., Natsume, M., Hirayama, M., Adachi, T., and Kaminogawa, S., Characterization of a water-soluble polysaccharide fraction with immunopotentiating activity from *Bifidobacterium adolescentis* M101-4, *Biosci. Biotechnol. Biochem.*, 61, 312, 1997.
81. Sreekumar, O. and Hosono, A., The antimutagenic properties of a polysaccharide produced by *Bifidobacetrium longum* and its cultured milk against some heterocyclic amines, *Can. J. Microbiol.*, 44, 1029, 1998.
82. Kitazawa, H., Ishii, Y., Uemura, J., Kawai, Y., Saito, T., Kaneko, T., Noda, K., and Itoh, T., Augmentation of macrophage functions by an extracellular phosphopolysaccharide from *Lactobacillus delbrueckii* sp. *bulgaricus*, *Food Microbiol.*, 17, 109, 2000.
83. Oba, T., Doesburg, K.K., Iwasaki, T., and Sikkema, J., Identification of biosynthetic intermediates of the extracellular polysaccharide viilian in *Lactococcus lactis* subspecies *cremoris* SBT 0495, *Arch. Microbiol.*, 171, 343, 1999.
84. Nakajima, H., Toba, T., and Toyoda, S., Enhancement of antigen-specific antibody production by extracellular slime products from slime-forming *Lactococcus lactis* subspecies *cremoris* SBT 0495 in mice, *Int. J. Food Microbiol.*, 25, 153, 1995.
85. Bekers, M., Laukevich, J., Karsakevich, A., Ventina, E., Kaminska, E., Upite, D., Vina, I., Linde, R., and Scherbaka, R., Levan-ethanol biosynthesis using *Zymomonas mobilis* cells immobilized by attachment and entrapment, *Proc. Biochem.*, 36, 979, 2001.
86. Abraham, E. and Robinson, A., Oral immunization with bacterial polysaccharide and adjuvant enhances antigen-specific pulmonary secretory antibody response and resistance to pneumonia, *Vaccine*, 9, 757, 1991.
87. Calazans, G.M.T., Lima, R.C., de Franca, F.P., and Lopes, C.E., Molecular weight and antitumour activity of *Zymomonas mobilis* levans, *Int. J. Biol. Macromol.*, 27, 245, 2000.
88. Bekers, M., Laukevich, J., Upite, D., Kaminska, E.,Vigants, A., Viesturs, U., Pankova, L., and Danilevich, A., Fructooligosaccharides and levan producing activity of *Zymomonas mobilis* extracellular levansucrase, *Proc. Biochem.*, 38, 701, 2002.
89. Duncan, C.J.D., Pugh, N., Pasco, S.P., and Ross, S.A., Isolation of a galactomannan that enhances macrophage activation from the edible fungus *Morchella esculenta*, *J. Agric. Food. Chem.*, 50, 5683, 2002.

90. Mizuno, M., Shiomi, Y., Minato, K., Kawakami, S., Ashida, H., and Tsuchida, H., Fucogalactan isolated from *Sarcodon aspratus* elicits release of tumor necrosis factor-α and nitric oxide from murine macrophages, *Immunopharmacology*, 46, 113, 2000.
91. Zhang, J., Wang, G., Li, H., Zhuang, C., Mizuno, T., Ito, H., Suzuki, C., Okamoto, H., and Li, J., Antitumor polysaccharides from Chinese mushroom "Yuhuahgmo", the fruiting body of *Pleurotus citrinopileatus*, *Biosci. Biotechnol. Biochem.*, 58, 1195, 1994.
92. Cho, S.M., Koshino, H., Yu, S.H., and Yoo, I.D., A mannofucogalactan, fomitellan A, with mitogenic effect from fruit bodies of *Fomitella fraxinea*, *Carbohydr. Polym.*, 37, 13, 1998.
93. Ukai, S., Kiho, T., Hara, C., Morita, M., Goto, A., Imaizumi, N., and Hasegawa, Y., Polysaccharides in fungi. XII. Antitumor activity of various polysaccharides isolated from *Dictyophora indusiata*, *Ganoderma japonicum*, *Cordyceps cicadae*, *Auricularia auricula-judae* and *Auricularia* sp., *Chem. Pharm. Bull.*, 31, 741, 1983.
94. Hara, C., Kumazawa, Y., Inagaki, K., Kaneko, M., Kiho, T., and Ukai, S., Mitogenic and colony stimulating factor-inducing activities of polysaccharide fractions from the fruit bodies of *Dictyophora indusiata* Fisch, *Chem. Pharm. Bull.*, 39, 1615, 1991.
95. Zhuang, C., Mizuno, T., Shimada, A., Ito, H., Suzuki, C., Mayuzumi, Y., Okamoto, H., Ma, Y., and Li, J., Antitumor protein-containing polysaccharides from a Chinese mushroom Fengweigu or Houbitake, *Pleurotus sajor-caju* (Fr.) Sing, *Biosci. Biotechnol. Biochem.*, 57, 901, 1993.
96. Zhuang, C., Mizuno, T., Ito, H., Shimura, K., Sumiya, T., and Kawade, M., Antitumor activity and immunological property of polysaccharides from the mycelium of liquid-cultured *Grifola fondosa*, *Nippon Shokuhin Kogyo Gakkaishi*, 41, 724, 1994.
97. Mizuno, T., Bioactive substances in *Heriium erinaceus* Pers (Yamabushitake), and its medical utilization, *Int. J. Med. Mushrooms*, 1, 105, 1999.
98. Khotimchenko, Y.S., Kovalev, V.V., Savchenko, U.V., and Ziganshina, O.A., Physical-chemical properties, physiological activity, and usage of alginates, the polysaccharides of brown algae, *Russ. J. Marine Biol.*, 27, S53, 2001.
99. Sabra, W., Zeng, A.P., and Deckwer, W.D., Bacterial alginate: physiology, product quality and process aspects, *Appl. Microbiol. Biotechnol.*, 56, 315, 2001.
100. Annison, G. and Couperwhite, I., Influence of calcium on alginate production and composition in continuous cultures of *Azotobacter vineladii*, *Appl. Microbiol. Biotechnol.*, 25, 55, 1986.
101. Skjak-Braek, G., Alginate: biosynthesis and some structure-function relationships relevant to biomedical and biotechnological applications, *Biochem. Soc. Trans.*, 20, 27, 1992.
102. Jansson, P.E., Kenne, L., and Lindberg, B., Structure of the extracellular polysaccharide from *Xanthomonas campestris*, *Carbohydr. Res.*, 45, 275, 1975.
103. Stokke, B.T., Christensen, B.E., and Smidsrod, O., Macromolecular properties of xanthan, in *Polysaccharides: Structural Diversity and Functional Versatility*, Dumitriu, S., Ed., Marcel Dekker, New York, 1998, p. 433.
104. Harvey, L.M. and McNeil, B., Thickeners of microbial origin, in *Microbiology of Fermented Foods*, 2nd ed., Vol. 1, Wood, B.J.B., Ed., Blackie Academic & Professional, London, 1988, p. 150.
105. Levrat-Verny, M.A., Behr, S., Mustad, V., Remesy, C., and Demigne, C., Low levels of viscous hydrocolloids lower plasma cholesterol in rats primarily by impairing cholesterol absorption, *J. Nutr.* 130, 243, 2000.

106. Castro, I.A., Tirapegui, J., and Benedicto, M.L., Effects of diet supplementation with three soluble polysaccharides on serum lipid levels of hypercholesterolemic rats, *Food Chem.*, 80, 323, 2003.
107. Sutherland, I.W., Bacterial exopolysaccharides: their nature and production, in *Surface Carbohydrates of the Procaryotic Cell*, Sutherlandk, I.W., Ed., Academic Press, London, 1997, p. 27.
108. Sutherland, I.W., Ed., *Biotechnology of Microbial Exopolysaccharides*, Cambridge University Press, Cambridge, U.K., 1990, chap. 2.
109. McNeil, B., Fungal biotechnology, in *Encyclopedia of Molecular Biology and Molecular Medicine*, Meyers, R., Ed., VCH, New York, 1996, p. 337.
110. Jann, K. and Jann, B., Biochemistry and expression of bacterial capsules, *Biochem. Soc. Trans.*, 19, 623, 1991.
111. Whitfield, C. and Valvano, M.A., Biosynthesis and expression of cell-surface polysaccharides, *Adv. Microb. Physiol.*, 35, 135, 1993.
112. Sharon, N. and Lis, H., Carbohydrates in cell recognition, *Sci. Am.*, 268, 82, 1993.
113. Cross, S.A., The biological significance of bacterial encapsulation, *Curr. Top. Microb. Immunol.*, 150, 87, 1990.
114. Roberts, I.S., Saunders, F.K., and Boulnois, G.J., Bacterial capsules and interactions with complements and phagocytes, *Biochem. Soc. Trans.*, 17, 462, 1989.
115. Veringa, E., Ferguson, D., Lambe, D., and Verhoef, J., The role of glycocalyx in surface phagocytosis of *Bacteroides* sp. in the presence and absence of clindamycin, *J. Antimicrob. Chemother.*, 23, 711, 1989.
116. Giavasis, I., Harvey, L.M., and McNeil, B., Gellan gum, *Crit. Rev. Biotechnol.*, 20, 177, 2000.
117. Rapp, P., 1,3-β-Glucanase, 1,6-β-glucanase and β-glucosidase activities of *Sclerotium glucanicum*: synthesis and properties, *J. Gen. Microbiol.*, 135, 2847, 1989.
118. Stephanopoulos, G., Aristidou, A., and Nielsen, J., Review of cellular metabolism, in *Metabolic Engineering: Principles and Methodologies*, Stephanopoulos, G., Aristidou, A., and Nielsen, J., Eds., Academic Press, New York, 1998, p. 21.
119. DeVuyst, L., De Vin, F., Vaningelgem, F., and Degeest, B., Recent developments in the biosynthesis and applications of heteropolysaccharides from lactic acid bacteria, *Int. Dairy J.*, 11, 687, 2001.
120. Kennedy, L., McDowell, K., and Sutherland, I., Alginases from *Azotobacter* species, *J. Gen. Microbiol.*, 138, 2465, 1992.
121. Kennedy, L. and Sutherland, I., Gellan lyases: novel polysaccharide lyases, *Microbiology*, 140, 3007, 1994.
122. Mattysse, A.G., White, S., and Lightfoot, R., Genes required for cellulose synthesis in *Agrobacterium tumefaciens*, *J. Bacteriol.*, 177, 1069, 1995.
123. Standal, R., Iversen, T.G., Coucheron, D.H., Fjaervik, E., Blatny, J.M., and Valla, S., A new gene required for cellulose production and a gene encoding cellulolytic activity in *Acetobacter xylinum* are colocalized with the *bcs* operon, *J. Bacteriol.*, 176, 665, 1994.
124. Degeest, B., Vaningelgem, F., and De Vuyst, L., Microbial physiology, fermentation kinetics, and process engineering of heteropolysaccharide production by lactic acid bacteria, *Int. Dairy J.*, 11, 747, 2001.
125. Cutfield, S.M., Davies, G.J., Murshudov, G., Anderson, B.F., Moody, P.C.E., Sullivan, P.A., and Cutfield, J.F., The structure of the exo-β-(1,3)-glucanase from *Candida albicans* in native and bound forms: relationship between a pocket and groove in family 5 glycosyl hydrolases, *J. Mol. Biol.*, 294, 771, 1999.

126. Minato, K., Mizuno, M., Terai, H., and Tsuchida, H., Autolysis of lentinan, an antitumor polysaccharide, during storage of *Lentinus elodes*, Shiitake mushroom, *J. Agric. Food Chem.*, 47, 1530, 1999.
127. Basco, R.D., Cueva, R., Andaluz, E., and Larriba, G., *In vivo* processing of the precursor of the major exoglucanase by KEX2 endoprotease in the *Saccharomyces cerevisiae* secretory pathway, *Biochim. Biophys. Acta*, 1310, 110, 1996.
128. Agrawal, P.B. and Pandit, A.B., Isolation of α-glucosidase from *Saccharomyces cerevisiae*: cell disruption and adsorption, *Biochem. Eng. J.*, 15, 37, 2003.
129. Hernandez, L.F., Espinosa, J.C., Fernandez-Gonzalez, M., and Briones, A., β-Glucosidase activity in a *Saccharomyces cerevisiae* wine strain, *Int. J. Food Microbiol.*, 80, 171, 2003.
130. Bekers, M., Linde, R., Danilevich, A., Kaminska, E., Upite, D., Vigants, A., and Scherbaka, R., Sugar beet diffusion juice and syrup as media for ethanol and levan production by *Zymomonas mobilis*, *Food Biotechnol.*, 63/64, 1595, 1999.
131. De Vuyst, L. and Degeest, B., Heteropolysaccharides from lactic acid bacteria, *FEMS Microbiol. Rev.*, 23, 153, 1999.
132. Fu, J.F. and Tseng, Y.H., Construction of lactose utilising *Xanthomonas campestris* and production of xanthan from whey, *Appl. Environ. Microbiol.*, 56, 919, 1990.
133. Schilling, B., *Sclerotium rolfsii* ATCC 15205 in continuous culture: economical aspects of scleroglucan production, *Bioproc. Eng.*, 22, 57, 2000.
134. De Vuyst, L., Van Loo, J., and Vandamme, E.J., Two-step fermentation process for improved xanthan production by *Xanthomonas campestris* NRRL-B-1459, *J. Chem. Technol. Biotechnol.*, 39, 263, 1987.
135. Taurhesia, S. and McNeil, B., Production of scleroglucan by *S. glucanicum* in batch and supplemented batch cultures, *Enz. Microbiol. Technol.*, 16, 223, 1994.
136. Shu, C.H., Chen, Y.C., and Hsu, Y.C., Effects of citric acid on cell growth and Schizophyllan formation in the submerged culture of *Schizophyllum commune*, *J. Chin. Inst. Chem. Eng.*, 33, 315, 2002.
137. Farina, J., Sineriz, F., Molina, O., and Perotti, N., High scleroglucan production by *Sclerotium rolfsii*: influence of medium composition, *Biotechnol. Lett.*, 20, 825, 1998.
138. Harvey, L.M., Production of Microbial Polysaccharides by the Continuous Culture of Fungi, Ph.D. thesis, University of Strathclyde, Glasgow, 1984.
139. Kim, M.K., Lee, I.Y., Ko, J.H., and Park, Y.H., Higher intracellular levels of uridinemonophosphate under nitrogen-limited conditions enhance metabolic flux of curdlan synthesis in *Agrobacterium species*, *Biotechnol. Bioeng.*, 62, 317, 1999.
140. Kennedy, J.F., Jones, P., Barker, S.A., and Banks, G.T., Factors affecting microbial growth and polysaccharide production during the fermentation of *Xanthomonas campestris* cultures, *Enz. Microb. Technol.*, 4, 39, 1982.
141. Larsson, C., von Stockar, U., Marison, I., and Gustafsson, L., Metabolic uncoupling in *Saccharomyces cerevisiae*, *Thermochim. Acta*, 251, 99, 1995.
142. Aguilar-Uscanga, B. and François, J.M., A study of the yeast cell wall composition and structure in response to growth conditions and mode of cultivation, *Lett. Appl. Microbiol.*, 37, 268, 2003.
143. Lee, K.M., Lee, S.Y., and Lee, H.Y., Bistage control of pH for improving exopolysaccharide production from mycelia of *Ganoderma lucidum* in an air-lift fermentor, *J. Biosci. Bioeng.*, 88, 646, 1999.
144. Fang, Q.H. and Zhong, J.J., Submerged fermentation of higher fungus *Ganoderma lucidum* for production of valuable bioactive metabolites: ganoderic acid and polysaccharide, *Biochem. Eng. J.*, 10, 61, 2002.

145. Cerning, J., Bouillanne, C., Landon, M., and Desmazeaud, M.J., Isolation and characterization of exopolysaccharides from slime-forming mesophilic lactic acid bacteria, *J. Dairy Sci.*, 75, 692, 1992.
146. Grobben, G.J., Chin-Joe, I., Kitzen, V.A., Boels, I.C., Boer, F., Sikkema, J., Smith, M.R., and De Bont, J.A.M., Enhancement of exopolysaccharide production by *Lactobacillus delbrueckii* subsp. *bulgaricus* NCFB 2772 with a simplified defined medium, *Appl. Environ. Microbiol.*, 64, 1333, 1998.
147. Grobben, G.J., Smith, M.R., Sikkema, J., and De Bont, J.A.M., Influence of fructose and glucose on the production of exopolysaccharides and the activities of enzymes involved in the sugar metabolism and the synthesis of sugar nucleotides in *Lactobacillus delbrueckii* subsp. *bulgaricus* NCFB 2772, *Appl. Microbiol. Biotechnol.*, 46, 279, 1996.
148. Wang, Y. and McNeil, B., The effect of temperature on scleroglucan synthesis and organic acid production by *Sclerotium glucanicum*, *Enz. Microb. Technol.*, 17, 893, 1995.
149. Casas, J.A., Santos, V.E., and Garcia-Ochoa, F.G., Xanthan gum production under several operational conditions: molecular structure and rheological properties, *Enz. Microb. Technol.*, 26, 282, 2000.
150. Shu, C.H. and Yang, S.T., Effects of temperature on cell growth and xanthan production in batch cultures of *Xanthomonas campestris*, *Biotechnol. Bioeng.*, 35, 454, 1990.
151. Mozzi, F., Oliver, G., Savoy De Giori, G., and Font De Valdez, G., Influence of temperature on the production of exopolysaccharides by thermophilic lactic acid bacteria, *Milchwissenshaft*, 50, 80, 1995.
152. Kang, K. and Cotrell, I., Polysaccharides, in *Microbial Technology*, 2nd ed., Peppler, H. and Perlman, D., Eds., Academic Press, New York, 1979, p. 417.
153. Wang, Y. and McNeil, B., pH effects on exopolysaccharide and oxalic acid production in cultures of *Sclerotium glucanicum*, *Enz. Microb. Technol.*, 17, 124, 1995.
154. Kapteyn, J.C., Ter Riet, B., Vink, E., Blad, S., De Nobel, H., Van Den Ende, H., and Kli, F.M., Low external pH induces HOG1-dependent changes in the organization of the *Saccharomyces cerevisiae* cell wall, *Mol. Microbiol.*, 39, 469, 2001.
155. Pena, C., Trujillo-Roldan, M.A., and Galindo, E., Influence of dissolved oxygen tension and agitation rate on alginate production and its molecular weight in cultures of *Azotobacter vinelandii*, *Enz. Microb. Technol.*, 27, 390, 2000.
156. Gassem, M.A., Schmidt, K.A., and Frank, J.F., Exopolysaccharide production from whey lactose by fermentation with *Lactobacillus delbrueckii* ssp. *bulgaricus*, *J. Food Sci.*, 62, 171, 1997.
157. Peters, H.U., Herbst, H., Hesselink, P., Lunsdorf, H., Schume, A., and Deckwer, W.D., The influence of agitation rate on xanthan production by *Xanthomonas campestris*, *Biotechnol. Bioeng.*, 34, 1393, 1989.
158. Rau, U., Gura, E., Olzewski, E., and Wagner, F., Enhanced glucan formation of filamentous fungi by effective mixing, oxygen limitation and fed-batch processing, *Ind. Microbiol.*, 9, 19, 1992.
159. Lawford, H. and Rousseau, J., Effect of oxygen on the rate of β-1,3-glucan microbial exopolysaccharide production, *Biotechnol. Lett.*, 11, 125, 1989.
160. Christensen, B.E., Myhr, M., and Smidsrod, O., The degradation of xanthan by hydrogen peroxide in the presence of ferrous ions. Comparison to acid hydrolysis, *Carbohydr. Res.*, 280, 85, 1996.
161. Yang, F.C. and Liau, C.B., The influence of environmental conditions on polysaccharide formation by *Ganoderma lucidum* in submerged cultures, *Process Biochem.*, 33, 547, 1998.

162. Tang, Y.J. and Zhong, J.J., Role of oxygen in submerged fermentation of *Ganoderma lucidum* for production of *Ganoderma* polysaccharide and ganoderic acid, *Enz. Microb. Technol.*, 32, 478, 2003.
163. Hromadkova, Z., Ebringerova, A., Sasinkova, V., Sandula, J., Hribalova, V., and Omelkova, J., Influence of the drying method on the physicochemical properties and immunomodulatory activity of the particulate (1→3)-β-D-glucan from *Saccharomyces cerevisiae*, *Carbohydr. Polym.*, 51, 9, 2003.
164. Krizkova, L., Durackova, Z., Sandula, J., Sasinkova, V., and Krajcovic, J., Antioxidative and antimutagenic activity of yeast cell wall mannans *in vitro*, *Mutat. Res.*, 497, 213, 2001.
165. Mau, J.L., Lin, H.C., and Chen, C.C., Antioxidant properties of several medicinal mushrooms, *J. Agric. Food. Chem.*, 50, 6072, 2002.
166. Mizuno, T., Inagaki, R., Kanao, T., Hagiwara, T., and Nakamura, T., Antitumor activity and some properties of water-insoluble heteroglycans from "Himematsutake", the fruiting body of *Agaricus blazei* Murill, *Agric. Biol. Chem.*, 54, 2897, 1990.
167. Morin, A., Screening of polysaccharide-producing microorganisms, factors influencing the production, and recovery of microbial polysaccharides, in *Polysaccharides: Structural Diversity and Functional Versatility*, Dumitriu, S., Ed., Marcel Dekker, New York, 1998, p. 275.
168. Lee, I.Y., Seo, W.T., Kim, G.J., Kim, M.K., Park, C.S., and Park, Y.H., Production of curdlan using sucrose or sugar cane molasses by two-step fed-batch cultivation of *Agrobacterium* species, *J. Ind. Microbiol. Biotechnol.*, 18, 255, 1997.
169. Sandula, J., Kogan, G., Kacuracova, M., and Machova, E., Microbial (1→3)-β-D-glucans, their preparation, physicochemical characterization and immunomodulatory activity, *Carbohydr. Polym.*, 38, 247, 1999.
170. Sandula, J., Machova, E., and Hribalova, V., Mitogenic activity of particulate yeast β-(1→3)-D-glucan and its water-soluble derivatives, *Int. J. Biol. Macromol.*, 17, 323, 1995.
171. Cross, G.G., Jennings, H.J., Whitfield, D.M., Penney, C.L., Zacharie, B., and Gagnon, L., Immunostimulant oxidized β-glucan conjugates, *Int. Immunopharm.*, 1, 539, 2001.
172. Furue, H., Biological characteristics and clinical effects of sizofilan (SPG), *Med. Actual.*, 23, 335, 1987.
173. Zhu, D., Recent advances on the active components in Chinese medicines, *Abstr. Chin. Med.*, 1, 251, 1987.
174. Chihara, G., Hamuro, J., Maeda, Y.Y., Shiio, T., and Suga, T., Antitumor metastasis-inhibitory activities of lentinan as an immunomodulator: an overview, *Cancer Detect. Prev.*, 1, 423, 1987.
175. Ikekawa, T., Beneficial effects of edible and medicinal mushrooms in health care, *Int. J. Med. Mushrooms*, 3, 291, 2001.
176. Aoki, T., Lentinan, in *Immune Modulation Agents and Their Mechanisms*, Fenichel, R.L. and Chirigos, M.A., Eds., Marcel Dekker, New York, 1984, p. 63.
177. Reizenstein, P. and Mathe, G., Immunomodulating agents, *Immunol. Ser.*, 35, 347, 1984.
178. Miyakoshi, H. and Aoki, T., Acting mechanisms of lentinan in human. I. Augmentation of DNA synthesis and immunoglobulin production of peripheral mononuclear cells, *Int. J. Immunopharm.*, 6, 365, 1984.
179. Arinaga, S., Karimine, N., Takamuku, K., Nanbara, S., Nagamatsu, M., Ueo, H., and Akiyoshi, T., Enhanced production of interleukin 1 and tumor necrosis factor by peripheral monocytes after lentinan administration in patients with gastric carcinoma, *Int. J. Immunopharm.*, 14, 43, 1992.

180. Wasser, S.P. and Weis, A.L., Medicinal properties of substances occurring in higher Basiomycetes mushrooms: current perspectives, *Int. J. Med. Mushrooms*, 1, 31, 1999.
181. Kraus, J. and Franz, G., β(1→3) Glucans: anti-tumor activity and immunostimulation, in *Fungal Cell Wall and Immune Responses*, NATO ASI Series H53, Latge, J.P. and Boucias, D., Eds., Springer, Berlin, 1991, p. 431.
182. Hamuro, J. and Chihara, G., Lentinan, a T-cell orientated immunopotentiator: its experimental and clinical applications and possible mechanism of immune modulation, in *Immunomodulation Agents and Their Mechanisms*, Fenichel, R.L and Chirigos, M.A., Eds., Dekker, New York, 1985, p. 409.
183. Jamas, S., Easson, J., Davidson, D., and Ostroff, G., Use of Aqueous Soluble Glucan Preparations to Stimulate Platelet Production, U.S. Patent 5,532,223, 1996.
184. Suzuki, T., Sakurai, T., Hashimoto, K., Oikawa, S., Masuda, A., Ohsawa, M., and Yadomae, T., Inhibition of experimental pulmonary metastasis of Lewis lung carcinoma by orally administered β-glucan in mice, *Chem. Pharm. Bull.*, 39, 1606, 1991.
185. Otterlei, M., Ostgaard, K., Skjak-Braek, G., Smidsrod, O., Soon-Shiong, P., and Espevik, T., Induction of cytokine production from human monocytes stimulated with alginate, *J. Immunother.*, 10, 286, 1991.
186. Kitazawa, H., Itoh, T., Tomioka, Y., Mizugaki, M., and Yamaguchi, T., Induction of IFNγ and IL-1α production in macrophages stimulated with phosphopolysaccharide produced by *Lactococcus lactis* spp. *cremoris*, *Int. J. Food Microbiol.*, 31, 99, 1996.
187. Mattila, P., Suonpaa, K., and Piironen, V., Functional properties of edible mushrooms, *Nutrition*, 16, 694, 2000.
188. Markova, N., Kussovski, V., Radoucheva, T., Dilova, K., and Georgieva, N., Effects of intraperitoneal and intranasal application of lentinan on cellular response in rats, *Int. Immunopharm.*, 2, 1641, 2002.
189. Jagodzinski, P., Wiaderkiewicz, R., Kursawski, G., Kloczewiak, M., Nakashima, H., Hyjek, E., Yamamoto, N., Uryu, T., Kaneko, Y., Osner, M., and Kosbor, D., Mechanism of the inhibitory effect of curdlan sulphate on HIV-1 infection *in vitro*, *Virology*, 202, 735, 1994.
190. Marchetti, M., Pisani, S., Petropaolo, V., Seganti, L., Nicoletti, R., Degener, A., and Orsi, N., Antiviral effect of a polysaccharide from *Sclerotium glucanicum* towards herpes simplex virus type 1 infection, *Planta Med.*, 62, 303, 1996.
191. Mastromarino, P., Petruzziello, R., Macchia, S., Rieti, S., Nicoletti, R., and Orsi, N., Antiviral activity of natural and semisynthetic polysaccharides on early steps of rubella virus infection, *J. Antimicrob. Chemother.*, 39, 339, 1997.
192. Hetland, G. and Sandven, P., β-1,3-Glucan reduces growth of *Mycobacterium tuberculosis* in macrophage cultures, *FEMS Immunol. Med. Microbiol.*, 33, 41, 2002.
193. Liang, J., Melican, D., Cafro, L., Palace, G., Fisette, L., Armstrong, R., and Patchen, M.L., Enhanced clearance of a multiple antibiotic resistant *Staphylococcus aureus* in rats treated with PGG-glucan is associated with increased leukocyte counts and increased neutrophil oxidative burst activity, *Int. J. Immunopharm.*, 20, 595, 1998.
194. Hotta, H., Hagiwara, K., Tabata, K., Ito, W., and Homma, M., Augmentation of protective immune-responses against Sendai virus infection by fungal polysaccharide schizophyllan, *Int. J. Immunopharm.*, 15, 55, 1993.
195. Kakumu, S., Ishikawa, T., Wakita, T., Yoshioka, K., Ito, Y., and Shinagawa, T., Effect of sizofiran, a polysaccharide, on interferon gamma, antibody production and lymphocyte proliferation specific for hepatitis-B virus antigen in patients with chronic hepatitis-B, *Int. J. Immunopharm.*, 13, 969, 1991.

196. Seal, C.J. and Mathers, J.C., Comparative gastrointestinal and plasma cholesterol responses of rats fed on cholesterol-free diets supplemented with guar gum and sodium alginate, *Br. J. Nutr.*, 85, 317, 2001.
197. Klingberg-Knol, E.C., Festen, H.P., and Meuwissen, S.G., Pharmacological management of gastro-oesophageal reflux disease, *Drugs*, 49, 695, 1995.
198. Sutton, A., Reduction of strondium absorption in man by the addition of alginate to the diet, *Nature*, 216, 1005, 1967.
199. Hikino, H., Konno, C., Mirin, Y., and Hayashi, T., Isolation and hypoglycemic activity of ganoderans A and B, glycans of *Ganoderma lucidum* fruit bodies, *Planta Med.*, 4, 339, 1985.
200. Song, T.Y. and Yen, G.C., Antioxidant properties of *Antrodia camphorata* in submerged culture, *J. Agric. Food Chem.*, 50, 332, 2002.
201. Kacew, S., Kim, H.S., and Lee, B.M., *In vitro* chemopreventive effects of plant polysaccharides (*Aloe barbadensis Miller*, *Lentinus elodes*, *Ganoderma lucidum* and *Coriolous versicolor*), *Carcinogenesis*, 20, 1637, 1999.
202. Franz, G. and Alban, S., Structure-activity relationship of antithrombotic polysaccharide derivatives, *Int. J. Biol. Macromol.*, 17, 311, 1995.
203. Cleary, J.A., Kelly, G.E., and Husband, A.J., The effect of molecular weight and β-1,6-linkages on priming of macrophage function in mice by (1,3)-β-D-glucan, *Immunol. Cell Biol.*, 77, 395, 1999.
204. Kojima, T., Tabata, K., Itoh, W., and Yanaki, T., Molecular weight dependence of the antitumor activity of schizophyllan, *Agric. Biol. Chem.*, 50, 231, 1986.
205. Gao, Q.P., Seljelid, R., Chen, H.Q., and Jiang, R., Characterisation of acidic heteroglycans from *Tremella fuciformis* Berk with cytokine stimulating activity, *Carbohydr. Res.*, 288, 135, 1996.
206. Kimura, Y., Watanabe, K., and Okuda, H., Effects of soluble sodium alginate on cholesterol excretion and glucose tolerance in rats, *J. Ethnopharmacol.*, 54, 47, 1996.
207. Maeda, Y.Y., Watanabe, S.T., Chihara, C., and Rokutanda, M., Denaturation and renaturation of a β-1,6; 1,3-glucan, lentinan, associated with expression of T-cell-mediated responses, *Cancer Res.*, 48, 671, 1988.
208. Yanaki, T., Ito, W., and Tabata, K., Correlation between antitumor activity of schizophyllan and its triple helix, *Agric. Biol. Chem.*, 509, 2415, 1986.
209. Saito, H., Yoshioka, Y., Uehara, N., Aketagawa, J., Tanaka, S., and Shibata, Y., Relationship between conformational and biological response for (1→3)-β-D-glucans in the activation of coagulation factor G from lumilus amebocyte lysate and host-mediated antitumor activity. Demonstration of single-helix conformation as a stimulant, *Carbohydr. Res.*, 217, 181, 1991.
210. Gomaa, K., Kraus, J., Rosskopf, F., Roper, H., and Franz, G., Antitumor and immunological activity of a β-1→3/1→6 glucan from *Glomerella cingulata*, *J. Cancer Res. Clin. Oncol.*, 118, 136, 1992.
211. Kulicke, W.M., Lettau, A.I., and Thielking, H., Correlation between immunological activity, molar mass, and molecular structure of different (1→3)-β-D-glucans, *Carbohydr. Res.*, 297, 135, 1997.
212. Suzuki, T., Ohno, N., Saito, K., and Yadomae, T., Activation of the complement system by (1,3)-beta-D-glucans having different degrees of branching and different ultrastructures, *J. Pharmacobiodyn.*, 15, 277, 1992.
213. Itoh, W., Sugawara, I., Kimura, S., Tabata, K., Hirata, A., Kojima, T., Mori, S., and Shimada, K., Immunopharmacological study of sulphated schizophyllan (SPG). I. Its action as a mitogen and anti-HIV agent, *Int. J. Immunopharmacol.*, 12, 225, 1990.

214. Paulik, S., Mojisova, J., Durove, A., Benisek, Z., and Huska, M., The immunomodulatory effect of the soluble fungal glucan (*Pleurotus ostreatus*) on the delayed hypersensitivity and phagocytic ability of blood leucocytes in mice, *J. Vet. Med.*, 43, 129, 1996.
215. Yoshida, I., Kiho, T., Usui, S., Sakushima, M., and Ukai, S., Polysaccharides in fungi. XXXVII. Immunomodulating activities of carboxymethylated derivatives of linear (1→3)-alpha-D-glucans extracted from the fruiting bodies of *Agrocybe cylindracea* and *Amanita muscaria*, *Biol. Pharm. Bull.*, 19, 114, 1996.
216. Ohno, N., Kurachi, K.m and Yadomae, T., Physicochemical properties and antitumor activities of carboxymethylated derivatives of glucan from *Sclerotinia sclerotiorum*, *Chem. Pharm. Bull.*, 36, 1016, 1988.
217. Yokoyama, W.H., Knuckles, B.E., Wood, D., and Inglett, G.E., Food processing reduces size of soluble cereal beta-glucan polymers without loss of cholesterol-reducing properties, *Bioact. Comp. Food ACS Symp. Ser.*, 816, 105, 2002.
218. Guillon, F. and Champ, M., Structural and physical properties of dietary fibres, and consequences of processing on human physiology, *Food Res. Int.*, 33, 233, 2000.
219. Pilnik, W. and Rombouts, F.M., Polysaccharides and food processing, *Carbohydr. Res.*, 142, 93, 1985.
220. Giavasis, I., Harvey, L.M., and McNeil, B., Simultaneous and rapid monitoring of biomass and biopolymer production by *Sphingomonas paucimobilis* using Fourier-transform near infrared spectroscopy, *Biotechnol. Lett.*, 25, 957, 2003.
221. Vaccari, G., Dosi, E., Campi, A.L., Gonzalez-Vara, A., Matteuzzi, D., and Mantovani, G., A near-infrared spectroscopy technique for the control of fermentation processes: an application to lactic acid fermentation, *Biotechnol. Bioeng.*, 43, 913, 1993.
222. Von Stockar, U., Duboc, P., Menoud, L., and Marison, I.W., On-line calorimentry as a technique for process monitoring and control in biotechnology, *Thermochim. Acta*, 300, 225, 1997.
223. Welman, A.D. and Maddox, I.S., Exopolysaccharides from lactic acid bacteria: perspectives and challenges, *Trends Biotechnol.*, 21, 269, 2003.
224. Lamothe, G.T., Jolly, L., Mollet, B., and Stingele, F., Genetic and biochemical characterization of exopolysaccharide biosynthesis by *Lactobacillus delbrueckii* subsp. *bulgaricus*, *Arch. Microbiol.*, 178, 218, 2002.

6 Chitosan as a Dietary Supplement and a Food Technology Agent

Riccardo A.A. Muzzarelli and Corrado Muzzarelli

CONTENTS

6.1 INTRODUCTION

Chitin is the most abundant organic compound of nitrogen. At least 10 gigatons (1.10^{13} kg) of chitin are synthesized and degraded each year in the biosphere. Chitin is therefore important for making nitrogen available to countless living organisms: it is widely distributed among invertebrates. Nitrogen fixation from the atmosphere is made by *Rhizobium* in root nodules of leguminous plants, the major step being the formation of lipo-chitin oligomeric forms. Alpha-chitin is found in the calyces of hydrozoa, the egg shells of nematodes and rotifers, the radulae of mollusks, and the cuticles of arthropods, and beta-chitin is in the shells of brachiopods and mollusks, the cuttlefish bone, the squid pen, and pogonophora tubes. Chitin is found in exoskeletons, peritrophic membranes, and cocoons of insects. The septation apparatus in budding yeast is based on a chitin septum. Chitin is ubiquitous in the fungi: the chitin in the fungal walls varies in crystallinity, degree of covalent bonding to other wall components, mainly glucans, and degree of acetylation.

In the areas of fisheries, chitin is recovered to exploit renewable resources and alleviate waste problems. Today chitins and chitosans from different animals are commercially available, mainly from shrimp, but also from squid, lobster, and crab. Chitin is obtained from the shells by removing calcium carbonate, pigments, proteins, and lipids immediately after peeling the shrimp to be deacetylated to chitosans; minor quantities are transformed into O-carboxymethyl chitin, glycol chitin, and 6-oxychitin.

Chitin isolates differ from each other in many respects, including degree of acetylation, typically close to 0.90; elemental analysis, with nitrogen content typically close to 7%; N/C ratio, 0.146 for fully acetylated chitin; molecular size; and polydispersity. The average molecular weight of chitin *in vivo* is probably in the order of the megadaltons, but chitin isolates have lower molecular weights due to partial random depolymerization occurring during the chemical treatments and depigmentation steps. Polydispersity may vary depending on such treatments as powder milling and blending of various chitin batches.

Isolated chitin is a highly ordered copolymer of 2-acetamido-2-deoxy-β-D-glucose, the major component, and 2-amino-2-deoxy-β-D-glucose. Chitobiose, O-(2-amino-2-deoxy-β-D-glucopyranosyl)-(1,4)-2-amino-2-deoxy-D-glucose, is the structural unit of native chitin (Fig. 6.1). Bound water is also a part of the structure.[1–7]

Chitin is degraded *in vivo* either during digestion by many animals or during the molting: chitinases permit the resorbing of most of the N-acetylglucosamine, so that it can be reused within hours for a new, larger exoskeleton built under the action of chitin synthase. In the environment chitin is degraded by bacteria present in the sea floor.

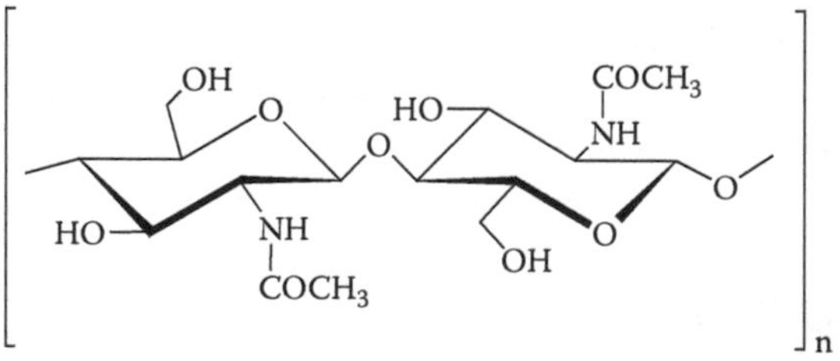

FIGURE 6.1 This formula shows the structural unit of the polysaccharide chitin, the repeating unit being N-acetylglucosamine. *In vitro*, approximately one repeating unit out of ten is deacetylated and linked to a protein. Extended *in vitro* deacetylation yields chitosan.

Chitosan is the only largely available cationic polysaccharide, while most polysaccharides are neutral or anionic, for instance, alginates and pectins. Chitosan indicates a family of deacetylated chitins. In general, chitosans have a nitrogen content higher than 7% and a degree of acetylation lower than 0.40. The removal of the acetyl group is a harsh treatment usually performed with concentrated NaOH. Protection from oxygen, with a nitrogen purge or by addition of sodium borohydride to the alkali solution, is necessary to avoid undesirable reactions such as depolymerization and generation of reactive species. The acetyl groups in the acid-soluble fractions are randomly distributed, while the insoluble fractions contain relatively long sequences of acetylated units.

The presence of a prevailing number of 2-amino-2-deoxyglucose units in a chitosan allows the polymer to be brought into the solution by salt formation. Chitosan is a primary aliphatic amine that can be protonated by acids, the pK of the chitosan amine being 6.3. Certain salts are water soluble, for instance, hydrochloride, formate, acetate, lactate, malate, citrate, glyoxylate, pyruvate, glycolate, and ascorbate.

Under particular conditions chitin and chitosan give hydrophilic highly water swellable hydrogels: gel formation is also promoted by cross-linking agents or organic solvents, particularly for chitosan derivatives. Chemical and physical gels are produced, thermally reversible and not reversible.[8–11]

Chitin and chitosan are not present in human tissues, but acetylglucosamine and chitobiose are found in glycoproteins and glycosaminoglycans. Since chitosan is biodegradable, nontoxic, nonimmunogenic, and biocompatibile in animal tissue, much research has been directed toward its use in medical applications such as drug delivery, artificial skin, and blood anticoagulants. These biopolymers offer a wide range of unique applications, including formation of biodegradable films, immobilization of enzymes, preservation of foods from microbial deterioration, clarification and deacidification of fruits and beverages, emulsion formation, thickening, color stabilization, and dietary supplementation. Most recent pertinent reviews are available.[12–16]

6.2 CHITIN AS A FOOD COMPONENT

Crustaceans, insects, rodents, frogs, snails, earthworms, spiders, scorpions, centipedes, and millipedes have been used as food sources for millennia. Insects utilized in this fashion are more extensive than one can imagine (Table 6.1). One usually thinks only of ants, bees, wasps, beetles, and caterpillars, commonly referred to as "bush tucker" by the Australian aborigines. A closer examination of world insect consumption shows that more than 2000 edible species have been utilized as a food source to date, although this is stated to be underreported, and continued research will increase this number considerably.[17–23] Nevertheless, it should be said that of the 15 million plants, animals, and microbes on Earth, more than 90% of the world's food supply comes from just 15 crop species and 8 livestock species.

What is important is that chitinases are present in plant foods, and that human and bacterial chitinolytic enzymes occur in the human digestive apparatus. The bibliographic research allows confirmation of the presence of chitinolytic enzymes

TABLE 6.1
Consumption of Insects as Food in the World

Continent	Species	Countries	Most Popular Insects Cooked and Eaten
Africa	524	35	*Gonimbrasia belina* (grub)
Central and South America	679	23	*Rhynchophorus* spp. (snout beetle)
			Aegiale hesperiaris (butterfly)
			Atta spp. (ant)
Asia	349	18	*Locusta* spp.
			Bombyx mori (silkworm)
Australia	152	14	*Rhynchophorus ferrugineus papuanus*

in a lot of plant organs and the making of a list of the vegetables and fruits that man uses that could help digest chitin. Although the level of chitinases is often correlated to the presence of stress, like infections, these enzymes also occur in healthy plant tissue and, above all, in fruits under maturation, which are most often used in human nutrition.[23–25]

In the excreta of healthy people there is at least a bacterial species able to hydrolyze chitin to N-acetyl glucosamine. *Clostridium paraputrificum* in the human colon is able to synthesize and secrete chitinases and beta-N-acetyl glucosaminidase, which could take part in the digestion of chitin. Similarly, chitotriosidase, a chitinase produced by activated macrophages, is able to hydrolyze chitin[26–30] (see below).

An inherited deficiency in chitotriosidase activity is frequently reported in plasma of Caucasian subjects, whereas in the African population this deficiency is rare. The study of Musumeci et al.[31] compares chitotriosidase activity in colostrum of 53 African women and 50 Caucasian women. Elevated chitotriosidase was found in the colostrum of African women on the first day after delivery (1230 ± 662 nmol/ml h^{-1}), which decreased with time. The chitotriosidase activity on the first day after delivery in the colostrum of Caucasian women, however, was significantly lower (293 ± 74 nmol/ml h^{-1}) and decreased to 25 ± 20 and 22 ± 19 nmol/ml h^{-1} on the second and third days, respectively. The chitotriosidase activity in the plasma of African women was also higher (101 ± 80 nmol/ml h^{-1}) than that of Caucasian women (46 ± 16 nmol//ml h^{-1}), but no correlation was found between the plasma and colostrum activities. The elevated chitotriosidase activity in the colostrum of African women suggests the presence of activated macrophages in human milk, consistent with the genetic characteristics of the African population and their chitin-eating habits. Actually, a chitin-rich diet induces the secretion of human chitinases.

Chitosan is present in certain fermented cheeses in Northern Europe, and high glucosamine contents have been documented in Oriental fermented foods.[32,33]

6.3 CHARACTERISTICS OF DIETARY CHITOSANS

A number of European countries as well as the U.S. and some Oriental countries (Japan, South Korea) have approved the sale of chitosan-based nutraceuticals as over-the-counter products for the control of weight, hypercholesterolemia, and hypertension.[34,35] Chitosan, in this context, is generally regarded as safe, being a partially deacetylated chitin, i.e., a polysaccharide widely present in nature, in particular in human food such as certain cheeses and marine animals, including crustaceans and squids.[36]

The chitosans actually used to prepare dietary supplements come from crustaceans, mainly shrimp. The freshly caught crustaceans are peeled in the canning and freezing factories, and while the meat is canned or packed, the shells are collected and treated by chemical or microbiological means to extract chitin. Due to seasonal fluctuations and varieties of shrimp caught during fishing activities, the chitosans resulting from the chitin extraction are inherently different in terms of certain characteristic properties.

The exposure of chitin to acids, alkali, surfactants, and solvents for the extraction of carotenoids, and the submission of the chitosan flakes to drying, milling, sieving, and other operations introduce more diversity among various lots of chitosan. Thermal drying, for instance, may introduce a limited degree of cross-linking, while the degree of deacetylation of the final chitosan may depend on the grain size of the chitin powder exposed to hot alkali during the deacetylation process.

At this date, there is no chitosan standard for any application. In other words, even though various grades are available, such as technical, food, and medical grades, no clear recommendation has been made and accepted for adopting a chitosan standard in food applications. As a consequence, the chitosans currently used to manufacture chitosan tablets for human consumption are chosen mainly on the basis of their constant supply, economical convenience, and cost.

In the eyes of today's producers, the characteristics of chitosan are a secondary aspect, and, in general, only the degree of acetylation, the viscosity, and the microbiological contamination are certified. The producers do not declare the degree of crystallinity, the polydispersity of the molecular weight, and the presence of aminoacids and metals. Therefore, it is difficult to predict the performances of the dietary supplements.

A further aspect of uncertainty is introduced by the preparation of the tablets: at this stage, the tendency is to adopt the most usual tabletting process, regardless of the consequences that certain excipients have on the chitosan activity *in vivo*. For example, tabletting involves the use of a binder to hold together the poorly compressible chitosan powder; a popular binder, magnesium stearate, a slightly soluble compound, once coated on the chitosan powder deeply alters the capacity of chitosan to react or effectively contact other compounds.

6.4 MODERN APPLICATIONS OF CHITOSAN IN FOOD SCIENCES

The data so far available clearly show that chitosan may be either bactericidal or bacteriostatic, or even be a growth promoter depending on the bacterial strain. For each strain, the characteristics of chitosan may be more or less significant. While general statements are therefore to be avoided, it appears that chitosans are certainly active against most human pathogens and food spoiling microbes.

Chitosan is not an antimicrobial per se, but its performances can occasionally be superior to biocides. For example, the antimicrobial efficacy of 0.5, 1.0, and 2.0% chitosan and a commercial biocide based on hydrogen peroxide was determined at 20°C against *Listeria monocytogenes*, *Salmonella enterica serovar*, *Typhimurium*, *Staphylococcus aureus*, and *Saccharomyces cerevisiae* adhered to stainless steel. Dried films of *S. aureus* were most sensitive to chitosan but relatively resistant to the biocide. By contrast, yeast films were least sensitive to chitosan.[37]

6.4.1 Antibacterial Activity

The antibacterial activity of chitosan was originally documented by Muzzarelli et al.,[38] who published electron micrographs showing the alterations produced in the bacterial cell wall and organelles. Those results were brilliantly confirmed more than a decade later by Helander et al.,[39] who studied the mode of antimicrobial action of chitosan on Gram-negative bacteria, with special emphasis on its ability to bind to and weaken the barrier function of the outer membrane. Chitosan (250 ppm) at pH 5.3 induced significant uptake of the hydrophobic probe 1-N-phenylnaphthylamine in *Escherichia coli*, *Pseudomonas aeruginosa*, and *Salmonella typhimurium*. Chemical and electrophoretic analyses of cell-free supernatants of chitosan-treated cell suspensions showed that interaction of chitosan with *E. coli* and the salmonellae involved no release of lipopolysaccharide or other membrane lipids. Highly cationic mutants of *S. typhimurium* were more resistant to chitosan than the parent strains. Electron microscopy showed that chitosan caused extensive cell surface alterations and covered the outer membrane with vesicular structures. Chitosan thus appeared to bind to the outer membrane, explaining the loss of the barrier function. This property makes chitosan useful for food protection. It was also found that the antibacterial activity of quaternized chitosan against *E. coli* is stronger than that of chitosan.[40]

New interest has emerged in partially hydrolyzed chitosan and chitosan oligosaccharides. Enzymatic preparation methods captured interest due to safe and nontoxic conditions, and production has been developed into a continuous process. Many of the biological activities reported for chitosan oligosaccharides, such as antimicrobial, anticancer, antioxidant, and immunostimulant effects, depend on their physicochemical properties. In a review, Kim and Rajapakse[41] have summarized different enzymatic preparation methods of chitosan oligosaccharides and their biological activities.

Chitosan activity is synergistically enhanced by traditional preservatives such as benzoic acid, acetic acid, and sulfite. For fresh pork sausages, two pilot-scale trials showed that 0.6% chitosan combined with low sulfite (170 ppm) retarded the growth of spoilage organisms more effectively (3 to 4 log cfu/g) than high levels (340 ppm) of sulfite alone at 4°C for up to 24 days.[42] Trials in real foods showed that dipping of standard and skinless pork sausages in chitosan solutions (1.0%) reduced the native microflora (total viable counts, yeasts and molds, and lactic acid bacteria) by approximately 13 log cfu g^{-1} for 18 days at 7°C. Chitosan treatment increased the shelf life.[43–45] The combined use of chitosan and sulfite permitted the slowing down of deterioration of chilled pork sausages.[42] In the case of the preservation of herring and Atlantic cod, chitosan as an edible invisible film enhanced the quality of seafood during storage.[46]

Recent investigations point out that chitosan in emulsions might be particularly more effective than in aqueous systems; for example, Jumaa et al.[47] found that lipid emulsions containing 0.5% chitosan conformed to the requirements of the preservation efficacy test for topical formulations according to the European Pharmacopoeia.

6.4.2 Antifungal Activity

The use of chitosan to control postharvest fungal decay has attracted much attention due to problems associated with chemical agents, consumer reluctance against fungicide-treated produce, and an increasing number of fungicide-tolerant postharvest pathogens. Chitosan reduces the *in vitro* growth of numerous fungi with the exception of Zygomycetes, i.e., the fungi containing chitosan as a major cell wall component.

Tripathi and Dubey[48] reviewed the exploitation of some natural products, such as flavor compounds, acetic acid, jasmonates, glucosinolates, propolis, fusapyrone and deoxyfusapyrone, chitosan, essential oils, and plant extracts, for the management of fungal rotting of fruits and vegetables, capable of prolonging shelf life.

The antifungal effect of chitosan on *in vitro* growth of common postharvest fungal pathogens in strawberry fruits consists of the marked reduction of the radial growth of *Botrytis cinerea* and *Rhizopus stolonifer*, with a greater effect at higher concentrations. Signs of infection in chitosan-coated fruits appeared after 5 days of storage at 13°C compared with 1 day for the control treatment. After 14 days of storage, chitosan coating at 15 mg/ml reduced decay of strawberries caused by the same fungi by more than 60%, and coated fruits ripened normally and did not show any apparent sign of phytotoxicity. Similarly, the preservative effect of chitosan was observed on low-sugar candied kumquat. The growth of *Aspergillus niger* was inhibited by the addition of chitosan (0.1 to 5 mg/ml) to the medium (pH 5.4). Cuero et al.[49] observed that N-carboxymethylchitosan reduced aflatoxin production in *Aspergillus flavus* and *Aspergillus parasiticus* by more than 90%, while fungal growth was reduced to less than one half.

Table grape treated with 1% chitosan showed increased phenylalanine ammonia-lyase activity, besides a direct activity against *Botrytis cinerea*.[50] Chitosan coatings reduced the incidence of molds occurring on apples over 12 weeks. The combination of hypobaric and chitosan treatments was found to be a valid strategy for decreasing the decay of sweet cherries.[51] The effect of glycol chitosan applied as a coating to act as a biocontrol treatment of postharvest diseases of apple and citrus fruits was evaluated under simulated commercial packinghouse conditions by El-Ghaouth et al.[52]

Devlieghere et al.[53] studied the antimicrobial effects of chitosan coatings on decay of minimally processed strawberries and lettuce. Several bacteria and yeasts were exposed to chitosan concentrations varying from 40 to 750 mg/l. Generally, Gram-negative bacteria seemed to be very sensitive to chitosan minimum inhibitory concentration (MIC 0.006% (w/v)), while the sensitivity of Gram-positive bacteria was variable and that of yeast was 0.01% (w/v). A chitosan coating was formed by dipping the products in a chitosan–lactic acid/Na–lactate solution; the pH was adjusted to the pH of the products. These products were equilibrium modified atmosphere packaged, stored at 7°C, and during storage sensorially and microbiologically evaluated. The microbiological load on the chitosan-treated samples was lowered for both products. The antimicrobial effect of chitosan on lettuce disappeared after 4 days of storage, while on the strawberries, it lasted 12 days. The treatment of mandarins and oranges with a chitosan coat produced excellent results in terms of percentage of weight loss and visual appearance.[54]

A study carried out on chitosan coating for the inhibition of *Sclerotinia sclerotiorum* rot of carrot showed that the incidence of rotting was reduced from 88 to 28% by coating carrot roots with 2% chitosan.[55–57] Carrot slices coated with a starch and chitosan mixture showed reductions in mesophilic aerobes, mold and yeast, and psychrotrophic amounting to 1.34, 2.50, and 1.30 log cycles, respectively. The presence of 1.5% chitosan in the coatings inhibited the growth of total coliforms and lactic acid bacteria throughout the storage period (10°C for 15 days). The use of edible antimicrobial yam starch and chitosan coating was deemed to be a viable alternative for controlling microbial growth in minimally processed carrot.

Coating fruits and vegetables with chitosan or its derivatives has positive advantages for long-term storage of these foods, particularly fruits of exotic origin such as mango.[58] The preliminary treatment of the plants provided further advantages: *Pseudomonas fluorescens*, *Bacillus subtilis*, and *Saccharomyces cerevisiae* were evaluated for their potential to attack the mango (*Mangifera indica* L.) anthracnose pathogen *Colletotrichum gloeosporioides* Penz. under endemic conditions. The plant growth-promoting rhizobacteria *P. fluorescens* amended with chitin sprayed at fortnightly intervals gave the maximum induction of flowering, a yield attribute in the preharvest stage; consequently, reduced latent symptoms were recorded at the postharvest stage. An enormous induction of the defense-mediating lytic enzymes chitinase and beta-1,3-glucanase was recorded.[59]

Mangosteen, an economically important fruit of Thailand, has a short shelf life. Kungsuwan et al.[60] showed that the most effective chitosan concentration for aerial spraying was 2%, which could extend the shelf life of mangosteens to more than 23 days.

Extension of the storage life and better control of decay of peaches, pears, and kiwi fruits by application of chitosan film have been documented.[61] Cucumbers, bell peppers, strawberries, and tomatoes could be stored for long periods after coating with chitosan. These results may be attributed to decreased respiration rates, inhibition of fungi development, and delayed ripening due to the reduction of ethylene and carbon dioxide evolution.

A series of O-acyl chitosans with a degree of substitution between 0.02 and 0.28 were synthesized by reaction of alkanoic acid derivatives with chitosan in the presence of H_2SO_4 as a catalyst. O-Decanoyl chitosan (mole ratio of 1:2 chitosan to decanoic acid) was the most active compound against *Botrytis cinerea* and O-hexanoyl chitosan displayed the highest activity against *Pyricularia grisea*. Some derivatives also repressed spore formation at rather high concentrations (1.0, 2.0, and 5.0 g l^{-1}).[62]

When administered to a plant, chitosan has a dual function, i.e., direct interference of fungal growth and activation of several defense processes that include accumulation of chitinases, which degrades fungal cell walls' synthesis of proteinase inhibitors, lignification, and induction of callous synthesis. The microbial transport systems seem to be highly affected by the presence of chitosan.[63,64] Chitosan induced the accumulation of the antifungal phytoalexin pisatin in pea pods.[65,66]

Saprolegnia parasitica is responsible for infection of fish and eggs in aquaculture facilities and grows on injured, stressed, or infected fish. Electron microscopy observation provided evidence of ultrastructural alteration, damaged fungal structure, hyphal distortion, and retraction. The antifungal action of chitosan could be modulated by proper chemical modification and put to use in protecting aquacultured fish.[67]

The quantitative determination of chitin, a constituent of the fungal cell walls, offers the advantage that it reflects the total amount of mycelium. Bishop et al.[68] used chitin to further evaluate the detection of mold in tomato products, ketchup, paste, and puree. Variations were observed in chitin content among different fungal species, depending upon age and growth conditions. Insect contamination did not change the glucosamine level significantly except in cases of extremely high contamination.

In addition to its direct antimicrobial activity, chitosan induces a series of defense reactions correlated with enzymatic activities. Chitosan increases the production of glucanohydrolases, phenolic compounds, and synthesis of specific phytoalexins with antifungal activity, and reduces macerating enzymes such as polygalacturonases and pectin metil esterase. For some horticultural and ornamental commodities, chitosan increased the harvested yield. Due to its ability to form a semipermeable coating, chitosan extends the shelf life of treated fruit and vegetables by minimizing the rate of respiration and reducing water loss. As a nontoxic biodegradable material, as well as an elicitor, chitosan has the potential to become a new class of plant protectant, thus assisting toward the goal of sustainable agriculture.[69]

6.4.3 Edible Films and Textural Agents

Edible films can provide supplementary and sometimes essential means of controlling physiological, morphological, and physicochemical changes in food products. High-density polyethylene film, a common packaging material used to protect foods, has disadvantages like fermentation due to the depletion of oxygen and condensation of water, which promotes fungal growth.

Due to their filmogenicity, chitin and chitosan are satisfactorily used as food wraps. Semipermeable chitosan films modify the internal atmosphere, decrease the transpiration, and delay the ripening of fruits.[70,71] For the preparation of chitosan/pectin-laminated films and chitosan/methylcellulose films, several approaches have been used, including simple coacervation. Chitosan films are tough, flexible, and tear resistant; moreover, they have favorable permeation characteristics for gases and water vapor.

Chitosan is also suitable as a texturizing agent for perishable foods: for instance, high-viscosity chitosan solutions were used to prepare tofu, a widely consumed Oriental food, for which the organoleptic properties did not vary appreciably, while shelf life was extended.[72–74]

6.4.4 Control of Enzymatic Browning in Fruits

Mechanical injury during postharvest handling and processing causes browning of fruits and vegetables with loss of quality and value. Polyphenol oxidase is responsible for this phenomenon that affects color, taste, and nutritional value of fruits and vegetables.[75] Dark-colored pigments are generated from *o*-quinones, under the effect of polyphenol oxidase activity. Concern over the adverse health effects of sulfite, the most effective browning inhibitor, has stimulated a search for surrogate anti-browning compounds. The effect of a chitosan film on the enzymatic browning of litchi fruit (*Litchi chinensis* Sonn.) was studied by Zhang and Quantick,[76] who reported that chitosan film coating delayed changes of contents of anthocyanins, flavonoids, and total phenolics. It also delayed the increase in polyphenol oxidase activity and partially inhibited the increase in peroxidase activity.

Manually peeled litchi fruits were treated with aqueous solutions of 1, 2, or 3% of chitosan, placed into trays overwrapped with plastic film, and then stored at –1°C. Application of chitosan coating retarded weight loss and the decline in sensory quality, with higher contents of total soluble solids, titratable acid, and ascorbic acid, and suppressed the increase of polyphenol oxidase and peroxidase. Application of a chitosan coating effectively maintained quality attributes and extended shelf life of the peeled fruit.[77]

6.4.5 Clarification and De-acidification of Fruit Juices

Processing of clarified fruit juices commonly involves the use of clarifying agents, including gelatin, bentonite, tannins, potassium caseinate, and polyvinyl pyrrolidone. Chitosan is a de-hazing agent used to control acidity in fruit juices, besides being a good clarifying agent for grapefruit juice, with or without pectinase treatment, and apple, lemon, and orange juices, as well as a fining agent for apple juice, which can afford zero turbidity products with as little as 0.8 kg/m^3 of chitosan. No impact on the biochemical parameters of the juices was found.[78] Apple juice can be protected from fungal spoilage with the aid of modest additions of chitosan glutamate.[79] Spagna et al.[80] observed that chitosan has a good affinity for polyphenolic compounds, such as catechins, proanthocyanidins, cinnamic acid, and their derivatives, which can change the color of white wines due to their oxidative products. By adding chitosan to grapefruit juice (15 g/l), the total acid content (citric, tartaric, malic, oxalic, and ascorbic acid) was sharply reduced.

6.4.6 Recovery of Solids from Food Processing Wastes

Complying with water quality regulations is one of the major endeavors of the food industry. Effluents from food processing plants are characterized by high chemical oxygen demand, biochemical oxygen demand, and total suspended solids. Recovery of suspended solids by coagulation and decanting may also be convenient in view of their utilization.[81]

Chitosan as a coagulating agent for waste treatment systems is particularly effective in removing proteins from butchery and fishery wastes: the coagulated by-products serve as animal feed, chitosan being digestible by most animals. Similarly, plant proteins can be recovered from waters used to prepare vegetables for canning. In the cheese manufacture, for example, Fernandez and Fox[82] reported the use of chitosan to remove proteins and peptides from whey, while Ausar et al.[83] precipitated casein with chitosan. Altieri et al. used chitosan to prolong mozzarella cheese shelf life by taking advantage of the growth inhibition of spoilage microorganisms such as coliforms; in this context chitosan was found to stimulate lactic acid bacteria.[84]

In fact, the presence of chitosan together with growth of milk fermentative bacteria was found useful in cheese making. In nutrient broth, all chitosans showed a dose-dependent inhibition of *Streptococcus thermophilus* and *Lactobacillus delbrueckii* ssp. *bulgaricus* growth. Chitosan of high and low molecular weight, but not chitosan oligosaccharides, showed a dose-dependent inhibition of *Propionibacterium freudenreichii*. The effect of chitosan on milk fermentative processes depended not only on its molecular weight and concentration, but also on the presence of casein micelles or milk fat, which could prevent the inhibitory activity of these biopolymers on bacterial growth.[85]

6.5 CHITOSAN AS A NUTRACEUTICAL

6.5.1 Hypercholesterolemia

Atherosclerotic diseases, such as coronary heart disease and stroke, are the major cause of adult mortality in developed countries. Although the etiology of these diseases is clearly multifactorial, it is now well accepted that elevated serum cholesterol concentrations play a causal role in the development of atherosclerosis. Although there are a number of drugs available on the market that effectively lower serum cholesterol, such as the statins, these drugs are expensive and are not without risk. The recent withdrawal of Baycol from the U.S. market by the Food and Drug Administration due to development of rhabdomyolysis in some individuals taking the drug, resulting in a number of deaths, illustrates this point rather dramatically.

A physicochemical characteristic of chitosan potentially related to its hypocholesterolemic effect is its molecular weight. In dilute acid, chitosans develop a high viscosity, the degree of which mainly depends on their molecular weight. Intestinal contents' viscosity has been demonstrated to be a key characteristic of cholesterol-lowering polysaccharides.[86–89] The hypocholesterolemic effect of chitosan preparations with widely varying *in vitro* viscosities was examined in cholesterol-fed rats.[90] When fed at either 2 or 5% in the diet, all chitosan preparations, regardless of their *in vitro* viscosity, were equally effective at reducing hepatic cholesterol concentrations. In a similar study, broiler chickens were fed three chitosan preparations of different *in vitro* viscosities at a dietary concentration of 1.5%.[91] All three preparations significantly and equally reduced plasma cholesterol. More recently, the viscosity of native (i.e., undiluted) intestinal contents from rats fed meals containing 7.5% chitosan was measured and found to be very low.[87] From these studies it is clear that chitosan molecular weight does not explain the hypocholesterolemic effect.

When discussing the role of chitosan as a cholesterol-lowering nutraceutical and considering possible mechanisms for its action, one should keep in mind that animal studies might not be predictive of results in humans because of the occurrence of chitinases in the digestive systems of many animals.

6.5.1.1 Cholesterol Lowering in Animals

The first report of the cholesterol-lowering ability of chitosan appears to be that of Sugano et al.,[92] who found that a diet containing 5% chitosan reduced liver cholesterol concentration by one half or more in cholesterol-fed rats. This effect was confirmed by Kobayashi et al.[93] In a 4-week study, a 4% chitosan diet reduced both liver and serum cholesterol dramatically in rats fed diets containing 1% cholesterol + 0.1% bile salts.[94] Chitosan has also been shown to reduce plasma cholesterol in cholesterol-fed broiler chickens at dietary concentrations of 1.5 to 3.0%.[91,95] Thus, the ability of chitosan to reduce cholesterol in animal models is well established.

Three low molecular weight chitosans obtained by enzymatic hydrolysis of a high molecular weight chitosan had low viscosity and were water soluble. The water-soluble 46-kDa chitosan was the most effective at inhibiting pancreatic lipase activity *in vitro* and plasma triacylglycerol elevation after the oral lipid tolerance test. Chitosan (300 mg kg^{-1}, twice daily) prevented increases in body weight, various white adipose tissue weights, and liver lipids (cholesterol and triacylglycerol) in mice fed a high-fat diet, and further increased the fecal bile acid and fat. The lipid-lowering effects of chitosan may be mediated by increases in fecal fat or bile acid excretion resulting from the binding of bile acids, and by a decrease of the absorption of triacylglycerols and cholesterol in the small intestine as a result of the inhibition of pancreatic lipase activity. Chitosan did not cause liver damage with the elevation of glutamic oxaloacetic transaminase and glutamic pyruvic transaminase, or kidney damage with the elevation of blood nitrogen urea. It was concluded that chitosan is a safe functional food ingredient.[96]

The apoprotein E-deficient mouse has extreme hypercholesterolemia, atherosclerosis develops rapidly, and the lesions appear histologically similar to those seen in humans.[97] Apoprotein E-deficient mice fed diets containing 5% chitosan for approximately 6 months had a serum cholesterol that was only 64% of that of the control group.[98] The atherosclerotic lesion area in the aortic arch was reduced by half and in the total aorta by 42% in mice fed chitosan. Chitosan consumption reduces adiposity in animal models, based on the increase in fecal fat. Mice fed high-fat diets and dietary concentrations of chitosan had a reduced adipose tissue weight, with the reduction being dose related. However, these animals also experienced a reduced body weight gain and lower liver weight, indicating that the reduction in tissue weight was not completely specific to adipose tissue, making interpretation of the study somewhat challenging. In contrast, apoE-deficient mice fed 5% chitosan actually had a greater rate of body weight gain than the control group, and had equivalent weights of liver and epididymal and uterine horn fat pads.[98] In mice fed *ad libitum* for 9 weeks, chitosan prevented the increase of body weight, hyperlipidemia, and fatty liver.

6.5.1.2 Cholesterol Lowering in Humans

Chitosan was first shown to reduce serum cholesterol in humans in 1993, when adult males fed chitosan-containing meals for 2 weeks (3 g/day for week 1, 6 g/day for week 2) experienced a significant decrease of 6% in total cholesterol.[99] The subjects also demonstrated a 10% increase in HDL cholesterol. However, in a 28-day study in overweight subjects given a daily dose of approximately 0.6 g/day of chitosan, no reduction in total cholesterol was detected.[34] The failure to find a cholesterol reduction in this study is likely due to the very modest dose of chitosan used. Two studies have reported serum cholesterol reductions with chitosan treatment. Obese women consuming 1.2 g of microcrystalline chitosan for 8 weeks demonstrated significant reductions in low-density lipoprotein (LDL) cholesterol, although not total serum cholesterol.[100] Eighty-four female subjects with mild to moderate hypercholesterolemia receiving 1.2 g of chitosan per day experienced a significant decrease in total serum cholesterol.[101]

Metso et al. observed 83 middle-aged men and women without severe disease and with a total cholesterol of 4.8 to 6.8 mmol/l and triglycerides below 3.0 mmol/l. They concluded that treatment with microcrystalline chitosan had no effect on the concentrations of plasma lipids or glucose in healthy middle-aged men and women with moderately increased plasma cholesterol concentrations.[102]

For oral administration to humans, chitosan is generally recognized as safe.[13,103] Twenty-one overweight normocholesterolemic subjects were fed a supplement containing equal amounts of glucomannan and chitosan for 28 days. The observed serum cholesterol reduction was mediated by increased fecal steroid excretion and was not linked to fat excretion. Greater fecal excretion of neutral sterols and bile salts was observed. The topic has been reviewed by Muzzarelli[104,105] and Pittler and Ernst.[106]

6.5.1.3 Mechanism of Cholesterol Lowering

Bile salts are formed from cholesterol in the liver and are secreted into the duodenum by the enterohepatic circulation: the bile salt pool is maintained stable, the newly ingested cholesterol compensating the excreted quantities. However, if bile salts are sequestered by any suitable compound, some cholesterol is oxidized to produce more bile salts.[107] Bile is produced at the rate of 700 to 1200 ml/day, bile salts accounting for 1.24 to 1.72% and cholesterol for 0.86 to 1.76 g/l; the average pH is 7.3. For the various intestinal tracts, the pH values are as follows: duodenum, 4.7 to 6.5; upper jejunum, 6.2 to 6.7; lower jejunum, 6.2 to 7.3; ileum, 6.1 to 7.3; and colon, >7.3. It is worth noting that these values tend to keep the bile salts in solution, while depressing the chitosan solubility (chitosan pK = 6.3).[108]

The uptake of bile salts into chitosan–alginate gel beads was observed by Murata et al.[109]. The presence of weak acids (orotic, citric, folic, and ascorbic) did not hinder the uptake; rather, chitosan orotate salt was found to enhance it. Various chitosans and heavily modified Chitopearl® chitosans were studied by Murata et al.[110] (see also Kumar et al.[13]) and found to have capacities in the range of 0.53 to 1.20 mmol taurocholate/g of chitosan (various degrees of acetylation) and 0.2 to 0.9 mmol taurocholate for Chitopearl products; the capacity of Questran® was ca. 1.0 mmol taurocholate/g. Higher capacities were observed for taurodeoxycholate (1.8 mmol taurodeoxycholate/g of chitosan). Of course, these capacities depend on the initial bile salt concentration, grain size, and other parameters. One millimole taurocholate corresponds to 515 mg; thus, the weight ratio taurocholate:chitosan orotate is 0.515:1.000, which appears to be a high ratio, notwithstanding the uncertainties related to the description of the experimental conditions.

Thongngam and McClements[111,112] focused on the binding of Na taurocholate to chitosan and provided thermodynamic data by isothermal titration calorimetry. At 30°C, Na taurocholate binds strongly to chitosan to form an insoluble complex containing ca. 4 mmol Na taurocholate/g of chitosan at saturation — that means a taurocholate:chitosan molar ratio of ca. 2:3 and a weight ratio of ca. 2:1. Ionic strength had scarce influence; the enthalpy changes went from endothermic (at 10°C) to exothermic (at 40°C), indicating the importance of changes of hydrophobic interactions leading to the formation of micelle-like clusters within the chitosan structure. The binding capacities of sodium glycocholate to chitosan, diethylaminoethyl chitosan, quaternized diethylaminoethyl chitosan, and cholestyramine were 1.42, 3.12, 4.06, and 2.78 mmol/g, respectively. The capacity of dialkylaminoalkyl chitosans increased with the number of carbons in the alkyl groups, indicating that hydrophobic interaction plays a major role in the sequestration of bile acids;[113] similarly, the capacity of 6-oxychitosan for cholic acid decreases with increasing degree of oxydation, i.e., loss of cationicity.[114]

Among the chitosan salts that instantly form from chitosan and bile acids *in vitro*, taurocholate appears to possess infrared spectral characteristics that qualify it as a real ionic, water-insoluble, poorly crystalline, hydrophobic chitosan salt. In this chitosan taurocholate and homologues, chitosan accounts for no more than one half by weight; therefore, the hydrophobic nature prevails when contacting lipids. Butter and oils are collected to high extents with no discrimination of their components, including cholesterol and tocopherols. For these chitosan salts, the lipid uptake is much higher than for plain chitosan. The salts studied in the present work are scarcely hydrolyzed by a variety of hydrolases, and it is presumed that their complexes with lipids would be even more resistant to enzymatic attack.

If the above findings are extrapolated from the *in vitro* model to the physiological environment, it seems reasonable to speculate that:

1. Chitosan glycocholate and chitosan taurocholate insoluble salts subtract bile salts from the circulation, thus forcing the organism to replete the bile pool at the expense of cholesterol.
2. The activity of lipases on triglycerides is depressed as a consequence of the poor emulsification of lipids due to the lowered availability of taurocholate, the emulsifier. It is known that the pancreatic lipases require a certain dimension of the oil droplets in the emulsion to hydrolyze triglycerides: now when the bile salts become scarce, inadequate emulsions are formed and then limited hydrolysis of triglycerides takes place. Ample information on digestive lipases supports these views.[115–119] Lipases work thanks to the presence of bile salts that in one case activate the bile salts-dependent lipases and in the other case provide the emulsion necessary

to the pancreatic lipases for enzymatic activity. Moreover, as soon as the bile salt availability decreases due to chitosan ingestion, the bile salts-dependent lipases are poorly activated, and assimilation of lipids by the organism decreases sharply.

3. While pectinase is one of the most representative bacterial enzymes in the intestine, the resistance of the chitosan taurocholate, glycocholate, and taurodeoxycholate salts to hydrolysis by this enzyme, as well as by other hydrolases, would certainly be an indication of the capacity of these hydrophobic salts to be excreted, presumably with accompanying adsorbed lipids.

From the cited literature, it appears that chitosan would be most effective in the reduction of cholesterol and body weight provided that chitosan formulations lend themselves to prompt reaction with bile salts. Unfortunately, little attention has been paid so far to the composition of the tablets and to the questionable need to administer chitosan in the form of tablets.

6.5.2 Current Views on Side Aspects

One remark against the use of chitosan has been the worry that chitosan could deplete iron from the organism, based on the well-known chelating ability of chitosan, which was studied *in vitro* for a number of transition metals ions.[1] However, many chelating compounds are introduced into the organism with food, such as alginate or, more commonly, citric acid. The latter in particular is a very strong chelating agent, superior to chitosan, from which it can strip off the chelated metal ions. Citric acid is naturally present in fruits and vegetables and is largely added to foods, drinks, and candies, but no one has ever expressed worries about its capacity to remove iron from living tissues. On the contrary, chitosan may contain relatively high trace levels of various metals coming from the equipment used to manufacture chitosan, especially iron and copper that are not declared.

There is no deep knowledge on chitosanolytic microbes in the human digestive tract. Bacteriostatic properties of chitosan can affect the microflora of the digestive tract. It was proved that chitosan derivatives inhibit several pathogens, but the effect on common strict anaerobes in the digestive tract is unknown. There is no information about the possibility of different effects of chitosan on adherent bacteria and the anaerobic bacterial population in the colon lumen. There is a scattered knowledge of chitosan and chitosan derivatives digestion in the human digestive tract and their favorable effect on prevention and treatment of Crohn's disease and ulcerative colitis. Chitosan has been proposed and tested as a suitable vehicle for the delivery of drugs to the colon in the treatment of the mentioned diseases.[120]

As for the treatment of celiac intolerance, important scientific indications exist in the literature concerning the capacity of chitosan to bind the toxic fraction of gluten.[121,122] The recently acquired status of being generally considered a safe substance will foster the extension of chitosan applications in the food area, particularly foods for celiac patients.

6.5.3 Overweight

NiMhurchu et al.[123] conducted two systematic reviews of randomized controlled trials to determine whether chitosan, a popular over-the-counter weight-loss supplement, is an effective treatment for overweight and obese people. Included were randomized controlled trials of chitosan with a minimum duration of 4 weeks in adults who were overweight or obese or had hypercholesterolemia at baseline.

Fourteen published trials including a total of 1131 participants met the inclusion criteria in a study by NiMhurchu et al.[124] Analyses including all trials indicated that chitosan preparations result in a significantly greater weight loss (weighted mean difference = −1.7 kg, 95% confidence interval (CI) = −2.1 to −1.3 kg, $p < 0.00001$), decrease in total cholesterol (−0.2 mmol/l, 95% CI = −0.3 to −0.1, $p < 0.00001$), and decrease in systolic (−5.9 mmHg, 95% = CI −7.3 to −4.6, $p < 0.0001$) and diastolic (−3.4 mmHg, 95% CI = −4.4 to −2.4, $p < 0.00001$) blood pressure compared with placebo. There were no clear differences between intervention and control groups in terms of frequency of adverse events or in fecal fat excretion. However, the quality of many studies was suboptimal, and analyses restricted to studies that met allocation concealment criteria, were larger, or were of longer duration showed that such trials produced substantially smaller decreases in weight and total cholesterol. It was concluded that there is some evidence that chitosan is more effective than placebo in the short-term treatment of overweight and obese persons. However, many trials have been of poor quality and results have been variable. Results obtained from high-quality trials indicate that the effect of chitosan on body weight is minimal and unlikely to be of clinical significance.

The same research team[125] conducted a 24-week randomized, double-blind, placebo-controlled trial, with a total of 250 participants (82% women, mean (s.d.) body mass index = 35.5 (5.1) kg/m^2, mean age = 48 (12) years). The chitosan group lost more body weight than the placebo group (mean (s.e.) = −0.4 (0.2) kg (0.4% loss) vs. +0.2 (0.2) kg (0.2% gain), $p = 0.03$) during the 24-week intervention. Similar small changes occurred in circulating total and LDL cholesterol and glucose ($p < 0.01$). There were no significant differences between groups for any of the other measured outcomes. Therefore, in this 24-week trial, chitosan treatment did not result in a clinically significant loss of body weight compared with placebo.

Studies of the effect of chitosan on human adiposity suggest that results may differ depending on whether the subjects are eating *ad libitum* or are on a weight-loss diet. Overweight subjects consuming 2.4 g/day of chitosan for 28 days, as well as their normal diet, showed no change in body weight during the trial.[88]

A 6-week study was conducted on obese adults who lost more body weight (−2.3 vs. 0.0 kg), body fat (−1.1 vs. 0.2%), and absolute fat mass (−2.0 vs. 0.2 kg) than a placebo group. The combination of glucomannan, chitosan, fenugreek *Gymnema sylvestre*, and vitamin C was therefore proven effective.[126] Gades and Stern tested the fat-trapping capacity of a chitosan product in 12 men and 12 women. Fecal fat excretion increased with chitosan by 1.8 ± 2.4 g/day in males, but did not increase with chitosan in females. Therefore, chitosan failed to meet fat-trapping claims.[127]

Chitosan greatly increases fecal fat excretion when consumed in sufficient amounts.[128] Therefore, chitosan may be useful for promoting weight loss. However, chitosan can accelerate weight loss when subjects follow a low-calorie diet, but will be ineffective in those consuming their normal diets.

6.5.4 Osteoarthritis

Osteoarthritis is the most common form of arthritis and an important medical problem in the aging population. Currently, osteoarthritis is treated with a variety of pharmacologic therapies, including acetaminophen, nonsteroidal anti-inflammatory drugs (NSAIDs), cyclo-oxygenase (COX)-2 inhibitors, intra-articular steroids, viscosupplements, vitamins, and capsaicin.[129] Most of these are directed at pain relief and do not address the more fundamental issue of correcting the underlying degenerative disorder of connective tissue. For this reason, a great deal of interest is directed toward the use of glucosamine and N-acetyl-D-glucosamine (NAG) for treatment of osteoarthritis. These compounds are fundamental building blocks of the structural matrix of connective tissue in joints (e.g., glycosaminoglycans (GAGs), chondroitin, and hyaluronic acid). Glucosamine, occurring in the connective and cartilage tissues, contributes to maintaining the strength, flexibility, and elasticity of these tissues. Glucosamine is not only necessary for the synthesis of GAGs, but also stimulates their synthesis and prevents their degradation. Moreover, these compounds have been shown to offer protection from oxidative damage. The subject of glucosamine in nutraceuticals has been reviewed by Deal and Moskowitz,[130] Laverty et al.,[131] and Anderson et al.[132]

Glucosamine is widely used to relieve symptoms from osteoarthritis. Its safety and effects on glucose metabolism were critically evaluated by Anderson et al.[132] The LD_{50} of oral glucosamine in animals is ca. 8 g/kg, with no adverse effects at 2.7 g/kg for 12 months. Because altered glucose metabolism can be associated with parenteral administration of large doses of glucosamine in animals and with high concentrations in *in vitro* studies, the clinical importance of these effects should be critically evaluated. Oral administration of large doses of glucosamine in animals has no documented effects on glucose metabolism. *In vitro* studies demonstrating effects of glucosamine on glucose metabolism have used concentrations that are 100 to 200 times higher than tissue levels expected with oral glucosamine administration in humans. The reviewed clinical trial data included 3063 human subjects. Fasting plasma glucose values decreased slightly for subjects after oral glucosamine for 66 weeks. There were no adverse effects of oral glucosamine administration on blood, urine, or fecal parameters. Side effects were significantly less common with glucosamine than with placebo or nonsteroidal anti-inflammatory drugs. In contrast to the latter, no serious or fatal side effects have been reported for glucosamine. Therefore, glucosamine is safe under current conditions of use and does not affect glucose metabolism, notwithstanding the worries expressed by Laverty et al.[131] during their studies on horses.

Glucosamine sulfate is a commonly used compound for the treatment of osteoarthritis. Largo et al.[133] studied whether or not glucosamine sulfate could modify the NFkappaB activity and the expression of COX-2, a NFkappaB-dependent gene. Using human osteoarthritic chondrocytes in culture stimulated with interleukin-1 beta (IL-1beta), the effects of glucosamine sulfate on NFkappaB activation, nuclear translocation of NFkappaB/Rel family members, COX-1 and COX-2 expressions and syntheses, and prostaglandin E2 (PGE2) concentration were studied. Glucosamine sulfate significantly inhibited NFkappaB activity in a dose-dependent manner, as well as the nuclear translocation of p50 and p65 proteins. Glucosamine sulfate also inhibited the gene expression and the protein synthesis of COX-2 induced by IL-1beta, while no effect on COX-1 synthesis was seen. Glucosamine sulfate also inhibited the release of PGE2 to conditioned media of chondrocytes stimulated with IL-1beta. It can be affirmed, therefore, that glucosamine sulfate inhibits the synthesis of proinflammatory mediators in human osteoarthritic chondrocytes stimulated with IL-1beta through a NFkappaB-dependent mechanism. Glucosamine sulfate has a role as a symptom-releaving and structure-modifying drug in the treatment of osteoarthritis.

According to a study by Persiani et al.,[134] glucosamine is rapidly absorbed after oral administration and its pharmacokinetics are linear in the dose range of 750 to 1500 mg, but not at 3000 mg. Plasma levels increased over 30-fold from baseline and peaked at about 10 μM with the standard 1500 mg once-daily dosage. Glucosamine distributed to extravascular compartments and its plasma concentrations were still above baseline up to the last collection time. The glucosamine elimination half-life was only tentatively estimated to average 15 h.

Glucosamine is bioavailable after oral administration of crystalline glucosamine sulfate, persists in circulation, and its pharmacokinetics support once-daily dosage. Steady-state peak concentrations at the therapeutic dose of 1500 mg were in line with those found to be effective in selected *in vitro* mechanistic studies. As a consequence of this study, pharmacokinetic, efficacy, and safety data are now available for a glucosamine formulation, in agreement with previous data.[135]

A problem with the use of oral glucosamine or its derivatives has been its relatively short half-life in blood, and thus a sustained release form of glucosamine has long been sought. The potential use of polymeric forms such as chitin or chitosan was explored by Talent et al. and Rubin et al.[136–138] The bioavailability of glucosamine was evaluated in normal healthy volunteers who ingested 1 to 1.5 g each day of NAG or chitin (Poly-NAG®).

The resulting serum levels of glucosamine and NAG from subjects who ingested Poly-NAG were equal to or greater than those from subjects who ingested NAG. When subjects stopped ingesting NAG, the levels of serum NAG decreased. In contrast, the serum NAG levels remained elevated for subjects who had been ingesting Poly-NAG. Thus, the serum half-life of glucosamine was longer in the Poly-NAG ingestors. The efficacy of Poly-NAG in treating patients with osteoarthritis was assessed in a double-blind, placebo-controlled pilot study that showed that Poly-NAG-treated patients had lessened pain compared to placebo.

A number of studies led to the formulation of chitosan suitable for commercial exploitation, mainly based on the association of chitosan with chondroitin sulfate, a chondroprotective agent, and manganese ascorbate, a promoter of collagen formation.[139–143]

As a consequence, a great deal has been published on the use of glucosamine sulfate and chondroitin sulfate for the treatment of osteoarthritis, and insight has been gained on the mechanism by which these monomers, oligomers, or polymers may function. In contrast, glucosamine did not affect the body weights, platelet counts, and bleeding time in guinea pigs after administration. These observations suggest that glucosamine is likely to exert an inhibitory action on platelets in guinea pigs receiving an average of 400 mg of glucosamine/animal/day by suppressing platelet aggregation, ATP release, and thromboxane A(2) production. Thus, glucosamine could be expected as a novel and safe antiplatelet agent.[144]

Chen et al.[145] investigated whether glucosamine sulfate could modulate the proinflammatory cytokine-induced expression of the gene for intercellular adhesion molecule-1, an inflammatory protein in human retinal pigment epithelial cells. They demonstrated the potentially important property of glucosamine sulfate in reducing intercellular adhesion molecule-1-mediated inflammatory mechanisms in the eye.

In a study on 32 patients with low back pain, an analgesic effect and improvement in quality of life were found with the use of a glucosamine complex.[146] The glucosamine complex was well tolerated.

Glucosamine and chondroitin sulfate were used to treat osteoarthritis in 1583 patients with symptomatic knee osteoarthritis. For patients with moderate to severe pain at baseline, the rate of response was significantly higher with combined therapy than with placebo (79.2% vs. 54.3%, $p = 0.002$).[147]

Chitosan as a vector system is expected to be useful for direct gene therapy for joint disease.[148] This study first sought to confirm that foreign genes can be transferred to articular chondrocytes in primary culture. Next, chitosan–DNA nanoparticles containing the interleukin-1 receptor antagonist (IL-1Ra) or interleukin-10 (IL-10) gene were injected directly into the knee joint cavities of osteoarthritic rabbits to clarify the *in vivo* transfer availability of the chitosan vectors. Clear expression of IL-1Ra was detected in the knee joint synovial fluid of the chitosan IL-1Ra-injected group. No expression was detected in the chitosan IL-10-injected group; this demonstrates that the transfection efficiency of chitosan–DNA nanoparticles was closely related to the type of the gene product. A significant reduction was also noted in the severity of histologic cartilage lesions in the group that received the chitosan IL-1Ra injection. This approach may therefore represent a promising future treatment for osteoarthritis.

6.5.4.1 Depolymerization of Chitosan

It has long been rationalized that several enzymes, including nonspecific lipases, lysozyme, and N-acetyl-glucosaminidase from the flora of the gut,[4] account for this degradation. Zhang and Neau[149] have described the *in vitro* degradation of chitosan by bacterial enzymes from the cecal and colonic contents in rats. A mammalian chitinase (chitotriosidase) from human macrophages is now being well characterized.[28,150] Its function has been suggested to be in the defense against chitin-containing pathogens.[151] Human cartilage glycoprotein 39 (HC-gp39) is a glycoprotein secreted by articular chondrocytes, synoviocytes, and macrophages. Increased levels of HC-gp39 have been demonstrated in synovial fluids of patients with rheumatoid or osteoarthritis. Recklies et al.[152] identified a second mammalian chitinase relatively abundant in the gastrointestinal tract. This chitinase is extremely acid stable and exhibits a pH optimum around pH 2. It is referred to as acidic mammalian chitinase and is capable of hydrolyzing artificial chitin-like substrates as well as animal and fungal chitin. Suzuki et al.[153] described the cellular expression of this gut chitinase in the parotid gland, von Ebner's gland, and gastric chief cells. Thus, it is now clear that there are enzymes in the gut promoting the degradation of chitin/chitosan to its oligomers and monomers and their absorption into the bloodstream.

6.5.4.2 Chitosan Depolymerization Products

Over the last decade, there has been a marked increase in the numbers of papers on glucosamine that deal with osteoarthritis and a parallel increase in the osteoarthritis literature focusing on the use of glucosamine.[154–162]

While the majority of recent studies have found positive results, there are a few exceptions. Eighty patients were studied in a randomized, placebo-controlled, double-blind trial of glucosamine sulfate (1.5 g daily for 6 months).[163] The assessment of pain used found no significant differences compared to the placebo group. Past clinical trials on glucosamine treatment of osteoarthritis suffered from nonstandard classification criteria in patient selection, small size sampling, short-duration studies, poor or absent radiological assessment, heterogeneous patients, and nonstandardized clinical outcomes measures.[164,165]

On the other hand, well-documented and carefully controlled studies have been completed. Setnikar and Rovati[166] reviewed absorption, distribution, metabolism, and excretion of glucosamine sulfate, and Towheed et al.[167] reviewed all randomized, controlled trials evaluating the effectiveness and toxicity of glucosamine in osteoarthritis. They identified 16 randomized, controlled trials with evidence that glucosamine is both safe and effective in treatment of osteoarthritis. In 13 of these in which glucosamine was compared to a placebo, glucosamine was found to be superior in all but one. In four studies in which glucosamine was compared to an NSAID, glucosamine was superior in two and equivalent in two. An update included 20 studies with 2570 patients. Pooled results from studies using a non-Rotta preparation or adequate allocation concealment failed to show benefit in pain and function (according to WOMAC, a validated health status questionnaire), while those studies evaluating the Rotta preparation showed that glucosamine was superior to a placebo in the treatment of pain and functional impairment resulting from symptomatic osteoarthritis. Two randomized controlled trials using the Rotta preparation showed that glucosamine was able to slow radiological progression of osteoarthritis of the knee over a 3-year period. Glucosamine was as safe as a placebo in terms of the number of subjects reporting adverse reactions.[167]

Ruane and Griffiths[168] compared oral glucosamine with ibuprofen for the relief of joint pain in a mini-review of double-blind, randomized, controlled trials. Studies on 218 participants compared 1.2 g of ibuprofen with 1.5 g of glucosamine sulfate daily and found glucosamine and ibuprofen to be of similar efficacy. They concluded that glucosamine provides cartilage-rebuilding properties not seen with simple analgesics, and that glucosamine could be used as an alternative to anti-inflammatory drugs and analgesics or as a useful adjunct to standard analgesic therapy. Similarly, Hochberg[169] reviewed the American College of Rheumatology recommendations for the medical management of patients with lower-limb osteoarthritis and suggested glucosamine as a first-line agent for patients with knee osteoarthritis with mild to moderate pain.

Glucosamine sulfate and ibuprofen reduced pain levels. Glucosamine sulfate combines efficacy with good tolerability.[170] Reginster et al.[171] did a randomized, double-blind, placebo-controlled trial in which 212 patients with knee osteoarthritis were assigned 1500 mg of oral glucosamine sulfate or placebo once daily for 3 years. The 106 patients on placebo had a progressive joint-space narrowing, with a mean joint-space loss after 3 years of –0.31 mm. There was no significant joint-space loss in the 106 patients on glucosamine sulfate: –0.06 mm. Similar results were reported with minimum joint-space narrowing. Symptoms worsened slightly in patients on placebo compared with the improvement observed after treatment with glucosamine sulfate.

The results by Braham et al.[172] suggest that glucosamine supplementation can provide some degree of pain relief and improved function in persons who experience regular knee pain, which may be caused by prior cartilage injury or osteoarthritis. The trends in the results also suggest that, at a dosage of 2000 mg per day, the majority of improvements are present after 8 weeks.

Radiological evidence supports the efficacy of glucosamine and its derivatives. Pavelka et al.[173] conducted a randomized, placebo-controlled trial of over 200 patients with mild to moderately severe knee osteoarthritis. They evaluated treatment (1.5 g of glucosamine sulfate per day or placebo) by radiological analyses over a 3-year period. Their conclusions were that long-term treatment with glucosamine sulfate retarded the progression of knee osteoarthritis, possibly determining disease modification. Similarly, Bruyere et al.[174] reported the results of a 3-year prospective, placebo-controlled study of 212 patients with different degrees of knee osteoarthritis. They concluded that patients with the less severe knee osteoarthritis experienced the most dramatic disease progression in terms of joint-space narrowing.

The oral ingestion of Poly-NAG was followed by the appearance of oligomers and monomers of glucosamine and N-acetyl glucosamine in the blood.[175,176] Significant clinical improvement was found in the symptom assessment of the subjects taking oral Poly-NAG. It was concluded that oral ingestion of the polymer resulted in degradation to monomers of glucosamine and its derivatives, which were absorbed and functioned in the chondroprotective processes in a manner similar to that of free glucosamine.

A portion of orally administered chondroitin sulfate or dermatan sulfate was absorbed in polymeric form, apparently by pinocytosis.[177] These studies also rationalized clinical studies that concluded that oral chondroitin sulfate provided pain relief to patients suffering from osteoarthritis.[178]

6.5.4.3 Mechanism of Action

The general assumption has been that glucosamine and its derivatives primarily function to directly provide substrates for the synthesis of new GAGs in the joints, largely based on observations in animal studies utilizing radiolabeled tracers.[8] The most obvious potential mechanisms in which oral Poly-NAG might function include the provision of monomers (e.g., glucosamine) for GAG synthesis. Glucosamine has also been shown to function as an activator of GAG synthesis, inhibitor of GAG degradation, and antioxidant. Glucosamine sulfate has also been shown to modify chondrocyte metabolism by acting on protein kinase C, cellular phospholipase A2, protein synthesis, and possibly collagenase.[179]

By enhancing the protective metabolic response of chondrocytes to stress, GlcN and chondroitin sulfate may improve their ability for repair and regeneration. These compounds function as biological response modifiers that boost natural protective responses of tissue under adverse environmental conditions.[180]

Chondrocytes, the cells of cartilage, consume glucose as a primary substrate for ATP production in glycolysis, utilize glucosamine sulfate and other sulfated sugars as structural components for extracellular matrix synthesis, and are dependent on hexose uptake and delivery to metabolic and biosynthetic pools. Data from several laboratories suggest that chondrocytes express multiple isoforms of the GLUT/SLC2A family of glucose/polyol transporters.[181]

Studies on intact articular cartilage explants showed that GlcN was taken up by the chondrocytes and incorporated selectively into the hexosamine, but not the hexuronic acid, components of the glycosaminoglycan chains of articular cartilage proteoglycan.[182]

With respect to effects on inflammation, glucosamine has been reported to inhibit human peripheral blood neutrophil functions, including superoxide generation, phagocytosis, granule enzyme release, chemotaxis, CD11b expression, actin polymerization, and p38 mitogen-activated protein kinase (MAPK) phosphorylation.[183] These studies explain its anti-inflammatory actions in arthritis. Shikhman et al.[184] proposed that glucosamine and NAG inhibit IL-1beta and TNF-alpha-induced NO production in normal human articular chondrocytes via inhibition of the inducible NO synthase expression. In addition, they found that NAG also supressed the production of IL-1beta-induced cyclooxygenase-2 and IL-6. Gouze et al.[185] also found that glucosamine reversed the decrease in proteoglycan synthesis induced by interleukin-1 beta, and that the glucosamine modulation of IL-1-induced activation of rat chondrocytes is at the receptor level inhibiting the NFkappaB pathway.

Hua et al.[183] evaluated the effects of glucosamine on neutrophil functions using human peripheral blood neutrophils. Glucosamine (0.01 to 1 m*M*) dose dependently suppressed the superoxide anion generation induced by formyl-Met-Leu-Phe (fMLP) or complement-opsonized zymosan and inhibited the phagocytosis of complement-opsonized zymosan or immunoglobulin G (IgG)-opsonized latex particles. Furthermore, glucosamine inhibited the release of the granule enzyme lysozyme from phagocytosing neutrophils and suppressed neutrophil chemotaxis toward zymosan-activated serum. In addition, glucosamine inhibited fMLP-induced upregulation of CD11b (significantly), polymerization of actin, and phosphorylation of p38 MAPK. In contrast, N-acetyl-glucosamine did not affect these neutrophil functions (superoxide generation, phagocytosis, granule enzyme release, chemotaxis, CD11b expression, actin polymerization, and p38 MAPK phosphorylation) at the concentrations examined (1 to 10 m*M*). These observations suggest that glucosamine suppresses the neutrophil functions, thereby possibly exhibiting anti-inflammatory actions in arthritis.

6.5.4.4 Product Quality

Blakeley and Ribeiro[186,187] investigated the decision-making and self-treatment practices of older users of glucosamine products. Advice from friends was a major factor in the decision to take glucosamine. Cost was a major factor influencing the type and brand name of the product used. Ninety-two percent managed their own treatment. The majority (88%) used glucosamine regularly; only one third took the recommended daily dose. Sixty-seven percent perceived the product to be helpful in relieving symptoms.

The quality of the products varies widely, however: Russell et al.[188] analyzed in a coded, blinded manner 14 commercially available capsules or tablets of GLS, plus one herbal mixture as a control. The amount of free base varied from 41 to 108% of the content stated on the label; the amount of glucosamine varied from 59 to 138% even when expressed as sulfate. The content should be best expressed in terms of free base. No analogous investigation is available for chitosans.

6.6 CONCLUSIONS

Chitosan is effective in lowering serum cholesterol concentration and hypertension in subjects with a restricted diet. Of course, the quality and chemical form of chitosan should be adequate to the scope.

Chitosan, glucosamine, and its sulfate salt appear to provide an effective therapy in the prevention and treatment of osteoarthritis. The fact that chitosan provides an effective sustained release of glucosamine can now be appreciated. Mechanism studies are beginning to reveal the pathways of involvement of glucosamine and its derivatives in preventing inflammatory cascades and oxidative damage, in addition to its direct effects on GAG synthesis and degradation.[189–191]

The existence of many new and encouraging biological approaches to cartilage repair justifies the future investment of time and money in this research area, particularly given the extremely high socioeconomic importance of such therapeutic strategies in the prevention and treatment of these common joint diseases and traumas. Moreover, clinical epidemiological and prospective trials are urgently needed for an objective, scientific appraisal of current therapies and future novel approaches.[192] Permissible claims for dietary supplements should comply with criteria such as those provided by the Food and Drug Administration.

ACKNOWLEDGMENTS

Thanks are due to Mrs. Maria Weckx for her valuable assistance.

REFERENCES

1. Muzzarelli, R.A.A., *Natural Chelating Polymers*, Pergamon, Oxford, 1973.
2. Muzzarelli, R.A.A., *Chitin*, Pergamon, Oxford, 1977.
3. Muzzarelli, R.A.A., Jeuniaux, C., and Gooday, G.W., Eds., *Chitin in Nature and Technology*, Plenum, New York, 1986.
4. Jollès, P. and Muzzarelli, R.A.A., Eds., *Chitin and Chitinases*, Birkhauser, Basel, 1999.
5. Muzzarelli, R.A.A., Ed., *Chitin Enzymology*, Vol. 1, Atec, Grottammare, Italy, 1993.
6. Muzzarelli, R.A.A., Ed., *Chitin Enzymology*, Vol. 2, Atec, Grottammare, Italy, 1996.
7. Muzzarelli, R.A.A., Ed., *Chitin Enzymology*, Vol. 3, Atec, Grottammare, Italy, 2001.
8. Muzzarelli, R.A.A., Ed., *Chitosan per os: From Dietary Supplement to Drug Carrier*, Atec, Grottammare, Italy, 2000.
9. Muzzarelli, R.A.A. and Muzzarelli, C., Native and modified chitins in the biosphere, in *Nitrogen Containing Macromolecules in Biosphere and Geosphere*, Stankiewicz, A., Ed., American Chemical Society, Philadelphia, 1998.
10. Muzzarelli, R.A.A., Stanic, V., and Ramos, V., Enzymatic depolymerization of chitins and chitosans, in *Carbohydrate Biotechnology Protocols*, Bucke, C., Ed., Humana Press, Totowa, NJ, 197, 1999.
11. Muzzarelli, R.A.A., Chitin, in *The Polysaccharides,* Aspinall, G.O., Ed., Academic Press, New York, 3, 418, 1985.
12. Muzzarelli, R.A.A. and Muzzarelli, C., Chitosan chemistry: relevance to the biomedical sciences, in *Advances in Polymer Science*, Heinze, T., Ed., Springer Verlag, Berlin, 151, 2005.
13. Kumar, M.N.V.R., Muzzarelli, R.A.A., Muzzarelli, C., Sashiwa, H., and Domb, A.J., Chitosan chemistry and pharmaceutical perspectives, *Chem. Rev.*, 104, 6017, 2004.
14. Dutta, P.K., Dutta, J., and Tripathi, V.S., Chitin and chitosan: chemistry, properties and applications, *J. Sci. Ind. Res.*, 63, 20, 2004.
15. Shalaby, S.W., DuBose, J.A., and Shalaby, M., Chitosan-based systems, *Absorb. Biodegrad. Polym.*, 77, 2004.
16. Agullo, E., Rodriguez, M.S., Ramos, V., and Albertengo, L., Present and future role of chitin and chitosan in food, *Macromol. Biosci.*, 3, 521, 2003.
17. Paoletti, M.G., *Ecological Implications of Minilivestock: Potential of Insects, Rodents, Frogs and Snails*, Science Publishers, Enfield, NH, 2005.
18. Backwell, L.R. and D'Errico, F., Evidence for termites foraging by Swartkrans early hominids, *Proc. Natl. Acad. Sci. U.S.A.*, 98, 1358, 2001.
19. DeFoliart, G.R., The human use of insects as food and animal feed, *Bull. ESA*, 13, 22, 1989.
20. DeFoliart, G.R., Insects as human food, *Crop Prot.*, 11, 395, 1992.
21. DeFoliart, G.R., Insects as food: why the Western attitude is important, *Ann. Rev. Entomol.*, 44, 21, 1999.
22. Gawaad, A.A.A. and Brune, H., Insect protein as a possible source of protein to poultry, *Z. Tierphysiol. Tierrenahrung Futtermittelkde*, 42, 216, 1979.
23. Taira, T., Ohnuma, T., Yamagami, T., Aso, Y., Ishiguro, M., and Ishihara, M., Antifungal activity of rye (*Secale cereale*) seed chitinase: the different binding manner of class I and class II chitinases to the fungal cell walls, *Biosci. Biotechnol. Biochem.*, 66, 970, 2002.
24. Taira, T., Toma, N., and Ishihara, M., Purification, characterization and antifungal activity of chitinases from pineapple (*Ananas comosus*) leaf, *Biosci. Biotechnol. Biochem.*, 69, 189, 2004.

25. Truong, N.H., Park, S.M., Nishizawa, Y., Watanabe, T., Sasaki, T., and Itoh, Y., Structure, heterologous expression, and properties of rice (*Oryza sativa* L.) family 19 chitinases, *Biosci. Biotechnol. Biochem.*, 67, 1063, 2003.
26. Gardiner, T., Dietary N-acetylglucosamine (GlcNac): absorption, distribution, metabolism, excretion, and biological activity, *Glycosci. Nutrition*, 1, 1, 2000.
27. Gianfrancesco, F. and Musumeci, S., The evolutionary conservation of the human chitotriosidase gene in rodents and primates, *Cytogenet. Genome Res.*, 105, 54, 2004.
28. Renkema, G.H., Boot, R.G., Au, F.L., Donker-Koopman, W.E., Strijland, A., Muijsers, A.O., Hrebicek, M., and Aerts, J.M.F.G., Chitotriosidase, a chitinase, and the 39-kDa human cartilage glycoprotein, a chitin-binding lectin, are homologues of family 18 glycosyl hydrolases secreted by human macrophages, *Eur. J. Biochem.*, 251, 504, 1998.
29. Renkema, G.H., Boot, R.G., Muijsers, A.O., Donker-Koopman, W.E., and Aerts, J.M.F.G., Purification and characterization of human chitotriosidase, a novel member of the chitinase family of proteins, *J. Biol. Chem.*, 270, 2198, 1995.
30. Tsukada, H., Miyake, T., Ueda, S., Shirahase, H., Matsunaga, T., Sakai, M., and Uchino, H., Hexosaminidase activity in human gastric mucosa, *Nippon Shokakibyo Gakkai Zasshi*, 79, 1391, 1982.
31. Musumeci, M., Malaguarnera, L., Simpore, J., Barone, R., Whalen, M., and Musumeci, S., Chitotriosidase activity in colostrum from African and Caucasian women, *Clin. Chem. Lab. Med.*, 43, 198, 2005.
32. Sparringa, R.A. and Owens, J.D., Glucosamine content of tempe mould, *Rhizopus oligosporus*, *Int. J. Food Microbiol.*, 47, 153, 1999.
33. Sakamoto, H., Characteristics and applications of glucosamines to food, *Food Style*, 21, 58, 1997.
34. Pittler, M.H. et al., Randomized, double-blind trial of chitosan for body weight reduction, *Eur. J. Clin. Nutr.*, 53, 379, 1999.
35. Ernst, E. and Pittler, M.H., Chitosan as a treatment for body weight reduction? A meta-analysis, *Perfusion*, 11, 461, 1998.
36. Shahidi, F., Arachchi, J.K.V., and Jeon, Y.J., Food applications of chitin and chitosan, *Trends Food Sci. Technol.*, 10, 37, 1999.
37. Knowles, J. and Roller, S., Efficacy of chitosan, carvacrol, and a hydrogen peroxide-based biocide against foodborne microorganisms in suspension and adhered to stainless steel, *J. Food Prot.*, 64, 1542, 2001.
38. Muzzarelli, R.A.A., Tarsi, R., Filippini, O., Giovanetti, E., Biagini, G., and Varaldo, P.E., Antimicrobial properties of N-carboxybutyl chitosan, *Antimicrob. Agents Chemother.*, 34, 2019, 1990.
39. Helander, I.M., Nurmiaho, E.L., Ahvenainen, R., Rhoades, J., and Roller, S., Chitosan disrupts the barrier properties of the outer membrane of Gram-negative bacteria, *Int. J. Food Microbiol.*, 71, 235, 2001.
40. Jia, Z.S., Shen, D.F., and Xu, W.L., Synthesis and antibacterial activities of quaternary ammonium salt of chitosan, *Carbohydr. Res.*, 333, 1, 2001.
41. Kim, S.K. and Rajapakse, N., Enzymatic production and biological activities of chitosan oligosaccharides, *Carbohydr. Polym.*, 62, 357, 2005.
42. Roller, S., Sagoo, S., Board, R., O'Mahony, T., Caplice, E., Fitzgerald, G., Fogden, M., Owen, M., and Fletcher, H., Novel combination of chitosan, carnocin and sulphite for the preservation of chilled pork sausages, *Meat Sci.*, 62, 165, 2002.
43. Sagoo, S., Board, R., and Roller, S., Chitosan inhibits growth of spoilage microorganisms in chilled pork products, *Food Microbiol.*, 19, 175, 2002.

44. Lin, K.W. and Chao, J.Y., Quality characteristics of reduced-fat sausage as related to chitosan molecular weight, *Meat Sci.*, 59, 343, 2001.
45. Jo, C., Lee, J.W., Lee, K.H., and Byun, M.W., Quality properties of pork sausage prepared with chitosan oligomer, *Meat Sci.*, 59, 369, 2001.
46. Jeong, Y.J., Kamil, J.Y.V.A., and Shahidi, F., Chitosan as an edible invisible film for quality preservation of herring and Atlantic cod, *J. Agric. Food Chem.*, 50, 5167, 2002.
47. Jumaa, M., Furkert, F.H., and Muller, B.W., A new lipid emulsion formulation with high antimicrobial efficacy using chitosan, *Eur. J. Pharm. Biopharm.*, 53, 115, 2002.
48. Tripathi, P. and Dubey, N.K., Exploitation of natural products as an alternative strategy to control postharvest fungal rotting of fruit and vegetables postharvest, *Biology Technol.*, 32, 235, 2004.
49. Cuero, R.G., Antimicrobial action of exogenous chitosan, in *Chitin and Chitinases*, Jollès, P. and Muzzarelli, R.A.A., Eds., Birkhauser, Basel, 315, 1999.
50. Romanazzi, G., Nigro, F., Ippolito, A., Di Venere, D., and Salerno, M., Effects of pre- and postharvest chitosan treatment to control storage grey mold of table grapes, *J. Food Sci.*, 67, 1, 2002.
51. Romanazzi, G., Nigro, F., and Ippolito, A., Short hypobaric treatments potentiate the effect of chitosan in reducing storage decay of sweet cherries, *Postharvest. Biol. Technol.*, 29, 73, 2003.
52. El-Ghaouth, A., Smilanick, J.L., Brown, G.E., Ippolito, A., Wisniewski, M., and Wilson, C.L., Application of *Candida saitoana* and glycolchitosan for the control of postharvest diseases of apple and citrus fruit under semicommercial conditions, *Plant Dis.*, 84, 243, 2000.
53. Devlieghere, F., Vermeulen, A., and Debevere, J., Chitosan, antimicrobial activity, interactions with food components and applicability as a coating on fruit and vegetables, *J. Food Microbiol.*, 21, 703, 2004.
54. Galed, G., Fernández-Valle, M.E., Martínez, A., and Heras, A., Application of MRI to monitor the process of ripening and decay in citrus treated with chitosan solutions, *Magn. Reson. Imaging*, 22, 127, 2004.
55. Cheah, L.H., Page, B.B.C., and Shepherd, R., Chitosan coating for inhibition of *Sclerotinia* rot of carrots, *New Zealand J. Crop Hort. Sci.*, 25, 89, 1997.
56. Molloy, C., Chedah, L.H., and Koolaard, J.P., Induced resistance against *Sclerotinia sclerotiorum* in carrots treated with enzymatically hydrolysed chitosan, *Postharvest Biol. Technol.*, 33, 61, 2004.
57. Durango, A.M., Soares, N.F.F., and Andrade, N.J., Microbiological evaluation of an edible antimicrobial coating on minimally processed carrots, *Food Control*, 17, 336, 2006.
58. Srinivasa, P.C., Baskaran, R., Ramesh, M.N., Prashanth, K.V.H., and Tharanathan, R.N., Storage of mango packed using biodegradable chitosan film, *Eur. Food Res. Technol.*, 215, 504, 2002.
59. Vivekananthan, R., Ravi, M., Ramanathan, A., and Samiyappan, R., Lytic enzymes induced by *Pseudomonas fluorescens* and other biocontrol organisms mediate defence against the anthracnose pathogen in mango, *World J. Microbiol. Biotechnol.*, 20, 235, 2004.
60. Kungsuwan, A., Ittipong, B., Inthuserd, P., and Dokmaihom, S., Shelf life extension of mangosteen by chitosan coating, in *Proceedings of Conference on Advances in Seafood Byproducts*, 241, 2003.
61. Du, J., Gemma, H., and Iwahori, S., Effects of chitosan coating on the storage of peach, Japanese pear and kiwifruit, *J. Jpn. Soc. Hort. Sci.*, 66, 15, 1997.

62. Badawy, M.E.I., Rabea, E.I., Rogge, T.M., Stevens, C.V., Steurbaut, W., Hofte, M., and Smagghe, G., Fungicidal and insecticidal activity of O-acyl chitosan derivatives, *Polym. Bull.*, 54, 279, 2005.
63. Winkelmann, G., Ed., *Microbial Transport Systems*, Wiley-VCH, Weinheim, 2001.
64. Hirano, S. and Nagano, N., Effects of chitosan pectic acid, lysozyme and chitinase on growth of several phytopathogens, *Agric. Biol. Chem.*, 53, 3065, 1989.
65. Walker-Simmons, M., Hadwiger, L., and Ryan, C.A., Chitosan and pectic polysaccharides both induce the accumulation of the antifungal phytoalexin pisatin in pea pods and antinutrient proteinase inhibitors in tomato leaves, *Biochem. Biophys. Res. Commun.*, 110, 194, 1983.
66. Hadwiger, L.A. and Beckman, J.M., Chitosan as a component of pea *Fusarium solani* interactions, *Plant Physiol.*, 66, 205, 1980.
67. Muzzarelli, R.A.A., Miliani, M., Cartolari, M., Tarsi, R., Tosi, G., and Muzzarelli, C., Fungistatic activity of modified chitosans against *Saprolegnia parasitica, Biomacromolecules*, 2, 165, 2001.
68. Bishop, R.H., Duncan, C.L., Evancho, G.N., and Young, H., Estimation of fungal contamination on tomato products by chemical assay for chitin, *J. Food Sci.*, 47, 437, 1982.
69. Bautista-Baños, S., Hernández-Lauzardo, A.N., Velázquez-del Valle, M.G., Hernández-López, M., Barka, E.A., Bosquez-Molina, E., and Wilson, C.L., Chitosan as a potential natural compound to control pre- and postharvest diseases of horticultural commodities, *Crop Prot.*, 25, 108, 2006.
70. El-Ghaouth, A., Arul, J., Ponnampalam, R., and Boulet, M., Chitosan coating effects on storability and quality of fresh strawberries, *J. Food Sci.*, 56, 1618, 1991.
71. El-Ghaouth, A., Ponnampalam, R., Castaigne, F., and Arul, J., Chitosan coating to extend the storage life of tomatoes, *Hort. Sci.*, 27, 1016, 1992.
72. Kim, M. and Han, J., Evaluation of physico-chemical characteristics and microstructure of tofu containing high viscosity chitosan, *Int. J. Food Sci. Technol.*, 37, 277, 2002.
73. No, H.K., Park, N.Y., Lee, S.H., Hwang, H.J., and Meyers, S.P., Antibacterial activities of chitosans and chitosan oligomers with different molecular weights on spoilage bacteria isolated from tofu, *J. Food Sci.*, 67, 1511, 2002.
74. No, H.K., Park, N.Y., Lee, S.H., and Meyers, S.P., Antibacterial activity of chitosans and chitosan oligomers with different molecular weights, *Int. J. Food Microbiol.*, 74, 65, 2002.
75. Lattanzio, V., Cardinali, A., and Palmieri, S., The role of phenolics in the post-harvest physiology of fruits and vegetables: browning reaction and fungal diseases, *Ital. J. Food Sci.*, 1, 3, 1994.
76. Zhang, D. and Quantick, C., Effects of chitosan coating on the enzymatic browning and assay during post-harvest storage of litchi fruit, *Post-harvest Biol. Technol.*, 12, 195, 1997.
77. Dong, H.Q., Cheng, L.Y., Tan, J.H., Zheng, K.W., and Jiang, Y.M., Effects of chitosan coating on quality and shelf life of peeled litchi fruit, *J. Food Eng.*, 64, 355, 2004.
78. Chatterjee, S., Chatterjee, S., Chatterjee, B.P., and Guha, A.K., Clarification of fruit juice with chitosan, *Process Biochem.*, 39, 2229, 2004.
79. Roller, S. and Covill, N., The antifungal properties of chitosan in laboratory media and apple juice, *Int. J. Food Microbiol.*, 47, 67, 1999.
80. Spagna, G., Pifferi, P.G., Rangoni, C., Mattivi, F., Nicolini, G., and Polmonari, R., The stabilization of white wines by absorption of phenolic compounds on chitin and chitosan, *Food. Res. Int.*, 29, 241, 1996.

81. No, H.K. and Meyers, S.P., Crawfish chitosan as a coagulant in recovery of organic compounds from seafood processing streams, *J. Agric. Food Chem.*, 37, 580, 1989.
82. Fernandez, M. and Fox, P.F., Fractionation of cheese nitrogen using chitosan, *Food Chem.*, 58, 319, 1997.
83. Ausar, S.F., Characterization of casein micelle precipitation by chitosans, *J. Dairy Sci.*, 84, 361, 2001.
84. Altieri, C., Scrocco, C., Sinigaglia, M., and DelNobile, M.A., Use of chitosan to prolong mozzarella cheese shelf life, *J. Dairy Sci.*, 88, 2683, 2005.
85. Ausar, S.F., Passalacqua, N., Castagna, L.F., Bianco, I.D., and Beltramo, D.M., Growth of milk fermentative bacteria in the presence of chitosan for potential use in cheese making, *Int. Dairy J.*, 12, 899, 2002.
86. Gallaher, D.D., Hassel, C.A., and Lee, K.J., Relationships between viscosity of hydroxypropyl methylcellulose and plasma cholesterol in hamsters, *J. Nutr.*, 123, 1732, 1993.
87. Gallaher, D.D., Hassel, C.A., and Lee, K.J., Cholesterol reduction by glucomannan and chitosan is mediated by changes in cholesterol absorption and bile acid and fat excretion in rats, *J. Nutr.*, 130, 2753, 2000.
88. Gallaher, D.D., Gallaher, C.M., Mahrt, G.J., Carr, T.P., Hollingshead, C.H., Hesslink, R., Jr., and Wise, J., A glucomannan and chitosan fiber supplement decreases plasma cholesterol and increases cholesterol excretion in overweight normocholesterolemic humans, *J. Am. Coll. Nutr.*, 21, 428, 2002.
89. Gallaher, D.D., Chitosan, cholesterol lowering and caloric loss, *Agro-Food Ind. High Technol.*, 14, 32, 2003.
90. Sugano, M., Watanabe, S., Kishi, A., Izume, M., and Ohtakara, A., Hypocholesterolemic action of chitosans with different viscosity in rats, *Lipids*, 23, 187, 1988.
91. Razdan, A. and Pettersson, D., Hypolipidaemic, gastrointestinal and related responses of broiler chickens to chitosans of different viscosity, *Br. J. Nutr.*, 76, 387, 1996.
92. Sugano, M., Fujikawa, T., Hiratsuji, Y., and Hasegawa, Y., Hypocholesterolemic effects of chitosan in cholesterol-fed rats, *Nutr. Rep. Int.*, 18, 531, 1978.
93. Kobayashi, T., Otsuka, S., and Yugari, Y., Effect of chitosan on serum and liver cholesterol levels in cholesterol-fed rats, *Nutr. Rep. Int.*, 19, 327, 1979.
94. Nagyvary, J.J., Falk, J.D., Hill, M.L., Schmidt, M.L., Wilkins, A.K., and Bradbury, E.L., The hypolipidemic activity of chitosan and other polysaccharides in rats, *Nutr. Rep. Int.*, 20, 677, 1979.
95. Razdan, A., Pettersson, D., and Pettersson, J., Broiler chicken body weights, feed intakes, plasma lipid and small-intestinal bile acid concentrations in response to feeding of chitosan and pectin, *Br. J. Nutr.*, 78, 283, 1997.
96. Sumiyoshi, M. and Kimura, Y., Low molecular weight chitosan inhibits obesity induced by feeding a high-fat diet long-term in mice, *J. Pharm. Pharmacol.*, 58, 201, 2006.
97. Plump, A.S., Smith, J.D., Hayek, T., Aalto-Setala, K., Walsh, A., Verstuyft, J.G., Rubin, E.M., and Breslow, J.L., Severe hypocholesterolemia and atherosclerosis in apolipoprotein E-deficient mice created by homologous recombination in ES cells, *Cell*, 71, 343, 1992.
98. Ormrod, D.J., Holmes, C.C., and Miller, T.E., Dietary chitosan inhibits hypercholesterolaemia and atherogenesis in the apolipoprotein E-deficient mouse model of atherosclerosis, *Atherosclerosis*, 138, 329, 1998.
99. Maezaki, Y., Keisuke, T., Nakagawa, Y., Kawai, Y., Akimoto, M., Tsugita, T., Takekawa, W., Terada, A., Hara, H., and Mitsuoka, T., Hypocholesterolemic effect of chitosan in adult males, *Biosci. Biotechnol. Biochem.*, 57, 1439, 1993.

100. Wuolijoki, E., Hirvela, T., and Ylitalo, P., Decrease in serum LDL cholesterol with microcrystalline chitosan, *Meth. Find. Exp. Clin. Pharmacol.*, 21, 357, 1999.
101. Bokura, H. and Kobayashi, S., Chitosan decreases total cholesterol in women: a randomized, double-blind, placebo-controlled trial, *Eur. J. Clin. Nutr.*, 57, 721, 2003.
102. Metso, S., Ylitalo, R., Nikkila, M., Wuolijoki, E., Ylitalo, P., and Lehtimaki, T., The effect of long-term microcrystalline chitosan therapy on plasma lipids and glucose concentrations in subjects with increased plasma total cholesterol: a randomised placebo-controlled double-blind crossover trial in healthy men and women, *Eur. J. Clin. Pharmacol.*, 59, 741, 2003.
103. Harrison, T.A., The FDA regulation of labelling claims for nutraceuticals in the U.S., *Agro-Food Ind. High Technol.*, 13, 8, 2002.
104. Muzzarelli, R.A.A., Chitosan-based dietary foods, *Carbohydr. Polym.*, 29, 309, 1996.
105. Muzzarelli, R.A.A., Management of hypercholesterolemia and overweight by oral administration of chitosans, in *New Biomedical Materials: Applied and Basic Studies*, Chapman, D. and Haris, P.I., Eds., IOS Press, Amsterdam, 135, 1998.
106. Pittler, M.H. and Ernst, E., Dietary supplements for body-weight reduction: a systematic review, *Am. J. Clin. Nutr.*, 79, 529, 2004.
107. Brady, T., *Nutritional Biochemistry*, Academic Press, New York, 1999.
108. Muzzarelli, R.A.A., Orlandini, F., Tosi, G., and Muzzarelli, C., Chitosan taurocholate capacity to bind lipids and to undergo enzymatic hydrolysis: an *in vitro* model, *Carbohydr. Polym.*, 66, in press, 2006.
109. Murata, Y., Toniwa, S., Miyamoto, E., and Kawashima, S., Preparation of alginate gel beads containing chitosan nicotinic acid salt and the functions, *Eur. J. Pharm. Biopharm.*, 48, 49, 1999.
110. Murata, Y., Kojima, N., and Kawashima, S., Functions of a chitosan-orotic acid salt in the gastrointestinal tract, *Biol. Pharmacol. Bull.*, 26, 687, 2003.
111. Thongngam, T. and McClements, D.J., Isothermal titration calorimetry study of the interaction between chitosan and bile salt (sodium taurocholate), *Food Hydrocolloids*, 19, 813, 2005.
112. Thongngam, T. and McClements, D.J., Influence of pH, ionic strength and temperature on self-association and interactions of sodium dodecyl sulfate in the absence and presence of chitosan, *Langmuir*, 21, 79, 2005.
113. Lee, J.K., Kim, S.U., and Kim, J.H., Modification of chitosan to improve its hypocholesterolemic capacity, *Biosci. Biotechnol. Biochem.*, 63, 833, 1999.
114. Yoo, S.H., Lee, J.S., Park, S.Y., Kim, Y.S., Chang, P.S., and Lee, H.G., Effects of selective oxidation of chitosan on physical and biological properties, *Int. J. Biol. Macromol.*, 35, 27, 2005.
115. Lombardo, D., Bile salt-dependent lipase: its patho-physiological implications, *Biochim. Biophys. Acta*, 1533, 1, 2001.
116. Miled, N., Canaan, S., Dupuis, L., Roussel, A., Riviere, M., Carriere, F., de Caro, A., Cambillau, C., and Verger, R., Digestive lipases: from three-dimensional structure to physiology, *Biochimie*, 82, 973, 2000.
117. Mukherjee, M., Human digestive and metabolic lipases, a brief review, *J. Mol. Catalysis B Enzymatic*, 22, 369, 2003.
118. Wickman, M., Wilde, P., and Fillery-Travis, A., A physico-chemical investigation of two phosphatidylcholine/bile salt interfaces: implications for lipase activation, *Biochim. Biophys. Acta*, 1580, 110, 2002.
119. Cajal, Y., Svendsen, A., DeBolos, J., Patkar, S.A., and Alsina, M.A., Effect of the lipid interface on the catalytic activity and spectroscopic properties of a fungal lipase, *Biochimie*, 82, 1053, 2000.

120. Tozaki, H., Odoriba, T., Okada, N., Fujita, T., Terabe, A., Suzuki, T., Okabe, S., Muranishi, S., and Yamamoto, A., Chitosan capsules for colon-specific drug delivery: enhanced localization of 5-aminosalicylic acid in the large intestine accelerates healing of TNBS-induced colitis in rats, *J. Control Rel.*, 82, 51, 2002.
121. DeVincenzi, M., Maialetti, F., and Muzzarelli, R.A.A., Modified chitosan binds prolamine peptides toxic in the coeliac disease, in *Advances in Chitin Sciences*, Domard, A., Jeuniaux, C., Muzzarelli, R.A.A., and Roberts, G.A.F., Eds., Jacques André Publ., Lyon, France, 1996.
122. Muzzarelli, R.A.A. and DeVincenzi, M., Chitosans as dietary food additives, in *Applications of Chitin*, Goosen, M.F.A., Ed., Technomic, Lancaster, PA, 115, 1996.
123. NiMhurchu, C., Dunshea-Mooij, C., Bennett, D., and Rodgers, A., Effect of chitosan on weight loss in overweight and obese individuals: a systematic review of randomized controlled trials, *Obesity Rev.*, 6, 35, 2005.
124. NiMhurchu, C., Dunshea-Mooij, C.A.E., Bennett, D., and Rodgers, A., Chitosan for overweight or obesity, *Cochrane Database Syst. Rev.*, 3, 1136, 2005.
125. NiMhurchu, C., Poppitt, S.D., McGill, A.T., Leahy, F.E., Bennett, D.A., Lin, R.B., Ormrod, D., Ward, L., Strik, C., and Rodgers, A., The effect of the dietary supplement, chitosan, on body weight: a randomised controlled trial in 250 overweight and obese adults, *Int. J. Obesity*, 28, 1149, 2004.
126. Woodgate, D.E. and Conquer, J.A., Effects of a stimulant free dietary supplement on body weight and fat loss in obese adults, *Curr. Ther. Res.*, 64, 248, 2003.
127. Gades, M.D. and Stern, J.S., Chitosan supplementation and fat absorption in men and women, *J. Am. Dietetic Assoc.*, 105, 72, 2005.
128. Gades, D.M. and Stern, J.S., Chitosan supplementation and fecal fat excretion in man, *Obesity Res.*, 11, 683, 2003.
129. Goldberg, S.H., Von Feldt, J.M., and Lonner, J.H., Pharmacologic therapy for osteoarthritis, *Am. J. Orthop.*, 12, 673, 2002.
130. Deal, C.L. and Moskowitz, R.W., Nutraceuticals as therapeutic agents in osteoarthritis, *Rheum. Dis. Clin. North Am.*, 25, 379, 1999.
131. Laverty, S., Sandy, J.D., Celeste, C., Vachon, P., Marier, J.F., and Plaas, A.H.K., Synovial fluid levels and serum pharmacokinetics in a large animal model following treatment with oral glucosamine at clinically relevant doses, *Arthritis Rheum.*, 52, 181, 2005.
132. Anderson, J.W., Nicolosi, R.J., and Borzelleca, J.F., Glucosamine effects in humans: a review of effects on glucose metabolism, side effects, safety considerations and efficacy, *Food Chem. Toxicol.*, 43, 187, 2005.
133. Largo, R., Alvarez-Soria, M.A., Díez-Ortego, I., Calvo, E., Sánchez-Pernaute, O., Egido, J., and Herrero-Beaumont, G., Glucosamine inhibits IL-1 beta-induced NF kappa B activation in human osteoarthritic chondrocytes, *Osteoarthr. Cartil.*, 11, 290, 2003.
134. Persiani, S., Roda, E., Rovati, L.C., Locatelli, M., Giacovelli, G., and Roda, A., Glucosamine oral bioavailability and plasma pharmacokinetics after increasing doses of crystalline glucosamine sulfate in man, *Osteoarthr. Cartil.*, 13, 1041, 2005.
135. James, C.B. and Uhl, T.L., A review of articular cartilage pathology and the use of glucosamine sulfate, *J. Athl. Train.*, 36, 413, 2001.
136. Talent, J.M. and Gracy, R.W., Pilot study of oral polymeric N-acetyl-D-glucosamine as a potential treatment for patients with osteoarthritis, *Clin. Ther.*, 18, 1184, 1996.
137. Rubin, B.R., Talent, J.M., Kongtawelert, P., Pertusi, R.M., Forman, M.D., and Gracy, R.W., Oral polymeric N-acetyl-D-glucosamine and osteoarthritis, *J. Am. Osteopath. Assoc.*, 101, 339, 2001.

138. Rubin, B.R., Talent, J.M., Pertusi, R.M., Forman, M.D., and Gracy, R.W., Oral polymeric N-acetyl-D-glucosamine and osteoarthritis, in *Chitosan per os: From Dietary Supplement to Drug Carrier*, Muzzarelli, R.A.A., Ed., Atec, Grottammare, Italy, 187, 2000.
139. Williams, S.K. and Bynum, S.A., Glucosamine Composition and Method, U.S. Patent 5,679,344, 1997.
140. Henderson, R., Aminosugar and GAG Composition for the Treatment and Repair of Connective Tissue, Patent Application WO 98/27988, 1998.
141. Leffler, C.T., Glucosamine, chondroitin and Mn ascorbate for degenerative joint disease of the knee or low back, *Mil. Med.* 164, 85, 1999.
142. Buckwalter, J.A., Callaghan, J.J., and Rosier, R.N., From oranges and lemons to glucosamine and chondroitin sulfate: clinical observations stimulate basic research, *J. Bone Joint Surg. Am.*, 83A, 1266, 2001.
143. de los Reyes, G.C., Koda, R.T., and Lien, E.J., Glucosamine and chondroitin sulfates in the treatment of osteoarthritis: a survey, *Prog. Drug Res.*, 55, 81, 2000.
144. Lu-Suguro, J.F., Hua, J., Sakamoto, K., and Nagaoka, I., Inhibitory action of glucosamine on platelet activation in guinea pigs, *Inflamm. Res.*, 54, 493, 2005.
145. Chen, J.T., Liang, J.B., Chou, C.L., Chien, M.W., Shyu, R.C., Chou, P.I., and Lu, D.W., Glucosamine sulfate inhibits TNF-alpha and IFN-gamma-induced production of ICAM-1 in human retinal pigment epithelial cells *in vitro*, *Invest. Ophthalmol. Vis. Sci.*, 47, 664, 2006.
146. Tant, L., Gillard, B., and Appelboom, T., Open-label, randomized, controlled pilot study of the effects of a glucosamine complex on low back pain, *Curr. Ther. Res. Clin. Exp.*, 66, 511, 2005.
147. Clegg, D.O., Reda, D.J., Harris, C.L., Klein, M.A., O'Dell, J.R., Hooper, M.M., Bradley, J.D., Bingham, C.O., III, Weisman, M.H., Jackson, C.G., Lane, N.E., Cush, J.J., Moreland, L.W., Schumacher, H.R., Jr., Oddis, C.V., Wolfe, F., Molitor, J.A., Yocum, D.E., Schnitzer, T.J., Furst, D.E., Sawitzke, A.D., Shi, H., Brandt, K.D., Moskowitz, R.W., and Williams, H.J., Glucosamine, chondroitin sulfate, and the two in combination for painful knee osteoarthritis, *N. Engl. J. Med.*, 354, 795, 2006.
148. Zhang, X.L., Yu, C.L., Xu, S., Zhang, C., Tang, T.T., and Dai, K.R., Direct chitosan-mediated gene delivery to the rabbit knee joints *in vitro* and *in vivo*, *Biochem. Biophys. Res. Commun.*, 341, 202, 2006.
149. Zhang, H. and Neau, S.H., *In vitro* degradation of chitosan by bacterial enzymes from rat cecal and colonic contents, *Biomaterials*, 13, 2761, 2002.
150. Boot, R.G., Renkema, G.H., Strijland, A., Zonneveld, A.J., and Aerts, J.M.F.G., Cloning of a cDNA encoding chitotriosidase, a human chitinase produced by macrophages, *J. Biol. Chem.*, 270, 26252, 1995.
151. Boot, R.G., Blommaart, E.F., Swart, E., Ghauharali-van der Vlugt, K., Bijl, N., Moe, C., Place, A., and Aerts, J.M., Identification of a novel acidic mammalian chitinase distinct from chitotriosidase, *J. Biol. Chem.*, 276, 6770, 2001.
152. Recklies, A.D., White, C., and Ling, H., The chitinase 3-like protein human cartilage glycoprotein 39 (HC-gp39) stimulates proliferation of human connective-tissue cells and activates both extracellular signal-regulated kinase- and protein kinase beta-mediated signalling pathways, *Biochem J.*, 365, 119, 2002.
153. Suzuki, M., Fujimoto, W., Goto, M., Morimatsu, M., Syuto, M., and Iwanaga, T., Cellular expression of gut chitinase mRNA in the gastrointestinal tract of mice and chickens, *Histochem. Cytochem.*, 8, 1081, 2002.
154. McCarty, M.F., The neglect of glucosamine as a treatment for osteoarthritis, *Med. Hypotheses*, 42, 323, 1994.

155. Kanurnyy, I.I., Usage of derivatives of D-glucosamine in emotional hypertension due to stress, *Eur. Psychiatry*, 17, Supp. 1, 161, 2002.
156. Sutton, L., Rapport, L., and Lockwood, B., Glucosamine: con or cure, *Nutrition*, 18, 534, 2002.
157. Meininger, C.J., Kelly, K.A., Lee, H., Haynes, T.E., and Wu, G., Glucosamine inhibits inducible nitric oxide synthesis, *Biochem. Biophys. Res. Commun.*, 279, 234, 2000.
158. Ma, L., Immuno suppressive effects of glucosamine, *J. Biol. Chem.*, 277, 39343, 2002.
159. Wu, G., Glutamine metabolism to glucosamine is necessary for glutamine inhibition of endothelial nitric oxide synthesis, *Biochem. J.*, 353, 245, 2001.
160. Reichelt, A., Foster, K.K., Fischer, M., Rovati, L.C., and Setnikar, I., Efficacy and safety of intramuscular glucosamine sulfate, *Drug Res.*, 44, 75, 1994.
161. Ebube, N.K., Mark, W., and Hahm, H., Preformulation studies and characterization of proposed condroprotective agents: glucosamine·HCl and chondroitin sulfate, *Pharm. Dev. Technol.*, 7, 457, 2002.
162. Oegema, T.R., Effect of oral glucosamine on cartilage and meniscus, *Arthritis Rheum.*, 46, 2495, 2002.
163. Hughes, R. and Carr, A., A randomized, double-blind, placebo-controlled trial of glucosamine sulphate as an analgesic in osteoarthritis of the knee, *Rheumatology*, 41, 279, 2002.
164. Bellamy, N. and Lybrand, S., Glucosamine therapy: does it work? *Med. J. Aust.*, 175, 399, 2001.
165. Chard, J. and Dieppe, P., Glucosamine for osteoarthritis: magic, hype or confusion? *Br. Med. J.*, 322, 1439, 2001.
166. Setnikar, I. and Rovati, L.C., Absorption, distribution, metabolism and excretion of glucosamine sulfate, *Drug Res.*, 51, 699, 2001.
167. Towheed, T.E., Anastassiades, T.P., Shea, B., Houpt, J., Welch, V., and Hochberg, M.C., Glucosamine therapy for treating osteoarthritis, *Cochrane Database Syst. Rev.*, 1, CD002946, 2001.
168. Ruane, R. and Griffiths, P., Glucosamine therapy compared to ibuprofen for joint pain, *Br. J. Community Nurs.*, 7, 148, 2002.
169. Hochberg, M.C., What a difference a year makes: reflections on the ACR recommendations for the medical management of osteoarthritis, *Curr. Rheumatol. Rep.*, 3, 473, 2001.
170. Bijlsma, J.W., Glucosamine and chondroitin sulfate as a possible treatment for osteoarthritis, *Ned. Tijdschr. Geneeskd.*, 146, 1819, 2002.
171. Reginster, J.Y., Deroisy, R., Rovati, L.C., Lee, R.L., Lejeune, E., Bruyere, O., Giacovelli, G., Henrotin, Y., Dacre, J.E., and Gossett, C., Long-term effects of glucosamine sulphate on osteoarthritis progression: a randomised, placebo-controlled clinical trial, *Lancet*, 357, 251, 2001.
172. Braham, R., Dawson, B., and Goodman, C., The effect of glucosamine supplementation on people experiencing regular knee pain, *Br. J. Sport Med.*, 37, 45, 2003.
173. Pavelka, K., Gatterova, J., Olejarova, M., Machacek, S., Giacovelli, G., and Rovati, L.C., Glucosamine sulfate use and delay of progression of knee osteoarthritis: a 3-year, randomized, placebo-controlled, double-blind study, *Arch. Intern. Med.*, 162, 2113, 2002.
174. Bruyere, O., Honore, A., Ethgen, O., Rovati, L.C., Giacovelli, G., Henrotin, Y.E., Seidel, L., and Reginster, J.Y., Correlation between radiographic severity of knee osteoarthritis and future disease progression. Results from a 3-year prospective, placebo-controlled study evaluating the effect of glucosamine sulfate, *Osteoarthr. Cartil.*, 11, 1, 2003.

175. Sherman, W.T. and Gracy, R.W., Treatment of Osteoarthritis by Administered Poly-N-Acetyl-D-Glucosamine, Patent Application WO 98/25631, 1998.
176. Gracy, R.W., Chitosan and glucosamine derivatives in the treatment of osteoarthritis, *Agro-Food Ind. High Technol.*, 74(5), 53, 2003.
177. Ronca, G. and Conte, A., Metabolic fate of partially depolymerized shark chondroitin sulfate in man, *Int. J. Clin. Pharm. Res.*, 13, 27, 1993.
178. McCarty, M.F., Russell, A.L., and Seed, M.O., Sulfated glycosaminoglycans and glucosamine may synergize in promoting synovial hyaluronic acid synthesis, *Med. Hypotheses*, 54, 798, 2000.
179. Piperno, M., Reboul, P., Hellio, L.E., Graverand, M.P., Peschard, M.J., Annefeld, M., Richard, M., and Vignon, E., Glucosamine sulfate modulates dysregulated activities of human osteoarthritic chondrocytes *in vitro*, *Osteroarthr. Cartil.*, 8, 207, 2000.
180. Lippiello, L., Glucosamine and chitosan sulfate: biological response modifiers of chondrocytes under simulated conditions of joint stress, *Osteroarthr. Cartil.*, 11, 335, 2003.
181. Mobasheri, A., Vannucci, S.J., Bondy, C.A., Carter, S.D., Innes, J.F., Arteaga, M.F., Trujillo, E., Ferraz, I., Shakibaei, M., and Martin-Vasallo, P., Glucose transport and metabolism in chondrocytes: a key to understanding chondrogenesis, skeletal development and cartilage degradation in osteoarthritis, *Histol. Histopathol.*, 17, 1239, 2002.
182. Noyszewski, E.A., Wroblewski, K., Dodge, G.R., Kudchodkar, S., Beers, J., Sarma, A.V.S., and Reddy, R., Preferential incorporation of glucosamine into the galactosamine moieties of chondroitin sulfates in articular cartilage explants, *Arthritis Rheum.*, 44, 1089, 2001.
183. Hua, J., Sakamoto, K., and Nagaoka, I., Inhibitory actions of glucosamine, a therapeutic agent for osteoarthritis, on the functions of neutrophils, *J. Leukoc. Biol.*, 71, 632, 2002.
184. Shikhman, A.R., Kuhn, K., Alaaeddine, N., and Lotz, M., N-Acetylglucosamine prevents IL-1beta-mediated activation of human chondrocytes, *J. Immunol.*, 166, 5155, 2001.
185. Gouze, J.N., Bianchi, A., Becuwe, P., Dauca, M., Netter, P., Magdalou, J., Terlain, B., and Bordji, K., Glucosamine modulates IL-1-induced activation of rat chondrocytes at a receptor level, and by inhibiting the NF-kappa B pathway, *FEBS Lett.*, 510, 166, 2002.
186. Blakeley, J.A. and Ribeiro, V., A survey of self-medication practices and perceived effectiveness of glucosamine products among older adults, *Complement Ther. Med.*, 10, 154, 2002.
187. Blakeley, J.A. and Ribeiro, V.E.S., Glucosamine and osteoarthritis, *Am. J. Nurs.*, 104, 54, 2004.
188. Russell, A.S., Aghazadeh-Habashi, A., and Jamali, F., Active ingredient consistency of commercially available glucosamine sulfate products, *J. Rheumatol.*, 29, 2407, 2002.
189. Hungerford, D.S. and Jones, L.C., Glucosamine and chitosan-sulfate are effective in the management of osteoarthritis, *J. Arthroplasty*, 18, 5, 2003.
190. Nolte, R.M. and Klimkiewicz, J.J., Pharmacologic treatment alternatives, *Sport Med. Arthroscopy Rev.*, 11, 102, 2003.
191. Dodge, G.R. and Jimenez, S.A., Glucosamine sulfate modulates the levels of aggrecan and matrix metallo-proteinase-3, *Osteoarthr. Cartil.*, 11, 424, 2003.
192. Hunziker, E.B., Articular cartilage repair: basic science and clinical progress. A review of the current status and prospects, *Osteoarthr. Cartil.*, 10, 432, 2002.

7 Arabinoxylans: Technologically and Nutritionally Functional Plant Polysaccharides

Marta S. Izydorczyk and Costas G. Biliaderis

CONTENTS

7.1 INTRODUCTION

Arabinoxylans are nonstarch polysaccharides present in various tissues of cereal grains. Although they are minor constituents of cereals, they are the predominant matrix polysaccharides in gramineous cell walls. They have been speculated to play a role in cross-linking of cellulose microfibrils, and therefore may create a rigid network within cell wall architecture as well as regulate cell expansion.[1] Arabinoxylans are the pentose-containing carbohydrate polymers, and hence are often referred to as pentosans. Arabinoxylans were first identified in wheat flour by Hoffman and Gortner in 1927,[2] and they have been of interest to cereal chemists and technologists ever since because of their technological importance. Arabinoxylans have been shown to significantly affect cereal-based processes such as milling, brewing, and bread making. Furthermore, arabinoxylans offer nutritional benefits of soluble and insoluble fiber, and because of the presence of phenolic moieties in their molecular structures, they may also have some antioxidant properties.[3] Arabinoxylans might constitute a significant portion of human dietary fiber intake because they occur in a wide variety of cereal crops. In addition, some agriculture by-products with little current economic value, such as corn husk, brewer's grain, flaxseed cake, or banana peels, are potential sources of arabinoxylans.[4–6]

7.2 ARABINOXYLANS AS CONSTITUENTS OF AGRICULTURAL CROPS

Arabinoxylans have been identified in all major cereal grains, including wheat, barley, oats, rye, rice, sorghum, maize, and millet,[7] as well as in other plants, such as psyllium,[8] pangola grass, bamboo shoots, and rye grass. Recent studies on the water-soluble mucilage extracted from flaxseeds have demonstrated that heterogeneous, high molecular weight arabinoxylans are a major component of the extracted polysaccharides.[5] Arabinoxylans have also been isolated from corn husk and banana peels.[6] In cereal grains, arabinoxylans are localized mainly in the cell walls of starchy endosperm and the aleurone layer, in the bran tissues, and in the husk of some cereals. However, depending on the genus and species, the amount of arabinoxylans in a particular tissue may vary (Table 7.1). For example, the cell walls of wheat and rye starchy endosperm and the aleurone layer are built up mainly of arabinoxylans (~60 to 70%), yet the total arabinoxylan content in bran of these grains is higher than that of endosperm.[9] In barley, the aleurone cell walls are built up mainly of arabinoxylans (60 to 70%), whereas the starchy endosperm cell walls contain only about 20 to 40% of these polymers and a much greater amount of β-glucans. The fine molecular structure of arabinoxylans may also vary depending on the specific tissue from which these polymers are derived. The outer layers of cereal grains (husk and bran) appear to contain acidic arabinoxylans (glucuronoarabinoxylans) containing glucuronic acid in addition to arabinose and xylose residues.

The level of arabinoxylans in cereals depends on genetic and environmental factors. Among the cereal grains, rye has the highest content of arabinoxylans, followed by wheat and barley (Table 7.1). It has been shown that translocation of wheat, with the short arm of the 1B chromosome of wheat replaced by the short

TABLE 7.1
Content of Total and Water-Soluble Arabinoxylans in Various Grains and Grain Tissues

Source	Total Arabinoxylans (%)	Water-Soluble Arabinoxylans (%)	Reference
Barley			
Whole grain	6.11	0.35	Hashimoto et al.[147]
Whole grain	3.4–4.1	—	Izydorczyk et al.[148]
Whole grain	—	0.40–0.88	Oscarsson et al.[82]
Pearled grain	4.45	0.27	Hashimoto et al.[147]
Pearlings	14.14	0.54	Hashimoto et al.[147]
Pearled flour	—	0.3–1.08	Dervilly-Pinel et al.[55]
Wheat			
Whole grain	5.77	0.59	Hashimoto et al.[149]
Whole grain	—	0.38–0.83	Saulnier et al.[46]
Bran	19.38	0.88	Hashimoto et al.[149]
Flour	1.37–2.06	0.54–0.68	Izydorczyk et al.[23]
Durum wheat	4.07-6.02	0.37–0.56	Lempereur et al.[13]
Rye			
Whole grain	7.6	—	Bengtsson and Åman[79]
Whole grain	8–12.1	2.6–4.1	Hansen et al.[15]
Bran	—	1.7	Figueroa-Espinoza et al.[31]
Flour	3.2–3.64	2.2–2.65	Cyran et al.[80]
Oats			
Whole grain	2.73	0.17	Hashimoto et al.[147]
Hulls	8.79	0.40	Hashimoto et al.[147]
Bran	3.50	0.33	Hashimoto et al.[147]
Pearled grain	3.00	0.15	Hashimoto et al.[147]
Rice			
Whole grain	2.64	0.06	Hashimoto et al.[147]
Hulls	8.36–9.24	0.11–0.12	Hashimoto et al.[147]
Bran	4.84–5.11	0.35–0.77	Hashimoto et al.[147]
Sorghum			
Whole grain	1.8	0.08	Hashimoto et al.[147]
Pearlings	5.4	0.35	Hashimoto et al.[147]
Corn			
Bran	29.86	0.28	Hashimoto et al.[147]
Soybean		1.33	
Hulls	13.10		Hashimoto et al.[147]

arm of the 1R chromosome of rye (1B/1R gene), increases the content of water-soluble arabinoxylans, but does not affect the amount of total arabinoxylans in comparison with standard wheat.[10] Other studies have proposed that the content of water-extractable arabinoxylans in rye is controlled by many factors scattered throughout the genome, and while chromosomes 2R, 5R, and 6R are responsible for increased arabinoxylan content, chromosome 3R is responsible for its reduced level.[11,12] Significant genetic and environmental variations in arabinoxylan content have been reported for durum wheat and barley.[13,14] Hansen et al.[15] examined the effect of harvest year on the content and composition of dietary fiber in seven rye varieties grown in Denmark and found that yearly variations in the content of total and water-extractable arabinoxylans were higher (27 to 55% of total variance) than those associated with genotype effects (14 to 19% of total variance). Yearly differences in the content of total arabinoxylans were also found among five rye varieties grown in Finland.[16] It was also found that a cold and wet season resulted in small rye kernels with a high arabinoxylan content. Coles et al.[17] conducted drought studies using a mobile rain shelter and found a positive relationship between the amount of arabinoxylan accumulated in wheat and the dry conditions. Also, the exposure of plants to UV light increased the degree of cross-linking in arabinoxylans, but the effect on the total content of these polymers was not clearly established. Wakabayashi et al.[18] investigated the changes in the amount and composition of arabinoxylans in cell walls of wheat coleoptiles grown under continuous hypergravity conditions. It was found that the amount of arabinoxylans per unit length of coleoptiles increased under hypergravity conditions. In addition, the amount of the acidic arabinoxylans (glucuronoarabinoxylans) and ferulic acid–cross-linked arabinoxylans increased as a result of continuous hypergravity.

7.3 EXTRACTION, ISOLATION, AND PURIFICATION OF ARABINOXYLANS

7.3.1 Aqueous Extraction

The most common approach to isolating arabinoxylans from various plant materials involves aqueous or alkali extraction of these polymers either from whole grains or from specific plant tissues. Once isolated from the cell wall matrix, arabinoxylans are water soluble; however, in the intact wall, these polymers are cross-linked with other wall constituents to form a structural fabric that is not soluble in an aqueous environment.[19] Some of the cross-links involved are noncovalent and, while individually weak, they may confer strength and insolubility if present in large numbers (e.g., hydrogen bonds). Arabinoxylan chains can also be covalently cross-linked to each other or to other cell wall constituents. As a consequence of these cross-links, a certain portion of arabinoxylans cannot be easily extracted from the plant materials with water and requires harsher treatments with alkali solutions to liberate them from the networks of covalent and noncovalent bonds, as well as physical entanglements. Various methods and procedures for obtaining highly purified water-soluble arabinoxylans from common cereals for analytical purposes have been published.[20–23] Extractions are usually conducted in water or in buffers. After extraction, purification procedures usually involve inactivation of endogenous enzymes in aqueous extracts and the use of hydrolytic enzymes to eliminate contaminating proteins and starch from the preparations. Crowe and Rasper[21] and Izydorczyk et al.[23] obtained nearly protein-free pentosan extracts from wheat flour after adsorption of contaminating proteins in the water extracts on various clays.

Recently, Faurot et al.[24] developed a large-scale procedure for isolation of water-soluble and insoluble pentosans from wheat flour (Figure 7.1). The procedure was based on mixing flour and water (50 kg/250 l), separating the soluble arabinoxylans from the insoluble residue by centrifugation, and heat treating the supernatant. The insoluble residue was then treated with Alcalase and Termamyl to solubilize the initially water-unextractable polymers. The yields of water-soluble and -insoluble arabinoxylans ranged from 100 to 200 g and from 250 to 350 g, respectively. Both preparations, however, were contaminated with either soluble proteins or starch, but the enrichment factor ranged from 98 to 68 for water-soluble and 27 to 19 for water-insoluble arabinoxylans compared to their initial content in wheat flour. A pilot-scale isolation of water-extractable arabinoxylans from rye[25] involved a heat treatment of the ground whole meal (130°C, 90 min) to inactivate the endogenous enzymes, followed by stirring of the rye whole meal (10 kg) and deionized water (100 l) at room temperature (90 min) in a pilot-scale brewing vessel. After decanting of the supernatant, the water-extractable arabinoxylans were purified by treatments with a heat-stable α-amylase, protein coagulation, and partial concentration by heat evaporation. Subsequently, arabinoxylans were precipitated by addition of ethanol. The precipitated material contained about 54% arabinoxylans, 22% proteins, 4.4% arabinogalactan peptide, and 4.7% β-glucans. An additional purification step, based on a treatment of the dissolved material with montmorillonite clay, increased the arabinoxylan content to about 93%, while leaving almost no protein in the extract.[25] Gruppen et al.[26] reported a large-scale isolation of highly purified arabinoxylan-enriched cell wall material from wheat endosperm based on dough kneading in combination with wet sieving. Mares and Stone[27] and Selvedran and Dupont[28] developed a method for preparation of cell wall material enriched in arabinoxylans based on wet sieving of wheat flour in aqueous ethanol to remove starch granules, followed by sonication or removal of starch and intracellular proteins by organic solvents to improve the purity of the preparations.

7.3.2 Strategies to Extract Arabinoxylans from Agricultural By-products

A considerable effort has recently been directed toward extraction and purification of arabinoxylans from agricultural by-products. Brewer's (spent) grain, wheat or rye bran, sugar beet pulp, corncobs, and banana peels constitute abundant low-value by-products of the food industry. These materials are potential sources of arabinoxylans as they are rich in noncellulosic polysaccharides, with arabinoxylans being a primary component. Bataillon et al.[29] combined delignification (with 37% sodium chlorite) and alkali extraction (with 43% sodium hydroxide) to obtain arabinoxylan preparations from de-starched wheat bran. The extracted arabinoxylans were purified by a microfiltration and dried by an atomization system (Figure 7.2); the yield and purity of this preparation were 13 and 75%, respectively.[29] In cereal grains, the extractability of arabinoxylans from the outer layers of the kernel is lower than that from the starchy endosperm, and physical and chemical treatments are often needed to ease their solubilization. However, Bergmans et al.[30] observed that autoclaving, alkaline peroxide treatment, or chlorite delignification of wheat bran was not effective in

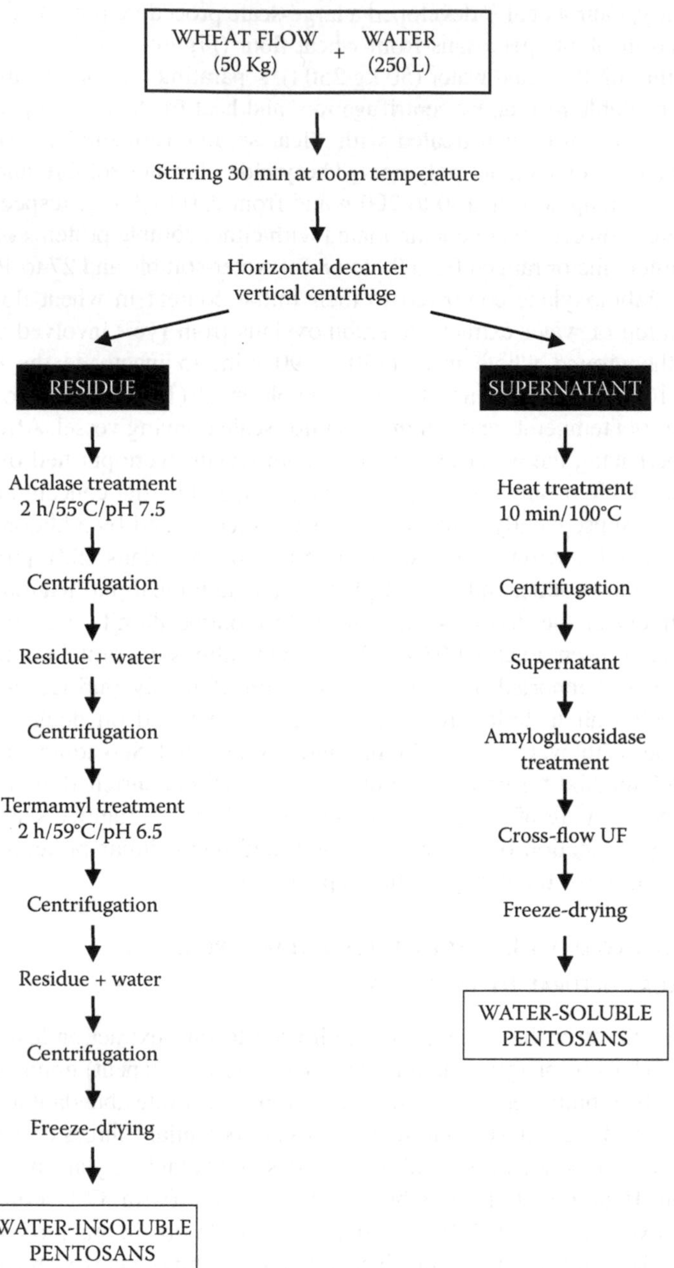

FIGURE 7.1 Flow scheme for extraction procedure and purification of water-extractable and water-insoluble arabinoxylans from wheat flour. (Adapted from Faurot, A.-L. et al., *Lebensm. Wiss. Technol.*, 28, 436, 1995. With permission.)

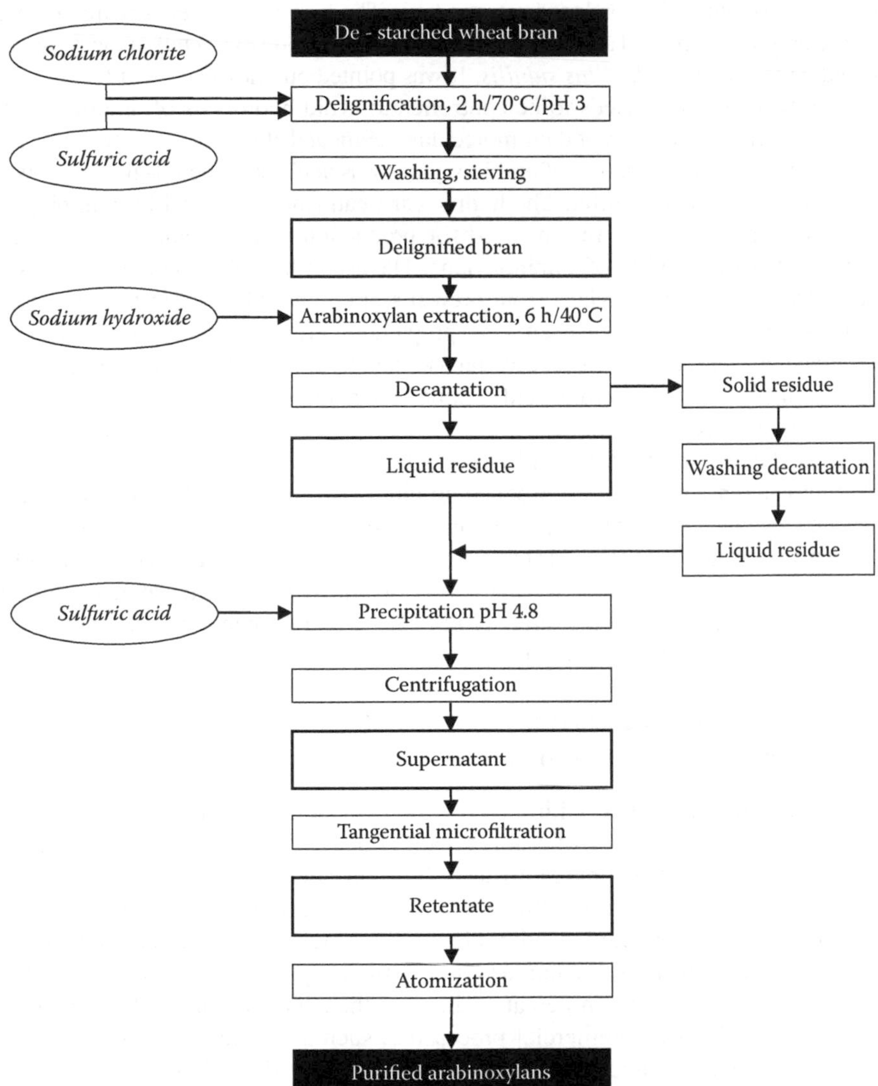

FIGURE 7.2 Flow scheme for extraction of arabinoxylans from wheat bran. (Adapted from Bataillon, M. et al., *Ind. Crop Products*, 8, 37, 1998. With permission.)

increasing the extract yield in the subsequent saturated barium hydroxide extraction of glucuronoarabinoxylans. Consequently, researchers have turned to various hydrolyzing enzymes, such as β-glucanases, arabinofuranosidases, endoxylanases, and ferulic acid esterases to facilitate the liberation of arabinoxylans from the cell wall materials. Figueroa-Espinoza et al.[31] combined various physical treatments (extrusion and high shear) with the use of different endoxylanases (from *Aspergillus niger*, *Talaromyces emersonii*, and *Bacillus subtilis*) to investigate the release of high molecular weight arabinoxylans from rye bran. The best results were obtained when the rye bran was extruded at high temperature (~140°C) and extracted in the presence of endoxylanase from *Bacillus subtilis*. It was pointed out, however, that the amount of hydrolytic enzymes needs to be controlled to avoid hydrolysis of arabinoxylans and a consequent reduction of their molecular weight and ability to form gels. Further improvement in solubilization of arabinoxylans was achieved when β-glucanase was added to the extraction buffer. The high-shear treatments, achieved by pumping of rye bran slurry through a reactor used for production of microemulsions, did not improve the extractability of arabinoxylans. The use of hydrolytic enzymes to assist the extraction of arabinoxylans from brewer's grain and wheat bran has also been explored by Faulds et al.[4] An enzyme preparation from the thermophilic fungus *Humicola insolens*, containing endoxylanases and feruloyl esterases, proved efficient in releasing ferulic and diferulic acid residues and solubilizing arabinoxylans present in these by-products.

Lu and coworkers[32] utilized a by-product generated during processing of wheat flour into starch and gluten to produce arabinoxylan-rich fiber. The arabinoxylan-enriched residue left after flour processing was simply collected on a sieve (75 μm), washed with water, and spray dried to a powder. Another product, containing 60% of arabinoxylan-enriched dietary fiber, was obtained from wastewater, remaining after starch extraction, after a series of enzymatic and fermentative treatments followed by cross-flow ultrafiltration and spray drying.[33]

7.3.3 Arabinoxylan-Enriched Fractions Obtained by Physical Grain Fractionation

It is prudent to predict that with the growing demand for functional and designer foods, physical fractionation of grain to obtain fractions enriched in specific dietary fiber components will become more desirable than isolation of a single constituent. This approach will ensure that the preparations are enriched in the component of interest but are not depleted of other dietary fiber ingredients and phytochemicals (polyphenols, phytosterols, vitamins, etc.) naturally present in grain tissues. The location of arabinoxylans in cereal grains and their interactions with other grain constituents influence commercial processing, such as milling and isolation procedures aiming at obtaining arabinoxylan-enriched fractions. Various approaches to milling whole rye grain have recently been investigated to obtain fractions with improved functionality and dietary fiber characteristics.[25,34] As a consequence of variable concentration of arabinoxylan in various grain tissues, milling leads to fractions differing in arabinoxylan content. In the study by Glitsø and Knudsen,[34] the whole rye grain was passed twice through a dehuller and the outer layers were ground in a fine grinder yielding the pericarp/testa-enriched fractions. Part of the residual rye material was passed through three corrugated roller mills and sieved to obtain particles smaller than 1200 μm, thus producing the fraction enriched in aleurone cells. Another part of the rye material, remaining after dehulling, was passed through one smooth roller mill retrieving particles smaller than 1200 μm and through another smooth roller mill retrieving particles smaller than 250 μm, thus producing the endosperm-enriched fraction.[34] The fractions varied in the

TABLE 7.2
Composition, Extractability of Arabinoxylans (% Total), and Arabinose-to-Xylose Ratio of Whole Rye and Enriched Rye-Milling Fractions

	Whole Rye	Pericarp/Testa Fraction	Aleurone Fraction	Endosperm Fraction
Total dietary fiber, %	15.1	73.3	28.3	6.5
Cellulose, %	1.3	13.6	2.0	0.4
β-Glucans, %	1.5	0.46	3.3	0.75
Lignin, %	1.5	11.0	3.9	0.2
Total AX, %	9.0	39.5	17.1	4.2
Water-extractable AX, %	41	14	27	70
Ara/Xyl (WE)	0.64	1.17	0.57	0.71
Ara/Xyl (WUE)[b]	0.63	1.02	0.35	0.83
Ferulic acid, mg/100 g	93	659	193	17

Note: AX = arabinoxylans; WE = water-extractable arabinoxylans; WUE = water-unextractable arabinoxylans.

Source: Adapted from Glitsø, L.V. and Knudsen, K.E.B., *J. Cereal Sci.*, 29, 89, 1999.

content and composition of dietary fiber; a wide range of arabinose-to-xylose (Ara/xyl) ratios and the proportions of soluble-to-insoluble arabinoxylans implied that variations in arabinoxylans' structure were obtained (Table 7.2). Fractionation of barley grain via pearling and roller milling was also shown to afford fractions enriched in specific components of dietary fiber (Figure 7.3).[35] The roller milling of whole or pearled barley involved several break and sizing passages using corrugated rolls, and reduction passages with smooth frosted or corrugated rolls. It was shown that addition of impact milling between grinding passages enabled better separation of the bran and endosperm cell walls and resulted in fiber-rich fractions containing between 15 and 25% of β-glucans and 8 and 12% of arabinoxylans, depending on the barley genotype and certain preprocessing practices, such as pearling and tempering of grain.[35] Most recently, Harris et al.[36] described a large-scale procedure for the production of aleurone-rich and pericarp-rich fractions from hard Australian wheat. The pericarp fraction was obtained after wheat grain, conditioned to 16% moisture, was abraded using a debranning machine to remove 2.5% of bran by weight. The debranned wheat was put onto the first break of the pilot mill. The bran and pollard produced by this milling were mixed to form the aleurone-rich fraction, which included the seed coat with its cuticle. Both fractions were enriched in arabinoxylans, but the concentration of ester-linked ferulic acid and β-glucans was greater in the aleurone than in the pericarp-rich fraction. Consistent with the results of Glitsø and Knudsen,[34] it was found that the arabinoxylans in the walls of the pericarp-rich fraction were more highly substituted (Ara/Xyl = 0.93) than those in the walls of the aleurone-rich fraction (Ara/Xyl = 0.59).[36]

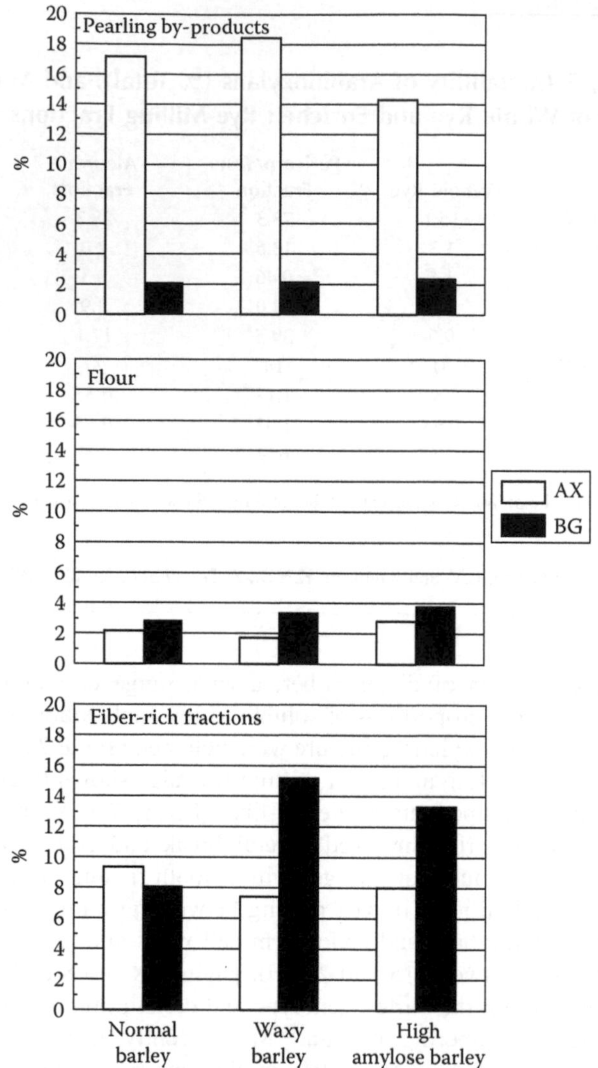

FIGURE 7.3 Content of arabinoxylans and β-glucans in barley pearling by-products (obtained by abrading barley grain to 10%), flour, and fiber-rich fractions (obtained via roller milling of whole barley grain). (Adapted from Izydorczyk, M.S. et al., *Cereal Chem.*, 80, 637, 2003.)

7.3.4 Production of Xylooligosaccharides

Recent advances in the area of prebiotic activity of oligosaccharides have prompted developments of various procedures for obtaining partially degraded arabinoxylans from agricultural biomass. It has been shown that xylooligosaccharides have prebiotic properties and improve the intestinal function by enhancing the growth of healthy *Bifidobacteria*, while suppressing the growth of *Clostridium* and having bacteriostatic effects against *Vibrio anguillarum*.[37–39] The breakdown of arabinoxylans into xylooligosaccharides can be carried out by direct enzymatic or acid conversion of raw materials or by hydrolysis of isolated polysaccharides. Endo-β-(1→4)-xylanases that cut internal β-(1→4) linkages in the arabinoxylan's backbone belong to glycoside hydrolase families (GHFs) 10 and 11.[40] Endoxylanases of GHF 10 (from *Cryptococcus albidus* or *Aspergillus aculeatus*) and GHF 11 (from *Trichoderma reesei* or *Bacillus subtilis*) have been reported to have different substrate specificities, and therefore may produce different oligosaccharides.[40,41] A variety of strategies, including adsorption and chromatographic separation, have been employed for purification of xylooligosaccharides from the hydrolysis liquors.[42,43] Vegas et al.[44] treated rice husk with hot compressed water to cause hydrolytic degradation of arabinoxylans. Purification of the concentrated autohydrolysis liquors via ethyl acetate extraction and ion exchange led to concentrates with arabinoxylan-derived oligomers with a potential application as functional food ingredients.

7.4 MOLECULAR STRUCTURE OF ARABINOXYLANS

7.4.1 Monosaccharide Residues and Glycosidic Linkages in Arabinoxylan Structures

Arabinoxylans consist of linear (1→4)-β-D-xylopyranosyl chains to which α-L-arabinofuranosyl residues are attached as side branches. Arabinose residues can be attached to xylose units at the O-2, O-3, or both O-2,3 positions, resulting in four structural elements in the molecular structure of arabinoxylans: monosubstituted Xyl*p* at O-2 or O-3, disubstituted Xyl*p* at O-2,3, and unsubstituted Xyl*p* (Figure 7.4). The relative amount and the sequence of distribution of these structural elements vary depending on the source of arabinoxylans. The majority of arabinofuranosyl residues in arabinoxylans are present as monomeric substituents; however, a small proportion of oligomeric side chains consisting of two or more arabinosyl residues linked via 1→2, 1→3, and 1→5 linkages have been reported.[7] The molecular structure of arabinoxylans from rice, sorghum, finger millet, and maize bran is more complex than that from wheat, rye, and barley, since the side branches contain, in addition to arabinose residues, small amounts of xylopyranose, galactopyranose, and α-D-glucuronic acid or 4-O-methyl-α-D-glucuronic residues (Figure 7.5).[45–48] Glucuronopyranosyl residues constitute about 4% of the arabinoxylans from barley husk[49] and are also present in arabinoxylans from wheat bran.[50]

A recent investigation of the molecular structure of water-soluble wheat endosperm arabinoxylans using atomic force microscopy has confirmed the generally linear structure of arabinoxylans as described above. However, it also revealed that a small fraction (~15%) of the polymers might, in fact, be branched.[51] It was reported that the branches were composed of β-(1→4)-linked xylose residues, and they appeared to be randomly located along the chain. The likelihood of their presence increased with the increasing length of the molecules. Only about 1% of the branched chains contained more than one branch.

The physiologically active gel-forming polysaccharides of psyllium husk (*Plantago ovata* Forsk) have recently been found to consist of neutral arabinoxylans,

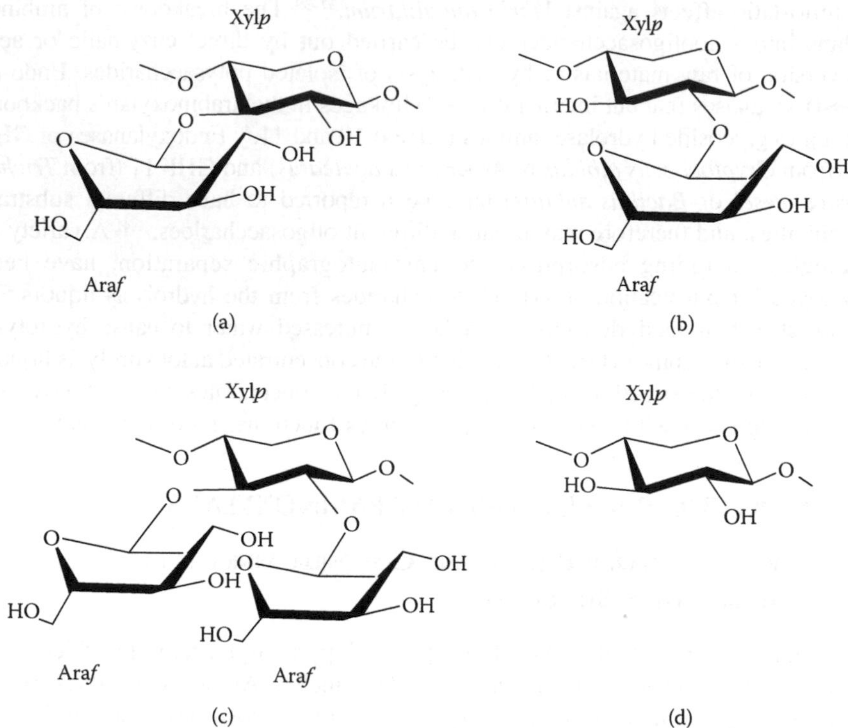

FIGURE 7.4 Structural elements present in arabinoxylans: (a) monosubstituted Xyl*p* at O-3, (b) monosubstituted Xyl*p* at O-2, (c) disubstituted Xyl*p* at O-2,3, (d) unsubstituted Xyl*p*.

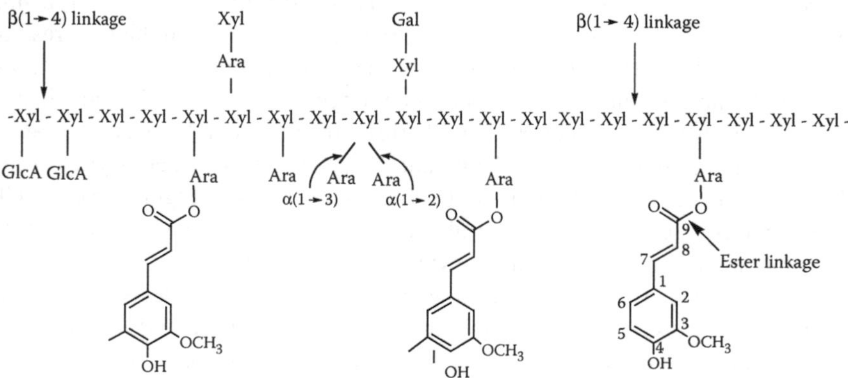

FIGURE 7.5 General structure of arabinoxylans.

containing mainly arabinose (22.6%) and xylose (74.6%) residues, with only traces of other sugars. However, their structure is significantly different from that of arabinoxylans in common cereals. Psyllium arabinoxylans, despite their low Ara/Xyl ratio (0.30), are highly branched polymers with the main chain of densely substituted

(1→4)-linked xylopyranosyl residues, some carrying single xylose units at position O-2 and others bearing trisaccharide branches (Ara*f*-α-(1→3)-Xyl*p*-β-(1→3)-Ara*f*) at position O-3.[8] This unique molecular structure is associated with some unusual physicochemical and physiological properties of psyllium arabinoxylans, such as strong gelling potential and low fermentability by the intestinal microflora.

7.4.2 Ferulic Acid Residues and Intermolecular Cross-Linking

An unusual feature of the structure of arabinoxylans is the presence of ferulic acid residues covalently linked via an ester linkage to O-5 of the arabinose residue (Figure 7.5). Digestion of wheat aleurone cell walls and detailed analysis of the hydrolyzates led to identification of two feruloylated arabinoxylosides: (5-O-feruloyl-α-L-arabinofuranosyl)-(1→3)-β-D-xylopyranosyl-(1→4)-β-D-xylanopyranose (FAXX) and β-D-xylanopyranosyl-(1→4)-(5-O-feruloyl-α-L-arabinofuranosyl)-(1→3)-β-D-xylopyranosyl-(1→4)-β-D-xylanopyranose (XFAXX).[52] Ferulic acid is usually reported in its trans-isomeric form; however, the exposure of grain to UV light at the later stages of development may cause trans to cis isomerization, resulting in up to 30 to 40% of cis-ferulic acid, especially in arabinoxylans originating from the outer grain tissues. Smith and Hartley[53] estimated that the cell walls of wheat bran contained approximately 34.0 μmol of feruloyl groups per gram of walls, whereas the endosperm cell walls contained 5.6 μmol/g of walls. Rattan et al.[54] reported that arabinoxylans from flours of several Canadian wheat varieties contained 0.63 to 1.37 mg of ferulic acid per gram of isolated and purified polymers. Dervilly-Pinel et al.[55] compared the contents of ferulic acid in arabinoxylans isolated from the endosperm of wheat, barley, rye, and triticale. It was reported that the amounts ranged from 18 to 60 ferulic acid residues per 10,000 xylose residues, and that purified arabinoxylans from wheat and barley contained more ferulic acid than those from rye and triticale. A greater concentration of ferulic acid was found in the wheat aleurone preparations (0.71%) than in the pericarp preparations (0.31%).[36,56]

Ferulic acid residues can act as cross-linking agents between polysaccharides or between polysaccharides and lignin. The cross-linking is effected by ferulate dimerization by either photochemical or, more importantly, free radical coupling reactions of ferulate–polysaccharide esters. Ferulate esters dimerize via phenoxy radicals to form dehydrodiferulate esters. Three electron-delocalized phenoxy radicals can be induced at position 4-O, C-5, or C-8 of ferulic acid residue, giving rise to at least five known diferulate esters coupled via 8-5', 8-O-4', 5-5', 8-8', and 4-O-5' linkages (Figure 7.6). Bunzel and coworkers[57] have recently isolated and measured the amount of different diferulates in cereals. The distribution patterns of various diferulates in water-insoluble arabinoxylans were similar among the different cereals. The 8-5'-coupled diferulate predominated, followed by the 8-O-4'-linked dimer. The diferulate distribution patterns in water-soluble arabinoxylans in various cereals were very different from those in water-insoluble fiber. The amount of 8-8' dimers increased substantially, whereas the amounts of 5-5' and 8-O-4' dimers decreased. It has to be pointed out that the identification of dehydrodiferulic acids released upon saponification of plant material does not provide sufficient evidence of

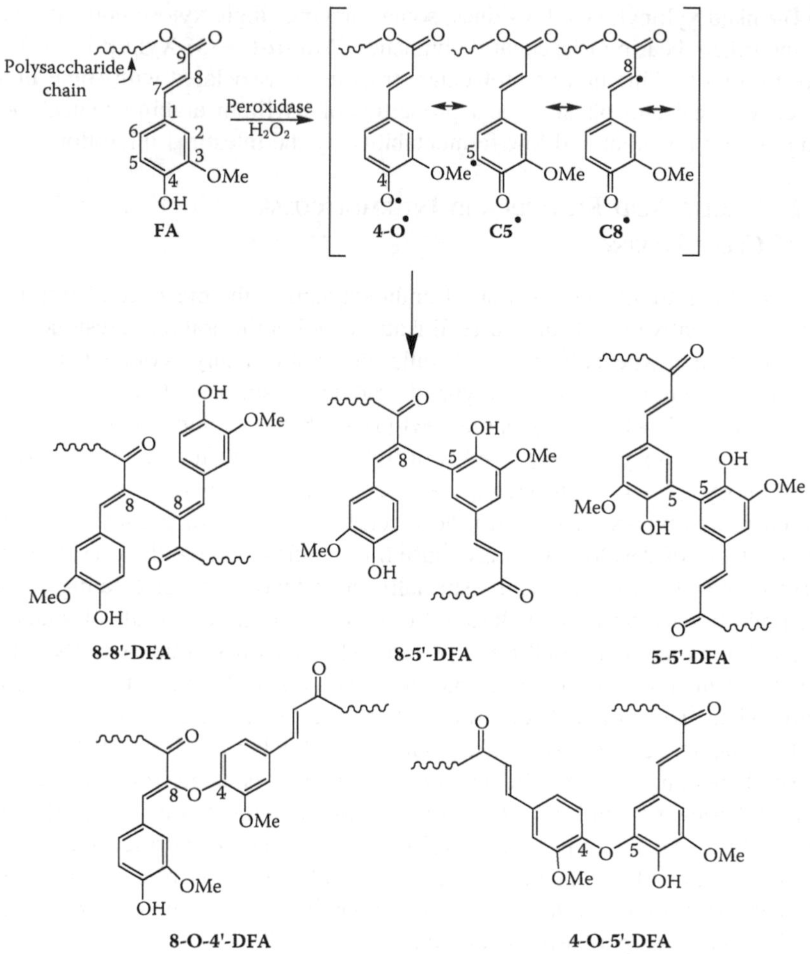

FIGURE 7.6 Three electron-delocalized phenoxy radicals generated from ferulic acid (FA) by a free radical-generating system (peroxidase–H_2O_2) and molecular structure of five known dehydrodiferulic acids.

polysaccharide cross-linking, since these diferulate bridges can theoretically be formed intramolecularly. Evidence for polysaccharide cross-linking via dehydrodiferulates was provided, however, by isolation and identification of feruloylated saccharide fragments. The isolation of 5-5'-diferuloyl saccharides[58] and diarabinosyl ester of 8-O-4'-dehydrodiferulate[59] provided more satisfactory evidence for intermolecular cross-linking, although the molecular modeling experiments did not exclude the possibility that 5-5'-diferulate may be tethered to the same arabinoxylan chains as long as the arabinose residues bearing ferulate units are at least three xylose residues apart.[60] Fry et al.[61] speculated that free radical polymerization of ferulates does not stop at the dimer stage *in vivo*, but proceeds to form higher oligomers. Indeed, in 2003 the first 4-0-8'/5-'5"-coupled dehydrotriferulic acid was

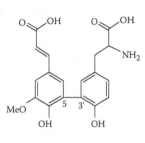

8-O-4′/8′-O-4"dehydro-triferulic acid

8-8′/4′-O-8"dehydro-triferulic acid

4-O-8′/5′-5"dehydro-triferulic acid

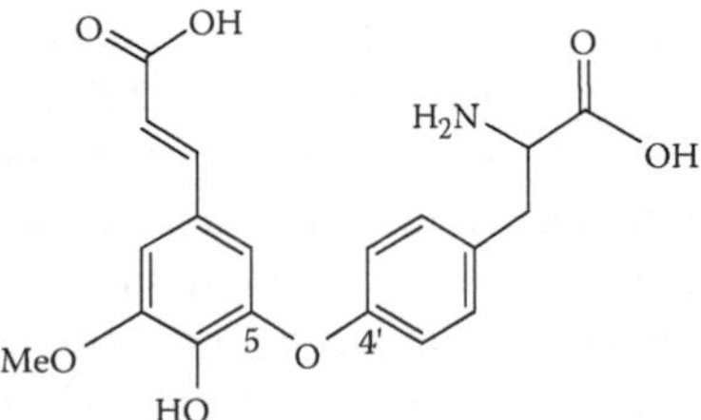

5-3′-dehydro-ferulic acid-tyrosine

4-O-3′-dehydro-ferulic acid-tyrosine

5-O-4′-dehydro-ferulic acid-tyrosine

FIGURE 7.7 Structures of new ferulic acid dehydrotrimers and speculative structures of dehydroferulic acid–tyrosine heterodimers.

isolated from maize (Figure 7.7).[62,63] It was suggested, however, that this trimer originates from initial intramolecular 5-5' linkage, further coupled with ferulate on another chain, rather than from the interaction of three ferulates on three separate chains.[62,64] With the isolation and identification of another two trimers coupled by 8-O-4'/8'-O-4" and 8-8'/4'-O-8" linkages, it became clear that higher oligomers of ferulate may play a significant role in cross-linking of cell wall polysaccharides (Figure 7.7).[65]

Cross-linking of cell wall polymers via diferulate bridges is of importance to plant physiologists, food chemists, and technologists. Thermal stability of cell adhesion and maintenance of crispness of plant-based foods (e.g., water chestnut after cooking),[66] gelling properties of cereal arabinoxylans and sugar beet pectins,[67] insolubility of cereal dietary fibers,[57] and limited cell wall degradability by ruminants[68] are related to the formation of ferulate cross-links. Bunzel et al.[57] estimated the degree of cross-linking in arabinoxylans from the ratio of diferulic acid to xylose residues and proposed that the diferulate bridges are partly responsible for the insolubility of these polymers. It was found that the degree of cross-linking in insoluble arabinoxylans was 8 to 39 times higher than in their water-soluble counterparts. Theoretically, in the presence of proteins, ferulic acid residues could be linked to the N-terminal of the protein amino group or to tyrosine.[69] Recent studies by Piber and Koehler[70] provided evidence for a covalent linkage between arabinoxylans and proteins. Dehydroferulic acid–tyrosine dimers were isolated from wheat and rye dough preparations and identified based on mass spectrometric data (Figure 7.7).

7.4.3 Structural Heterogeneity and Polydispersity of Arabinoxylans

Arabinoxylans from various cereals and different plant tissues share the same general molecular structures; however, they differ drastically in fine structural features, which may affect their physicochemical properties. These differences are reflected in the degree of polymerization, in the ratio of arabinose to xylose residues, in the relative proportions and sequence of various glycosidic linkages, in the pattern of substitution of the xylan backbone with arabinose residues, and in the presence and amount of other substituents, such as feruloyl groups or glucuronic acid residues. Since arabinoxylans are not under strict genetic control, even polymers isolated from a single plant or tissue exhibit structural microheterogeneity. To obtain a better insight into the structural characteristics of these polymers, arabinoxylans have been fractionated into more homogeneous populations by chromatographic or chemical means. Perhaps the most successful fractionation practices utilize a stepwise precipitation with alcohol or ammonium sulfate. The fractionation is based on differential solubilities of arabinoxylans with different molecular weights and structures in solutions containing various amounts of ethanol or ammonium sulfate. Arabinoxylans precipitating at increasing concentrations of ethanol or ammonium sulfate exhibit an increasing ratio of Ara/Xyl but decreasing weight-average molecular weight (Figure 7.8).

The ratio of Ara/Xyl residues indicates a degree of branching in these polysaccharides. Depending on the origin of arabinoxylans, the ratio of Ara/Xyl may vary from 0.3 to 1.1, although some minor fractions with the Ara/Xyl ratio outside this common range have also been reported (e.g., corncob arabinoxylans with Ara/Xyl = 0.07,[71] rye bran arabinoxylans with Ara/Xyl = 0.14,[72] water-soluble wheat arabinoxylans with Ara/Xyl = 1.28,[73] water-soluble rye arabinoxylans with Ara/Xyl = 1.42[74]). The ratio of Ara/Xyl in arabinoxylans from wheat endosperm generally varies from 0.50 to 0.71, and it is usually slightly lower than in arabinoxylans from wheat bran (0.82 to 1.07).[23,54,75–77] Antoine and coworkers[56] reported Ara/Xyl ratios of 0.33, 0.37, and 1.13 for arabinoxylans in the walls of aleurone, intermediate, and pericarp fractions, respectively. A study of the water-soluble and -insoluble arabi-

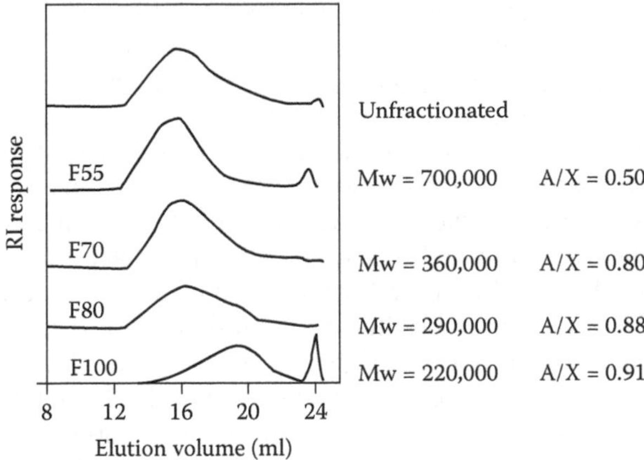

FIGURE 7.8 High performance size exclusion chromatography (HPSEC) profiles and molecular weight of arabinoxylan fractions obtained by sequential precipitation of arabinoxylan solution with increasing saturation level of ammonium sulfate. F55, F70, F80, F100, fractions obtained with 55, 70, 80, and 100% saturation of ammonium sulfate; Mw, weight-average molecular weight of arabinoxylan fractions determined with multiangle light-scattering detector after elution from the size exclusion column.

noxylans from wheat endosperm found a lot of similarities between these two types of polymers,[78] but significant differences between the structures of water- and alkali-extractable arabinoxylans from bran were also found.[77] The water-extractable arabinoxylans from wheat bran were found to have a lower degree of substitution (Ara/Xyl = 0.45) and lower molecular weights (20- and 5-kDa fractions), compared to their alkali-extractable counterparts (Ara/Xyl = 0.82; 100- to 120-kDa and 5- to 10-kDa fractions). In rye, the ratio of Ara/Xyl falls between 0.48 and 0.78,[9,79] and a distinction can be made between the structures of arabinoxylans from the rye starchy endosperm and those from the bran.

The ratio of Ara/Xyl indicates the degree of branching in arabinoxylans; however, it does not reveal detailed structural features of these polymers. The relative amounts of unsubstituted, monosubstituted (at O-3 or O-2), and doubly substituted xylose residues, as well as the sequences of these four structural elements, are better indicators of the molecular structures of cereal arabinoxylans. However, the relative proportions of the four structural elements in arabinoxylan chains may be related to the Ara/Xyl ratio, and some trends have been reported. For example, in water-extractable rye and wheat arabinoxylans, a higher Ara/Xyl ratio was associated with a higher content of 2-monosubstituted and disubstituted xylose residues, and a lower content of 3-monosubstituted and unsubstituted xylose residues.[9,78,80] Compared to rye, water-extractable arabinoxylans from wheat exhibit a higher proportion of unsubstituted xylose residues (50 to 80% vs. 22 to 54%), a somewhat lower proportion of monosubstituted xylose residues (<20% vs. up to 40%), and a higher proportion of arabinose residues in 2-, 3-, or 5-linked short chains.[78,81]

The barley grain water-soluble arabinoxylans generally contain 47 to 65% unsubstituted, 20 to 25% monosubstituted, and 19 to 26% disubstituted xylose residues.[82–84] In barley, more arabinoxylans are found in the aleurone than in the endosperm cell walls. Recent studies have identified significant differences in the molecular structure between arabinoxylans isolated from the pearling by-products (PBPs) (enriched in the aleurone layer) and fiber-rich fractions (FRFs) (enriched in the endosperm cell walls) of hull-less barley samples with various starch types.[85] In general, the water-extractable arabinoxylans from PBPs were more substituted than those from the FRFs. This was observed for all barley types and evidenced by a higher Ara/Xyl ratio and a lower content of unsubstituted Xyl*p* residues in PBPs than in FRFs. The mode of substitution varied, however, depending on the barley type and tissue (Figure 7.9). While the water-extractable arabinoxylans from PBPs of high-amylose and normal barley had almost the same amount of mono- and disubstituted Xyl*p* residues, those from waxy barley had twice as many singly as doubly substituted xylose

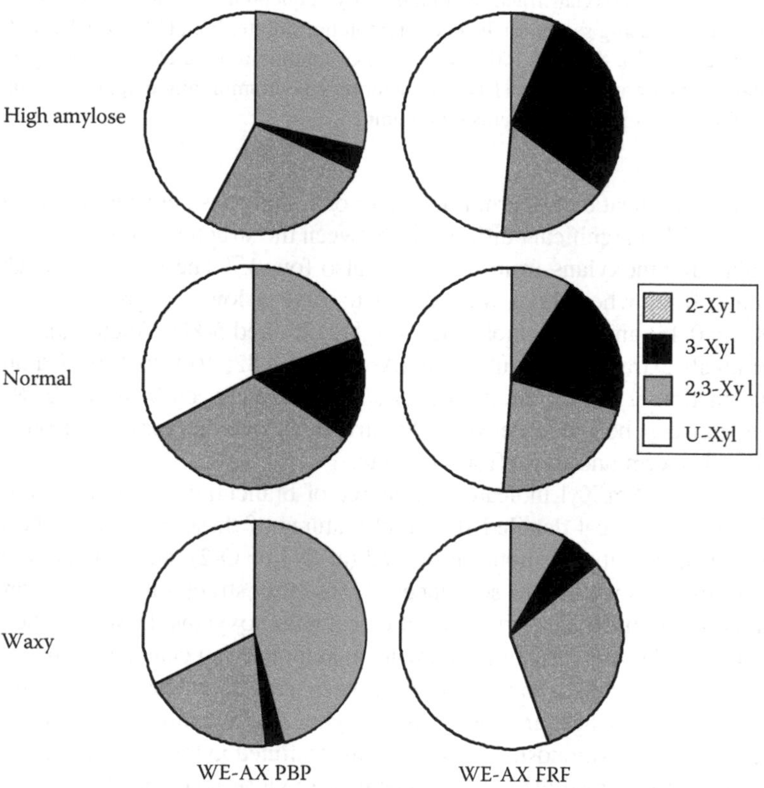

FIGURE 7.9 Substitution patterns in water-extractable (WE) arabinoxylans from pearling by-products (PBPs, 10% abraded) and fiber-rich fractions (FRFs, obtained by roller milling of 10% pearled grain) of barley with high-amylose, normal, and waxy starch characteristics. (Adapted from Izydorczyk, M.S. et al., *Cereal Chem.*, 80, 645, 2003.)

residues. An interesting feature of arabinoxylans from PBPs, common to all barley types, was a very high content of Xyl*p* residues substituted at the O-2 position.

It is generally agreed that cereal arabinoxylans are highly heterogeneous, consisting of a range of structures with different degrees and patterns of substitution. However, despite the structural heterogeneity of arabinoxylans, most studies point to a nonrandom distribution of Ara*f* residues along the xylan backbone.[73,86,87] Detailed structural analysis of isolated and purified arabinoxylans from various cereals resulted in tentative structural models for these polysaccharides. For the water-extractable arabinoxylans from wheat endosperm, it was shown that the highly substituted regions (or entire chains) of the fraction obtained with 100% saturation of ammonium sulfate are enriched in Xyl*p* residues doubly substituted at C(O)-2,3, Xyl*p* monosubstituted at C(O)-2, and short arabinose side chains.[88] Furthermore, these regions contain sequences of up to four contiguously substituted Xyl*p* residues, with three contiguously substituted Xyl*p* residues occurring most frequently. More recently, Dervilly-Pinel et al.[73] isolated homogeneous fractions of wheat arabinoxylans using both graded ethanol precipitation and size exclusion chromatography (SEC) fractionation and found blocks up to six contiguously substituted xylosyl residues in highly branched populations of these polymers. In contrast, the less substituted fractions of wheat arabinoxylans, obtainable by precipitation at a relatively low concentration of ammonium sulfate or ethanol, are built up mainly of less densely substituted regions, containing sequences of contiguously (at least up to six, but possibly more) unsubstituted xylose residues. The branched Xyl*p* in these regions are more frequently mono- than disubstituted and occur mostly as single or blocks of two substituted residues.[88]

Bengtsson et al.[89] proposed that in rye arabinoxylans, the mono- and disubstituted xylose residues were present in different polymers or in different regions of the same polymer chain; it was hypothesized that the major polymer structure (arabinoxylans I) contained only un- and monosubstituted (~46%) xylose residues, whereas the minor structure (arabinoxylans II) contained un- and disubstituted (57%) Xyl*p*. The ratio of arabinoxylans I/II could vary from 1.1 to 2.8 for different rye varieties grown in different countries.[90] Vinkx et al.[74] isolated a highly branched rye arabinoxylan (Ara/Xyl = 1.42) with a very low proportion of 3-monosubstituted Xyl*p*. However, this polymer differed from arabinoxylans II, hypothesized by Bengtsson et al.,[89] in that it contained a relatively high level of 2-monosubstituted Xyl*p* (14%) in addition to disubstituted Xyl*p* residues (60%). It was therefore concluded that a range of polymer structures exists in rye arabinoxylans, rather than two classes as initially suggested.[80,91]

7.4.4 Molecular Weight

The molecular weight (Mw) of arabinoxylans varies depending on their origin and the method used for its determination. The early studies using sedimentation techniques[92,93] reported relatively low Mw values (65,000 to 66,000) for water-extractable wheat arabinoxylans. However, very high Mw values (800,000 to 5,000,000) were reported when gel filtration chromatography was used, and the molecular weight of arabinoxylans was estimated by comparing their elution volume with that of standards with known Mw.[20,94] The latter approach overestimates the molecular weight and highlights the difficulties in accurately measuring the molecular weight of polymers with different conformations than those of commonly used gel filtration standards (pullulans, dextrans). More recent studies, utilizing high-performance size exclusion chromatography, combined with a multiangle light-scattering detector, have determined that the weight-average molecular weight of wheat arabinoxylans ranges from 220,000 to 700,000 (Figure 7.8).[78,95] Higher molecular weight values were reported for alkali-extractable arabinoxylans from hull-less barley (850,000 to 2,430,000).[85] The size exclusion chromatography profiles of arabinoxylans indicate a very broad distribution of molecular weights. High ratios of weight-average molecular weight to number-average molecular weight (Mw/Mn) reported for alkali-extractable wheat arabinoxylans (1.3 to 2.5), as well as for water-extractable wheat (4.1) and rye (8.5) arabinoxylans,[7] indicate the inherent polydisperse nature of these polymers. Warrand et al.[5] reported that the main fraction of the mucilage extracted from seeds of *Linum usitassinum* contained highly polydispersed arabinoxylans with three different populations of these polymers with molecular weights of 5×10^6 (~10%), 1×10^6 (40%), and 0.2×10^6 (~50%).

7.5 BIOSYNTHESIS OF ARABINOXYLANS

Relatively little is known about the exact mechanism of arabinoxylan synthesis. Like other polysaccharides, arabinoxylans are products of synthases and glycosyl transferases and, as such, are secondary gene products.[96] Arabinoxylans and other cell wall polysaccharides (except cellulose) are synthesized within the cell in the Golgi apparatus and endoplasmic reticulum.[97] The immediate donors of monosaccharides for synthesis of arabinoxylans are UDP-D-Xyl*p* and UDP-L-Ara*f*, formed from UDP-D-Glc*p* by the action of appropriate epimerases.[61] Some generalizations can be made concerning the mechanism of arabinoxylan polymerization (Figure 7.10). It is believed that distinct glycosyl transferases are necessary for each of the different monosaccharide and glycosidic linkages existing in arabinoxylan chains. The process of polymerization can be divided into three steps: chain initiation, elongation, and termination. Available evidence suggests that the first sugar donation is not to a free monosaccharide, but to a protein or lipid primer. Tailward growth, i.e., addition of the new residue to the nonreducing end of the chain, has been generally accepted as the direction of chain elongation. Recently, a β-(1→4)-xylosyl transferase, isolated from the microsomal membranes of the developing barley endosperm, has been shown to transfer xylose from uridine 5'-diphosphoxylose (UDP-Xyl) into an exogenous xylooligosaccharide chain (derivatized at the reducing end).[98] Repeated attachment of xylose residues occurred at the nonreducing end of the pyridylaminated-xylotriose chain through β-(1→4) linkages. During the stepwise addition of the xylose residues to the growing polymer chain, a stage must be reached that involves the addition of the branching points. Based on the *in vivo* studies on xyloglucans and glucuronoxylans,[99] it is assumed that arabinose residues are incorporated simultaneously with the polymerization of the xylan backbone. However, it is not clear whether separate arabinosyl transferases are required for substitution of arabinose at C-2 and C-3

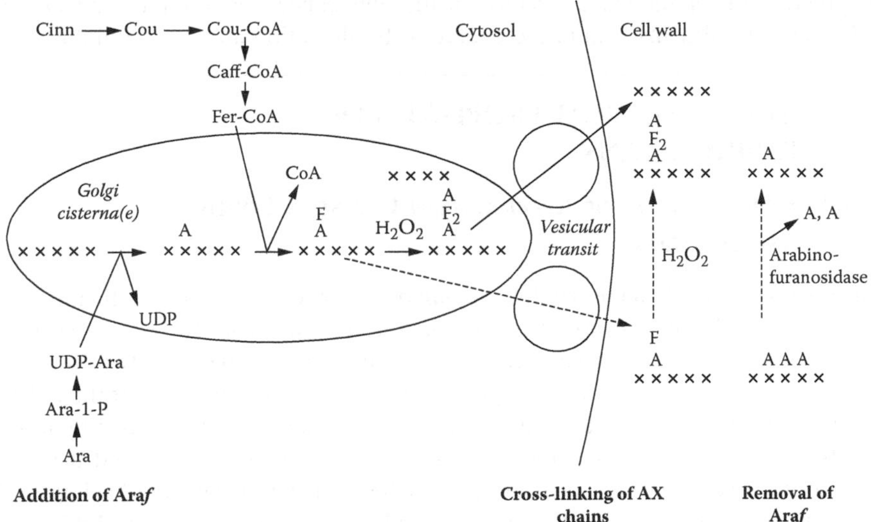

FIGURE 7.10 Biosynthesis of arabinoxylans. X, xylose residues; A, arabinose residues; F, ferulic acid residue; F2, diferulic acid; Cinn, cinnamic acid; Cou, coumaric acid; caff, caffeic acid; Fer, ferulic acid. (Adapted from Fry, S.C. et al., *Planta*, 211, 679, 2000. With permission.)

of the xylose residues. The attachment of feruloyl groups to arabinoxylans may occur by transacylation, and the polysaccharides are feruloylated co-instantaneously with the polymerization processes within the endomembrane system.[61,97] Fry and coworkers[61] showed that in maize cell cultures, the coupling of the feruloyl groups, leading to cross-linking of arabinoxylans, can occur within 1 min of the attachment of the feruloyl group to the polymer chain. However, cross-linking of the feruloylated arabinoxylans can also occur after their deposition in the cell walls. It is also thought that when arabinoxylans are initially deposited into walls, the xylan backbone is heavily substituted with arabinosyl residues. Subsequently, arabinosyl residues are removed by the action of arabinofuranohydrolases.[100] These postdeposition processes, debranching and cross-linking, lead to changes in physicochemical properties of arabinoxylans, such as solubility or capability to interact with other cell wall polysaccharides, thereby allowing the plant to control the tissue cohesion, cell expansion, and permeability of the cell walls to metabolites and pathogens.

The mechanism of chain termination, which directly controls the length of arabinoxylan chains, is the least known. It has been suggested that the rates of vesicle movement and fusion with plasma membrane play some role in determining the degree of polymerization (DP) of cell wall polysaccharides;[97] however, no evidence exists to support this proposition. Because of the numerous and complex events involved in biosynthesis of arabinoxylans, the formation of these polymers is not strictly regulated and may depend on several factors. As a result, arabinoxylans show a high degree of microheterogeneity and belong to the class of polydispersed polysaccharides. Their polydispersity can be reflected in the degree of polymerization of individual chains, in the abundance, distribution, and DPs of side-chain substituents, and in the degree of feruloylation and cross-linking.

7.6 PHYSICOCHEMICAL PROPERTIES OF ARABINOXYLANS

7.6.1 CONFORMATION OF ARABINOXYLAN CHAINS IN SOLIDS AND SOLUTIONS

The conformation of an unsubstituted xylan bears some resemblance to that of other β-(1→4)-linked polysaccharides, such as cellulose or mannan. However, the single hydrogen bond between two adjacent xylosyl residues, compared to two hydrogen bonds between two adjacent glycosyl residues in cellulose, has an important effect on the capacity of the xylan chain to form cooperative intramolecular hydrogen bonds, and hence on its conformation. As a result, xylans form twisted threefold ribbon-like strands that are more flexible than the rigid twofold helices of cellulose.[101] The β-(1→4)-linked xylan in the unsubstituted form aggregates into insoluble complexes stabilized by numerous intermolecular H bonds. The presence of side groups protruding from the xylan backbone drastically suppresses the interchain linking system, thus making the polymer partially soluble in water, but apparently does not substantially change the basic backbone conformation.[101] Andrewartha and coworkers[92] postulated that the presence of arabinosyl substituents stiffens the chain by maintaining a more extended xylan backbone.

The origin of the viscous behavior of arabinoxylans was associated with the formation of rod-like structures in solution.[92] However, the evidence presented by other workers indicates that arabinoxylans behave in solution as semiflexible coils. Ebringerová et al.[71] reported the Mark–Houwink exponent 'a' of 0.50 ($[\eta] = KM_w^a$) for corncob arabinoxylans with an unusually low degree of substitution (Ara/Xyl = 0.07), a value characteristic for molecules in unperturbed coil-shaped structures. Recent studies, based on the light-scattering measurements, also indicate that in solution, arabinoxylans behave as locally stiff, semiflexible random coils.[95,102,103] Dervilly-Pinel et al.[102] reported that the conformational parameters, such as the exponent 'a' (0.74) and the hydrodynamic parameter v (0.47) ($R_g^{0.5} = KM_w^v$) determined for the water-soluble wheat arabinoxylans, were typical for a random coil conformation. However, the persistence length (q = 6 to 8 nm) representing the chain rigidity indicated that arabinoxylan chains are semiflexible, in comparison with very flexible polysaccharides (q = 1.70), such as amylose or pullulans.

The solubility of arabinoxylans is closely related to the presence of the arabinosyl substituents along the xylan backbone. Andrewartha and coworkers[92] prepared a series of arabinoxylan chains with various degrees of branching by partially removing the arabinofuranosyl side branches with α-L-arabinofuranosidase. At Ara/Xyl ~ 0.43, the solubility of arabinoxylans abruptly declined. The amount of arabinose substituents as well as their distribution along the xylan backbone affects the potential of arabinoxylan chains to interact with each other or with other polysaccharides. It is prudent to predict that the presence of segments of unsubstituted xylose residues in the polymer chains will increase the potential of arabinoxylans to form intermolecular aggregates. This may lead to either an increase in viscosity or precipitation of polymer chains if the interactions are numerous. Izydorczyk and MacGregor[104] provided empirical evidence of noncovalent interactions between sparsely substituted arabinoxylan chains (Ara/Xyl = 0.18 to 0.32) and cellulose-like fragments of β-glucans. In the plant cell wall material, the noncovalent topological associations between β-glucans and arabinoxylans might contribute to poor water solubility or to low enzymic digestibility of these polymers.

7.6.2 Viscosity of Arabinoxylan Solutions

As a result of the high molecular weight and locally stiff, semiflexible random coil conformation, arabinoxylans exhibit a very high viscosity in aqueous solutions. An intrinsic viscosity of 5.9 dl/g was reported for rye arabinoxylans, whereas for wheat arabinoxylans the values ranged from 2.75 to 5.48 dl/g depending on cultivars.[9,81] In dilute solutions, the zero-shear-rate specific viscosity $(\eta_{sp})_0$ increases linearly with increasing arabinoxylan concentration (c) with a slope of log $(\eta_{sp})_0$ vs. log c≈1. Above the so-called critical concentration (c*), marking the onset of physical entanglements and coil overlap, the concentration dependence of $(\eta_{sp})_0$ increases and the slope increases to ≈3.7 to 3.9 (Table 7.3). The critical concentration values for arabinoxylans from wheat endosperm are relatively low, ranging from 0.2 to 0.4% (w/v) depending on the chain length and fine molecular structure of these polymers (Table 7.3).[105,106]

The apparent viscosity of aqueous solutions of arabinoxylans is strongly dependent on their concentration and the rate of shear at which the viscosity measurements are taken. At low shear, arabinoxylan solutions behave like Newtonian fluids and exhibit very little shear rate dependence (Figure 7.11a). The apparent viscosity increases with the polymer concentration. With the increasing shear rates, arabinoxylans display a reduction in the apparent viscosity, commonly known as shear thinning. The molecular size of arabinoxylans is an important determinant of the solution behavior of these polymers. Izydorczyk and Biliaderis[105] reported that the high molecular weight fractions of wheat arabinoxylans (with intrinsic viscosities of 8.5 and 6.2 dl/g) exhibited weakly elastic properties in solutions (Figure 7.11b). The dynamic rheological measurements indicated that with increasing polymer concentration, the viscoelastic behavior of arabinoxylan fractions changed from that of viscous solution (G" > G' at all frequencies) to weakly elastic (G' > G" at higher frequencies) (Figure 7.11b).

Warrand et al.[107] showed that arabinoxylans have a tendency to form macrostructures in aqueous solutions via chain aggregation and physical entanglements, and emphasized the importance of hydrogen bonds in stabilizing these aggregates. This aggregation tendency may be responsible for pseudo-gel behavior of arabinoxylan solutions under some conditions. For example, it was shown that 2% (w/v) solutions of high molecular weight arabinoxylans isolated from flaxseed mucilage exhibited preponderantly elastic properties, with the elastic modulus, G', exceeding the viscous modulus, G" (Figure 7.12).[107] These weak gel properties were, however, greatly diminished in the presence of chaotropic salts, known for weakening the hydrogen bonds between solute molecules. The behavior of arabinoxylans in

TABLE 7.3
Values for Intrinsic Viscosity, Critical Concentration (c*), Coil Overlap, and Slopes of Dilute and Entangled Domains of Arabinoxylan Fractions Obtained by Stepwise Precipitation with Ammonium Sulfate or Size Exclusion Chromatography Fractionation

Arabinoxylan Fraction	[η] (dl/g)	c* (g/100 ml)	Coil Overlap (c*[η])	Slope: Dilute Domain	Slope: Entangled Domain
F60[a]	4.70	0.26	1.24	1.13	2.19
F70	4.20	0.31	1.30	1.12	2.04
F80	3.16	0.38	1.20	1.07	2.00
F95	1.90	nd	nd	nd	nd
F1[b]	8.5	0.17	1.44	1.1	3.9
F2	6.2	0.20	1.24	1.1	3.7
F3	4.3	0.28	1.20	1.1	3.8
F4	3.8	0.29	1.10	1.1	3.8
F5	3.4	0.30	1.02	1.1	3.9

Note: nd = no data.

[a]Fractions obtained by stepwise precipitation of isolated and purified water-soluble wheat flour arabinoxylans with ammonium sulfate; the numbers 60 to 95 indicate the saturation of ammonium sulfate.
[b]Isolated and purified water-soluble wheat flour arabinoxylans were fractionated by size exclusion chromatography.

Source: Adapted from Izydorczyk, M.S. and Biliaderis, C.G., *J. Agric. Food Sci.*, 40, 561, 1992; Izydorczyk, M.S. and Biliaderis, C.G., *Carbohydr. Polym.*, 17, 237, 1992.

solutions and their viscosity-building properties are probably the most important characteristics responsible for the functional properties of arabinoxylans in the human digestive tract.

7.6.3 Oxidative Cross-Linking

Arabinoxylan solutions possess a unique capacity to form hydrogels in the presence of free radical-generating agents, such as peroxidase–H_2O_2, laccase, linoleic acid–lipoxygenase, ammonium persulfate, or ferric chloride.[108–112] Covalent cross-linking of arabinoxylan chains through dimerization of ferulic acid substituents is responsible for this unusual property of arabinoxylans (Figure 7.6). Although five dimers of ferulic acid (identical to those found in the cell walls) have been identified in gelled arabinoxylans, the 8-5' and 8-O-4' forms are preponderant, suggesting that both the aromatic ring and the double bond in the structure of ferulic acid serve as cross-linking sites.[112] Recently, a very small amount of a trimer of ferulic acid (4-O-8'/5'-5") has been detected in the laccase cross-linked wheat arabinoxylan gels.[64]

Small-amplitude oscillatory measurements have been useful in following the development of three-dimensional networks in solutions of arabinoxylans undergoing oxidative gelation. Upon addition of a free radical-generating agent to a solution

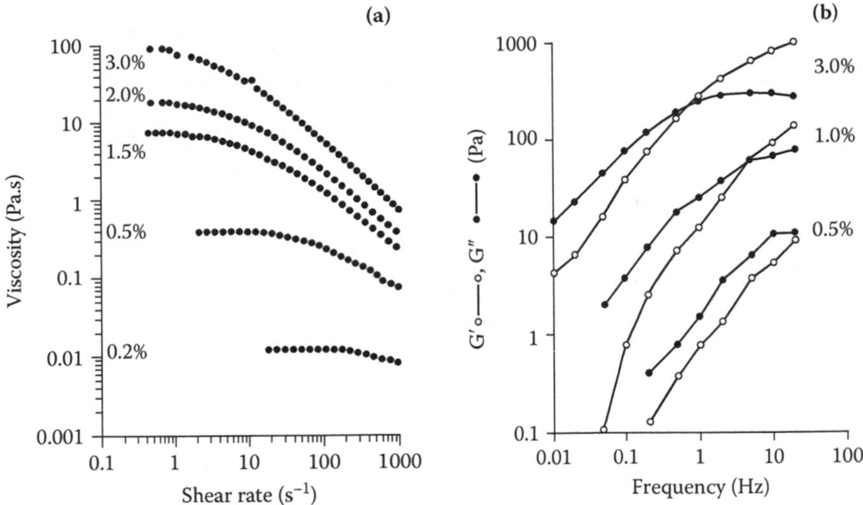

FIGURE 7.11 Effects of shear rate and polymer concentration of the apparent viscosity(a) and viscoelastic properties (b) of arabinoxylan solutions. (Adapted from Izydorczyk, M.S. and Biliaderis, C.G., *J. Agric. Food Sci.*, 40, 561, 1992. With permission.)

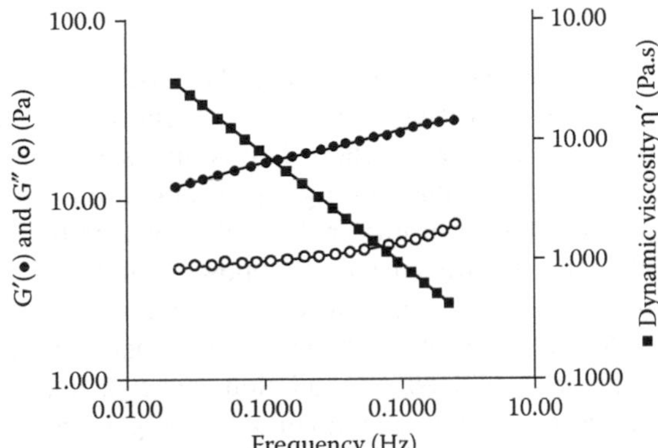

FIGURE 7.12 Frequency dependence of the storage modulus, G' (•), loss modulus, G" (○), and dynamic viscosity (■) of the solution (20 g/l, 25°C) of arabinoxylan isolated from flaxseed mucilage. (Adapted from Warrand, J. et al., *Biomacromolecules*, 6, 1871, 2005. With permission.)

of arabinoxylans, there is usually a rapid rise in the elastic modulus (G') followed by a plateau region with very little further increase of G' (Figure 7.13). This behavior has been attributed to the initial formation of covalent cross-linkages between ferulic acid residues of adjacent arabinoxylan chains that, once created in sufficient amount, could impede chain mobility and thus prevent further formation of cross-links.[7,54,55] The central role of feruloyl groups in gelation of arabinoxylan solutions is evidenced by disappearance of ferulic acid residues with a simultaneous formation of ferulic acid dimers and trimers during the initial stage of the gelation process.[64,110] Carvajal-Millan et al.[112] prepared arabinoxylan samples having the same molecular structure

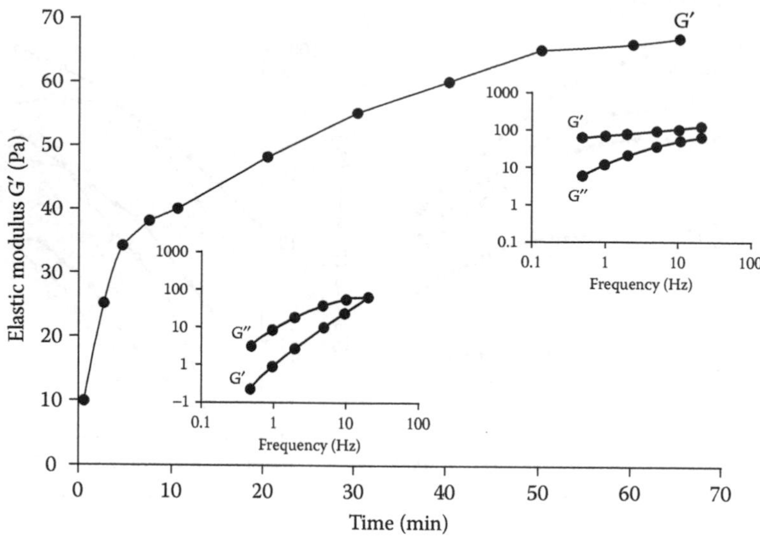

FIGURE 7.13 Development of storage modulus, G', with time of a water-soluble arabinoxylan (2.3% (w/v)) isolated from wheat flour and treated with horseradish peroxidase (0.11 purpurogallin units/ml) and H_2O_2 (3 mg/ml). Insets represent the molecular spectra of arabinoxylans before (left) and after (1 h) (right) addition of the oxidant.

but varying ferulic acid content, and showed that the elasticity of arabinoxylan gels increased linearly with the increasing ferulic acid content (in the range of 1.4 to 2.4 μg of ferulic acid per mg of arabinoxylan). It was also shown that arabinoxylans with ferulic acid content lower than 0.4 μg/mg of arabinoxylan (i.e., less than 1 ferulic acid residue per 2000 Xyl*p*) are not capable of forming gel structures.[112] Izydorczyk and Biliaderis[7] and Rattan et al.[54] proposed that in addition to the pivotal role of ferulic acid in the gelation of arabinoxylans, the molecular structure (amount and distribution of Ara*f* residues along the xylan backbone) and molecular weight of arabinoxylan chains also affected their gelation capacity. Rattan et al.[54] found a positive relationship between the elastic modulus of the gels from wheat arabinoxylans and the intrinsic viscosity and ferulic acid content in native polymers. It was also established that the rate and extent of gelation were dependent on the concentration of arabinoxylans and oxidizing agents.[54,110] Recently, Vansteenkiste et al.[113] and Carvajal-Millan et al.[112] have shown that cross-linking densities of gels, determined from swelling experiments, were higher than those calculated theoretically from the amount of dimers and trimers formed during gelation; these results suggest that in addition to the covalent cross-links, concomitant formation of noncovalent bonds between adjacent arabinoxylan chains might also occur and contribute to the final gel structure of arabinoxylans.

7.6.4 Physicochemical Properties of Arabinoxylan Gels

Arabinoxylan gels have neutral taste and odor, very high water absorption capacity (up to 100 g of water per gram of dry polymer), and are not susceptible to changes in pH or electrolyte concentrations.[7] These properties, together with the macroporous texture of gels (mesh sizes varying from 200 to 400 nm) and the dietary fiber nature of arabinoxylans, give them potential to be used as matrices with controlled releases of active agents in the food, cosmetic, and pharmaceutical industries. Vansteenkiste et al.[113] demonstrated that proteins embedded in the arabinoxylan gel network are protected against enzymic hydrolysis. Carvajal-Millan et al.[114] modified the rheological properties of arabinoxylan gels by altering either the initial ferulic acid content or the concentration of polymers before gelation. Consequently, the differences in rheological properties of gels were shown to affect their capacity to load and release proteins with various molar masses. The possibility to modulate protein release from arabinoxylan gels makes them useful for controlled delivery of therapeutic proteins.

7.7 ARABINOXYLANS AS TECHNOLOGICALLY FUNCTIONAL FOOD INGREDIENTS

The role of arabinoxylans in the bread-making processes and their effects on the final bread product have been studied extensively over the last few decades. The results reported by earlier studies were sometimes contradictory, primarily due to utilization of polymers with varying degrees of purity and composition, different levels of supplementation, and various baking procedures utilized by different workers. Clear distinction has to be made between water-soluble arabinoxylans and those that are not readily extractable with water. When added to wheat flour, water-soluble arabinoxylans compete for water with other flour constituents. As a consequence, dough consistency is increased. Michniewicz et al.[115] and Wang et al.[116] reported also that water-soluble arabinoxylans negatively affect gluten yield and increase its resistance against extension. These effects, however, can be corrected by adding more water or adding xylanase before dough mixing. Biliaderis et al.[117] used two different preparations of arabinoxylans to demonstrate the effect of molecular size of these polymers on baking absorption and showed that the high molecular weight arabinoxylans increased the farinograph water absorption to a greater extent than their low molecular weight counterparts. Water-soluble arabinoxylans are believed to increase the viscosity of the dough aqueous phase, and therefore to have a positive effect on the dough structure and its stability, especially during the early baking processes, when a relatively high pressure is generated inside the gas cells. The increased stability of the film surrounding the gas cells is useful in prolonging the oven rise and preventing coalescence. These phenomena lead to a higher loaf volume and improve crumb structure. McCleary[118] convincingly demonstrated the beneficial role of arabinoxylans on the loaf volume and appearance when wheat flours, treated with an excessive amount of purified endoxylanase, were shown to produce breads with low loaf volume and soggy texture. The overall effect of arabinoxylans on the bread-making process is, however, dependent on the concentration of these polymers in the dough system. A higher than optimum amount of arabinoxylans may cause viscosity buildup and hinder their beneficial effects.[117,119] The optimum concentration of arabinoxylans may depend on their molecular structure (size) and the baking characteristics of the wheat flours.[117] Despite a number of investigations, the mechanism of interactions between arabinoxylans and gluten has not been fully determined. Wang et al.[116,120] observed that water-soluble pentosans lower the extensibility of dough, and that this effect is related to the ferulic acid content of arabinoxylans. The authors proposed that covalent cross-linking between protein and arabinoxylans might be partly responsible for changes in the rheological properties of gluten. However, two separate studies by Hilhorst et al.[121] and Labat et al.[122] reported no evidence for covalent complexes between arabinoxylans and protein in mixed, oxidized, or overmixed doughs. Labat et al.[122] proposed that free ferulic acid esters are likely to react with proteins during mixing and, therefore, these free ferulic acid residues, rather than feruloylated arabinoxylans, might be involved in gluten breakdown during dough overmixing. Santos and coworkers[123] reported that the presence of water-soluble pentosans reinforced the gluten network and decreased the irreversible changes occurring during heating the gluten.

In contrast to water-soluble arabinoxylans, their insoluble counterparts, which are not readily extractable from the cell walls with water, destabilize the dough structure and have a negative effect on the loaf volume and other bread characteristics.[124,125] They also absorb a large amount of water, thus depleting the pool available for proper gluten development and film formation. Water-insoluble arabinoxylans that are present in dough as discrete cell wall fragments can form physical barriers for the gluten network during dough development. The resulting gluten has lower extensibility and a lower rate of aggregation, and therefore a different network structure.[125] During fermentation, water-insoluble arabinoxylans decrease the film stability. Their presence in dough results in lower loaf volume and coarser and firmer crumbs. However, Courtin and Delcour[124] showed that the negative effects of water-insoluble arabinoxylans in bread could be reversed by using endoxylanases with specificity toward the water-insoluble arabinoxylans. These enzymes catalyze the conversion of detrimental water-insoluble arabinoxylans to the high molecular weight water-soluble counterparts. The addition of optimal doses of endoxylanases during the bread-making processes has been found to positively affect the dough handling and bread properties. The addition of xylanases has been reported to improve dough consistency, fermentation stability, oven rise, loaf volume, and crumb structure and softness.[124–128] An excessive amount of endoxylanases or addition of endoxylanases with specificity toward water-soluble arabinoxylans may lead to too extensive degradation of these polymers and to negative effects on bread loaves. The use of xylanase in combination with peroxidase prevents extensive degradation of arabinoxylans by cross-linking them into larger aggregates.[121]

Another functional property of arabinoxylans may be associated with their role in bread staling. Bread staling is a complex phenomenon involving loss of aroma, deterioration of crust characteristics, and increase in crumb firmness. Biliaderis et al.[117] measured bread staling by monitoring the crumb firmness of breads fortified with water-soluble arabinoxylans. It was shown that over a 7-day storage period, the arabinoxylan-fortified breads exhibited lower crumb firmness than the controls. The observed effects were attributed to a higher moisture content of breads substituted with arabinoxylans and to the plasticizing effects of water. Contrary to the observed decrease in crumb firmness, starch retrogradation, as measured by calorimetry and x-ray diffraction, increased in the presence of arabinoxylans. The enhanced kinetics of chain ordering were attributed to a higher water content in the breads and a correspondingly greater mobility of the starch molecules. As established by Zeleznak and Hoseney,[129] starch retrogradation increases with increasing moisture content between 20 and 45% moisture in the system. Gudmudson et al.[130] showed that the rate of amylopectin crystallization in starch gels containing arabinoxylans is either increased or decreased, depending on the final water content of starch; retrogradation was minimal in starch gels with a moisture content below 20% and increased significantly in gels with a moisture content between 20 and 30%.

The recent revival of interest in incorporating arabinoxylans into food products is due to the beneficial effects of these polymers on human health. However, as pointed out above, the addition of arabinoxylans to food products may have some negative effects on their quality, depending on the amount and properties of arabinoxylans. The food industry is therefore faced with a challenge to create foods that are both healthy and appealing in terms of taste, aroma, and appearance. Several attempts to enrich various food products with arabinoxylans have been documented. Lu et al.[32] incorporated between 7 and 14% of arabinoxylan-rich fiber into bread and reported that the product's palatability was equal to that of 50% whole wheat bread. Utilization of barley pearling by-products enriched in dietary fiber for making pasta has been reported by Marconi et al.[131] Substitution of 50% of durum wheat semolina with pearling by-products resulted in a darker pasta with good cooking characteristics as regards stickiness, firmness, and cooking losses. The high dietary content, including arabinoxylans, has made this product a healthy alternative for health-conscious consumers. Izydorczyk et al.[132] incorporated barley fiber-rich fractions enriched in arabinoxylans and β-glucans into noodles and baked products. Appropriate baking procedures, protein content of wheat flour, and addition of specific hydrolytic enzymes in the bread formula resulted in satisfactory crumb texture and loaf volume of bread baked with partial replacement (15%) of wheat flour with the barley fiber-rich fractions.[133] Izydorczyk et al.[132] showed that yellow alkaline and white salted noodles enriched in barley fiber-rich fractions offer consumers convenience due to shorter cooking time, acceptable cooking quality, and increased and diversified nutritional value.

7.8 ARABINOXYLANS AS NUTRITIONALLY FUNCTIONAL FOOD INGREDIENTS

Arabinoxylans as part of dietary fiber have many potential physiological effects along the entire human gastrointestinal tract. These effects are dependent on a complex mixture of molecular and physical properties of arabinoxylan preparations as well as on the site, rate, and extent of their digestion and fermentation in the gut. In the large bowel, the undigested arabinoxylans may change the composition of the microbial flora, affect the activity of bacterial enzymes, and influence the end products of bacterial fermentation, thus having an effect on colonic health. It is generally agreed that arabinoxylans and various xylooligosaccharides enhance the growth of potentially health-promoting bacteria, the so-called probiotics, although some inconsistencies as to the effectiveness of arabinoxylans in proliferation of specific bacterial species are reported in the literature. Jaskari et al.[37] and Crittenden et al.[38] reported that xylooligosaccharide preparations (containing mainly β-(1→4)-Xyl oligosaccharides ranging in size from DP 2 to DP 5) support the growth of many *Bifidobacterium* and *Bacteroides* species, as well as *Lactobacillus brevis*, but are not fermented by *Escherichia coli*, enterococci, *Clostridium* sp., and the majority of *Lactobacillus* sp. Katrien et al.[134] generated various arabino-xylooligosaccharides with DPs of 5 to 10 containing mainly doubly branched xylose residues and observed that they were fermented completely by *Bifidobacterium adolescentis*, *Bifidobacterium longum*, and *Bacteroides vulgatus*. The same authors reported that intact arabinoxylans from wheat were fermented by *Bifidobacterium longum* and *Bacteroides ovatus*, but glucuronoarabinoxylans from sorghum could only be fermented by *Bacteroides ovatus*. It appears that the presence of arabinosyl substitution affects the ability of microorganisms to ferment the intact polysaccharide since arabinoxylans can support the growth of some bacterial species, whereas unsubstituted xylans are not fermented by any of the probiotic bacteria. Unsubstituted xylans are known to form insoluble aggregates and hindered accessibility of the bacterial enzymes to the substrates. Water-insoluble arabinoxylans are also not likely to be digested in the large intestine. The degree of branching and the distribution of arabinose along the xylan backbone may additionally influence the degradation of arabinoxylans. When four types of rye bread containing either whole rye, pericarp/testa, aleurone, or endosperm-enriched fractions (Table 7.2), and therefore differing in arabinoxylans structure, were fed to pigs, substantial differences in the rate and extent of digestibility were found.[135] The pericarp/testa diet exhibited the lowest digestibility and the endosperm the highest (Figure 7.14). The endosperm fractions contained the highest amount of water-soluble arabinoxylans and a large content of unsubstituted xylose residues. Mono- and disubstituted xylose, terminal xylose, and nonterminal

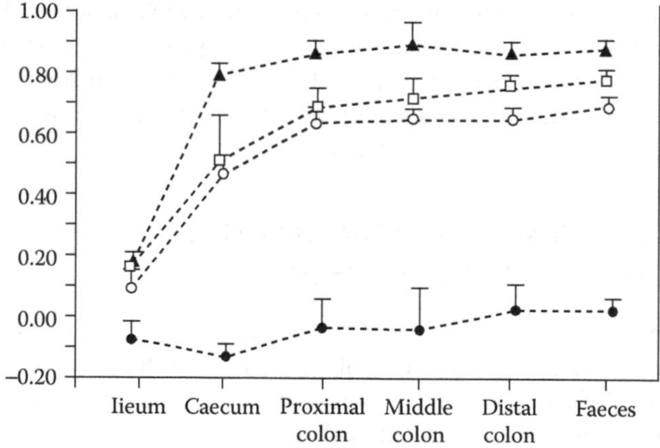

FIGURE 7.14 Digestibility of arabinoxylans in intestinal segments of pigs receiving diets based on (○) whole rye, (•) pericarp/testa, (□) aleurone, and (▲) endosperm. Values are means and standard deviations of five pigs. (Adapted from Glitsø, L.V. et al., *J. Sci. Food Agric.*, 79, 961, 1999. With permmission.)

arabinose residues were concentrated in the unfermented fractions, collected from the intestinal material of pigs, showing that these structural features were difficult to digest. Harris et al.[36] also reported that the wheat-lignified pericarp-rich fractions were not degraded, whereas the wheat aleurone-rich fractions were partially degraded when fed as a dietary supplement to rats. Hopkins and coworkers[137] compared the degradation of cross-linked (via H_2O_2–peroxidase) and non-cross-linked arabinoxylans by the intestinal microbiota of children and reported that the non-cross-linked arabinoxylans were utilized more quickly than their more viscous cross-linked counterparts. The degradation of non-cross-linked arabinoxylans resulted in initial production of arabinose, followed by release of xylose residues, whereas the xylan backbone of the cross-linked arabinoxylans appeared to be more resistant to degradation, probably because the dense structure of the covalently bound chains restricted access of xylanolytic enzymes to their target sites. Treatment of the pericarp-enriched fractions with alkali considerably increased the fermentability and suggested that in the lignified tissues, the solubilization of arabinoxylans by breaking the alkali-labile cross-links was a more important determinant of arabinoxylans' degradability than arabinoxylans' structure itself (Table 7.4).[136] Aura et al.[138] investigated the effects of various preprocessing of rye bran on the fermentation rate of arabinoxylans. Rye bran was extruded and treated with xylanase, and a part of xylanase-treated rye bran was separated into a soluble rye bran extract and an insoluble residue. Xylanase treatment of extruded rye bran made the cell wall polysaccharides more accessible to colonic bacteria, and thus increased the initial

TABLE 7.4
Amount of Total Fermented Carbohydrates (CHO) and Amount of Short-Chain Fatty Acids (SCFAs) Produced during *In Vitro* Fermentation of Ileal Effluent, Collected from Cannulated Pigs Fed Rye Bread Diets, with Fecal Inocula from Pigs

Diet[a]	Treatment	g CHO fermented (in 48 h)	mmol SCFA (in 48 h)	Molar Distribution of SCFAs			
				Acetate	Propionate	Butyrate	b-SCFAs
Whole rye		1.64	18	40	41	17	0.9
Pericarp/testa		–0.23	8	46	32	18	4.5
Aleurone		1.82	19	42	39	18	1.1
Endosperm		1.37	20	40	37	20	2.9
Pericarp/testa	NaOH	2.64	20	52	33	15	nd
Pericarp/testa	$NaClO_2$	0.68	7	54	36	11	nd
Endosperm	NaOH	1.54	18	49	31	20	nd
Endosperm	$NaClO_2$	1.27	12	47	21	32	0.4

Note: nd = no data; b-SCFAs = branched short-chain fatty acids (isobutyrate, isovalerate).

[a]Rye bread diets based on either whole rye or on one of three rye-milling fractions enriched in pericarp/testa, aleurone, or endosperm, as shown in Table 7.2, fed to pigs.

Source: Adapted from Glitsø, L.V. et al., *J. Sci. Food Agric.*, 80, 1211, 2000. With permission.

rate of fermentation. The soluble rye bran extract showed the fastest fermentation rate and the highest extent of fermentation as determined from the consumption of neutral sugars (arabinose, xylose, and glucose) and the production of short-chain fatty acids (SCFAs). The molar proportions of acetic, propionic, and butyric acids after 24 h of fermentation were 60:21:19 for the extruded rye bran, 66:18:16 for the xylanase-treated rye bran, and 67:16:17 for the soluble rye extract. The authors reported that the Ara/Xyl ratio increased during fermentation, indicating that the nonsubstituted xylan was preferentially consumed during fermentation.

In addition to prebiotic properties, the effects of arabinoxylans on lipid metabolism and mineral absorption have recently been investigated. The study involving diets containing water-soluble arabinoxylans extracted from maize bran with alkaline solutions revealed striking effects of these polymers on cecal fermentation as well as on lipid metabolism and mineral balance in rats.[139] It was reported that the arabinoxylan diets caused an enlargement of the cecum and cecal walls and induced significant proliferation of the beneficial microflora. The authors postulated that fermentation of arabinoxylans and the resulting production of SCFAs were involved in a decrease of serum cholesterol and improve the calcium and magnesium adsorption. One of the interesting effects of corncob arabinoxylan was production of a relatively high percentage of propionate rather than butyrate. Propionate is believed to affect various metabolic pathways; for example, it inhibits cholesterogenesis and lipogenesis.[140] Hopkins et al.[137] investigated *in vitro* breakdown of arabinoxylans by the intestinal microbiota, collected from fecal samples from children, and also found a relatively high production of propionate together with acetate. These results support the contention that the species of genus *Bacteroides* are particularly active in arabinoxylan metabolism. This contrasts with starch digestion, which is considerably more bifidogenic and associated with significant production of acetate, butyrate, and lactate (Table 7.5). However, the studies by Grasten et al.[141] reported that in healthy humans, the pentosan bread diet increased the concentration of butyrate, whereas the inulin bread diet increased the concentration of acetate and propionate in feces. Fecal levels of SCFAs only partly reflect the SCFA production in the colon, because they can be absorbed rapidly in the proximal colon. However, the authors claimed that a high fecal concentration of SCFAs also indicates their high level in the distal colon, the site of most colon tumors. In addition, the authors postulated that butyrate has beneficial effects on maintaining normal functions in colonocytes, and implied that pentosans are beneficial in the human large intestine.[141]

The unfermented arabinoxylans can absorb water and increase fecal bulk, thus resulting in the dilution of intestinal contents and lowering the concentration of putative carcinogens, such as the secondary bile acids. Also, fecal bulking results in decreased time of transit through the colon, and therefore in reduction of the exposure time to irritants or carcinogens. As indicated above, the fine molecular structure of arabinoxylans may have an important effect on their fermentability, and thus on their specific physiological functions and efficacy to treat various bowel disorders. Edwards et al.[142] postulated that terminal arabinose residues are particularly susceptible to fermentation, and since the distribution of these residues along the xylan backbone is nonrandom, there will be regions of these polymers that are more or less fermentable in the gut. The portion of arabinoxylans that is resistant

TABLE 7.5
Production of Fermentation Acids in pH-Controlled Fecal Incubation Mixtures Containing Arabinoxylans, Ferulic Acid Cross-Linked Arabinoxylans, and Starch[a]

Substrate	Incubation Period (h)	Production (m*M*) of:					
		Acetate	Propionate	Butyrate	Total SCFAs	Lactate	Succinate
Arabinoxylans							
	12	45.5 (11.8)	6.8 (1.3)	2.3 (1.1)	54.3 (13.4)	ND	8.0 (2.2)
	24	62.1 (11.5)	15.9 (2.6)	4.1 (2.0)	82.0 (14.7)	ND	8.7 (2.4)
	48	72.0 (11.8)	22.5 (2.7)	9.7 (4.6)	104 (15.9)	ND	6.8 (2.1)
Cross-linked arabinoxylans							
	12	43.4 (11.6)	7.2 (1.2)	1.7 (0.6)	50.9 (13.2)	ND	6.5 (1.8)
	14	55.4 (11.4)	14.9 (3.0)	2.6 (0.5)	72.9 (13.6)[c]	ND	6.1 (1.4)
	48	55.8 (8.7)	17.7 (3.3)	3.4 (0.6)	73.4 (10.9)	ND	5.1 (1.6)
Starch							
	12	64.0 (13.8)	7.3 (1.2)	10.0 (2.4)	81.4 (15.2)[c]	0.7 (0.7)	9.0 (1.7)
	14	75.0 (12.0)	11.6 (1.9)	11.6 (1.9)	98.2 (12.2)	0.6 (0.6)	7.3 (1.8)
	48	81.4 (13.3)	13.4 (2.2)	13.1 (2.3)	108 (13.8)	0.3 (0.3)	5.8 (1.5)

Note: ND = not detected.

[a]The values are means (n = 10). The values in parentheses are standard errors of the means. The concentrations of minor SCFAs were <0.1 m*M*.

Source: Adapted from Hopkins, M.J. et al., *Appl. Environ. Microbiol.*, 69, 6354, 2003. With permmission.

to fermentation will retain its native structure and will likely increase the fecal bulk, thereby alleviating the symptoms of certain bowel disorders, such as constipation. For example, the psyllium arabinoxylans, because of their more complex and more substituted structure, are thought to be more efficient in inducing the laxative effects than their wheat bran counterparts. Marlett and Fischer[143] conducted a series of human clinical studies to test the hypothesis that a gel-forming, nonfermented fraction of psyllium seed husk is the active component responsible for the laxative and cholesterol-lowering properties of psyllium. The nonfermented fraction of psyllium appeared to be a highly branched arabinoxylan with both arabinose and xylose side chains. However, as pointed out by the authors, psyllium arabinoxylans must possess unique structural features that hinder their fermentation by typical colonic microflora and distinguish them from the extensively fermented arabinoxylans from wheat or oats. In 2004, Fischer et al.[8] identified the physiologically active gel-forming fraction of psyllium seed husk to be a neutral arabinoxylan branched with single xylose units as well as with trisaccharides consisting of Ara*f*-α-(1→3)-Xyl*p*-β-(1→3)-Ara*f*.

Another physiological role of arabinoxylans may be associated with the management of diabetes in humans. Lu et al.[32] showed that addition of arabinoxylan-rich fiber to bread eaten at breakfast lowered postprandial glucose and insulin responses in healthy humans. Similar results were obtained by Zunft and coworkers.[33] The mechanism by which arabinoxylans affect the postprandial glucose response is not clear, but it has been postulated that due to their viscosity-generating properties, arabinoxylans impair mixing of the food mass and can markedly affect the degree of contact of food substrates with the enzymes that digest them in the small intestine. In this way, arabinoxylans may slow down the rate of gastric emptying and reduce small intestine motility, which results in delayed glucose absorption.

The beneficial role of arabinoxylans in the human diet may also be associated with the presence of ferulic acid covalently bound to these polymers. Recent studies have shown that ferulic acid has strong anti-inflamatory properties, inhibits chemically induced carcinogenesis in rats, and plays a role as an antioxidant, inhibiting lipid peroxidation and low-density lipoprotein (LDL) oxidation and scavenging oxygen radicals.[144] Katapodis et al.[3] evaluated the antioxidant activity of feruloylated oligosaccharides derived from partial hydrolysis of wheat flour arabinoxylans. The researchers found that feruloyl arabinoxylotrisaccharide (FAX_3) had profound antioxidant activity in 2,2-diphenyl-1-picrylhydrazyl reduction assay and inhibited the copper-mediated oxidation of human LDL. This antioxidant activity of FAX_3 may be important in preventing or reducing the progression of arteriosclerosis by inhibiting the peroxidation of lipids. The rate of release of ferulic acid from arabinoxylans may affect its bioavailability and bioactivity, and human studies are needed to evaluate which form of ferulic acid, free or bound, would be more effective. Adam et al.[144] showed that the cereal matrix severely limits ferulic acid bioavailability in rats. Rondini et al.,[145] on the other hand, demonstrated that plasmas of rats fed with wheat bran, where ferulic acid is mainly bound to arabinoxylans, showed a better antioxidant activity than those of the pure ferulic acid-supplemented group. The contribution of other bran components acting as antioxidants cannot, however, be disregarded in the latter study. Recent studies on the cell culture and animal models have indicated that arabinoxylans show positive oxidative burst activity in murine macrophage cells *in vitro* and tend to increase the body weight and reduce the attachment of pathogen Salmonella to ileal tissue in broiler chicks undergoing mild heat stress *in vivo*.[6] Deters et al.[146] showed that water-soluble and gel-forming polysaccharides from psyllium seed husk promote proliferation of human epithelial cells (skin keratinocytes and fibroblasts) via enhanced growth factor receptors and energy production.

REFERENCES

1. Carpita, N.C., Structure and biogenesis of the cell wall of grasses, *Annu. Rev. Plant Physiol. Plant Mol. Biol.*, 47, 445, 1996.
2. Hoffman, W.F. and Gortner, R.A., The preparation and analysis of the various proteins of wheat flour with special reference to the globulin, albumin, and proteose fractions, *Cereal Chem.*, 4, 221, 1927.
3. Katapodis, P. et al., Enzymatic production of a feruloylated oligosaccharide with antioxidant activity from wheat flour arabinoxylans, *Eur. J. Nutr.*, 42, 55, 2003.
4. Faulds, C.B. et al., Arabinoxylan and mono- and dimeric ferulic acid release from brewer's grain and wheat bran by feruloyl esterases and glycosyl hydrolases from *Humicola insolens*, *Appl. Microbiol. Biotechnol.*, 64, 644, 2004.
5. Warrand, J. et al., Flax (*Linum usitatissimum*) seed cake: a potential source of high molecular weight arabinoxylans? *J. Agric. Food Chem.*, 53, 1449, 2005.
6. Zhang, P. et al., Effects of arabinoxylans on activation of murine macrophages and growth performance of broiler chicks, *Cereal Chem.*, 81, 511, 2004.
7. Izydorczyk, M.S. and Biliaderis, C.G., Cereal arabinoxylans: advances in structure and physicochemical properties, *Carbohydr. Polym.*, 28, 33, 1995.
8. Fischer, M.H. et al., The gel-forming polysaccharide of psyllium husk (*Plantago ovata* Forsk), *Carbohydr. Res.*, 339, 2009, 2004.
9. Vinkx, C.J.A. and Delcour, J.A., Rye (*Secale cereale* L.) arabinoxylans: a critical review, *J. Cereal Sci.*, 24, 1, 1996.
10. Selanere, M. and Andersson, R., Cell wall composition of 1B/1R translocation wheat grains, *J. Sci. Food Agric.*, 82, 538, 2002.
11. Cyran, M., Rakowska, M., and Miazga, D., Chromosomal location of factors affecting content and composition of non-starch polysaccharides in wheat-rye addition lines, *Euphytica*, 89, 153, 1996.
12. Boros, D., Lukaszewski, A.J., Anil, A., and Ochodzki, P., Chromosome location of genes controlling the content of dietary fibre and arabinoxylans in rye, *Euphytica*, 128, 1, 2002.
13. Lempereur, I., Rouau, X., and Abecassis, J., Genetic and agronomic variation in arabinoxylan and ferulic acid contents of durum wheat (*Triticum durum* L.) grain and its milling fractions, *J. Cereal Sci.*, 25, 103, 1997.
14. Henry, R.J., Genetic and environmental variation in the pentosan and -glucan contents in barley, and their relation to malting quality, *J. Cereal Sci.*, 4, 269, 1986.
15. Hansen, H.B., Rasmussen, C.V., Knudsen, K.E.B., and Hansen, Å., Effects of genotype and harvest year on content and composition of dietary fibre in rye (*Secale cereale* L) grain, *J. Sci. Food Agric.*, 83, 76, 2003.

16. Saastamoinen, M., Plaami, S., and Kumpulainen, J., Pentosan and beta-glucan content of Finnish winter rye varieties as compared with rye grain from several countries, *J. Cereal Sci.*, 10, 199, 1989.
17. Coles, G.D. et al., Environmentally-induced variation in starch and non-starch polysaccharide content in wheat, *J. Cereal Sci.*, 26, 47, 1997.
18. Wakabayashi, K., Soga, K., Kamisaka, S., and Hoson, T., Increase in the level of arabinoxylans-hydroxycinnamate network cell walls of wheat coleoptiles grown under continuous hypergravity conditions, *Physiol. Plant.*, 125, 127, 2005.
19. Fry, S.C., Primary cell wall metabolism: tracking the careers of wall polymers in living plant cells, *New Physiol.*, 161, 641, 2004.
20. Fincher, G.B. and Stone, B.A., A water-soluble arabino-galactan peptide from wheat endosperm, *Aust. J. Biol. Sci.*, 27, 117, 1974.
21. Crowe, N.L. and Rasper, V.F., The ability of chlorine and chlorine-related oxidants to induce oxidative gelation in wheat flour pentosans, *J. Cereal Sci.*, 7, 283, 1988.
22. Michniewicz, J., Biliaderis, C.G., and Bushuk, W., Water-insoluble pentosans of wheat: composition and some physical properties, *Cereal Chem.*, 67, 434, 1990.
23. Izydorczyk, M.S., Biliaderis, C.G., and Bushuk, W., Comparison of the structure and composition of water-soluble pentosans from different wheat varieties, *Cereal Chem.*, 68, 139, 1991.
24. Faurot, A.-L. et al., Large scale isolation of water-soluble and water-insoluble pentosans from wheat flour, *Lebensm. Wiss. Technol.*, 28, 436, 1995.
25. Delcour, J.A., Rouseu, N., and Vanhaesendonck, I.P., Pilot scale isolation of water-extractable arabinoxylans from rye, *Cereal Chem.*, 76, 1, 1999.
26. Gruppen, H. et al., Mild isolation of water-insoluble cell wall material from wheat flour. Composition of fractions obtained with emphasis on non-starch polysaccharides, *J. Cereal Sci.*, 9, 247, 1989.
27. Mares, D.J. and Stone, B.A., Studies on wheat endosperm. Properties of wall components and studies on their organization in the wall, *Aust. J. Biol. Sci.*, 26, 813, 1973.
28. Selvedran, R.R. and DuPont, M.S., An alternative method for the isolation and analysis of cell wall material from cereals, *Cereal Chem.*, 57, 278, 1980.
29. Bataillon, M., Mathaly, P., Nunes Cardinali, A.-P., and Duchiron, F., Extraction and purification of arabinoxylans from destarched wheat bran in a pilot scale, *Ind. Crop Products*, 8, 37, 1998.
30. Bergmans, M.E.F., Beldman, G., Gruppen, H., and Voragen, A.G.J., Optimisation of the selective extraction of (glucurono) arabinoxylans from wheat bran: use of barium and calcium hydroxide solution at elevated temperatures, *J. Cereal Sci.*, 23, 235, 1996.
31. Figueroa-Espinoza, M.-C. et al., Enzymatic solubilization of arabinoxylans from native, extruded, and high-shear-treated rye bran by different endo-xylanases and other hydrolytic enzymes, *J. Agric. Food Chem.*, 52, 4240, 2004.
32. Lu, Z.X. et al., Arabinoxylan fiber, a by-product of wheat flour processing, reduces the postprandial glucose response in normoglycemic subjects, *Am. J. Clin. Nutr.*, 71, 1123, 2000.
33. Zunft, H.J. et al., Reduction of postprandial glucose and insulin response in serum of healthy subjects by an arabinoxylans concentrate isolated from wheat starch plant process water, *Asia Pac. J. Clin. Nutr.*, 13 (Suppl.), S147, 2004.
34. Glitsø, L.V. and Knudsen, K.E.B., Milling of whole grain rye to obtain fractions with different dietary fibre characteristics, *J. Cereal Sci.*, 29, 89, 1999.
35. Izydorczyk, M.S. et al., Roller milling of Canadian hull-less barley: optimization of roller milling conditions and composition of mill streams, *Cereal Chem.*, 80, 637, 2003.

36. Harris, P.J., Chavan, R.R., and Ferguson, L.R., Production and characterization of two wheat-bran fractions: an aleurone-rich and a pericarp-rich fraction, *Mol. Nutr. Food Res.*, 49, 536, 2005.
37. Jaskari, J. et al., Oat β-glucan and xylan hydrolysates as selective substrates for *Bifidobacterium* and *Lactobacillus* strains, *Appl. Microbiol. Biotechnol.*, 49, 175, 1998.
38. Crittenden, R. et al., *In vitro* fermentation of cereal dietary fibre carbohydrates by probiotic and intestinal bacteria, *J. Sci. Food Agric.*, 82, 781, 2002.
39. Izumi, Y. and Kojo, A., Long-Chain Xylooligosaccharide Compositions with Intestinal Function-Improving and Hypolipemic Activities, and Their Manufacture, JP Patent 2003048901, 2003.
40. Biely, P. et al., Endo-β-1,4-xylanase families: differences in catalytic properties, *J. Biotechnol.*, 57, 151, 1997.
41. Trogh, I. et al., Enzymic degradability of hull-less barley flour alkali-solubilized arabinoxylan fractions by endoxylanases, *J. Agric. Food Chem.*, 53, 7243, 2005.
42. Pellerin, P. et al., Enzymatic production of oligosaccharides from corncob xylan, *Enzyme Microb. Technol.*, 13, 617, 1991.
43. Sun, H.J., Yoshiba, S., Park, N.H., and Kusakabe, I., Preparation of (1→4)-β-xylooligosaccharides from an acid hydrolysate of cotton-seed xylan: suitability of cotton-seed xylan as a starting material for the preparation of (1→4)-β-xylooligosaccharides, *Carbohydr. Res.*, 37, 657, 2002.
44. Vegas, R., Alonso, J.L., Dominguez, H., and Parajo, J.C., Processing of rice husk autohydrolysis liquors for obtaining food ingredients, *J. Agric. Food Chem.*, 52, 7311, 2004.
45. Shibuya, N., Misaki, A., and Iwasaki,T., The structure of arabinoxylan and arabino-glucuronoxylan isolated from rice endosperm cell wall, *Agric. Biol. Chem.*, 47, 2223, 1983.
46. Saulnier, L., Peneau, N., and Thibault, J.F., Variability in grain extract viscosity and water soluble arabinoxylan content in wheat, *J. Cereal Sci.*, 22, 259, 1995.
47. Subba Rao, M.V.S.S.T. and Muralikrishna, G., Structural analysis of arabinoxylans isolated from native and malted finger millet (*Eleusine coracana*, ragi), *Carbohydr. Res.*, 339, 2457, 2004.
48. Chilkunda, D., Nandini, D., Paramahans, V., and Salimath, V., Structural features of arabinoxylans from sorghum having good roti-making quality, *Food Chem.*, 74, 417, 2001.
49. MacGregor, A.W. and Fincher, G.B., Carbohydrates of the Barley Grain, in *Barley Chemistry and Technology*, MacGregor, A.W. and Bhatty, R.S., Eds., American Association of Cereal Chemists, St. Paul, MN, 1993, p. 73.
50. Schooneveld-Bergmans, M.E.F., van Dijk, Y.M., Beldman, G., and Voragen, A.G.J., Physicochemical characteristics of wheat bran glucuronoarabinoxylans, *J. Cereal Sci.*, 29, 49, 1999.
51. Adams, E.L., Kroon, P.A., Williamson, G., and Morris, V.J., Characterization of heterogeneous arabinoxylans by direct imaging of individual molecules by atomic force microscopy, *Carbohydr. Res.*, 338, 771, 2003.
52. Rhodes, D.I., Sadek, M., and Stone, B.A., Hydroxycinnamic acids in walls of wheat aleurone cells, *J. Cereal Sci.*, 36, 67, 2002.
53. Smith, M.M. and Hartley, R.D., Occurrence and nature of ferulic acid substitution of cell-wall polysaccharides in graminaceous plants, *Carbohydr. Res.*, 118, 65, 1983.

54. Rattan, O., Izydorczyk, M.S., and Biliaderis, C.G., Structure and rheological behaviour of arabinoxylans from Canadian bread wheat flours, *Lebensm. Wiss. Technol.*, 27, 550, 1994.
55. Dervilly-Pinel, G. et al., Water-extractable arabinoxylans from pearled flours of wheat, barley, rye, and triticale. Evidence for the presence of ferulic acid dimers and their involvement in gel formation, *J. Cereal Sci.*, 34, 207, 2001.
56. Antoine, C. et al., Individual contribution of grain outer layers and their cell wall structure to the mechanical properties of wheat bran, *J. Agric. Food Chem.*, 51, 2026, 2003.
57. Bunzel, M. et al., Diferulates as structural components in soluble and insoluble cereal dietary fiber, *J. Sci. Food Agric.*, 81, 653, 2001.
58. Saulnier, L. et al., Isolation and structural determination of two 5-5'-diferuloyl oligosaccharides indicate that maize heteroxylans are covalently cross-linked by oxidatively coupled ferulates, *Carbohydr. Res.*, 320, 82, 1999.
59. Allerdings, E. et al., Isolation and structural identification of diarabinosyl 8-O-4-dehydrodiferulate from maize bran insoluble fibre, *Phytochemistry*, 66, 113, 2005.
60. Hatfield, R.D. and Ralph, J., Modelling the feasibility of intramolecular dehydrodiferulate formation in grass walls, *J. Sci. Food Agric.*, 79, 425, 1999.
61. Fry, S.C., Willis, S.C., and Paterson, A.E., Intraprotoplasmic and wall-localized formation of arabinoxylan-blound diferulates and larger ferulate coupling-products in maize cell-suspension cultures, *Planta*, 211, 679, 2000.
62. Bunzel, M. et al., Isolation and identification of a ferulic acid dehydrotrimer from saponified maize bran insoluble fibre, *Eur. Food Res. Technol.*, 217, 128, 2003.
63. Rouau, X. et al., A dehydrotrimer of ferulic acid from maize bran, *Phytochemistry*, 63, 899, 2003.
64. Carvajal-Millan, E. et al., Storage stability of laccase induced arabinoxylan gels, *Carbohydr. Polym.*, 59, 181, 2005.
65. Funk, C. et al., Isolation and structural characterization of 8-O-4/8-O-4 and 8-8/8-O-4-coupled dehydrotriferulic acids from maize bran, *Phytochemistry*, 66, 363, 2005.
66. Parker, C.C. et al., Thermal stability of texture of water chestnut may be dependent on 8,8'-diferulic acid (aryltetralyn form), *J. Agric. Food Chem.*, 51, 2034, 2003.
67. Oosterveld, A. et al., Formation of ferulic acid dehydrodimers through oxidative cross-linking of sugar beet pectin, *Carbohydr. Res.*, 300, 179, 1997.
68. Grabber, J.H., Hatfield, R.D., and Ralph, J., Diferulate cross-links impede the enzymatic degradation of non-lignified maize walls, *J. Sci. Food Agric.*, 77, 193, 1998.
69. Neukon, H. and Markwalder, H.U., Oxidative gelation of wheat flour pentosans: a new way of cross-linking polymers, *Cereal Food World*, 23, 374, 1978.
70. Piber, M. and Koehler, P., Identification of dehydro-ferulic acid–tyrosine in rye and wheat: evidence for a covalent cross-link between arabinoxylans and proteins, *J. Agric. Food Chem.*, 53, 5276, 2005.
71. Ebringerova, A., Hromadkova, Z., Alföldi, J., and Berth, G., Structural and solution properties of corn cob heteroxylans, *Carbohydr. Polym.*, 19, 99, 1992.
72. Ebringerová, A., Hromádková, Z., and Berth, G., Structural and molecular properties of a water-soluble arabinoxylan-protein complex isolated from rye bran, *Carbohydr. Res.*, 264, 97, 1994.
73. Dervilly-Pinel, G., Tran, V., and Saulnier, L., Investigation of the distribution of arabinose residues on the xylan backbone of water-soluble arabinoxylans from wheat flour, *Carbohydr. Polym.*, 55, 171, 2004.

74. Vinkx, C.J.A., Delcour, J.A., Verbruggen, M.A., and Gruppen, H., Rye water-soluble arabinoxylans also vary in their 2-monosubstituted xylose content, *Cereal Chem.*, 72, 227, 1995.
75. Cleemput, G. et al., Heterogeneity in the structure of water-soluble arabinoxylans in European wheat flours of variable bread-making quality, *Cereal Chem.*, 70, 324, 1993.
76. Brillouet, J.M. and Joseleau, J.P., Investigation of the structure of a heteroxylan from the outer pericarp (besswing bran) of wheat kernel, *Carbohydr. Res.*, 159, 109, 1987.
77. Maes, C. and Delcour, J.A., Structural characterization of water-extractable and water-unextractable arabinoxylans in wheat bran, *J. Cereal Sci.*, 35, 315, 2002.
78. Gruppen, H. et al., Characterization by ^{1}H-NMR spectroscopy of enzymatically-derived oligosaccharides from alkali-extractable wheat flour arabinoxylans, *Carbohydr. Res.*, 233, 45, 1992.
79. Bengtsson, S. and Aman, P., Isolation and chemical characterization of water-soluble arabinoxylans in rye grain, *Carbohydr. Polym.*, 12, 267, 1990.
80. Cyran, M., Courtin, C.M., and Delcour, J.A., Structural features of arabinoxylans extracted with water at different temperatures from two rye flours of diverse bread-making quality, *J. Agric. Food Chem.*, 51, 4404, 2003.
81. Izydorczyk, M.S. and Biliaderis, C.G., Structural heterogeneity of wheat endosperm arabinoxylans, *Cereal Chem.*, 70, 641, 1993.
82. Oscarsson, M., Andersson, R., Salomonsson, A.-C., and Åman, P., Chemical composition of barley samples focusing on dietary fiber components, *J. Cereal Sci.*, 24, 161, 1996.
83. Izydorczyk, M.S., Macri, L.J., and MacGregor, A.W., Structure and physicochemical properties of barley non-starch polysaccharides. I. Water-extractable β-glucans and arabinoxylans, *Carbohydr. Polym.*, 35, 249, 1998.
84. Dervilly, G. et al., Isolation and characterization of high molar mass water-soluble arabinoxylans from barley and barley malt, *Carbohydr. Polym.*, 47, 143, 2002.
85. Izydorczyk, M.S., Jacobs, M., and Dexter, J.E., Distribution and structural variation of non-starch polysaccharides in milling fractions of hull-less barley with variable amylose content, *Cereal Chem.*, 80, 645, 2003.
86. Gruppen, H., Kormelink, F.J.M., and Voragen, A.G.J., Water-unextractable cell wall material from wheat flour. 3. A structural model for arabinoxylans, *J. Cereal Chem.*, 18, 111, 1993.
87. Vietor, R.J., Angelino, S.A.G.F., and Voragen, A.G.J., Structural features of arabinoxylans from barley and malt cell wall material, *J. Cereal Sci.*, 15, 213, 1992.
88. Izydorczyk, M.S., Studies on Structure and Physicochemical Properties of Wheat Endosperm Arabinoxylans, Ph.D. thesis, University of Manitoba, Canada, 1993.
89. Bengtsson, S., Åman, P., and Andersson, R., Structural studies on water-soluble arabinoxylans in rye grain using enzymic hydrolysis, *Carbohydr. Polym.*, 17, 277, 1992.
90. Bengtsson, S., Andersson, R., Westerlund, E., and Åman, P., Content, structure, and viscosity of soluble arabinoxylans in rye grain from several countries, *J. Sci. Food Agric.*, 58, 331, 1992.
91. Vinkx, C.J.A., Reynaert, H.R., Grobet, P.J., and Delcour, J.A., Physicochemical and functional properties of rye nonstarch polysaccharides. V. Variability in the structure of water-soluble arabinoxylans, *Cereal Chem.*, 70, 311, 1993.
92. Andrewartha, K., Philips, D.R., and Stone, B.A., Solution properties of wheat flour arabinoxylans and enzymatically modified arabinoxylans, *Carbohydr. Res.*, 77, 191, 1979.

93. Girhammar, U., Nakamura, M., and Nair, B.M., Water-soluble pentosans from wheat and rye: chemical composition and some physical properties in solution, in *Gums and Stabilisers for the Food Industry*, Phillips, G.O., Wedlock, P.A., and Williams, D.J., Eds., IRL Press, Oxford, 1986, p. 123.
94. Fincher, G.B. and Stone, B.A., Cell walls and their components in cereal grain technology, in *Advances in Cereal Science and Technology*, Vol. VIII, Pomeranz, Y., Ed., American Association of Cereal Chemists, St. Paul, MN, 1986, p. 207.
95. Dervilly, G., Saulnier, L., Roger, P., and Thibault, J.-F., Isolation of homogeneous fractions from wheat water-soluble arabinoxylans. Influence of the structure on their macromolecular characteristics, *J. Agric. Food. Chem.*, 48, 270, 2000.
96. Bacic, A., Harris, P.J., and Stone, B.A., Structure and function of plant cell walls, in *The Biochemistry of Plants*, Vol. 3, Preiss, J., Ed., Academic Press, New York, 1988, p. 297.
97. Delmer, D.P. and Stone, B.A., Biosynthesis of plant cell walls, in *Plant Biotechnology*, Preiss, J., Ed., Academic Press, San Diego, 1988, p. 373.
98. Urahara, T. et al., A β-(1→4)-xylosyltranferase involved in the synthesis of arabinoxylans in developing barley endosperm, *Physiol. Plant.*, 122, 169, 2004.
99. Brett, C. and Waldron, K., *Physiology and Biochemistry of Plant Cell Walls*, 2nd ed., Chapman & Hall, New York, 1996.
100. Lee, R., Burton, R.A., Hrmova, M., and Fincher, G.B., Barley arabinoxylan arabinofuranohydrolases: purification, characterization and determination of primary structures from cDNA clones, *Biochem J.*, 356, 181, 2001.
101. Atkins, E.D.T., Three-dimensional structure, interactions and properties of xylans, in *Xylans and Xylanase*, Visser, J., Beldman, G., Kusters-van Somerom, M.A., and Voragen, A.G.J., Eds., Elsevier Science, London, 1992, p. 39.
102. Dervilly-Pinel, G., Thibault, J.-F., and Saulnier, L., Experimental evidence for a semi-flexible conformation for arabinoxylans, *Carbohydr. Res.*, 330, 365, 2001.
103. Picout, D.R. and Ross-Murphy, S.B., On the chain flexibility of arabinoxylans and other β-(1→4) polysaccharides, *Carbohydr. Res.*, 337, 1781, 2002.
104. Izydorczyk, M.S. and MacGregor, A.W., Evidence of intermolecular interaction of beta-glucans and arabinoxylans, *Carbohydr. Polym.*, 41, 417, 2000.
105. Izydorczyk, M.S. and Biliaderis, C.G., Effect of molecular size on physical properties of wheat arabinoxylans, *J. Agric. Food Sci.*, 40, 561, 1992.
106. Izydorczyk, M.S. and Biliaderis, C.G., Influence of structure on the physicochemical properties of wheat arabinoxylans, *Carbohydr. Polym.*, 17, 237, 1992.
107. Warrand, J. et al., Contribution of intermolecular interactions between constitutive arabinoxylans to the flaxseeds mucilage properties, *Biomacromolecules*, 6, 1871, 2005.
108. Durham, R.K., Effect of hydrogen peroxide on relative viscosity measurements of wheat and flour suspensions, *Cereal Chem.*, 2, 297, 1925.
109. Geissman, T. and Neukom, H., Composition of the water-soluble wheat flour pentosans and their oxidative gelation, *Lebensm. Wiss. Technol.*, 6, 59, 1973.
110. Izydorczyk, M.S., Biliadersis, C.G., and Bushuk, W., Oxidative gelation studies of water-soluble pentosans from wheat, *J. Cereal Sci.*, 11, 153, 1991.
111. Figueroa-Espinoza, M.-C. and Rouau, X., Oxidative cross-linking of pentosans by fungal laccase and horseradish peroxidase: mechanism of linkage between feruloylated arabinoxylans, *Cereal Chem.*, 75, 259, 1998.
112. Carvajal-Millan, E. et al., Arabinoxylan gels: impact of the feruloylation degree on their structure and properties, *Biomacromolecules*, 6, 309, 2005.

113. Vansteenkiste, E., Babot, C., Rouau, X., and Micard, V., Oxidative gelation of feruloylated arabinoxylan as affected by protein. Influence on protein enzymatic hydrolysis, *Food Hydrocolloids*, 18, 557, 2004.
114. Carvajal-Millan, E., Guilbert, S., Morel, M.-H., and Micard, V., Impact and the structure of arabinoxylan gels on their rheological and protein transport properties, *Carbohydr. Polym.*, 60, 431, 2005.
115. Michniewicz, J., Biliaderis, C.G., and Bushuk, W., Effect of added pentosans on some physical and technological characteristics of dough and gluten, *Cereal Chem.*, 68, 252, 1991.
116. Wang, M., Vliet von, T., and Hamer, R.J., How gluten properties are affected by pentosans, *J. Cereal Sci.*, 39, 395, 2004.
117. Biliaderis, C.G., Izydorczyk, M.S., and Rattan, O., Effect of arabinoxylans on bread-making quality of wheat flours, *Food Chem.*, 5, 165, 1995.
118. McCleary, B.V., Enzymatic modification of plant polysaccharides, *Int. J. Biol. Macromol.*, 8, 349, 1986.
119. Delcour, J.A., Vanhamel, S., and Hoseney, R.C., Physicochemical and functional properties of rye nonstarch polysaccharides. II. Impact of a fraction containing water-soluble pentosans and proteins on gluten-starch loaf volume, *Cereal Chem.*, 68, 72, 1991.
120. Wang, M., Hamer, R.J., Vliet von, T., and Oudgenoeng, G., Interaction of water extractable pentosans with gluten protein: effect of dough properties and gluten quality, *J. Cereal Sci.*, 36, 25, 2002.
121. Hilhorst, R. et al., Effects of xylanase and peroxidase on soluble and insoluble arabinoxylans in wheat bread dough, *J. Food Sci.*, 67, 497, 2002.
122. Labat, E., Rouau, X., and Morel, M.-H., Effect of flour water-extractable pentosans on molecular associations in gluten during mixing, *Lebensm. Wiss. Technol.*, 35, 185, 2002.
123. Santos, D.M.J., Monteiro, S.R., and da Silva, J.A.L., Small strain viscoelastic behaviour of wheat gluten-pentosan mixtures, *Eur. Food Res. Technol.*, 221, 398, 2005.
124. Courtin, C. and Delcour, J.A., Arabinoxylans and endoxylanases in wheat flour bread-making, *J. Cereal Sci.*, 35, 225, 2002.
125. Wang, M. et al., Effect of water unextractable solids on gluten formation and properties: mechanistic considerations, *J. Cereal Sci.*, 37, 55, 2003.
126. Rouau, X., Hayek, E.I., and Moreau, M.L., Effect of an enzyme preparation containing pentosanases on the bread-making quality of flours in relation to changes in pentosan properties, *J. Cereal Sci.*, 19, 259, 1994.
127. Hilhorst, R. et al., Baking performance, rheology, and chemical composition of wheat dough and gluten affected by xylanase and oxidative enzymes, *J. Food Sci.*, 64, 808, 1999.
128. Jiang, Z. et al., Improvement of the breadmaking quality of wheat flour by the hyperthermophilic xylanase B from *Thermotoga maritime*, *Food Res. Int.*, 38, 37, 2005.
129. Zeleznak, K. and Hoseney, R.C., The role of water in the retrogradation of wheat starch gels and bread crumbs, *Cereal Chem.*, 63, 402, 1986.
130. Gudmudson, M., Eliasson, A.-C., Bengtsson, S., and Åman, P., The effects of water-soluble arabinoxylans on gelatinization and retrogradation of starch, *Starch/Starke*, 43, 5, 1991.
131. Marconi, E., Graziano, M., and Cubadda, R., Composition and utilization of barley pearling by-products for making functional pastas rich in dietary fiber and β-glucans, *Cereal Chem.*, 77, 133, 2000.

132. Izydorczyk, M.S. et al., The enrichment of Asian noodles with fiber-rich fractions derived from roller milling of hull-less barley, *J. Sci. Food Agric.*, 85, 2094, 2005.
133. Dexter, J.E., Izydorczyk, M.S., Preston, K.R., and Jacobs, M., The enrichment of bread with a fibre-rich fraction derived from roller milling of hull-less barley, in *Proceedings of the ICC-SA Bread and Cereals Symposium. Advances in Cereal Science and Technology. World Perspectives*, Fowler, A., Ed., 2005 (CD-ROM).
134. Katrien, M.J. et al., Fermentation of plant cell wall derived polysaccharides and their corresponding oligosaccharides by intestinal bacteria, *J. Agric. Food. Chem.*, 48, 1644, 2000.
135. Glitsø, L.V. et al., Degradation of rye arabinoxylans in the large intestine of pigs, *J. Sci. Food Agric.*, 79, 961, 1999.
136. Glitsø, L.V., Jensen, B.B., and Knudsen, K.E.B., *In vitro* fermentation of rye carbohydrates including arabinoxylans of different structures, *J. Sci. Food Agric.*, 80, 1211, 2000.
137. Hopkins, M.J. et al., Degradation of cross-linked and non-cross-linked arabinoxylans by the intestinal microbiota in children, *Appl. Environ. Microbiol.*, 69, 6354, 2003.
138. Aura, A.-M. et al., Processing of rye bran influences both the fermentation of dietary fibre and the bioconversion of lignans by human faecal flora *in vitro*, *J. Sci. Food Agric.*, 85, 2085, 2005.
139. Lopez, H.W. et al., Effects of soluble corn bran arabinoxylans on cecal digestion, lipid metabolism, and mineral balance (Ca, Mg) in rats, *J. Nutr. Biochem.*, 10, 500, 1999.
140. Demingé, C. et al., Effect of propionate on fatty acid and cholesterol synthesis and acetate metabolism in isolated rat hepatocytes, *Br. J. Nutr.*, 74, 209, 1995.
141. Grasten, S. et al., Effects of wheat pentosan and inulin on the metabolic activity of fecal microbiota and on bowel function in healthy humans, *Nutr. Res.*, 23, 1503, 2003.
142. Edwards, S., Chaplin, M.F., Blackwood, A.D., and Dettmar, P.W., Primary structure of arabinoxylans of ispaghula husk and wheat bran, *Proc. Nutr. Soc.*, 62, 217, 2003.
143. Marlett, J.A. and Fischer M.H., The active fraction of psyllium seed husk, *Proc. Nutr. Soc.*, 62, 207, 2003.
144. Adam, A. et al., The bioavailability of ferulic acid is governed primarily by the food matrix rather than its metabolism in intestine and liver of rats, *J. Nutr.*, 132, 1962, 2002.
145. Rondini, L. et al., Bound ferulic acid from bran is more available than the free compound in rat, *J. Agric. Food Chem.*, 52, 4338, 2004.
146. Deters, A.M., Schroder, J.K.R., Smiatek, T., and Hensel, A., Ispaghula (*Plantago ovata*) seed husk polysaccharides promote proliferation of human epithelial cells (skin keratinocytes and fibroblasts) via enhanced growth factor receptors and energy production, *Planta*, 71, 33, 2005.
147. Hashimoto, S., Shogren, M.D., and Pomeranz, Y., Cereal pentosans: their estimation and significance. III. Pentosans in abraded grains and milling products of wheat and milled wheat products, *Cereal Chem.*, 64, 39, 1987.
148. Izydorczyk, M.S. et al., Variation in total and soluble β-glucan content in hulless barley: effects of thermal, physical, and enzymatic treatments, *J. Agric. Food Chem.*, 48, 982, 2000.
149. Hashimoto, S., Shogren, M.D., and Pomeranz, Y., Cereal pentosans: their estimation and significance. I. Pentosans in wheat and milled wheat products, *Cereal Chem.*, 64, 30, 1987.

8 Carbohydrates and the Risk of Cardiovascular Disease

Gunilla Önning

CONTENTS

8.1 INTRODUCTION

Cardiovascular disease (CVD) is the major contributor to the global burden of disease among the noncommunicable diseases, and the World Health Organization (WHO) currently attributes one third of all global deaths to CVD.[1] There are large differences in the incidence of coronary heart disease (CHD) between countries. In Finland, the incidence of CHD was found to be 198/100,000 in men aged 40 to 59 and initially free of CVD, while the corresponding figure for Japan was 15/100,000.[2] In Europe, there are also large differences in CVD incidence between countries, and such international comparisons have played an important part in the search of causes of these diseases. Important causes are unhealthy dietary habits, low physical activity, and high tobacco consumption. As summarized by WHO,[1] unhealthy dietary practices include high consumption of saturated fats, salt, and refined carbohydrates, as well as low consumption of fruit and vegetables, and these tend to cluster together.

Carbohydrates provide a large part of the energy for almost all human diets, but there are marked differences between countries concerning the kind of carbohydrate-rich foods consumed. The more developed a country is, the more refined carbohydrates are consumed. Many agencies have issued recommendations to prevent CVD by an increased intake of unrefined foods rich in dietary fiber and other bioactive nutrients (cereals, fruits, vegetables). In this chapter the effects of different available carbohydrates and dietary fibers on CVD are evaluated. First, the development of CVD and different risk factors are briefly presented for readers not familiar with this subject. Then results from observational or intervention studies on the relationship between carbohydrates and CVD are discussed, including different mechanisms of action. Finally, different health claims and recommendations regarding carbohydrate intake and CVD are presented.

8.2 CARDIOVASCULAR DISEASES

Myocardial infarction, angina, and stroke are major cardiovascular diseases. The diseases are caused by atherosclerotic plaques developed in the blood vessels, making the vessels narrower and stiffer. It is uncertain how this process starts, but one important factor is low-density lipoprotein (LDL). LDL is rich in cholesterol, and if there is no need of cholesterol supply in the cells, its uptake is low and it continues to circulate in the blood. It can also enter the endothelium of the blood vessels and be converted to oxidized LDL (LDLox), which can start an inflammatory process. Briefly, macrophages engulf the LDLox, leading to the formation of the first stage of atherosclerosis — a fatty streak (Figure 8.1). The inflammation can become permanent, and later smooth muscle cells and fibrous tissues will proliferate and surround the lipids (the second stage). The third stage is represented by the development of unstable plaques that are prone to rupture and lead to the formation of luminal thrombosis. Plaque rupture is responsible for most acute coronary syndromes.

Many different factors influence the development of atherosclerosis. The factors are normally divided into those possible to modify and nonmodifiable ones (Table 8.1). The atherosclerotic process outlined above takes many years to develop, and most of the CVDs occur in middle-aged or older subjects. Women that have not entered menopause are somewhat protected from atherosclerosis. Thus, the incidence of CHD is three to four times higher in men than in women during middle age, but only two times higher in the elderly.[3] It is also well documented that a sedentary lifestyle increases the risk for CVD. Data from the Framingham study indicate that active subjects were only one half to one third as likely to develop CHD as sedentary subjects.[4] In men, physical exercise reduces CVD and also the overall mortality. Cardiovascular events occur more frequently in smokers than in nonsmokers, and myocardial infarctions are more likely to be fatal in smokers.[3] The relationship of smoking to CVD risk is dose dependent and observed in both women and men.[5] Patients with diabetes mellitus have a 2.5-fold increase in cardiovascular disease risk for males and a 6-fold increase for women. Other examples of risk factors are

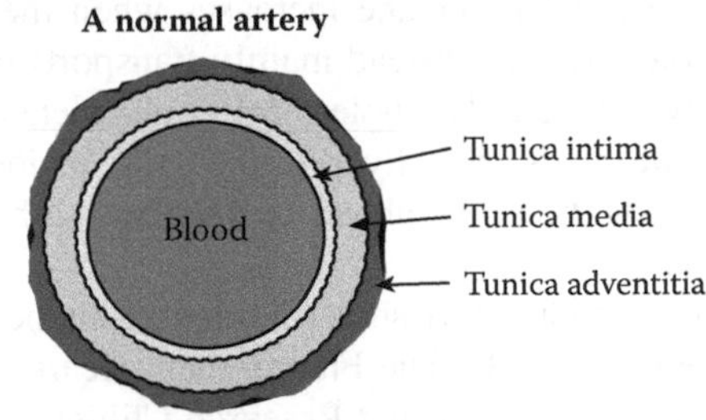

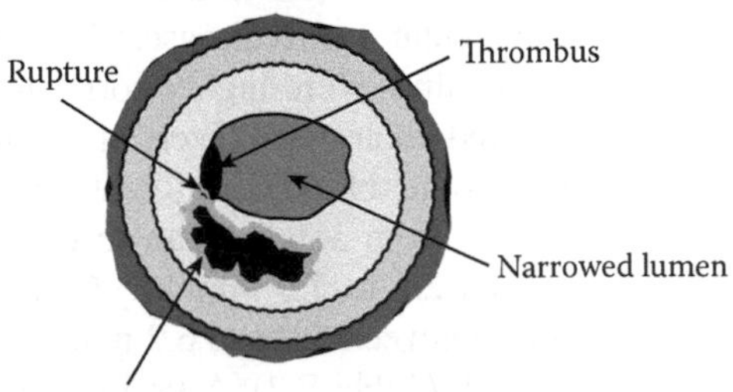

FIGURE 8.1 Development of atherosclerosis.

TABLE 8.1
Factors Linked to the Development of Atherosclerosis

Modifiable	Nonmodifiable
Smoking	Age
Obesity	Gender
Diet	Genetic disposition
Alcohol	
Negative stress	
Physical activity	
Diabetes	

obesity (upper-body location), family history of CVD, high consumption of alcohol, and negative stress.

8.2.1 Blood Lipids, Apolipoproteins, Blood Pressure, and Hemostatic Factors

Chylomicrons transport lipids absorbed after a meal to different tissues and the chylomicron remnants are taken up in the liver. Very low density lipoprotein (VLDL) also mainly transports triglycerides. It is produced in the liver when there is a need for lipids in other tissues. Triglycerides are delivered to the cells and VLDL is transformed to another lipoprotein — LDL. LDL has a longer half-life and binds to

LDL receptors in cells, and its uptake increases when the cells need cholesterol. High-density lipoprotein (HDL) instead mainly transports cholesterol from peripheral tissues to the liver, where the cholesterol is channeled to the bile or used for the production of bile acids. The liver is also the major organ for cholesterol biosynthesis, and its synthesis is subject to feedback inhibition if HDL delivers cholesterol to the liver.

Many observational studies have shown a relationship between serum cholesterol levels and the incidence of CVD. The Framingham Heart Study, the Multiple Risk Factor Intervention Trial, and the Lipid Research Clinics trials found a direct association between total cholesterol levels and the rate of new-onset cardiovascular heart disease in both men and women that were initially free of CVD, and also in subjects with established cardiovascular heart disease.[5] Moreover, in the Seven Country Study, median cholesterol values were highly correlated with mortality in cardiovascular heart disease. Elevated cholesterol levels play a role in the development of mature coronary plaque. No other blood constituent varies so much between different populations as cholesterol. From south Japan to east Finland, the mean serum cholesterol value ranges from 2.6 to 7.0 mmol/l.[2] Cholesterol levels above 5.2 mmol/l are considered elevated, between 5.2 and 6.2 mmol/l are borderline high and levels above 6.2 mmol/l are high (Table 8.2). A prolonged reduction in serum cholesterol with 0.6 mmol/l was associated with an almost 30% reduction in the risk of coronary heart disease.[6]

The strong relationship between blood cholesterol and risk of CVD is due to elevations of LDL cholesterol. Several lines of evidence support the link between LDL cholesterol and CVD.[7] The level of LDL cholesterol correlates with cross-cultural variance in CVD risk, and prospective cohort studies show a positive relation between serum LDL cholesterol and CHD mortality. Furthermore, oxidatively modified forms of LDL have been isolated from plaques in postmortem samples in amounts directly proportional to the concentration of LDL. Recent evidence indicates that elevated LDL cholesterol levels contribute to plaque instability, and thus lowering of the LDL cholesterol stabilizes the plaques and reduces the risk for CVD.[5] There is another rationale for long-term lowering of the LDL cholesterol since it also slows development of the atherosclerotic plaques. In five studies, treatment with LDL cholesterol-lowering drugs like statins led to a 28% reduction in the LDL cholesterol level and a 30% decrease in cardiovascular heart incidence.[8] Observational and clinical trials suggest that each 0.026 mmol/l increment in LDL cholesterol

TABLE 8.2
Relation between Blood Lipid Levels and an Increased Risk for Cardiovascular Diseases[5]

Level	Increased Risk
Cholesterol	>5.2 mmol/l
LDL cholesterol	>3.4 mmol/l
HDL cholesterol	<1.0 mmol/l
Triglycerides	>2.3 mmol/l

causes a 1% increase in coronary risk.[9] LDL cholesterol is one of the best markers for risk of CVD, and it is easily measured in health control programs. It is thus also the main target of lipid-lowering therapy, and a decrease to a value below 3.4 mmol/l of LDL cholesterol reduces the risk for CVD (Table 8.2).

A high level of HDL cholesterol is related to a decreased risk of CVD. Data from the Framingham study indicate that every 0.26 mmol/l increase in HDL cholesterol results in a 33 to 50% reduction in risk for cardiovascular heart disease.[10] HDL transports cholesterol from the artery wall and other organs to the liver and in that way reduces the size of the atherosclerotic plaque.

The mean HDL levels are higher in women than in men, and other factors that influence the HDL levels are weight loss (increase), physical activity (increase), weight gain (reduce), and smoking (reduce). An HDL cholesterol level below 1.0 mmol/l leads to an increased risk of CVD. At a low HDL concentration the HDL particles become smaller and denser, and this type of particles is more strongly associated with increased risk for CHD.[11] In the elderly, low concentrations of HDL are more predictive of vascular disease risk than elevated LDL cholesterol levels.[12]

Another parameter often calculated is the total cholesterol/HDL cholesterol ratio. It has been shown to be a powerful predictor of coronary heart disease risk, but usually not as good as the LDL cholesterol level.[5]

Elevated triglyceride levels in the blood are associated with a substantial increase in risk for CVD. If the triglyceride level is higher than 1.5 mmol/l, it can lead to an increased formation of small, dense LDL particles and a low HDL cholesterol level.[13] Small, dense LDL particles can more easily infiltrate the intima than larger ones, and thus increase the risk for development of atherosclerosis. The triglyceride level has been recommended to be used as a predictor for CVD, especially in women, based on results in the Framingham study.[14] A meta-analysis by Hokanson and Austin[15] showed that raised triglyceride levels were positively associated with the risk of CHD. Triglyceride levels are influenced by many factors, like alcohol intake, diet, weight changes, and physical activity. The independent impact of elevated triglyceride levels on CVD risk has earlier been questioned,[3] except in the presence of diabetes. There is increasing consensus on triglyceride levels being a valuable predictor for CVD according to a recent review.[7] A triglyceride concentration over 2.3 mmol/l is regarded as supernormal, but lower triglyceride levels also negatively influence the pattern of LDL and HDL particles.

Apolipoproteins in lipoproteins are important for receptor recognition and enzyme regulation. The apolipoproteins are categorized into A, B, C, D, and E classes, with additional subclasses. The dominating apolipoprotein in LDL is B, as it is also in chylomicrons and VLDL. The level of apolipoprotein B is positively related to CVD since it is in these potentially atherogenic lipoproteins. It is also raised in subjects with increased triglyceride levels. Different apolipoproteins A are the major proteins in HDL, and a low level of apolipoprotein A-I is associated with an increased risk for coronary heart disease. This marker is naturally dependent of the HDL level. In the AMORIS study, a high apolipoprotein B and a low apolipoprotein A-I level were shown to be good predictive factors for fatal myocardial infarction.[16] Apolipoprotein E is important for the cholesterol clearance in the blood, and its level is higher in subjects that have had a myocardial infarction than in controls. There are also different genetic variants (polymorphism) of apolipoprotein E containing ε4, ε3, or ε2 alleles. Subjects with the ε4 allele have higher total and LDL cholesterol levels, and some studies have shown that these subjects are more responsive to dietary changes than those without an ε4-allele.[17] More research is needed to relate different polymorphic forms of apolipoproteins to the response to diet changes.

One important risk factor of CVD is hypertension. It is defined as a systolic blood pressure of 140 mmHg or a diastolic blood pressure of 90 mmHg. An increased blood pressure affects the artery wall, and this can lead to smooth muscle cell proliferation and a narrower vessel lumen. This contributes to an increased risk for myocardial infarction, stroke, and peripheral vascular disease. Systolic blood pressure is a good predictor of CHD in subjects older than 60 years, and therapy to decrease the blood pressure in subjects with hypertension reduces cardiovascular morbidity and mortality.[7,18]

High levels of plasma fibrinogen are positively correlated with CVD and are identified as a major independent risk marker for CVD. Fibrinogen increases the platelet aggregation and fibrin formation and influences plasma viscosity. Other hemostatic parameters that have been investigated in relation to CVD risk are platelet functions and the level of plasminogen activator inhibitor-1 as a marker of fibrinolysis. However, all these hemostatic factors are relatively unaffected by the diet, and thus not very suitable markers for the effects of diet.[7]

8.3 DIET AND CARDIOVASCULAR DISEASES

In a newly published WHO report,[1] different dietary factors related to cardiovascular disease were evaluated. Among the most investigated components are fatty acids, and one major conclusion from human studies is that saturated fatty acids raise the total and LDL cholesterol levels. Among the saturated fatty acids, myristic and palmitic acids have the largest negative effects, while stearic acid has not been shown to elevate the cholesterol levels. Trans fatty acids seem to be even more atherogenic than saturated fatty acids since they decrease the HDL cholesterol concentration, besides increasing the LDL cholesterol level. On the contrary, monounsaturated and n-6 polyunsaturated fatty acids lower the total and LDL cholesterol levels, and the long n-3 polyunsaturated fatty acids lower the triglyceride concentration. In the last decade, plant sterols have been incorporated in large amounts in foods since they have cholesterol-lowering effects. Other dietary components that may decrease the risk for CVD are antioxidants, folate, and flavonoids, while sodium increases the risk, especially for stroke. Dietary fiber was briefly mentioned in the WHO report to reduce total and LDL cholesterol, and probably also to decrease the risk for cardiovascular diseases. In this chapter the main focus is on dietary fiber and its relation to CVD, but also the roles of other carbohydrates are discussed.

8.3.1 MONO- AND DISACCHARIDES

A diet with a high content of available carbohydrates can increase the triglyceride level in the blood.[19] Fructose is considered the most hypertriglyceridemic monosaccharide, since fructose promotes triglyceride synthesis and VLDL production in the liver. Since fructose is a component of the most common disaccharide, sucrose, an increased intake of sucrose can lead to higher triglyceride levels. The hypertriglyceridemic effect of sucrose is only seen when substantial quantities are consumed (>140 g/day) and when a relatively high proportion of dietary fat is derived from saturated fatty acids.[20] However, more long-term studies are needed to assess at which intake level these negative effects on the triglyceride metabolism arise. Some subjects are more sensitive to sucrose than others, and men are more sensitive than women. The effect of sucrose on VLDL metabolism is variable, and it is more apparent in individuals with elevated blood triglyceride and insulin concentrations or in obese subjects, especially with abdominal obesity.

It has been suggested that the glycemic index (GI) is a better predictor of the metabolic effects of a diet than the sucrose content.[21] A diet with a low GI gives a slower absorption of the carbohydrates than a diet with a high GI. The health benefits with a low-GI diet include reduced insulin demand, improved blood glucose control, and reduced blood lipid levels, and all these factors play important roles in the prevention or management of coronary heart disease and other chronic diseases.[22] The GI value of the diet is influenced by type of starch, nature of monosaccharide components, content of viscous fiber, cooking/food processing and storage, particle size, ripeness, alpha-amylase inhibitors, organic acids, and nutrient–starch interactions. In observational studies, however, no strong relations between the GI and cardiovascular diseases have been observed.[22] By using the glycemic load concept (GI × carbohydrate content), it was found that a high glycemic load increased the risk for CHD in one observational study on women followed for 10 years.[23] Also, Fried and Rao[19] found that a high dietary glycemic load was associated with higher triglyceride levels and a greater risk of coronary disease in women. Intervention studies comparing low- and high-GI diets have shown more consistent results.[24,25] Significant reductions in total cholesterol (–8.8%), LDL cholesterol (–9.1%), and triglycerides (–19.3%) were seen in healthy subjects given a low-GI diet, compared with a high-GI diet.[24] In diabetic subjects given either a low- or high-GI diet for 24 days, significant reductions in total and LDL cholesterol, apolipoprotein B, and the plasminogen activator inhibitor-1 level were observed after the low-GI diet.[25]

To conclude, short-term studies indicate that a high intake of sucrose can increase the triglyceride level, but more long-term studies are needed to assess if and at which level this negative effect arises. More studies are needed to evaluate if GI or the glycemic load is a better predictor of metabolic effects than the sucrose intake.

8.3.2 Oligosaccharides

Oligosaccharides include inulin and oligofructose and are also named fructans. They are composed of linear chains of fructose units linked to a terminal sucrose molecule. Oligosaccharides are water soluble, and the water-binding capacity and gelling tendency increase with the number of hexose units.[26] In plants, inulin is used as an energy reserve and osmoregulator. Inulin occurs, for instance, in onions, garlic, banana, leeks, and asparagus. Different oligofructoses (OFSs) can be obtained by enzymatic hydrolysis of inulin.[27] Intake of high doses of OFS can reduce serum triglycerides in experimental animals. Equivalent doses cannot be used in humans due to gastrointestinal side effects, and the few human studies conducted have produced unclear results, as will be discussed below. One study in hyperlipidemic subjects showed significantly lower total and LDL cholesterol concentrations after a diet rich in inulin (18 g/day) compared with the control diet.[28] The study was a randomized, double-blind, crossover trial with two 6-week treatment periods, separated with a 6-week washout period. However, the blood lipids increased significantly after giving the control diet, and the authors therefore concluded that the study could not be used to evaluate the lipid-lowering effects of inulin. Brighenti et al.[29] observed lower cholesterol and triglyceride levels in men consuming 9 g of inulin added to a rice breakfast cereal for 4 weeks. No effect on the blood lipids was observed when Luo et al.[30] gave a higher dose (20 g of OFS/day) for 4 weeks. Compared to a sucrose diet, no significant changes in cholesterol, triglycerides, or apolipoproteins in subjects with normal lipid values were found. In another study on subjects with normal lipid levels, inulin (14 g/day) was added to a low-fat spread for 4 weeks, but no significant differences in blood lipids, compared to a placebo, were detected.[31] The study had a background diet high in fat, and a triglyceride-lowering effect of inulin would be more likely if the diet were high in carbohydrates.[27] In a study on subjects with type 2 diabetes,[32] an intake of 15 g of OFS/day for 20 days resulted in no significant differences in lipid values in comparison with the placebo period. However, the length of the study may be of importance. Jackson et al.[33] conducted a study on hyperlipidemic subjects that consumed 10 g of inulin daily for 8 weeks. The triglyceride level was significantly lower than in the placebo group at 8 weeks, but not after 4 weeks. Williams and Jackson[34] reported that from nine studies on OFS and blood lipids, three showed reductions in triglycerides and four in total and LDL cholesterol, but in three no effects were observed. Additional studies are needed to evaluate the effects of different doses of oligosaccharides on lipid metabolism.

The main mechanism for the effects of oligosaccharides on blood lipids is probably related to the fermentation of oligosaccharides in the colon and the production of short-chain fatty acids. The pattern of fermentation varies with the type of oligosaccharides and the duration of the oligosaccharide supply. Butyrate seems mainly to serve as a fuel for the colonic mucosa, while acetate and propionate can be absorbed and modulate the lipid metabolism in the body. In the liver, acetate may be lipogenic and cholesterogenic, whereas propionate inhibits the synthesis of triglycerides.[26] The relative importance of these mechanisms remains to be established.

8.3.3 Polysaccharides: Dietary Fiber

8.3.3.1 Observational Studies

Inverse associations between the intake of whole-grain foods or dietary fiber and CVD morbidity and mortality have been demonstrated in several observational studies.[35] Some of theses studies will be evaluated in the following, starting with whole-grain foods. In one study, the dietary habits of postmenopausal women from

Iowa were investigated in 1986 and the incidence of CVD was followed for 9 years.[36] The mean weekly intake was 6.4 slices of dark bread, 1.9 cups of whole-grain breakfast cereals, and 2.5 servings of other whole-grain foods, and the intake of whole-grain foods ranged from 1.5 to 22.5 servings per week. A striking inverse relation between the whole-grain intake and the incidence of ischemic heart disease (IHD) was found. The age- and energy-adjusted relative risks from the lowest to the highest intake by quintiles were 1.0, 0.84, 0.58, 0.45, and 0.60. After adjustment of potentially confounding variables, the risk of IHD death was reduced by about one third in those eating >1 serving of a whole-grain product per day, compared with those who reported rarely eating any whole-grain products. In contrast, no association between the total refined-grain intake and risk of IHD death was found.

The association between the intake of whole-grain bread and CVD mortality was investigated among Norwegian subjects.[37] The men reported a mean intake of 6.3 slices of bread per day using 23% whole-grain flour, while the women consumed 3.9 slices per day containing 26% whole-grain flour. A whole-grain bread score was calculated and ranged from 0.05 (one slice per day made with 5% whole-grain flour) to 5.4 (nine slices per day made with 60% whole-grain flour). In total, five whole-grain intake categories were formed and the hazard ratios for total mortality in CVD were inverse and graded across the categories (from 1 to 0.63).

Several studies have dealt with the association between dietary fiber intake and risk for cardiovascular disease. The intake of dietary fiber and risk of coronary heart disease were followed for 10 years among women in the Nurses' Health Study.[38] In total, 68,782 women were included in the analysis and the age-adjusted relative risk for major CHD events was 0.53 for women in the highest quintile of total dietary fiber intake (median, 22.9 g/day) compared with women in the lowest quintile (median, 11.5 g/day). Among different sources of dietary fiber, only cereal fiber was strongly associated with a reduced risk of CHD. Male health professionals have been followed in a similar study.[39] Their food habits were assessed using a 131-item food frequency questionnaire. During 6 years of follow-up there were 511 nonfatal cases of myocardial infarction (MI) and 229 coronary deaths. The age-adjusted relative risk of myocardial infarction for the top quintile of total dietary fiber intake was 0.64 (median intake, 28.9 g/day) compared with men in the lowest quintile (median intake, 12.4 g/day). Cereal fiber was most strongly associated with a reduced risk of MI compared with other fiber sources, like vegetables and fruits. Thus, the same associations were found for male health professionals as for female health professionals. In one of the earliest studies in the field, the CHD incidence in men working at London transport or banks was also found to be lower in those eating more cereal fibers.[40] Elderly people were the focus in another study from the U.S., and the aim was to assess whether fiber consumption from fruit, vegetable, and cereal sources was associated with the incidence of CVD.[41] Using a 99-item food frequency questionnaire, the average intake of the different fibers was calculated: cereal fiber, 4.2 g/day; fruit fiber, 5.2 g/day; vegetable fiber, 6.9 g/day; and total fiber, 16.2 g/day. The main foods contributing to cereal fiber intake were dark breads and bran cereals; to fruit fiber intake, apples, oranges, and bananas; and to vegetable fiber intake, beans, broccoli, peas, corn, and cauliflower. During 9 years, the incidence of combined stroke, ischemic heart disease death, and nonfatal myocardial infarction was followed. Neither fruit fiber, vegetable fiber, nor total fiber was associated with the incidence of CVD, while cereal fiber consumption, adjusted for several parameters, was inversely associated with incident CVD, with 21% lower risk in the highest quintile intake compared with the lowest quintile. Thus, all these studies show that the intake of fiber, especially cereal fiber, is associated with a lower risk for CVD.

Observational studies have also evaluated the relation between intake of soluble fiber relative to insoluble fiber and its association with CVD risk. In the study on female health professionals, an inverse relation between both soluble and insoluble fiber and the risk of CVD and MI was found, but after multivariate adjustments the associations were no longer significant.[42] Pietinen et al.[43] investigated smoking men included in the Alpha-Tocopherol, Beta Carotene (ATBC) Cancer Prevention Study (21,930 subjects). They filled in a 276-food-item dietary questionnaire and the incidence of CHD was followed for 6 years. The median intake of dietary fiber in the highest quintile was 34.8 g/day and in the lowest quintile 16.1 g/day. In age- and treatment-adjusted analysis, water-soluble fiber was more strongly associated with reduced coronary death than water-insoluble fiber, and cereal fiber also had a stronger association than vegetable and fruit fiber. A 10-g larger daily intake of fiber appeared to lower the risk of coronary death by 17%. A study was conducted in Italy comparing subjects who had had a nonfatal acute myocardial infarction with a control group.[44] Compared with the lowest tertile, the odds ratio (OR) of risk of acute myocardinal infarction in the highest tertile was 0.72 for total fiber, 0.64 for soluble fiber (significant), 0.77 for insoluble fiber, 0.71 for cellulose, 0.82 for vegetable fiber, 0.64 for fruit fiber (significant), and 1.1 for cereal fiber. In this population, cereal fiber was derived chiefly from refined grains, and this may explain the lack of protection found for this type of fiber.

In a meta-analysis, Anderson[45] evaluated the strength of the association between dietary fiber-rich foods and risk for CHD. For total dietary fiber, 82% and, for whole grains, 75% of the studies in the meta-analysis showed a significant inverse association to the risk for CHD (Figure 8.2). For cereals, fruits, and vegetables, the association was less pronounced. In the meta-analysis, the risk ratio for CHD comparing the highest to the lowest groups of intake was 0.73 for dietary fiber and 0.71 for intake of whole grains. Thus, dietary fiber was associated with a significant reduction of the risk for CHD of approximately 27%. Also, fruit and vegetable consumption was associated with a significant reduction of CHD risk (23 and 14%, respectively), while for cereals, no significant association was found. This may be related to the fact that most of the cereals consumed were refined and low in dietary fiber.

8.3.3.2 Intervention Studies

Many intervention studies have been made investigating the effects of dietary fibers on different variables related to CVD. The main interest has been focused on effects of soluble fibers, and thus on foods rich in soluble fibers. First, some studies investigating mixtures of dietary fibers or food fiber concentrates will be reviewed, followed by studies using only one fiber preparation (psyllium, oats, barley, guar gum, pectin).

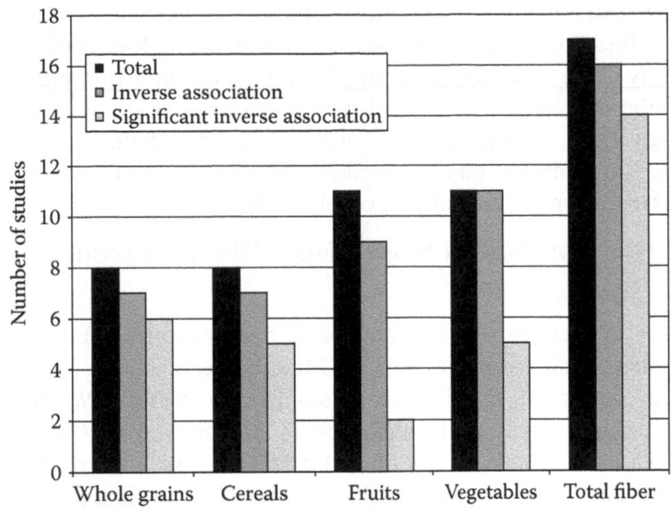

FIGURE 8.2 Number of observational studies reporting an association between dietary fiber-rich foods and risk for CHD. (Adapted from Anderson, J.W., *Proc. Nutr. Soc.*, 62, 135, 2003.)

8.3.3.2.1 Effects of Mixtures of Dietary Fibers

In one study, subjects with increased plasma cholesterol values were given a daily supplement of 15 g of soluble fiber from psyllium, pectin, guar gum, and locust bean gum during 6 months.[46] The fibers were mixed in water and consumed with each of three major daily meals. In comparison with the control group given acacia gum, the total and LDL cholesterol values were significantly lower in the test group. After 8 weeks, the reductions in comparison with baseline were 6.4 and 10.5%, respectively, and about the same reductions were found at weeks 16 and 24. In another study, a combination of soluble fibers from psyllium, oats, and barley was given to men with hypercholesterolemia.[47] They consumed the fibers as a breakfast cereal (50 g containing 12 g of soluble fiber) for 6 weeks. In comparison with a control group given a breakfast cereal based on wheat, the total cholesterol and LDL cholesterol levels fell significantly in the test group, with 3.2 and 4.4%, respectively. In yet another study, hypercholesterolemic subjects consumed soluble fibers (gum arabic:pectin = 4:1) mixed in apple juice for 12 weeks using doses of 0, 5, or 15 g/day.[48] No differences in blood lipid levels were found between the treatments, and even increases in blood lipids were observed. A combination of soluble and insoluble fibers has also been investigated.[49] The soluble fibers were guar gum and pectin, and the insoluble fibers were soy fiber, pea fiber, and corn bran, and they were given to hypercholesterolemic subjects consuming 20 g of the fibers daily for 15 to 36 weeks. The supplement was mixed with a beverage and was consumed before breakfast and dinner. Intake of the fiber supplement significantly decreased the levels of LDL cholesterol (–9.7%), total cholesterol (–6.5%), and apolipoprotein B (–12.6%) compared with the placebo at week 15. There was a tendency to smaller reductions toward the end of the treatment period. Another fiber combination of oat beta-glucan and psyllium (4 servings per day, 8 g of soluble fiber) and a control diet were compared in a randomized crossover study of subjects with increased plasma cholesterol values.[50] Compared with the control diet, the high-fiber diet reduced total cholesterol by 2.1% and, applying the Framingham cardiovascular disease risk equation to the data, gave a calculated reduction in risk of 4.2%. The reduction in risk, although small, is likely to be significant on a population basis. Thus, many studies using supplementation by a combination of soluble fibers in daily doses of 5 to 20 g have shown positive effects on blood lipid levels.

8.3.3.2.2 *Effects of Different Single Dietary Fiber-Rich Foods*

8.3.3.2.2.1 *Psyllium*

Psyllium has been extensively investigated in relation to its effects on CVD, and over 40 human studies were found in a Medline search using the search terms *psyllium* and *cholesterol*. Psyllium seed husk is derived from *Plantago ovata* and contains a mixture of polysaccharides such as arabinoxylans; 100 g of seed husk contains about 71% soluble dietary fiber and 15% insoluble fiber.[51] Psyllium preparations have been commercially available for over 60 years, and their physiological properties are thought to be due to the gel-forming ability.[3] In a long-term study, men and women with hypercholesterolemia were given an American Heart Association (AHA) Step I diet for 8 weeks and thereafter randomly assigned to receive either 10.2 g of psyllium (n = 197) or a cellulose placebo (n = 51) daily for 26 weeks.[52] The fibers were mixed in liquid and were taken immediately before breakfast and dinner. Serum total and LDL cholesterol were 4.7 and 6.7% lower, respectively, in the psyllium group than in the placebo group. In a study by Bell et al.,[53] psyllium was incorporated into a breakfast cereal also containing oat bran, sugar beet fiber, and wheat bran. The cereal was consumed once daily for 6 weeks by subjects with increased blood cholesterol values. The daily intake of soluble fiber from the cereal was 5.8 g, whereof 50% was from psyllium. Compared to a control group that consumed cornflakes, the total cholesterol and LDL cholesterol levels were reduced significantly in the psyllium group. The effects of psyllium (15 g/day) have also been studied in diabetes type 2 patients for 6 weeks.[54] In comparison with cellulose placebo treatment, the total and LDL cholesterol in the test group were reduced significantly, 2.7 and 7.2%, respectively. Furthermore, significant increases of HDL cholesterol and a decrease in triglycerides were observed in the test group. As in the study of Anderson et al.,[52] the fiber preparations were mixed in water and taken before regular meals three times per day. A dose–response study was made by Davidson et al.[55] using psyllium seed husk given in daily doses of 0, 3.4, 6.8, or 10.2 g for 24 weeks. The fibers were included in different foods like ready-to-eat cereals, bread, pasta, and snack bars. A change in LDL cholesterol (–5.3% in comparison to control) after 24 weeks consumption was only shown for the group that took the highest dose of psyllium husk — 10.2 g/day. The reduction in LDL cholesterol was more pronounced in the beginning of the intervention (week 4) for all groups. Davidson et al.[55] also investigated the lipid-lowering effect of psyllium in hypercholesterolemic children (6 to 18 years). They were given psyllium for 6 weeks and, after a 6-week washout period, a control cereal. Consumption of psyllium gave a 7% reduction in LDL cholesterol compared with the control cereal. Williams et al.[56] have also used psyllium as a treatment for hypercholesterolemic children.

The children consumed 6.4 g of soluble fiber from psyllium per day during 12 weeks and, compared to a control group, their total cholesterol and LDL cholesterol levels decreased. A meta-analysis of eight controlled trials showed that psyllium supplementation was well tolerated and resulted in an additional 7% reduction in LDL cholesterol in hypercholesterolemic individuals already consuming a low-fat diet.[57] An earlier meta-analysis of data from 12 studies of subjects with increased cholesterol levels who consumed psyllium in a low-fat diet showed similar results.[58] The total cholesterol and LDL cholesterol were reduced an average of 5 and 9%, respectively, in comparison with a control diet. Gender, age, and menopausal status did not affect the response.

Thus, many existing studies have shown that psyllium seed husk can reduce the total and LDL cholesterol levels significantly, and meta-analysis of the studies has confirmed this. More dose–response studies are needed to evaluate how much psyllium is needed to be consumed to get a significant cholesterol-lowering effect.

8.3.3.2.2.2 Oats

The cholesterol-lowering properties of oats have also been studied extensively. More than 50 human studies were found in a Medline search by including human studies conducted after 1980 and using the search terms *oats* and *cholesterol*. It is generally agreed that the main cholesterol-lowering compound in oats is the soluble fiber component, beta-glucan. Beta-glucans are composed of β-(1-4)-linked glucose units separated every two to three units by a single β-(1-3)-linked glucose. The beta-glucans in the oat kernel are distributed throughout the endosperm and are located in the endosperm cell walls. In different oat varieties harvested in Sweden in three different years, the beta-glucan content varied from 3.5 to 5.7% of the dry matter.[59] Oat bran is defined by the American Association of Cereal Chemists as a material fractionated so that the oat bran fraction is not more than 50% of the starting material, with a total beta-glucan content of at least 5.5% (dry weight basis) and a total dietary fiber content of at least 16.0% (dry weight basis), and with at least one third of the total dietary fiber being soluble fiber.[60]

To further purify the beta-glucans, different techniques have been used. In an EU project (QLK1-CT-2000-00535), concentrates with a content of 25% oat beta-glucans have been prepared (R. Öste and A. Öste Triantafyllou, EP 1124441). The manufacturing process includes dry or wet milling of the flakes at 60°C followed by an enzymatic reaction step using beta-amylase. In this way, maltose and beta-limit dextrins, the main carbohydrate species in the final product, are formed from starch. After the enzymatic step, insoluble fibers can be optionally separated using a decanting step. However, the beta-glucans are somewhat degraded during the concentration process. In the starting material (oat bran), 30% of the beta-glucan had a molecular weight under 200,000, but in the final concentrate, all beta-glucans had a molecular weight below 200,000. The molecular weight distribution of beta-glucans in oat-based foods has also been investigated by Åman et al.[61] Oats, rolled oats and oat bran concentrates, extruded flakes, macaroni, muffins, and porridge contained beta-glucans with a high average molecular weight, while pasteurized apple juice, fresh pasta, and a tea cake contained degraded beta-glucans.

The molecular weight of the beta-glucans may be of importance for the cholesterol-lowering effects. In one study, an intake of 5.9 g of beta-glucan from oat bran incorporated in bread and cookies did not have any significant effect on blood lipids, while an intake of 5 g of beta-glucans mixed in orange juice significantly lowered LDL cholesterol compared with a control group.[62] The molecular weight of the beta-glucans was lower in the bread than in the cookies and the preparation mixed with orange juice. However, the molecular weight is probably not the only important factor for the cholesterol-reducing potential. In another trial, an oat drink containing beta-glucans of rather low molecular weight (peak molecular weight, 82,400; Biliaderis, unpublished) was compared with a control drink low in beta-glucans.[63] The intake of oat drink (3.8 g of beta-glucans/day) resulted in significantly lower total cholesterol (6%) and LDL cholesterol (6%) levels compared to the control drink, and thus had the expected quantitative cholesterol-lowering effects, compared with less processed products. One explanation for the results could be that the beta-glucans were in a soluble form. In another human study, the solubility of the beta-glucans used was measured.[64] After the addition of an oat bran concentrate to food products like bread, tea cake, muesli, muffins, macaroni, pasta, and apple drink, the solubility was analyzed after extraction of the ground sample in water at 37°C for 2 h. It was found that the solubility of the beta-glucans in the products was rather low (about 50%), but the daily dose of soluble beta-glucans consumed by hypercholesterolemic subjects (2.7 g) was still high enough to decrease the blood cholesterol levels significantly compared to a control diet. The molecular weight and the solubility of the beta-glucans used in human studies have seldom been documented, and this makes it difficult to compare results from different studies in relation to the doses used. However, several investigators have addressed the task to make overall evaluations on the relation between oat beta-glucan intake and its cholesterol-lowering effects. A large meta-analysis of oat products and their lowering effects of plasma cholesterol was made by Ripsin et al.[65] The studies included in the analysis varied in study design, oat products, doses of oats, control products, and subjects with different initial cholesterol levels, gender, and age, and the influence of these parameters was assessed in the meta-analysis. The criteria for inclusion in the meta-analysis were that the study was controlled and randomized and that the control product should have a very low soluble fiber content. Moreover, the trial should also include a dietary assessment and measurement of the body weight. Twelve trials were included in the calculation of the summary effect size.[65] Most of the trials used a parallel design, and the length of the treatment phase varied between 18 days and 12 weeks. The summary effect size for a change in total cholesterol was found to be –0.13 mmol/l. The initial cholesterol level was highly predictive of the reduction in total cholesterol level, while age and gender could not predict the response to oats. The dose–response effect was also evaluated, and after dividing the material in an intake of <3 and 3 g of soluble fiber, the interaction was statistically significant. Mälkki[66] also evaluated the dose–response effect of oats. She identified 53 clinical trials, and 37 of them showed significant reductions in blood cholesterol levels after consumption of oat products, while in 10 studies no significant effects were detected. The dose–response effect was not very obvious.

It is known that the cholesterol-lowering effect is larger in subjects with increased cholesterol levels. Ripsin et al.[65] indicated that if the initial cholesterol levels were over 5.9 mmol/l, the reduction was larger. Önning[67] compared 17 studies on hyperlipidemic subjects. The total cholesterol level at the end of the intervention period in the control group was between 5.9 and 7.4 mmol/l. The change in total cholesterol in the oat group in comparison with the control group varied from 0 to –13% and in LDL cholesterol from 0 to –16.5%. The studies with the largest reductions incorporated the oats in hot cereals, muffins, and beverages, while in the studies with small reductions the oats were given in cold cereals and also in bread. Thus, there are several problems concerning the estimate of the lipid-lowering effects of different oat products, but incorporation of the oats in a hydrated form seems to have a more positive effect. Additional studies are needed to relate the chemical and physicochemical characterization of the oat product to the cholesterol-lowering capacity.

8.3.3.2.2.3 Barley

Barley usually contains beta-glucans in about the same amount as oats (3.5 to 5.9% of the dry matter), but higher amounts have been detected in some varieties (up to 11%).[68] Barley may have cholesterol-reducing effects similar to those of oats, but very few human studies have been done. In one animal study, hamsters consumed barley in doses of 0, 25, 50, and 75% of the diet, and the total cholesterol concentration was reduced but no dose–response relationship was found.[69] No differences in the cholesterol-lowering effects between barley and oats were observed in another study on hamsters, where the cereals were given in three doses: 2, 4, and 8 g/100 g of diet.[70] Chicken has also been fed beta-glucans from barley, and the effect on the blood cholesterol concentration was followed.[71] A nonwaxy (Franubet) and a waxy (Washonupana) starch genotype with the same content of beta-glucans were compared, and it was shown that only the waxy genotype had an effect on the blood cholesterol. This was probably due to the waxy genotype having a higher viscosity when mixed with water, a greater average degree of polymerization, and lower endogenous beta-glucanase activity.[72]

Six studies were found in a Medline search on the effect of barley on blood cholesterol by including human studies done after 1980 and using the search terms *barley* and *cholesterol.* Keogh et al.[73] gave a highly concentrated barley beta-glucan preparation (75% beta-glucans) to hypercholesterolemic men in a crossover trial. The beta-glucan intake was 8 to 12 g/day, depending on body weight, and the preparation was incorporated in different foods taken throughout the day (bread, waffles, muffins, cakes, in dishes). The diet was controlled during the whole study period (7-day diet rotation) and, in the control diet, the beta-glucans were exchanged with glucose. Eighteen subjects completed the study, and the average intake of beta-glucans was 10 g/day. The total cholesterol level fell 1.3% and the LDL cholesterol level fell 3.8% over the 4-week beta-glucan period, but no significant differences between the beta-glucan and the control period were found. The lack of effect could be due to structural changes in the beta-glucans during the isolation process or during the incorporation of the preparation in foods (freezing, storage, baking). In an earlier study, however, barley fiber was more effective in reducing the blood cholesterol level than wheat fiber.[74] Twenty-one men with mild hypercholesterolemia consumed barley fiber incorporated in bread, muesli, spaghetti, and biscuits (total of about 8 g of beta-glucan/day) for 4 weeks. In comparison with the wheat foods, the barley foods gave significantly lower total cholesterol (6%) and LDL cholesterol (7%) levels, while the HDL cholesterol and triglyceride levels did not differ. A similar result was also obtained in a recent study by Behall et al.[75] Eighteen men with increased cholesterol levels consumed a controlled diet (AHA Step I) for 2 weeks, and then the same diet for 5 weeks in a Latin square design, but about 20% of the energy intake was exchanged with brown rice/whole wheat, $^1/_2$ barley, and $^1/_2$ brown rice/whole wheat or barley (<0.4, 3, and 6 g soluble fiber per 2800 kcal, respectively). The barley (flakes, flour, pearled) was incorporated into foods like pancakes, spice cake, no-bake cookies, hot cereal, toasted flakes, steamed pilaf, and muffins. The total cholesterol and LDL cholesterol levels were significantly lower after the high-soluble-fiber diet than after the low- or medium-fiber diets. Newman et al.[76] found that a diet with high amounts of barley (corresponding to 42 g of dietary fiber daily) gave decreased cholesterol levels for subjects that had high pretreatment cholesterol levels, but not for subjects with average pretreatment cholesterol levels. Barley flour that contained 10.7% beta-glucan was incorporated in breads, cookies, muffins, and bars, and the subjects were asked to consume one serving of three test products daily and otherwise follow their normal meal pattern. In another study in Japan, replacement of 30% of the carbohydrates in the diet with barley gave significantly lower total and LDL concentrations in young female subjects.[77] Thus, most of the studies published on barley and blood lipids indicate that barley, like oats, has positive effects on the blood lipid pattern.

8.3.3.2.2.4 Guar Gum

Guar gum is the ground endosperm of seeds from the guar plant (*Cyamopsis tetragonoloba*), and the main component is galactomannan. Guar gum is the natural gum that produces the highest viscosity, and many human studies on its effect on lipid metabolism have been conducted. A Medline search using the search terms *guar gum* and *cholesterol* showed that nearly 70 studies have been published since 1980. Its effects on glucose, lipid metabolism, and blood pressure were investigated in healthy men by Landin et al.[78] The dose of guar gum was 10 g and it was taken three times a day for 6 weeks. In comparison with a placebo, the guar gum decreased the blood cholesterol and triglyceride levels and blood pressure significantly. Thirty grams of guar gum was taken daily for 6 weeks in a double-blind, placebo-controlled, crossover study.[79] The subjects had primary hyperlipidemia, and their total cholesterol, LDL cholesterol, and intermediate-density lipoprotein levels were decreased about 10% when they consumed guar gum. The effect of guar gum on LDL metabolism seemed to be related to an increased LDL apolipoprotein B fractional catabolism. Modified guar gum has also been studied, and in one study partially depolymerized guar gum decreased the total cholesterol levels by 10%, which is a reduction similar to that found earlier for high molecular weight guar gum.[80] The effects of solid or liquid guar gum and preparations with high or medium viscosity on lipid metabolism were followed in hypercholesterolemic subjects.[81] Both solid and liquid guar gum preparations lowered the total and LDL cholesterol, but the high-viscosity preparation gave a larger reduction in blood lipid levels than the medium-viscosity preparation. In yet another study, hypercholesterolemic men were given granulated guar gum (15 g/day) or placebo for 12 weeks in a crossover trial.[82] After 6 weeks' consumption of guar gum, the total and LDL cholesterol levels were significantly decreased compared with the placebo, but after 12 weeks the difference was no longer statistically significant. The hypocholesterolemic effect of guar gum seems to decrease during prolonged dietary supplementation. However, most studies on guar gum have noted significant reductions in serum cholesterol (mean, 11.2%) and LDL cholesterol (mean, 17.7%).[3] The doses given in the studies varied between 6 to 26 g/day, but in most of the studies a dose of over 10 g/day was used. Other gums (xanthan, locust bean, karaya, acacia, arabic) also significantly lower the cholesterol levels, but the effects are slightly less than those reported with guar gum.

8.3.3.2.2.5 Pectin

Pectic substances are a complex group of acidic polysaccharides consisting mainly of galacturonic acid, rhamnose, arabinose, and galactose residues.[83] Also for pectin, several human studies have been done and about 25 studies were found in a Medline search by using the search terms *pectin* and *cholesterol*. One observational study especially focused on pectin. Here, the influence of pectin on the progression of atherosclerosis was studied in a cohort of subjects free of heart disease aged 40 to 60 years (n = 573).[84] A significant inverse association between the intima-media thickness of the common carotid arteries and the intake of pectin was found. The ratio of total to HDL cholesterol was inversely related to the intake of pectin, and also to total fiber and viscous fiber intake. Pectin can be included in the diet as a supplement, but also as fruits, which often contain much pectin. In another study, subjects with hypertension were given guava fruits before meals during 12 weeks, and the effect on the blood lipids and blood pressure was followed.[85] In comparison with a group that was not given guava, the total cholesterol, HDL cholesterol, triglycerides, and blood pressure decreased significantly. An intake of another pectin-rich fruit (100 g of prunes daily) and its effect on blood lipids and bile acid excretion were investigated in men with mild hypercholesterolemia.[86] LDL cholesterol in plasma and the concentration of lithocholic acid in feces were significantly lower after the prune diet in comparison with the control diet (grapefruit juice). Consumption of apple powder (52 g daily) for 4 weeks by subjects with type 2 diabetes did, however, not decrease the total cholesterol and LDL cholesterol levels significantly.[87] The apple powder was added to bread and seven slices were consumed per day. The apple powder would be expected to contain approximately 10.9% pectic substances, which corresponds to a daily intake of 5.7 g. When pectin isolated from grapefruit was given to hypercholesterolemic subjects, the plasma cholesterol and LDL cholesterol decreased 7.6 and 10.8%, respectively.[88] In another study on male hyperlipidemic subjects, the total cholesterol and LDL cholesterol levels were reduced 2 and 4%, respectively, by a pectin supplement, but the reductions were not significant compared to those after a cornflakes control diet.[89] The pectin was added to a ready-to-eat breakfast cereal consumed once daily for 6 weeks. The breakfast cereal also contained oat bran, sugar beet fiber, and wheat bran. The daily intake of soluble fiber in the pectin group was 6 g, whereof about 3 g was from pectin. Challen et al.[90] gave healthy volunteers 36 g of pectin per day during 3 weeks, which resulted in significantly lower serum cholesterol concentrations (5.18 mmol/l) in comparison with a control (5.73 mmol/l), but had no effects on the platelet aggregation, platelet fatty acid concentration, blood clot lysis time, and bleeding times. The influence of pectins on the fibrin network was also investigated. Dietary pectin influenced the fibrin network in hypercholesterolemic men (more permeable, lower tensile strength) but did not change the fibrinogen concentration.[91] Moreover, male hyperlipidemic subjects were given either pectin (15 g/day) or acetate (6.8 g/day) and the effects on the fibrin network were followed.[92] Both supplements gave significant and similar changes on the fibrinogen network (more permeable, lower tensile strength), and thus the effect of pectin on the fibrin network could partially be mediated by acetate.

Several studies have also been done with different kinds of pectins. Dongowski and Lorentz[93] gave diets containing pectin with different degrees of methylation (34.5, 70.8, and 92.6%) to rats for 3 weeks. The concentration of bile acids in the plasma decreased when pectin was given, and with increasing degree of methylation more bile acids were excreted with the feces. A study on hamsters indicated that the viscosity of the pectin was important for the cholesterol-lowering effect.[94]

According to Anderson et al.,[3] most clinical studies on pectin noted a significant reduction in serum cholesterol after pectin supplementation, with a mean decrease of 12.4%. The doses given were 2 to 50 g/day, with most of the studies using a dose of 15 g/day.

8.3.3.3 Cholesterol-Lowering Mechanisms of Dietary Fiber

Several mechanisms have been proposed for the cholesterol-lowering effects of dietary fiber. The main hypothesis is that they decrease the intestinal uptake of bile acids. Bile acids (mainly cholic acid and chenodeoxycholic acid) are produced in the liver from cholesterol and they are excreted into the bile. When food enters the intestine, bile acids flow into the duedenum and mix with the contents. By active transport, over 90% of the bile acids are reabsorbed from the intestine and transported back to the liver (enterohepatic circulation). Decreasing the bile acid reuptake will lead to an increased use of cholesterol for bile acid synthesis, and in this way, the blood cholesterol values will be reduced. Several types of dietary fiber have been shown to increase the fecal excretion of chenodeoxycholic acid. In subjects with an ileostomy, intake of oat bran bread led to an increased excretion of chenodeoxycholic acid and total bile acids in comparison with a wheat flour bread.[95] The median (range) bile acid excretions were 851 (232 to 1550) and 606 (101 to 980) mg/day in the oat bran and wheat flour periods, respectively. However, barley bread did not increase the bile acid excretion significantly in comparison to the wheat flour bread, but instead the excretion of total cholesterol was significantly higher (583 and 494 mg/day, respectively). In another study on ileostomy patients, either pectin (15 g) or wheat bran (16 g) was given and excretion of bile acids and cholesterol from the small intestine was studied.[96] Pectin significantly increased the bile acid excretion

35% and the net cholesterol excretion 14%, but an intake of wheat bran did not change these parameters.

The composition of the bile can also be influenced by a dietary fiber intake. Hillman et al.[97] gave fiber components like pectin (12 g), cellulose (15 g), or lignin (12 g) to healthy subjects daily for 4 weeks, and the effect on the biliary composition was followed. The fiber components gave different effects on the bile acid composition, probably mainly related to changes in the colonic metabolism of the fiber components. Pectin significantly decreased the mean percentage of cholic acid from 42.8 to 39%, and cellulose increased the mean percentage of chenodeoxycholic acid from 33.6 to 35.4%. Lignin had no significant effect on the bile acid composition. The effects on the blood lipids were also followed, but no significant changes were observed in these normolipidemic subjects.[98]

In animals, the main outcome of fiber action is a lowering of hepatic cholesterol pools as a result of more cholesterol being diverted to bile acid synthesis and lower cholesterol delivery to the liver through chylomicron remnants.[99] The activity of major regulatory enzymes of cholesterol homeostasis are increased (3-hydroxy-3-methylglutaryl coenzyme A reductase and Cyp 7) — the former to compensate for sterol loss in the liver and the latter to produce more bile acids to balance the depletion of bile acid pools. In addition, acyl–coenzyme A cholesterol:acyltransferase activity is decreased because of low availability of free cholesterol. Furthermore, hepatic LDL receptors are upregulated to reestablish a cholesterol balance in the liver. In addition, the VLDL apolipoprotein B synthesis rate is decreased and the conversion of VLDL to LDL is reduced. The combination of all of these mechanisms is the probable cause of the consistent hypocholesterolemic effect induced by soluble fiber.[99] Furthermore, soluble fiber can decrease the gastric emptying, prolong glucose absorption, and increase insulin sensitivity. These changes can also have an effect on the liver metabolism of lipids (cholesterol, fatty acids). Dietary fiber can also interfere with the dietary fat absorption by binding the bile acids necessary for micelle formation. Furthermore, some fibers are fermented in the colon and the short-chain fatty acids produced may influence the cholesterol metabolism in the liver, but at present only a few human studies on this point exist and more studies are needed to estimate this effect.

The method of fiber administration seems also to have an influence on the cholesterol-lowering effect. Psyllium fibers (7.3 to 7.6 g daily) were taken either with meals or between meals for 2 weeks.[100] In comparison with a wheat bran control, the total, LDL, and HDL cholesterol were reduced 8, 11, and 7%, respectively, if the psyllium was taken with meals, while when it was taken between meals no cholesterol-lowering effect was observed. It is therefore important to include the soluble fiber in each meal, and in most of the controlled studies, it is taken with a meal at least twice daily.

8.3.3.4 Cholesterol-Lowering Effects of Dietary Fiber in Comparison with Other Food Components

Brown et al.[101] evaluated the cholesterol-lowering effect of different soluble dietary fibers and made a meta-analysis of controlled trials (25 trials of oat products, 17 of psyllium, 7 of pectin, 18 of guar gum). The intake of soluble fiber was associated with a small but significant decrease in total cholesterol (–0.045 mmol/l/g of fiber) and LDL cholesterol (–0.057 mmol/l/g of fiber). The different soluble fibers' effect on the cholesterol levels did not differ significantly (Table 8.3). As a comparison, other meta-analyses and also studies on barley are included in Table 8.3. Brown et al.[101] concluded that increasing the intake of soluble fiber can only make a limited contribution to dietary therapy to lower cholesterol. Jenkins and Kendall[102] compared the ability of different food components to reduce the LDL cholesterol (Table 8.4). Consumption of 5 to 10 g of viscous fiber daily can reduce the LDL cholesterol by approximately 5%. The same reduction can also be obtained with an intake of 1 to 3 g of plant sterols or 25 g of soy protein daily. Changing the amount of saturated

TABLE 8.3
Doses of Fiber Preparations Used in Dietary Studies and Resulting Lowering Effect of Mean Blood Cholesterol (mmol/l/g)

Dietary Fiber Source	Amount, Mean (Range), g/day	Blood Cholesterol-Lowering Effect, mmol/l/g	Reference
Psyllium	9.1 (4.7–16.2)[a]	–0.028	Brown et al.,[101] meta-analysis
	10.2[a]	–0.023	Anderson et al.,[57] meta-analysis
Oats	5.0 (1.5–13.0)[b]	–0.037	Brown et al.,[101] meta-analysis
	3 (1.2–7.6)[b]	–0.043	Ripsin et al.,[65] meta-analysis
Barley	8[c]	–0.016	McIntosh et al.[74]
	9.9 (8–12)[c]	–0.008	Keogh et al.[73]
	6[b]	–0.065	Behall et al.[75]
Guar gum	17.5 (6.6–30.0)[b]	–0.026	Brown et al.,[101] meta-analysis
Pectin	4.7 (2.2–9.0)[b]	–0.070	Brown et al.,[101] meta-analysis

[a]Psyllium.
[b]Soluble fiber.
[c]Beta-glucan.

TABLE 8.4
Ability of Different Food Components to Lower the Blood LDL Cholesterol Level

Component	Approximate LDL Cholesterol Reduction
5–10 g of viscous fiber per day	5%
1–3 g of plant sterols per day	5%
25 g of soy protein per day	5%
<7% of the energy intake from saturated fat	10%

Source: Adapted from Jenkins, D.J.A. and Kendall, C.W.C., *J. Am. Coll. Nutr.*, 18, 559, 1999.

fat to below 7% of the energy intake will lead to lower LDL cholesterol levels by approximately 10%. A reduction of LDL cholesterol by 5% is estimated to have clinical relevance on a population basis, but a combination of different dietary components known to improve the LDL cholesterol values can probably give an even better result. Intake of a diet low in saturated fat and high in plant sterols (1.2 g/1000 kcal), soy protein (16.2 g/1000 kcal), viscous fibers (8.3 g/1000 kcal), and almonds (16.6 g/1000 kcal) lowered the LDL cholesterol similarly to statins (35%), a very efficient cholesterol-lowering agent.[103] The viscous fibers used were psyllium husk and beta-glucans from oats and barley. No significant differences in blood pressure, HDL cholesterol, or triglycerides were seen in comparison with subjects only consuming a low-saturated-fat diet. The calculated coronary artery disease risk was significantly reduced by 32% (Framingham 10-year cardiovascular disease risk).

8.4 HEALTH CLAIMS RELATED TO CARBOHYDRATES AND CVD

The European guideline on cardiovascular disease prevention includes recommendations on the intake of carbohydrates and dietary fibers.[104] The consumption of whole-grain cereals and bread should be encouraged, and saturated fat should be replaced partly by complex carbohydrates.

The WHO[1] lists several factors related to dietary fiber that probably decrease the risk for developing cardiovascular diseases, i.e., nonstarch polysaccharides and whole-grain cereals. According to the WHO, more convincing evidence exists for the intake of fruit and vegetables and a decreased risk of developing CVD.

However, the U.S. Food and Drug Administration (FDA) has approved several health claims related to dietary fiber and coronary heart disease (Table 8.5).[105] A general claim is related to fruits, vegetables, and grain products that contain fiber, particularly soluble fiber, and risk of coronary heart disease. The food without fortification should have a content of 0.6 g of soluble dietary fiber per reference amount and be low in saturated fat, cholesterol, and total fat. In 1999, the FDA also approved a health claim for whole-grain foods: "Diets rich in whole-grain foods and other plant foods and low in total fat, saturated fat, and cholesterol may reduce the risk of heart disease and some cancers." Whole-grain foods should contain 51% or more whole-grain ingredients by weight per reference amount, with a dietary fiber content of 3.0 g/50 g or 1.7 g/35 g, and the food should also be low in fat.

Enough evidence of the cholesterol-lowering effects of psyllium had been gathered to authorize a health claim from the FDA in 1998, stating that psyllium can reduce the risk of heart disease if consumed in an otherwise healthy diet.[106] The FDA based this health claim on a review of 21 human studies. It found no reliable data to establish a dose–response relationship, but an intake of 10.2 g daily of psyllium seed husk as part of a diet low in saturated fat and cholesterol seemed to have significant effects on blood total and LDL cholesterol levels. This amount of psyllium husk contains about 7 g of soluble fiber, and a food should provide at least 1.7 g of soluble fiber from psyllium per serving to be able to use the health claim.

TABLE 8.5
Health Claims Approved by the U.S. Food and Drug Administration on the Relation between Intake of Dietary Fiber and Coronary Heart Disease[105]

Health Claim	Requirements
Fruit, vegetables, and grain products that contain fiber, particularly soluble fiber, and risk of coronary heart disease	A fruit, vegetable, or grain product that contains fiber Low in saturated fat, cholesterol, and total fat At least 0.6 g of soluble fiber per reference amount (RA) (without fortification)
Whole-grain foods and risk of heart disease and certain cancers	Contains 51% or more whole-grain ingredients by weight per RA Low in total fat Dietary fiber content at least: 3.0 g/RA of 55 g 2.8 g/RA of 50 g 2.5 g/RA of 45 g 1.7 g/RA of 35 g
Soluble fiber from certain foods and risk of coronary heart disease	Low in saturated fat, cholesterol, and total fat Containing at least 0.75 g of whole-oat soluble fiber per RA or psyllium seed husk containing at least 1.7 g of psyllium husk soluble fiber per RA

A similar health claim also exists for oats.[107] The FDA reviewed 37 oat trials in 1996, and 17 studies showed a positive effect of oat bran and oatmeal on total and LDL cholesterol. The amount of oat bran or oat meal given in the studies ranged from 34 g (2.5 g of soluble fiber) to 123 g (10.3 g of soluble fiber). Five studies showed equivocal results in reducing serum cholesterol, 4 had a too short study period, and 11 showed no effect on serum lipid levels. But the overall conclusion was that oats could lower serum cholesterol levels, specifically LDL cholesterol, without any significant change in the HDL fraction. The FDA thus authorized a health claim stating: "Diets high in oat bran or oat meal and low in saturated fat and cholesterol may reduce the risk of heart disease." To be able to use the health claim, a product must contain at least 0.75 g of beta-glucan soluble fiber per serving, and the recommended daily intake is 3 g. In 2005, barley was added as an eligible source of beta-glucan soluble fiber. In other countries, similar health claims on the relations between a high intake of dietary fiber or whole-grain foods and a decreased risk for coronary heart disease have been issued.

8.5 SUMMARY AND CONCLUSIONS

Cardiovascular diseases like myocardial infarction, angina, and stroke are major contributors to the global burden of disease. The diseases are caused by the development of atherosclerosis in the blood vessels, and one major factor influencing the atherosclerotic process is the carbohydrate content in the diet. A diet with a high content of available carbohydrates may increase the triglyceride level in the blood, and this in turn can enhance the development of atherosclerosis. Oligosaccharides like inulin and oligofructose have been shown in animal studies to reduce the triglyceride level in the blood and also the total and LDL cholestesterol concentrations, and thus decrease the risk for CVD. Studies in humans consuming lower oligosaccharide doses than in the animal studies have shown divergent results on the blood lipids and more studies are needed. Concerning the intake of dietary fiber, several observational studies indicate that a high intake is linked to a decreased risk for CVD. There are also indications from observational studies that soluble fibers are more efficient than insoluble fibers, and different intervention studies have confirmed this relation. Supplementation of a diet with a combination of soluble fibers in daily doses of 5 to 20 g has shown positive effects on the blood lipid pattern; the total cholesterol and LDL cholesterol levels especially are decreased. Psyllium husk is rich in soluble fiber, and meta-analyses of studies indicate that it can reduce the LDL cholesterol level 7 to 9%. The dose–response effect is not clear, but the FDA suggests that a daily intake of 10.2 g of psyllium husk is needed (7 g of soluble fiber) in a diet low in saturated fat and cholesterol to reduce the risk of heart disease. Oats are also rich in soluble fibers (beta-glucans), and a meta-analysis of studies gave a summary effect size for change in total cholesterol of –0.13 mmol/l. The dose–response effect has also been evaluated, and an intake of over 3 g of soluble fiber from oats seems to be needed to decrease the blood cholesterol levels significantly, and thus reduce the risk for CVD. Barley, like oats, is rich in beta-glucans, and most of the few studies published indicate that barley has positive effects on the blood lipid pattern. Guar gum is the natural gum that produces the highest viscosity, and human studies have shown that it reduces the total and LDL cholesterol levels significantly. Most of the studies used relatively high doses of guar gum (>10 g/day). Pectin has also been investigated thoroughly and shown to have significant effects on blood cholesterol levels. The main mechanism behind the lipid-lowering effect of soluble fiber is probably that the lipids decrease the uptake of bile acids in the intestine. A meta-analysis of 67 trials investigating either psyllium, oats, pectin, or guar gum showed that intake of soluble fiber was associated with a small but significant decrease in total and LDL cholesterol. It is estimated that an intake of 5 to 10 g of viscous fiber per day can reduce the LDL cholesterol level by 5%, a reduction similar to that of a daily intake of 1 to 3 g of plant sterols and 25 g of soy protein. Several health claims exist on the relation between intake of dietary fiber and related components and a decreased risk for coronary heart disease. But more information is needed to clarify what happens when dietary fiber components are purified or processed in other ways and the effect this has on CHD. The physiological effects of dietary fiber might be affected by factors such as fiber degradation (which probably decreases the physiological effects) and/or fiber solubilization via processing (which probably increases the physiological effects).

REFERENCES

1. WHO Technical Report Series 916, *Diet, Nutrition and the Prevention of Chronic Diseases*, Geneva, 2003.

2. Keys, A., Diet, in *Seven Countries: A Multivariate Analysis of Death and Coronary Heart Disease*, Harvard University Press, Cambridge, MA, 1980, p. 248.
3. Anderson, J.W. et al., Dietary fiber and coronary heart disease, *Food Sci. Nutr.*, 29, 95, 1990.
4. Dawber, T.R. et al., Physical activity and cardiovascular disease, in *The Framingham Study*, Harvard University Press, Cambridge, MA, 1980, p. 157.
5. National Cholesterol Education Program (NCEP), Expert panel on detection, evaluation and treatment of high blood cholesterol in adults, third report, *Circulation*, 206, 3143, 2002.
6. Law, M.R., Wald, N.J., and Thompson, S.G., By how much and how quickly does reduction in serum cholesterol concentration lower risk of ischaemic heart disease, *BMJ*, 308, 267, 1994.
7. Mensink, R.P. et al., PASSCLAIM-diet-related cardiovascular disease, *Eur. J. Nutr.*, 42 (Suppl. 1), 6, 2003.
8. LaRosa, J.C., He, J., and Vupputuri, S., Effects of statins on risk of coronary disease: a meta-analysis of randomized controlled trials, *JAMA*, 282, 2340, 1999.
9. Mensink, R.P. and Katan, M.B., Effect of dietary fatty acids on serum lipids and lipoproteins. A meta-analysis of 27 trials, *Arterioscler. Thromb.*, 12, 911, 1992.
10. Gordon, T. et al., High density lipoprotein as a protective factor against coronary heart disease: the Framingham study, *Am. J. Med.*, 62, 707, 1977.
11. Stampfer, M.J. et al., A prospective study of cholesterol, apolipoproteins, and the risk of myocardial infarction, *N. Engl. J. Med.*, 325, 373, 1991.
12. Cuchel, M. et al., Atherogenic lipid profile in elderly patients with ischaemic cerebrovascular disease, *Lancet*, 356, 401, 2000.
13. Griffin, B.A. et al., Role of plasma triglyceride in the regulation of plasma low density lipoprotein (LDL) subfractions: relative contribution of small, dense LDL to coronary heart disease risk, *Atherosclerosis*, 106, 241, 1994.
14. Castelli, W.P., The triglyceride issue: a view from Framingham, *Am. Heart J.*, 112, 432, 1986.
15. Hokanson, J.E. and Austin, M.A., Plasma triglyceride level is a risk factor for cardiovascular disease independent of high-density lipoprotein cholesterol level: a meta-analysis of population-based prospective studies, *J. Cardiovasc. Risk*, 3, 213, 1996.
16. Walldius, G. et al., High apolipoprotein B and a low apolipoprotein A-I, and improvement in the prediction of fatal myocardial infarction (AMORIS study): a prospective study, *Lancet*, 358, 2026, 2001.
17. Sarkkinen, E. et al., Effect of apolipoprotein E polymorphism on serum lipid response to the separate modification of dietary fat and dietary cholesterol, *Am. J. Clin. Nutr.*, 68, 1215, 1998.
18. Franklin, S.S. et al., Does the relation of blood pressure to coronary heart disease risk change with aging? *Circulation*, 103, 1245, 2001.
19. Fried, S.K. and Rao, S.P., Sugars, hypertriglyceridemia, and cardiovascular disease, *Am. J. Clin. Nutr.*, 78, 873S, 2003.
20. Garrow, J., *Human Nutrition and Dietetics*, 10 rev. ed., Elsevier, Edinburgh, 1999.
21. Hung, T. et al., Fat versus carbohydrate in insulin resistance, obesity, diabetes and cardiovascular disease, *Curr. Opin. Clin. Nutr. Metab. Care*, 2, 165, 2003.
22. Augustin, L.S. et al., Glycemic index in chronic disease: a review, *Eur. J. Clin. Nutr.*, 56, 1049, 2002.
23. Liu, S. et al., A prospective study of dietary glycemic load, carbohydrate intake, and risk of coronary heart disease in US women, *Am. J. Clin. Nutr.*, 71, 1455, 2000.

24. Jenkins, D.J. et al., Metabolic effects of a low-glycemic-index diet, *Am. J. Clin. Nutr.*, 46, 968, 1987.
25. Järvi, A.E. et al., Improved glycemic control and lipid profile and normalized fibrinolytic activity on a low-glycemic index diet in type 2 diabetic patients, *Diabetes Care*, 22, 10, 1999.
26. Delzenne, N.M., Oligosaccharides: state of the art, *Proc. Nutr. Soc.*, 62, 177, 2003.
27. Williams, C.M., Effects of inulin on lipid parameters in humans, *J. Nutr.*, 129, 1471S, 1999.
28. Davidson, M.H. and Maki, K.C., Effects of dietary inulin on serum lipids, *J. Nutr.*, 129, 1474S, 1999.
29. Brighenti, F. et al., Effect of consumption of a ready-to-eat breakfast cereal containing inulin on the intestinal milieu and blood lipids in healthy male volunteers, *Eur. J. Clin. Nutr.*, 53, 726, 1999.
30. Luo, J. et al., Chronic consumption of short-chain fructooligosaccharides by healthy subjects decreased basal hepatic glucose production but had no effect on insulin-stimulated glucose metabolism, *Am. J. Clin. Nutr.*, 63, 939, 1996.
31. Pedersen, A., Sandström, B., and Van Amelsvoort, J.M.M., The effect of ingestion of inulin on blood lipids and gastrointestinal symptoms in healthy females, *Br. J. Nutr.*, 78, 215, 1997.
32. Alles, M.S. et al., Consumption of fructooligosaccharides does not favorably affect blood glucose and serum lipid concentrations in patients with type 2 diabetes, *Am. J. Clin. Nutr.*, 69, 64, 1999.
33. Jackson, K.G. et al., The effect of the daily intake of inulin on fasting lipid, insulin and glucose concentration in middle-aged men and women, *Br. J. Nutr.*, 82, 23, 1999.
34. Williams, C.M. and Jackson, K.G., Inulin and oligofructose: effects on lipid metabolism in human studies, *Br. J. Nutr.*, 87, 261S, 2002.
35. Pereira, M.A. and Pins, J.J., Dietary fiber and cardiovascular disease: experimental and epidemiologic advances, *Curr. Atheroscler. Rep.*, 2, 494, 2002.
36. Jacobs, D.R. et al., Whole-grain intake may reduce the risk of ischemic heart disease death in postmenopausal women: the Iowa women's health study, *Am. J. Clin. Nutr.*, 68, 248, 1998.
37. Jacobs, D.R., Meyer, H.E., and Solvoll, K., Reduced mortality among whole grain bread eaters in men and women in the Norwegian County Study, *Eur. J. Clin. Nutr.*, 55, 137, 2001.
38. Wolk, A. et al., Long-term intake of dietary fiber and decreased risk of coronary heart disease among women, *JAMA*, 281, 1998, 1999.
39. Rimm, E.B. et al., Vegetable, fruit, and cereal fiber intake and risk of coronary heart disease among men, *JAMA*, 275, 447, 1996.
40. Morris, J.N., Marr, J.W., and Clayton, D.G., Diet and heart: a postscript, *Br. Med. J.*, 2, 1307, 1977.
41. Mozaffarian, D. et al., Cereal, fruit, and vegetable fiber intake and the risk of cardiovascular disease in elderly individuals, *JAMA*, 289, 1659, 2003.
42. Liu, S. et al., A prospective study of dietary fiber intake and risk of cardiovascular disease among women, *J. Am. Coll. Cardiol.*, 39, 49, 2002.
43. Pietinen, P. et al., Intake of dietary fiber and risk of coronary heart disease in a cohort of Finnish men. The alpha-tocopherol, beta-carotene cancer prevention study, *Circulation*, 94, 2720, 1996.
44. Negri, E. et al., Fiber intake and risk of nonfatal acute myocardial infarction, *Eur. J. Clin. Nutr.*, 57, 464, 2003.

45. Anderson, J.W., Whole grains protect against atherosclerotic cardiovascular disease, *Proc. Nutr. Soc.*, 62, 135, 2003.
46. Jensen, C.D., Haskell, W., and Whittam, J.H., Long-term effects of water-soluble dietary fiber in the management of hypercholesterolemia in healthy men and women, *Am. J. Cardiol.*, 79, 34, 1997.
47. Roberts, D.C. et al., The cholesterol lowering effect of a breakfast cereal containing psyllium fibre, *Med. J. Aust.*, 161, 660, 1994.
48. Davidson, M.H. et al., A low-viscosity soluble fiber fruit juice supplement fails to lower cholesterol in hypercholesterolemic men and women, *J. Nutr.*, 128, 1927, 1998.
49. Knopp, R.H. et al., Long-term blood cholesterol-lowering effects of a dietary fiber supplement, *Am. J. Prev. Med.*, 17, 18, 1999.
50. Jenkins, D.J. et al., Soluble fiber intake at a dose approved by the US Food and Drug Administration for claim of health benefits: serum lipid risk factors for cardiovascular disease assessed in a randomized controlled crossover trial, *Am. J. Clin. Nutr.*, 75, 834, 2002.
51. Lee, S.C. et al., Determination of soluble and insoluble dietary fiber in psyllium-containing cereal products, *J. AOAC*, 78, 724, 1995.
52. Anderson, J.W. et al., Long-term cholesterol lowering effects of psyllium as an adjunct to diet therapy in the treatment of hypercholesterolemia, *Am. J. Clin. Nutr.*, 71, 1433, 2000.
53. Bell, L.P. et al., Cholesterol-lowering effects of soluble-fiber cereals as part of a prudent diet for patients with mild to moderate hypercholesterolemia, *Am. J. Clin. Nutr.*, 52, 1020, 1990.
54. Rodriguez-Morán, M., Guerrero-Romero, F., and Lazcano-Burciaga, G., Lipid- and glucose-lowering efficacy of plantago psyllium in type II diabetes, *J. Diabetes Complications*, 12, 273, 1998.
55. Davidson, M.H. et al., Long-term effects of consuming foods containing psyllium seed husk on serum lipids in subjects wih hypercholesterolemia, *Am. J. Clin. Nutr.*, 67, 367, 1998.
56. Williams, C.L. et al., Soluble fiber enhances the hypocholesterolemic effect of the step I diet in childhood, *J. Am. Coll. Nutr.*, 14, 251, 1995.
57. Anderson, J.W. et al., Cholesterol-lowering effects of psyllium intake adjunctive to diet therapy in men and women with hypercholesterolemia: meta-analysis of 8 controlled trials, *Am. J. Clin. Nutr.*, 71, 472, 2000.
58. Olson, B.H. et al., Psyllium-enriched cereals lower blood total cholesterol and LDL cholesterol, but not HDL cholesterol, in hypercholesterolemic adults: results of a meta-analysis, *J. Nutr.*, 127, 1973, 1997.
59. Asp, N.-G., Mattsson, B., and Önning, G., Variation in dietary fibre, β-glucan, starch, protein, fat and hull content of oats grown in Sweden 1987–1989, *Eur. J. Clin. Nutr.*, 46, 31, 1991.
60. AACC Committee Adopts Oat Bran Definition, *Cereal Foods World*, December 1989.
61. Åman, P., Rimsten, L., and Andersson, R., Molecular weight distribution of beta-glucan in oat-based foods, *Cereal Chem.*, 81, 356, 2004.
62. Kerckhoffs, D.A.J.M., Hornstra, G., and Mensink, R.P., Cholesterol-lowering effect of beta-glucan from oat bran in mildly hypercholesterolemic subjects may decrease when beta-glucan is incorporated into bread and cookies, *Am. J. Clin. Nutr.*, 78, 221, 2003.
63. Önning, G. et al., Consumption of oat milk for 5 weeks lowers serum cholesterol and LDL cholesterol in free-living men with moderate hypercholesterolemia, *Ann. Nutr. Metab.*, 43, 301, 1999.

64. Lia Amundsen, Å., Haugum, B., and Andersson, H., Changes in serum cholesterol and sterol metabolites after intake of products enriched with an oat bran concentrate within a controlled diet, *Scand. J. Nutr.*, 47, 68, 2003.
65. Ripsin, C.M. et al., Oat product and lipid lowering. A meta-analysis, *JAMA*, 267, 3317, 1992.
66. Mälkki, Y., Oat fibers: production, composition, physico-chemical properties, physiological effects, safety and food applications, in *Handbook of Dietary Fibre*, Cho, S.S. and Dreher, M., Eds., Marcel Dekker, New York, 2001, p. 497.
67. Önning, G., The use of cereal beta-glucans to control diabetes and cardiovascular disease, in *Functional Foods, Cardiovascular Disease and Diabetes*, Arnoldi, A., Ed., Woodhead Publishing Ltd., Cambridge, U.K., 2004, p. 401.
68. Oscarsson, M. et al., Effects of cultivar, nitrogen ferilization rate and environment on yield and grain quality of barley, *J. Sci. Food Agric.*, 78, 359, 1998.
69. Ranhotra, G.S. et al., Dose response to soluble fiber in barley in lowering blood lipids in hamster, *Plant Foods Hum. Nutr.*, 52, 329, 1998.
70. Delaney, B. et al., β-Glucan fractions from barley and oats are similarly antiatherogenic in hypercholesterolemic Syrian golden hamsters, *J. Nutr.*, 133, 468, 2003.
71. Fadel, J.G. et al., Hypocholesterolemic effects of β-glucans in different barley diets fed to broiler chicks, *Nutr. Rep. Int.*, 35, 1049, 1987.
72. Bengtsson, S. et al., Chemical studies on mixed-linked beta-glucans in hull-less barley cultivars giving different hypocholesterolaemic responses in chickens, *J. Sci. Food Agric.*, 52, 435, 1990.
73. Keogh, G.F. et al., Randomized controlled cross-over study of the effect of a highly beta-glucan-enriched barley on cardiovascular disease risk factors in mildly hypercholesterolemic men, *Am. J. Clin. Nutr.*, 78, 711, 2003.
74. McIntosh, G.H. et al., Barley and wheat foods: influence on plasma cholesterol concentrations in hypercholesterolemic men, *Am. J. Clin. Nutr.*, 53, 1205, 1991.
75. Behall, K.M., Scholfield, D.J., and Hallfrisch, J., Lipids significantly reduced by diets containing barley in moderately hypercholesterolemic men, *J. Am. Coll. Nutr.*, 23, 55, 2004.
76. Newman, R.K. et al., Hypocholesterolemic effect of barley foods on healthy men, *Nutr. Rep. Int.*, 39, 749, 1989.
77. Li, J. et al., Effects of barley intake on glucose tolerance, lipid metabolism, and bowel function in women, *Nutrition*, 19, 926, 2003.
78. Landin, K. et al., Guar gum improves insulin sensitivity, blood lipids, blood pressure, and fibrinolysis in healthy men, *Am. J. Clin. Nutr.*, 56, 1061, 1992.
79. Turner, P.R. et al., Metabolic studies on the hypolipidaemic effect of guar gum, *Atherosclerosis*, 81, 145, 1990.
80. Blake, D.E. et al., Wheat bread supplemented with depolymerized guar gum reduces the plasma cholesterol concentration in hypercholesterolemic human subjects, *Am. J. Clin. Nutr.*, 65, 107, 1997.
81. Superko, H.R. et al., Effects of solid and liquid guar gum on plasma cholesterol and triglyceride concentrations in moderate hypercholesterolemia, *Am. J. Cardiol.*, 62, 51, 1988.
82. Aro, A. et al., Effects of guar gum in male subjects with hypercholesterolemia, *Am. J. Clin. Nutr.*, 39, 911, 1984.
83. Selvendran, R.R. and Robertson, J.A., The chemistry of dietary fibre: an holistic view of the cell wall matrix, in *Dietary Fibre: Chemical and Biological Aspects*, Southgate, A.T., Waldron, K., Johson, I.T., Fenwick, G.R., Eds., Royal Society of Chemistry, Cambridge, U.K., 1990, p. 27.

84. Wu, H. et al., Dietary fiber and progression of atherosclerosis: the Los Angeles atherosclerosis study, *Am. J. Clin. Nutr.*, 78, 1085, 2003.
85. Singh, R.B. et al., Effects of guava intake on serum total and high density lipoprotein cholesterol levels and on systemic blood pressure, *Am. J. Cardiol.*, 70, 1287, 1992.
86. Tinker, L.F. et al., Consumption of prunes as a source of dietary fiber in men with mild hypercholesterolemia, *Am. J. Clin. Nutr.*, 53, 1259, 1991.
87. Mahalko, J.R. et al., Effect of consuming fiber from corn bran, soy hulls, or apple powder on glucose tolerance and plasma lipids in type II diabetes, *Am. J. Clin. Nutr.*, 39, 25, 1984.
88. Cerda, J.J. et al., The effects of grapefruit pectin on patient risk for coronary heart disease without altering diet or lifestyle, *Clin Cardiol.*, 11, 589, 1988.
89. Bell, L.P. et al., Cholesterol-lowering effects of soluble fiber cereals as part of a prudent diet for patients with mild to moderate hypercholesterolemia, *Am. J. Clin. Nutr.*, 52, 1020, 1990.
90. Challen, A.D., Branch, W.J., and Cummings, J.H., The effect of pectin and wheat bran on platelet function and haemostasis in man, *Hum. Nutr. Clin. Nutr.*, 37, 209, 1983.
91. Veldman, F.J. et al., Dietary pectin influences fibrin network structure in hypercholesterolaemic subjects, *Thromb. Res.*, 86, 183, 1997.
92. Veldman, F.J. et al., Possible mechanisms through which dietary pectin influences fibrin network architecture in hypercholesterolaemic subjects, *Thromb. Res.*, 93, 253, 1999.
93. Dongowski, G. and Lorentz, A., Intestinal steroids in rats are influenced by the structural parameters of pectin, *J. Nutr. Biochem.*, 15, 196, 2004.
94. Terpstra, A.H. et al., Dietary pectin with high viscosity lowers plasma and liver cholesterol concentration and plasma cholesterol ester transfer protein activity in hamsters, *J. Nutr.*, 128, 1944, 1998.
95. Lia, Å. et al., Oat β-glucan increases bile acid excretion and a fiber-rich barley fraction increases cholesterol excretion in ileostomy subjects, *Am. J. Clin. Nutr.*, 62, 1245, 1995.
96. Bosaeus, I. et al., Effect of wheat bran and pectin on bile acid and cholesterol excretion in ileostomy patients, *Hum. Nutr. Clin Nutr.*, 40, 429, 1986.
97. Hillman, L.C. et al., Effects of the fibre components pectin, cellulose, and lignin on bile salt metabolism and biliary lipid composition in man, *Gut*, 27, 29, 1986.
98. Hillman, L.C. et al., The effects of the fiber components pectin, cellulose and lignin on serum cholesterol levels, *Am. J. Clin. Nutr.*, 42, 207, 1985.
99. Fernandez, M.-L., Soluble fiber and nondigestible carbohydrate effects on plasma lipids and cardiovascular risk, *Curr. Opin. Lipidol.*, 12, 35, 2001.
100. Wolever, T.M. et al., Psyllium reduces blood lipids in men and women with hyperlipidemia, *Am. J. Med. Sci.*, 307, 269, 1994.
101. Brown, L. et al., Cholesterol-lowering effects of dietary fiber: a meta-analysis, *Am. J. Clin. Nutr.*, 69, 30, 1999.
102. Jenkins, D.J.A. and Kendall, C.W.C., Plant sterols, health claims and strategies to reduce cardiovascular disease risk, *J. Am. Coll. Nutr.*, 18, 559, 1999.
103. Jenkins, D.J.A. et al., The effect of combining plant sterols, soy protein, viscous fibers, and almonds in treating hypercholesterolemia, *Metabolism*, 52, 1478, 2003.
104. De Backer, G. et al., European guidelines on cardiovascular disease prevention in clinical practise, *Eur. Heart J.*, 24, 1601, 2003.
105. FDA, Health claims, Appendix C, in *A Food Labelling Guide*, available online at http://www.cfsan.fda.gov/~dms/flg-6c.html.

106. FDA, Health claims; Soluble fiber from certain foods and coronary heart disease. Psyllium, *Fed. Reg.*, 63, 8103, 1998.
107. FDA, Food labelling: health claims; oats and coronary heart disease, *Fed. Reg.*, 61, 296, 1996.

9 Carbohydrates and Obesity

Gail Woodward-Lopez, Dana E. Gerstein, Lorrene D. Ritchie, and Sharon E. Fleming

CONTENTS

9.1 INTRODUCTION

Today, obesity represents one of the leading public health concerns worldwide. Since the mid-1970s, rates of obesity have risen precipitously in the U.S.; the prevalence of obesity has doubled among adults,[1] and the prevalence of overweight has tripled among children.[2] Excess body fat is not merely an aesthetic concern. Overweight and obesity increase the risk of numerous chronic diseases, including type 2 diabetes mellitus, cardiovascular disease, dyslipidemia, hypertension, gall bladder disease, osteoarthritis, and several forms of cancer.[3] Overweight and obesity can also impact quality of life due to decreased mobility, reduced self-esteem, and discrimination.[4,5]

Overweight status is typically assessed on the basis of a common measurement known as *body mass index* (BMI). BMI adjusts the weight of an individual by height using the following formula: [weight (in kg)/height (in m)]^2. For individuals aged 2 to 20 years, overweight is identified using national growth charts (developed and released in 2000 by the National Center for Health Statistics and the Centers for Disease Control). Currently in the U.S., an individual between 2 and 20 years of age with a BMI greater than or equal to the 95th percentile for a specific age and gender is considered *overweight*; an individual with a BMI between the 85th and 95th percentiles is considered at risk for becoming overweight. For individuals over the age of 20 years, being *overweight* is defined as having a BMI of 25 and *obesity* is defined as having a BMI of 30.

To maintain body weight (i.e., not gain or lose weight), total caloric energy from food and beverages must be equivalent to total energy expenditure (energy used for metabolic processes, growth, and physical activity). Weight gain occurs when a *positive energy balance* persists over time; energy intake is greater than energy expenditure. Weight loss occurs when a *negative energy balance* persists over time; energy intake is less than energy expenditure. Therefore, changes in body weight occur if uncompensated changes occur in energy intake, energy expenditure, or both.

Much debate has occurred recently concerning the role that the macronutrient composition of the diet plays in the development of obesity. The macronutrients (carbohydrates, fats, and proteins), because of their unique properties, have the potential to differentially impact energy balance. The macronutrients are not equal in terms of their potential impact on satiety and properties of foods such as palatability and energy density. Likewise, they are not treated the same by the body metabolically. In this chapter, we discuss the role of dietary carbohydrates in the development of energy imbalance. To do so, we examine four lines of evidence:

Mechanisms: Metabolism and properties of carbohydrates that have a potential to impact energy balance.
Secular trends: Changes in dietary intake over the same time period during which obesity has risen most steeply.
Observational studies: Studies that examine the associations between carbohydrate intake and adiposity in free-living populations.
Intervention studies: The impact on overweight of interventions that manipulate carbohydrate intake.

Each of these lines of evidence has strengths and weaknesses, but taken as a whole, they provide a more complete picture regarding the state of knowledge on the relationship between carbohydrate intake and obesity.

9.2 MECHANISMS BY WHICH DIETARY CARBOHYDRATES AFFECT ENERGY BALANCE

For body weight changes to occur, energy intake and energy expenditure cannot be equivalent. Thus, if a dietary component such as carbohydrates is to consistently influence body weight, it follows that this component must influence one or more factors that influence the consumption or expenditure side of the energy balance equation. In this section, we review the evidence for an association between carbohydrates and several key factors that have the potential to influence energy balance. These factors include satiation and satiety, energy density of foods, palatability and taste preference, the fat:carbohydrate ratio of the diet, metabolic fuel partitioning, and the glycemic index.

9.2.1 Satiation and Satiety

To begin to understand the potential role of carbohydrates in energy regulation and the development of overweight, it is necessary to understand the role carbohydrates play in appetite and hunger control. The term *satiation* will refer to the sensation of fullness experienced during a meal or eating episode that controls meal size.[6] *Satiety* will refer to the sensation of fullness experienced between meals or eating episodes that inhibits the resumption of eating and extends the duration between meals.[6] Because protein exerts a distinctly high level of satiation and satiety, most studies maintain constant protein levels when studying the effects of the other macronutrients, carbohydrates and fat, on satiation and satiety.[7,8] When lean subjects were allowed to eat from a range of high-fat or high-sucrose foods, passive overconsumption occurred only when high-fat foods were consumed.[9,10] The term *passive overconsumption* indicates that an individual does not deliberately intend to ingest excess calories, but rather signals for satiation do not function effectively to control meal size. Therefore, although dietary carbohydrates do not exert the powerful satiation and satiety effects of dietary protein, they have a greater effect than dietary fat.[7,11–13]

Not all carbohydrates, however, exert the same effect on satiation and satiety. It has been hypothesized that sugars and other simple carbohydrates may encourage consumption beyond the sensation of fullness, because of their relatively high palatability and rapid rate of digestion. The rapid absorption of simple carbohydrates into the bloodstream has been shown to trigger a release of insulin in excess of need, resulting in a rapid and steep decline in blood sugar, and hence a more rapid and intense resumption of hunger (see *glycemic index* below). A high sugar intake has also been shown to lead to insulin resistance, *de novo* hepatic fatty acid synthesis, and visceral fat deposition in rats.[14] Both animal and human studies suggest that high consumption of simple carbohydrates can lead to excess energy intake.[8,15]

Dietary fiber, on the other hand, appears to protect individuals from passive overconsumption and subsequent positive energy imbalance by promoting increased satiation and satiety in comparison to digestible complex carbohydrates and simple sugars.[10,16–18] The mechanisms by which dietary fiber promotes satiation and satiety include increased chewing, increased gastric distention, delayed gastric emptying, decreased rate of nutrient absorption and digestion, and unique effects on gut hormones.

The increased chewing time required for foods that are naturally high in dietary fiber may promote satiation by reducing the rate of ingestion as well as distending the stomach as a result of increased gastric juice and saliva secretion. The ingestion of high-fiber foods, especially those foods that are high in soluble fiber, results in a viscous gel formation in the stomach, which is thought to increase gastric distention and ultimately reduce the rate of gastric emptying. As a result, nutrient digestion and absorption are also delayed, which tends to increase satiety.[16–18] Postprandial blood glucose concentration also tends to be lower after high-fiber than after low-fiber meals or foods. As a result, insulin secretion is reduced, as is the likelihood of reactive hypoglycemia during the postabsorptive period, consequences which may also promote satiety. Finally, dietary fiber ingestion increases the secretion of such gut hormones as cholecystokinin, glucagon-like peptide-1 (GLP-1), peptide YY, and neurotensin, which may alter glucose homeostasis or act independently of glycemic response as satiety factors.[17,18] For example, when GLP-1 is provided exogenously, it slows gastric emptying and reduces hunger in the presence of fiber.[17] It is likely that dietary fiber's unique ability to promote satiety and satiation through a number of different mechanisms may be influential in preventing excess calorie intake, and ultimately positive energy imbalance or overweight development.

9.2.2 Energy Density

Energy density refers to the calorie content of a given weight of food (calories/gram). Comparisons of results from studies examining energy density are difficult because of differing or poorly defined measures of energy density.[19] However, energy density may be an important determinant of energy intake and thus energy balance.[20] Clinical studies have demonstrated that total caloric intake is markedly affected by manipulations in energy density, independent of the percentages of energy from macronutrients.[21,22] It has also been found that obese subjects consume a greater proportion of foods high in energy density than lean subjects.[19,23]

The energy density of a food is, in part, a function of its macronutrient composition.[24] Both dietary proteins and carbohydrates have a relatively low energy density, providing only 4 calories per gram. Dietary fat, on the other hand, is much more energy dense, providing 9 calories per gram. Alcohol is also fairly energy dense, providing 7 calories per gram. Yao and Roberts[20] concluded that the two most significant determinants of dietary energy density are water and fat content. Figure 9.1 illustrates the association of energy density (kcal/100 g) with fat, water, and fiber contents (g/100 g) of 200 commonly consumed foods.

Dietary fat is a major determinant of energy density as a result of its high caloric density compared to carbohydrates or proteins, and the wide variation in the proportion in which it is found in commonly consumed foods.[20] Because carbohydrates have a naturally lower energy density than fat, they are less likely to lead to passive overconsumption and subsequent positive energy imbalance. However, it should not be assumed that all foods high in carbohydrates are low in energy density. For instance, commercially available low-fat foods or virtually fat-free foods (specifically those with large amounts of added sugars or other concentrated carbohydrates) can have considerable energy density.[8] In general, nonprocessed foods that are naturally high in carbohydrates and low in dietary fat (e.g., fruits and whole grains) are low in energy density and unlikely to lead to passive overconsumption and a subsequent positive energy imbalance.

The impact of water on the energy density of foods is due to its zero-energy content and the wide variation in the proportion in which it is found in commonly consumed foods. It has been shown in experimental studies that so-called wet carbohydrates (high-carbohydrate foods with high water content, such as soups and fruits) have a higher satiety value than dry carbohydrates, such as pretzels and bagels. An exception to this rule, however, is beverages. Sugars in liquid form have a particularly weak appetite suppressant effect. Calorie intake in the form of beverages is not proportionately compensated for by a reduction in subsequent intake of solid foods and could therefore lead to excess calorie intake.[25]

Dietary fiber has the potential to influence energy density because of its minimal energy content.[20] However, in actuality, the influence of fiber on energy density is modest because the fiber content of foods does not vary widely, and there is an upper limit to the amount of fiber found in foods typically consumed by humans. As seen in Figure 9.1,[20] fiber content was not significantly related to energy density when studied in 200 commonly consumed foods. Therefore, although dietary fiber can contribute to a reduction in the energy density, this effect is dwarfed by the much larger impact of water and fat on energy density of commonly eaten foods.

9.2.3 Palatability and Taste Preference

Palatability, a subjective measure of the pleasantness of food, has consistently been shown to influence food choice and ultimately dietary intake.[26] Palatability and energy density are inextricably linked;[27] energy-dense foods are generally highly palatable. Highly palatable energy-dense foods, such as those foods high in dietary fat and low in water, are associated with increased caloric intake during single meals and with increased intake at subsequent meals.[20] Conversely, carbohydrate-rich foods, which are naturally low in energy density, such as foods high in fiber, tend to be less palatable. However, as previously noted, foods high in carbohydrates are

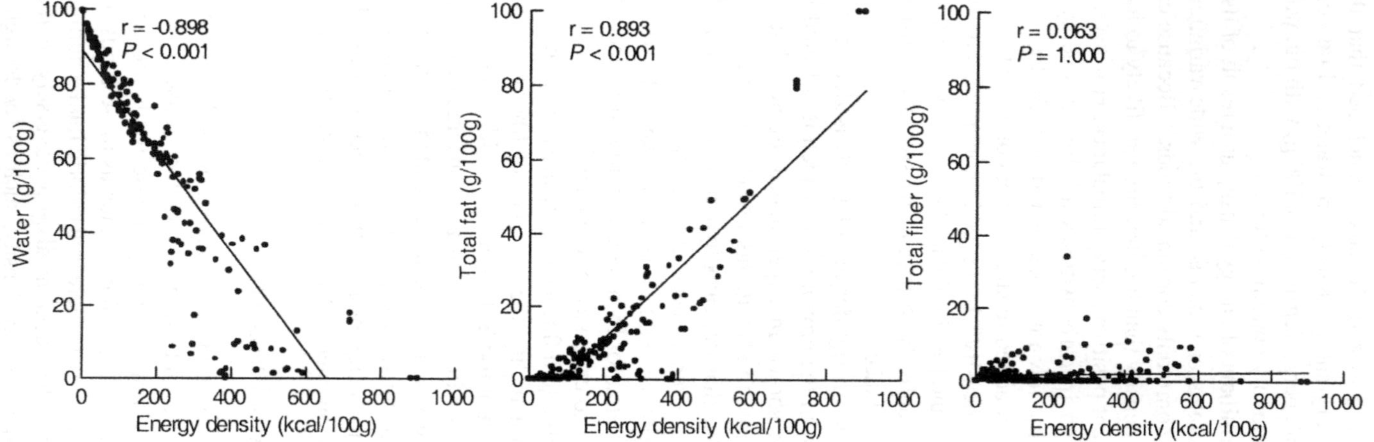

FIGURE 9.1 The association of energy density (kcal/100 g) with fat, water, and fiber contents (g/100 g) of 200 common foods randomly selected from the Fred Hutchinson Cancer Research Center Food Frequency Questionnaire (FHCRC/Block FFQ, version 06.10.88). Nutrient contents of the food were calculated using standard food composition tables (Minnesota Nutrition Data System, software developed by the Nutrition Coordination Center, University of Minnesota, Minneapolis; Food Database version 11A; Nutrient Database version 26, 1996). (From Yao, M. and Roberts, S.B., *Nutr. Rev.*, 59, 247, 2001. Permission for use granted by the International Life Sciences Institute.)

not always low in energy density. There are a number of high-carbohydrate foods that are both energy dense and highly palatable (e.g., desserts such as cakes, cookies, doughnuts, and ice cream). Consuming these foods could lead to passive overconsumption, and ultimately to a positive energy imbalance.

Taste is usually the number one reason given for eating a specific food, and a decrease in good taste is often given as a reason for terminating or reducing food intake.[20] The sense of taste, termed *taste preference*, mediates the relationships among metabolic status, food acceptance, and actual food consumption.[28] Taste preferences appear to have both genetic and acquired components and tend to vary by gender. Women tend to prefer foods such as chocolate, ice cream, doughnuts, cookies, and cake (foods with a high sweet–fat combination), and men tend to prefer salty, meaty foods, such as meatloaf and steak (foods with high protein and fat content).[29] Taste preferences also tend to vary by weight status; lean subjects have stronger taste preferences for sweet foods,[19] whereas obese subjects tend to have stronger preferences for fat-rich foods.[30]

9.2.4 Dietary Fat: Carbohydrate Ratio

There is a high degree of intercorrelation between the percentages of energy derived from dietary fats and carbohydrates in the diet. Observational studies have identified this significant inverse correlation in a number of populations.[31,32] Intervention trials have also illustrated this inverse relationship, whereby reductions in dietary fat intake have been typically accompanied by increases in the percentage of energy derived from carbohydrates, while the percentage of energy derived from protein remains rather stable.[33] The reciprocal relationship between the percentages of energy from dietary fats and carbohydrates has been coined the *fat–sugar seesaw*.[8]

A high dietary fat:carbohydrate ratio has been associated with greater total energy intake and greater body weight.[33,34] The dominant mechanism by which this ratio influences positive energy imbalance is through properties of dietary fat that lead to excess calorie intake and not through properties of dietary carbohydrates. Dietary fat's energy density leads to passive overconsumption,[9] and reductions in dietary fat have been accompanied by decreases in total energy intake.[33] Therefore, it can be concluded that a low-carbohydrate diet may increase the risk of positive energy imbalances or overweight development by way of its correlation with a high dietary fat intake.

9.2.5 Metabolic Fuel Partitioning

Use of carbohydrates and proteins as fuel sources by the body varies in accordance with the amounts consumed in a given meal. In other words, the consumption of carbohydrates promotes their oxidation. Carbohydrate consumption also promotes carbohydrate storage. *Glycogen*, the storage form of carbohydrate in animals, is stored in both the liver and muscle cells. The storage capacity for carbohydrates is limited and, for optimal functioning, is normally maintained within a relatively narrow range (i.e., ~200 to 500 g in adults). In comparison, body fat, the other form of stored energy in the body, is present at highly differing levels among individuals, and the capacity to store fat is relatively unlimited.[35] Following meal consumption, the carbohydrate component of the meal is utilized first for energy, and the dietary fat is taken up predominantly by the adipose tissue and stored until insulin levels fall during the postprandial period. The amount of carbohydrates consumed in a meal determines the extent to which carbohydrates suppress oxidation of dietary fat and promote its storage. *De novo lipogenesis*, the conversion of carbohydrates into fat, occurs when the body's total glycogen stores are considerably raised from their usual 4 to 6 g/kg of body weight to >8 to 10 g/kg of body weight. This requires deliberate and sustained overconsumption of large amounts of carbohydrates for 2 to 3 days.[35] It should be noted that dietary protein can similarly be converted to fat. In summary, increased dietary carbohydrate intake increases the rate of carbohydrate oxidation and storage, and an elevated rate of carbohydrate oxidation suppresses fat oxidation and promotes fat storage.[36]

The use of fat as body fuel is determined primarily by the gap between total energy expenditure and the amount of energy ingested in the form of carbohydrates and proteins.[35] Since the fraction of total dietary energy provided by proteins is relatively small and relatively constant, and because the body spontaneously maintains a nearly constant protein content by adjusting amino acid oxidation to amino acid intake, fat oxidation is regulated primarily by events pertaining to the body's carbohydrate economy.[35,36]

9.2.6 Glycemic Index

The *glycemic index* (GI) is a classification of the blood glucose-raising potential of carbohydrate-containing foods.[37] It is computed by calculating the area under the glycemic response curve during a 2-h period after consumption of 50 g of carbohydrates from a test food; the value is expressed relative to the effect of a standard, which is either glucose or white bread. Many dietary factors, such as starch chemistry, fiber content, fat content, and physical form of food (e.g., liquid vs. solid form), influence the GI of a food.[38] There is concern that the consumption of high-GI foods increases insulin output from the pancreas. Chronically high insulin output leads to a number of deleterious effects on the body, such as high blood triglycerides, increased fat deposition in adipose tissue, increased fat synthesis in the liver, and a more rapid return of hunger after a meal.

Some scientists have hypothesized that the consumption of high-GI foods is associated with a positive energy imbalance and the rising prevalence of overweight and obesity.[39,40] They further suggest that the reduction in dietary fat, as advocated by the federal government and various other official medical and health agencies, has led to a compensatory rise in carbohydrate consumption, and the carbohydrates that tend to replace fat in low-fat diets are typically high in GI.[39] Studies that have examined the relationship between the glycemic index and energy imbalance have been limited by their short-term duration. These studies have investigated the associations between GI and hunger, satiation, and satiety and have had inconsistent results.[40] Futhermore, higher dietary fiber content is often associated with low GI in foods and meals. It is often difficult to assess if the positive metabolic effects are a result of the high dietary fiber content or low-glycemic-index nature of a meal, as these two characteristics are often present together in foods.[41] Long-term clinical trials are necessary to understand the effects of GI on body weight regulation.

9.2.7 Summary of Mechanisms

In conclusion, our current understanding regarding the properties and metabolism of carbohydrates suggests that a higher proportion of carbohydrates in the diet would be protective against a positive energy imbalance, and therefore obesity development, but not all carbohydrates are alike. Different types of carbohydrates have distinct properties and distinct impacts on various factors that impact energy balance. Nonbeverage foods with wet carbohydrates (those with high water content), high fiber content, and low energy density have properties that would tend to protect against obesity development. Foods with more concentrated and palatable forms of carbohydrates, especially in beverage form, would be more "obesigenic."

9.3 SECULAR TRENDS: CHANGES IN CARBOHYDRATE INTAKE DURING THE TIME PERIOD THAT OBESITY RATES HAVE RISEN

9.3.1 Total Carbohydrates

As the prevalence of overweight and obesity has risen in the U.S. since the 1970s, so has the average per capita intake of carbohydrates. To estimate intake, we examine both national food supply and intake data. *National food supply* refers to the amount of food available for human consumption in the U.S., whereas *intake data* refers to the amount actually eaten based on a random sample of self-reported intakes. Both methodologies have their strengths and weaknesses. Food supply data are not subject to recall bias, but do not completely account for waste. Food intake data, on the other hand, are subject to many biases on the part of the individual, but have the advantage that they directly measure the variable of interest, i.e., intake. The increase in carbohydrate intake can be illustrated on the basis of both national food supply (Figure 9.2) and intake data (Figure 9.3). Since the late 1970s, total carbohydrate intake has increased in absolute terms as well as in proportion to total energy intake. According to data from the National Food Consumption Survey (NFCS) and the Continuing Survey of Food Intake by Individuals (CSFII), the percent of calories consumed as carbohydrates has increased among adults from about 40% in the late 1970s to about 50% in the mid-1990s.[42] Over this same time period, carbohydrate intake has increased from 47 to 54% of calories among adolescents.[43]

9.3.2 Sugars and Refined Carbohydrates

Of course, total carbohydrates are not a single entity but are a class of dietary compounds, some of which, on the basis of their chemical and metabolic properties,

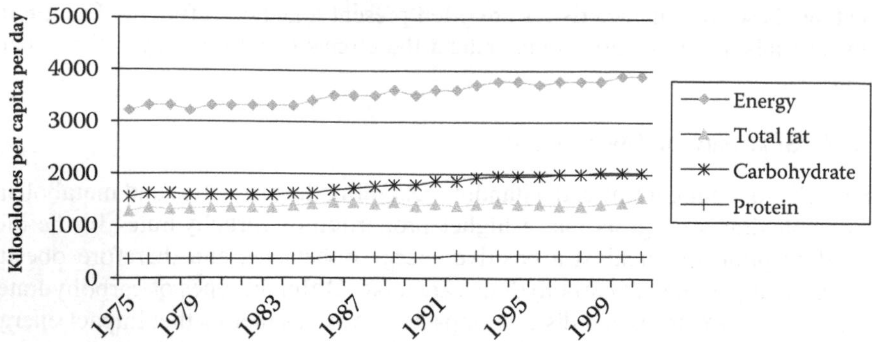

FIGURE 9.2 Energy sources in U.S. food supply. (From Lisa Bente, USDA Center for Nutrition and Policy Promotion, personal communication, 2003.)

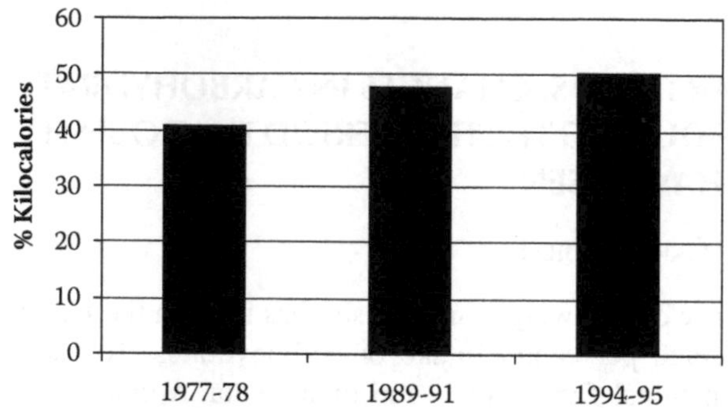

FIGURE 9.3 Trend in carbohydrate intake among adults (NFCS and CSFII). (From Enns, C. et al., *Fam. Econ. Nutr. Rev.*, 10, 27, 1997.)

may have quite different and even opposite effects on energy balance. Sugars, other refined carbohydrates, and fiber are three types of carbohydrates of particular interest. Food supply data suggest that Americans on average have increased their intake of all of these types of carbohydrates. Since the early 1970s, added sugars in the food supply have increased by 20% (Figure 9.4) and intake of foods that are classified as sugars and sweets have increased by 12%.[42] The rise in the intake of soda has been particularly precipitous (Figure 9.5), up from about 25 gallons per capita per year in the mid-1970s to almost 40 gallons per capita in the late 1990s. The intake of soda in the 1990s is therefore equivalent to about 400 calories per day, or 20% of the daily energy requirement for a sedentary adult. According to a 1994–1996 national survey, Americans get about 43% of their added sugars from soft drinks and fruit drinks.[44]

9.3.3 Fiber

Concomitantly, fiber intake has increased by 5 g or 26% (Figure 9.6), and intakes of total flour and cereal products have increased by 48%.[45] Nevertheless, a very

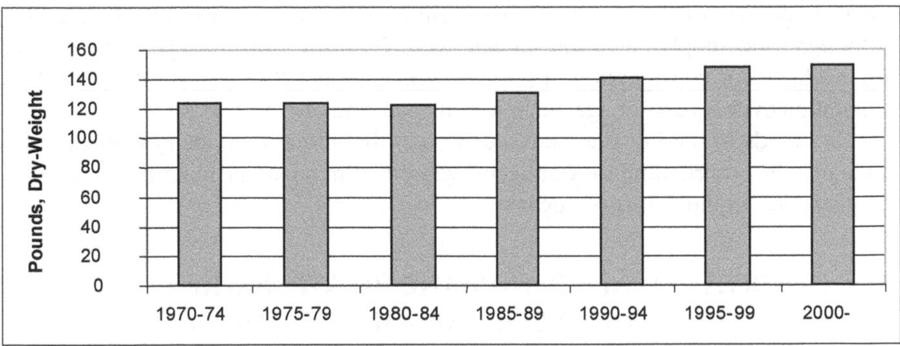

FIGURE 9.4 Per capita annual average of total caloric sweeteners. (From USDA, Nutrient content of the U.S. food supply, available online at http://147.208.9.134.)

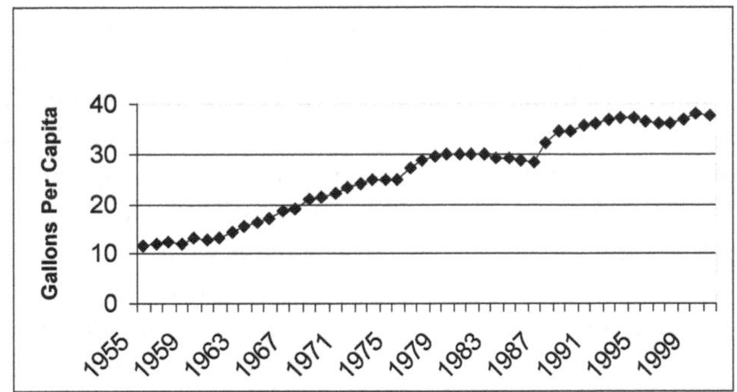

FIGURE 9.5 Per capita annual carbonated soft drink consumption in the U.S. (From USDA/ERS, Food availability spreadsheets, available online at http://www.ers.usda.gov/data/FoodConsumption/FoodAvailSpreadsheets.htm.)

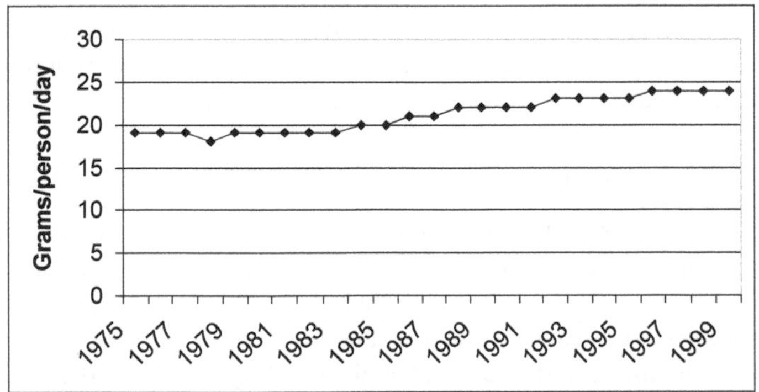

FIGURE 9.6 Fiber in U.S. food supply. (From USDA, Nutrient content of the U.S. food supply, available online at http://147.208.9.134.)

small proportion of the total flour and grain intake is in the form of whole grains. Although food supply data indicate that over 10 servings of grain products per day are available for consumption in the U.S., in the mid-1990s two thirds of the U.S. population over 2 years of age ate less than 1 serving of whole grain foods per day.[45] Secular trends in food intake therefore suggest that more than one type of carbohydrate may be contributing to Americans' weight gain, most particularly refined flour products and liquid forms of sugar.

9.4 OBSERVATIONAL STUDIES: THE RELATIONSHIP BETWEEN THE INTAKE OF CARBOHYDRATES AND OVERWEIGHT

The observational studies described in this section compare the carbohydrate intake and adiposity of free-living individuals in the absence of any intervention designed to influence the subjects' intake or weight. In other words, they are merely observations of the associations between naturally occurring phenomena. Although causation cannot be determined from these studies, they can be used to assess differences in intake by leaner vs. heavier individuals and groups. These studies are particularly useful in describing the type of dietary intake most likely to be protective against the development of obesity, but cannot as appropriately be used to characterize diets that would be most effective for weight loss. Data from nationally representative studies have the advantage of being generalizable to the U.S. population as a whole when they include representative samples of all age, ethnic, and socioeconomic groups. Since age, ethnicity, and socioeconomic status are powerful predictors of adiposity in the U.S., results from one of these groups cannot be generalized to another.

For the purposes of this review, studies are restricted to those that were identified through a key word search of the PubMed database, published between 1992 and March 2003, and conducted among populations in industrialized countries. Studies that examined the relationship between some measure of adiposity and dietary intake of total carbohydrates, fiber, starch, sugars, and complex and simple carbohydrates are included. We begin with a discussion of the longitudinal studies and follow with a discussion of studies that were cross-sectional in design.

9.4.1 LONGITUDINAL STUDIES

In comparison to cross-sectional studies, longitudinal studies are considered more compelling in pointing to causal relationships between dietary intake and adiposity, because they examine change over time. Thirteen longitudinal studies that examined the relationship between carbohydrates and adiposity were identified in the literature published within the last decade (Table 9.1). Of these 13 studies, 2 examined associations in adults,[46,47] 1 followed adolescents into adulthood from age 13 to 28 years,[48] and the other 10 studied children of various ages.[49–58]

Of the 11 studies that reported on the association between total carbohydrates and weight change, 2 found a significant negative association between total carbohydrate intake and adiposity,[51,54] 1 found a significant negative association only with regard to subscapular skinfolds,[53] and the other 8 found no significant association.[46,48,49,52,55–58] In a well-controlled study, Eck et al.[51] followed 187 white children ages 3 to 4 years in Tennessee for 1 year. They found that children at risk for overweight (as defined by parental weight status) gained marginally but significantly more weight than their low-risk counterparts and ate a relatively larger percentage of calories from carbohydrates. Lee[54] followed 5-year-old white girls for 2 years and found that larger absolute intakes of carbohydrates were associated with lower BMI; however, few control variables were included in the analysis. All of the longer-term studies (from 4 to 15 years in length)[48,49,52,55–57] and the two studies conducted among adults[46,49] failed to find an association between carbohydrate intake and adiposity. Robertson et al.[49] took a slightly different approach by examining the association between adiposity take-off (defined as a 1.5 or greater standard deviation increase in skinfolds over a 1-year period) and carbohydrate intake in children ages 3 through 7, but also found no significant association.

The relation of dietary fiber intake to adiposity was reported in two longitudinal studies.[47,50] The CARDIA study[47] followed black and white adults in five U.S. cities for 10 years and showed a significant negative relationship between fiber intake and change in BMI after controlling for a number of possible confounders. The Growing Up Today study[50] followed a large cohort of children ages 9 to 14 for 1 year and showed no significant relationship between fiber intake and change in BMI. Although several potential covariates were controlled for, there was no reported control for socioeconomic status or physical activity, two factors that can impact overweight risk.

Sucrose intake was not found to be associated with weight change in adults in the one study that examined this relationship.[46] In the CARDIA study,[47] however, the glycemic index (defined as low fiber and low fat) was significantly and positively associated with increases in BMI among whites.

Given that none of the adult studies found a significant association between total carbohydrates and adiposity, it appears that carbohydrate intake may have a limited influence on adiposity in adults. However, due to the limited number of longitudinal studies conducted among adults, it is not possible to draw firm conclusions. Likewise, among children it appears that the percent of calorie intake from total carbohydrates has a limited influence on weight change in free-living populations; less than half of the studies found a significant relationship between these two variables. However, these studies suffered from several methodological limitations, including relatively small sample sizes and a limited number of control variables, as well as being conducted among primarily white populations. Furthermore, dietary intake data for children are notoriously inaccurate due to the difficulty children have in recalling or recording their intake and the incomplete knowledge parents have of their children's intake. Unfortunately, there have also been too few longitudinal studies conducted to draw conclusions about the impact of sugar or fiber intake on weight changes, although fiber appears to be protective against weight gain. Further study is merited.

TABLE 9.1
Summary of Longitudinal Studies That Examined the Relationship between Carbohydrate Intake and Adiposity (in Descending Order by Sample Size, Adults First)

Study Characteristics			Association with Adiposity[a]		
Study	Time Frame	Subjects	Total CHO[b]	Fiber	Sugars
Ludwig (1999),[47] CARDIA, 5 U.S. cities	10 years	2909 adults, black and white, ages 18–30		–	
Parker (1997),[46] Pawtucket Heart Health Study, New England	4 years, 1986–1987 and 1991–92	176 men, 289 women	0		0 Sucrose[c]
Kemper (1999),[48] Amsterdam Growth and Health Longitudinal Study	15 years	83 males, 98 females, age 13, nonobese at start	0		
Berkey (2000),[50] Growing Up Today, U.S.	1 year	6149 girls, 4620 boys, ages 9–14		0	
Magarey (2001),[53] Adelaide Nutrition Study, Australia	13 years	243 children, age 2 at start	0 BMI/TC skinfolds – SS skinfolds		
Alexy (1999),[58] DONALD Study, Germany	2 years	105 boys, 100 girls, age 3 at start	0		
Lee (2001),[54] Pennsylvania	2 years	192 white girls, age 5 at start, and their mothers	– (g)		
Eck (1992),[51] Tennessee	1 year	187 white children	–		
Scaglioni (2000),[52] Italy	5 years, began 1991	67 girls, 80 boys, birth to 5 years	0		
Maffeis (1998),[56] Italy	4 years	112 white children, mean age = 8.7 at start	0		
Rolland-Cachera (1995),[55] France	4 years	112 children, age 10 months at start	0		
Carruth (2001),[57] Tennessee	6 years	53 white children, age 2 at start	0 Mean longitudinal intakes of total CHO (g)		
Robertson (1999),[49] SCAN Study, Texas	Case-control, 4 years	15 subjects and 33 controls, ages 3–7, mixed race	0[d]		

Note: CHO = carbohydrate; BMI = body mass index; TC = triceps; SS = subscapular.

[a]Association with adiposity: – indicates an inverse and significant association with adiposity; + indicates a direct and significant association with adiposity; 0 indicates the association was not significant; blank space indicates no association was reported. Findings were the same for all age, gender, and ethnic groups unless otherwise indicated.
[b]All studies defined carbohydrate intake in terms of percent of total calories unless otherwise indicated by g (grams).
[c]Definition of independent variable is indicated if different from that indicated in the column heading.
[d]Outcome variable was adiposity take-off (increase of 1.5 standard deviations in skinfolds in 1 year).

9.4.2 Cross-Sectional Studies

Cross-sectional studies that examined the relationship between carbohydrate intake and adiposity are both much more numerous and consistent in their results than longitudinal studies. We identified 47 cross-sectional studies published since 1991 (Table 9.2) that evaluated the relationship between adiposity and total carbohydrate intake (38 studies), fiber intake (13) studies), or sugars and simple vs. complex carbohydrates (15 studies).

Significant negative associations were reported between total carbohydrate intake and BMI or some other measure of adiposity for 71% of 17 studies conducted among adults.[23,31,47,59–68] Of the remaining five studies, two reported no significant association,[19,69] two reported mixed results,[31,61] and only one study reported a positive association with BMI.[70] Of the 21 studies to examine the relationship between carbohydrate intake and adiposity in children, 7 found a significant negative association[54,71–76] and 2 found a significant negative association among boys, but no significant association among girls.[77,78] The remainder of the studies found no significant association between these two variables.[52,79–89] Two of the studies[60,64] that reported significant negative associations between adult BMI and total carbohydrate intake used nationally representative data sets (National Health and Nutrition Examination Survey (NHANES) and CSFII). Twenty-nine of the studies examined populations outside the U.S. (in Europe, Japan, and Canada). All seven of the studies conducted among adults in the U.S.[31,47,60,61,64,66,67] found a significant negative association between total carbohydrate intake and adiposity, although one study[31] found no association when the dependent variable was continuous (as opposed to categorical). Among children, four of the seven studies conducted in the U.S.[54,74–76] found significant negative associations between BMI and carbohydrate intake. The 39 cross-sectional studies varied widely in the extent to which they controlled for potentially confounding variables, yet even after controlling for an array of potential confounders, significant and negative associations between total carbohydrate intake and adiposity in adults were still observed in several studies. Taken as a whole, these results establish that intakes of a relatively larger percent of calories from carbohydrates are consistently associated with lower BMI, at least among adults. However, Ludwig et al.[47] reported that the association between total carbohydrate intake and adiposity was weak and substantially attenuated when adjusted for fiber, suggesting that fiber may explain most of the association among the cohort studied.

Fewer studies have examined different types of carbohydrates and their association with some measure of adiposity. We identified 13 cross-sectional studies that reported on the association between dietary fiber and adiposity, of which 8 reported a significant negative association[47,65–67,72,74,84,90,91] and 1 reported a significant negative association among females but not males.[77] Only one study (conducted in Sweden) reported a significant positive association.[70] These studies were conducted in Europe, the U.S., Canada, and Japan. Six were conducted among children, over half of which found a significant negative association.[72,74,77,84] The studies among children, however, were less consistent in their associations and the study designs tended to be weaker. Outcome measures were not the same for all studies; they included both measured and self-reported BMI, as well as height for weight based on underwater

TABLE 9.2
Summary of Cross-Sectional Studies That Examined the Relationship between Carbohydrate Intake and Adiposity (in Descending Order by Sample Size, Adults First)

Study Characteristics		Association with Adiposity[a]		
Study	**Subjects**	**Total CHO[b]**	**Fiber**	**Sugars**
Gonzalez (2000),[59] EPIC Study, Spain	14,374 men, 23,289 women, ages 29–69, diverse SES	–		
Lewis (1992),[15] Nationwide Food Consumption Survey (NFCS), 1977–78	30,770 adults, children over 4, nationally representative			0 Added sugars
Trichopulou (2002),[69] EPIC study, Greece	27,862 adults, ages 25–82, healthy volunteers	0		
Kromhout (2001),[90] Seven Countries Study, Europe, Japan, and U.S., 1958 and 1964	12,763 men, middle aged, diverse SES		–	
Bolton-Smith (1994),[63] Scottish Health Heart Study, 1984–1976	5768 men, 5858 women, ages 25–65	– (p value not reported)		– Total sugar – Extrinsic sugar 0 Intrinsic sugar, lactose, starch
Stam-Moraga (1999),[62] Belgian Interuniversity Research on Nutrition and Health, 1979–1984	5837 men, 5243 women, ages 25–74, nationally representative	–		– Total sugar,[d] strongest association
Kennedy (2001),[64] Continuous Survey of Intake of Individuals (CSFII), 1994–96	9786 adults, nationally representative	–		
Yang (2003),[60] NHANES III	3754 men, 4074 women, ages 25–64, nationally representative	–		
Appleby (1998),[91] Oxford Vegetarian Study, 1980–84	1914 men, 3378 women, nonsmokers, about half nonmeat eaters		–	
Slattery (1992),[61] CARDIA, 5 U.S. cities	5115 adults, white and black, ages 18–30, diverse SES study baseline	– Men		

Ludwig (1999),[47] CARDIA, 5 U.S. cities	2902 adults, ages 18–30, black and white, study endpoint	–	–	
Gibson (1996),[92] Dietary and Nutritional Survey of British Adults, 1986–87	1087 men, 1110 women, ages 16–64			– (weak) High-sugar/high-fat – High-sugar/low-fat consuming men[c] – Sugar among high-fat consumers
Macdiarmid (1998),[29] 1986–87, Britain	1239–1853 healthy, nondieting adults, ages 16–64			–/0 Total sugar males/females –/+ High-fat/sweet males/females
Ruidavets (2002),[68] France	330 men, ages 45–64	– CHO and polysaccharides		0 Oligosaccharides
Wamala (1997),[70] Stockholm Female Coronary Risk Study	300 healthy women, ages 30–65	+	+	+ Sucrose
Tucker (1992),[31] Western U.S.	205 adult females, mostly white, mean age = 34.6	– (categorical) 0 (continuous)		
Nelson (1996),[66] Utah	203 men, mostly white, mean age = 41	–	–	0 Simple CHO – Complex CHO
Alfieri (1995),[65] Ontario	150 adults, mostly women, ages 18–65	–	– R2 = 17%	
Miller (1994), [67] Indiana	46 men, 32 women	–	–	0 Total sugars + Added sugars

TABLE 9.2 (continued)
Summary of Cross-Sectional Studies That Examined the Relationship between Carbohydrate Intake and Adiposity (in Descending Order by Sample Size, Adults First)

Study Characteristics		Association with Adiposity[a]		
Study	**Subjects**	**Total CHO[b]**	**Fiber**	**Sugars**
Westerterp (1996),[23] The Netherlands	34 obese women, 34 nonobese women, ages 20–50, case-control	–		
Cox (1999),[19] England	41 lean and 35 obese nondieting healthy adults	0		
Lluch (2000),[142] Stanislas Family Study, France	1320 members of 387 families, ages 11–65	–		
Gibson (1993), as reported by Hill (1995),[14] Britain	897 boys, 747 girls, ages 10–11 and 14–15			– Extrinsic sugar (both age groups for boys; 1 group for girls)
Lewis (1992),[15] USDA NFCS	4682 children, ages 4–10			– Added sugars (g/kg of body weight)
Davies (1997),[79] NDNS, U.K.	1444 British children, ages 1.5–4.5	0		
Bao (1996),[83] Bogalusa Heart Study, Louisiana	1419 children, age 10, 35% black, 65% white	0		
Rodriguez-Artalejo (2002),[87] Spain	557 boys, 555 girls, ages 6–7	0	0	
Guillaume (1998),[73] Belgium	955 children, ages 6–12	–		
Maffeis (2000),[78] Italy	278 boys, 252 girls, ages 7–11	– Males 0 Females		
Stewart (1999),[80] FRESH, Maryland	468 children, grades 2–5 (mean age = 8.9)	0		

Garaulet (2000),[77] Spain	192 girls, 139 boys, ages 14–18, Spanish	– Males 0 Females	0 Males – Females	
Tucker (1997),[74] Utah	253 children, ages 9–10	–	– Association found only when no variables controlled for	
Hanley (2000),[84] Canada	242 children, ages 2–19, native Canadians	0	– Fiber	0 Simple sugars 0 Starch
Lee (2001),[54] Pennsylvania	192 white girls, age 5 (cross-sectional results from longitudinal study)	– (g)		
Gills (2002),[93] Canada	181 children, ages 4–16			+ Total sugar (g)
Dennison (1997),[89] New York	168 predominantly white children, 94 age 2, 74 age 5	0 (g)		
Scaglioni (2000),[52] Italy	80 white boys, 67 white girls, age 5 (cross-sectional results from longitudinal study)	0		
McGloin (2002),[94] Northern Ireland	66 boys, 48 girls, ages 6–8, predominantly white			0 total sugar
Maffeis (1998),[56] Italy	112 white children, mean age = 8.7, Italian (cross-sectional results from longitudinal study)	0		
Maffeis (1996),[81] Italy	82 white children, 30 obese, 52 nonobese	0		
Atkin (2000),[86] Great Britain	39 boys, 38 girls, ages 1.5–4.5	0		
Ortega (1995),[71] Spain	37 boys, 27 girls, ages 15–17, Spanish	–	0	
Gazzaniga (1993),[76] Iowa	25 girls, 23 boys, ages 9–11	–		
Bandini (1999),[75] Massachusetts	11 obese girls, 10 obese boys, 12 nonobese girls, 10 nonobese boys, ages 12–18 Case-control	–		
Koivisto (1994),[82] Sweden	15 overweight children, 24 normal-weight children, ages 3–7 Case-control	0 (g)		0 Sucrose

TABLE 9.2 (continued)
Summary of Cross-Sectional Studies That Examined the Relationship between Carbohydrate Intake and Adiposity (in Descending Order by Sample Size, Adults First)

Study Characteristics		Association with Adiposity[a]		
Study	**Subjects**	**Total CHO[b]**	**Fiber**	**Sugars**
Rocandio (2001),[72] Spain	16 boys, 16 girls, age 11	–	– Mean fiber intake (g)	
Francis (1999),[88] Texas	12 pairs white nonobese children (12 with obese mother, 12 with nonobese mother)	0		

Note: CHO = carbohydrate; SES = socioeconomic status.

[a]Association with adiposity: – indicates an inverse and significant association with adiposity; + indicates a direct and significant association with adiposity; 0 indicates the association was not significant; blank space indicates no association was reported. Findings were the same for all age, gender, and ethnic groups unless otherwise indicated.
[b]All studies defined carbohydrate intake in terms of percent of total calories unless otherwise indicated by g (grams).
[c]Characteristics of subgroup to which the results apply are indicated when they do not apply to all groups studied.
[d]Definition of independent variable is indicated if different from that indicated in the column heading.

weighing and on skinfold measurements. Several studies controlled for numerous potential confounders. No study included a nationally representative sample from the U.S., but the CARDIA study[47] did include blacks and whites from five U.S. cities. One study in Canada[65] found that fiber accounted for 17% of the variance in BMI, and another[47] found that dietary fiber accounted for a larger proportion of the variation in BMI than total carbohydrates, fats, or proteins. These studies provide strong evidence that a higher fiber intake is consistently associated with lower adiposity.

Fourteen cross-sectional studies were identified that examined the relationship between sugar or simple carbohydrates and adiposity.[14,15,29,62,63,66–68,70,82,84,92–94] Of those that looked at total sugar or simple carbohydrates, two found significant negative associations with adiposity,[62,63] one found a negative association for males but none for females,[29] and five found no significant association.[66–68,84,94] One found a positive association, but only when intake was measured as an absolute value.[93] Of the seven that looked at added sugars or sucrose, three found a negative association,[14,15,63] two found a positive association,[67,70] and two found no significant association with adiposity.[15,82] Three studies analyzed the relationship between starch[63,84] or complex carbohydrates[66] and adiposity. The former two found no significant association, whereas the latter found a negative association. Only one study looked at intrinsic sugar (sugar naturally present in foods as opposed to being added during processing or preparation) and found no significant relationship with adiposity.[63] All of the studies were conducted in the U.S., Canada, or Europe. The only study of children in the U.S.[15] examined a nationally representative data set (NFCS) and reported a negative association between added sugars and BMI, as determined by self-reported height and weight. Using the same data set, but including adults and children, Lewis[15] found no association between added sugars and adiposity. The only study of native Canadian children found no association between total sugar intake and adiposity.

An additional study looked at intake of high-sweet and -fat foods in Britain and found that those with a high-sugar, low-fat intake had the lowest BMI of all groups.[92] Because sugar was also negatively associated with adiposity among high-fat consumers, Gibson[92] concludes that sugar and fat have antagonistic rather than synergistic influences on adiposity, with sugar having a protective effect. Macdiarmid et al.[29] found that high-fat/high-sweet foods were negatively associated with adiposity in males and positively associated with adiposity in females. These two studies suggest that the relationship between sugar intake and adiposity is a complex one, is not independent of the relative intake of other nutrients, and may have distinct impacts on adiposity among males and females.

When interpreting these results, it must be kept in mind that underreporting of intake can be particularly problematic with foods that are high in sugar (or fat), given that these foods are considered less desirable from a nutritional standpoint.[32] Macdiarmid et al.,[29] for example, found that an observed negative association between the intake of high-fat and -sweet products was reversed (and became significant) when low-energy reporters were excluded from the analysis. Overweight individuals may be the most likely to underreport these "forbidden" foods.[95] This may explain in part the conflicting results when examining the association between sugar intake and adiposity in studies using approaches that rely on self-reporting.

In summary, findings from cross-sectional studies suggest that both greater intakes of total sugar/simple carbohydrates (three negative associations and six nonsignificant results,) and complex carbohydrates (one negative association, two nonsignificant) may be protective and no conclusion can be made regarding added sugars (three negative, two positive, and two nonsignificant associations). But, given the inconsistent definitions of *sugar*, wide variations in the number of potential confounders controlled for, and lack of data from a more recent nationally representative sample, as well as the mixed results, it is not possible to draw a firm conclusion regarding the relationship between intake of simple carbohydrates and obesity.

9.4.3 Summary of Observational Studies (Longitudinal and Cross-Sectional)

When taken as a whole, the evidence from observational studies is fairly compelling that intakes characterized by a higher percentage of calories from carbohydrates are protective against the development of overweight. The cross-sectional data clearly establish that higher intakes of total carbohydrates (percent of total calories) are associated with lower adiposity in all age and ethnic groups studied. The lack of significant findings among longitudinal studies, however, suggests that the protective effect of total carbohydrate intake may be fairly limited. Fiber appears to account for some, but not all, of the protective effects of total carbohydrate intake, as fiber intake is quite consistently associated with lower adiposity. Given the limited number of studies relating fiber intake to adiposity, this is a promising area for further study. Despite commonly held concerns regarding sugar intake, the evidence relating sugar or simple carbohydrate intake to obesity is conflicting and therefore inconclusive. More studies, especially of a longitudinal nature, that examine the impact of different types of carbohydrates on obesity are clearly merited.

9.5 TREATMENT STUDIES: THE ROLE OF DIETARY CARBOHYDRATES IN WEIGHT LOSS AND PREVENTION OF WEIGHT REGAIN

Weight loss and subsequent weight maintenance may be, at least in part, metabolically different from prevention of weight gain. Consequently, weight loss and prevention of weight regain may be more difficult to achieve than prevention of weight gain in the first place, and the influence of diet on these three metabolic states may differ. For this reason, the influence of dietary carbohydrates on weight loss will be reviewed separately from the influence of dietary carbohydrates on the prevention of weight gain. We will begin with a discussion of what amount (high or low) and what type of carbohydrates are optimal for weight loss and maintenance, followed by a brief overview of health outcomes associated with low-calorie, high- and low-carbohydrate diets. Modifying carbohydrate intake to prevent weight gain and the development of obesity will then be addressed.

Interpreting the research in this area is complicated by the fact that when carbohydrate intake changes, a concurrent change in fat intake typically occurs. As discussed previously, a low-carbohydrate diet is usually associated with a higher fat intake and vice versa,[8] and when subjects reduce their carbohydrate intake, they typically increase their intake of fat.[96] Protein intake tends to remain relatively constant both within an individual and between individuals, whereas intakes of the other two macronutrients may change. Although the debate about the optimal ratio of macronutrients has focused on carbohydrates vs. fats, there is evidence to suggest that a high-protein diet may be beneficial for weight loss and maintenance,[97,98] possibly by promoting increased postprandial thermogenesis[99] and satiety.[7,12] Further complicating interpretation of clinical studies is the fact that carbohydrate consumption varies considerably, most certainly in a laboratory setting, but also among and between individuals in a free-living population. Further, carbohydrates are heterogeneous as a group, and not all types may be equally advantageous (or disadvantageous) for weight loss and prevention of weight gain.

At any given time it is estimated that approximately one third of adults in the U.S. are attempting to lose weight, the majority by altering diet composition.[100] Paradoxically, both low- and high-carbohydrate diets have been recommended and adopted for weight loss. The conventional approach, advocated by the majority of medical groups and health professionals, involves consuming a hypocaloric diet with a high carbohydrate-to-fat ratio.[101,102] However, a relatively low-carbohydrate, high-fat, and high-protein hypocaloric diet has also been advocated in the popular press. The Atkins diet in particular, which initially gained popularity in the 1970s, has enjoyed resurgence in interest in recent years.[103]

9.5.1 Low- vs. High-Carbohydrate Diets for Weight Loss

Total energy, rather than carbohydrate content, has been shown to be the most important determinant of weight loss. Kennedy et al.[64] concluded that weight loss is relatively independent of macronutrient composition, but is dependent instead on a reduction in total energy intake. This conclusion followed a systematic review of over 200 weight-loss studies of various designs published between 1956 and 2000. In this review, weight loss following consumption of a relatively low carbohydrate intake (ranging from 0 to 85 g/day) was assessed in 20 studies. In the majority of these 20 studies, weight changes were assessed over a relatively short term of approximately 6 weeks in duration. Within this period, weight loss ranging between 2.8 and 12.0 kg occurred in all studies. A similar number (22) of studies of diets relatively high in carbohydrates (6 of which were very low fat, <10% kcal, and 16 moderate in fat, 20 to 30% kcal) caused weight loss to the same extent as reported for the studies of low-carbohydrate diets.

Others have similarly concluded that weight loss in obese subjects is not related to dietary carbohydrate content.[104] Bravata et al.[104] recently reviewed 107 English-language studies published between 1966 and 2003 that manipulated dietary carbohydrate in a total of 3268 adult subjects in an outpatient setting. When studies of heterogeneous design (e.g., randomized and controlled as well as nonrandomized and noncontrolled trials) were considered altogether (as well as when segregated by energy content of the diet, diet duration, and baseline weight of subjects), weight loss was greater with lower-carbohydrate (60 g/day carbohydrate, mean weight loss of 16.9 kg weight) than with higher-carbohydrate diets (>60 g/day carbohydrate, mean weight loss of 1.9 kg). However, when studies of similar design were compared, no difference in weight loss was observed. For example, considering only randomized controlled and randomized crossover trials, mean weight loss did not significantly differ between the lower-carbohydrate diets (3.6 kg (95% confidence interval (CI), 1.2 to 6.0 kg)) and the higher-carbohydrate diets (2.1 kg (95% CI, 1.6 to 2.7 kg)). Rather, the degree of weight loss varied as a function of energy intake and diet duration: the lowest energy intakes for the longest periods of time resulted in the greatest weight loss. Although studies of duration less than 4 days were excluded in this analysis, the majority of studies had a relatively short duration, averaging 50 and 73 days for the lower- and higher-carbohydrate diets, respectively. It must also be noted that a minority of studies involved the consumption of few carbohydrates (<20 g/day), typical of many of the popular diet regimens, such as Atkins. Similarly, very high carbohydrate/very low fat diets were not included.

More recently, results have been published from several randomized controlled clinical trials of subjects on *ad libitum* diets involving a relatively longer duration of intervention (6 to 12 months). These results lend support to the hypothesis that a low-carbohydrate diet is superior to a high-carbohydrate diet for weight loss.[105–107] For example, in one 6-month study, subjects on the *ad libitum* low-carbohydrate Atkins diet (<20 g/day) lost significantly more weight and body fat (8.5 and 4.8 kg, respectively) than subjects following a conventional high-carbohydrate (approximately 54% of calories) and low-fat (approximately 28% of calories) hypocaloric diet (3.9 and 2.0 kg, respectively), a difference that remained significant even when an intention-to-treat analysis including all randomized subjects (i.e., dropouts) was performed.[105] In another trial, unique in that an attempt was made to mimic the approach used by most dieters (i.e., a copy of Dr. Atkins' book was provided to subjects who otherwise received minimal diet counseling or support), obese adults on the low-carbohydrate diet lost significantly more weight at 6 months (7.0 vs. 3.2% change) than adults on the high-carbohydrate, low-calorie diet. However, by 12 months this gap in weight loss had narrowed (4.4 vs. 2.5% change) and was no longer statistically significant.[106]

A variety of advantages have been proposed, but not proven, for a low-carbohydrate diet for weight loss. Interestingly, subjects on the low-carbohydrate diets in these three studies were not given instructions on calorie restriction, but reduced their calorie intake nonetheless to approximately the same degree as subjects instructed on the hypocaloric, high-carbohydrate diets. Proponents of popular low-carbohydrate diets frequently argue that low-carbohydrate (and high-fat and -protein) diets are more palatable, more satiating, and more acceptable to consumers than the conventional counterpart. The purported, but yet to be proven, mechanistic advantage of this low-carbohydrate (<10% of daily calorie intake at inception) diet for weight loss is its ability to promote lipid oxidation, satiety, and energy expenditure, attributed at least in part to the induction of ketosis. However, data from clinical trials suggest that ketosis is not a likely explanation for weight loss when consuming a low-carbohydrate diet. For example, ketosis does not appear to persist after the first few months on the diet, and presence of urinary ketones has not been correlated with degree of weight loss.[105,106] It has also been hypothesized that the greater limitation in food choices, simplicity, or novelty of a very low carbohydrate diet may be responsible for a greater reduction in calorie intake than in the more conventional higher-carbohydrate diet, which allows for intake of a wider variety of foods.[106,107] Similar limitations in food choices may explain the effectiveness of other less conventional diets, including ones that are very low in fat and very high in carbohydrates.[108] Finally, there was a trend in all studies toward lower attrition among subjects consuming a low-carbohydrate diet in comparison to a higher-carbohydrate one. For example, in one study the attrition rate over the course of the 6-month study was 26% in the high-carbohydrate group and 15% in the low-carbohydrate group.[105]

In summary, it is premature to endorse a low-carbohydrate diet over a conventional low-fat, high-carbohydrate diet for weight loss. Low-calorie intake and negative energy balance can be achieved on a variety of carbohydrate distributions. Regardless of diet composition, diets that are intended to reduce body weight generally achieve only small weight loss, and high attrition is the rule rather than the exception. Further, there are established benefits of dietary fat restriction for reduction of serum cholesterol and risk of cardiovascular disease. Finally, the short-term nature of the majority of studies precludes assessment of risk vs. benefit over extended periods.

9.5.2 Low- vs. High-Carbohydrate Diets for Prevention of Weight Regain

For successful weight loss to be maintained, it is critical to determine whether dietary composition is a critical element and, if so, which dietary composition is most likely to prevent excess energy intake over a prolonged period in a free-living population in an *ad libitum* setting. Unfortunately, most weight-loss studies have been of short duration and have not included long-term follow-up. Data from the National Weight Registry indicate that 90% of long-term weight maintainers consume a low-fat, high-carbohydrate, low-calorie diet and regularly participate in physical activity.[109] This registry is the largest U.S. database available of adults who have lost weight (averaging 30 kg) and maintained a minimum of 13.6 kg weight loss over a period of 5 years. Whether observations of individuals in this registry represent the experience of the general American population of weight maintainers is not known. In no controlled studies has the influence of low- vs. high-carbohydrate diets on maintenance of weight loss been evaluated. Before a low- or high-carbohydrate diet can be advocated, additional studies are warranted in which isocaloric diets of varying carbohydrate contents are provided for extended periods, and during which hunger ratings, diet acceptability, and health-related outcomes in addition to adiposity are monitored.

9.5.3 Type of Carbohydrate for Weight Loss

Based on a review of weight-loss studies, Howarth et al.[17] reported that adults on relatively higher fiber diets (fixed or *ad libitum*) consistently lost more weight than those consuming lower-fiber diets. This effect was observed even when energy intake was held constant. The amount of weight loss in general was modest (averaging 1.9 kg over 4 months in *ad libitum* studies and 1.3 kg over 3 months in studies in which fiber intake was fixed), but similar in magnitude to that observed when conventional low-fat diets are consumed *ad libitum.* Levine and Billington[110] came to a similar conclusion after finding that 26 of the 38 studies identified reported a decrease in body weight following a high-fiber regimen. There is also evidence that fiber may play a role in the prevention of weight regain following weight loss,[111,112] although this has been investigated to a limited extent. No differences were observed between types of fibers (soluble vs. insoluble vs. mixed or fiber supplements vs. high-fiber foods) in the meta-analysis by Howarth et al.[17] Similarly, Saltzman and Roberts[16] have argued that results are mixed with respect to the impact of soluble fiber on weight loss, with the numbers of studies reporting no effect equal to the numbers of studies reporting a positive impact on weight loss.

There is evidence that reducing added sugar intake may also be a beneficial weight-loss strategy. Vermunt et al.[113] recently performed a systematic review of weight reduction trials that focused on replacing added sugar with either low-energy artificial sweeteners (e.g., aspartame) or complex carbohydrates. The cumulative results suggested that removing added sugar from the diet was beneficial in reducing energy intake and body weight.[113] The authors have cautioned, however, that the studies addressing added sugars and weight loss were limited in number (<10) and relatively short term (<1 year) and that the results could not be extended to weight maintenance or the prevention of weight regain.

The glycemic response induced by carbohydrate-rich foods may also play a role in weight loss,[114] though the few clinical trials that have been performed to date have yielded conflicting results.[115,116] In one randomized controlled 6-month study, overweight or obese subjects consumed low-fat *ad libitum* diets that were high in either complex carbohydrates (one fourth of dietary fat replaced by complex carbohydrates) or simple sugars (one fourth of dietary fat replaced by simple sugars). Both groups lost significantly greater amounts of weight (1.8 and 0.9 kg, respectively) than the higher-fat (30 to 40% of energy), lower-carbohydrate control group, even though none of the subjects were encouraged to reduce their energy intake. Dietary fiber intake did not differ significantly between groups, a fact that was not entirely unexpected given that prepackaged foods (rather than fruits and vegetables) were provided to study participants.[117] Interestingly, in a subset of 46 subjects with symptoms of the metabolic syndrome at baseline, significantly greater weight loss was achieved on the diet high in complex carbohydrates (–4.25 kg) than on both the diet high in simple sugars (–0.29 kg) and the control diet (+1.03 kg).[118] Complex and simple carbohydrates may have differential effects depending on the individual's metabolic predisposition.

9.5.4 Other Health Considerations of a High-Carbohydrate Diet

In addition to weight outcomes, other health and nutrition indicators associated with dietary macronutrient composition must be considered. The merits of high- and low-carbohydrate intakes with respect to cardiovascular disease, diabetes, cancer, and gastrointestinal health are discussed in detail in other chapters. What follows is a brief description of the effects of high- vs. low-carbohydrate diets on the general health of persons undertaking weight loss.

Low-fat, high-carbohydrate diets, particularly those low in saturated and trans fats, have been shown to produce biologically meaningful reductions in blood cholesterol levels, without adverse effects on blood glucose.[64] For example, on the basis of a meta-analysis of 37 intervention studies, low-fat, high-carbohydrate diets were shown to reduce blood concentrations of total, low-density lipoprotein (LDL), and high-density lipoprotein (HDL) cholesterol as well as triacylglycerol.[64,119] However, it is not clear the extent to which some of the beneficial effects were due to weight loss, because weight loss, regardless of how achieved, is frequently accompanied by improved glycemic control, blood pressure, and blood lipid concentrations. In controlled feeding studies in which body weight was maintained, low-fat diets have been associated with decreases in HDL cholesterol *and* increases in triacylglycerol, both of which are risk factors for cardiovascular disease.[120] Unfortunately, the majority of weight-loss studies performed specifically to test the effects of different carbohydrate intakes have focused on weight and have not included other measures of health.[64] In the recent systematic review of carbohydrate intervention studies that examined health measures, the majority of which included only overweight or obese adult subjects, alterations in carbohydrate intake were not significantly associated with any adverse changes in serum lipid or glucose concentrations, or systolic blood pressure.[104] However, the authors caution that because of limited sample size, larger and longer-term studies are warranted.

9.5.5 Other Health Considerations of a Low-Carbohydrate Diet

A concern voiced by health professionals is that low-carbohydrate diets can result in excess consumption of total fat, saturated fat, and cholesterol with adverse effects on blood lipids. According to several systematic reviews, however, the most consistent metabolic effect of a short-term low-carbohydrate diet that produces weight loss is a reduction or no change in serum triglycerides.[64,104] Effects on blood glucose and insulin concentrations and blood pressure were less consistent, while the effects on lipoproteins were the most mixed.[64,104] However, it has been noted that studies of low-carbohydrate weight-loss diets have been relatively short term (i.e., the majority shorter than 90 days) and have involved adults who are relatively young (i.e., younger than 60 years old) and, in most instances, healthy (i.e., free from hyperlipidemia, hypertension, or diabetes). The effects of low-carbohydrate diets on the health of children and individuals with hyperlipidemia, diabetes, or other metabolic disorders merit additional study. Assessments of additional health outcomes, such as renal function, bone health, cardiovascular function, and quality of life, are needed. Further, the lowest-carbohydrate diets (e.g., 20 g/day) have been investigated in very few studies, few of which included measures of blood pressure or blood metabolites. It has also been suggested that low-carbohydrate diets may result in ketosis, high urinary nitrogen loads, impaired liver and renal function, and, in children, myocardial dysfunction.[121] Again, these potential consequences have not been adequately studied. It remains possible that, in the short term at least, the effect of weight loss on blood lipids may override effects due to the macronutrient composition of the diet.[105] If so, very low carbohydrate diets may only have a positive impact on blood lipids during the weight-loss phase. Decades of research suggest that prolonged high-fat intake, especially of saturated fat, increases the risk of heart disease.[122,123]

Finally low-carbohydrate diets have been criticized for their lack of adequate dietary fiber and antioxidant vitamins.[105] In an analysis of U.S. national survey data (1994 to 1996 CSFII), diet quality, as assessed by the USDA Healthy Eating Index (a composite rating based on the U.S. Food Guide Pyramid and Dietary Guidelines), was higher for high- than for low-carbohydrate diets.[64] The high-carbohydrate group ate more low-fat foods, grain products, fruits, and fruit juices and had higher intakes per 1000 kcal of vitamin A, vitamin C, folate, calcium, magnesium, and iron, and lower intakes of sodium and alcoholic beverages.[96] On the other hand, adults consuming lower-carbohydrate intakes had higher micronutrient densities of vitamin B_{12} and zinc and lower contents of sweets and sugars in their diets.[96]

9.6 PREVENTION STUDIES: THE ROLE OF DIETARY CARBOHYDRATES IN PREVENTING WEIGHT GAIN

9.6.1 Low- vs. High-Carbohydrate Diet for Prevention of Weight Gain

It has been estimated that about one third of adults in the U.S. are attempting to maintain their weight and avoid weight gain.[100] Intervention studies to prevent weight gain in nonoverweight individuals are, by their nature, more difficult to design and conduct than treatment studies of individuals who are already overweight[124]. Therefore, it is not surprising that few such studies have been undertaken. A systematic review of the literature between 1992 and March 2003 revealed 16 prevention trials that targeted or measured changes in carbohydrate intake. Unfortunately, none of these studies were specifically designed to determine the relative effectiveness of low- vs. high-carbohydrate diets for the prevention of weight gain, and none evaluated the effectiveness of very low carbohydrate diets. The majority of these studies aimed to reduce the risk of chronic disease (cancer, heart disease, etc.) and improve health rather than prevent weight gain per se. Furthermore, all of the studies aimed to measure the effectiveness of a diet consistent with a prudent diet, compared to a control group consuming their usual diet.

All but three of the studies were randomized controlled trials. Two of these three were not randomized;[137,140] the other was randomized but did not include a control group.[125] The interventions among adults involved counseling and education in a variety of settings, including worksite, clinic, home, and community. None mentioned the use of other environmental, mass media or institutional change strategies. All but two of the interventions among children were school based. The two that were not included a clinic-based, family-oriented counseling program[126] and a parent-focused behavioral intervention.[125]

None of the seven identified adult intervention studies (Table 9.3) targeted total carbohydrate intake, although one did aim to increase grain consumption[127] and another[128] aimed to increase the intake of complex carbohydrates. Both of these studies observed reductions in the weight of subjects relative to controls. Six did measure and detect increases in percent of calories from carbohydrate relative to controls.[127,129–133] Three of these six studies detected a significant decrease in the weight of subjects relative to controls,[127,129,131] all three of which targeted a reduction in dietary fat in addition to an increase in fiber or complex carbohydrates. Two of them also targeted an increase in fruits and vegetables.[127,129] Two other interventions[132,133] that only targeted fruit and vegetable intake (thereby resulting in an increased percent of energy as carbohydrate), however, did not have a significant impact on weight. It appears, therefore, that increases in percent calories from carbohydrates among adults can, but do not always, have a protective effect on weight gain. This effect may be larger when the increase in proportion of calories from carbohydrates comes from whole grains as well as from fruits and vegetables.

As with adults, none of the nine prevention trials conducted among children (Table 9.4) specifically aimed to change the intake of total carbohydrates, although a few interventions aimed to decrease certain high-carbohydrate foods. Generally speaking, however, when fat intake decreases (as a percent of energy intake), carbohydrate intake increases; therefore, our conclusions regarding the impact of carbohydrate intake could be the reverse of those regarding fat intake: a relative increase in carbohydrate intake (as a percent of energy intake) is sometimes, but not always, associated with lower adiposity. However, the only three studies[126,134,135] that actually measured and detected an increase in the intake of total percent of calories from carbohydrates relative to controls did not observe any impact of the intervention on any measure of adiposity. Of the seven studies that *aimed* to increase total carbohydrates, only two observed an impact on adiposity,[138,139] and this impact was only observed with regard to skinfolds and not BMI. Again, there are many possible reasons why these interventions were not effective. Therefore, it is not possible to come to a conclusion regarding the independent impact of carbohydrate intake on these results.

9.6.2 Type of Carbohydrates for Prevention of Weight Gain

Five prevention trials among adults were identified that resulted in a significant increase in dietary fiber intake relative to controls.[129–132,136] Two[129,131] detected a decrease in weight or BMI relative to controls. The others found no significant impact. However, only two aimed to increase fiber intake per se,[129,130] one of which favorably impacted weight. Another aimed to increase complex carbohydrates[131] and also favorably impacted weight. Both the trial that aimed only to increase fruit and vegetable intake[132] and the trial that only targeted a decrease in dietary fat[136] failed to significantly impact weight relative to controls. It is difficult to draw conclusions

TABLE 9.3
Prevention Trials among Adults That Targeted or Impacted Carbohydrate Intake (in Descending Order by Sample Size)

Author, Year *Study Name* Location	Study Population, Sample Size, Age, Ethnicity	Study Design and Strategies	Length of Intervention and Timing of Follow-Up Measure	Relevant Target Behaviors: Targeted	Relevant Target Behaviors: Reported Change	Results: Impact on Measure of Adiposity: Direction of Change Relative to Control Group	Results: Impact on Measure of Adiposity: Control Variables	Comments
Lanza, 2001 *Polyp Prevention Trial* U.S.	2079 adults with large-bowel adenomatous polyps, 35–89 years old, 88–91% white	Randomized controlled trial; individualized instruction and counseling program to prevent the recurrence of adenomatous polyps	4 years	↓ Dietary fat ↑ Intake of fruits and vegetables ↑ Dietary fiber	↓ Dietary fat (% and g/day) ↑ Intake of fruits and vegetables (servings/MJ and g/day) ↑ Fiber (g/MJ) 0 Change in total calories ↑ Carbohydrates (%) ↑ Protein (%; men) ↑ Calcium (men), ↑ Whole grains (g/day) ↓ High-fat foods (g/day; red/processed meats, high-fat dairy and desserts) ↑ Low-fat alternatives (g/day)	↓ (Weight)	Gender	No significant differences at baseline between groups for gender, age, minority race, education, marital status, BMI, current smoking, current aspirin use, or vigorous and moderate physical activity

Rock, 2001 Women's Healthy Eating and Living (WHEL) Study California, Arizona, Texas, and Oregon	1010 women, 18–70 years old, 86% non-Hispanic white	Randomized controlled trial; individualized dietary counseling; tested effect of low-fat, high-vegetable diet on patients at risk for breast cancer recurrence	1 year	↑ Fruit and vegetable intake ↑ Fiber intake ↓ Percentage of calories from dietary fat	↑ Fruit and vegetable intake (servings/1000 kcal) ↑ Fiber intake (g/1000 kcal) ↓ Percentage of calories from dietary fat ↑ Percentage of calories from carbohydrates	0 (Weight)	Age, stage of cancer at initial diagnosis, overweight status, menopausal status	No significant differences in age, ethnicity, education level, stage at diagnosis, or BMI between intervention and comparison groups at baseline *Change in vegetable intake was inversely associated with weight change
Bhargava, 2002 *Women's Health Trial: Feasibility Study in Minority Populations* Georgia, Alabama, and Florida	926 postmenopausal women, 50–70 years old, 28% black, 16% Hispanic, 54% white	Randomized control trial; group dietary counseling; weight gain prevention intervention	1 year	↓ Energy intakes from fat to ~20% calories ↓ Intake of saturated fat ↑ Consumption of fruits, grain products, and vegetables	↓ Fat intake (g) (saturated fat, monounsaturated fat, and polyunsaturated fat) ↓ Energy intake ↑ % total calories from carbohydrates	↓ (Weight) ↓ (Waist circumference) ↓ (Hip circumference)		Baseline values were similar across groups. *In both intervention and control groups, weight change was explained by changes in carbohydrate and saturated, monounsaturated, and polyunsaturated fats

TABLE 9.3 (continued)
Prevention Trials among Adults That Targeted or Impacted Carbohydrate Intake (in Descending Order by Sample Size)

Author, Year *Study Name* Location	Study Population, Sample Size, Age, Ethnicity	Study Design and Strategies	Length of Intervention and Timing of Follow-Up Measure	Relevant Target Behaviors		Results: Impact on Measure of Adiposity		Comments
Boyd, 1997 *Canadian Diet and Breast Cancer Prevention Study Group* Toronto, Hamilton, London, and Windsor in Ontario	786 women with mammographic dysplasia, aged 30–65 years, race/ethnicity not specified	Randomized controlled trial; intensive individual dietary counseling to determine whether adoption of low-fat, high-carbohydrate diet would reduce the area of radiologically dense breast tissue	2 years	↓ Intake of dietary fat to 15% of total calories ↑ Intake of complex carbohydrates	↓ Mean percentage of calories derived from fat (fell from 33 to 21%) Protein intake as a percent of total calories was unchanged ↑ Carbohydrate intake (rose from 50 to 61% of calories) ↑ Intake of total dietary fiber (from 17.2 to 20.3 g/day)	↓ (Weight)		Baseline characteristics were similar across (intervention and control) groups
Smith-Warner, 2000 *Minnesota Cancer Prevention Research Unit Intervention Study* Minneapolis, MN	201 adults with adenomatous large-bowel polyps, 30–74 years old, 99% white	Randomized controlled trial; individual diet counseling to prevent colon cancer	1 year	↑ Intake of fruits and vegetables	↑Intake of fruits/vegetables (servings/day) 0 Change energy intake or protein (%) ↓ Dietary fat (%) ↑ Carbohydrate (%) ↑ Fiber (g/day)	0 (Weight) 0 (BMI)	Baseline value, gender	No significant differences at baseline between groups for age, gender, household income, education, marital status, employment, ethnicity, smoking, BMI, alcohol intake, or use of nutrient supplements

Cox, 1998 Reading and Glasgow, U.K.	125 adults, 16–65 years old	Randomized controlled trial; education to promote fruit and vegetable consumption	8 weeks	↑ Intake of fruits and vegetables	↑ Intake of fruits and vegetables (g/day; attenuated but remained after 1-year follow-up) 0 Difference in calories ↓ (in both intervention and control groups), fat (%; except significant ↓ in subgroup with baseline fat of >35%), or starch (%) ↑ CHO (%) and total sugars (%)	0 (Weight) (Both intervention and control groups gained weight)		No significant differences at baseline between groups for age, gender, occupation, household income, employment status, or BMI
Simon, 1997 Detroit, MI, and Wichita, KS	133 women at high risk for developing breast cancer, aged 18–67 years, 89% Caucasian, 9% African American, 2% Hispanic	Randomized Controlled trial; combination of education, goal setting, evaluation, feedback, and participant self-monitoring, included both intensive individual counseling sessions and group meetings; weight reduction was not encouraged	3 months of intensive intervention, 12 months follow-up measurements	↓ Dietary fat intake to 15% of total calories	↓ Mean percent caloric intake from fat (from 36 to 18%); this change was maintained at 12 months ↓ Mean caloric intake at 12 months compared to baseline ↑ Dietary fiber intake at 3 months; increase was maintained at 12 months	0 (Weight and percent body fat)	Study site	No *p* values were reported for dietary changes but they appeared to be significant; women in the low-fat diet group lost an average of 3 lb and women in the nonintervention group lost an average of 5 lb; % body fat ↓ slightly for both groups

TABLE 9.4
Prevention Trials among Children That Targeted or Impacted Carbohydrate Intake (in Descending Order by Sample Size)

				Relevant Targeted Behaviors		Results: Impact on Measure of Adiposity		
Author, Year *Study Name* Location	Study Population, Sample Size, Age, Ethnicity	Study Design and Strategies	Length of Intervention and Timing of Follow-Up Measure	Targeted	Reported Change	Direction of Change Relative to Control Group	Control Variables	Comments
Luepker, 1996 *Child and Adolescent Trial for Cardiovascular Health (CATCH)* San Diego, Houston, New Orleans, and Minneapolis	5106 children (school-level data), 4019 children (individual-level data), grades 3–5, 69% white, 13% African American, 14% Hispanic	Randomized controlled trial; school-based multicomponent CVD risk reduction program (education, PE, school lunch, and home programs for half of the families)	3 school years	↓ Dietary fat ↑ Physical activity	*School level:* ↓ Fat (%) in school lunch menus ↓ Dietary energy in school lunch menus ↑ Physical activity intensity in PE *Individual level:* ↓ Rise in dietary energy (MJ/d) ↓ Dietary fat (%) ↑ CHO intake (%) 0 Change in protein intake (%) ↑ vigorous physical activity	0 (BMI) 0 (TSF or SSF)	*School level:* Observation days within semester and lessons within observation days, location of the lesson, specialty of the teacher *Individual level:* Baseline value, gender, ethnicity, CATCH field site, random effect of school with site and intervention group	No significant differences at baseline between groups at school-level for environmental, behavioral, psychosocial, and risk factor data Not explicitly stated in paper, but prevention of obesity was not a goal of CATCH, but rather, it aimed to avoid growth retardation

Nader, 1999 *CATCH* San Diego, Houston, New Orleans, and Minneapolis	3714 children, grades 6–8 (73% of original cohort)	Randomized controlled trial; school-based multicomponent CVD risk reduction program (education, PE, school lunch, and home programs for half of the families)	3-year follow-up (after 3-year intervention)	↓ Dietary fat ↑ Physical activity	↓ Dietary fat (%) ↑ CHO (%) 0 Change protein (%) ↑ Vigorous physical activity ↓ (Note: dietary energy at end of 3-year intervention gone by 3-year follow-up)	0 (BMI) 0 (TSF or SSF)	Age, site, gender, ethnicity, intraclass correlation within school and among students, gender × ethnicity interaction	Not explicitly stated in paper, but prevention of obesity was not a goal of CATCH, but rather, it aimed to avoid growth retardation
The Writing Group for the DISC Collaborative Research Group, 1995 *The Dietary Intervention Study in Children* Baltimore, MD, Chicago, IL, Iowa City, IA, Newark, NJ, New Orleans, LA, and Portland, OR	663 children with elevated LDL cholesterol, 8–10 years, race/ethnicity not specified	Randomized controlled trial; clinic-based, personalized, family-oriented dietary counseling program to reduce LDL cholesterol	3 years	↓ Dietary fat energy and other nutrients at RDA	↓ Total fat (%) ↓ Total energy (kJ/d) ↑ Protein (%) ↑ Carbohydrate (%) 0 Change in diet, Ca, Zn, Fe, vitamins A and C 0 Change in serum ferritin, Zn, retinol	0 (BMI) 0 (Weight) 0 (Sum TSF, SSF, and suprailiac skinfold) 0 (WHR)	Gender, baseline value	No significant differences at baseline between groups for age, gender(?) (appears so, but not explicitly stated), anthropometry, blood lipid levels, and blood pressure; small differences in dietary intake, with intervention group having slightly lower % energy from PUFA and slightly higher intakes of vitamin B_6 and Zn Intervention group had a slightly higher proportion with household income <$20K

TABLE 9.4 (continued)
Prevention Trials among Children That Targeted or Impacted Carbohydrate Intake (in Descending Order by Sample Size)

Author, Year *Study Name* Location	Study Population, Sample Size, Age, Ethnicity	Study Design and Strategies	Length of Intervention and Timing of Follow-Up Measure	Relevant Targeted Behaviors		Results: Impact on Measure of Adiposity		Comments
Resnicow, 1992 *Know Your Body, comprehensive school health education* program New York	1209 children, 6–13 years old, predominantly Hispanic population	Nonrandomized control trial; school-based comprehensive school health education program (classroom curriculum, schoolwide health activities, and environmental modifications)	1 ½-year intervention and 3-year follow-up measure	↑ Health knowledge ↑ Fiber content of foods served in school ↓ Fat content of foods served in school ↑ Vegetable and heart-healthy indices	0 Significant differences in dietary indices ↑ Health knowledge at 3-year follow-up ↑ Number of servings of vegetables and heart-healthy foods ↓ Number of servings of meat and desserts	0 (BMI)	Age, ethnicity, gender, baseline values	At baseline, intervention students did not significantly differ with regard to sex, total cholesterol, systolic blood pressure, BMI, health attitudes, and self-efficacy; they were significantly younger, more likely to be Hispanic, and had significantly lower health knowledge and fruit intake scores than control students

Vandongen, 1995 West Australia	869 children, 10–12 years old, considered to be representative sample of the socioeconomic mix of the community in West Australia, race/ethnicity not specified	Randomized controlled trial; school-based nutrition and fitness program to improve cardiovascular risk factors	9 months	↑ Consumption of fruit, vegetables, whole-grain bread and cereal relative to other foods ↓ Consumption of fatty, sugary, and salty foods relative to other foods	↓ % Total energy from sugar (boys only) ↓ % Total energy from fat (girls only) ↓ % Total energy from saturated fat (girls only) 0 Change in total energy (boys and girls) ↑ % Total energy from protein (boys only) ↑ Fiber intake, g/day (boys and girls)	↓ (Triceps skinfolds, boys and girls) 0 (Subscapular skinfolds, boys and girls) 0 (% Body fat, boys and girls) 0 (BMI, boys and girls)	Gender, baseline values	
Burke, 1998 Western Australia	720 children, 11 years old, race/ethnicity not specified	Randomized controlled trial; school- and home-based physical enrichment program for children at higher risk of cardiovascular disease	20-week (2 school terms) intervention and 6-month follow-up measure	↑ Duration and frequency of physical activity ↓ Consumption of fat, sugar, and salt ↑ Fiber intake	↑ Physical fitness (boys and girls) 0 Change in physical activity (boys and girls) ↓ TV watching (boys only) 0 Dietary change (boys and girls)	0 (BMI, boys and girls) ↓ (Subscapular skinfolds, girls only) ↓ (Triceps skinfolds, boys and girls)	Gender, baseline values	At baseline, there were no significant differences in dietary variables, time spent in leisure-time physical activity, or hours of TV watching
Donnelly, 1996 Rural Nebraska	44 subjects, 64 controls, grades 3–5, subsample of 11 subjects and 25 controls with 22% body fat, race/ethnicity not specified	Nonrandomized controlled trial; school-based multicomponent program (nutrition education, modified school lunches, and enhanced physical activity program)	2 school years	↓ Intake of fat ↓ Intake of cholesterol ↓ Intake of sodium ↑ Intake of fiber ↑ Physical fitness and knowledge and awareness of diet in health	0 Significant change in total energy, % kcal from fats, carbohydrates, proteins, or fiber ↓ Intake of sodium	0 (Weight, BMI, and body fat % — underwater weighing) 0 (Attenuation of obesity — subjects with body fat of 22%)	Schools were matched for ethnicity/SES and baseline characteristics (grade, height, weight, BMI, mile-run time)	

TABLE 9.4 (continued)
Prevention Trials among Children That Targeted or Impacted Carbohydrate Intake (in Descending Order by Sample Size)

Author, Year *Study Name* Location	Study Population, Sample Size, Age, Ethnicity	Study Design and Strategies	Length of Intervention and Timing of Follow-Up Measure	Relevant Targeted Behaviors		Results: Impact on Measure of Adiposity		Comments
Sahota, 2001 *Active Programme Promoting Lifestyle in Schools (APPLES)* Leeds, U.K.	203–303 children, 8–10 years old, some ethnic minority children in sample	Randomized controlled trial; school-based, multidisciplinary intervention to reduce risk factors for obesity	12 months	Influence of diet and physical activity behaviors ↑ Consumption of fruits and vegetables ↓ Consumption of foods high in fat ↓ Consumption of foods and drinks high in sugars	↑ Vegetable intake ↓ Fruit intake in obese children ↑ Intake of foods and drinks high in sugar in overweight children ↑ Sedentary behavior in the overweight children	0 (BMI)	Gender, age, baseline BMI	No significant differences were found between the intervention and comparison pupils for any of the measures at baseline

Epstein, 2001 *Childhood Weight Control and Prevention Program* Buffalo, NY	27 intervention families (at least 1 obese (BMI > 85th percentile) parent + a nonobese (BMI < 85th percentile) child, 6–11 years old)	Randomized noncontrolled trial; parent-focused behavioral intervention on parent and child eating changes, and weight control treatment for parents	6-month intervention and 1-year follow-up measure	↑ Fruit and vegetable intake ↓ High-fat and high-sugar food intake	↑ Fruit and vegetable intake (servings/day) ↓ High-fat and high-sugar foods (servings/day)	0 (Percentage of overweight)		No significant differences at baseline between groups for gender, age, weight, % overweight, family history of obesity and chronic diseases (except for more HTN in families in fat/sugar group), food habits, child feeding practices, and confidence in making choices Correlations showed no differences by age or gender in outcomes

regarding the independent impact of dietary fiber on adiposity given the limited number of studies and the variability in the combination of target behaviors in each intervention. However, it appears that dietary fiber is sometimes, but not always, effective in preventing weight gain. Interventions may be more effective when they explicitly target fiber intake from a variety of sources, including whole grains.

Four programs specifically targeted an increase in fiber intake among children.[137–140] Only one, however, measured and detected an increase in fiber intake among subjects compared to controls.[138] Therefore, once again it is hard to arrive at a conclusion regarding the impact of a reduction in fiber intake, when most of the studies were unable to produce a detectable change. Although half of these four studies did observe an impact on adiposity,[138,139] this impact was observed only with regard to skinfolds, and not BMI. All of these programs aimed to reduce fat or sugar as well as increase fiber, and two of them also targeted physical activity,[139,140] again making it difficult to attribute program impact (or lack thereof) to any specific targeted behavior. We can conclude only that increases in fiber intake may be part of an effective strategy to prevent overweight in children.

Only one study examined sugar intake among adults.[133] This study, which aimed to increase fruit and vegetable intake, resulted in an increase in total sugar intake and did not have a significant impact on weight of subjects relative to controls. Based on this limited data, no conclusions can be drawn regarding sugar intake among adults for the prevention of weight gain.

Four studies were identified that specifically targeted the reduction of sugar or foods high in sugar among children.[125,138,139,141] Three of these were school-based randomized controlled trials, and the other[125] was a family-focused behavior intervention that did not include a control group. The only study that detected a decrease in the intake of high-sugar foods did not observe an impact on adiposity.[125] Another study[141] actually detected an *increase* in the intake of high-sugar foods and beverages relative to controls and observed no impact on BMI. Although half of these trials demonstrated an impact on adiposity (in regard to skinfolds, but not BMI), all of these interventions targeted multiple dietary changes, and some also aimed to alter physical activity, making it impossible to determine the independent effect of sugar intake on adiposity. It appears that a reduction in sugar intake can be part of an effective strategy to prevent increases in adiposity, but interventions reported to date have been largely ineffective in reducing sugar intake.

9.6.3 Summary of Treatment and Prevention Studies

The ratios of the dietary macronutrients (carbohydrates, fats, and proteins) ideal for weight loss or weight maintenance remain unknown and may not be the same for both purposes. Limitations of weight-loss intervention studies performed to date include relatively short duration, failure to account for physical activity, few intention-to-treat analyses (particularly critical with high dropout rates), lack of control for other macronutrients, no accounting for type of carbohydrates (e.g., simple vs. complex/fiber or high vs. low glycemic index) or energy density, and few studies of subjects with hyperlipidemia, hypertension, or impaired glucose tolerance. Prevention trials have not focused on carbohydrate intake per se, and, therefore, it is not possible to determine the independent impact of carbohydrate intake on adiposity from these studies.

Diets of quite varied macronutrient composition have proven to be successful in producing weight loss in the short term. None have a good track record with maintenance of weight loss in the long term. Prevention trials published to date have also been limited in duration. Although preliminary results suggest that low-carbohydrate/high-fat diets may not be detrimental to health in the short term when accompanied by weight loss, their safety in the long term remains unknown. It is well established that long-term consumption of high-fat diets, particularly diets high in saturated fat, increases the risk of cardiovascular disease and several types of cancer. Because of this, it would be prudent to avoid very high fat diets for long-term weight maintenance until their safety is established. At this point, the most prudent recommendation for weight loss is to reduce energy intake from carbohydrates and fats and increase energy expenditure with regular physical activity. The exact proportions of intake from the macronutrients may be more a question of individual preference than scientific rationale. For weight maintenance, energy intake from carbohydrates, fats, and proteins must be balanced with energy expenditure. Particular attention should be placed on reducing foods rich in saturated and trans fatty acids and increasing foods rich in complex carbohydrates and dietary fiber.

In terms of prevention of weight gain, all trials include multiple behavior changes and are therefore of limited utility for assessing the independent impact of carbohydrates. The results from currently available studies suggest that increases in the percent of calories from carbohydrates that result from decreases in the intake of fat (i.e., a prudent diet) can be, but are not always, sufficient for preventing weight gain. Increases in fiber, especially from whole grains, appear to be a particularly promising strategy, although it is hard to arrive at a firm conclusion given the limited number of studies. Studies that examined the impact of sugar intake on adiposity are even more limited and further research is merited before definitive conclusions can be drawn.

9.7 CONCLUSIONS

Although many issues remain unresolved regarding how diet influences obesity, it is certain that weight gain occurs when energy intake exceeds energy expenditure, weight loss occurs when energy expenditure exceeds energy intake, and weight maintenance results from a balance between energy intake and expenditure. A minor imbalance over time can have a dramatic effect on body weight. As an example, if energy intake exceeds energy expenditure by 1% per day, an annual weight gain of ~2.5 lb results. In this case, a person with an ideal body weight at 20 years old would be 25 lb overweight at 30 years of age. Thus, factors that upset this energy balance may be subtle and difficult to identify with certainty, even in very well controlled and carefully conducted studies. In a society where food is abundant, where a significant proportion of food is consumed outside of the home, and where food providers fiercely compete for clientele, it is our challenge to identify modifiable factors that can offset the temptation to consume food in the absence of hunger.

Via several mechanisms, the macronutrient composition of a food can influence how much of that food and energy is consumed. For example, dietary carbohydrates do not exert the powerful satiation and satiety effects of dietary protein, but they do have a greater effect than dietary fat. The relative satiety values of simple vs. complex carbohydrates are less clear. By influencing the energy density of food, macronutrient composition can also influence energy intake. The water and fat contents of diets are the most significant determinants of energy density, and they decrease and increase energy density of foods, respectively. Macronutrient composition can also influence consumption by impacting the palatability of food. Foods high in carbohydrates, especially fiber, may be less palatable than foods high in fat, thereby reducing consumption. Some high-carbohydrate foods are both highly palatable and energy dense, and their consumption can lead to passive overconsumption. Palatability is, to an unknown extent, influenced by taste preferences, and preferences are influenced by genetics, familiarity, gender, and weight status. Although it may be possible in a confined research or clinical setting to eliminate availability of foods that are highly palatable, and thus reduce risk of passive overconsumption, this is less likely to be achieved in a free-living environment.

The relationship between total carbohydrate content of a diet and adiposity has been well studied using several approaches. Secular trend data indicate that intakes of total carbohydrates and energy have increased over the period that obesity has increased. It appears that the intake of all types of carbohydrates has increased, but sugar in the form of sweetened beverages and grain products in their refined form have increased the most. Results of longitudinal observational studies indicate that total carbohydrate intake has a limited, if any, influence on weight change, however. Although most cross-sectional observational studies show significant negative associations between total carbohydrate intake (after correcting for energy intake) and adiposity, results of intervention studies indicate, at least in the short term, that diets low in carbohydrates may be superior to diets high in carbohydrates for weight loss. However, it has not been determined with certainty if it is the carbohydrate levels in these diets that have contributed to their success. These diets are distinct from the control diets in many aspects, notably palatability and variety. Studies designed to determine why these diets are successful are needed. Trials designed to test the relative efficacy of high- vs. low-carbohydrate diets for the prevention of weight gain are lacking. Available data suggest that increasing the proportion of calories from carbohydrates can be part of a successful weight maintenance strategy in the short term. Longer-term studies that examine the independent impact of carbohydrate intake are needed. It should also be determined with greater certainty whether, as some current data might suggest, low-carbohydrate diets are more likely to be adhered to over the long term than high-carbohydrate diets, thereby promoting greater weight loss, especially in the light of strong epidemiological evidence suggesting the contrary. Adherence to diet may be the single most important area for further study, as weight loss has been found to be highly dependent on energy consumption and relatively independent of macronutrient composition, making adherence to a low-energy diet the factor most likely to influence weight loss. Finally, it must be determined whether very low carbohydrate diets are not only effective, but also safe over the long term.

Secular trend data indicate that Americans have also increased their intakes of refined carbohydrates and dietary fiber over the period that obesity has increased. There are too few longitudinal studies to draw conclusions regarding the impact of either refined carbohydrates or fiber, although fiber appears to be protective. Most cross-sectional studies report a significant negative association between dietary fiber and adiposity, with fiber accounting for as much as 17% of the variance in BMI. In weight-loss studies, high-fiber diets have been found to be more conducive to weight loss than low-fiber diets. The available prevention trials also suggest that fiber contributes to successful weight maintenance. Paradoxically, intakes of refined carbohydrates have also been significantly and negatively associated with adult adiposity in many cross-sectional studies, though the only nationally representative sample from the U.S. failed to find a significant relationship. Results of several studies suggest that low-energy diets providing complex carbohydrates may be more conducive to weight loss and weight maintenance than low-energy diets with simple sugars, but this needs further investigation. If fiber is confirmed to be protective against weight gain and conducive to weight loss, it will be important to learn whether high-fiber diets will also promote maintenance of weight following weight reduction.

Based on the limited number and quality of studies available to date, the intake of simple carbohydrates appears to have a limited, if any, impact on adiposity. If these findings are confirmed, the value of further studies evaluating the role of the glycemic index in adiposity will be questioned, as sugar content is one of the main factors influencing the glycemic index of food. If a high glycemic index continues to be implicated in weight gain, however, long-term clinical trials will be needed to determine whether the glycemic index of a food influences body weight regulation.

Currently, it is not certain that either the proportion or type of carbohydrate in a diet influences the ability of a diet to prevent weight gain following weight reduction. Studies that address this need will be expensive to conduct and time-consuming, but they are nonetheless of high priority. As there is reason to suggest that prevention of weight gain prior to obesity and maintenance of weight loss following obesity/overweight may not be completely synonymous metabolic states, very basic metabolic studies are needed to determine whether dietary characteristics will be the same or different for these two conditions.

ACKNOWLEDGMENTS

This chapter was supported in part by Cooperative Agreement U48/CCU909706-10 from the Centers for Disease Control and Prevention; the College of Natural Resources, University of California, Berkeley; and the American Dietetic Association. Its contents are solely the responsibility of the authors and do not necessarily represent the official views of the funders.

REFERENCES

1. Flegal, K.M. et al., Prevalence and trends in obesity among US adults, 1999–2000, *JAMA*, 288, 1723, 2002.
2. Ogden, C.L. et al., Prevalence and trends in overweight among US children and adolescents, 1999–2000, *JAMA*, 288, 1728, 2002.
3. Must, A. et al., The disease burden associated with overweight and obesity, *JAMA*, 282, 1523, 1999.
4. French, S.A., Story, M., and Perry, C.L., Self-esteem and obesity in children and adolescents: a literature review, *Obes. Res.*, 3, 479, 1995.
5. Gortmaker, S.L. et al., Social and economic consequences of overweight in adolescence and young adulthood, *N. Engl. J. Med.*, 329, 1008, 1993.
6. Blundell, J.E. and King, N.A., Overconsumption as a cause of weight gain: behavioural-physiological interactions in the control of food intake (appetite), *Ciba Found. Symp.*, 201, 138, 1996.
7. Blundell, J.E. and MacDiarmid, J.I., Fat as a risk factor for overconsumption: satiation, satiety, and patterns of eating, *J. Am. Diet. Assoc.*, 97, S63, 1997.
8. Stubbs, R.J., Mazlan, N., and Whybrow, S., Carbohydrates, appetite and feeding behavior in humans, *J. Nutr.*, 131, 2775S, 2001.
9. Green, S.M., Burley, V.J., and Blundell, J.E., Effect of fat- and sucrose-containing foods on the size of eating episodes and energy intake in lean males: potential for causing overconsumption, *Eur. J. Clin. Nutr.*, 48, 547, 1994.
10. Rolls, B.J., Carbohydrates, fats, and satiety, *Am. J. Clin. Nutr.*, 61, 960S, 1995.
11. Rolls, B.J. and Hammer, V.A., Fat, carbohydrate, and the regulation of energy intake, *Am. J. Clin. Nutr.*, 62, 1086S, 1995.
12. Johnstone, A.M., Stubbs, R.J., and Harbron, C.G., Effect of overfeeding macronutrients on day-to-day food intake in man, *Eur. J. Clin. Nutr.*, 50, 418, 1996.
13. Stubbs, J., Ferres, S., and Horgan, G., Energy density of foods: effects on energy intake, *Crit. Rev. Food Sci. Nutr.*, 40, 481, 2000.
14. Hill, J.O. and Prentice, A.M., Sugar and body weight regulation, *Am. J. Clin. Nutr.*, 62, 264S, 1995.
15. Lewis, C.J. et al., Nutrient intakes and body weights of persons consuming high and moderate levels of added sugars, *J. Am. Diet. Assoc.*, 92, 708, 1992.
16. Saltzman, E. and Roberts, S.B., Soluble fiber and energy regulation, current knowledge and future directions, *Adv. Exp. Med. Biol.*, 427, 89, 1997.
17. Howarth, N.C., Saltzman, E., and Roberts, S.B., Dietary fiber and weight regulation, *Nutr. Rev.*, 59, 129, 2001.
18. Pereira, M.A. and Ludwig, D.S., Dietary fiber and body-weight regulation, observations and mechanisms, *Pediatr. Clin. North Am.*, 48, 969, 2001.
19. Cox, D.N. et al., Sensory and hedonic associations with macronutrient and energy intakes of lean and obese consumers, *Int. J. Obes. Relat. Metab. Disord.*, 23, 403, 1999.
20. Yao, M. and Roberts, S.B., Dietary energy density and weight regulation, *Nutr. Rev.*, 59, 247, 2001.
21. Bell, E.A. et al., Energy density of foods affects energy intake in normal-weight women, *Am. J. Clin. Nutr.*, 67, 412, 1998.
22. Bell, E.A. and Rolls, B.J., Energy density of foods affects energy intake across multiple levels of fat content in lean and obese women, *Am. J. Clin. Nutr.*, 73, 1010, 2001.

23. Westerterp-Plantenga, M.S. et al., Energy intake adaptation of food intake to extreme energy densities of food by obese and non-obese women, *Eur. J. Clin. Nutr.*, 50, 401, 1996.
24. Rolls, B.J. and Bell, E.A., Dietary approaches to the treatment of obesity, *Med. Clin. North Am.*, 84, 401, 2000.
25. Mattes, R.D., Dietary compensation by humans for supplemental energy provided as ethanol or carbohydrate in fluids, *Physiol. Behav.*, 59, 179, 1996.
26. Nasser, J., Taste, food intake and obesity, *Obes. Rev.*, 2, 213, 2001.
27. Drewnowski, A., Energy density, palatability, and satiety: implications for weight control, *Nutr. Rev.*, 56, 347, 1998.
28. Drewnowski, A., Human preferences for sugar and fat, in *Appetite and Body Weight Regulation: Sugar, Fat, and Macronutrient Substitutes*, Fernstrom, J. and Miller, G.D., Eds., CRC Press, Boca Raton, FL, 1994, p. 137.
29. Macdiarmid, J.I. et al., The sugar-fat relationship revisited: differences in consumption between men and women of varying BMI, *Int. J. Obes. Relat. Metab. Disord.*, 22, 1053, 1998.
30. Rissanen, A. et al., Acquired preference especially for dietary fat and obesity: a study of weight-discordant monozygotic twin pairs, *Int. J. Obes. Relat. Metab. Disord.*, 26, 973, 2002.
31. Tucker, L.A. and Kano, M.J., Dietary fat and body fat: a multivariate study of 205 adult females, *Am. J. Clin. Nutr.*, 56, 616, 1992.
32. Lafay, L. et al., Does energy intake underreporting involve all kinds of food or only specific food items? Results from the Fleurbaix Laventie Ville Sante (FLVS) study, *Int. J. Obes. Relat. Metab. Disord.*, 24, 1500, 2000.
33. Lissner, L. and Heitmann, B.L., The dietary fat:carbohydrate ratio in relation to body weight, *Curr. Opin. Lipidol.*, 6, 8, 1995.
34. Drewnowski, A. et al., The fat-sucrose seesaw in relation to age and dietary variety of French adults, *Obes. Res.*, 5, 511, 1997.
35. Flatt, J.P., Use and storage of carbohydrate and fat, *Am. J. Clin. Nutr.*, 61, 952S, 1995.
36. Stubbs, R.J., Prentice, A.M., and James, W.P., Carbohydrates and energy balance, *Ann. N.Y. Acad. Sci.*, 819, 44, 1997.
37. Wolever, T.M. et al., Determination of the glycaemic index of foods: interlaboratory study, *Eur. J. Clin. Nutr.*, 57, 475, 2003.
38. Bjorck, I. et al., Food properties affecting the digestion and absorption of carbohydrates, *Am. J. Clin. Nutr.*, 59, 699S, 1994.
39. Ludwig, D.S., Dietary glycemic index and obesity, *J. Nutr.*, 130, 280S, 2000.
40. Roberts, S.B., High-glycemic index foods, hunger, and obesity: is there a connection? *Nutr. Rev.*, 58, 163, 2000.
41. Bjorck, I. and Elmstahl, H.L., The glycaemic index: importance of dietary fibre and other food properties, *Proc. Nutr. Soc.*, 62, 201, 2003.
42. Enns, C., Goldman, J., and Cook, A., Trends in food and nutrient intakes by adults: NFCS 1977-78, CSFII 1989-91, and CSFII 1994-95, *Fam. Econ. Nutr. Rev.*, 10, 2, 1997.
43. Siega-Riz, A.M., Popkin, B.M., and Carson, T., Differences in food patterns at breakfast by sociodemographic characteristics among a nationally representative sample of adults in the United States, *Prev. Med.*, 30, 415, 2000.
44. Johnson, R.K. and Frary, C., Choose beverages and foods to moderate your intake of sugars: the 2000 dietary guidelines for Americans — what's all the fuss about? *J. Nutr.*, 131, 2766S, 2001.

45. Putnam, J., Allshouse, J.E., and Kantor, L.S., US per capita food supply trends: more calories, refined carbohydrates and fats, *Food Rev.*, 25, 2, 2002.
46. Parker, D.R. et al., Dietary factors in relation to weight change among men and women from two southeastern New England communities, *Int. J. Obes. Relat. Metab. Disord.*, 21, 103, 1997.
47. Ludwig, D.S. et al., Dietary fiber, weight gain, and cardiovascular disease risk factors in young adults, *JAMA*, 282, 1539, 1999.
48. Kemper, H.C. et al., Lifestyle and obesity in adolescence and young adulthood: results from the Amsterdam Growth and Health Longitudinal Study (AGAHLS), *Int. J. Obes. Relat. Metab. Disord.*, 23, S34, 1999.
49. Robertson, S.M. et al., Factors related to adiposity among children aged 3 to 7 years, *J. Am. Diet. Assoc.*, 99, 938, 1999.
50. Berkey, C.S. et al., Activity, dietary intake, and weight changes in a longitudinal study of preadolescent and adolescent boys and girls, *Pediatrics*, 105, E56, 2000.
51. Eck, L.H. et al., Children at familial risk for obesity: an examination of dietary intake, physical activity and weight status, *Int. J. Obes. Relat. Metab. Disord.*, 16, 71, 1992.
52. Scaglioni, S. et al., Early macronutrient intake and overweight at five years of age, *Int. J. Obes. Relat. Metab. Disord.*, 24, 777, 2000.
53. Magarey, A.M. et al., Does fat intake predict adiposity in healthy children and adolescents aged 2–15 y? A longitudinal analysis, *Eur. J. Clin. Nutr.*, 55, 471, 2001.
54. Lee, P.A., Guo, S.S., and Kulin, H.E., Age of puberty: data from the United States of America, *Apmis*, 109, 81, 2001.
55. Rolland-Cachera, M.F. et al., Influence of macronutrients on adiposity development: a follow up study of nutrition and growth from 10 months to 8 years of age, *Int. J. Obes. Relat. Metab. Disord.*, 19, 573, 1995.
56. Maffeis, C., Talamini, G., and Tato, L., Influence of diet, physical activity and parents' obesity on children's adiposity: a four-year longitudinal study, *Int. J. Obes. Relat. Metab. Disord.*, 22, 758, 1998.
57. Carruth, B.R. and Skinner, J.D., The role of dietary calcium and other nutrients in moderating body fat in preschool children, *Int. J. Obes. Relat. Metab. Disord.*, 25, 559, 2001.
58. Alexy, U. et al., Macronutrient intake of 3- to 36-month-old German infants and children: results of the DONALD Study. Dortmund Nutritional and Anthropometric Longitudinally Designed Study, *Ann. Nutr. Metab.*, 43, 14, 1999.
59. Gonzalez, C.A. et al., Types of fat intake and body mass index in a Mediterranean country, *Public Health Nutr.*, 3, 329, 2000.
60. Yang, E.J. et al., Carbohydrate intake and biomarkers of glycemic control among US adults: the third National Health and Nutrition Examination Survey (NHANES III), *Am. J. Clin. Nutr.*, 77, 1426, 2003.
61. Slattery, M.L. et al., Associations of body fat and its distribution with dietary intake, physical activity, alcohol, and smoking in blacks and whites, *Am. J. Clin. Nutr.*, 55, 943, 1992.
62. Stam-Moraga, M.C. et al., Sociodemographic and nutritional determinants of obesity in Belgium, *Int. J. Obes. Relat. Metab. Disord.*, 23, 1, 1999.
63. Bolton-Smith, C. and Woodward, M., Dietary composition and fat to sugar ratios in relation to obesity, *Int. J. Obes. Relat. Metab. Disord.*, 18, 820, 1994.
64. Kennedy, E.T. et al., Popular diets: correlation to health, nutrition, and obesity, *J. Am. Diet. Assoc.*, 101, 411, 2001.
65. Alfieri, M.A. et al., Fiber intake of normal weight, moderately obese and severely obese subjects, *Obes. Res.*, 3, 541, 1995.

66. Nelson, L.H. and Tucker, L.A., Diet composition related to body fat in a multivariate study of 203 men, *J. Am. Diet. Assoc.*, 96, 771, 1996.
67. Miller, W.C. et al., Dietary fat, sugar, and fiber predict body fat content, *J. Am. Diet. Assoc.*, 94, 612, 1994.
68. Ruidavets, J. et al., Eating frequency and body fatness in middle-aged men, *Int. J. Obes.*, 26, 1476, 2002.
69. Trichopoulou, A. et al., Lipid, protein and carbohydrate intake in relation to body mass index, *Eur. J. Clin. Nutr.*, 56, 37, 2002.
70. Wamala, S.P., Wolk, A., and Orth-Gomer, K., Determinants of obesity in relation to socioeconomic status among middle-aged Swedish women, *Prev. Med.*, 26, 734, 1997.
71. Ortega, R.M. et al., Relationship between diet composition and body mass index in a group of Spanish adolescents, *Br. J. Nutr.*, 74, 765, 1995.
72. Rocandio, A.M., Ansotegui, L., and Arroyo, M., Comparison of dietary intake among overweight and non-overweight schoolchildren, *Int. J. Obes. Relat. Metab. Disord.*, 25, 1651, 2001.
73. Guillaume, M., Lapidus, L., and Lambert, A., Obesity and nutrition in children. The Belgian Luxembourg Child Study IV, *Eur. J. Clin. Nutr.*, 52, 323, 1998.
74. Tucker, L.A., Seljaas, G.T., and Hager, R.L., Body fat percentage of children varies according to their diet composition, *J. Am. Diet. Assoc.*, 97, 981, 1997.
75. Bandini, L.G. et al., Comparison of high-calorie, low-nutrient-dense food consumption among obese and non-obese adolescents, *Obes. Res.*, 7, 438, 1999.
76. Gazzaniga, J.M. and Burns, T.L., Relationship between diet composition and body fatness, with adjustment for resting energy expenditure and physical activity, in preadolescent children, *Am. J. Clin. Nutr.*, 58, 21, 1993.
77. Garaulet, M. et al., Difference in dietary intake and activity level between normal-weight and overweight or obese adolescents, *J. Pediatr. Gastroenterol. Nutr.*, 30, 253, 2000.
78. Maffeis, C. et al., Distribution of food intake as a risk factor for childhood obesity, *Int. J. Obes.*, 24, 75, 2000.
79. Davies, P., Diet composition and body mass index in pre-school children, *Eur. J. Clin. Nutr.*, 51, 443, 1997.
80. Stewart, K. et al., Dietary fat and cholesterol intake in young children compared with recommended levels, *J. Cardiopulmonary Rehabil.*, 19, 112, 1999.
81. Maffeis, C., Pinelli, L., and Schutz, Y., Fat intake and adiposity in 8- to 11-year-old obese children, *Int. J. Obes. Relat. Metab. Disord.*, 20, 170, 1996.
82. Koivisto, U., Fellenius, J., and Sjoden, P.O., Relations between parental mealtime practices and children's food intake, *Appetite*, 22, 245, 1994.
83. Bao, W., Srinivasan, S.R., and Berenson, G.S., Persistent elevation of plasma insulin levels is associated with increased cardiovascular risk in children and young adults. The Bogalusa Heart Study, *Circulation*, 93, 54, 1996.
84. Hanley, A.J. et al., Overweight among children and adolescents in a Native Canadian community: prevalence and associated factors, *Am. J. Clin. Nutr.*, 71, 693, 2000.
85. Maffeis, C. et al., Energy intake and energy expenditure in prepubertal males with asthma, *Eur. Respir. J.*, 12, 123, 1998.
86. Atkin, L.-M. and Davies, P.S.W., Diet composition and body composition in preschool children, *Am. J. Clin. Nutr.*, 72, 15, 2000.
87. Rodriguez-Artalejo, F. et al., Dietary patterns among children aged 6–7 y in four Spanish cities with widely differing cardiovascular mortality, *Eur. J. Clin. Nutr.*, 56, 141, 2002.

88. Francis, C. et al., Body composition, dietary intake, and energy expenditure in nonobese, prepubertal children of obese and nonobese biological mothers, *J. Am. Diet. Assoc.*, 99, 58, 1999.
89. Dennison, B., Rockwell, H., and Baker, S., Excess fruit juice consumption by preschool-aged children is associated with short stature and obesity, *Pediatrics*, 99, 15, 1997.
90. Kromhout, D. et al., Physical activity and dietary fiber determine population body fat levels: the Seven Countries Study, *Int. J. Obes. Relat. Metab. Disord.*, 25, 301, 2001.
91. Appleby, P.N. et al., Low body mass index in non-meat eaters: the possible roles of animal fat, dietary fibre and alcohol, *Int. J. Obes. Relat. Metab. Disord.*, 22, 454, 1998.
92. Gibson, S.A., Are high-fat, high-sugar foods and diets conducive to obesity? *Int. J. Food Sci. Nutr.*, 47, 405, 1996.
93. Gills, L. et al., Relationship between juvenile obesity, dietary energy and fat intake and physical activity, *Int. J. Obes.*, 26, 458, 2002.
94. McGloin, A. et al., Energy and fat intake in obese and lean children at varying risk of obesity, *Int. J. Obes.*, 26, 200, 2002.
95. Fogelholm, M. et al., Determinants of energy imbalance and overweight in Finland 1982 and 1992, *Int. J. Obes. Relat. Metab. Disord.*, 20, 1097, 1996.
96. Bowman, S.A. and Spence, J.T., A comparison of low-carbohydrate vs. high-carbohydrate diets: energy restriction, nutrient quality and correlation to body mass index, *J. Am. Coll. Nutr.*, 21, 268, 2002.
97. Skov, A.R. et al., Randomized trial on protein vs. carbohydrate in *ad libitum* fat reduced diet for the treatment of obesity, *Int. J. Obes. Relat. Metab. Disord.*, 23, 528, 1999.
98. Layman, D.K. et al., A reduced ratio of dietary carbohydrate to protein improves body composition and blood lipid profiles during weight loss in adult women, *J. Nutr.*, 133, 411, 2003.
99. Johnston, C.S., Day, C.S., and Swan, P.D., Postprandial thermogenesis is increased 100% on a high-protein, low-fat diet versus a high-carbohydrate, low-fat diet in healthy, young women, *J. Am. Coll. Nutr.*, 21, 55, 2002.
100. Serdula, M.K. et al., Prevalence of attempting weight loss and strategies for controlling weight, *JAMA*, 282, 1353, 1999.
101. National Institutes of Heatlh, Clinical guidelines on the identification, evaluation, and treatment of overweight and obesity in adults: the evidence report. National Institutes of Heatlh, *Obes. Res.*, 6, 464, 1998.
102. American Diabetes Association, Position of the American Dietetic Association: weight management, *J. Am. Diet. Assoc.*, 97, 71, 1997.
103. Atkins, R., *Dr. Atkins' New Diet Revolution*, Avon Books, New York, 1998.
104. Bravata, D.M. et al., Efficacy and safety of low-carbohydrate diets: a systematic review, *JAMA*, 289, 1837, 2003.
105. Brehm, B.J. et al., A randomized trial comparing a very low carbohydrate diet and a calorie-restricted low fat diet on body weight and cardiovascular risk factors in healthy women, *J. Clin. Endocrinol. Metab.*, 88, 1617, 2003.
106. Foster, G.D. et al., A randomized trial of a low-carbohydrate diet for obesity, *N. Engl. J. Med.*, 348, 2082, 2003.
107. Samaha, F.F. et al., A low-carbohydrate as compared with a low-fat diet in severe obesity, *N. Engl. J. Med.*, 348, 2074, 2003.
108. Freedman, M.R., King, J., and Kennedy, E., Popular diets: a scientific review, *Obes. Res.*, 9, 1S, 2001.

109. Klem, M.L. et al., A descriptive study of individuals successful at long-term maintenance of substantial weight loss, *Am. J. Clin. Nutr.*, 66, 239, 1997.
110. Levine, A. and Billington, C., Eds., *Dietary Fiber: Does It Affect Food Intake and Body Weight? Appetite and Body Weight Regulation, Sugar, Fat and Macronutrient Substitutes*, CRC, Boca Raton, FL, 1994.
111. Ryttig, K.R. et al., A dietary fibre supplement and weight maintenance after weight reduction: a randomized, double-blind, placebo-controlled long-term trial, *Int. J. Obes.*, 13, 165, 1989.
112. Cairella, G., Cairella, M., and Marchini, G., Effect of dietary fibre on weight correction after modified fasting, *Eur. J. Clin. Nutr.*, 49, S325, 1995.
113. Vermunt, S.H. et al., Effects of sugar intake on body weight: a review, *Obes. Rev.*, 4, 91, 2003.
114. Brand-Miller, J.C. et al., Glycemic index and obesity, *Am. J. Clin. Nutr.*, 76, 281S, 2002.
115. Wolever, T.M. et al., Beneficial effect of a low glycaemic index diet in type 2 diabetes, *Diabet. Med.*, 9, 451, 1992.
116. Slabber, M. et al., Effects of a low-insulin-response, energy-restricted diet on weight loss and plasma insulin concentrations in hyperinsulinemic obese females, *Am. J. Clin. Nutr.*, 60, 48, 1994.
117. Saris, W.H. et al., Randomized controlled trial of changes in dietary carbohydrate/fat ratio and simple vs. complex carbohydrates on body weight and blood lipids: the CARMEN study. The Carbohydrate Ratio Management in European National Diets, *Int. J. Obes. Relat. Metab. Disord.*, 24, 1310, 2000.
118. Poppitt, S.D. et al., Long-term effects of *ad libitum* low-fat, high-carbohydrate diets on body weight and serum lipids in overweight subjects with metabolic syndrome, *Am. J. Clin. Nutr.*, 75, 11, 2002.
119. Yu-Poth, S. et al., Effects of the National Cholesterol Education Program's Step I and Step II dietary intervention programs on cardiovascular disease risk factors: a meta-analysis, *Am. J. Clin. Nutr.*, 69, 632, 1999.
120. Kris-Etherton, P.M. and Yu, S., Individual fatty acid effects on plasma lipids and lipoproteins: human studies, *Am. J. Clin. Nutr.*, 65, 1628S, 1997.
121. Best, T.H. et al., Cardiac complications in pediatric patients on the ketogenic diet, *Neurology*, 54, 2328, 2000.
122. Sacks, F.M. and Katan, M., Randomized clinical trials on the effects of dietary fat and carbohydrate on plasma lipoproteins and cardiovascular disease, *Am. J. Med.*, 113, 13S, 2002.
123. Ascherio, A., Epidemiologic studies on dietary fats and coronary heart disease, *Am. J. Med.*, 113, 9S, 2002.
124. Teufel, N.I. and Ritenbaugh, C.K., Development of a primary prevention program: insight gained in the Zuni Diabetes Prevention Program, *Clin. Pediatr. (Phila.)*, 37, 131, 1998.
125. Epstein, L.H. et al., Increasing fruit and vegetable intake and decreasing fat and sugar intake in families at risk for childhood obesity, *Obes. Res.*, 9, 171, 2001.
126. The Writing Group for the DISC Collaborative Research Group, Efficacy and safety of lowering dietary intake of fat and cholesterol in children with elevated low-density lipoprotein cholesterol. The Dietary Intervention Study in Children (DISC), *JAMA*, 273, 1429, 1995.
127. Bhargava, A. and Guthrie, J.F., Unhealthy eating habits, physical exercise and macronutrient intakes are predictors of anthropometric indicators in the Women's Health Trial: Feasibility Study in Minority Populations, *Br. J. Nutr.*, 88, 719, 2002.

128. Boyd, N.F. et al., Lack of effect of a low-fat high-carbohydrate diet on ovarian hormones in premenopausal women: results from a randomized trial, *IARC Sci. Publ.*, 156, 445, 2002.
129. Lanza, E. et al., Implementation of a 4-y, high-fiber, high-fruit-and-vegetable, low-fat dietary intervention: results of dietary changes in the Polyp Prevention Trial, *Am. J. Clin. Nutr.*, 74, 387, 2001.
130. Rock, C.L. et al., Reduction in fat intake is not associated with weight loss in most women after breast cancer diagnosis: evidence from a randomized controlled trial, *Cancer*, 91, 25, 2001
131. Boyd, N. et al., Effects at two years of a low-fat, high carbohydrate diet on radiologic features of the breast: results from a randomized trial, *J. Natl. Cancer. Inst.*, 89, 488, 1997.
132. Smith-Warner, S.A. et al., Increasing vegetable and fruit intake: randomized intervention and monitoring in an at-risk population, *Cancer Epidemiol. Biomarkers. Prev.*, 9, 307, 2000.
133. Cox, D. et al., Take Five, a nutrition education intervention to increase fruit and vegetable intakes: impact on consumer choice and nutrient intakes, *Br. J. Nutr.*, 80, 123, 1998.
134. Luepker, R.V. et al., Outcomes of a field trial to improve children's dietary patterns and physical activity. The Child and Adolescent Trial for Cardiovascular Health, CATCH collaborative group, *JAMA*, 275, 768, 1996.
135. Nader, P.R. et al., Three-year maintenance of improved diet and physical activity: the CATCH cohort. Child and Adolescent Trial for Cardiovascular Health, *Arch. Pediatr. Adolesc. Med.*, 153, 695, 1999.
136. Simon, M.S. et al., A randomized trial of a low-fat dietary intervention in women at high risk for breast cancer, *Nutr. Cancer*, 27, 136, 1997.
137. Resnicow, K. et al., A three-year evaluation of the Know Your Body program in inner-city schoolchildren, *Health. Educ. Q.*, 19, 463, 1992.
138. Vandongen, R. et al., A controlled evaluation of a fitness and nutrition intervention program on cardiovascular health in 10- to 12-year-old children, *Prev. Med.*, 24, 9, 1995.
139. Burke, V. et al., A controlled trial of health promotion programs in 11-year-olds using physical activity "enrichment" for higher risk children, *J. Pediatr.*, 132, 840, 1998.
140. Donnelly, J.E. et al., Nutrition and physical activity program to attenuate obesity and promote physical and metabolic fitness in elementary school children, *Obes. Res.*, 4, 229, 1996.
141. Sahota, P. et al., Randomised controlled trial of primary school based intervention to reduce risk factors for obesity, *BMJ*, 323, 1, 2001.
142. Lluch, A., Herbeth, B., Mejean, L., and Siest, G., Dietary intakes, eating style and overweight in the Stanislas Family Study, *Int. J. Obes.*, 24, 1493, 2000.

10 Dietary Carbohydrates and Risk of Cancer

Joanne Slavin

CONTENTS

10.1 INTRODUCTION

Large international differences in rates of cancer, as well as results from migratory studies that find individuals take on the cancer demographics of the population to which they migrate, suggest a strong role for environmental factors, including diet, on cancer incidence. Dietary carbohydrates may be protective against cancer, as is the case for many studies with dietary fiber. Other research suggests that intake of refined carbohydrates, such as sugars, may be linked to a higher risk of cancer, especially if intake of carbohydrates results in obesity. Finally, recent studies suggest that intake of food with a high glycemic index or glycemic load may be linked to increased risk of certain cancers.

The goal of this chapter is to first define and describe carbohydrates in foods that have been linked to cancer. Second, the dietary, metabolic, and physiological effects of food carbohydrates and links to cancer will be discussed. Finally, data to support the use of functional carbohydrates as protective substances in food products will be addressed.

10.2 CARBOHYDRATES IN FOODS

Carbohydrates are the predominant part of the Western diet. Generally, we consume at least 50% of our calories as carbohydrates. Carbohydrates are popular since they taste good, represent the core of many eating patterns (grains, breads, pasta), are inexpensive, and are convenient. The primary role of carbohydrates is to provide energy to the body, particularly the brain, which is the only carbohydrate-dependent organ in the body. Additionally, recommendations for carbohydrate intake for endurance athletes are higher, as much as 70% of calories as carbohydrates.

The Nutrition Facts label in the U.S. divides carbohydrates into total carbohydrates, sugars, and dietary fiber. Carbohydrates can also be subdivided based on the number of sugar units present. Monosaccharides consist of one sugar unit, such as glucose or fructose. Few monosaccharides occur naturally in foods. Disaccharides such as sucrose, lactose, and maltose are considered sugars. Sucrose and lactose are commonly consumed in foods, and maltose is a digestive breakdown product of starch. Sugars improve palatability of foods, but are also added to foods for viscosity, texture, and food preservation. Sugars enter our food supply in fruits and fruit juices, as sucrose in baked foods, and as sweeteners, usually isolated from corn. Corn syrups contain some trisaccharides and longer-chain carbohydrates, but they should still be considered sugars.

Attempts to further categorize sugars have been fraught with difficulty and dissent. In the U.K., sugars were separated into intrinsic and extrinsic sugars. Intrinsic sugars are present within the cell walls of plants (naturally occurring), while extrinsic sugars are typically added to foods. The U.S. Department of Agriculture (USDA) has defined "added sugars" for the purpose of analyzing the nutrient intake of Americans using nationwide surveys. Added sugars are sugars and syrups added to food during processing or preparation, including soft drinks, cakes, other desserts, and candy. Added sugars do not include naturally occurring sugars such as lactose in milk or fructose in fruits. Added sugars are not chemically different from naturally occurring sugars, but many foods and beverages that are sources of added sugars have low nutrient densities, and intakes of such foods need to be limited with the current obesity epidemic.

Oligosaccharides such as raffinose and stachyose are found in legumes, onions, and garlic and are generally 3 to 10 sugar units. Oligosaccharides are poorly digested in the small intestine, escape to the large intestine, and are fermented by the microflora. Polysaccharides include digestible carbohydrates such as starch and nondigestible carbohydrates termed dietary fiber. Even starch is much more complicated since it exists in different chain lengths and is also composed of amylose and amylopectin. Amylose is a straight-chain starch, while amylopectin is branch chained. Additionally, not all starch is digested, and the portion that is not digested, called resistant starch, is considered dietary fiber by new dietary fiber definitions. Glycogen, animal starch, is not consumed as food, but has a structure similar to that of amylopectin.

Glycemic response, glycemic index, and glycemic load have also been suggested as terms to assist in putting carbohydrates into categories that could improve food intake.[1] Glycemic index (GI) is defined as the area under the curve for the increase in blood glucose after the ingestion of a set amount of carbohydrates in an individual food in the 2-h post-ingestion period, compared with ingestion of the same amount of carbohydrates from a reference food (white bread or glucose) tested in the same individual under the same conditions using the initial blood glucose concentration as a baseline. The average glycemic load is derived the same way as the GI, but without dividing by the total amount of carbohydrate consumed. Glycemic load is an indicator of glucose response or insulin demand that is induced by total carbohydrate intake. The two main factors that influence GI are carbohydrate type and the rate of digestion, the latter factor being affected by grain granulation (whether grains are intact or ground into flour), and food firmness resulting from cooking, ripeness, and soluble fiber content. Intrinsic factors such as the amylose–amylopectin ratio and particle size, and extrinsic factors such as food preparation and processing all affect GI. When the glycemic indices of common foods are compared, they range from 126 in low-amylose white rice to 32 in fructose.

The Panel on Dietary Reference Intakes for Macronutrients was responsible for reviewing the research on dietary fiber and disease prevention and deciding whether to set a recommended intake level for dietary fiber. Prior to this report, there was no Recommended Dietary Allowance (RDA) for dietary fiber. The panel also found in its deliberations that there was no official definition of dietary fiber. Thus, a Panel on the Definition of Dietary Fiber was formed to review existing literature on dietary fiber and determine the best scientific definition.[2] New definitions for dietary fiber and recommendations for fiber intake were published in the Dietary Reference Intakes (DRIs).[3] *Dietary fiber* consists of nondigestible carbohydrates and lignin that are intrinsic and intact in plants. *Added fiber* consists of isolated, nondigestible carbohydrates that have beneficial physiological effects in humans. *Total fiber* is the sum of dietary fiber and added fiber. Two categories of fiber are described: *dietary fiber*, fiber in its natural state, and *functional fiber*, fiber that is isolated, manufactured, synthetic, or enzyme-produced. Functional fiber does not have to be plant based. Other important recommendations of the committee were that functional fiber must show a beneficial physiological effect to be classified as functional fiber. Additionally, the committee recommended phasing out the terms *soluble* and *insoluble dietary fiber*. Two properties, viscosity and fermentability, were recommended as meaningful alternative characteristics for the terms *soluble* and *insoluble fiber*.

All fiber is not created equal. Dividing dietary fiber into soluble and insoluble fiber was an attempt to assign physiological effects to chemical types of fiber. Scientific supports that soluble fibers lower serum cholesterol while insoluble fibers increase stool size are inconsistent at best. A meta-analysis testing the effects of pectin, oat bran, guar gum, and psyllium on blood lipid concentrations found that 2 to 10 g/day of viscous fiber was associated with small but significant decreases in total and LDL cholesterol concentrations.[4] Oat bran lowers serum lipids, while wheat bran does not.[5] Resistant starch, generally a soluble fiber, does not affect serum lipids.[6] Thus, not all soluble fibers are hypocholesterolemic agents, and other traits such as viscosity of fiber play a role and must be considered.

The insoluble fiber association with laxation is also inconsistent. Fecal weight increases 5.4 g/g of wheat bran fiber (mostly insoluble), 4.9 g/g of fruits and vegetables (soluble and insoluble), 3 g/g of isolated cellulose (insoluble), and 1.3

g/g of isolated pectin (soluble).[7] Many fiber sources are mostly soluble but still enlarge stool weight, such as oat bran and psyllium. Not all insoluble fibers are particularly good at relieving constipation, for example, isolated cellulose. The disparities between the amounts of soluble and insoluble fiber that are measured chemically and the physiological effects led a National Academy of Sciences panel to recommend that the terms *soluble* and *insoluble fibers* gradually be eliminated and replaced by specific beneficial physiological effects of a fiber, perhaps *viscosity* and *fermentability*.

The DRI committee used the new definitions for *dietary*, *functional*, and *total fiber* in their report. Additionally, they set adequate intakes (AIs) for total fiber in foods of 38 and 25 g/day for young men and women, respectively, based on the intake level observed to protect against coronary heart disease. AI is the recommended average daily intake level based on observed or experimentally determined approximations or estimates of nutrient intake by a group of apparently healthy people that are assumed to be adequate — used when an RDA cannot be determined. There was insufficient evidence to set a tolerable upper intake level (UL) for dietary fiber or functional fibers. The committee concluded that the recommended intake of dietary fiber should also provide protection against cancer, but there is not enough research to set a recommended fiber intake based on cancer prevention. The median intake of dietary fiber ranged from 16.5 to 17.9 g/day for men and 12.1 to 13.8 g/day for women. Thus, there is a large fiber gap to fill between usual intake of dietary fiber and recommended intakes.

Besides the AI for dietary fiber, the DRIs set a Recommended Dietary Allowance (RDA) for carbohydrates of 130 g/day for adults and children. Additionally, the committee recommended that carbohydrate intake as a percentage of calories range from 45 to 65% of kilocalories. The committee suggested a maximal intake of 25% of energy from added sugars, based on ensuring sufficient intakes of essential micronutrients that are for the most part present in relatively low amounts in foods and beverages that are major sources of added sugars in North American diets.

10.3 LINKS BETWEEN CANCER AND CARBOHYDRATE INTAKE

10.3.1 SUGARS AND DIGESTIBLE CARBOHYDRATES

Digestible carbohydrate intake is positively linked to cancer risk because of its role in promotion of obesity. Excess energy intake and obesity are related to higher incidences of many cancers, so it is difficult to isolate the adverse effects of any macronutrient source. Total sucrose intake and glycemic index were linked to increased risk of lung cancer in a case-control study in Uruguay.[8] Augustin et al.[9] reported a direct association between glycemic index and endometrial cancer in a case-control study. Ovarian cancer risk was also linked to glycemic index and glycemic load in an Italian case-control study.[10]

Results with breast cancer and sugar intake have been conflicting and not sufficient to determine a link.[11] Augustin et al.[12] reported that glycemic index and glycemic load were linked to increased breast cancer risk. Jonas et al.[13] found no association between postmenopausal breast cancer risk and dietary glycemic index and load among 63,307 U.S. women in the Cancer Prevention Study II Nutrition Cohort. Insulin resistance and insulin-like growth factors may play a role in development of breast cancer. Cho et al.[14] reported that associations between carbohydrate intake or glycemic load and breast cancer risk among young adult women differed by body weight, with more risk in women with a larger body mass index. The Health-Professional Follow-Up Study found a reduced risk of advanced prostate cancer with increased fructose intake.[15] A diet high in glycemic load may increase the risk of pancreatic cancer in women who already have an underlying degree of insulin resistance.[16]

More data have been published on the relationship between sugar consumption and digestive cancers. Augustin et al.[17] reported that high dietary glycemic index and glycemic load were associated with cancers of the upper aerodigestive tract. The World Cancer Research Fund and the American Institute for Cancer Research[18] reported an increase in colorectal polyps and colorectal cancer risk across intakes of sugar and sugar-rich foods. Satia-Abouta et al.[19] found that total energy intake was consistently associated with colon cancer risk, but associations with individual macronutrients varied by race and by adjustment for energy intake in a North Carolina case-control study. A Canadian prospective cohort study in women found no relationship between diets high in glycemic load, carbohydrates, or sugar and colorectal cancer risk.[20]

10.3.2 Dietary Fiber

Dietary fiber has physiological effects throughout the gastrointestinal tract that may explain its protectiveness against cancer. Although we generally think of dietary fiber as most active in the large intestine, it is known that fiber affects hormones in the upper digestive tract. These changes may alter satiety, slow digestion, and aid in weight maintenance. Additionally, insulin response has been considered most relevant in diabetes prevention, although it also has been linked to risk of colon cancer and breast cancer.

Potential mechanisms for the protective nature of dietary fiber against colon cancer are listed in Table 10.1. The fermentation of dietary fiber in the gut is considered its most important physiological effect. Fermentation of carbohydrate in the colon produces short-chain fatty acids (SCFAs) that help maintain the integrity of the gut.[21] More than 75% of dietary fiber in an average diet is broken down in the large intestine, resulting in the production of carbon dioxide, hydrogen, methane, and short-chain fatty acids, including butyrate, propionate, and acetate. Propionate and acetate are metabolized in colonic epithelial cells or peripheral tissue. Butyrate may regulate colonic cell proliferation and serve as an energy source for colonic cells. Propionate is transported to the liver and may suppress cholesterol synthesis, a potential explanation for how soluble dietary fiber lowers serum cholesterol.

According to calculations by Cummings and MacFarlane,[22] if approximately 20 g of fiber is fermented in the colon each day, approximately 200 m*M* SCFAs will be produced, of which 62% will be acetate, 25% propionate, and 16% butyrate.

TABLE 10.1
List of Mechanisms by which Fiber Can Protect against the Development of Cancer

- Increased stool bulk
    - Decrease of transit time
    - Dilution of carcinogens
- Binds with bile acids or other potential carcinogens
- Lower fecal pH
    - Inhibition of bacterial degradation of normal food constituents to potential carcinogens
- Changes in microflora
- Fermentation by fecal flora to short-chain fatty acids
    - Decrease in colonic pH
    - Inhibition of carcinogens
- Increase in lumenal antioxidants
- Peptide growth factors
- Alteration of sex hormone status
- Change in satiety resulting in lowered body weight
- Alterations in insulin sensitivity and glucose metabolism

Colonic absorption of SCFAs is concentration dependent with no evidence of a saturable process. The mechanism by which SCFAs cross the colonic mucosa is thought to be passive diffusion of the unionized acid into the mucosa cell. SCFAs are respiratory fuels for the colonic mucosa. In isolated human colonocytes, butyrate is actively metabolized to both CO_2 and ketone bodies, which accounts for about 80% of the oxygen consumption of colonocytes. Butyrate is almost completely consumed by the colonic mucosa, while acetate and propionate enter the portal circulation, extending the effects of dietary fiber beyond the intestinal tract.

Butyrate may be an important protective agent in colonic carcinogenesis.[23] Trophic effects on normal colonocytes *in vitro* and *in vivo* are induced by butyrate. In contrast, butyrate arrests the growth of neoplastic colonocytes and inhibits the preneoplastic hyperproliferation induced by some tumor promoters *in vitro*. Butyrate induces differentiation of colon cancer cell lines and regulates the expression of molecules involved in colonocyte growth and adhesion.

The effects of butyrate on colonic tumor cell lines *in vitro* seem to contradict what has been shown *in vivo*.[24] Butyrate appears to have two contrasting effects. It serves as the primary energy source for normal colonic epithelium and stimulates growth of colonic mucosa, yet in colonic tumor cell lines it inhibits growth and induces differentiation and apoptosis. Since SCFAs are volatile, they are quickly absorbed from the lumen. SCFAs acidify the gut, which may affect development of colon cancer because changes in gut pH will affect solubility of metabolites and activities of bacterial enzymes.[25]

Other mechanisms through which dietary fiber affects colon cancer have been examined in animal models. Reddy et al.[26] reported that the concentration of fecal secondary bile acids and fecal mutagenic activity were significantly lower during wheat bran supplementation compared to the control, whereas an oat bran diet supplemented at a level to achieve the same level of fiber had no impact on these measures.

Other studies have examined fiber's ability to increase fecal bulk and speed intestinal transit. Dietary fibers differ in their ability to hold water and their resistance to bacterial degradation in the gut. Pectin is effective in holding water, but is quickly fermented in the gut and cannot be found in feces. Wheat bran consistently has been found to have the most effect on stool bulk, probably because it is slowly fermented and survives transit through the gut. Milling of wheat bran may affect the laxative properties of the bran, with larger particle sizes causing larger increases in fecal weight.

Dietary fiber sources such as wheat bran are complex matrices, and attempts have been made to isolate the effects of chemical components of wheat bran. In an animal study, rats were fed wheat bran, dephytinized wheat bran, and phytic acid alone, and aberrant crypt foci were measured after treatment with azoxymethane.[27] Wheat bran without phytic acid was less protective than intact wheat bran, suggesting that the protective effects of wheat bran include fiber and phytic acid. Certain components of dietary fiber are more protective against colorectal cancer. Insoluble fibers have consistently been found to decrease cell proliferation, while soluble fibers may increase cell proliferation. Lu et al.[28] found that lignin, a component of insoluble dietary fiber, is a free radical scavenger. They suggest that the ability of dietary fiber to protect against colorectal cancer may be determined by the amount of lignin in dietary fiber as well as the free radical-scavenging ability of the lignin.

A usual criticism of animal studies in this area is the large amount of dietary fiber that is fed. Dietary fibers have been fed at levels of 30% of the diet and more. These levels of intake have no bearing on typical or recommended intakes in humans. Yet animal studies allow investigators to screen a wide range of different dietary fibers at many doses.

Two analyses, conducted as meta-analyses, have summarized the observational and case-control epidemiologic studies on dietary fiber and colorectal cancer. Trock et al.[29] analyzed 37 epidemiologic studies that examined the relationship between colorectal cancer and fiber, vegetables, grains, and fruits, either alone or in combination. Overall, 80% of the studies reported up to that time supported the protective role of dietary fiber in colorectal cancer. Howe et al.[30] conducted a combined analysis of data from 13 case-control studies in populations with different colorectal cancer rates and dietary practices. The risk of colorectal cancer decreased incrementally as dietary fiber intake increased. Consumption of more than 31 g of fiber/day was associated with a 50% reduction in risk of colorectal cancer compared to a diet incorporating <11 g/day. The authors estimate that the risk of colorectal cancer in the U.S. population could be reduced by about 31% from an average increase in fiber intake from food sources of about 13 g/day. In contrast, Giovannucci[31] recently concluded that more recent epidemiologic studies have not supported a strong influence of dietary fiber or fruits and vegetables on colorectal cancer.

Le Marchand et al.[32] report a protective role of fiber from vegetables against colorectal cancer that appears to be independent of its water solubility property and of the effects of other phytochemicals. High intake of vegetables, fruits, and grains was associated with a decreased risk of polyps in a case-control study.[33] Lubin et al.[34] found no significant protection against adenomatous polyps in a case-control study. They did find a significant interaction between water and fiber intake. They suggest that fiber and water increase the volume of colonic contents, as well as dilute and adsorb exogenous and endogenous toxic compounds present in the colonic contents. Also, the increased volume of bowel contents promotes peristalsis, reducing the duration of the contact of colonic contents with the mucosa.

Slattery et al.[35] examined eating patterns and risk of colon cancer in a population-based case-control study. The prudent patterns, which included vigorous exercise, smaller body size, and higher intakes of dietary fiber and folate, were associated with a lower risk of colon cancer. In contrast, the Western style was associated with increased risk of colorectal cancer. Mai et al.[36] reported that within a cohort of older women characterized by a relatively low fiber intake, there was little evidence that dietary fiber intake lowered the risk of colorectal cancer.

Peters et al.[37] assessed the relation of fiber intake and frequency of colorectal adenoma within the Prostate, Lung, Colorectal and Ovarian (PLCO) Cancer Screening Trial. High intakes of dietary fiber were associated with lower risk of colorectal adenoma. In this case-control study of over 38,000 subjects, subjects with the highest amounts of fiber in their diets (36 g/day) had the lowest incidence of colon adenomas. Their risk of having an adenoma detected by sigmoidoscopy was 27% less than that of the people who ate the least amount of fiber (12 g/day). Fiber from fruits and grain/cereals was significantly associated with lower adenoma risk, while fiber from vegetables and legumes was not.

The data from large cohort studies are not consistent. In the Health Professional Follow-up Study,[38] dietary fiber was inversely associated with risk of colorectal adenoma in men. All sources of fiber (vegetables, fruits, and grain) were associated with a decreased risk of adenoma. The Nurses' Health Study found no protective effect of dietary fiber on the development of colorectal cancer in women.[39] In the Iowa Women's Health Study, a weak and statistically nonsignificant inverse association was found between dietary fiber intake and risk of colon cancer.[40]

The European Prospective Investigation into Cancer and Nutrition (EPIC) is a prospective cohort study comparing the dietary habits of more than a half-million people in 10 countries with colorectal cancer incidence.[41] It found that people who ate the most fiber (those with total fiber from food sources averaging 33 g/day) had a 25% lower incidence of colorectal cancer than those who ate the least fiber (12 g/day). The investigators estimated that populations with low average fiber consumption could reduce colorectal cancer incidence by 40% by doubling their fiber intake.

Few epidemiologic studies have collected biomarkers of dietary fiber intake. Cummings et al.[42] collected data from 20 populations in 12 countries and found that average stool weight varied from 72 to 470 g/day and was inversely related to colon cancer risk. Dukas et al.[43] reported that in the Nurses' Health Study, women in the highest quintile of dietary fiber intake (median intake, 20 g/day) were less likely to experience constipation than women in the lowest quintile (median intake, 7 g/day).

It is known that different dietary fibers have different effects on stool weight. As summarized by Cummings,[7] wheat bran is most effective in increasing stool weight, with each gram of fiber fed as wheat bran increasing stool weight by 5.4 g.

In contrast, soluble fibers like pectin only increase stool weight by 1.2 g/g of fiber fed as pectin. Yet psyllium, a fiber that is at least 70% soluble, increases stool weight by 4.0 g/g of fiber fed as psyllium. Thus, the laxation properties of a fiber source cannot be predicted based on the solubility of the fiber. The relationship between stool weight and protection from colorectal cancer could be strong while the relationship between dietary fiber intake and colorectal cancer is weak because of all these inconsistencies in fiber measurement and physiological effect.

10.4 INSULIN AND COLON CANCER

Giovannucci[44] has proposed that the etiologies of insulin resistance and colorectal cancer are related. He suggests that diets high in fat and energy and low in complex carbohydrates and a sedentary lifestyle lead to insulin resistance and that the associated hyperinsulinemia, hypertriglyceridemia, and glycemia lead to increased colon cancer risk through the growth-promoting effect of insulin and the increased availability of energy. La Vecchia et al.[45] examined the relationship between diabetes mellitus and the risk of colorectal cancer in an Italian case-control study. They found that subjects with non-insulin-dependent diabetes have a slightly increased risk of colorectal cancer. Allowances for potential confounding factors, including body mass index, diet, and physical activity, could not explain the excess colorectal cancer risk among subjects with diabetes.

10.5 INTERVENTION STUDIES

Several randomized intervention studies with high fiber diets as a component of chemoprevention have been published. Alberts et al.[46] studied the effects of wheat bran fiber (an additional 13.5 g/day as wheat bran cereal) on rectal epithelial cell proliferation in patients with resection for colorectal cancers. They found that the wheat bran fiber cereal inhibited DNA synthesis and rectal mucosal cell proliferation in this high-risk group, which they argued should be associated with reduced cancer risk. They suggested that such a fiber regimen might be used as a chemopreventive agent for colorectal cancers. The study is weakened by poor compliance to the intervention over the 4 years of the study.

A double-blind, placebo-controlled randomized trial with supplements of fiber and calcium and measurement of labeling index in rectal biopsies was conducted.[47] It was concluded that 9 months of high-dose wheat bran fiber and calcium carbonate supplementation in study participants with a history of recently resected colorectal adenomas did not have a significant effect on cellular proliferation rates in rectal mucosal biopsies, comparing 3- and 9-month results to baseline results.

In a randomized trial of intake of fat, fiber, and beta-carotene to prevent colorectal adenomas, patients on the combined intervention of low fat and added wheat bran had zero large adenomas at both 24 and 48 months, a statistically significant finding.[48] The Toronto Polyp Prevention trial demonstrated no significant difference between low-fat/high-fiber and high-fat/low-fiber dietary groups with regard to the recurrence of adenomatous polyps.[49]

Two widely publicized intervention studies do not support the protective properties of dietary fiber against colon cancer.[50,51] The studies found no significant effect of high fiber intakes on the recurrence of colorectal adenomas. Both papers described well-planned dietary interventions to determine whether high fiber food consumption could lower colorectal cancer risk, as measured by a change in colorectal adenomas, a precursor of most large-bowel cancers. Perhaps the fiber interventions were not long enough, the fiber dose was not high enough, and recurrence of adenoma is not an appropriate measure of fiber's effectiveness in preventing colon cancer. Yet the results from the studies are clear. Increasing dietary fiber consumption over 3 years did not alter recurrence of adenomas. Bonithon-Kopp et al.[52] found that the addition of 3.5 g/day psyllium (ispaghula husk) decreased polyp recurrence, with an adjusted odds ratio of 1.67 ($p = 0.042$) for the psyllium fiber intervention on polyp recurrence.

10.6 BREAST CANCER

Limited epidemiologic evidence has been published on fiber intake and human breast cancer risk. Since the fat and fiber content of the diet are generally inversely related, it is difficult to separate the independent effects of these nutrients, and most research has focused on the fat and breast cancer hypothesis. International comparisons show an inverse correlation between breast cancer death rates and consumption of fiber-rich foods. An interesting exception to the high-fat diet hypothesis in breast cancer was observed in Finland, where intake of both fat and fiber is high and the breast cancer mortality rate is considerably lower than in the U.S. and other Western countries, where the typical diet is high in fat. The large amount of fiber in the rural Finnish diet may modify the breast cancer risk associated with a high-fat diet.

A meta-analysis of 12 case-control studies of dietary factors and risk of breast cancer found that high dietary fiber intake was associated with reduced risk of breast cancer.[53] Dietary fiber intake has also been linked to a lower risk of benign proliferative epithelial disorders of the breast.[54] Not all studies find a relationship between dietary fiber intake and breast cancer incidence, including a prospective cohort study reported by Willett and colleagues.[55] Jain et al.[56] also found no association among total dietary fiber, fiber fractions, and risk of breast cancer. Still, nutrition differences, including dietary fiber intake, appear to be important variables that contribute to the higher rate of breast cancer experienced by younger African American women.[57]

Considerable evidence suggests that both breast and colon cancers are hormone-mediated diseases.[58] Few studies have examined the effects of dietary fiber on hormone metabolism while fat content of the diet was held constant. Rose and colleagues[59] reported that when wheat bran was added to the usual diet of premenopausal women, it significantly reduced serum estrogen concentrations, whereas neither corn bran nor oat bran had an effect. Dietary fiber intake was increased from about 15 to 30 g/day in this study, an increase similar to that recommended by the National Cancer Institute. Goldin and associates[60] reported that a high-fiber, low-fat diet significantly decreased serum concentrations of estrone, estrone sulfate, testosterone, and sex hormone binding globulin in premenopausal women. Dietary fiber also caused prolongation of the menstrual cycle by 0.72 day and the follicular phase by 0.85 day, changes thought to reduce overall risk of developing breast cancer.

10.7 OTHER CANCER SITES

Few studies have been reported in this area. Studies of dietary fiber and endometrial cancer have reported both increases and decreases in risk.[61–63] Ovarian cancer risk is decreased with higher intakes of dietary fiber.[64,65] No significant relationships have been reported between dietary fiber intake and risk of prostate cancer.[66,67]

10.8 ADDITION OF CARBOHYDRATES TO FUNCTIONAL FOODS TO REDUCE CANCER RISK

Given the state of knowledge in carbohydrates and cancer risk, it is difficult to make research-based recommendations for specific carbohydrate fractions to add to functional foods. In general, any carbohydrate that is less digestible may be useful since it can reduce calorie intake and potentially aid in weight loss or weight maintenance. Since there are no accepted biomarkers to support the protective properties of food components on cancer prevention, it is impossible to design clinical trials in this area.

The DRI committee concluded that recommended intake levels for dietary fiber could not be set based on the prevention of colon cancer, breast cancer, or other cancers. They suggest that future research trials should evaluate the protectiveness of fiber against colon cancer in subsets of the population by applying genotyping and phenotyping to individuals. Additionally, they suggest increased validation of intermediate markers, such as polyp recurrence, and assessment of functional markers, such as fecal bulk, so future studies can determine if dietary treatments such as carbohydrates have beneficial physiological effects.

Some ongoing studies are examining the effect of prebiotics, probiotics, and synbiotics on cancer prevention.[68] Resistant starch has also been examined as an active carbohydrate source in cancer prevention, although results are inconsistent.[69] Stronger protective support has been found for whole foods high in dietary fiber, such as cereal fiber, fruits, and whole grains. Dietary fiber may be just part of the protective puzzle, with other components including antioxidants, phenolic compounds, and associated substances also providing protection against colorectal cancer.

Other cancer sites are equally elusive as to their connection with dietary fiber intake. Since fiber intake is linked to lower body mass index, it will be protective against breast and prostate cancer. Also, breast cancer may be prevented by high fiber intakes, especially if the fibers consumed are high in phytoestrogens that alter sex hormone metabolism.

Despite many years of research and nutrition education, dietary fiber intakes are not increasing. We must continue to promote consumption of foods high in complex carbohydrates, including resistant starch, oligosaccharides, and dietary fiber.[70] As many consumers depend on processed foods as the mainstay of their diets, efforts should be made to increase the fiber content of popular foods to assist consumers in obtaining recommended levels of unavailable carbohydrates. Differences in fiber composition must be considered since recent studies find that cereal fiber, and not vegetable fiber, is protective against cancer.

REFERENCES

1. Brand-Miller, J.C., Glycemic load and chronic disease, *Nutr. Rev.*, 61, S49, 2003.
2. Institute of Medicine of the National Academies, *Dietary Reference Intakes: Proposed Definition of Dietary Fiber*, National Academies Press, Washington, DC, 2001.
3. Institute of Medicine of the National Academies, *Dietary Reference Intakes: Energy, Carbohydrates, Fiber, Fat, Fatty Acids, Cholesterol, Protein and Amino Acids*, National Academies Press, Washington, DC, 2002.
4. Brown, L. et al., Cholesterol-lowering effects of dietary fiber: a meta-analysis, *Am. J. Clin. Nutr.*, 69, 30, 1999.
5. Anderson, J.W. et al., Lipid responses of hypercholesterolemic men to oat-bran and wheat-bran intake, *Am. J. Clin. Nutr.*, 54, 678, 1991.
6. Jenkins, D.J.A. et al., Physiological effects of resistant starches on fecal bulk, short chain fatty acids, blood lipids and glycemic index, *J. Am. Coll. Nutr.*, 17, 609, 1998.
7. Cummings, J.H., The effect of dietary fiber on fecal weight and composition, in *CRC Handbook of Dietary Fiber in Human Nutrition*, Spiller, G.A., Ed., CRC Press, Boca Raton, FL, 1993, p. 263.
8. De Stefani, E. et al., Dietary sugar and lung cancer: a case-control study in Uruguay, *Nutr. Cancer*, 31, 132, 1998.
9. Augustin, L.S.A. et al., Glycemic index and glycemic load in endometrial cancer, *Int. J. Cancer*, 105, 404, 2003.
10. Augustin, L.S.A. et al., Dietary glycemic index, glycemic load and ovarian cancer risk: a case-control study in Italy, *Ann. Oncol.*, 14, 78, 2003.
11. Burley, V.J., Sugar consumption and human cancer in sites other than the digestive tract, *Eur. J. Cancer Prev.*, 7, 253, 1998.
12. Augustin, L.S.A. et al., Dietary glycemic index and glycemic load, and breast cancer risk: a case-control study, *Ann. Oncol.*, 12, 1533, 2001.
13. Jonas, C.R. et al., Dietary glycemic index, glycemic load and risk of incident breast cancer in postmenopausal women, *Cancer Epidemiol. Biomarkers Prev.*, 12, 573, 2003.
14. Cho, E. et al., Premenopausal dietary carbohydrate, glycemic index, glycemic load, and fiber in relation to risk of breast cancer, *Cancer Epidemiol. Biomarkers Prev.*, 12, 1153, 2003.
15. Giovannucci, E. et al., Calcium and fructose intake in relation to risk of prostate cancer, *Cancer Res.*, 58, 442, 1998.
16. Michaud, D.S. et al., Dietary sugar, glycemic load and pancreatic cancer risk in a prospective study, *J. Natl. Cancer Inst.*, 94, 1293, 2002.
17. Augustin, L.S.A et al., Glycemic index and load and risk of upper aero-digestive tract neoplasms (Italy), *Cancer Causes Control*, 14, 657, 2003.
18. World Cancer Research Fund, American Institute for Cancer Research, *Food, Nutrition, and the Prevention of Cancer: A Global Perspective*, American Institute for Cancer Research, Washington, DC, 1997.
19. Satia-Abouta, A.J. et al., Associations of total energy and macronutrients with colon cancer risk in African Americans and Whites: results from the North Carolina colon cancer study, *Am. J. Epidemiol.*, 158, 951, 2003.
20. Terry, P.D. et al., Glycemic load, carbohydrate intake and risk of colorectal cancer in women: a prospective cohort study, *J. Natl. Cancer Inst.*, 95, 914, 2003.
21. Topping, D.L. and Clifton, P.M., Short-chain fatty acids and human colonic function: roles of resistant starch and nonstarch polysaccharides, *Physiol. Rev.*, 81, 1031, 2001.

22. Cummings, J.H. and MacFarlane, G.T., Colonic microflora: nutrition and health, *Nutrition*, 13, 476, 1997.
23. Valazquez, O.C., Lederer, H.M., and Rombeau, J.L., Butyrate and the colonocyte: implications for neoplasia, *Dig. Dis. Sci.*, 14, 727, 1996.
24. Hague, A., Singh, B., and Paraskeva, C., Butyrate acts as a survival factor for colonic epithelial cells: further fuel for the *in vivo* versus *in vitro* debate, *Gastroenterology*, 112, 1036, 1997.
25. Thornton, J.R., High colonic pH promotes colorectal cancer, *Lancet*, i, 1081, 1981.
26. Reddy, B. et al., Biochemical epidemiology of colon cancer: effect of types of dietary fiber on fecal mutagens, acid, and neutral sterols in healthy subjects, *Cancer Res.*, 49, 4629, 1989.
27. Jenab, M. and Thompson, L.U., The influence of phytic acid in wheat bran on early biomarkers of colon carcinogenesis, *Carcinogenesis*, 19, 1087, 1989.
28. Lu, F.J., Chu, L.H., and Gau, R.J., Free radical-scavenging properties of lignin, *Nutr. Cancer*, 30, 31, 1998.
29. Trock, B., Lanza, E., and Greenwald, P., Dietary fiber, vegetables, and colon cancer: critical review and meta-analyses of the epidemiologic evidence, *J. Natl. Cancer Inst.*, 82, 650, 1990.
30. Howe, G.R. et al., Dietary intake of fiber and decreased risk of cancers of the colon and rectum: evidence from the combined analysis of 13 case-control studies, *J. Natl. Cancer Inst.*, 84, 1887, 1992.
31. Giovannucci, E., Diet, body weight, and colorectal cancer: a summary of the epidemiologic evidence, *J. Women's Health*, 12, 173, 2003.
32. Le Marchand, L. et al., Associations of sedentary lifestyle, obesity, smoking, alcohol use, and diabetes with the risk of colorectal cancer, *Cancer Res.*, 57, 4787, 1997.
33. Witte, J.S. et al., Relation of vegetable, fruit, and grain consumption to colorectal adenomatous polyps, *Am. J. Epidemiol.*, 144, 1015, 1996.
34. Lubin, F. et al., Nutritional and lifestyle habits and water-fiber interaction in colorectal adenoma etiology, *Cancer Epidemiol. Biomarkers Prev.*, 6, 79, 1997.
35. Slattery, M.L. et al., Plant foods and colon cancer: an assessment of specific foods and their related nutrients (United States), *Cancer Causes Control*, 8, 575, 1997.
36. Mai, V. et al., Dietary fibre and risk of colorectal cancer in the Breast Cancer Detection Demonstration Project (BCDDP) follow-up cohort, *Int. J. Epidemiol.*, 32, 239, 2003.
37. Peters, L. et al., Dietary fibre and colorectal adenoma in a colorectal cancer early detection programme, *Lancet*, 361, 1491, 2003.
38. Giovannucci, E. et al., Relationship of diet to risk of colorectal cancer in men, *J. Natl. Cancer Inst.*, 84, 91, 1992.
39. Willett, W.C. et al., Relation of meat, fat, and fiber intake to the risk of colon cancer in a prospective study among women, *N. Engl. J. Med.*, 323, 1664, 1990.
40. Steinmetz, K.A. et al., Vegetables, fruit, and colon cancer in the Iowa Women's Health Study, *Am. J. Epidemiol.*, 139, 1, 1994.
41. Bingham, S.A. et al., Dietary fibre in food and protection against colorectal cancer in the European Prospective Investigation into Cancer and Nutrition (EPIC): an observational study, *Lancet*, 361, 1496, 2003.
42. Cummings, J.H. et al., Fecal weight, colon cancer risk and dietary intake of nonstarch polysaccharides (dietary fiber), *Gastroenterology*, 103, 1783, 1992.
43. Dukas, L., Willett, W.C., and Giovannucci, E.L., Association between physical activity, fiber intake, and other lifestyle variables and constipation in a study of women, *Am. J. Gastroenterol.*, 98, 1790, 2003.
44. Giovannucci, E., Insulin and colon cancer, *Cancer Causes Control*, 6, 164, 1995.

45. La Vecchia, C. et al., Diabetes mellitus and colorectal cancer risk, *Cancer Epidemiol. Biomarkers Prev.*, 6, 1007, 1997.
46. Alberts, D.S. et al., Effects of dietary wheat bran fiber on rectal epithelial cell proliferation in patients with resection for colorectal cancers, *J. Natl. Cancer Inst.*, 82, 1280, 1990.
47. Alberts, D.S. et al., The effect of wheat bran fiber and calcium supplementation on rectal mucosal proliferation rates in patients with resected adenomatous colorectal polyps, *Cancer Epidemiol. Biomarkers Prev.*, 6, 161, 1997.
48. MacLennan, R. et al., Randomized trial of intake of fat, fiber, and beta carotene to prevent colorectal adenomas, *J. Natl. Cancer Inst.*, 87, 1760, 1995.
49. McKeown-Eyssen, G.E. et al., A randomized trial of a low fat high fibre diet in the recurrence of colorectal polyps, *J. Clin. Epidemiol.*, 47, 525, 1994.
50. Schatzkin, A. et al., Lack of effect of a low-fat, high-fiber diet on the recurrence of colorectal adenomas, *N. Engl. J. Med.*, 342, 1149, 2000.
51. Alberts, D.S. et al., Lack of effect of a high-fiber cereal supplement on the recurrence of colorectal adenomas, *N. Engl. J. Med.*, 342, 1156, 2000.
52. Bonithon-Kopp, C. et al., Calcium and fibre supplementation in prevention of colorectal adenoma recurrence: a randomized intervention trial, *Lancet*, 356, 1300, 2000.
53. Howe, G.R. et al., Dietary factors and risk of breast cancer: combined analysis of 12 case-control studies, *J. Natl. Cancer Inst.*, 82, 561, 1990.
54. Baghurst, P.A. and Rohan, T.E., Dietary fiber and risk of benign proliferative epithelial disorders of the breast, *Int. J. Cancer*, 63, 481, 1995.
55. Willett, W.C. et al., Dietary fat and fiber in relation to risk of breast cancer. An 8-year follow-up, *JAMA*, 268, 2037, 1992.
56. Jain, T.P. et al., No association among total dietary fiber, fiber fractions, and risk of breast cancer, *Cancer Epidemiol. Biomarkers Prev.*, 11, 507, 2002.
57. Forshee, R.A. et al., Breast cancer risk and lifestyle differences among premenopausal and postmenopausal African-American women and white women, *Cancer*, 97 (1 Suppl.), 280, 2003.
58. Rose, D.P., Diet, hormones, and cancer, *Annu. Rev. Publ. Health*, 14, 17, 1993.
59. Rose, D.P. et al., High-fiber diet reduces serum estrogen concentrations in premenopausal women, *Am. J. Clin. Nutr.*, 54, 520, 1991.
60. Goldin, B.R. et al., The effect of dietary fat and fiber on serum estrogen concentrations in premenopausal women under controlled dietary conditions, *Cancer*, 74 (3 Suppl.), 1125, 1994.
61. Barbone, F., Austin, H., and Partridge, E.E., Diet and endometrial cancer: a case-control study, *Am. J. Epidemiol.*, 137, 393, 1993.
62. Goodman, M.T. et al., Association of soy and fiber consumption with the risk of endometrial cancer, *Am. J. Epidemiol.*, 146, 294, 1997.
63. McCann, S.E. et al., Diet in the epidemiology of endometrial cancer in western New York (United States), *Cancer Causes Control*, 11, 965, 2000.
64. McCann, S.E. et al., Intakes of selected nutrients and food groups and risk of ovarian cancer, *Nutr. Cancer*, 39, 19, 2001.
65. Risch, H.A. et al., Dietary fat intake and risk of epithelial ovarian cancer, *J. Natl. Cancer Inst.*, 86, 1490, 1994.
66. Andersson, S.O. et al., Energy, nutrient intake and prostate cancer risk: a population-based case-control study in Sweden, *Int. J. Cancer*, 68, 716, 1996.
67. Rohan, T.E. et al., Dietary factors and risk of prostate cancer: a case-control study in Ontario, Canada, *Cancer Causes Control*, 6, 145, 1995.

68. Van Loo, J. and Jonkers, N., Evaluation in human volunteers of the potential anti-carcinogenic activities of novel nutritional concepts: prebiotics, probiotics and synbiotics (the SYNCAN project QLK1-1999-00346), *Nutr. Metab. Cardiovasc. Dis.*, 11 (4 Suppl.), 87, 2001.
69. Van Gorkom, B.A. et al., Calcium or resistant starch does not affect colonic epithelial cell proliferation throughout the colon in adenoma patients: a randomized controlled trial, *Nutr. Cancer*, 43, 31, 2002.
70. Marlett, J.A., McBurney, M.I., and Slavin, J.L., Position of the American Dietetic Association: health implications of dietary fiber, *J. Am. Diet. Assoc.*, 102, 993, 2002.

11 The Role of Carbohydrates in the Prevention and Management of Type 2 Diabetes

Kaisa Poutanen, David Laaksonen, Karin Autio, Hannu Mykkänen, and Leo Niskanen

CONTENTS

11.1 INTRODUCTION

Increasing attention has been focused on the physical properties and physiological effects of carbohydrates in the prevention of type 2 diabetes. Type 2 diabetes, representing about 85% of diabetic patients, is defined as a metabolic disorder of multiple etiology, characterized by chronic hyperglycemia with disturbances of carbohydrate, fat, and protein metabolism resulting from defects in insulin secretion, insulin action, or both. Due to increasing obesity, physical inactitivity, and dietary factors, type 2 diabetes has emerged as a major epidemic of the 20th and 21st centuries. The global prevalence is estimated to increase from the current approximate 150 million cases to 300 million in 2025.[1] The costs of treatment of type 2 diabetes for the community are escalating due to the markedly increased risk of vascular complications. Microvascular complications, e.g., retinopathy, neuropathy, and nephropathy, are rather specific to diabetes. There is convincing evidence that treatment of hyperglycemia reduces the risk of microvascular complications, and treatment of cardiovascular risk factors prevents macrovascular complications. What is especially important is that recent intervention studies have shown that type 2 diabetes can be prevented by modest lifestyle changes, including increased dietary fiber intake, decreased total and saturated fat intake, increased physical activity, and moderate weight loss.[2]

Epidemiological evidence suggests that carbohydrates with a low glycemic index or load and high in cereal fiber content decrease the risk of type 2 diabetes. Intervention studies indicate that soluble fiber favorably affects postprandial insulin and glucose responses in both diabetic and nondiabetic individuals and, at least in type 2 diabetes patients, improves long-term glycemic control. Trial evidence on the role of the physical properties of carbohydrates in the development of type 2 diabetes and the more minor disturbances in glucose and insulin metabolism that precede it is more sketchy and inconsistent. Not only does the source of dietary carbohydrates per se affect glucose and insulin metabolism, but also the manner in which food is prepared.

This chapter reviews briefly glucose and insulin metabolism in relation to dietary carbohydrates, the pathogenesis of type 2 diabetes, and the metabolic disturbances that generally precede it. We will then focus on the role of the glycemic index and dietary fiber in reducing the risk of type 2 diabetes. Finally, we will address factors controlling the rate of glucose absorption from starchy foods, including food preparation and food structure.

11.2 GLUCOSE AND INSULIN METABOLISM IN RELATION TO DIETARY CARBOHYDRATES

Carbohydrates are the major source of energy in the Western diet. According to the population nutrient intake goals for preventing diet-related chronic diseases, 55 to 75% of the daily energy intake should come from carbohydrates.[3] However, this is often not reached, and in many countries increased carbohydrate intake is recommended. Nutritionally, carbohydrates can be classified as digestible and nondigestible, or available and nonavailable.[4] Mono- and disaccharides and starch, the most common digestible carbohydrate, are rapidly digested and absorbed in the small intestine. Dietary fiber is a nutritional definition for the nondigestible part of plant food and includes many oligosaccharides, cellulose, hemicellulose, pectins, gums, lignin, and associated plant substances.[5] For prevention of diet-related chronic diseases, an intake of at least 25 g/day of dietary fiber is recommended,[3] but in many countries this level of dietary fiber intake is not reached.

Consumption of digestible carbohydrates leads to elevation of the blood sugar level. The rate of gastric emptying influences blood glucose and insulin values.[6] Gastric emptying, in turn, is affected by particle size after chewing[6,7] and by the presence of viscous food polymers.[8] The current Western diet is often rich in rapidly digestible carbohydrates, which are absorbed quickly in the upper part of the small intestine and cause a rapid postprandial rise in blood glucose concentration. This normally reaches the peak value within 30 to 45 min, and then gradually decreases, returning to the fasting level within 90 to 180 min. The bloodstream carries the absorbed monosaccharides to the liver via the portal vein. As glucose is a key nutrient for many tissues, and the critical source of energy for some tissues (brain, kidney cortex, erythrocytes), the changes in plasma glucose are regulated within very narrow limits in healthy humans. The increase in plasma glucose after a meal is followed by a rapid discharge of insulin from pancreatic cells. This, in turn, suppresses the output of glucose from the liver and stimulates the uptake of glucose from plasma into the peripheral tissues.[9,10] Also, part of the glucose is taken up by the liver and converted to glycogen. The largest part of glucose is taken up by the muscle tissues.[11] The major part of absorbed fructose and galactose is taken up by liver and requires further metabolism before contributing to blood glucose. Their contribution to blood glucose level is small, but if excess fructose is consumed, hyperglycemia may result since fructose does not stimulate secretion of insulin.[12]

Glucose uptake and further metabolism in different target cells are controlled by insulin. Basal insulin secretion occurs in pulses of 10 to 12 min.[13] This provides a more efficient control of the glycemic response than a steady insulin infusion. Postprandial insulin secretion has two phases: a short acute first phase is seen within the first 10 min of the stimulus, followed by a second phase lasting until the stimulus ceases.[14] The glycemic response is the main stimulus to insulin secretion, but other factors, such as nutrients, hormones, and neurotransmitters, may potentiate or inhibit the secretory response to glucose.[15–17] For example, when ingested with glucose, arginine attenuates and prolongs the rise of blood glucose.[18]

11.3 PATHOGENESIS OF THE METABOLIC SYNDROME AND TYPE 2 DIABETES

Type 2 diabetes not only is a manifestation of the metabolic syndrome, but also, in most cases, is preceded by the metabolic syndrome.[19] The metabolic syndrome is a concurrence of disturbed glucose and insulin metabolism, overweight and abdominal fat distribution, dyslipidemia, and hypertension. The hallmark of the syndrome is resistance to the biological effects of insulin in target tissues, thus why it is also known as the insulin resistance syndrome. The metabolic syndrome as defined by the National Cholesterol Education Program Expert Panel (2001) and the World Health Organization[20] is common, present in about 24% of all adults in the U.S., and in somewhat smaller proportions in most European countries.

The metabolic syndrome constitutes a major threat to public health because of its association with a 5- to 10-fold increased risk of type 2 diabetes mellitus and a 2- to 3-fold higher risk of cardiovascular disease.[19,21–23] Cardiovascular risk factors such as dyslipidemia, hypertension, endothelial dysfunction, inflammation, hypercoagulability and impaired fibrinolysis, obesity, and abnormal insulin and glucose metabolism, all of which are part of or closely associated with the metabolic syndrome, contribute to this increased risk.

The pathogenesis of the metabolic syndrome is poorly understood. Skeletal muscle is a major determinant of whole-body glucose disposal. Defects in insulin signaling in muscle tissue contribute to lowered insulin-stimulated glucose uptake.[24] Furthermore, an abdominal distribution of fat is particularly deleterious. Abdominal obesity has been hypothesized to mediate its deleterious effects on carbohydrate and lipid metabolism through the increased lipolytic activity of especially omental fat, which drains directly into the portal-venous system.[25] This, in turn, results in higher nonesterified fatty acid concentrations, with consequent insulin resistance in the liver and skeletal muscle. In addition to abdominal subcutaneous and visceral fat, lipid accumulation in skeletal muscle and the liver has also been shown to be a powerful determinant of insulin sensitivity.[26]

As the metabolic syndrome becomes more severe, interplay among genetic susceptibility, insulin resistance, and dietary patterns may lead to progressive β-cell failure and impaired insulin secretion capacity. As β-cell function declines, impaired glucose tolerance (IGT) develops.[27] At this point, pulsatile and first-phase insulin secretion, essential for normal glucose tolerance, is impaired, and an exaggerated longer-lasting second phase occurs in compensation. On a yearly basis, in roughly 5 to 10% of persons with IGT, it converts to type 2 diabetes.[28] Type 2 diabetes is a heterogeneous clinical entity, however, and not all patients have features of metabolic syndrome; about 10 to 20% have insulin secretion deficiency as the major contributor to hyperglycemia.

Hyperglycemia begets further loss of pancreatic β-cell function. It has not been fully resolved whether this loss of pancreatic function results primarily from excessive secretion of insulin (i.e., β-cell exhaustion) or toxicity to β-cells because of hyperglycemia. By definition, high intake of carbohydrates with a high glycemic index (GI) produces high concentrations of plasma glucose and increased insulin demand, and may therefore hypothetically contribute to an increased risk of type 2 diabetes. The individual response to a given carbohydrate load is influenced by the degree of underlying insulin resistance and impaired insulin secretion, which is, in turn, determined primarily by the degree and type of adiposity, physical activity, genetics, and diet. Thus, it might be expected that the adverse metabolic effects of high-GI, low-fiber foods would be pronounced in sedentary, overweight, or genetically susceptible persons and be quite modest in the healthy university students that frequently participate in metabolic studies.

11.4 GLYCEMIC INDEX AND GLYCEMIC LOAD

The GI concept was developed over 20 years ago by Jenkins et al.[29] to classify foods based on their effects on the blood glucose level. The original aim was to aid diabetic patients in controlling their postprandial hyperglycemia. It ranks individual foods according to their postprandial rate of carbohydrate digestion and absorption. The GI is defined as the incremental area under the postprandial blood glucose curve (change in blood glucose level 3 h after a meal) after the consumption of 50 g of (digestible) carbohydrates from a test food, divided by the area under the corresponding curve after a meal containing a similar amount of the reference food, normally white bread or glucose.[30] The reference is given the value 100, and the lower the response, the smaller the GI. Using white wheat bread as a reference gives higher GI values than using glucose. Also, white rice has been suggested as the reference, especially for Asian populations.[31] Not only the reference carbohydrate, but also the reference group in which the carbohydrate has been measured pose problems in standardizing the values for GI. Values for the glycemic index have frequently been derived from (small) samples of individuals with normal glucose tolerance, impaired glucose tolerance, or type 1 or type 2 diabetes. The glucose tolerance of the individuals tested may affect the GI obtained, even when comparing with a standard reference.

Carbohydrates frequently have been divided into refined and unrefined or simple and complex carbohydrates. With respect to the glycemic index, however, these divisions are not very useful: unrefined potato products, which are classified as complex carbohydrates, frequently have GIs over 100, whereas pasta, containing refined carbohydrates, has a GI of around 53. Fruits, which contain simple carbohydrates in the form of disaccharides, also have a low GI. Similarly, whole-wheat (unrefined) bread has a GI similar to that of white (refined) bread. The GIs of starchy foods range from higher than 130 to lower than 30 (with white bread as the reference).[32] Many potato and bread products have high values, and unprocessed grains, pasta, and legumes have low values. Summary tables of the GIs of over 750 different food items are available.[32,33]

Not only the quality, but also the quantity of carbohydrates in a serving influence the postprandial glycemia. Glycemic load (GL) has been proposed to take into account both. GL is calculated by multiplying the GI of a food with the amount of total dietary carbohydrate per serving.[32] Assessing the total glycemic effect, GL has proved useful in epidemiological studies.[34–36] Table values are currently available for a range of food items.[32] There are also problems with respect to GL; reducing the total intake of carbohydrates or dietary GI decreases GL, but the long-term effects may not be the same.

One of the problems with the GI concept is that the insulin response is not directly proportional to the glucose response. The glycemic index has explained anywhere from about 49[16] to 79%[37] of the variability. This is in line with our experience in which, in healthy subjects with normal glucose tolerance, there was no difference in the glycemic responses between wheat or different rye breads, whereas insulin responses to rye breads where markedly different, so that the same amount of carbohydrate required less insulin.[38] Therefore, it seems that in healthy persons the use of GI to classify foods with respect to their effects on glucose metabolism is hampered with many difficulties, and insulin responses could be used instead. However, even if the total glucose responses did not differ between the different types of breads,[38] the shapes of the glucose curves showed interesting patterns: 3 h after ingestion of the wheat bread, plasma glucose levels had declined below the fasting level, whereas after rye bread ingestion glucose levels were still above the fasting level. The rapid decrease of blood glucose may increase the feeling of hunger and may cause more frequent snacking, which may be relevant with respect to weight gain.[39]

11.5 LOW-GI CARBOHYDRATES IN THE TREATMENT OF DIABETES

Current dietary guidelines usually recommend high intakes of carbohydrates. The main reason is that high-fat diets are thought to enhance the development of cardiovascular diseases and promote weight gain. Both suppositions have been questioned.[40–43] Regarding dietary fat, there is a general consensus that the type of fat is at least as important as the quantity of fat. On the other hand, reducing postprandial glucose is deemed beneficial in the dietary management of type 2 diabetes. In principle, this can be achieved by reducing the amount of dietary carbohydrates or reducing the GI of the diet. However, the American Diabetes Association guidelines state that with regard to the glycemic effect of carbohydrates, the total amount is more important than the source or type.[44] However, one can anticipate that we are encountering the same phenomenon with carbohydrates as with types of lipids in the diet; the quality is at least as important as the quantity.

11.5.1 Medium- to Long-Term Studies

In studies conducted mostly in patients with type 2 diabetes, the low-GI diets have shown an improvement in glycated hemoglobin (HbA_{1c}) of 10%, compared with high-GI foods.[45] This effect is clinically significant and comparable to the effects of monounsaturated fat with high-carbohydrate diets. Recently, a meta-analysis identified 14 studies, comprising 356 subjects, that met strict inclusion criteria and showed that low-GI diets reduced HbA_{1c} by 0.43% points over and above that produced by high-GI diets.[46] Taking both HbA_{1c} and fructosamine data together and adjusting for baseline differences, glycated proteins were reduced 7.4% more on the low-GI diet than on the high-GI diet. Therefore, choosing low-GI foods in place of conventional or high-GI foods has a small, but clinically significant, effect on medium-term glycemic control in patients with diabetes. The benefit is similar to that offered by pharmacological agents that also target postprandial hyperglycemia.

In a carefully conducted Swedish study[47] in patients with type 2 diabetes, the baseline diet for two study groups was in line with the current recommendations, but differed in GI. Although both diets improved insulin sensitivity and serum lipids, food modification by lowering GI about 30% resulted in lower low-density lipoprotein (LDL) cholesterol, glucose and insulin responses, and even a beneficial effect on plasminogen activator inhibitor-1. This implies improvement of impaired fibrinolytic capacity, a common phenomenon in subjects with metabolic syndrome. Low GI has been shown to have beneficial effects on LDL cholesterol levels in patients with type 2 diabetes during weight-loss diets.[48] The EURODIAB study showed that low GI was associated with better metabolic control, more favorable lipoprotein pattern, and lower waist circumference also in subjects with type 1 diabetes.[49]

11.6 LOW-GI CARBOHYDRATES AND RISK OF TYPE 2 DIABETES

11.6.1 Epidemiological Studies

In a large cohort of over 65,000 American nurses, the risk of diabetes over a 6-year follow-up was 40% higher in women in the highest fifth than those in the lowest fifth of the glycemic index and glycemic load, even after adjusting for baseline BMI and cereal fiber intake.[35] In updated analyses with 6 more years of follow-up, 5 more dietary assessments, and 3300 more cases, findings remained similar.[50] Dietary fiber per se is important, since products having a higher fiber content have less starch and a lower glycemic load, but the above associations were independent of and multiplicative to those of cereal fiber. Among the nurses, women with both a high glycemic load and low cereal fiber intake were 2.5 times more likely to develop diabetes than those in the respective lower and upper thirds. A similar apparent protective effect of the low glycemic index and glycemic load was seen in 43,000 American male health professionals.[34] In contrast, despite similar findings of an inverse association of cereal fiber and whole-grain intake, and no association of refined or total carbohydrate intake with incident cereal fiber with incident diabetes, no association of the glycemic index or load was found in a study of 36,000 Iowan women.[51] A possible explanation for the discrepancy may be related to the assessment of diet only at baseline in the Iowan women. Furthermore, although in both studies diabetes was self-reported, the latter study was in women without a medical background, and no confirmation was made of the diagnosis.

Several reviews on the glycemic index and risk of diabetes have recently been published.[39,52–56] Prospective epidemiological studies addressing the association of dietary factors with the metabolic syndrome have not yet been published, and there is an urgent need for controlled randomized studies in these groups of subjects. Large-cohort studies assessing the association of incidence of diabetes mellitus or cardiovascular disease nonetheless suggest that a diet that has a low glycemic index, is high in fiber, or is high in fruit and vegetable intake, may decrease the risk for obesity, type 2 diabetes, or cardiovascular disease and its risk factors.

In theory, low-GI foods may have a preventive effect on the development of type 2 diabetes by influencing the fundamental defects related to insulin resistance and impaired β-cell function. Furthermore, although evidence is limited and somewhat inconsistent, low-GI diets may also promote weight loss (see Chapter 9, this book), which strikingly influences insulin sensitivity.[57] Indirect evidence for the glycemic index and soluble fiber in the prevention of diabetes is provided by the glucosidase inhibitor acarbose, which delays intestinal absorption of carbohydrates.[58] Acarbose decreases postprandial insulin and glucose responses and improves glycemic control in diabetic patients. In 1429 individuals with IGT, acarbose has been shown to decrease the risk of diabetes by 25%, which was independent of the small weight loss (0.77 kg) that occurred in the acarbose group.

11.6.2 Postprandial Studies: Second-Meal Effects

The effect of low-GI foods on glucose and insulin balance has been shown to extend even to the next meal, so that foods eaten during dinner might influence the glycemic response at breakfast.[59–61] Some low-GI foods eaten at breakfast (e.g., pasta) have been reported to maintain a net increment in blood glucose and insulin at the time of lunch, thus reducing postprandial glycemia and insulinemia.[62] Recently it has been suggested that dietary fiber content not only slows down glucose absorption, but perhaps also contributes to the second-meal effect. An evening meal containing low-GI (53) barley meal rich in β-glucan improved glucose tolerance at breakfast, whereas an evening meal with pasta (GI = 54) did not have this effect.[61] Factors influencing the second-meal effect need to be studied further, but it is clearly beneficial if low-GI cereal foods are also rich in soluble fiber.

11.6.3 Medium- to Long-Term Studies

In a study by Frost et al.,[63] women were randomly assigned to consume high- or low-GI diets for 3 weeks. Insulin resistance measured *in vivo* with a short insulin tolerance test and in cultured adipocytes was greater in women consuming the high-GI diet. The adverse effects of the high-GI diet appeared to be due to an increased production of free fatty acids in the late postprandial state. In a study by Kiens and Richter,[64] an increase in insulin resistance was not found in seven healthy, lean young men after the subjects had consumed a high-GI diet. It should be noted, however, that subjects were quite healthy and had normal insulin sensitivity, underscoring the concept that metabolic effects of GI are difficult to observe in the healthy population.

A recent crossover study involving 11 overweight subjects showed that insulin sensitivity measured by the euglycemic hyperinsulinemic clamp improved after subjects consumed a whole-grain diet, compared with a refined-grain diet, for 6 weeks, independent of body weight.[65] In another uncontrolled study by the same group,[66] a 4-week low-glycemic diet also decreased an insulin sensitivity index calculated from an oral glucose tolerance test, but the glucose area under the curve only tended to decrease. In contrast, this group[67] found no effect on fasting glucose or insulin concentrations of a 12-week dietary intervention decreasing the GI of the diet (71 vs. 81) in 55 individuals participating in a randomized controlled trial.

Hypothetically, β-cell failure may be induced by high-GI foods by repeated postprandial hyperinsulinemia, leading to overstimulation and exhaustion of the β-cells.[68,69] In a crossover study of six healthy adults, Jenkins et al.[70] found that a low-GI diet containing mainly intact whole grains reduced C-peptide concentrations, a crude measure of insulin secretion, about 32% compared with a high-GI diet containing primarily refined-grain products. Recently, we showed that long-term ingestion of rye bread in elderly women enhanced the acute insulin response to an intravenous glucose tolerance test,[71] which could be a novel mechanism for the effects of low GI on glucose and insulin.

In this regard, it is interesting to address the recent carefully conducted study on six healthy men by Schenk et al.,[72] who showed by the stable-isotope technique that the low GI of high-fiber breakfast cereals, compared with high-GI breakfast cereals, was not due to lower systemic appearance of glucose, but instead to earlier disappearance of glucose, which was mainly due to early hyperinsulinemia. These findings support the study by Juntunen et al.[71] and suggest that one of the hitherto unanticipated mechanisms of foods with a low GI and high in whole grain may be to enhance the first phase of insulin secretion. Because the loss of the first phase of the insulin response to glucose is one of the early pathogenetic abnormalities in the development of type 2 diabetes, the potential importance of dietary modification is further strengthened and new developmental avenues are opened.

Low-GI foods may also have relevance with fuel oxidation. Postprandial rises in glucose and insulin concentrations increase carbohydrate oxidation and decrease fatty acid oxidation, but during later postprandial phases, increased release of counterregulatory hormones restores euglycemia and elevates free fatty acid (FFA) levels. Increased availability and oxidation of fatty acids may, in turn, decrease carbohydrate oxidation.[54] Recently, Wolever and Mehling[73] randomly assigned 24 subjects with IGT to a high-carbohydrate + high-GI, high-carbohydrate + low-GI, low-carbohydrate, or high-monounsaturated-fatty-acid (MUFA) diet for 4 months. Postprandial glucose levels were reduced by the same amount with high-GI and -MUFA diets, compared to the low-GI diet, but curiously, HbA_{1c} increased with the MUFA diet. The changes in FFA levels were not significant between the groups, but fasting insulin levels fell in high-GI and -MUFA groups. Weight changes were small, but significantly more weight was actually lost with the high-GI diet than with the other two diets, challenging the weight-controlling effects of a low-GI diet.

In summary, there are still methodological problems to be worked out for the glycemic index. Prospective epidemiological evidence supporting a role of the glycemic index is mainly limited to two cohort studies, although these are convincing in that they are both large and have repeated updating of dietary information. Medium- to long-term trial evidence is limited and conflicting. There is clearly a need for randomized, controlled, multicenter intervention studies comparing the effects of conventional and low-GI diets and fat-modified diets on insulin and glucose metabolism.

11.7 DIETARY FIBER IN THE TREATMENT OF HYPERGLYCEMIA IN TYPE 2 DIABETES

Soluble dietary fiber, such as oat β-glucan, psyllium, and guar gum, has been recommended in type 2 diabetic patients for improvement of especially postprandial insulin and glucose metabolism, in addition to antihyperlipidemic effects. It is well documented that viscous polysaccharides in the diet slow down the rate of carbohydrate digestion and absorption. A main reason for the slower responses is the hindered mixing of luminal contents, causing retarded diffusion and contact between gastrointestinal enzymes and their substrates and retarded transport through the gut.[74,75] An inverse highly significant linear relationship has been shown between the postprandial blood glucose and insulin responses and the viscosity of the liquid mixtures consumed.[76,77] The importance of viscosity for glycemic response was already emphasized by Jenkins et al.[78] Both soluble[79] and insoluble[80] dietary fibers have also been reported to increase mucin production, which increases viscosity beyond fiber alone. Insoluble dietary fiber does not have effects on glucose and insulin metabolism, at least in the short term,[81] despite epidemiological evidence that insoluble fiber may prevent diabetes independently of glycemic load or body mass index. We will not focus on the effects of dietary fiber on weight loss and weight control in type 2 diabetes, but the results have been conflicting.[82–84]

11.7.1 Postprandial Studies

Oat β-glucan is well known for its ability to reduce postprandial glucose and insulin responses after an oral glucose load in diabetic patients.[85–87] Both isolated oat gum[88] and an oat bran containing β-glucan[85,87] have been shown to be effective. In a study by Tappy et al.,[87] progressive decreases in blood glucose of 33 to 63% and in insulin of 33 to 41% relative to the control breakfast were achieved with the addition of 4 to 8.4 g of β-glucan to the oat bran breakfast. The efficacy is dependent on viscosity, however, and may be lost if viscosity is reduced in food processing due to, e.g., enzymatic breakdown. The viscosity of β-glucan, as measured in an assay simulating the physiological conditions, has also been emphasized in the case of oat bran-containing breakfast cereals.[87] It has also been pointed out that predicting the action of a polysaccharide on the basis of preingestion viscosity can be misleading, and that viscosity should be high at the site of absorption in the gut.[89]

The viscous polysaccharide psyllium[30,90,91] and guar gum[92–95] decrease postprandial glucose and insulin levels in patients with type 2 diabetes. In some studies, however, no effect of guar gum administration on postprandial glucose levels has been found, even at a dose of 5 g with meals.[96]

11.7.2 Medium- to Long-Term Studies

Medium- to long-term (i.e., lasting from weeks to months) studies of dietary soluble fiber other than guar gum on improving glycemic control in diabetes are not very common. A study comparing the effects over 6 months of barley bread, high in β-glucan, to white bread found that barley bread improved glycemic control compared with white wheat bread in 11 men with type 2 diabetes.[97] Insulin responses were increased, which hypothetically could reflect recovered β-cell function.

In men with diabetes and hypercholesterolemia participating in a crossover trial, 8 weeks of psyllium (15 g/day) decreased hemoglobin A_{1c} 6.1% (absolute change, 0.8%), with similar 6% decreases in fasting postprandial glucose.[98] A second-meal effect was noted, with the most marked decreases in postprandial glucose after lunch, 31% relative to the control. No changes in insulin sensitivity measured during a euglycemic hyperinsulinemic clamp were noted, however.

Improved fasting and postprandial glycemic control was found in 11 type 2 diabetic patients taking 21 g/day of guar gum or placebo in a randomized double-blinded crossover trial.[82] Small improvements in overall glycemic control and sizable improvements in postprandial glycemia after 4 weeks of treatment in a randomized controlled crossover trial were reported by Fuessl et al.[94] Guar gum decreased fasting blood glucose from 11.4 to 9.5 mmol/l in 19 obese patients with type 2 diabetes who were enrolled in a randomized double-blind crossover trial.[99] Guar gum (15 g/day) has also improved long-term glycemic control and postprandial glucose tolerance in 15 type 2 diabetic patients treated with guar gum over an 8-week period.[84] On the other hand, guar gum had no effect on fasting blood glucose or hemoglobin A_{1c} levels in a 3-month trial in type 2 diabetic patients on oral hypoglycemic medication,[100] or in an 8-week randomized crossover trial in type 2 diabetic patients treated by diet, oral agents, or lente insulin.[96] Guar gum has also decreased fasting glucose levels and hemoglobin A_{1c} in patients with type 1 diabetes participating in a randomized, double-blinded, placebo-controlled trial. In contrast, Ebeling et al.[101] found only improvements in postprandial glucose responses and a decrease in insulin doses without an improvement in overall glycemic control or insulin sensitivity. Bruttomesso et al.[102] found no benefits of adding 15 g/day of guar gum to meals.

In summary, dietary soluble fiber such as β-glucan, psyllium, and guar gum decreases glucose and insulin responses to carbohydrates if taken in sufficient amounts. A second-meal effect has also been described. Improved glycemic control with long-term use, over periods of weeks to months, has been suggested in isolated studies for β-glucan and psyllium, but more studies are needed. For the best studied dietary soluble fiber, guar gum, findings are conflicting.

11.8 DIETARY FIBER AND RISK OF TYPE 2 DIABETES

There are no trials focused on fiber intake alone in modifying the risk of diabetes in nondiabetic individuals. A multifactorial intervention including increasing fiber intake (mainly insoluble) as a component has been shown to prevent the development of type 2 diabetes.[2] In the Finnish Diabetes Prevention Study (DPS), 523 obese persons with IGT were randomized into an intervention or control group. During the trial the risk reduction of diabetes was 58%. When the results were analyzed according to the success score, none developed diabetes in either group if they achieved four or five intervention goals (weight loss at least 5%, physical activity at least 4 h/week, fiber intake > 15 g/1000 kcal, intake of total fat < 30% of energy, and intake of saturated fat <10% of energy). *Post hoc* analyses of individual intervention goals in the prevention of diabetes have not been reported.

11.8.1 Epidemiological Studies

The best epidemiological evidence for dietary fiber in the prevention of type 2 diabetes is ironically for insoluble dietary fiber. Two large-cohort studies, one in over 70,000 American nurses[35] and the other in over 42,000 American male health professionals,[34] showed that cereal fiber, but not fruit or soluble fiber, decreased the risk of diabetes during follow-up by 28 to 30% for the upper vs. lower thirds, even when taking into account the glycemic load of the diet and other potentially confounding factors. Findings in 36,000 women from Iowa confirmed these results.[51]

The mechanisms by which cereal fiber, which in the diets of these cohorts contains only small amounts of soluble fiber, such as β-glucan and arabinoxylan, may prevent diabetes independently of glycemic load and body weight is unclear, because insoluble fiber does not affect insulin and glucose metabolism, at least in the short term. It is possible that the association is due to other compounds in whole grains, but most of the apparent preventive effect of whole-grain foods against diabetes[51,103–105] seems to be mediated by cereal fiber or components closely associated with it. Epidemiological studies suggest that cereal fiber and whole-grain products also appear to prevent obesity and weight gain, in addition to independent effects on decreasing the risk for diabetes,[106] but trial evidence is conflicting. This topic will be covered in greater detail in the chapter on obesity (see Chapter 9, this book), but mechanisms that decrease obesity and weight gain are also critical in the prevention of type 2 diabetes. There is also currently controversy over whether soluble fiber can aid in weight loss.

Epidemiological evidence for soluble fiber in the prevention of diabetes is lacking. This may be because high quantities of soluble fiber, much more than normally consumed in Western diets, are required to alter postprandial insulin and glucose responses. Furthermore, dietary assessment in epidemiological studies is imprecise.

11.8.2 Postprandial Studies

Oat β-glucan at sufficient doses decreases postprandial insulin and glucose responses in healthy individuals.[76,85,88,107] Five grams or more of psyllium per dose also decrease postprandial glucose and insulin responses in nondiabetic subjects,[108,109] but doses of only 1.7[66] or 2.3 g[110] do not. Guar gum has decreased postprandial insulin levels, but not glucose concentrations, in nondiabetic individuals.[111,112] However, guar gum has decreased both postprandial insulin and glucose levels in several studies.[113–116] A low-GI (53) barley meal rich in dietary fiber eaten in the evening improved glucose tolerance at breakfast, whereas an evening meal with pasta (GI 54) had no effect, suggesting the importance of dietary fiber for second-meal effects.[61] Similar findings have been shown for breakfast and lunch.[117]

11.8.3 Medium- to Long-Term Studies

Longer-term studies on the effects of oat bran, psyllium, and guar gum on insulin and glucose metabolism in nondiabetic individuals are sparse. In 36 overweight and obese middle-aged and older nondiabetic glucose men, effectiveness improved, but insulin sensitivity and the acute first-phase insulin response as measured by the minimal model test remained unchanged in men who consumed oat cereal (14 g of dietary fiber, 5.5 g of β-glucan) for 12 weeks, compared with wheat cereal.[118] A 4-week β-glucan-enriched (8 to 12 g/day) barley diet did not have a significant effect on glucose tolerance in a crossover trial in 18 somewhat overweight hyperlipidemic men.[119] Oat bran (100 g/day) or oat bran containing β-glucan (10.3 g/day) did not affect fasting glucose or insulin levels in hypercholesterolemic individuals participating in a randomized crossover trial (n = 8)[120] or a randomized parallel trial (n = 52).[121] Psyllium at a dose of 3.4 g with meals did not lower fasting glucose levels in two 8-week randomized controlled trials with 26[122] and 75[123] mildly to moderately hypercholesterolemic patients.

In summary, soluble dietary fiber such as β-glucan, psyllium, and guar gum at sufficient doses (3.7 to 14 g/meal[124]) decreases postprandial insulin and glucose responses in healthy individuals. A second-meal effect has also been described. Longer-term studies on the effects of oat bran, psyllium, and guar gum on glucose and insulin metabolism are few and have shown no benefit. On the other hand, studies in groups in which such benefits might be expected to occur, e.g., in persons with IGT or the metabolic syndrome, have not been carried out. Furthermore, studies lasting longer than 1 or 2 months may be needed. The methodology in assessing insulin and glucose metabolism has often been restricted to fasting glucose and insulin determinations, and has not included oral and i.v. glucose tolerance tests or the euglycemic hyperinsulinemic clamp.

11.9 DEVELOPMENT OF CARBOHYDRATE FOODS AND PREVENTION OF TYPE 2 DIABETES

As already demonstrated above, a number of food factors affect the rate of glucose delivery and uptake in the human body. Some are related to the characteristics of the raw materials, and others to the processing conditions. As starch is a major carbohydrate source in the diet, special emphasis should be on factors controlling starch digestibility. The monomeric composition of the carbohydrate moiety also plays a role, and the GI of pure, low molecular weight carbohydrates decreases in the following order: glucose > sucrose > lactose > fructose. However, recently a cautionary note was given to the excessive use of fructose, and it was suggested that the primary dietary carbohydrates should be free glucose and starch.[12] The higher the levels of unavailable and complex carbohydrates in the diet, the better. Also, interactions with other food components, such as protein, lipids, and bioactive compounds, play a role in determining carbohydrate digestibility, and, e.g., arginine has been shown to attenuate the rise of blood glucose.[18] This is an area where research hitherto is rather limited.

11.9.1 Use of Dietary Fiber

As clearly shown on previous pages, dietary fiber is of utmost relevance when designing foods for improved glycemic control. It not only reduces the glycemic load, but may also influence absorption of glucose through creating viscous conditions in the small intestine, or influence glucose metabolism through the formation of short-chain fatty acids in colonic fermentation. Dietary fiber also may change food structure in such a way that it retards carbohydrate digestibility and release in the gastrointestinal tract. Dietary fiber is a group name for various poly- and oligosaccharides, lignin, and associated substances sharing the property of not being absorbed in the small intestine.[5] Dietary fiber content of a food may be increased either by choosing naturally dietary fiber-rich raw materials or by using commercial dietary fiber preparations as ingredients. Considering the large diversity of various dietary fiber carbohydrates available, it is obvious that care must be taken to achieve the desired physiological properties in the food produced.

Hydrolytic enzymatic reactions, catalyzed by endogenous and added enzymes, as well as mechanical and thermal energy, are the major causes for changed dietary fiber properties in food processing.[125–127] The major change in dietary fiber polysaccharides during processing is often depolymerization, leading to increased solubilization, but also reduction of viscosity. As discussed above, viscosity is an important determinant of soluble dietary fiber in retarding glycemic responses,[76,77,87] so careful processing is needed. Another challenge is to produce foods with a high enough dosage of soluble fiber to reach levels attenuating blood glucose levels. The dosages needed were discussed above.

11.9.2 Potential of Bioactive Components

Bioactive substances include a range of secondary plant metabolites that can evoke physiological, behavioral, or immunological effects. They often occur in the same parts of plants as dietary fiber and are also sometimes referred to as co-passengers or associated compounds of dietary fiber. There are many possible ways for bioactive substances to affect the metabolism of glucose, and hence they could provide one means of modulating glycemic responses of foods in the future. Much more research is needed, however, before practical applications are expected. Inhibition of starch hydrolysis in the small intestine is one of the rate-limiting steps where these compounds could play a role. Some polyphenols have been shown to inhibit amylase, and hence may indirectly affect glucose and insulin levels. Diacetylated anthocyanin was shown to have α-glucosidase inhibitory activity suppressing postprandial glucose in rats.[128] In rats, Touchi extract was shown to have inhibitory activity against rat intestinal α-glucosidase, reducing the postprandial blood glucose and insulin levels after ingestion of cooked rice in four diabetic subjects.[129] Intestinal glucose uptake, mainly performed by the sodium-dependent glucose transporter (SGLT1), is inhibited by, e.g., green tea polyphenols, especially polyphenols with galloyl residues, which could possibly play a role in dietary glucose uptake in the intestinal tract.[130] Tea has also been shown to increase insulin activity by about 15-fold *in vitro* in an epidymal fat cell assay.[131] Insulin-potentiating food factors are another interesting opportunity for future food design.

11.9.3 Use of Predictive *In Vitro* Methods

The rate of starch digestion can be evaluated *in vitro* using alimentary amylolytic enzymes.[132,133] An *in vitro* method based on chewing has been developed[134] to predict the metabolic responses to different starchy foods. Starch in the test foods is chewed *in vivo* prior to *in vitro* hydrolysis with pancreatic α-amylase in a dialysis tube for 3 h. The hydrolysis index (HI) is calculated as the area under the hydrolysis curve of the test food as a percentage of the reference (white wheat bread). A good correlation between the GI and HI ($r = 0.877$, $p < 0.001$) and between insulin index (II) and HI

($r = 0.647$, $p < 0.01$) was obtained for 17 different foods, but it was noted that the method is not suitable for ranking foods having differences in the rate of gastric emptying.[134] Another similar *in vitro* starch hydrolysis method was also shown to have a good correlation with the *in vivo* GI assay.[135] Starch hydrolysis in different foods at 90 min correlated well ($r = 0.909$, $p < 0.05$) with *in vivo* glycemic responses and was suggested as a simple way of predicting the GI.

Starch has also been classified as rapidly available glucose (RAG) and slowly available glucose (SAG) by using specific analytical methods.[136,137]

Gastrointestinal conditions are simulated by using both pepsin and a mixture of amylolytic enzymes, and adjusting pH, temperature, viscosity, and mechanical mixing.[136] SAG and fat content together accounted for 73% of the variance in GI, and RAG and protein content together 45% of the variance in II.[137]

In vitro methods are an important tool in development of foods with a slower rate of carbohydrate digestion and absorption. Clinical research is always needed to document and understand human responses, but choosing raw materials and especially development of food processing to produce a variety of foods with low GI, GL, and II requires screening methods to choose candidate products for the laborious, slow, and expensive human trials. These methods are especially useful in development of starchy foods, with large potential variation in rate and extent of starch digestibility and availability.

11.9.4 Tailoring of Food Structure

Food structure is one of the key factors affecting glucose and insulin responses. Food structure affects both enzymatic accessibility and the gastric emptying half-time.[6,7,138,139] In stomach, food pieces are subjected to the action of pepsin, acid conditions, and to the vigorous grinding action of gastric motility. Gastric emptying is affected by particle size.[7] The only exit from the stomach to the small intestine is the poylorus, which allows food pieces of smaller than 2 mm in diameter to exit. The gastric emptying half-time is longer (75 min) for spaghetti than for mashed potatoes (35 min).[6] Size reduction of foods that will leave the mouth as coherent and large particles will in the stomach take a longer time, and the blood sugar values will increase more gradually.

In whole-grain products and legumes, insoluble fiber in tissue particles retards the rate of starch hydrolysis. The less the tissue structures in cereal grains or vegetables have broken down during processing, the slower is the rate of starch hydrolysis in the upper intestinal tract. A linear relationship has been reported between the proportion of barley kernels in bread and the glycemic response in humans.[140] When barley kernels were milled to flour, the corresponding whole-meal bread produced equally high glucose and insulin responses, as did the white bread reference product. Similar results have also been shown with wheat, rye, and oats. The kernels in breads and whole-meal rye flour contain integrated cell structures even after baking to bread.[141] It is probable that integrated tissue structures originating from whole-meal flour in breads can decrease enzymatic hydrolysis due to limited accessibility, with no effects on gastric emptying times. The sizes of endosperm and aleurone tissue structures in whole-meal breads are in the range of

250 μm to 1 mm. These sizes do not influence the gastric emptying times, since particles smaller than 2 mm are easily passed from the stomach without size reduction.

In recent investigations, large differences were observed in the particle size of different breads after mastication and treatments that mimic the stomach phase.[142] Whole-meal rye bread residues remained as coherent, large particles (largest in the range of 2 to 20 mm in diameter), whereas white wheat breads broke down to very small particles (<3 mm in diameter). Also, pasta residues exist in large pieces after treatment mimicking stomach conditions.[142]

Some foods, such as high-amylose rice[143,144] and different types of rye breads,[38,71,145] have been shown to decrease insulin responses of healthy subjects, but have a small or no effect on glucose responses. In healthy men, consumption of barley-containing pasta with β-glucan also led to a more blunted insulin response than consumption of wheat pasta, although the plasma glucose did not differ significantly.[146] This probably reflects the good glycemic control of healthy subjects, and suggests that more emphasis should be put on studying effects on insulin responses instead of only glycemic responses.

Also, in the case of legumes, the tissue integrity and softness of the product seem to be important factors for starch digestibility and glycemic responses. Most legume products prepared by conventional cooking produce low glucose and insulin responses.[132,147,148] Great differences in GI (12 to 74) have been observed between different legume products due to variations in raw materials and the processing. Canning and mechanical disruption especially produce higher GI values.[149,150]

The physical state of starch is in itself a key determinant of glycemic response of starchy foods. Native starch granules are digested slowly by α-amylase and, during gelatinization, the *in vitro* rate of amylolysis increases remarkably.[151] It has been shown with wheat starch gels that the starch hydrolysis rate decreases markedly when amylose leaches out of the granule.[152] It is possible that amylose, when located near the granule surface, also retards hydrolysis of amylopectin. Lower glycemic and insulin indices and a higher content of resistant starch have been reported for a barley bread baked from a high-amylose barley genotype by using long-time/low-temperature baking conditions.[153] Retrograded amylose is found in a cooled, cooked potato, bread, and cornflakes. Resistant starch content can be increased by prolonging the wet stage after cooking and heating or freezing the cooked food.

Enzymes can be used for modification of carbohydrate and protein structure. In the former case, the solubility of arabinoxylans in wheat or rye-based products can be increased *in situ* by enzymic transformation of water-unextractable arabinoxylans to water-extractable arabinoxylans by using an endoxylanase with a strong selectivity for hydrolysis of water-unextractable arabinoxylans.[154] Bread-making processes for increased soluble fiber levels have been developed by replacement of 40% wheat flour by naked barley flour, using a *Bacillus subtilis* endoxylanase with a strong selectivity for solubilizing water-unextractable arabinoxylans to extractable ones.[155] This enzyme is not inhibited by wheat endogenous xylanase inhibitors.

Addition of acid in barley bread baking has reduced the rate of gastric emptying in reference to white wheat breads.[139] This has been shown to be due to the reduced starch digestion rate.[156] Acid in such concentrations will make the bread structure very firm and less porous. Porosity facilitates degradation of food in the mouth.[157]

It has been suggested that the presence of lactic acid during heat treatment promotes interactions between starch and gluten, hence reducing starch bioavailability.

In a preliminary study (only five breads), a positive correlation was found between bread hardness and insulin index (unpublished data). Bread hardness can be increased by cross-linking enzymes, such as transglutaminase.[158] Transglutaminase increases gluten–fiber interactions and elasticity of dough and, as a result, a harder and less porous crumb structure will be formed.[157]

11.10 CONCLUSIONS

Carbohydrates are a key nutrient in glucose and insulin metabolism and also play an important role in the prevention and management of type 2 diabetes. Strong epidemiological evidence supports a role for whole-grain products and insoluble dietary fiber in the prevention of type 2 diabetes. Ironically, in postprandial studies viscous, but not insoluble, dietary fiber has favorable effects on glucose and insulin metabolism, if taken in sufficient quantities. Findings from medium- to long-term studies of the role of dietary fiber in improving glucose and insulin metabolism in nondiabetic and diabetic individuals are less conclusive. There is also evidence for the importance of low-GI starchy foods in the prevention and treatment of type 2 diabetes. More randomized controlled trials are needed to establish the roles of dietary fiber and the glycemic index in the prevention and treatment of type 2 diabetes. It seems clear, however, that we are only beginning to understand the importance of the quality of dietary carbohydrates. There are several food factors influencing physiological functionality of carbohydrates in a food product. Food structure especially has a significant effect on the rates of digestion and absorption of starch in foods. In the future, we may also learn more about the effects of minor food constituents on carbohydrate metabolism, leading to new avenues for designing carbohydrous foods to assist in combating the rapidly expanding epidemic of type 2 diabetes.

REFERENCES

1. King, H., Aubert, R., and Herman, W., Global burden of diabetes, 1995–2025: prevalence, numerical estimates, and projections, *Diabetes Care*, 21, 1414, 1998.
2. Tuomilehto, J. et al., Prevention of type 2 diabetes mellitus by changes in lifestyle among subjects with impaired glucose tolerance, *N. Engl. J. Med.*, 344, 1343, 2001.
3. WHO/FAO, *Diet, Nutrition and the Prevention of Chronic Diseases*, WHO Technical Reports Series 916, Report of the Joint WHO/FAO Expert Consultation, Geneva, 2003, 149 pp.
4. Asp, N.-G., Classification and methodology of food carbohydrates as related to nutritional effects, *Am. J. Clin. Nutr.*, 61 (Suppl. 4), 930S, 1995.
5. De Vries, J.W., On defining dietary fibre, *Proc. Nutr. Soc.*, 62, 37, 2003.
6. Mourot, J. et al., Relationship between the rate of gastric emptying and glucose and insulin responses to starchy foods in young healthy adults, *Am. J. Clin. Nutr.*, 48, 1035, 1988.

7. Thomsen, C. et al., The glycaemic index of spaghetti and gastric emptying in non-insulin-dependendent diabetic patients, *Eur. J. Clin. Nutr.*, 48, 776, 1994.
8. Meyer, J.H. and Doty, J.E., GI transit and absorption of solid food: multiple effects of guar, *Am. J. Clin. Nutr.*, 48, 267, 1988.
9. Dinneen, S.F., The postprandial state: mechanisms of glucose intolerance, *Diabet. Med.*, 14 (Suppl. 3), S19, 1997.
10. Cherrington, A.D., Edgerton, D., and Sindelar, D.K., The direct and indirect effects of insulin on hepatic glucose production *in vivo*, *Diabetologia*, 41, 987, 1998.
11. Ferrannini, E. et al., Effect of insulin on the distribution and disposition of glucose in man, *J. Clin. Invest.*, 76, 357, 1985.
12. Elliot, S.S. et al., Fructose, weight gain, and the insulin resistance syndrome, *Am. J. Clin. Nutr.*, 76, 911, 2002.
13. Lang, D.A. et al., Cyclic oscillations of basal plasma glucose and insulin concentrations in human beings, *N. Engl. J. Med.*, 301, 1023, 1979.
14. Ward, W.K. et al., Pathophysiology of insulin secretion in non-insulin-dependent diabetes mellitus, *Diabetes Care*, 7, 491, 1984.
15. Lang, V. et al., Varying the protein source in mixed meal modifies glucose, insulin and glucagon kinetics in healthy men, has weak effects on subjective satiety and fails to affect food intake, *Eur. J. Clin. Nutr.*, 53, 959, 1999.
16. Holt, H.A., Brand-Miller, J.C., and Petocz, P., An insulin index of foods: the insulin demand generated by 1000kJ portions of common foods, *Am. J. Clin. Nutr.*, 66, 1264, 1997.
17. Van Loon, L.J. et al., Plasma insulin responses of different amino acid or protein mixtures with carbohydrate, *Am. J. Clin. Nutr.*, 72, 96, 2000.
18. Gannon, M.C., Nuttall, J.A., and Nuttall, F.Q., The metabolic response to ingested glycine, *Am. J. Clin. Nutr.*, 76, 1302, 2002.
19. Laaksonen, D.E. et al., Metabolic syndrome and development of diabetes mellitus: application and validation of recently suggested definitions of the metabolic syndrome in a prospective cohort study, *Am. J. Epidemiol.*, 156, 1070, 2002.
20. World Health Organization, *Definition, Diagnosis and Classification of Diabetes Mellitus and Its Complications*, Part 1, *Diagnosis and Classification of Diabetes Mellitus*, Report of a WHO consultation, Geneva, 1999.
21. Reaven, G.M., Banting lecture. Role of insulin resistance in human disease, *Diabetes*, 37, 1595, 1988.
22. Lakka, H.M. et al., The metabolic syndrome and total and cardiovascular disease mortality in middle-aged men, *JAMA*, 288, 2709, 2002.
23. Ridker, P.M. et al., C-reactive protein, the metabolic syndrome, and risk of incident cardiovascular events: an 8-year follow-up of 14,719 initially healthy American women, *Circulation*, 107, 391, 2003.
24. Kelley, D.E. and Mandarino, L.J., Fuel selection in human skeletal muscle in insulin resistance: a reexamination, *Diabetes*, 49, 677, 2000.
25. Bjorntorp, P., Metabolic implications of body fat distribution, *Diabetes Care*, 14, 1132, 1991.
26. Ravussin, E. and Smith, S.R., Increased fat intake, impaired fat oxidation, and failure of fat cell proliferation result in ectopic fat storage, insulin resistance, and type 2 diabetes mellitus, *Ann. N.Y. Acad. Sci.*, 967, 363, 2002.
27. Kahn, S.E., The relative contributions of insulin resistance and beta-cell dysfunction to the pathophysiology of type 2 diabetes, *Diabetologia*, 46, 3, 2003.

28. Edelstein, S.L., Knowler, W.C., Bain, R.P., Andres, R., Barrett-Connor, E.L., Dowse, G.K., Haffner, S.M., et al., Predictors of progression from impaired glucose tolerance to NIDDM: an analysis of six prospective studies, *Diabetes*, 46, 701, 1997.
29. Jenkins, D.J. et al., Glycemic index of foods: a physiological basis for carbohydrate exchange, *Am. J. Clin. Nutr.*, 34, 362, 1981.
30. Wolever, T.M. et al., The glycemic index: methodology and clinical implications, *Am. J. Clin. Nutr.*, 54, 846, 1991.
31. Sugiyama, M. et al., Glycemic index of single and mixed meal foods among common Japanese foods with white rice as a reference food, *Eur. J. Clin. Nutr.*, 57, 743, 2003.
32. Foster-Powell, K., Holt, S.H., and Brand-Miller, J., International table of glycemic index and glycemic load values, *Am. J. Clin. Nutr.*, 76, 5, 2002.
33. Foster-Powell, K. and Brand Miller, J., International tables of glycemic index, *Am. J. Clin. Nutr.*, 62, 871S, 1995.
34. Salmeron, J. et al., Dietary fiber, glycemic load, and risk of NIDDM in men, *Diabetes Care*, 20, 545, 1997.
35. Salmeron, J. et al., Dietary fiber, glycemic load, and risk of non-insulin-dependent diabetes mellitus in women, *J. Am. Med. Assoc.*, 277, 472, 1997.
36. Liu, S. et al., A prospective study of dietary glycemic load, carbohydrate intake, and risk of coronary heart disease in US women, *Am. J. Clin. Nutr.*, 71, 1455, 2000.
37. Björck, I., Liljeberg, H., and Ostman, E., Low glycaemic-index foods, *Br. J. Nutr.*, 83, S149, 2000.
38. Juntunen, K. et al., Postprandial glucose, insulin, and incretin responses to grain products in healthy subjects, *Am. J. Clin. Nutr.*, 75, 254, 2002.
39. Ludwig, D.S., The glycemic index: physiological mechanisms relating to obesity, diabetes, and cardiovascular disease, *JAMA*, 287, 2414, 2002.
40. Hu, F.B. and Willett, W.C., Optimal diets for prevention of coronary heart disease, *JAMA*, 288, 2569, 2002.
41. Willett, W.C. and Leibel, R.L., Dietary fat is not a major determinant of body fat, *Am. J. Med.*, 113 (Suppl. 9B), 47S, 2002.
42. Foster, G.D., Wyatt, H.R., and Hill, J.O., A randomized trial of a low-carbohydrate diet for obesity, *ACC Curr. J. Rev.*, 12, 29, 2003.
43. Samaha, F.F. et al., A low-carbohydrate as compared with a low-fat diet in severe obesity, *N. Engl. J. Med.*, 348, 2074, 2003.
44. Anonymous, Position of the American Dietetic Association: health implications of dietary fiber, *J. Am. Diet. Assoc.*, 102, 993, 2002.
45. FAO/WHO, Carbohydrates in Human Nutrition: A Summary of the Joint FAO/WHO Expert Consultation, FAO, Rome, 1997, available at http://www.fao.org/waicent/faoinfo/economic/esn/carboweb/carbo.htm, accessed March 27, 2002.
46. Brand-Miller, J. et al., Low-glycemic index diets in the management of diabetes: a meta-analysis of randomized controlled trials, *Diabetes Care*, 26, 2261, 2003.
47. Järvi, A.E. et al., Improved glycemic control and lipid profile and normalized fibrinolytic activity on a low-glycemic index diet in type 2 diabetic patients, *Diabetes Care*, 22, 10, 1999.
48. Heilbronn, L.K., Noakes, M., and Clifton, P.M., The effect of high- and low-glycemic index energy restricted diets on plasma lipid and glucose profiles in type 2 diabetic subjects with varying glycemic control, *J. Am. Coll. Nutr.*, 21, 120, 2002.
49. Buyken, A.E. et al., EURODIAB IDDM Complications Study Group. Glycemic index in the diet of European outpatients with type 1 diabetes: relations to glycated hemoglobin and serum lipids, *Am. J. Clin. Nutr.*, 73, 574, 2001.

50. Hu, F.B. et al., Diet, lifestyle, and the risk of type 2 diabetes mellitus in women, *N. Engl. J. Med.*, 345, 790, 2001.
51. Meyer, K.A. et al., Carbohydrates, dietary fiber, and incident type 2 diabetes in older women, *Am. J. Clin. Nutr.*, 71, 921, 2000.
52. Augustin, L.S. et al., Glycemic index in chronic disease: a review, *Eur. J. Clin. Nutr.*, *56*, 1049, 2002.
53. Jenkins, D.J. et al., High-complex carbohydrate or lente carbohydrate foods? *Am. J. Med.*, 113, 30S, 2002.
54. Jenkins, D.J. et al., Glycemic index: overview of implications in health and disease, *Am. J. Clin. Nutr.*, 76, 266S, 2002.
55. Ludwig, D.S., Dietary glycemic index and obesity, *J. Nutr.*, 130, 280S, 2002.
56. Willet, W., Manson, J., and Liu, S., Glycemic index, glycemic load, and risk of type 2 diabetes, *Am. J. Clin. Nutr.*, 76, 274S, 2002.
57. Uusitupa, M. et al., Long-term improvement in insulin sensitivity by changing life-styles of people with impaired glucose tolerance: 4-year results from the Finnish Diabetes Prevention Study, *Diabetes*, 52, 2532, 2003.
58. Chiasson, J.L. et al., Acarbose for prevention of type 2 diabetes mellitus: the STOP-NIDDM randomised trial, *Lancet*, 359, 2072, 2002.
59. Wolever, T.M. et al., Second-meal effect: low-glycemic-index foods eaten at dinner improve subsequent breakfast glycemic response, *Am. J. Clin. Nutr.*, 48, 1041, 1988.
60. Axelsen, M. et al., Suppression of nocturnal fatty acid concentrations by bedtime carbohydrate supplement in type 2 diabetes: effects on insulin sensitivity, lipids, and glycemic control, *Am. J. Clin. Nutr.*, 71, 1108, 1999.
61. Björck, I. and Liljeberg, H., The glycaemic index: importance of dietary fibre and other food properties, *Proc. Nutr. Soc.*, 62, 201, 2003.
62. Liljeberg, H. and Björck, I., Effects of a low-glycaemic index spaghetti meal on glucose tolerance and lipaemia at a subsequent meal in healthy subjects, *Eur. J. Clin. Nutr.*, 54, 24, 2000.
63. Frost, G. et al., The effect of low-glycemic carbohydrate on insulin and glucose response *in vivo* and *in vitro* in patients with coronary heart disease, *Metabolism*, 45, 669, 1996.
64. Kiens, B. and Richter, E.A., Types of carbohydrate in an ordinary diet affect insulin action and muscle substrates in humans, *Am. J. Clin. Nutr.*, 63, 47, 1996.
65. Pereira, M.A. et al., Effect of whole grains on insulin sensitivity in overweight hyperinsulinemic adults, *Am. J. Clin. Nutr.*, *75*, 846, 2002.
66. Frost, G.S. et al., Carbohydrate-induced manipulation of insulin sensitivity independently of intramyocellular lipids, *Br. J. Nutr.*, 89, 365, 2003.
67. Frost, G.S. et al., A prospective randomised trial to determine the efficacy of a low glycaemic index diet given in addition to healthy eating and weight loss advice in patients with coronary heart disease, *Eur. J. Clin. Nutr.*, 58, 121, 2004.
68. Ludwig, D.S. and Eckel, R.H., The glycemic index at 20 y, *Am. J. Clin. Nutr.*, 76, 264S, 2002.
69. Grill, V. and Bjorklund, A., Overstimulation and beta-cell function, *Diabetes*, 50, S122, 2001.
70. Jenkins, D.J. et al., Metabolic effects of a low-glycemic-index diet, *Am. J. Clin. Nutr.*, 46, 968, 1987.
71. Juntunen, K. et al., High-fiber rye bread and insulin secretion and sensitivity in healthy postmenopausal women, *Am. J. Clin. Nutr.*, 77, 385, 2003.
72. Schenk, S. et al., Different glycemic indexes of breakfast cereals are not due to glucose entry into blood but to glucose removal by tissue, *Am. J. Clin. Nutr.*, 78, 742, 2003.

73. Wolever, T.M. and Mehling, C., Long-term effect of varying the source or amount of dietary carbohydrate on postprandial plasma glucose, insulin, triacylglycerol, and free fatty acid concentrations in subjects with impaired glucose tolerance, *Am. J. Clin. Nutr.*, 77, 612, 2003.
74. Edwards, C.A., Johnson, I.T., and Read, N.W., Do viscous polysaccharides slow absorption by inhibiting diffusion or convection? *Eur. J. Clin. Nutr.*, 42, 307, 1988.
75. Lund, E.K. et al., Effect of oat gum on the physical properties of the gastrointestinal contents and on the uptake of C-galactose and cholesterol by rat small intestine *in vitro*, *Br. J. Nutr.*, 62, 91, 1989.
76. Wood, P.J. et al., Effect of dose and modification of viscous properties of oat gum on plasma glucose and insulin following an oral glucose load, *Br. J. Nutr.*, 72, 731, 1994.
77. Wood, P.J., Beer, M.U., and Butler, G., Evaluation of the role of concentration and molecular wight of oat β-glucan in determining the viscosity on plasma glucose and insulin following an oral glucose load, *Br. J. Nutr.*, 84, 19, 2000.
78. Jenkins, D.J. et al., Dietary fibres, fibre analogues, and glucose tolerance: importance of viscosity, *Br. Med. J.*, 1, 1392, 1978.
79. Begin, F. et al., Effect of dietary fibres on glycemia and insulinemia and on gastrintestinal function in rats, *Can. J. Physiol. Pharmacol.*, 67, 1265, 1989.
80. Satchithanandam, S. et al., Alteration of gastrointestinal mucin by fiber feeding in rats, *J. Nutr.*, 120, 1179, 1990.
81. Fontvieille, A.M. et al., *In vitro* and *in vivo* digestibility and metabolic effects of 3 wheat-flour products (white bread, french toast (rusk) and french toast bran-enriched) in normal subjects, *Diabetes Metab.*, 14, 92, 1988.
82. Aro, A. et al., Improved diabetic control and hypocholesterolaemic effect induced by long-term dietary supplementation with guar gum in type 2 (insulin-independent) diabetes, *Diabetologia*, 21, 29, 1981.
83. Beattie, V.A. et al., Does adding fibre to a low energy, high carbohydrate, low fat diet confer any benefit to the management of newly diagnosed overweight type II diabetics? *Br. Med. J.*, 296, 1147, 1988.
84. Groop, P.H. et al., Long-term effects of guar gum in subjects with non-insulin-dependent diabetes mellitus, *Am. J. Clin. Nutr.*, 58, 513, 1993.
85. Braaten, J.T. et al., High β-glucan oat bran and oat gum reduce postprandial blood glucose and insulin in subjects with and without type 2 diabetes, *Diabetic Med.*, 11, 312, 1994.
86. Pick, M.E. et al., Oat bran concentrate bread products improve long-term control of diabetes: a pilot study, *J. Am. Diet. Assoc.* 96, 1254, 1996.
87. Tappy, L., Gügolz, E., and Würsch, P., Effects of breakfast cereals containing various amounts of β-glucan fibers on plasma glucose and insulin responses in NIDDM subjects, *Diabetes Care*, 19, 831, 1996.
88. Braaten, J.T. et al., Oat gum lowers glucose and insulin after an oral glucose load, *Am. J. Clin. Nutr.*, 53, 1425, 1991.
89. Edwards, C.A. et al., Viscosity of food gums determined *in vitro* related to their hypoglycemia actions, *Am. J. Clin. Nutr.*, 46, 72, 1987.
90. Pastors, J.G. et al., Psyllium fiber reduces rise in postprandial glucose and insulin concentrations in patients with non-insulin-dependent diabetes, *Am. J. Clin. Nutr.*, 53, 1431, 1991.
91. Rodriguez-Moran, M., Guerrero-Romero, F., and Lazcano-Burciaga, G., Lipid- and glucose-lowering efficacy of *Plantago psyllium* in type II diabetes, *J. Diabetes Complications*, 12, 273, 1998.

92. Jenkins, D.J. et al., Decrease in postprandial insulin and glucose concentrations by guar and pectin, *Ann. Intern. Med.*, 86, 20, 1977.
93. McIvor, M.E. et al., Flattening postprandial blood glucose responses with guar gum: acute effects, *Diabetes Care*, 8, 274, 1985.
94. Fuessl, H.S. et al., Guar sprinkled on food: effect on glycaemic control, plasma lipids and gut hormones in non-insulin dependent diabetic patients, *Diabet. Med.*, 4, 463, 1987.
95. Gatenby, S.J. et al., Effect of partially depolymerized guar gum on acute metabolic variables in patients with non-insulin-dependent diabetes, *Diabet. Med.*, 13, 358, 1996.
96. Holman, R.R. et al., No glycemic benefit from guar administration in NIDDM, *Diabetes Care*, 10, 68, 1987.
97. Pick, M.E. et al., Barley bread products improve glycemic control of type 2 subjects, *Int. J. Food Sci. Nutr.*, 49, 71–78, 1998.
98. Anderson, J.W. et al., Cholesterol-lowering effects of psyllium intake adjunctive to diet therapy in men and women with hypercholesterolemia: meta-analysis of 8 controlled trials, *Am. J. Clin. Nutr.*, 71, 472, 1999.
99. Lalor, B.C. et al., Placebo-controlled trial of the effects of guar gum and metformin on fasting blood glucose and serum lipids in obese, type 2 diabetic patients, *Diabet. Med.*, 7, 242, 1990.
100. Uusitupa, M. et al., Metabolic and nutritional effects of long-term use of guar gum in the treatment of noninsulin-dependent diabetes of poor metabolic control, *Am. J. Clin. Nutr.*, 49, 345, 1989.
101. Ebeling, P. et al., Glucose and lipid metabolism and insulin sensitivity in type 1 diabetes: the effect of guar gum, *Am. J. Clin. Nutr.*, 48, 98, 1988.
102. Bruttomesso, D. et al., No effects of high-fiber diets on metabolic control and insulin-sensitivity in type 1 diabetic subjects, *Diabetes Res. Clin. Pract.*, 13,15, 1991.
103. Liu, S. et al., A prospective study of whole-grain intake and risk of type 2 diabetes mellitus in US women, *Am. J. Pub. Health*, 90, 1409, 2000.
104. Fung, T.T. et al., Whole-grain intake and the risk of type 2 diabetes: a prospective study in men, *Am. J. Clin. Nutr.*, 76, 535, 2002.
105. Montonen, J. et al., Whole-grain and fiber intake and the incidence of type 2 diabetes, *Am. J. Clin. Nutr.*, *77*, 622, 2003.
106. Liu, S., Whole-grain foods, dietary fiber and type 2 diabetes: searching for a kernel of truth, *Am. J. Clin. Nutr.*, 77, 527, 2003.
107. Hallfrisch, J., Schofield, D.J., and Behall, K.M., Diets containing soluble oat extracts improve glucose and insulin responses of moderately hypocholesterolemic men and women, *Am. J. Clin. Nutr.*, 61, 379, 1995.
108. Cherbut, C. et al., Involvement of small intestinal motility in blood glucose response to dietary fibre in man, *Br. J. Nutr.*, 71, 675, 1994.
109. Rigaud, D. et al., Effect of psyllium on gastric emptying, hunger feeling and food intake in normal volunteers: a double blind study, *Eur. J. Clin. Nutr.*, 52, 239, 1998.
110. Frape, D.L. and Jones, A.M., Chronic and postprandial responses of plasma insulin, glucose and lipids in volunteers given dietary fibre supplements, *Br. J. Nutr.*, 73,733, 1995.
111. Ellis, P.R. et al., Evaluation of guar biscuits for use in the management of diabetes: tests of physiological effects and palatability in non-diabetic volunteers, *Eur. J. Clin. Nutr.*, 42, 425, 1988.

112. Ellis, P.R., Dawoud, F.M., and Morris, E.R., Blood glucose, plasma insulin and sensory responses to guar-containing wheat breads: effects of molecular weight and particle size of guar gum, *Br. J. Nutr.*, 66, 363, 1991.
113. Gatti, E. et al., Effects of guar-enriched pasta in the treatment of diabetes and hyperlipidemia, *Ann. Nutr. Metab.*, 28, 1, 1984.
114. Jarjis, H.A. et al., The effect of ispaghula (Fybogel and Metamucil) and guar gum on glucose tolerance in man, *Br. J. Nutr.*, 51, 371, 1984.
115. Leclere, C.J. et al., Role of viscous guar gums in lowering the glycemic response after a solid meal, *Am. J. Clin. Nutr.*, 59, 914, 1994.
116. Heijnen, M.L., van Amelsvoort, J.M., and Weststrate, J.A., Interaction between physical structure and amylose:amylopectin ratio of foods on postprandial glucose and insulin responses in healthy subjects, *Eur. J. Clin. Nutr.*, 49, 446, 1995.
117. Liljeberg, H.G., Akerberg, A.K., and Bjorck, I.M., Effect of the glycemic index and content of indigestible carbohydrates of cereal-based breakfast meals on glucose tolerance at lunch in healthy subjects, *Am. J. Clin. Nutr.*, 69, 647, 1999.
118. Davy, B.M. et al., High-fiber oat cereal compared with wheat cereal consumption favorably alters LDL-cholesterol subclass and particle numbers in middle-aged and older men, *Am. J. Clin. Nutr.*, 76, 351, 2002.
119. Keogh, G.F. et al., Randomized controlled crossover study of the effect of a highly beta-glucan-enriched barley on cardiovascular disease risk factors in mildly hypercholesterolemic men, *Am. J. Clin. Nutr.*, 78, 711, 2003.
120. Kirby, R.W. et al., Oat-bran intake selectively lowers serum low-density lipoprotein cholesterol concentrations of hypercholesterolemic men, *Am. J. Clin. Nutr.*, 34, 824, 1981.
121. Uusitupa, M.I. et al., A controlled study on the effect of beta-glucan-rich oat bran on serum lipids in hypercholesterolemic subjects: relation to apolipoprotein E phenotype, *J. Am. Coll. Nutr.*, 11, 651, 1992.
122. Anderson, J.W. et al., Cholesterol-lowering effects of psyllium hydrophilic mucilloid for hypercholesterolemic men, *Arch. Intern. Med.*, 148, 292, 1988.
123. Bell, L.P. et al., Cholesterol-lowering effects of psyllium hydrophilic mucilloid. Adjunct therapy to a prudent diet for patients with mild to moderate hypercholesterolemia, *JAMA*, 261, 3419, 1989.
124. Liu, S. et al., A prospective study of dietary glycemic load, carbohydrate intake and risk of coronary heart disease in US women, *Am. J. Clin. Nutr.*, 71, 1455, 2000.
125. Thebaudin, J.Y. et al., Dietary fibres: nutritional and technological interest, *Trends Food Sci. Technol.*, 8, 41, 1997.
126. Guillon, F. and Champ, M., Structural and physical properties of dietary fibres, and consequences of processing on human physiology, *Food Res. Int.*, 33, 233, 2000.
127. Poutanen, K., Effect of processing on the properties of dietary fibre, in *Advanced Dietary Fibres*, McCleary, B., Ed., Blackwell Science, London, 2000, p. 262.
128. Matsui, T. et al., Anti-hyperglycemic effect of diacylated anthocyanin derived from *Ipomoea batatas* cultivar Ayamursaki can be achieved through the α-glucosidase inhibitory action, *J. Agric. Chem. Soc.*, 50, 7244, 2002.
129. Fujita, H., Yamagami, T., and Ohshima, K., Fermented soybean-derived water-soluble Touchi extract inhibits α-glucosidase and is antiglycemic in rats and humans after single oral treatment, *J. Nutr.*, 131, 1211, 2001.
130. Kobayashi, Y. et al., Green tea polyphenols inhibit the sodium-dependent glucose transporter of intestinal epithelial cells by a competitive mechanism, *J. Agric. Food Chem.*, 48, 5618, 2000.

131. Anderson, R.A. and Polansky, M.M., Tea enhances insulin activity, *J. Agric. Food Chem.*, 50, 7182, 2002.
132. Jenkins, D.J. et al., Rate of digestion of foods and postprandial glycaemia in normal and diabetic subjects, *Br. Med. J.*, 281, 14, 1980.
133. Bornet, F.R. et al., Insulin and glycemic responses in healthy humans to native starches processed in different ways: correlation with *in vitro* α-amylase hydrolysis, *Am. J. Clin. Nutr.*, 50, 315, 1989.
134. Granfeldt, Y. et al., An *in vitro* procedure based on chewing to predict metabolic response to starch in cereal and legume products, *Eur. J. Clin. Nutr.*, 46, 649, 1992.
135. Goni, I. et al., Analysis of resistant starch: a method for food and food products, *Food Chem.*, 56, 445, 1996.
136. Englyst, K.N. et al., Rapidly available glucose in foods: an *in vitro* measurement that reflects the glycemic response, *Am. J. Clin. Nutr.*, 69, 448, 1999.
137. Englyst, K.N. et al., Glycaemic index of cereal products explained by their content of rapidly and slowly available glucose, *Br. J. Nutr.*, 89, 329, 2003.
138. Read, N.W. et al., Swallowing food without chewing; a simple way to reduce postprandial glycaemia, *Br. J. Nutr.*, 55, 43, 1986.
139. Liljeberg, H. and Björck, I., Delayed gastric emptying rate as a potential mechanism for lowered glycemia after eating sourdough bread: studies in humans and rats using test products with added organic acids or an organic salt, *Am. J. Clin. Nutr.*, 64, 886, 1996.
140. Liljeberg, H. and Björck, I., Bioavailability of starch in bread products. Postprandial glucose and insulin responses in healthy subjects and *in vitro* resistant starch content, *Eur. J. Clin. Nutr.*, 48, 151, 1994.
141. Autio, K., Parkkonen, T., and Fabritius, M., Observing structural differences in wheat and rye breads, *Cereal Foods World*, 42, 702, 1997.
142. Liukkonen, K.-H., Autio, K., and Poutanen, K., An *in vitro* particle size method for estimating particle size reduction in the digestion of starchy foods and for predictiing insulin response (in press), 2006.
143. Van Amelsvoort, J.M. and Weststrate, J.A., Amylose-amylopectin ratio in a meal affects postprandial variables in male volunteers, *Am. J. Clin. Nutr.*, 55, 712, 1992.
144. Goddard, M.S., Young, G., and Marcus, R., The effect of amylose content on insulin and glucose responses to ingested rice, *Am. J. Clin. Nutr.*, 39, 388, 1984.
145. Leinonen, K. et al., Rye bread decreases postprandial insulin response but does not alter glucose response in healthy Finnish subjects, *Eur. J. Clin. Nutr.*, 53, 262, 1999.
146. Bourdon, I. et al., Postprandial lipid, glucose, insulin and cholecystokinin responses in men fed barley pasta enriched with beta-glucan, *Am. J. Clin. Nutr.*, 69, 55, 1999.
147. Brand, J.C. et al., Food processing and the glycemic index, *Am. J. Clin. Nutr.*, 42, 1192, 1985.
148. Wolever, T.M., Relationships between dietary fiber content and composition in foods and the glycemic index, *Am. J. Clin. Nutr.*, 51, 72, 1990.
149. Jenkins, D.J., Wolever, T.M., and Jenkins, A.L., Starchy foods and glycemic index, *Diabetes Care*, 11, 149, 1988.
150. Wolever, T.M. et al., Glycemic index of foods in individual subjects, *Diabetes Care*, 13, 126, 1990.
151. Holm, J. et al., Degree of starch gelatinization, digestion rate of starch *in vitro*, and metabolic responses in rats, *Am. J. Clin. Nutr.*, 47, 1010, 1998.
152. Slaughter, S.L., Ellis, P.R., and Butterworth, P.J., An investigation of the action of porcine pancreatic α-amylase on native and gelatinized starches, *Biochim. Biophys. Acta*, 1571, 55, 2002.

153. Åkerberg, A., Liljeberg, H., and Björck, I., Effects of amylose/amylopectin ratio and baking conditions on resistant starch formation and glycaemic indices, *J. Cereal Sci.*, 28, 71, 1998.
154. Courtin, C.M., Gelders, G.G., and Delcour, J.A., Use of two endoxylanases with different substrate selectivity for understanding arabinoxylan functionality in wheat flour breadmaking, *Cereal Chem.*, 78, 564, 2001.
155. Trogh, I. et al., Xylanase and naked barley flour technologies produce breads with increased soluble fibre, in *AACC Annual Meeting*, Portland, 28 September – 2 October 2003, Abstract book, 2003, p. 97.
156. Ostman, E.M., Liljeberg-Elmstahl, H.G., and Bjorck, I.M., Barley bread containing lactic acid improves glucose tolerance at a subsequent meal in healthy men and women, *J. Nutr.*, 132, 1173, 2002.
157. Autio, K. et al., Food structure and its relation to starch digestibility and glycaemic response, in *3rd International Conference of Food Rheology and Structure*, Fischer, P., Marti, I., and Windhab, E.J., Eds., 2003, p. 7.
158. Gerrard, J.A. et al., Dough properties and crumb strength of white pan bread by microbial transglutaminase, *J. Food Sci.*, 65, 472, 1998.

12 Carbohydrates and Mineral Metabolism

David D. Kitts

CONTENTS

12.1 INTRODUCTION

There is strong scientific evidence to show that an optimal dietary intake of fiber will assist in reducing the risk of many chronic diseases.[1] The recognition of setting a maximum limit for intake of dietary fiber has been based on the concern that certain dietary fibers may interfere with mineral (e.g., calcium, magnesium, iron, and zinc) intestinal absorption.[2–5] The link between reduced absorption efficiency of important minerals, such as calcium and magnesium, with subsequent utilization for bone health has been made with the demonstration that both ions have important roles in managing against osteoporosis[6] (albeit, it is also important to note that calcium deficiency does not always contribute to immediate symptoms of bone weakening). Osteoporosis is a condition where bones become extremely porous and fragile and are susceptible to fracture.[7] Bone constantly undergoes bone turnover, a coupled process that involves both resorption and formation. With remodeling, old bone is reabsorbed by osteoclasts that produce a cavity that is subsequently filled with a new bone matrix due to osteoblast activity. There is a tightly coupled balance between bone resorption and bone formation that is regulated by multiple local and systemic factors (e.g., growth factors and calcium metabolism hormones, such as vitamin D, parathyroid, estrogens, and calcitonin). Despite the wealth of knowledge on the etiology of osteoporosis, some controversy exists in the scientific literature as to the role of dietary fiber in positively or negatively influencing mineral metabolism homeostasis, which, in turn, may lead to a protection against or a predisposition to osteoporosis. Some insoluble fiber complexes that coexist with organic acids, such as phytic and oxalic acids, have been associated with reduced calcium bioavailability,[8] whereas other more soluble fibers, such as inulin and polydextrose, have no apparent adverse effect on calcium bioavailability, and may even enhance bioavailability under certain conditions.[9,10] Underlying mechanisms, such as the typical lowering of pH in the large intestine as a result of feeding oligosaccharides, that result in greater mineral solubilization have been associated with protection of bone mineral.[10] In addition, the role of certain soluble fibers to indirectly prevent bone loss has been proposed with findings that prebiotic substrate-enhancing specific gut bacteria (e.g., bifidobacteria) increase the metabolic transformation of isoflavone conjugates to corresponding aglycones, thereby generating a specific enhanced capacity to protect against decreased bone mineral density.[12] This chapter reviews the evidence for potential effects of different dietary fiber sources on calcium bioavailability and bone health.

12.2 CALCIUM AND MAGNESIUM HOMEOSTASIS

Calcium is present in abundant amounts in various body pools, with both bone and teeth containing approximately 99% of total body calcium in forms of calcium phosphate salt or hydroxyapatite crystal ($Ca_{10}(PO_4)_6(OH)_2$). The remaining 1% of calcium ion is present in blood, muscle, and extra- and intracellular fluids and has a critical role in nerve transmission, muscle contraction, and blood-clotting mechanisms. Magnesium is the second most abundant intracellular cation, with approximately half present in soft tissue and the other half found in bone. Both calcium and magnesium concentrations are maintained in a relatively narrow range in both extra- and intracellular compartments.[13] Calcium and magnesium homeostasis is regulated by three principal organ systems: the intestine, skeleton, and kidney. The intestine represents the sole port of entry of calcium and magnesium into the body and contributes significantly to mineral homeostasis by influencing both absorption and secretion activities (Table 12.1). The processes of intestinal calcium transport involve essentially a transmucosal transport, with the mucosal layer of the intestine representing the only barrier that calcium ions have to cross to reach the portal circulation. Thus, calcium movements from the intestinal lumen into the circulation, or from various body pools into the intestinal lumen, represent calcium transport. The rate of intestinal calcium transport, when plotted against the concentration of intraluminal calcium, gives a curvilinear fit,[14] indicating that

TABLE 12.1
Calcium Transport Mechanisms across Intestinal Barrier

	Intestinal Transport Mechanism	
Feature	**Cellular**	**Paracellular**
Intestinal compartment	Cytosolic component of enterocyte	Extracellular space between enterocytes
Active components	Sequestering by vitamin D-dependent, Ca binding proteins, facilitated diffusion, or endocytosis	Simple diffusion or osmotic properties between tight junctions of adjacent enterocytes
Regulation factors	Plasma mineral concentration, 1,25 $(OH)_2$ D_3	Presence of soluble mineral, co-nutrient (lactose)-induced generation of osmotic gradient

two process components — saturable and nonsaturable mechanisms — exist for the transport process. The two biochemical aspects that define the transport of calcium across the intestinal barrier involve both a cellular (e.g., facilitated) pathway, characterized by the involvement of biosynthesized and active proteins, and a paracellular (e.g., passive) pathway. With the cellular transport, calcium ions move into the absorption cell cytosolic compartment of the mucosal lining of the small intestine (e.g., enterocyte) by four coordinated steps. First, calcium enters into the cytosol via the apical membrane and is moved from the membrane to the basolateral membrane of the enterocyte before calcium extrusion takes place at the basolaeral membrane into the lateral space occupied by the lamina propria. From the lateral space, calcium diffuses into the portal circulation. To accomplish this process, the calcium ion must enter the enterocyte by diffusion or by binding to the apical membrane cytoskeletal proteins.[15] From here, the divalent ion is sequestered by vitamin D-dependent calcium binding proteins, calbindin (e.g., –9k in the rat, 32k in the chicken), or transported to the basolateral membrane by a facilitated diffusion.[16] Calcium is extruded into the lateral space by a calcium extrusion pump, which involves a transmembrane protein associated with the basolateral membrane.[17] Since the cellular pathway for both calcium and magnesium transport requires metabolic energy, it is considered an active, energy-requiring process.[18] Calcium and magnesium have been shown to produce an antagonistic effect on respective absorption. Classic *in vitro* experiments, using the inverted gut sac technique, have shown that magnesium and calcium compete with each other for intestinal absorption.[19] For example, feeding calcium supplements containing magnesium to rats resulted in a poorer absorption of calcium from the supplement than did control animals fed milk.[20] In human studies, the ingestion of excess calcium reduces intestinal absorption of magnesium.[21]

Paracellular uptake of calcium and magnesium in the intestine occurs by simple diffusion of the ion. Since mechanisms for both calcium and magnesium involve in part passive diffusion, the osmotic properties of the intestinal contents will influence the extent of absorption. A solvent drag effect occurs that involves the movement of soluble calcium ions from the intestinal lumen into the lateral space. The osmotic gradient developed between the lateral space and the intestinal lumen occurs by active extrusion of sodium and chloride ions into the lateral space. In the case of calcium, there is no apparent regulatory control for this mode of transport of calcium, albeit extracellular calcium is involved in the formation of tight junctions[22] and intracellular calcium has an important role for maintaining the integrity of the tight junctions.[23] Thus, calcium may modulate paracellular transport directly by influencing gap junction structure and function. Intestinal absorption of magnesium also occurs mainly in the distal parts of the intestine, such as the jejenum and the ileum, where passive diffusion contributes to a major portion of absorption.[24] Some of the more prominent co-nutrients, or digestion products that stimulate calcium transport, are lactose,[25] casein phosphopeptides,[26,27] bile salts, and some amino acids.[28,29] Some fats may also impair calcium bioavailability by forming insoluble soaps with the free ion, and sodium intake can also lead to increased urinary calcium loss and potential effects on bone density. Since bioavailability is influenced by so many factors, including absorption, transport, cellular organization, storage, and excretion of the nutrient, an absolute definition of the term *bioavailability* of calcium or magnesium is required when assessing the effect of different dietary constituents on absorption and utilization of this mineral.[26] To complicate matters, the complexity of the system likely makes a single method to evaluate calcium bioavailability unsatisfactory.

12.3 CARBOHYDRATES AND MINERAL BIOAVAILABILITY AND UTILIZATION

12.3.1 Nonfiber Carbohydrate Sources

Polyols, namely, lactitol, xylitol, maltitol, and isomalt, are carbohydrates commonly used in the low-sugar or sugar-free confectionary that produces products with a low glycemic index, reduced energy, and noncarcinogenicity. These sources of carbohydrate possess low digestibility in the small intestine, but are greatly degraded by fermentation in the large intestine. Malitol produces a beneficial effect on intestinal calcium bioavailabity in the rat,[30] as does xylitol.[31] Human studies have indicated that a 100 g/day intake of a polyol-containing diet for 1 month effectively improved magnesium absorption in young healthy males by 25%, but had no effect on calcium balance.[32]

Lactulose, analogous to lactose but synthesized through alkaline isomerization of lactose, is a good example of an osmotically active agent that has no effect on enhancing calcium bioavailability *in vitro*.[33] *In vivo* studies conducted in rats, however, have reported an increase in fractional calcium absorption that was maximum at 10% lactulose concentration. This effect was specific to lactulose, since feeding the component sugars (e.g., galactose and fructose) at equimolar concentrations had no effect on fractional calcium absorption.[31] A second example of a carbohydrate-based facilitated enhancement of calcium absorption occurs with transgalactooligosaccharides (TOSs), a mixture of glucose and galactose (or galactosyllactoses), which are obtained by a transgalactosylation reaction catalyzed by β-D-galactosidases derived from *Aspergillus oryzae*.[34,35] TOS, found in human and bovine milk and commercial yogurt, can improve apparent calcium absorption and retention in rats in a manner similar to that of lactose at dietary intake levels of 5 and 10%.[34] For both lactulose and TOS, the increase in calcium absorption is accompanied by a significant increase in both cecal wall and digesta weight, as well as a significant decrease in cecal pH. These responses of osmotically active carbohydrate are also related to an increased fluid uptake within the small intestine lumen, which occurs in response to maintain isotonicity. The resultant increased distension and permeability of the intracellular junction between enterocytes as a result of the solvent drag defines the mechanism for increased passive absorption of calcium in response to TOS feeding. In human studies, TOS feeding for 3 weeks at a level of 10 g/day resulted in a significant increase in fecal bifidobacteria.[35] Although studies that have focused on demonstrating the prebiotic potential of TOSs have not extended them to actually measure calcium bioavailability, other workers have attempted to link TOS feeding with increased calcium absorption. For example, feeding 15 g of TOS/day for 3 weeks to healthy young men failed to show enhanced uptake of calcium from 24-h urinary excretion data.[36] Extending the urinary collection period to 3 days to capture the impact of late colonic absorption was successful in showing that TOS feeding could enhance calcium absorption.[37]

12.3.2 Dietary Fiber Carbohydrate Sources

The characteristics of chemical composition and structural integrity of plant cell walls are critical factors for the different physicochemical properties of recovered fiber sources that influence gastrointestinal function. Due to the relative complexity of dietary fiber sources to otherwise simpler carbohydrates, an absolute extrapolation of mechanism of action explaining calcium bioavailability–fiber interactions is not completely warranted, albeit some similarities do exist. Dietary fiber, by definition, refers to the nondigestible carbohydrate and lignin plant materials that are unavailable for direct use by the host. Insoluble fiber sources, such as cellulose and wheat bran, have a limited effect on passage time and digestion or absorption activities in the stomach or small intestine, whereas the affinity to form viscous mixtures from soluble fiber sources (e.g., β-glucans) will change the rheological behavior of the stomach contents and result in a slower stomach emptying time.[39] Dietary fiber that eventually escapes the digestive processes of the small intestine becomes available for fermentation by microflora in the large intestine to produce a mixture of end products that include short-chain fatty acids (SCFAs) such as acetate, butyrate, and propionate. The intensity of the fermentation, and thus the rate of synthesis of the SCFAs, is influenced by the source of the fuel (e.g., prebiotic) for the proliferation of microflora.[39] A further characterization of dietary fiber can be made on the basis of relative solubility, which differentiates the fiber source into soluble (e.g., pectin, gums, and mucilages) and insoluble (e.g., cellulose and mucilages) forms. Depending on the definition of the dietary fiber source, it is important to recognize that the effect of fiber on calcium bioavailability should be regarded as fiber specific (Table 12.2 and Table 12.3). For example, soluble fibers contribute more to viscosity changes in the gastrointestinal tract and products of fermentation, while insoluble fibers may influence calcium absorption by altering intestinal transit time or the actual sequestering of ion.

TABLE 12.2
Soluble Fiber Sources That Enhance Mineral Bioavailablity and Utilization

Experimental Subject	Substrate (Dietary Conc.)	Effect	Reference
Rat	Inulin (10%)	Enhanced cecal weight, generation of SCFA, decreased pH, increased soluble calcium, magnesium	40
	(15%)	Increased soluble cecal calcium concentration, stimulated ODS activity, increased butyrate output	41, 42, 57
Human	Inulin (40 g/day)	Increased apparent magnesium balance	9
Rat	FOS (75 g/kg)	Enhanced calcium absorption	45, 46, 54
Mice	FOS (5% w/w)	Reduced fecal calcium loss	11
	FOS (5%) + isoflavone	Prevented femoral bone loss	12
Rat	FOS (6%, 14 days)	Generation of SCFA in bowel	57
Human	FOS (5 g/day, 10 days)	Enhanced magnesium absorption	50
Rat	GOS	Enhanced calcium absorption	34
Human	TOS (10 g/days, 2 days)	Enhanced intestinal bifidobacteria	35
	TOS (15 g/days, 1–3 weeks)	Increased urinary calcium output	37
Human	Amylomaize RS (25–50%, 21 days)	Increased SCFA production	9
Rat	Potato RS (35%, 21 days)	Increased cecum absorption area, lowered pH, increased calcium uptake	32
Human	Pectin (36 g/days, 5 weeks)	Increased bacterial population, increased calcium absorption	65
Rat	Pectin	Enlargement of cecum, SCFA production, drop in cecal pH	66
Rat	Psyllium (5–10%)	Increased fecal dry weight, lower apparent Ca absorption, lower Ca content in femur/tibia	90
Monkey		Increased hypertrophy in jejunum/ileum	
Rat	Guar gum hydrolysate (50-g diet/kg, 1–3 weeks)	Increased calcium absorption, lower cecal pH, higher SCFA	69–71

12.4 SOLUBLE FIBER SOURCES AND MINERAL BIOAVAILABILITY

12.4.1 INULIN AND FRUCTOOLIGOSACCHARIDES

Inulin and fructooligosaccharide (FOS) represent soluble, nondigestible fibers that contribute to the production of fermentation products in the large intestine. Feeding inulin to rats for 3 weeks, at four dietary concentrations ranging from 0 to 20%,

TABLE 12.3
Insoluble Fiber Sources and Mineral Bioavailablity and Utilization

Experimental Subject	Substrate (Dietary Conc.)	Effect	Reference
Rat	Cellulose (12%, 16 days)	Reduced calcium absorption	5
Rat	Wheat bran (15% diet)	No effect on mineral bioavailability	76
Rat	Cellulose (20%)	Lower duodenal Ca uptake, reduced bone calcium content	74
Human	Cellulose (16 g/day, 30 days)	Normal bowel function; increased fecal Ca, Mg excretion; negative Ca balance, n/c Mg balance	89
Human	Carboxymethylcellu lose (7.5 g/1000 kcal)	Negative Mg balance	67
	Karaya gum (19–27 g/days, 4 weeks)	No change in bowel transit time, increased fecal dry weight, positive effect on Mg, Ca	
Human	Wheat bran (18.5 g/day, 3× in 24 h)	No effect on mineral bioavailability	77

n/c = no change

increased cecal calcium absorption to an extent that was directly proportional to the amount of inulin consumed up to a 20% dietary level.[40] Inulin, which consists of one α-glucopyranosyl unit linked to β-fructosyl units, has a positive effect on calcium bioavailabilty regardless of the degree of polymerization (DP), which can vary between 3 and 50 DP, depending on the source. For example, feeding inulin products such as Raftiline® (average DP = 10) and Raftilose® (average DP = 4.8) at 10% diet for 24 days will produce similar rates of fecal calcium excretion, which corresponds to a significant increase in apparent calcium absorption.[41] It is important to note that in all of these studies, the level of dietary calcium was an important factor that may have influenced the action of inulin to improve calcium bioavailability. For example, despite feeding equal amounts of inulin to animals that also received 0.8% dietary calcium, significantly greater responses to the presence of inulin occurred in animals that were fed 0.3% calcium.[42]

Similar findings have been reported in rodents fed FOS. The administration of a 50 g of FOS/kg diet for 7 days exhibited a 28% increase in apparent calcium absorption over control animals.[43] This was confirmed by other workers who fed only 5% FOS, but for a relatively longer time (e.g., 12 days) and reported both greater true and apparent calcium absorption from calcium balance measurements and bioavailability estimates using ^{45}Ca tracers.[44] Enhanced calcium absorption in gastrectomized and sham-operated rats fed FOS for 4 weeks reduced postgastrectomy osteopenia to a significant extent, but did not completely prevent loss in femoral mineral density compared to sham-operated rates fed the control diet.[45] Extending the duration of FOS feeding to 6 weeks in the gastrectomized rats was successful at preventing postgastrectomy osteopenia.[46]

Studies conducted in human feeding trials with both inulin and FOS have produced less consistent results. For example, feeding healthy young men inulin-containing diets (e.g., 40 g/day) for 4 weeks produced significant increases in calcium absorption and apparent calcium balance.[9] This result was not confirmed by other workers who fed less inulin (e.g., 15 g/day) to healthy men and used dual-stable-isotope techniques to evaluate calcium absorption,[36] rather than the balance study protocol reported previously. Comparing the results and methodologies of both studies can explain the reason for these discrepancies, with marked differences in experimental design that included the amounts of inulin and FOS fed, as well as methods of assessing calcium bioavailability being particularly different. As shown above, measuring calcium absorption using radio-isotope methodology requires sufficient duration for urinary collection to account for calcium absorption from both the small and large intestines, respectively. As reported with earlier TOS findings, feeding 15 g of FOS for 9 days using a randomized, double-blind crossover design produced significant increases in true calcium absorption in response to feeding FOS only when urinary collections were extended to 36 h, from the original 24 h after isotope administration.[37] The importance of the amount of FOS made available in the diet to produce a positive effect was observed with a study conducted in 59 prepubertal girls fed only 8 g of FOS/day for 3 weeks.[47] Although these workers employed a similar randomized crossover design and measured true calcium absorption by the dual-isotope method, which included an extended urinary collection period to account for colonic calcium uptake, no significant effects of feeding FOS on enhancing calcium bioavailability were observed.

One explanation for a reduced effect of fermentable fibers on calcium bioavalability over longer experimental durations may involve downregulation of transcellular absorption in the proximate section of the small intestine in response to the long-term exposure to the fermentable carbohydrate-induced calcium absorption in the large intestine. This phenomona was shown by Chonan and Watanuki[48] with galactooligosaccharide feeding to ovariectomized rats that produced a positive effect on calcium absorption after days of short-term feeding, but no effect after 28 days of feeding. Similar results have been reported in both adolescent and postmenopausal women fed FOS, who exhibited an increased calcium bioavailability at 9 days,[39,49] but no effect when fed for 3 weeks.[38] A 5-week feeding study with short-chain FOS has been shown to enhance magnesium absorption in postmenopausal women.[50] The 11% increase in magnesium absorption attributed to feeding FOS diets was equivalent to 10 mg of Mg/day and paralleled a 10 mg of Mg/day urinary loss of magnesium, thereby indicating that the true net benefit of feeding FOS was negligible.

Rodent studies have shown that the daily intake of 5 g of FOS/100 g of diet for 40 days produced significant increases in both calcium and magnesium absorption, which corresponded to significant enhancement of trabecular bone volume at the neck of the femur.[51] More recently, the prebiotic action of FOS to enhance intestinal microorganism-derived deglycosylation of soy isoflavones, produces greater bioactivity for increase femoral bone mineral content in ovariectomized mice. This demonstrates the potential benefit of FOS for bone health.[12]

12.4.2 Resistant Starch

Resistant starch is another example of an incompletely digested carbohydrate that escapes the small intestine, but is metabolized or fermented by microbes in the large intestine. Feeding potato starch as a source of resistant starch to rats for 3 weeks has been shown to increase the cecal calcium absorption rate by more than fivefold, compared to control rats.[52] Reducing the dietary intake of potato starch by twofold (e.g., 150 g/kg) produced similar results with increased soluble calcium and significant increases in cecal weight and a decreased pH;[53] the latter effect was directly attributed to the production of short-chain fatty acids (SCFAs). An increase in calcium absorption in rats fed 150 g of resistant starch/kg of diet was shown to be equivalent to the feeding of 50 g of inulin/kg of diet. Moreover, feeding resistant starch to rats for 3 weeks in the form of 50% amylomaize starch produced an increased SCFA concentration together with a decreased cecal pH, which corresponded to increased soluble calcium and an enlarged cecum.[54] Similar results were obtained in adjacent studies with the feeding of 10% lactulose.

The underlying mechanism for enhanced mineral bioavailability from soluble carbohydrate sources such as inulin, FOS, some resistant starches, and polyols is linked to the extent of fermentation and resultant products of fermentation (Figure 12.1). The low digestibility of these carbohydrates in the small intestine ensures that the substrate will be available for prebiotic activity for indigenous microflora in the large intestine. Fermentation of carbohydrates by gut bacteria will result in the production of short-chain fatty acids,[55] which has been attributed to increased cecal weight, reduced pH, and increased soluble calcium and magnesium in this segment of the intestine.[42,56,57] The solubilization of divalent mineral ions in the cecum, due to acidic conditions of fermentation, will favor an increase in soluble ion concentrations that are required for enhanced paracellular absorption. Increases in cecal weight in response to these sources of nondigestible carbohydrates has been attributed to triggered cell proliferation — both cell number (hyperplasia) and cell size (hypertrophy) — by the generation of a butyrate fermentation product. Moreover, generation of butyrate has been associated with stimulation of calbindin–D9K expression and increased concentration of 1,25 $(OH)_2D_3$ receptor activity.[58] This activity may explain the observation that feeding fermentable carbohydrate for an extended time period can produce feedback inhibition, which downregulates calcium transcellular absorption in the upper part of the intestine,[9,36,59] albeit magnesium bioavailability is enhanced.[32] The generation of short-chain fatty acids with fermentation may have another distinctly different effect on enhancing calcium bioavailabity in the large intestine. For example, generation of protonated SCFAs that diffuse across the apical membrane eventually dissociates and increases intracellular hydrogen ion concentrations that are secreted from the cell in exchange for soluble calcium ion (Figure 12.2). It has been proposed that the flux of hydrogen ion outside the cell enables protonation of more SCFAs, with

Soluble fermentable fibre

Fermentation in large intestine

Production of SCFAs

Protonated SCFA uptake by enterocyte (e.g. acetate)

Build-up of H^+ (intracellular)

Ca/H exchange

Butyrate

Increased -calbindin — production

Lowering cecal- pH

Substrate for microbial proliferation

Polyamine production

Enhanced $(1,25)_2$ D_3 receptor activity

Increase mineral solubility

Increased gut absorption area (hyperplasia & hypertrophy)

Enhanced -paracellular— absorption (cecum)

Enhanced cellular absorption (cecum)

Increased mineral balance

Increased bone mass

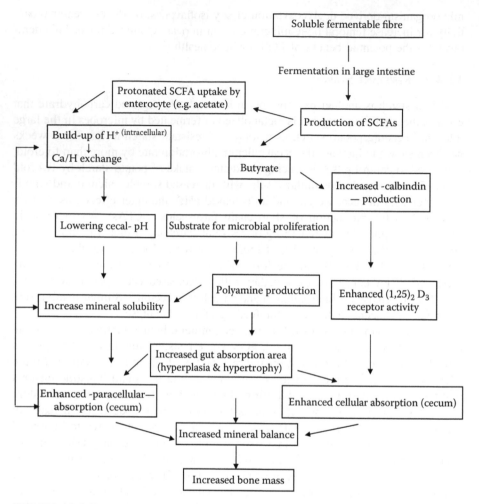

FIGURE 12.1 Proposed mechanisms of fiber-induced changes on mineral bioavailability.

Enterocyte Apical Membrane

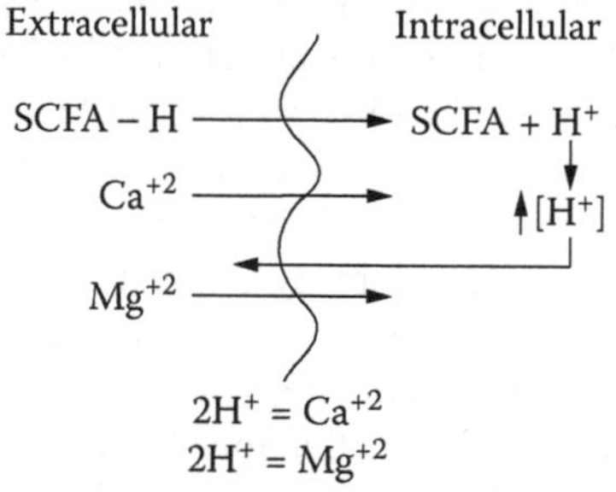

FIGURE 12.2 Schematic of SCFA-induced mineral influx to enterocyte.

the result that a Ca/H exchange cycle occurs.[60,61] This has been demonstrated *in vivo*, where graded amounts of SCFAs were reported to increase serum calcium concentrations, in contrast to the decrease in serum acetate response in the presence of graded amounts of calcium.[62] Thus, generation of SCFAs was active in stimulating calcium absorption in the cecum by forming a SCFA–Ca complex, which in turn facilitates the absorption of soluble calcium ions. Couple this effect with the proposed enhanced blood flow triggered by the presence of volatile fatty acids in the cecum,[63] and it is expected that the net result would be an increased uptake of calcium or magnesium in this section of the intestine.

Finally, it should also be considered that increased proliferation of certain microorganisms in the large intestine from fermentation products of low digestible carbohydrate could also result in increased phytase activity, thus reducing the potential for phytate-induced restriction of divalent minerals.[64]

12.4.3 Pectin

Pectin is a heteropolysaccharide that consists mainly of D-galactouronic acids and is linked with other sugar units, such as fucose, xylose, and galactose. It is a soluble and fermentable fiber that tends to form a viscous gel, as well as bind with divalent cations such as calcium. *In vitro* studies that have examined the effect of pectin on intestinal calcium absorption were conducted in simulated human gastrointestinal digestion procedures that involved dialysis and measurements of calcium release.[60] Calcium released from pectin was shown to be significantly reduced in the simulated small intestine as a result of the tendency of pectin to form a viscous gel upon ingestion and a capacity to bind with calcium due to the presence of the carboxylic acid side chain associated with uronic acid. Since the mineral binding is pH dependent (e.g., binding of calcium occurs at basic condition and is released at acidic conditions), it is a logical conclusion that calcium binding in the small intestine occurs in the basic environment. However, entry of pectin into the large intestine will result in partial fermentation, thereby releasing the calcium associated with pectin. With fermentation, the generation of SCFAs decreases the pH of the large intestine, thus facilitating the solubilization of inorganic calcium that leads to enhanced paracellular absorption. Therefore, the sum of both recoveries of calcium may not be different between pectin-containing and control diets. The workers attributed the reduced release of calcium in the simulated small intestinal to the tendency of pectin to bind to calcium and form a viscous gel that entraps free calcium ions. The presence of pectin carboxylic acid side chains that originate mainly from the uronic acid component also contributes to a pH-dependent binding affinity for soluble calcium in this section of the intestine, where paracellular absorption of calcium is the predominant mode of mineral uptake. The passage of pectin into the large intestine results in subsequent fermentation, and thus an induced release of pectin-bound calcium due to the lowering of gastrointestinal pH from the SCFA fermentation products. The increased solubility of calcium as a result of pH lowering explains the significant increase in calcium bioavailability observed in the large intestine-simulated model. A similar mechanism has been proposed for inulin.[52]

Presently, *in vivo* studies have not confirmed this finding. Experiments conducted in human trials with five medical students fed 36 g of pectin/day for 6 weeks, with 3 weeks of control period prior to treatment, produced no significant increase in calcium balance.[61] There was also no significant increase in the fiber content in the feces before and after pectin feeding, indicating that pectin was fully fermented. In addition, the fat, nitrogen, total dry matter, and bile acid contents of the excreted feces were increased by 80, 47, 28, and 35%, respectively, indicating an increase in bacterial mass in response to the fermentation of pectin.

12.4.4 Gums

Gums are soluble polysaccharides that include, in part, glucans, galactomannas, carrageenan, agar, and alginates. These substances are often used as thickening, emulsifying, and stabilizing agents in food materials. Although the effect of gums on calcium bioavailability is controversial, the overall consensus is that there is no effect of gums on enhancing intestinal calcium absorption. One *in vitro* study showed that alginate (AA), guar gum (GG), and locust bean gum (LBG) reduced calcium availability when used as a thickening agent in infant formula.[66] These workers used dialyzed whey-based infant formula treated with graded amounts of AA, GG, or LBG, in a preliminary intraluminal digestion step that was adapted to the conditions of infants younger than 6 months, using human milk as the reference. Calcium availability decreased in the presence of each fiber source, with the greatest effect associated with GG- and AA-added formula. Calcium availability from GG-added formula was significantly lower than that from LBG-added formula using a similar concentration of fiber (0.42 g/100 ml of formula). This significant decrease in calcium availability is possibly attributed to both the affinity of fiber to form nonabsorbable complexes with calcium and its capacity to create a viscous environment in the small intestine that would result in entrapped calcium being unavailable for absorption.

Again, the result of gums affecting calcium bioavailability *in vitro* may be quite different when examined *in vivo*. For example, refined fibers such as sodium–carboxymethylcellulose, LBG, or karaya gum produced no effect on calcium balance when fed at 7.5 g/1000 calories to 11 human subjects for 4 weeks.[67,68] The apparent calcium balance in non-insulin-dependent diabetic subjects has also been shown to be unaffected after their consumption of guar gum for 6 months.[68] Hara et al.[69] fed nephrectomized rats 50 g of a low-viscosity guar gum hydrolysate (GGH) soluble fiber, obtained by hydrolyzing guar gum with β-1,4-mannanase for 21 days. These workers found that GGH consumption helped to compensate the otherwise reduced proximal calcium absorption caused by nephrectomy by increasing calcium absorption in the large intestine. Femur calcium content, however, did not change. A significant difference in calcium absorption as well as femur calcium content between gastroectomized and sham-operated rats fed with GGH and a control diet was reported in a later study.[70] GGH-enhanced calcium absorption in the large intestine following GGH fermentation in the cecum will result in acidification of luminal contents, which eventually increases ionized calcium concentration. Calcium absorption in the proximal intestine can be unaffected due to the fact that GGH has low viscosity.

Gums, such as LBG and GG, generally form viscous gels within the intestine upon ingestion, and therefore it is possible that the change in viscosity restricts calcium bioavailability in the small intestine. Alternatively, fermentation of these particular sources of digestible fiber can release trapped calcium in the large intestine, thus enabling paracellular calcium absorption to occur and a compensation for the reduced calcium absorption occurring in the small intestine.[66]

In other studies, phosphorylated GGH (P-GGH) was shown to produce a positive effect on calcium solubilization that influenced calcium absorption based on inhibition of calcium precipitation by formation of an insoluble calcium phosphate salt.[71] Feeding rats a 50 g of P-GGH/kg diet significantly increased the apparent calcium absorption as well as the calcium content in the femur.

12.4.5 Polydextrose

Polydextrose (PD) is less fermentable and less viscous than GGH. Feeding 50 g of PD/kg to rats for 21 days produced a significant increase in the apparent calcium absorption and femur calcium concentration in both normal and gastrectomized rats, whereas the same dose of GGH enhanced only the apparent calcium absorption in gastrectomized rats only.[10] The fact that SCFA production was less in the cecum of the PD-fed rats than in the GGH-fed rats and the control animals, implies that fermentation in the cecum was not a major contributor to the observed enhanced calcium absorption. A similar increase in calcium absorption was obtained when 50 g/l of PD, but not GGH, was added to mucosal fluid of an inverted ileum sac system. These findings indicate that the different effect of PD on calcium absorption, compared to GGH, could be attributed to the fact that PD influenced calcium absorption in the small intestine rather than in the large intestine.

12.4.6 Psyllium

Psyllium is an example of a soluble fiber that has been found to reduce calcium bioavailability.[41,72] The strong affinity to form gels and bind to minerals in the small intestine, but relatively poor fermentation capacity in the colon, likely explains the negative effect of psyllium on calcium balance. *In vitro* studies have shown psyllium to release the least amount of calcium, compared to pectin and lactulose in a simulated small intestine, due to the viscous nature of the material in the gut and the fact that psyllium binds to calcium. Moreover, psyllium is not fermented to the same extent as pectin or lactulose, thereby causing more bound calcium to occur in the simulated large intestine condition. An *in vivo* study using rats confirmed the hypothesis that psyllium consumption reduces calcium bioavailability.[72] Feeding rats on 5 or 10% of psyllium for 4 weeks from three different fiber sources — purified psyllium (PP), Metamucil® (MET), and All-Bran® Bran Buds® (AB) — resulted in a decreased apparent calcium absorption at the 10% level. The negative influence of psyllium on calcium bioavailability was further evidenced by its effect on bone mineralization, where the decreased apparent calcium absorption was accompanied by a significant decrease in the calcium content in hind limb bones and mandibles of rats.

12.5 INSOLUBLE FIBER AND MINERAL BIOAVAILABILITY

12.5.1 CELLULOSE

Cellulose, which consists of many glucose units all linked together with a β-1,4 glycosidic bond to form a linear polysaccharide, is found in a higher concentration in fruits and vegetables than in cereals.[73] The presence of hydroxyl functional groups providing intramolecular hydrogen bonding capacity interactions with minerals has been shown *in vitro* with calcium. There is controversy, however, in regard to a similar effect on calcium bioavailability from *in vivo* studies. For example, no effect of feeding 7.5 g/1000 kcal of cellulose (ranged from 19.1 to 27.0 g/day) on mineral balance was observed in some studies,[68] while a decreased apparent mineral balance of calcium has been reported in others.[3-5,74,75] Rat studies that consisted of feeding 12% cellulose for 16 days have been shown to reduce the apparent calcium absorption relative to control levels.[5] Intestinal transit time was significantly decreased and fecal material content was significantly greater in animals fed cellulose than in controls. Other animal studies that fed 20% of cellulose to rats for 8 weeks also reported significant reductions in calcium binding activity in the duodenum and reduced calcium in the bone ash.[74]

In human studies, seven healthy women fed refined cellulose (i.e., 16 g/day of purified cellulose from wood pulp) for 30 days displayed an increased stool weight, decreased intestinal transit time, and increased frequency of defecation that corresponded to a mean negative calcium balance from −16 ± 76 to −199 ± 51 mg/day.[7] The fecal moisture content, however, was not significantly different in these subjects than in others fed the control diet, confirming that cellulose was not well hydrated. Taken together, both animal and human studies indicate that cellulose may negatively influence calcium absorption by increasing the bulkiness of intestinal contents and reducing intestinal transit time. The increase in bulkiness of contents upon cellulose ingestion may involve stimulation of peristaltic muscle movement in the intestine, thus increasing the time for transfer of food material and reducing intestinal transit time. The net effect could be a reduced frequency of important mucosal–mineral interactions that are necessary for optimal absorption. It is also possible that the bulkiness of the cellulose-containing intestinal contents acts to abrade the epithelial surface of the intestinal mucosa to an extent that results in an increased endogenous calcium excretion. This is evidenced by the reported decreased calcium binding activity occurring in the duodenum.[74]

12.5.2 WHEAT BRAN

Wheat bran is a widely consumed fiber that has been proposed to negatively influence calcium absorption.[2,3] One of the earlier studies that triggered interest to determine the effect of wheat bran on mineral absorption showed that feeding bread partly made from wheat whole meal resulted in a decreased calcium balance in two men.[4] It is important to recognize that the presence of phytate in this preparation of insoluble fiber may have been more of a factor for the reduced calcium absorption. This possibility was nicely shown in a study conducted in rats fed graded amounts of wheat bran for 2 weeks and measures of calcium bioavailability obtained using orally administered tracer ^{45}Ca. The results indicated that the addition of wheat bran up to 15% of the diet had no effect on altering ^{45}Ca absorption.[76] A good explanation for the inconsistent findings in rats, compared to studies conducted in pigs where reduced calcium bioavailability was observed when they were fed wheat bran, is the presence of active phytase enzyme in the rodent intestine. Earlier studies conducted in human subjects reported that feeding 16 g of wheat bran/day to eight ileostomy patients did not impair calcium absorption from the small intestine.[75] Other human studies conducted in six men, where the phytate content was controlled by adding sodium phytate to obtain an equal amount of phytate for each loaf of bread, concluded that wheat bran had no significant effect on calcium absorption.[77] Intrinsically ^{45}Ca-labeled wheat during plant growth propagation was used as a measure of interactions with calcium and other dietary constituents. This distinct experimental design produced results that had wheat products, but not wheat bran, showing no significant effect on calcium absorption in 26 healthy adult women.[78] It is noteworthy that leavening bread was found to improve the calcium absorption, which can be explained on the basis that yeasts can produce phytases, and thus reduce the phytic acid content during bread making.

Other *in vitro* studies have supported the hypothesis that wheat bran by itself has the potential to decrease mineral bioavailability, since the insoluble fraction of wheat bran has a large potential to bind calcium even after removal of phytate.[79] This is seen when calcium ions are added to an insoluble fiber residue of dephytinized wheat bran, and the calcium binding capacity is determined using flame atomic absorption spectroscopy after acid washing the wheat bran. A similar technique was used to determine the binding capacity of wheat bran, rice bran, and oat fiber for calcium and other minerals.[80] Wheat bran was found to bind significantly more calcium than rice bran and oat fiber, respectively, which suggests that wheat bran has either more specific calcium binding sites or a similar number of sites, but a higher affinity toward calcium than other fibers. An *in vitro* method that involves simulating gastrointestinal digestion estimated the effect of wheat and barley fiber on calcium bioavailability by measuring ^{45}Ca dialyzability in test meals.[81] Results indicated that ^{45}Ca dialyzability was reduced more by wheat fiber than barley fiber, relative to control. Again, it is important to note that wheat fiber contains a 3.2-fold higher content of phytate than the barley fiber. Cell culture studies utilizing Caco-2 cells (i.e., the human colon adenocarcinoma) as a model test system to determine the effect of barley hull and wheat bran, as well as the dephytinized barley and wheat fiber on calcium absorption, showed that wheat bran significantly decreased (e.g., 17%) calcium transport across the cells and uptake of calcium (e.g., 24%), in comparison to the fiber-free control. However, this inhibitory effect on calcium uptake was removed when the wheat bran was dephytinized,[82] again indicating that the presence of phytate, rather than wheat bran itself, had a major role in inhibiting calcium absorption.

12.5.3 Maillard Reaction Products

The Maillard reaction is a nonenzymatic browning reaction that involves the condensation between a reducing sugar and a free -amino group derived from amino acid, peptide, or protein.[83] The reaction contributes to the overall development of brown color and flavor during the heat treatment of foods. The initial condensation forms a labile glycosyl–amino product, which rearranges to Amadori or Heyns compounds and subsequently degrades to form deoxysones and Strecker degradation products, which are precursors for higher molecular weight polymers, referred to as melanoidins. Little is known about the structure of melanoidins, with no single melanoidin presently isolated for identification. Melanoidins have an effect similar to that of nondigestible fiber, in terms of reducing calcium absorption.[84] Nonsoluble melanoidins produced from the final stage of the Maillard reaction bind to calcium, thereby decreasing solubility and thus hindering paracellular absorption in the small intestine.[85,86] The calcium-sequestering potential of melanoidins has been described to be related to the number of acidic donor groups in the food pigment derived from both model browning reactions and coffee brew.[87] As is the case with other examples of nondigestible fibers, the melanoidin–calcium ion complex that reaches the large intestine may be transformed to reverse the reduced bioavailabilities of divalent ions if colonic activities related to the complexes compensate for reduced bioavailability in the small intestine. Other studies have indicated that smaller molecular and soluble premelanoidins may also bind to calcium and retain the solubility of this ion, thereby making it available for absorption, as evidenced by increased urinary calcium excretion.[84]

In addition to having melanoidins sequester calcium ions, which results in reduced solubility for paracellular absorption, melanoindins may also influence enterocyte metabolism. Studies using Caco-2 cells have indicated that enterocyte metabolism is not affected by the presence of the Maillard reaction product (MRP), but in fact may act to facilitate the transport of soluble calcium.[85] Potential enhancement of calcium absorption in the large intestine by MRPs is again most likely due to the fermentation of browning products. It has been proposed that the effect of MRPs on microflora present in the large intestine is comparable to other forms of nondigestible carbohydrates, since MRPs can stimulate the nonspecific growth of anaerobic bacteria,[88] such as lactobacilli, and associated increase in lactic acid concentration, which will contribute to a lowering of gut pH.[86] Moreover, simulated upper-gut digestion using both peptic and pancreatic activity has not led to a production of low molecular weight products, thus reflecting the relative nondigestibility of MRPs. Despite the potential for MRPs to reduce calcium absorption, due to the *in vitro* evidence for sequestering activity, there is less indication that the net effect of MRPs to decrease calcium absorption occurs *in vivo*. Although apparent calcium absorption may indeed be decreased in the small intestine, some degree of compensation resulting from an increased apparent calcium absorption in the large intestine is important in evaluating the overall effect of MRPs on calcium bioavailability.

Moreover, since consumption of MRP is also associated with an increase in calcium urinary excretion, there is no evidence for reduced change in calcium retention.

12.6 CONCLUSIONS

This chapter has attempted to explain how different fiber sources pose potentially different effects on mineral bioavailability, with special emphasis on calcium and magnesium. Soluble fibers that do not form a viscous milieu in the small intestine and are available for fermentation in the large intestine may actually enhance calcium bioavailability by increasing the amount of calcium absorbed in the colon via passive absorption. Some examples may include inulin, oligofructose, FOS, resistant starch, and sugar-like compounds, such as lactulose. Similarly, some soluble fibers, such as pectin and gum, which provide a viscous gut environment yet are fermentable, generally do not adversely influence calcium absorption. On the other hand, soluble fibers such as psyllium, which have a tendency to form viscous gels and entrap calcium and are not available for fermentation, will likely reduce calcium bioavailability to some degree. Insoluble fibers may also adversely affect calcium bioavailability by influencing intestinal transit time. However, phytate, a nonfiber constituent in the mixture, may be a contributing factor that decreases calcium bioavailability, as evidenced by the fact that dephytinized wheat bran does not significantly affect calcium absorption. Finally, some MRPs may also be fermented, which could enhance calcium balance. More work is needed to evaluate this possibility, especially in light of the potential for increased urinary calcium excretion resulting from ingestion of MRPs.

REFERENCES

1. Fraser, G.E., Associations between diet and cancer, ischemic heart disease and all-cause mortality in non-Hispanic, white, Californian Seventh Day Adventists, *Am. J. Clin. Nutr.*, 70, 532S. 1999.
2. Rheinhiold, J.G., Fardji, B., and Ismail-Beigi, F., Decreased absorption of calcium, magnesium, zinc and phosphorous by humans due to increased fibre and phosphorous consumption as wheat bran, *J. Nutr.*, 106, 493, 1976.
3. Donagelo, C.M. and Eggum, B.O., Comparative effects of wheat bran and barley husk on nutrient utilization in rats. 2. Zinc, calcium, phosphorous, *Br. J. Nutr.*, 56, 269, 1986.
4. Reinhold, J.G. et al., Decreased absorption of calcium, magnesium, zinc and phosphorus by humans due to increased fiber and phosphorus consumption as wheat bread, *Nutr. Rev.*, 49, 204, 1991.
5. Ward, A.T. and Reichert, R.D., Comparison of the effect of cell wall and hull fibre from canola and soybean on the bioavailability for rats of minerals, protein and lipid, *J. Nutr.*, 116, 233, 1986.
6. Seelig, M.S., Increased need for magnesium with the use of combined estrogen and calcium for osteoporosis treatment, *Magnesium Res.*, 3, 197, 1990.
7. Geinoz, G. et al., Relationship between bone mineral density and dietary intakes in the elderly, *Osteoporos. Int.*, 3, 242, 1993.

8. Wisker E. et al., Calcium, magnesium, zinc and iron balances in young women: effects of a low-phytate barley-fiber concentrate, *Am. J. Clin. Nutr.*, 54, 553, 1991.
9. Coudray, C. et al., Effect of soluble or partly soluble dietary fibres supplementation on absorption and balance of calcium, magnesium, iron and zinc in healthy young men, *Eur. J. Clin. Nutr.*, 51, 375, 1997.
10. Hara, H., Suzuki, T., and Aoyama, Y., Ingestion of the soluble dietary fibre, polydextrose, increases calcium absorption and bone mineralization in normal and total-gastrectomized rats, *Br. J. Nutr.*, 84, 655, 2000.
11. Scholz-Ahrens, K.E. and Schrezenmeir, J., Inulin, oligofructose and mineral metabolism: experimental data and mechanism, *Br. J. Nutr.*, 87, S179, 2002.
12. Ohta, A. et al., A combination of dietary fructooligosaccharides and isoflavone conjugates increase femoral bone mineral density and equal production in ovariectomized mice, *J. Nutr.*, 132, 2048, 2002.
13. Bronner, F. and Stein, W.D., Calcium homeostasis: an old problem revisited, *J. Nutr.*, 125, 1987S, 1995.
14. Pansu, D., Bellation, D., and Bronner, F., Effect of calcium intake on saturable and nonsaturable components of duodenal calcium transport, *Am. J. Physiol.*, 240, G32, 1981.
15. Fullmer, C.S., Intestinal calcium absorption. Calcium entry, *J. Nutr.*, 122, 644, 1992.
16. Stein, W.D., Facilitated diffusion of calcium across the rat intestinal epithelial cell, *J. Nutr.*, 122, 651, 1992.
17. Wasserman, R.H. et al., Vitamin D and mineral deficiencies increase the plasma membrane calcium pump of chicken intestine, *Gasteroenterology*, 102, 886, 1992.
18. Hendrix, J.Z., Aleock, N.W., and Archibald, R.M., Competition between calcium, strontinium and magnesium for absorption in the isolated rat intestine, *Clin. Chem.*, 9, 734, 1963.
19. Greger, J.L. et al., Mineral utilization by rats fed various commercially available calcium supplements or milk, *J. Nutr.*, 117, 717, 1987.
20. Greger, J.L., Smith, S.A., and Snedeker, S.M., Effect of dietary calcium and phosphorous levels on the utilization of calcium, phosphorous, magnesium, manganese and selenium in adult rats, *Nutr. Res.*, 1, 315, 1981.
21. Anderson, J.M., Balda, M.S., and Fanning, A.S., The structure and regulation of tight junctions, *Curr. Opin. Cell Biol.*, 5, 772, 1993.
22. Palant, C.E. et al., Calcium regulation of tight junction permeability and structure in Necturus gallbladder, *Am. J. Physiol.*, 245, C203, 1983.
23. Brink, E.J. and Beynen, A.S., Nutrition and magnesium absorption, a review, *Prog. Food Nutr. Sci.*, 16, 125, 1992.
24. Yuan, Y.V., Kitts, D.D., and Nagasawa, T., The effect of lactose and fermentation products on paracellular calcium absorption and femur biomechanics in rats, *Can. J. Inst. Food Sci. Technol.*, 24, 74, 1991.
25. Kitts, D.D., Effect of casein, casein phosphopeptides and calcium intake on ileal 45Ca disappearance and temporal systolic blood pressure in spontaneously hypertensive rats, *Br. J. Nutr.*, 68, 765, 1992.
26. Kitts, D.D. and Kwong, W.Y., Calcium bioavailability of dairy products, in *Handbook of Functional Dairy Products*, Shortt, C. and O'Brien, J., Eds., CRC Press, New York, 2004, chap. 9.
27. Pappenheimer, J.R., Physiological regulation of transepithelial impedence in the intestinal mucosa of rats and hamsters, *J. Membrane Biol.*, 100, 137, 1987.
28. Pappemheimer, J.R., Paracellular intestinal absorption of glucose, creatinine and mannitol in normal animals: relation to body size, *Am. J. Physiol.*, 259, G290, 1990.

29. Torre, M., Rodriguez, A.R., and Saura-Calixto, F., Effects of dietary fiber and phytic acid on mineral availability, *Crit. Rev. Food Sci. Nutr.*, 30, 1, 1991.
30. Goda, T. et al., Effect of malitol intake on intestinal calcium absorption in the rat, *J. Nutr. Sci. Vitaminol.*, 38, 277, 1992.
31. Brommage, R. et al., Intestinal calcium absorption in rats is stimulated by dietary lactulose and other resistant sugars, *J. Nutr.*, 123, 2186, 1993.
32. Coudray, C. et al., Two polyol, low digestible carbohydrates improve the apparent absorption of magnesium but not calcium in healthy young men, *J. Nutr.*, 133, 90, 2003.
33. Trinidad, T.P., Wolever, T.M.S., and Thompson, L.U., Availability of calcium for absorption in the small intestine and colon from diets containing available and unavailable carbohydrates: an *in vitro* assessment, *Int. J. Food Sci. Nutr.*, 47, 83, 1996.
34. Chonan, O. and Watanaki, M., Effect of galactooligosaccharides on calcium absorption in rats, *J. Nutr. Sci. Vitaminol.*, 41, 95, 1995.
35. Bouhnik, Y. et al., Administration of transgalacto-oligosaccharides increases fecal bifidobacteria and modifies colonic fermentation metabolism in healthy humans, *J. Nutr.*, 127, 444, 1997.
36. Van den Heuvel, E.G. et al., Nondigestible oligosaccharides do not interfere with calcium and nonheme-iron absorption in young, healthy men, *Am. J. Clin. Nutr.*, 67, 445, 1998.
37. Van den Heuvel, E.G. et al., Oligofructose stimulates calcium absorption in adolescents, *Am. J. Clin. Nutr.*, 69, 544, 1999.
38. Johansen, H.N. et al., Effects of varying content of soluble dietary fibre from wheat flour and oat milling fractions or gastric emptying in pigs, *Br. J. Nutr.*, 75, 339, 1996.
39. Gibson, G.R. and Roberford, M.B., Dietary modulation of the human colonic microbiota, *J. Nutr.*, 125, 1401, 1995.
40. Levrat, M.A., Remesy, C., and Demigne, C., High propionic acid fermentations and mineral accumulation in the cecum of rats adapted to different levels of inulin, *J. Nutr.*, 121, 1730, 1991.
41. Delzenne, N. et al., Effect of fermentable fructo-oligosaccharides on mineral, nitrogen and energy digestive balance in the rat, *Life Sci.*, 57, 1579, 1995.
42. Remesy, C. et al., Cecal fermentations in rats fed oligosaccharides (inulin) are modulated by dietary calcium level, *Am. J. Physiol.*, 264, G855, 1993.
43. Ohta, A. et al., Calcium and magnesium absorption from the colon and rectum are increased in rats fed fructooligosaccharides, *J. Nutr.*, 125, 2417, 1995.
44. Morohashi, T. et al., True calcium absorption in the intestine is enhanced by fructooligosaccharide feeding in rats, *J. Nutr.*, 128, 1815, 1998.
45. Ohta, A. et al., Dietary fructooligosaccharides prevent osteopenia after gastrectomy in rats, *J. Nutr.*, 128, 106, 1998.
46. Ohta, A. et al., Dietary fructooligosaccharides prevent postgastrectomy anemia and osteopenia in rats, *J. Nutr.*, 128, 485, 1998.
47. Griffin, I.J., Davila, P.M., and Abrams, S.A., Non-digestible oligosaccharides and calcium absorption in girls with adequate calcium intakes, *Br. J. Nutr.*, 87, S187, 2002.
48. Chonan, O. and Watanuki, M., Effect of galactooligosaccharides on calcium absorption in rats, *J. Nutr. Sci. Vitaminol.*, 41, 95, 1995.
49. Van den Heuvel, E.G. et al., Lactulose stimulates calcium absorption in postmenopausal women, *J. Bone Miner. Res.*, 14, 1211, 1999.
50. Tahiri, M. et al., Five week intake of short chain fatty fructooligosaccharides increases intestinal absorption and status of magnesium in postmenopausal women, *J. Bone Miner. Res.*, 16, 2152, 2001.

51. Takahara, S. et al., Fructooligosaccharide consumption enhances femoral bone mineral volume and mineral composition in rats, *J. Nutr.*, 130, 1792, 2000.
52. Demigne, C., Levrat, M.A., and Remesy, C., Effects of feeding fermentable carbohydrates on the cecal concentrations of minerals and their fluxes between the cecum and blood plasma in the rat, *J. Nutr.*, 119, 25, 1989.
53. Younes, H., Demigné, C., and Remesy, C., Acidic fermentation in the caecum increases calcium absorption of calcium and magnesium in the large intestine of the rat, *Br. J. Nutr.*, 75, 301, 1996.
54. Younes, H. et al., Effects of two fermentable carbohydrates (inulin and resistant starch) and their combination on calcium and magnesium balance in rats, *Br. J. Nutr.*, 86, 479, 2001.
55. LeBlay, G. et al., Prolonged intake of fructooligosaccharides induce a short term elevation in lactic acid producing bacteria and a persistent increase in cecal butyrate in rats, *J. Nutr.*, 129, 2231, 1999.
56. Remesy, C. et al., Fibre fermentation in the cecum and its physiological consequences, *Nutr. Res.*, 12, 1235, 1992.
57. Campbell, J.M., Fahey, G.C., and Wolf, B.W., Selected undigestible oligosaccharides affect large bowel mass, cecal fecal short chain fatty acids, pH and microflora in rats, *J. Nutr.*, 127, 13, 1997.
58. Bronner, F., Calcium absorption: a paradigm for mineral absorption, *J. Nutr.*, 128, 917, 1998.
59. Ohta, A. et al., Dietary fructooligosaccharides change the intestinal mucosal concentrations of calbindin-D9k in rats, *J. Nutr.*, 128, 934, 1998.
60. Trinidad, T.R., Wolever, T.M.S., and Thompson, L.U., Effect of acetate and propionate on calcium absorption from the rectum and distal colon of humans, *Am. J. Clin. Nutr.*, 63, 574, 1996.
61. Trinidad, T.R., Wolever, T.M.S., and Thompson, L.U., Effect of calcium concentration, acetate and propionate on calcium absorption in human distal colon, *Nutrition*, 15, 529, 1999.
62. Trinidad, T.R., Wolever, T.M.S., and Thompson, L.U., Interactive effects of calcium and short chain fatty acids on absorption in the distal colon of man, *Nutr. Res.*, 13, 417, 1993.
63. Krietys, R.R. and Granger, D.N., Effect of volatile fatty acids on blood flow and oxygen uptake by the dog colon, *Gastroenterology*, 80, 962, 1981.
64. Lopez, W.H. et al., Intestinal fermentation lessens the inhibitory effects of phytic acid on mineral utilization in rats, *J. Nutr.*, 128, 1192, 1998.
65. Cummings, J.H. et al., The digestion of pectin in the human gut and its effect on calcium absorption and large bowel function, *Br. J. Nutr.*, 41, 477, 1979.
66. Bosscher, D., Van Caillie-Bertrand, M., and Deelstra, H., Effect of thickening agents, based on soluble dietary fiber, on the availability of calcium, iron, and zinc from infant formulas, *Nutrition*, 17, 614, 2001.
67. Behall, K.M. et al., Mineral balance in adult men: effect of four refined fibers, *Am. J. Clin. Nutr.*, 46, 307, 1987.
68. Behall, K.M., Effect of soluble fibers on plasma lipids, glucose tolerance and mineral balance, *Adv. Exp. Med. Biol.*, 270, 7, 1990.
69. Hara, H. et al., Increases in calcium absorption with ingestion of soluble dietary fibre, guar-gum hydrolysate, depend on the caecum in partially nephrectomized and normal rats, *Br. J. Nutr.*, 76, 773, 1996.
70. Hara, H. et al., Ingestion of guar gum hydrolysate, a soluble fiber, increases calcium absorption in totally gastrectomized rats, *J. Nutr.*, 129, 39, 1999.

71. Watanabe, O., Hara, H., and Kasai, T., Effect of a phosphorylated guar gum hydrolysate on increased calcium solubilization and the promotion of calcium absorption in rats, *Biosci. Biotechnol. Biochem.*, 64, 160, 2000.
72. Luccia, B.H. and Kunkel, M.E., Psyllium reduces relative calcium bioavailability and induces negative changes in bone composition in weanling Wistar rats, *Nutr. Res.*, 22, 1027, 2002.
73. Torre, M., Rodriguez, A.R., and Saura-Calixto, F., Effects of dietary fiber and phytic acid on mineral availability, *Crit. Rev. Food Sci. Nutr.*, 30, 1, 1991.
74. Oku, T., Konishi, F., and Hosoya, N., Mechanism of inhibitory effect of unavailable carbohydrate on intestinal calcium absorption, *J. Nutr.*, 112, 410, 1982.
75. Sandberg, A.S. et al., The effect of wheat bran on the absorption of minerals in the small intestine, *Br. J. Nutr.*, 48, 185, 1982.
76. Bagheri, S.M. and Gueguen, L., Effects of wheat bran on the metabolism of calcium-45 and zinc-65 in rats, *J. Nutr.*, 112, 2047, 1982.
77. Andersson, H. et al., The effects of breads containing similar amounts of phytate but different amounts of wheat bran on calcium, zinc and iron balance in man, *Br. J. Nutr.*, 50, 503, 1983.
78. Weaver, C.M. et al., Human calcium absorption from whole-wheat products, *J. Nutr.*,121, 1769, 1991.
79. Bergman, C.J., Gualberto, D.G., and Weber, C.W., Mineral binding capacity of dephytinized insoluble fiber from extruded wheat, oat and rice brans, *Plant Foods Hum. Nutr.*, 51, 295, 1997.
80. Idouraine, A., Khan, M.J., Kohlhepp, E.A., and Weber, C.W., *In vitro* mineral binding capacity of three fiber sources for Ca, Mg, Cu and Zn by two different methods, *Int. J. Food Sci. Nutr.*, 47, 285, 1996.
81. Kennefick, S. and Cashman, K.D., Investigation of an *in vitro* model for predicting the effect of food components on calcium availability from meals, *Int. J. Food Sci. Nutr.*, 51, 45, 2000.
82. Kennefick, S. and Cashman K.D., Inhibitory effect of wheat fibre extract on calcium absorption in Caco-2 cells: evidence for a role of associated phytate rather than fibre per se, *Eur. J. Nutr.*, 39, 12, 2000.
83. Jing, H. and Kitts, D.D., Chemical and biochemical properties of casein-sugar Maillard reaction products, *Food Chem. Toxicol.*, 40, 1007, 2002.
84. Seiquer, I. et al., Effects of heat treatment of casein in the presence of reducing sugars on calcium bioavailability: *in vitro* and *in vivo* assays, *J. Agric. Food Chem.*, 49, 1049, 2001.
85. Andrieux, C. and Sacquet, E., Effects of Maillard's reaction products on apparent mineral absorption in different parts of the digestive tract. The role of microflora, *Reprod. Nutr. Dev.*, 24, 379, 1984.
86. O'Brien, J. and Morrissey, P.A., Nutritional and toxicological aspects of the Maillard browning reaction in foods, *Crit. Rev. Food Sci. Nutr.*, 28, 211, 1989.
87. Rendleman, J.A., Complexation of calcium by melanoidin and its role in determining bioavailability, *J. Food Sci.*, 52, 1699, 1987.
88. Ames, J.M. et al., The effect of a model melanoidin mixture on fecal bacterial populations *in vitro*, *Br. J. Nutr.*, 82, 489, 1999.
89. Slavin, J.L. and Marlett, J.A., Influence of refined cellulose on human bowel function and calcium and magnesium balance, *Am. J. Clin. Nutr.*, 33, 1932, 1980.
90. Pauline, I., Mehta, T., and Hargis, A., Intestinal structural changes in African green monkeys after long term psyllium or cellulose feeding, *J. Nutr.*, 117, 253, 1987.

13 Dietary Carbohydrates as Mood and Performance Modulators

Larry Christensen

CONTENTS

13.1 INTRODUCTION

There seems to be a persistent belief that the food we eat somehow has an impact on our behavior. Throughout recorded history there has been the assumption that the food we eat affects both our physical and mental well-being. The Egyptians and Greeks believed that food contributed to a person's health.[1] For example, foods such as salt were supposed to stimulate passion.[2] This belief of a connection between the food a person eats and his or her behavior is alive and operating today, but with different assumptions. For example, many psychotherapists consider a diet therapy helpful for a variety of disorders, such as major depression,[3] and over 40% of parents or guardians of children with attention deficit disorder believe that sugar is the cause of their child's disorder.[4] Ryan and Lakshman[5] believe that research on psychiatric disorders has neglected the contribution that nutrition may have to treatment regimes. They believe that dietitians must have appropriate information to dispense to patients when they are discharged into the community.

This belief that nutrition has an important contribution to make to psychiatric disorders is probably a function of the scientific evidence that has accumulated over the past several decades revealing that nutrition has a significant effect on behavior. The macronutrient that has received the most attention is carbohydrates, and the behavioral effect that has received the most attention is the effect of carbohydrates on affect or mood, although there is also a literature on the effect of food, much of which is targeted at carbohydrates, on performance.

The attention that carbohydrates has received seems to have been promoted by a number of factors. During the decade of the 1970s, considerable attention within the popular press was devoted to hypoglycemia, or low blood sugar, and the mood and behavioral manifestations of this disorder. Individuals with hypoglycemia report a variety of symptoms, including nervousness, irritability, impaired concentration, and depression,[6] leading some[7,8] to believe that hypoglycemia is the cause of a variety of psychiatric and behavioral disorders, such as delinquency and violent criminal behavior, and that sugar consumption is the cause of hypoglycemia. While hypoglycemia does result in the development of numerous psychological symptoms, there is little evidence to support the notion that it is caused by sugar consumption or that it is the cause of various behavioral or psychiatric disorders.[6] In spite of this lack of relationship, the presumed adverse effect of consumption of simple carbohydrates has persisted and transferred to other behavioral disorders, such as attention deficit hyperactivity disorder.[4,9]

The research program that seems to have stimulated the most interest and research on the effect of carbohydrates on mood and performance is the work of Fernstrom and Wurtman at the Massachusets Institute of Technology demonstrating the possibility that foods might produce behavioral and mood alterations through their effect on brain neurotransmitters.[10–12] Many of the studies to be reviewed in this chapter are the result of hypotheses formulated from this program of research. More recently, some of the research investigating the relationship between carbohydrates, mood, and performance has shifted to a focus on the involvement of the endogenous opiate peptide system,[13] as well as the learned preferences for certain foods as a result of the reinforcement received from their psychoactive involvement.[14] Also, applied psychologists have been interested in the effect that various macronutrients can have on performance in the workplace or in the classroom.[15]

In this chapter I will focus on the research investigating the relationship between carbohydrates and behavior in both healthy and distressed individuals. I will first look at several metabolic mechanisms that have been hypothesized to explain carbohydrates' effects on behavior. In reviewing these mechanisms, I will focus on the two that have received considerable attention and stimulated most of the research in this area. Other mechanisms have been proposed.[16,17] However, they have received little attention or have not been supported and will not be discussed here. In discussing these mechanisms, I will also present data that call into question the most frequently proposed mechanism. I will then focus on the relationship between carbohydrates and both distressed and healthy individuals and conclude the chapter with a review of the effect of carbohydrates on performance.

13.2 HYPOTHESIZED METABOLIC DETERMINANTS MEDIATING THE BEHAVIORAL EFFECT OF CARBOHYDRATES

During the past 30 years there has been a dramatic growth in our knowledge regarding the metabolic effect of food. Information has accumulated indicating that consumption of certain types and amounts of foods can affect neurotransmitters in the brain. Of particular importance is the fact that the neurotransmitters that are affected are also ones that have been linked to specific behaviors, such as sleep and mood.

There are several dozen substances that are believed to act as neurotransmitters in the mammalian central nervous system. These substances fall into the three chemical groups of (1) monoamines, such as serotonin, the catecholamines (dopamine and norepinephrine), and acetylcholine; (2) peptides, such as thyrotropin-releasing hormone, somatostatin, endorphins, and substance P; and (3) nonessential amino acids, such as glutamine, aspartate, and glycine. The investigation of the relationship between carbohydrate intake and neurotransmitters has focused on the monoamine serotonin and on the aPD peptides because these are the substances that have been demonstrated to be related to diet and particularly to carbohydrates.

13.2.1 The Relationship between Carbohydrates and Serotonin

The biosynthesis of serotonin, 5-HT or 5-hydroxytryptamine, is accomplished by converting tryptophan to 5-hydroxytryptophan, 5-HTP, by the reaction of the enzyme tryptophan hydroxylase. 5-HTP is then converted to serotonin by the enzyme 5-HTP decarboxylase. Tryptophan is therefore necessary for the synthesis of 5-HT, and this amino acid must be obtained from the diet. If the diet contains little tryptophan, the precursor for synthesis of serotonin will not be available and the synthesis of serotonin will decline. This effect was clearly demonstrated by Fernstom and Wurtman[18] in an experiment involving feeding rats a corn diet for 5 weeks. Because corn has very little tryptophan, the rats were deprived of it, resulting in a decline in plasma levels of tryptophan and brain levels of tryptophan and serotonin. Similarly, increasing the availability of tryptophan in the diet should result in increased plasma tryptophan levels and brain levels of tryptophan and serotonin. This is exactly what has been found in both rats[18] and humans.[19] A sixfold elevation in cortical tryptophan levels was revealed in excised brain tissue obtained from neurological patients following infusion of tryptophan.[19] Additionally, cerebral spinal fluid 5-HIAA (the principal metabolite of serotonin) has been shown to increase following a tryptophan load.[20]

While it is quite logical that increasing the dietary intake of tryptophan should increase plasma and brain tryptophan levels and, consequently, the synthesis of central serotonin, it is counterintuitive to think that consumption of a carbohydrate-rich meal would have a similar effect, because such a meal contains little, if any, tryptophan. However, this is exactly what Wurtman and his colleagues identified. Interest in the effect of a meal composed of a single nutrient on brain serotonin levels was prompted by studies indicating that both insulin and a carbohydrate meal have little effect on plasma tryptophan but decreased all other amino acids.[21,22] Additionally, both insulin and the carbohydrate meal increased brain tryptophan and serotonin levels.[21] This meant that insulin, whether administered exogenously or secreted endogenously, resulted in an increase in both plasma tryptophan and brain tryptophan and serotonin levels. Subsequent research[23] revealed that plasma tryptophan has the characteristic of loosely binding to circulating albumin. When insulin is secreted, nonesterified fatty acid molecules, which are typically bound to albumin, dissociate themselves and enter adipocytes. This dissociation permits tryptophan to be loosely bound to albumin and protects it from being taken up by peripheral cells. The net effect is that there is little change in total plasma tryptophan levels following insulin secretion, although the plasma levels of many of the other amino acids decrease.

The sparing of tryptophan from being taken up by peripheral cells is important because the system by which tryptophan is transported across the blood–brain barrier is competitive because it also transports the other large neutral amino acids.[24] Because it is competitive, anything that increases the plasma level of tryptophan relative to the other large neutral amino acids would increase the amount of tryptophan transported into the brain. Because a carbohydrate-rich and protein-poor meal has, by definition, little protein, consumption of such a meal would preclude a rise in plasma levels of amino acids. However, the carbohydrate component would stimulate the secretion of insulin, which would cause an uptake of plasma amino acids by peripheral cells. It would also cause nonesterified fatty acid molecules to enter adipocytes, leaving albumin in an unbound state. The unbound albumin would bind loosely to tryptophan, sparing it from entering peripheral cells. This would increase the ratio of plasma tryptophan to the other large neutral amino acids and increase the amount of tryptophan that enters the brain's extracellular space. There would therefore be more of the precursor of serotonin available for synthesis, leading to an increased synthesis of central serotonin. Because the enzyme catalyzing the hydroxylation of tryptophan to 5-HTP is only half saturated with its substrate,[24] increasing the availability of tryptophan should increase the synthesis of 5-HT. It has been well established that increasing brain tryptophan levels increases the level of brain serotonin as well as its major metabolite, 5-hydroxyindoleacetic acid.[12]

This demonstrated effect of consumption of a carbohydrate-rich, protein-poor meal on the synthesis of central serotonin has been the driving force behind a large portion of the research investigating the mood- and performance-altering effects of carbohydrates. It is also one of the most frequently used explanations for a demonstrated behavioral effect following the consumption of a carbohydrate-rich and protein-poor meal. There are, however, a number of serious limitations surrounding the use of this idea as an explanation for a carbohydrate-mediated effect on behavior.

Teff et al.[25] demonstrated that in humans, meals containing as little as 4% protein could counteract a carbohydrate-induced rise in the plasma tryptophan ratio. In rats, the protein content of a meal has to exceed 6% to negate a carbohydrate-induced rise in brain tryptophan level.[26] This is a significant finding because it is difficult to consume a meal or even a carbohydrate snack that contains less than 4% protein. Therefore, unless a person consumes pure sucrose or glucose, sufficient protein would be consumed to negate a rise in the ratio of tryptophan to the other large neutral amino acids (LNAAs), the essential requirement for a rise in brain tryptophan levels, and an increase in the synthesis of central serotonin. Even if a meal or snack contained such a small amount of protein that a rise in the tryptophan/LNAA ratio did occur, the evidence suggests that the rise would be so small as to have little effect on central serotonin. Apparently the plasma tryptophan/LNAA ratio must rise at least 50 to 100% to produce a change in brain 5-HT synthesis.[27,28] A rise of that magnitude does not occur even when a pure carbohydrate, such as glucose, is administered.[29] Most studies have found a rise in the tryptophan/LNAA ratio of <25%, and only one study[25] found a rise as high as 47%, which still does not meet the criteria of a 50 to 100% rise needed to increase the synthesis of central serotonin.

Fernstrom[30] has appropriately pointed out that an animal must fast prior to consuming a carbohydrate-rich and protein-poor meal to elevate the plasma tryptophan/LNAA ratio sufficiently to stimulate serotonin synthesis. However, humans eat, on average, five to seven times a day.[31] This means that humans seldom meet the criteria of fasting prior to consuming a meal regardless of its composition. Rather, the frequency of consumption of food by humans suggests that the consumption of one meal or snack at one time influences the physiologic response to a subsequent meal or snack. Fernstrom and Fernstrom[26] revealed that consumption of a carbohydrate meal had no effect on brain tryptophan or 5-HTP levels if it was consumed 2 h after a meal containing 12% protein. Also, the physiologic effect of a carbohydrate meal on cortical and hypothalamic 5-HTP levels was reversed by the ingestion, 2 h later, of a meal containing a moderate amount of protein (12%).

Taken together, these results reveal that both a prior and a subsequent protein-containing meal can affect the physiologic response to a carbohydrate meal. If a 3-h intermeal interval existed, the physiologic effect of a meal was unaffected by either a subsequent or prior meal. However, this effect was observed in rats and the rate of metabolism in rats is greater than that which exists in humans.[32] Therefore, the intermeal interval would probably have to be longer than 3 h to avoid a sequential effect in humans. Because an intermeal interval longer than 3 h does not typically exist with humans,[31] and most meals and snacks comprise at least 10% protein, a carbohydrate-rich meal or snack would seem to have little effect on brain tryptophan levels or the synthesis of central serotonin. As Young[33] has stated, "At the moment the weight of the evidence indicates that it [the effect of a carbohydrate meal on synthesis of central serotonin] is not an important phenomenon in humans" (p. 900).

While the evidence seems to overwhelmingly suggest that any behavioral effect created by a carbohydrate-rich meal is not due to its effect on central serotonin function, this does not preclude an effect occurring under extreme physiologic or pathologic conditions that would lead to an increased use of central serotonin or to the depletion of serotonin.[16] Under such conditions, a specific meal condition that alters the availability of a precursor may also alter neurotransmitter levels[34] and, as a result, have a behavioral effect.

13.2.2 The Relationship between Carbohydrates and Endogenous Opioids

Opioids comprise families of peptides, including the endorphins, enkephalins, dynorphins, and many nociceptins, and a variety of receptor subtypes. These peptides are found in a variety of networks in the brain, especially in areas that are involved in regulating emotions, responses to pain and stress, endocrine regulation, and food intake.[35] Because of this extensive distribution of opioids throughout the brain and the regions they innervate, it has been hypothesized that these peptides play a role in many biological and psychological processes, such as food intake, reinforcement, pain modulation, affect and emotion, and response to stress.[35,36] I will focus on the effect that endogenous opioids exert on food intake and affect.

Considerable evidence has accumulated revealing that endogenous opioids influence food intake. In general, it has consistently been demonstrated that endogenous aPD antagonists decrease food intake, whereas agonists increase food intake (see Mercer and Holder[37] for a review). Although there is considerable agreement that opioids are involved in food intake, the precise nature of the relationship has not been determined. However, there is increasing evidence and agreement that opioids regulate the palatability or hedonic value of food,[36] or what Berridge[38] has called the liking value of food. The suggestion that endogenous opioids mediate the reward value or liking of food[38] is based on studies demonstrating that aPD antagonists selectively decrease and aPD agonists selectively increase the palatability of sweet-tasting food.[35–38] For example, Fantino et al.[39] has demonstrated that the aPD antagonist naltrexone reduces subjective ratings of sweetness. Drewnowski et al.[40] revealed that the aPD antagonist naloxone reduced hedonic preference for sugar–fat mixtures among female binge eaters and controls, but did not affect perceived hunger. Similarly, it has been demonstrated that animals show a decrease in the palatability of sweet substances following the administration of an aPD antagonist and an increase in the palatability of sweet foods following administration of an aPD agonist.[35–37]

If opioids mediate the liking or palatability of food, then the consumption of palatable foods should activate the aPD systems. There are several lines of evidence to suggest that this effect does occur. Studies (see Mercer and Holder[37] for a review) have revealed that consumption of sweet high-fat foods results in the release of beta-endorphins in the hypothalamus and alters pain perception in both rats and humans. For example, sucrose reduces infants' crying that is provoked by some standard painful hospital procedures. Additionally, electrical stimulation of areas of the brain that are involved in reward are also those brain areas that are involved in feeding. Electrical stimulation of the lateral hypothalamus, for example, is not only rewarding, but also induces feeding. However, aPD antagonists inhibit the electrical induced feeding, probably by reducing the ability of the electrical stimulation to enhance the reward value of the food consumed (see Carr[41] for a review).

The research provides strong evidence indicating that endogenous opioids exert a powerful affective state that is summarized by the terms *palatability* or *liking*. This positive state is strong enough to entice animals to engage in learning to obtain intravenous opiates, suggesting that opiates play an important role in brain reinforcement.[35,36] Because food palatability or the liking component of food is mediated by aPD activity, several investigators[37,40,42] have hypothesized that endogenous aPD peptides mediate behaviors such as food cravings and binge eating. While it may seem logical that a brain substrate that modulates our liking for food would induce cravings and binge eating, there are apparently two components involved in the seeking out and consumption of food. Food reward includes both liking and wanting.[38] The liking component is mediated by endogenous opioids and represents the affective reaction to the consumption of food. It is the reaction to having consumed a food. The wanting component refers to the craving or incentive to seek out and obtain the food. These are two different components and are primarily controlled by different brain substrates.

While the liking or affective component of food reward is determined by endogenous opioids, the wanting component seems to be affected by dopaminergic neural systems, as these systems play a role specific to motivation because massive depletion of dopamine does not influence the affective reaction to food — only the appetite for or the motivation to attain the food. Similarly, neither dopamine agonists or antagonists alter the affective or liking response to food. Rather, these drugs seem to alter the wanting or motivation to seek out and eat the available food (see Berridge[38] for a review). Therefore, the dopaminergic neural system influences the wanting or motivation to eat specific foods and the aPD peptide system influences the hedonic response to the consumption of the food. The dopaminergic neural system thus seems to be the brain substrate that accounts for the desire or cravings for specific foods.[39] Endogenous opioids therefore mediate the liking for or palatability of a food, and the dopaminergic neural systems mediate the wanting or desire for a specific food. These two systems would seem to work together and represent the neural substrate explaining our desire for certain foods and our hedonic reaction to or liking for the food consumed.

13.3 CARBOHYDRATES AND MOOD IN EMOTIONALLY DISTRESSED INDIVIDUALS

There has been considerable attention devoted to the relationship between stress and eating. Perhaps the most widely assumed relationship is that stress increases eating in all exposed individuals. However, this model is far too simplistic, as the research (see Greeno and Wing[43] for a review) has demonstrated a far more complex picture. Some individuals consume more when stressed and some consume less. The picture that is emerging is that there are individual differences in the stress–eating relationship. While dietary restraint seems to be an important variable in the stress–eating relationship, it is important to understand the types of stress that may affect eating and the types of foods that will be selected.[43]

There is some evidence to indicate that stress increases snacking, and the snacking is composed of carbohydrates, particularly foods that are sweet and have a higher fat content.[44,45] This relationship between stress and consumption of foods that are sweet carbohydrate- and fat-rich foods is even more apparent when the focus of attention is on individuals experiencing emotional distress. Westover and Marangell[46] found a 0.948 correlation between the prevalence of major depression and the per capita consumption of sugar. Christensen and Somers[47] found no difference between the nutrient intake of depressed and nondepressed individuals. However, depressed individuals consumed significantly more carbohydrates than nondepressed individuals, and this increased consumption of carbohydrates came primarily from an increased consumption of sugar. This altered preference for sweet carbohydrate-rich foods is consistent with reports[48] that many depressed individuals express a preference for sweet carbohydrate- and fat-rich foods.

13.3.1 Carbohydrate Cravings

Cravings, an intense desire or longing for a particular substance,[49] are hypothetical constructs that frequently appear in the addictions literature. This is a term that has also found its way into the literature on food and is typically used to describe in intense desire or urge to seek out and consume particular foods.[50,51] Up to 97% of women and 68% of men have reported cravings for specific foods.[52] However, the prevalence of food cravings depends on the definition used, and as the definition becomes more stringent, the percentage of identified food cravers can decline to as little as 4%.[53] Regardless of how stringent a definition is used, all studies revealed that some individuals experience food cravings.

Several studies have investigated factors affecting food cravings, and these studies have identified that a variety of factors, such as age, gender, prior meal consumed, mood, and type of food, affect food cravings. Several studies[50,54] have revealed that food cravings decline with age and that food cravings can be triggered by the repeated consumption of the same food.[55] Studies have also investigated the type of food craved and repeatedly revealed that sweet foods, such as chocolate, cake, and ice cream, are most frequently craved, followed by starch foods, such as pasta and breads.[53,56] Interestingly, it is women that not only demonstrate more food cravings,[54] but also more frequently report cravings for sweet foods.[51,56] This craving for sweet foods or, more specifically, foods that are sweet and rich in carbohydrates led to the origination of the term *carbohydrate craving*.[57] Since that time, the term *carbohydrate craver* has been used repeatedly in the literature to refer to individuals expressing an intense desire for sweet carbohydrate-rich foods. This may be an unfortunate use of the term because the foods that are desired by so-called carbohydrate cravers contain considerable amounts of fat, and the most frequently craved foods are not just any carbohydrate, but sweet carbohydrate-rich foods that are also fat rich.[40,58]

Given that carbohydrate cravers have a desire for sweet carbohydrate- and fat-rich foods, one might assume that these individuals would consume more daily energy than noncarbohydrate cravers. There is some evidence to suggest that this is not the case. Hill et al.[59] revealed that there was no significant difference in the daily energy consumed of food cravers and food noncravers. While this study focused on food cravers rather than carbohydrate cravers, the most commonly craved food is sweet carbohydrate- and fat-rich foods,[57] so most of the participants in the study would probably have been carbohydrate cravers. The failure to find a difference between the energy consumption of cravers and noncravers may be due to the fact that food cravers consume most of the craved food as snacks[60] rather than at meals and typically in the afternoon or evening.[60]

Why is it that the most frequently craved food is a sweet carbohydrate- and fat-rich food? One of the obvious answers is that this is the most pleasant tasting food,[40] so it would be the one that is most rewarding and, therefore, craved. However, carbohydrate cravings (I will continue to use this label to refer to the desire for sweet foods even though it is recognized that the food craved is sweet carbohydrate- and fat-rich foods) seem to occur most frequently during certain mood states, which would suggest that the desire or craving for the carbohydrate food is more than just its pleasant taste. This is not to imply that mood is the only factor contributing to the experience of carbohydrate cravings, as research[61] has demonstrated that cravings are influenced by many factors, such as environmental cues and thoughts. Mood seems to be a significant factor contributing to the experience of carbohydrate cravings and is the factor that has been most frequently investigated with respect to carbohydrate cravings.

There are several studies that have demonstrated that carbohydrate cravings are correlated with various indices of emotional distress. Hill et al.[59] has revealed that food cravings are correlated with emotional eating or with an increased appetite stimulated by the presence of a negative mood. This is consistent with studies revealing that many cravers, especially carbohydrate cravers, report that cravings are associated with feelings of fatigue, especially in women,[60] or being anxious and depressed.[51] Cravings for chocolate are correlated with depression and guilt[62] and a variety of other measures of distress, such as tension/anxiety, confusion, and fatigue.[51] It is especially important to realize that these correlations with measures of a dysphoric mood occur only with carbohydrate cravers. Cravings for other foods, such as protein, are not correlated with dysphoric mood states.[51] The interesting thing about the correlation between carbohydrate cravings and dysphoric mood states is that this occurs primarily with cravers of sweet carbohydrate-rich foods, such as desserts, and not with non-sweet carbohydrate-rich foods, such as pasta.[51] This relationship between a dysphoric mood state and carbohydrate cravings occurs with about half of all carbohydrate cravers.[51,62]

Why should individuals with a dysphoric mood state be motivated to consume sweet carbohydrate-rich foods? The evidence suggests that the sweet carbohydrate- and fat-rich food snacks are consumed for the ameliorative effect that they have on the dysphoric mood state.[51,60] Carbohydrate cravers tend to report feeling anxious, tired, depressed, and hungry prior to consuming the food that they crave, and following the consumption of this food, they report positive moods such as satisfied, happy, and relaxed.[51] Additionally, experimental studies[63] have demonstrated that the induction of a mood state such as depression results in an increase in the effort both humans and animals will expend to receive a sweet carbohydrate- and fat-rich reward.

While all studies[62] do not demonstrate an improvement of mood following consumption of a sweet carbohydrate-rich food, and the lifetime experience of a psychiatric diagnosis between cravers and noncravers[56] does not differ, the bulk of the evidence does support the finding of a positive mood-altering effect following consumption of a sweet carbohydrate- and fat-rich snack in carbohydrate cravers.

Further evidence indicating that consumption of a sweet carbohydrate- and fat-rich snack ameliorates a dysphoric mood comes from studies investigating the effect of a carbohydrate-rich snack in individuals with a variety of psychiatric diagnoses.

13.3.2 Carbohydrates and Seasonal Affective Disorder

Seasonal affective disorder (SAD) is defined as a "condition characterized by recurrent depressive episodes."[64] The initial description of this disorder revealed that the depressive episodes occurred in the winter, with the depressive symptoms beginning in the fall and continuing throughout the winter months. As spring approaches, the depressive symptoms begin to remit and the individual with this disorder is symptom-free during the summer months.

Awareness of the fact that some individuals have seasonal depression has existed for centuries. Hippocrates[65] recorded the existence of seasonal depressions, and eight centuries later, Posidonius observed that "melancholy occurs in Autumn, whereas Mania in Summer" (cited in Roccatagliata[66]). While there was continued awareness of the existence of a seasonal depression and Greco-Roman physicians even treated it with sunlight, seasonal depressions were not systematically described or investigated until the 1980s. Interest in seasonal depression was stimulated by the report of an engineer who documented his own seasonal mood swings for 15 years. He even postulated that light may have an influence on his mood state.[67] The initial investigation, conducted by Rosenthal and his colleagues, consisted of several pilot studies designed to document the validity of a seasonal variation in depression. One of the studies focused on a woman who consistently developed depression every winter since she was an adolescent. When she moved north, her depression started earlier in the fall, was more severe, and persisted longer into the spring. On several occasions she took a winter vacation to Jamaica, and within 2 days she experienced a complete elimination of her depression. She was treated with light therapy, the current standard treatment for SAD, early in the morning to determine whether exposure to a full spectrum of light would ameliorate her symptoms. This pilot study and others[64,68] led to the classification of SAD as a psychiatric disorder.

SAD is characterized by many of the typical symptoms of depression, including dysphoria, concentration difficulties, a decline in energy, irritability, anxiety, decreased libido, social withdrawal, and being overrepresented by the female gender.[67,69,70] Some individuals believe that patients with SAD differ from classically depressed individuals in that they experience increased fatigue, sleep duration, appetite, and weight gain.[67] However, others have found a difference only in sleep duration and carbohydrate cravings[69] or carbohydrate cravings and increased appetite.[70] These studies reveal that the symptom most consistently identified as distinguishing SAD from other depressive disorders is carbohydrate cravings. This is a symptom that has been identified in 60 to 88% of individuals with SAD (see Takahashi et al.[71] for a review). However, there is some evidence that the carbohydrate cravings are more prevalent in females than in males.[71] Also, most of the studies investigating SAD are conducted on females, probably because this is the group that is most affected by this disorder. Consequently, most of what we know about SAD is from investigations with a predominantly female population.

Also, most of the research has been conducted on individuals with winter depression. However, not all individuals with SAD have winter depression. Some have a summer depression and experience an amelioration in their depression during the winter months.[72,73] This is an important distinction, because these two groups of individuals seem to exhibit a different set of symptoms. The individuals with winter SAD most frequently exhibit carbohydrate cravings and an increase in appetite with an accompanying increase in food consumption and weight gain.[72–74] However, individuals with summer SAD are less likely to report carbohydrate cravings and more likely to report an appetite decrease and weight loss.[72,73] These are important differences, suggesting that there may be two distinct groups of SAD patients. Most of what we know relates to individuals with winter SAD, in terms of both treatment and characteristics, as most studies have focused on winter SAD. This may be because the prevalence of winter depression seems to be much greater than summer depression. Rosen et al.[75] revealed that the prevalence rates for winter depression in three northeastern cities ranged from 4.7 to 9.7%, whereas that for a southern city was 1.4%. The corresponding rates for summer depression ranged from 0.5 to 3.1% for three northeastern cities and was 1.2% for a southern city. It is therefore important to keep this point in mind, because the information presented seems to be most applicable to individuals with winter SAD.

The carbohydrate cravings that have been identified as the most distinguishing characteristic of individuals with winter SAD are correlated with an increase in appetite and food consumption.[74] Not only are these variables correlated, but it has been demonstrated that SAD patients retrospectively report changes in food consumption with the seasons of the year.[76] Meal consumption during the summer was less than during any other season, and carbohydrate consumption was the least during the summer and the greatest during the winter months. This increase in the consumption of carbohydrates came primarily from sweet snack foods consumed in the afternoon and evening.[77] About 85% of the total snack carbohydrate consumption came from sweet foods such as cakes. The increased consumption of total carbohydrates, including both starch and sweet foods, declined during the second half of the day with therapy and as the SAD patients' depression ameliorated.

The studies conducted with SAD patients clearly demonstrate that they increase their cravings for and consumption of carbohydrates, especially sweet snack foods, during the afternoon and evening as the winter months approach and as they become depressed. As the summer months approach or with therapy, these patients show a reduction of both carbohydrate cravings and intake. Anecdotally, some of these patients report that they eat the carbohydrates to energize themselves or to perk up.[64] This is consistent with the evidence documenting that SAD patients state not only that they ingest carbohydrates in response to a variety of symptoms of depression, but also that they do so to alleviate these symptoms.[78] Individuals with SAD therefore may be using carbohydrates to combat their feelings of depression and the anergy that accompanies it.

The use of carbohydrates to combat some of the symptoms of depression has been experimentally demonstrated by Rosenthal and his colleagues.[79] These investigators compared the mood-altering effect of a carbohydrate-rich and protein-rich meals on 16 patients with a diagnosis of SAD and 16 matched control individuals.

This study revealed that following consumption of the carbohydrate-rich meal, the SAD patients demonstrated a slight decrease in fatigue and a temporary decline in vigor. The normal controls, however, experienced an increase in fatigue and a continued decline in vigor for 3 h after meal consumption. Following a protein-rich meal, both SAD and normal controls experienced an increase in fatigue and a decline in vigor. This study has demonstrated that a carbohydrate-rich meal affects SAD patients differently than normal controls and provides some suggestive evidence that these individuals may be using carbohydrates to combat the anergy they may be feeling. However, this is only one study, so the results must be viewed as tentative until further evidence confirms the mood-altering effect of carbohydrate consumption. Given that SAD patients have cravings for sweet carbohydrate-rich snacks, consume more of these snacks when depressed, and report that they feel better when consuming them, as well as the suggestive experimental evidence from the Rosenthal et al.[79] study, the hypothesis that these individuals are consuming sweet carbohydrate-rich snacks to self-medicate seems to have some validity.

13.3.3 Carbohydrates and Mood in Obese Individuals

Obesity is an increasingly significant problem in many countries. In the U.S., a recent random-digit telephone survey of 195,005 adults aged 18 or older revealed that the prevalence of obesity increased from 19.8% in 2000 to 20.9% in 2001, an increase of 5.6%.[80] This is a trend that has been occurring for decades.[81] The significance of such a continuing increase in weight is apparent in the data demonstrating that being overweight and obese is significantly related to a variety of negative health consequences, such as diabetes, high blood pressure, high cholesterol, asthma, arthritis, and poor health status.[80]

Obesity is typically defined as excessive storage of energy in the form of fat.[81] While obesity is a complex multifaceted problem with a variety of antecedent and predisposing factors, overeating is, to many health practitioners and the general public, the primary cause.[82] A focus on the overconsumption of calories is counterproductive and probably will contribute little to our understanding of the cause and treatment of obesity. However, dietary intake is a significant variable. One of the variables that has received attention is the type of food preferred by obese individuals. Obese men list fat and fat/protein foods (meat dishes) as the type of foods they most prefer, whereas obese women list carbohydrate/fat foods and foods that are sweet, such as doughnuts, cookies, and cakes.[83] It is this desire for sweet carbohydrate- and fat-rich food that primarily exists in obese women that has been of interest to a number of researchers.

The desire for sweet carbohydrate- and fat-rich foods has been expressed as a carbohydrate craving. The hypothetical construct of carbohydrate craving in obese individuals is somewhat controversial, with some researchers believing that it is a meaningful construct[84] and others[58,85] considering it a useless construct and an impediment to scientific progress in obesity research. In spite of the controversy over its validity, it is a construct that continues to appear in the literature and one that is expressed by many obese individuals.

Although the construct of carbohydrate craving among the obese is controversial, the fact still remains that the obese, particularly obese women, express a preference for sweet carbohydrate- and fat-rich foods. Interestingly, foods from a fat food group were the least preferred, as were foods from a sugar food group.[83] Consequently, it is the combination of these two food groups that produces a highly desired food. It is also interesting to note that the consumption of sweet carbohydrate- and fat-rich foods occurs primarily in the form of snacks consumed in the afternoon and evening.[86,87]

While there is consistent evidence that obese individuals, particularly obese women, demonstrate a preference for and consumption of sweet carbohydrate- and fat-rich foods, there is some disagreement as to the frequency with which these individuals report craving such foods. Wurtman et al.[87] reported significant difficulty in finding overweight noncarbohydrate cravers, whereas other studies[85,86] seem to have little difficulty finding such individuals. The difference is probably due to the way in which these different studies identify so-called carbohydrate cravers, because up to 80% of obese women list sweet carbohydrate- and fat-rich foods as one of their preferred foods.[83] The evidence therefore indicates that a significant percentage of obese individuals, particularly women, experience a craving or intense desire for sweet carbohydrate- and fat-rich foods and consume these foods primarily as snacks in the afternoon and evening when they experience a craving for such foods.[86]

One of the interesting questions is why obese individuals would demonstrate a preference for and consumption of sweet carbohydrate- and fat-rich foods primarily in the afternoon and evening. It has been reported that overweight individuals reporting carbohydrate cravings suffer from transient depression[88] and state that they feel restless, tense, and unable to concentrate prior to consumption of the snack and calm, relaxed, and able to concentrate after snack consumption.[89] Such statements suggest that these individuals are engaged in a form of self-medication where food is used to induce a positive mood state. The notion that emotions induce eating in obese individuals has received significant attention, and the literature suggests that emotional eating is very prevalent in the obese and that the major determinant of emotional eating is its ability to reduce negative emotions such as anger, depression, and loneliness.[90]

The hypothesis that obese individuals engage in a form of self-medication with sweet carbohydrate- and fat-rich foods is rather controversial and has received conflicting support. Lieberman et al.[84] found that obese individuals who reported carbohydrate cravings and consumed carbohydrate-rich snacks during a 3-day period experienced a decline in depression 2 h after consumption of a carbohydrate-rich lunch. Obese noncarbohydrate cravers experienced an increase in depression and fatigue and a decrease in alertness 2 h after the consumption of the carbohydrate-rich lunch. This study clearly supports the idea that obese carbohydrate cravers experience a positive mood alteration following a carbohydrate-rich meal, whereas obese noncarbohydrate cravers do not. However, neither Toornvliet et al.[85] nor Reid and Hammersley[91] found any mood-enhancing effect from a carbohydrate-rich meal.

The failure to find an effect in these two studies may be due to the nature of the way in which the research participants were selected. Reid and Hammersley used individuals who were obese, but no attempt was made to determine if they were carbohydrate cravers. Toornvliet et al.[85] distinguished carbohydrate cravers from noncarbohydrate cravers based on their assessment of 3-day food records. Lieberman et al.[84] distinguished carbohydrate cravers from noncarbohydrate cravers based on self-report and actual selection of carbohydrate-rich snack foods. Also, Lieberman et al.[84] reported difficulty in finding noncarbohydrate cravers, whereas Toornvliet et al.[85] identified more noncarbohydrate cravers than cravers from their assessment of food records. This would strongly suggest that significant differences existed in the characteristics of the samples studied. Also, Corsica[92] has demonstrated that, when dysphoric, women who report craving carbohydrates selectively choose a carbohydrate-rich beverage vs. a nutrient-balanced beverage. Additionally, when the participants consumed the carbohydrate-rich beverage, they reported a greater improvement in depression than when they consumed a protein-rich beverage. The evidence therefore seems to be tilted in the direction of supporting the idea that obese carbohydrate cravers self-medicate with sweet carbohydrate- and fat-rich snacks to experience an improvement in their mood.

13.3.4 Carbohydrates, Mood, and Premenstrual Symptoms

The waxing and waning in many women of a variety of physical and psychological symptoms over the course of the menstrual cycle has been known and documented since the time of Hippocrates. However, the importance and the distress accompanying these symptoms remained in relative obscurity with, prior to about 1980, much of the research focused on documenting and verifying that affective symptoms actually do show cyclical changes across the menstrual cycle.[93] The premenstrual syndrome attracted considerable attention in the late 1970s and early 1980s when two women attributed the murder they committed to their premenstrual symptoms.[94] Since that time, this syndrome has generated considerable controversy and debate at every level, from its definition to its diagnostic criteria, and even with respect to the label used to summarize this condition. About four decades ago, the syndrome was referred to as premenstrual tension. Since that time, the term *premenstrual syndrome* has frequently been used, although the label *premenstrual dysphoric disorder* is used in the most recent edition of the *Diagnostic and Statistical Manual of Mental Disorders*.[95] For purposes of the present chapter, I will use the most common label of premenstrual syndrome (PMS).

Regardless of the label used, most definitions and diagnostic criteria emphasize the presence of symptoms only during the late luteal phase, with a change in severity from the follicular to the luteal phase and an impairment in function during the late luteal phase of the menstrual cycle.[96] There is also little knowledge regarding the etiology of PMS other than the fact that there is a high probability of involvement of the hypothalamo-pituitary-gonadal system, with environmental and psychological factors contributing to the severity and development of the symptoms.[96]

Although controversy has existed over the label, definition, and diagnostic criteria surrounding PMS, there is consistent evidence indicating that females, both with and without PMS, demonstrate an alteration in energy intake, with energy intake significantly increasing during the luteal phase of the menstrual cycle. The mean intake difference between the follicular and luteal phases ranged between 87 and

674 kcal/day, representing a 4 to 35% increase in the luteal phase.[97] While there seems to be agreement on the fact that there is an increase in energy intake during the luteal phase of the menstrual cycle, this increase in energy intake does not seem to be specific to a particular macronutrient in either animal or human studies, although a preference for carbohydrate, protein, and fat has been proposed by different studies (see Buffenstein et al.[96] for a summary of these studies).

Not only is there an increase in energy intake during the luteal phase of the menstrual cycle, but there is also, as might be expected, an increase in appetite.[98] However, the increase in appetite was greater in women with premenstrual symptoms. Additionally, appetite is correlated with depression in women with premenstrual symptoms, but not in women without premenstrual symptoms, and this correlation exists only during the luteal phase. Depression accounted for 65% of the variance in appetite. This suggests that the existence of premenstrual symptoms, such as depression, may influence appetite and food consumption, as it has been demonstrated that depression increases during the luteal phase in women with premenstrual symptoms.[99] This suggestion is supported by research demonstrating that negative affect is correlated with carbohydrate intake during the perimenstrual period.[100]

A number of studies have also indicated that food cravings, specifically cravings for carbohydrate-rich foods, increase during the luteal phase of the menstrual cycle.[101–103] While some studies have not found such an effect,[103,104] the difference between those studies that have identified an increase in carbohydrate cravings in the luteal phase and those that have not seems to be a function of the research sample used or the time in which cravings are assessed. Studies[104,106] that have not identified an increase in carbohydrate cravings have typically used normal participants or have tested for cravings in the morning,[105]which is wrong time to assess for cravings, as it has been identified that carbohydrate cravings and consumption of carbohydrate-rich foods occur primarily in the afternoon and evening[106] and in the form of snack foods.[106,107] Most of the evidence[101,106,109,110] indicates that the preference is for sweet carbohydrate-rich foods, although there is one study[107] indicating that there is no preference for sweet vs. starch type foods.

The literature reveals that cravings for and consumption of carbohydrate-rich foods in the luteal phase of the menstrual cycle occur primarily for women experiencing premenstrual symptoms, particularly when the symptom of depression is present. It has been suggested that the increase in appetite and the cravings for and increased consumption of carbohydrate-rich foods may be due to an attempt to ameliorate the heightened level of depression experienced during the luteal phase,[98,108] and suggests that these women engage in a form of self-medication. There are several studies demonstrating just this effect with women experiencing severe premenstrual symptoms. Wurtman et al.[107] compared the mood-altering effect of a carbohydrate-rich meal in a group of women with severe premenstrual symptoms to that in a group of women without symptoms during both the follicular and luteal phases of the menstrual cycle. Results of this study revealed that the carbohydrate meal significantly decreased the depression, tension, anger, confusion, and fatigue experienced by the women with severe premenstrual symptoms, and this effect existed only during the luteal phase of the menstrual cycle. There was no change in mood states during the follicular phase or in women without premenstrual symptoms during the luteal phase. The carbohydrate-rich meal had a positive effect only in the women with severe premenstrual symptoms and only in the luteal phase when these symptoms were prominent.

The ameliorative effect of a carbohydrate-rich meal on mood was replicated in a study[111] investigating the effect of a specially formulated carbohydrate-rich beverage in women with severe premenstrual symptoms. This beverage, when administered at a time when the selected women were experiencing a significant worsening of their premenstrual symptoms, was effective in improving these individuals' overall level of affect and reducing their anger and depression. These two studies reveal that a carbohydrate-rich meal or snack is capable of ameliorating some of the affective symptoms experienced by women with severe premenstrual symptoms and support the self-medication hypothesis. It should be emphasized that this ameliorative effect and the appetite increase and carbohydrate cravings that wax and wane with the various phases of the menstrual cycle are most prominent and occur most reliably in women with the most severe symptoms.

13.3.5 Carbohydrates and Major Depression

Depression is a serious mental disorder affecting about 1 in every 20 individuals at some time in their life.[112] To be diagnosed with major depression, a person must have a depressed mood or loss of interest or pleasure in nearly all activities as well as a number of other symptoms, such as a change in appetite or weight, sleep, and psychomotor activity; feeling fatigued, worthless, or guilty; having difficulty concentrating, thinking, or making decisions; or having suicidal thoughts, plans, or suicide attempts.[95] While these are the typical and diagnostic features of major depression, it has also been documented that about 25% of individuals with major depression experience carbohydrate cravings.[48,69,113] Depressed individuals also report that their preference for sweet carbohydrate- and fat-rich foods increases as they become depressed,[48] and they also consume more carbohydrates, particularly from sugars,[47] as revealed in Table 13.1. This should not be surprising, as it has been demonstrated that food cravings tend to lead to consumption of the craved food.[103]

TABLE 13.1
Mean Daily Carbohydrate Intake in Depressed and Nondepressed Individuals (Mean ± Standard Deviation)

	Depressed		Nondepressed		
	M	SD	M	SD	t-ratio
Carbohydrate (g)	330	140	232	118	2.86[a]
Sugars (g)	169	165	89	63	2.43[a]
Sucrose (g)	91	150	29	25	2.19[a]

[a] $p < 0.05$.

Adapted from data taken from Christensen, L. and Somers, S., *Int. J. Eating Disord.*, 20, 105, 1996.

Why would the depressed individuals report a change in preference for sweet carbohydrate-rich foods as they became depressed and then proceed to consume such foods? These individuals report consuming such foods to combat their depressive symptoms,[78] which essentially means that they are engaging in a form of self-medication. The interesting component is that they report using caffeine in the same manner, so both caffeine and sweet carbohydrate-rich foods seem to be used for the ameliorative effect that they have on depressive symptoms.

While the self-medication hypothesis has not been investigated in individuals with major depression, several studies have focused on the effects of carbohydrates on depressive symptoms. Moorhouse et al.[114] investigated the effect of administering a carbohydrate-rich and protein-poor, or a protein-rich and carbohydrate-poor breakfast, mid-morning, mid-afternoon, and evening snack to alcohol-dependent individuals and controls for two consecutive days. These investigators found that the alcohol-dependent individuals with carbohydrate cravings demonstrated an increase in depression after consumption of the carbohydrate-rich breakfast and snacks, whereas the protein-rich breakfast and snacks had no deleterious effect. These investigators hypothesized that "clinical populations who crave sugar are paradoxically sensitive to its adverse effect on mood" (Moorhouse et al.,[114] p. 641).

This hypothesis is supported by a number of studies[115–117] that my students and I have conducted. In the first series of studies,[115,117] we found that eliminating added sucrose and caffeine from the diet of selected emotionally distressed individuals ameliorated their distress. The constant theme running through these studies was that the individuals who benefit from eliminating added sucrose and caffeine were depressed. Therefore, we conducted a study[116] investigating the effect of eliminating added sucrose and caffeine from the diet of individuals experiencing a current episode of major depression. Table 13.2 reveals that the experimental group who eliminated added sucrose and caffeine not only experienced a significant amelioration in their depression, but this decline in depression was also apparent at the 3-month follow-up.

TABLE 13.2
Mean Depression Scores after 3-Week Dietary Elimination of Added Sucrose and Caffeine (Experimental) or Red Meat and Artificial Sweeteners (Control) and 3-Month Follow-Up of Individuals Who Demonstrated a Significant Amelioration in Depression in the Experimental Group (Mean ± Standard Deviation)

	3-Week Dietary Intervention				3-Month Follow-Up	
	Experimental		Control			
Depression Measure	M	SD	M	SD	M	SD
MMPI-D	63.30	14.63	81.20	15.87	55.83	15.82
BDI	9.50	8.29	23.10	9.12	5.43	6.27
SCL-90 Depression Scale	34.90	5.04	50.10	6.38	29.57	8.71

Note: MMPI-D = MMPI Depression Scale; BDI = Beck Depression Inventory; SCL-90 = Symptom Checklist-90. Higher scores indicate increased severity of depression.

Adapted from data from Christensen, L. and Burrows, R., *Behav. Ther.*, 21, 183, 1990.

Although these studies have revealed that some individuals experiencing a current episode of major depression can benefit from eliminating added sucrose and caffeine from their diet, it is important to emphasize that this ameliorative effect does not exist with all depressed individuals. In most of our studies,[115,117] we have identified the sensitive individuals using a single-case design followed by double-blind challenges, which is very time-consuming. However, given that Moorhouse et al.[114] revealed that only carbohydrate-craving alcohol-dependent men experienced an increase in depression following the high-carbohydrate breakfasts and snack foods, it is possible that individuals with major depression and reporting carbohydrate cravings would represent the sensitive population. It is also possible that these individuals represent atypical depressed individuals, as some of the features of these individuals are increased appetite, hypersomnia, and leaden paralysis.[95] These are also characteristic symptoms of individuals who profit from eliminating caffeine and added sucrose from their diet.

13.4 THE EFFECT OF CARBOHYDRATES ON HEALTHY ADULTS

For at least the past 50 years there has been a continuing interest in the effect that eating a meal may have on one's mood and performance. For example, there seems to be a general belief that consumption of breakfast is very important and will enhance performance. This belief seems to have its genesis in a series of studies[118] conducted in the mid-20th century by Tuttle and colleagues. More recently, interest has focused on the effect of carbohydrates on mood and performance. This interest in carbohydrates was stimulated by the demonstration that carbohydrate consumption can alter the synthesis of central serotonin in rats and the plasma tryptophan–large amino acid ratio in rats and humans.[12,15,18,21] This interest was given further emphasis by the demonstration[119] that consumption of a carbohydrate-rich and protein-poor meal vs. a protein-rich and carbohydrate-poor meal made females sleepier, males calmer, and people over 40 less tense and calmer. Not only did this study demonstrate that the carbohydrate macronutrient may have an effect on behavior, but also that there was a physiological reason for this effect. Since that time, a number of studies have been conducted trying not only to verify the presumed effect that carbohydrates have on mood and performance, but also to determine if the presumed reason for this effect was the effect that carbohydrate consumption may have on the synthesis of central serotonin. I will focus only on studies investigating the effect of carbohydrates on mood and performance.

13.4.1 Carbohydrates and Mood in Healthy Adults

The studies that have investigated the effect of carbohydrates on mood in healthy adults are summarized in Table 13.3. From this table you can see that some studies have focused on the effect of carbohydrate consumption in the morning, some have focused on both morning and afternoon, and some have limited their attention to the afternoon. If attention is focused on effects of carbohydrate consumption in the morning, most of the studies[120–125] indicate a positive effect, including a lowering of

TABLE 13.3
Summary of Studies Investigating the Effect of Carbohydrates on Mood

Study	Participants	Meal Type and Time Given	Outcome Measures	Time of Assessment	Outcome Affected
Effect of Consuming Carbohydrates in the Morning					
Fischer et al., 2002	Male students	CHO rich vs. balanced vs. protein rich at 7 A.M.	Rating 20 mood states on 7-point bipolar scales	Before and hourly for 3 h after breakfast	No effect of meals on ratings of any mood
Fischer et al., 2001	Male students	CHO vs. protein vs. fat at 7 A.M.	Ratings of mood states on 7-point bipolar scales, POMS	Before and hourly for 3 h after breakfast	Lower POMS depression scale score after CHO vs. protein
Benton et al., 2001	Female adults	Fast vs. high vs. low CHO at 10 A.M., each crossed with CHO snack or no snack at 11:30	Analogue-scale assessment of six mood states	Before, 15 min, and 1 h after breakfast and 15 min and 1 h after snack	More confused and anxious after high CHO; snack improved mood only in high-CHO condition
Lloyd et al., 1996	Mostly female adults	Low, medium, and high fat crossed with low, medium and high CHO and no breakfast	Ratings of 16 different mood states on 100-mm line	Before and at 30, 90, and 150 min after breakfast	Decline in fatigue after low-fat high-CHO breakfast
Reid and Hammersley, 1995	Male and female adults	Sugar or saccharin-sweetened orange beverage or water	POMS	At 9 A.M. before consuming beverage and 30 and 60 min after	Increase in energy 30 min after consuming the sugar beverage
Smith et al., 1994	Male and female students	High-CHO, high-protein or no breakfast	Introversion–extroversion, anxiety, levels of psychiatric symptoms	Before 8 A.M. or 8:30 breakfast and 1.5 and 2.5 h after breakfast	Less content, interested, sociable, and outgoing after CHO and no breakfast vs. protein
Wells et al., 1998	Male and female	Gastric infusion of sucrose, lipid, or saline	POMS, SSS, VAS, UWIST	Before and at half-hour intervals after 10:15 A.M. gastric infusion	Less dreamy and sleepy 30 min after sucrose, but more sleepy 2 h after sucrose and less dreamy after sucrose at 3.5 h

TABLE 13.3 (continued)
Summary of Studies Investigating the Effect of Carbohydrates on Mood

Study	Participants	Meal Type and Time Given	Outcome Measures	Time of Assessment	Outcome Affected
Consumption of Carbohydrates in the Morning and Afternoon					
Spring et al., 1983	Male and female adults	High CHO or high protein	POMS, SSS, analogue ratings of mood	2 h after eating breakfast at 7:15 to 8:30 A.M. and 2 h after eating lunch at 11 A.M. and 1 P.M.	Overall, females sleepier after CHO meal; people 40 and older are less tense and calmer after CHO breakfast; males are calmer after CHO meal
Deijen et al., 1989	Primarily female college students	High-protein food until 1:30 P.M. and high-CHO food after 1:30 P.M. or *ad libtum* diet	POMS	10:30 A.M. and 6 P.M. 1.5 to 2 h after meal and after being on the diet for 3 weeks	Diet group had higher anger in the morning
Thayer, 1987	Primarily females	Sugar snack vs. exercise	AD/ACL	1 h after abstaining from food and at a time that was convenient to them	Morning snack increased energy, but afternoon snack increased and then decreased energy; snack increased tension
Effect of Consuming Carbohydrates at Lunch or in the Afternoon					
Markus et al., 2000	High- and low-stress-prone students	CHO rich vs. protein rich at 11 A.M. or noon	POMS	1.5 h after lunch and before and after stress induction	Stress increased depression in high-stress group after protein-rich but not CHO-rich meal
Markus et al., 1998	High- and low-stress-prone students	CHO rich vs. protein rich at 11 A.M. or 1 P.M.	POMS	1.5 h after lunch and before and after stress induction	Stress increased depression and decreased vigor in low-stress group on both diets, but only on protein-rich diet in high-stress group

Wells et al., 1995	Male adults	High fat, low CHO or low fat, high CHO, both with moderate protein	Ratings of alertness, cheerfulness, and calmness	Hourly starting at 9 A.M., with lunch given at 12:45 P.M.	2.5 h after high-CHO lunch alertness greater vs. high-fat, low-CHO condition
Reid and Hammersley, 1999	Male and female adults	Saccharin, sucrose, maize oil, or sucrose and maize oil added to yogurt at 11 A.M.	Ratings of 10 mood states using visual analogue scales	Immediately after and 60 and 120 min after eating yogurt at 11	Sucrose group felt more fatigue at 120 min and was calmer at 60 min than saccharin group
Lloyd et al., 1994	Mostly females	Low fat, high CHO; medium fat, medium CHO; and high fat, low CHO	Visual analogue ratings of 16 mood states	30 min before and 30, 90, and 150 min after 12:30 P.M. lunch	Less drowsy, uncertain, and muddled after medium-fat and -CHO lunch, but less tense after low-fat, high-CHO lunch
Christensen and Redig, 1993	Female college students	Sugar rich, starch rich, or protein rich at 1 P.M.	SSS, POMS, AD-ACL	Before and at 30, 60, 90, 120, and 180 min after 1 P.M. lunch	No effect on mood
Pivonka and Grunewald, 1990	Female college students	Sugar or aspartame Kool-Aid or water at 3:30 P.M.	SSS, visual analogue ratings of 32 adjectives and POMS	Before and 1 h after consuming beverage at 3:30 P.M.	Greater sleepiness after sugar Kool-Aid
Smith et al., 1988	Male and female adults	High-protein, high-starch, or high-sugar lunch	Visual analogue ratings of 18 adjectives	Before and 1.25 h after lunch	No effect on mood
Spring et al., 1986	Female adults	CHO-rich, protein-rich, balanced, or no lunch	SSS, POMS, visual analogue ratings	Before and 4 times after lunch	Fatigue greater 2 h after CHO than after no lunch
Spring et al., 1986	Male adults	Starch or protein lunch	SSS, POMS	Before and hourly for 5 h after noon lunch	No effect on mood

Note: POMS = Profile of Mood States; SSS = Stanford Sleepiness Scale; VAS = Visual Analogue Scales; UWIST = Mood Adjective Checklist; AD-ACL = Activation/Deactivation Adjective Check List.

depression, a decline in fatigue, an increase in energy, less dreamy and sleepy 30 min and 3.5 h after carbohydrate consumption, but sleepier at 2 h after carbohydrate consumption, and a decline in tension and an increase in calmness in people over 40 years of age. However, at least one study[126] found no effect of a carbohydrate breakfast when compared with a protein and balanced breakfast. Two studies[127,128] found primarily negative effects, indicating that a carbohydrate load in the morning decreased feelings of contentedness, interest, sociability, and outgoingness, but confusion and anxiety also decreased. However, some of these effects declined with a carbohydrate snack.

These contradictory findings may be due to the discrepancy in the procedures followed in the various studies. A review of Table 13.3 reveals that the meal comparisons ranged from comparing carbohydrate, protein, and balanced meals, sometimes including a fast, to various combinations of carbohydrates and fats, to gastric infusion of single macronutrients. Similarly, the type of outcome measures varied from using standardized mood assessment measures to analogue ratings, and the time of consumption of the selected foods and assessment of outcome measures varied considerably. When these study variations are considered, it is perhaps understandable that there are variations in the outcome of these studies. In spite of these variations, the majority of the studies suggest that a carbohydrate load consumed in the morning has a beneficial effect on mood, and the beneficial effect seems to be one of improved energy or decreased fatigue and a general improvement in mood. Because most studies indicate a positive mood-altering effect from consumption of a carbohydrate load in the morning, it seems safe to conclude that such a positive effect will occur in most individuals. However, this positive effect should not be promoted as a definite outcome of morning carbohydrate consumption, as some individuals will obviously not feel such an effect, and some may have a negative reaction. Additionally, it is difficult to identify the time in which the effect will occur, as the studies varied not only the time of morning in which carbohydrates were administered, but also the amount of time that elapsed before testing the effect of the carbohydrate.

Table 13.3 also summarizes the results of studies assessing the effect of consuming carbohydrates at lunch or in the afternoon. In reviewing these studies, it is again apparent that there is considerable variation in terms of the macronutrients varied, the time in which the macronutrients were consumed, and the time of assessment of mood following consumption of the different macronutrients. In spite of these variations, there seem to be some consistencies emerging. When either a carbohydrate-rich meal or sucrose is compared to a condition in which no macronutrient is consumed (no lunch, water, saccharin), the participants consuming the sucrose or carbohydrate-rich meal report being sleepier or more fatigued.[129–131] However, if the effect of a carbohydrate-rich meal is compared with that of a protein-rich meal, there seems to be no significant difference between them on mood.[132–134] There is also little effect on mood if the research participants consume a high-protein meal in the morning and a high-carbohydrate meal in the afternoon for an extended period of time, such as 3 weeks.[135] However, this failure to find an effect of a carbohydrate-rich vs. a protein-rich meal may be due to the heterogeneous nature of the research participants, as it has also been demonstrated that high- and low-stress-prone individuals respond differently to these two meals when stressed. Two studies[136,137] have demonstrated that only low-stress-prone research participants experienced a rise in depression following a stressful task when consuming a carbohydrate-rich meal. The carbohydrate-rich lunch protected the high-stress-prone participants from experiencing a similar rise in depression.

Several studies have also compared the effect of varying the carbohydrate and fat contents of a meal. In one of these studies,[138] the medium-fat and medium-carbohydrate lunch had the positive effects of less drowsiness, uncertainty, and muddledness, and in the other study,[139] it was the low-fat and high-carbohydrate meal that created the positive effect of greater alertness. However, in both studies, the meal that had the positive effect was the meal type that was most comparable to the participants' typical meal. Therefore, it may be that any deviation from a person's typical meal pattern has a negative effect.

These studies seem to suggest that a carbohydrate-rich lunch will increase feelings of sleepiness and fatigue compared to not eating lunch. However, there is no difference when comparing a protein- vs. carbohydrate-rich lunch, so it may be the consumption of any lunch that makes the primary difference. This effect may be particularly noticeable if the lunch deviates from what is typically consumed.

13.4.2 Carbohydrates and Performance in Healthy Adults

Table 13.4 summarizes the results of studies investigating the effect of carbohydrate consumption on a variety of performance measures. From this table you can see that there is considerable variation in the macronutrient content of the meals provided in the various studies, as well as the type of tasks chosen and the time of assessment of the effect of the meals. This variation from study to study has certainly contributed to the variation in results across studies. In spite of this, there seem to be some trends emerging. In reviewing the studies[120,121,126–128,133] that have investigated the effect of a carbohydrate load in the morning, typically consumed at breakfast time, it is apparent that most of the studies have investigated some form of reaction time and some type of memory or recall task. These studies have revealed that the consumption of a carbohydrate breakfast either has no effect or slows reaction time. The slowed reaction time seems to occur in tasks requiring speed of reaction following some choice decision. Where the measure is some simple measure of speed of reaction without a choice, the carbohydrate breakfast seems to have no effect when compared to breakfasts composed primarily of other macronutrients, such as protein or fat.

When short-term memory is the task, consumption of a carbohydrate breakfast seems to result in worse or slower performance. However, if the memory task involves recall, the type of meal consumed seems to have no effect. While it may seem contradictory that a carbohydrate breakfast would result in worse short-term memory but have no effect on recall, this differential effect is probably due to the type of task involved in what was labeled short-term memory vs. recall. The recall task involved remembering a list of words, whereas the short-term memory task involved recognizing a defined sequence of colored circles.

TABLE 13.4
Summary of Studies Investigating the Effect of Carbohydrates on Performance

Study	Participants	Meal Type	Outcome Measure	Time of Effect	Outcome Affected
Effect of Carbohydrates on Performance in the Morning					
Fischer et al., 2002	Male students	CHO rich vs. balanced vs. protein rich at 7 A.M.	Choice reaction time, combined short-term memory and peripheral attention, task and multitask test	Before and hourly for 3 h	Longer decision time, less accurate short-term memory, slower central reaction time after CHO meal
Fischer et al., 2001	Male students	CHO vs. protein vs. fat at 7 A.M.	Simple and choice reaction time, combined short-term memory and peripheral attention task	Before and hourly for 3 h after breakfast	Reaction time worse for protein and CHO; short-term memory and focused attention tasks better after fat vs. CHO and protein
Benton et al., 2001	Female adults	Fast vs. high vs. low CHO at 10 A.M. crossed with snack or no snack at 11:30	Recall of words presented on a tape recorder	Before, 15 min and 1 h after breakfast, and 15 min and 1 h after snack	No meal effect before snack, but better memory 15 min but not 1 h after CHO snack
Lloyd et al., 1996	Mostly females	Low, medium, and high fat crossed with low, medium, and high CHO and no breakfast	Visual processing, tapping, recall, reaction time	Before and at 30, 90, and 150 min after breakfast	No effect of meal type on any of the performance measures
Smith et al., 1994	Male and female students	High-CHO vs. high-protein or no breakfast	Reaction time, serial response, repeated digits	Before 8 A.M. or 8:30 breakfast and 1.5 and 2.5 h after breakfast	No effect on any task
Smith, 1988	Adults	CHO of cereal and toast or no breakfast	Logical reasoning and search task	Before, right after, and 1.5 h after breakfast	No effect on either task
Effect of Carbohydrates on Performance in the Afternoon					
Markus et al., 1998	High- and low-stress-prone students	CHO rich vs. protein rich at 11 A.M. or 1 P.M.	Memory scanning task	1.5 h after lunch and right after stress induction	Reaction time in responding faster on CHO-rich diet

Wells et al., 1995	Male adults	High fat, low CHO or low CHO, high fat, both with moderate protein	Sustained attention task	Hourly between 9 A.M. and 5 P.M. with lunch at 12:45 P.M.	Response speed faster in low-fat high CHO condition
Kelly et al., 1994	Male adults	Study 1: High CHO, low CHO, high fat, low fat at lunch; study 2: high, medium, and low CHO and fat at lunch and breakfast	Digit symbol, number recogition, repeated acquisition, reinforcement of low-rate schedule	At 9:30, 1:30, and 7 P.M. for both studies	No effect of type of meal on any of the psychomotor tasks
Spring et al., 1983	Male and female adults	High CHO or high protein	Reaction time and dichotic shadowing	2 h after eating breakfast at 7:15 to 8:30 A.M. and 2 h after eating lunch at 11 A.M. and 1 P.M.	Shadowing less accurate after CHO meal due largely to older individuals eating a CHO lunch
Lloyd et al., 1994	Mostly females	Low fat, high CHO; medium fat, medium CHO; high fat, low CHO	Visual information processing, tapping, recall, and reaction time	30 min before and 30, 90, and 150 min after 12:30 P.M. lunch	Faster reaction time after medium-fat, medium-CHO lunch
Deijen et al., 1989	Primarily female college students	High-protein food until 1:30 P.M. and high-CHO food after 1:30 P.M. or *ad libitum* diet	CPT, pattern comparison, symbol–digit substitution, memory scanning, finger tapping	10:30 A.M. and 6 P.M., 1.5 to 2h after meal and after being on the diet for 3 weeks	Faster finger tapping and slower memory scanning in diet group in the morning
Smith et al., 1988	Male and female adults	High-protein, high-starch, or high-sugar lunch	Focused attention and search test meals	Before and 1.25 h after lunch	Slower reactions to peripheral stimuli after CHO
Spring et al., 1986	Female adults	CHO-rich, protein-rich, balanced, or no lunch	Digit–symbol substitution, letter cancellation, and test of addition	Before and 4 times after lunch	No meal effect on any performance measure
Spring et al., 1986	Male adults	Starch or protein lunch	Auditory reaction time, digit–symbol substitution, dichotic listening	Before and hourly for 5 h after noon lunch	Slower reaction time at 1:45 P.M. and impaired digit–symbol substitution after CHO lunch

Many other performance tasks were used as outcome measures, ranging from visual processing tasks to logical reasoning tasks. The studies did not demonstrate any meal effect on any of these tasks. Therefore, to date, there seems to be little effect of consumption of a carbohydrate breakfast relative to consumption of either a protein or high-fat breakfast on most of the performance measures investigated. This does not mean that consumption of breakfast does not affect performance, but only that consumption of a carbohydrate-rich breakfast seems to have little beneficial or detrimental effect compared with a protein-rich or fat-rich breakfast, at least on the performance tasks currently investigated.

The studies[124,131,136,138–141] that have investigated the effect of consuming a carbohydrate-rich meal at lunchtime have produced a mixed bag, probably due to the variation in types of tasks investigated. Perhaps the most consistent finding is that consumption of a carbohydrate-rich lunch results in faster speed of responding to a task that requires some type of sustained attention. However, there may be a slower reaction time to an auditory attention task as well as to stimuli seen in the periphery of the visual field. There seems to be no effect of consumption of carbohydrates on tasks such as number recognition, digit–symbol substitution, addition, or letter cancellation.

13.5 SUMMARY AND CONCLUSIONS

The literature focusing on the mood- and performance-altering effects of carbohydrate consumption has attracted considerable attention over the last three decades. The primary driving force behind this interest was the hypothesis that carbohydrate consumption, particularly in the absence of protein, caused a rise in the synthesis of central serotonin. More recently, this hypothesis has been questioned and seriously challenged on several fronts, leading some investigators[16,33] to conclude that this is probably not the mechanism by which carbohydrates affect behavior. The failure of studies to continue to demonstrate support for the carbohydrate–serotonin hypothesis, and the apparent general lack of acceptance of this hypothesis, has resulted in a decline in interest and research in this area, probably because of the absence of a metabolic theory that would explain a carbohydrate-induced effect on mood and performance.

It has been hypothesized[35,36,38,40] that endogenous aPD peptides may represent the underlying metabolic substrate for carbohydrate cravings and liking of sweet carbohydrates, and there is some support for this hypothesis. This hypothesis makes even more sense when it is considered that peptides are found in brain areas that are involved with regulating emotions, responses to pain and stress, endocrines, and food intake.[35] However, the precise relationship between endogenous aPD peptides, carbohydrate consumption, and mood and performance has yet to be delineated.

Even if it is demonstrated that endogenous aPD peptides contribute to the mood- and performance-altering effect of carbohydrate consumption, there is much more to the picture than just a metabolic connection. There are also psychological processes that are involved in any behavioral reaction foods, such as carbohydrates. For example, the sight or thought of food has no intrinsic motivational value in and of itself. It is a neutral stimulus that creates no response. As Berridge[38] has stated, food merely represents "an aggregation of visual shapes and colors like the sight of any object" (p. 15). Food does not become a desired commodity until there is some value placed on it. This intrinsic value of a carbohydrate comes through learning or associating carbohydrates with some desired outcome. Perhaps the most dramatic example of the influence of psychological processes on the behavioral reaction to food comes from the strong and specific aversions that can develop when food consumption occurs in close temporal contiguity with gastrointestinal illness.[142] The relationship between macronutrients, such as carbohydrates, psychological processes, such as the learning that occurs when food is paired with various outcomes, and underlying metabolic processes must be investigated and integrated into an overall theory to gain a more complete understanding of how macronutrients influence mood and performance. Currently, we have minimal information in this arena.

In spite of not having a well-supported theory or hypothesis to guide research on the mood- and performance-altering effects of carbohydrate consumption, progress has been made in this area. Almost two decades ago, Spring et al.[17] concluded that consumption of a carbohydrate-rich and protein-poor meal often induces fatigue in healthy adults, particularly women. While there continues to be support for this conclusion, the more recent research has revealed that making such a broad generalization is inappropriate, and the effect of an unbalanced carbohydrate-rich meal depends on whether it is consumed in the morning or afternoon. When a carbohydrate-rich meal is consumed in the morning, typically at breakfast time, there seems to be a positive mood-altering effect of being more energetic and less fatigued and having an improved mood. With regard to performance, a carbohydrate-rich breakfast has been demonstrated to slow reaction time to tasks requiring a choice and results in worse or slower performance in tasks requiring memory for and recognition of some specified sequence. However, most of the tasks investigated across studies did not demonstrate an effect of any specific type of breakfast eaten. The most logical conclusion seems to be that consumption of a carbohydrate-rich breakfast has little effect on most performance tasks.

When a carbohydrate-rich lunch is consumed, individuals are sleepier and more fatigued compared to a no-lunch condition. However, there is no difference in effect between a carbohydrate-rich and a protein-rich lunch. The effect of type of meal seems to also depend on whether a person deviates from what is typically eaten. As the lunch meal deviates from what is typically eaten, a more negative mood-altering effect seems to occur. With regard to performance, about the only consistent result that seems to have occurred is that a carbohydrate-rich lunch increases the speed of reaction to tasks that require some type of sustained attention.

The research indicates that there is minimal effect on both mood and performance in healthy adults. However, the picture is quite different with respect to emotionally distressed individuals, especially individuals experiencing some level of depression. While many individuals experience and report cravings for sweet carbohydrate- and fat-rich foods, this experience is most common and is strongest in emotionally distressed individuals, particularly in individuals experiencing symptoms of depression. Interestingly, this craving for and consumption of carbohydrate- and fat-rich food is in the form of snacks eaten primarily in the afternoon and evening. Cravings for and consumption of sweet carbohydrate- and fat-rich snacks are particularly apparent in individuals with seasonal affective disorder, individuals experiencing severe premenstrual symptoms, obese individuals experiencing depressive symptoms, and a proportion of individuals with major depression. It has even been demonstrated that individuals' preference for and consumption of these carbohydrate- and fat-rich foods become stronger as the depressive symptoms emerge. The evidence suggests that these individuals consume the carbohydrate- and fat-rich snack foods for the effect they have on ameliorating the depressive symptoms experienced.

It has been postulated[143] that there is a reciprocal and interactive relationship between emotional distress and carbohydrate cravings, as depicted in Figure 13.1. Emotional distress creates a learned preference and craving for carbohydrate- and fat-rich snack foods because the consumption of these snack foods is followed by a positive mood-altering effect. When a person experiences emotional distress such as depression, the emotional distress serves as a stimulus to generate the response of a desire to consume a carbohydrate- and fat-rich food so that some relief can be obtained from the emotional distress. This desire is expressed as a carbohydrate craving. Once the carbohydrate caving is initiated, the individual's attention is selectively directed toward the search for and consumption of sweet carbohydrate- and fat-rich snack foods. Once located, the snack food is consumed with the benefit of providing an enhancement in mood and a corresponding decline in the craving for carbohydrates. However, the positive benefit derived from consuming the sweet carbohydrate- and fat-rich snack is temporary. Therefore, the emotional distress returns, resulting in a reemergence of the carbohydrate cravings and a return of the cycle of carbohydrate cravings, resulting in a search for sweet carbohydrate- and fat-rich snacks to consume.

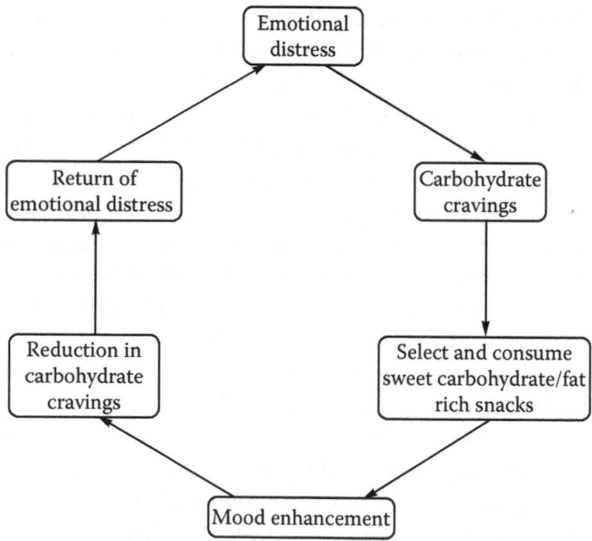

FIGURE 13.1 Interactive relationship between mood and carbohydrate cravings. (From Christensen, L., *Clin. Nutr.*, 20 (Suppl. 1), 164, 2001. With permission.)

This theoretical cycle provides a reasonable explanation of the literature demonstrating that emotionally distressed individuals, especially individuals with depressive symptoms, crave and consume carbohydrates. However, there is also research[117] demonstrating that the consumption of these carbohydrates may actually contribute to the emotional distress experienced. It seems contradictory that carbohydrates could ameliorate depressive symptoms and at the same time create symptoms of emotional distress. However, Moorhouse et al.[114] has pointed out that sugar seems to have a paradoxical effect in clinical populations who crave it. These individuals crave sugar, but the consumption of the sugar they crave contributes to the distress they are experiencing. The acute reaction to sugar consumption is one of providing a transitory amelioration in the distress symptoms in these populations, whereas the chronic effect is one of exacerbation of the distress symptoms.

This hypothesized acute vs. chronic effect of consumption of sweet carbohydrate- and fat-rich snacks on mood in sweet-craving emotionally distressed populations fits nicely with the literature. However, as with most hypotheses, this one could stand further verification.

REFERENCES

1. Cosman, M.P., A feast for Aesculapius: historical diets for asthma and sexual pleasure, *Annu. Rev. Nutr.*, 3, 1, 1983.
2. Darby, W.J., Ghalioungui, P., and Grivetti, L., *Food, the Gift of Osiris*, Academic Press, New York, 1977.
3. Burks, R. and Keeley, S., Exercise and diet therapy: psychotherapists' beliefs and practices, *Prof. Psychol. Res. Pract.*, 20, 62, 1989.
4. Bussing, R., Schoenberg, N.E., and Perwien, A.R., Knowledge and information about ADHD: evidence of cultural differences among African-American and White parents, *Soc. Sci. Med.*, 46, 919, 1998.
5. Ryan, C. and Lakshman, O.R., Nutrition and the psychiatric patient, *Psychiatr. Hosp.*, 18, 163, 1987.
6. Messer, S.C., Morris, T.L., and Gross, A.M., Hypoglycemia and psychopathology: a methodological review, *Clin. Psychol. Rev.*, 10, 631, 1990.
7. Schoenthaler, S.J., The effect of sugar on the treatment and control of antisocial behavior: a double-blind study of an incarcerated juvenile population, *Int. J. Biosoc. Res.*, 3, 1, 1982.
8. Virkkunen, M., Reactive hypoglycemic tendency among habitually violent offenders, *Nutr. Rev.*, 44 (Suppl.), 94, 1986.
9. Varley, C.K., Diet and the behavior of children with attention deficit disorder, *J. Am. Acad. Child Psychiatry*, 23, 182, 1984.
10. Fernstrom, J.D. and Wurtman, R.J., Brain serotonin content: increase following ingestion of carbohydrate diet, *Science*, 174, 1023, 1971.
11. Fernstrom, J.D. and Wurtman, R.J., Effect of chronic corn consumption on serotonin content of rat brain, *Nat. New Biol.*, 234, 62, 1971.
12. Fernstrom, J.D. and Wurtman, R.J., Brain serotonin content: physiological regulation by plasma neutral amino acids, *Science*, 178, 414, 1972.
13. Drewnowski, A., Food preferences and the aPD peptide system, *Trends Food Sci. Technol.*, 3, 97, 1992.

14. Rogers, P.J., Food, mood, and appetite, *Nutr. Res. Rev.*, 8, 243, 1995.
15. Spring, B., Effects of foods and nutrients on the behavior of normal individuals, in *Nutrition and the Brain*, Vol. 7, Wurtman, R.J. and Wurtman, J.J., Eds., Raven Press, New York, 1986, p. 1.
16. Christensen, L., The effect of carbohydrates on affect, *Nutrition*, 13, 503, 1997.
17. Spring, B., Chiodo, J., and Bowen, D.J., Carbohydrates, tryptophan, and behavior: a methodological review, *Psychol. Bull.*, 102, 234, 1987.
18. Fernstrom, J.D. and Wurtman, R.J., Brain serotonin content: physiological dependence on plasma tryptophan levels, *Science*, 173, 149, 1971.
19. Gillman, P.K. et al., Indolic substances in plasma, cerebrospinal fluid, and frontal cortex of human subjects infused with saline or tryptophan, *J. Neurochem.*, 37, 410, 1981.
20. Ashcroft, G.W. et al., 5-Hydroxytryptamine metabolism in affective illness: the effect of tryptophan administration, *Psychol. Med.*, 3, 326, 1973.
21. Fernstrom, J.D. and Wurtman, R.J., Brain serotonin content: increase following ingestion of a carbohydrate diet, *Science*, 174, 1023, 1971.
22. Fernstrom, J.D. and Wurtman, R.J., Elevation of plasma tryptophan by insulin in rat, *Metabolism*, 21, 337, 1972.
23. Madras, B.K. et al., Relevance of serum-free tryptophan to tissue tryptophan concentrations, *Metabolism*, 23, 1107, 1974.
24. Sved, A.F., Precursor control of the function of monoaminergic neurons, in *Nutrition and the Brain*, Wurtman, R.J. and Wurtman, J.J., Eds., Raven Press, New York, 1983, p. 224.
25. Teff, K.L. et al., Acute effect of protein or carbohydrate breakfasts on human cerebrospinal fluid monoamine precursor and metabolite levels, *J. Neurochem.*, 52, 235, 1989.
26. Fernstrom, M.H. and Fernstrom, J.D., Brain tryptophan concentrations and serotonin synthesis remain responsive to food consumption after ingestion of sequential meals, *Am. J. Clin. Nutr.*, 61, 312, 1995.
27. Ashley, D.V.M., Liardon, R., and Leathwood, P.D., Breakfast meal composition influences plasma tryptophan to large neutral amino acid ratios of healthy lean young men, *J. Neural Transm.*, 63, 271, 1985.
28. Curzon, G. and Sarna, G.S., Tryptophan transport into the brain: newer findings and older ones reconsidered, in *Progress in Tryptophan and Serotonin Research*, Schlossberger, H.G., Kochen, W., Linzen, B., and Steinhart, H., Eds., de Gruyter, Berlin, 1984, p. 145.
29. Martin-Du Pan, R. et al., Effect of various oral glucose doses on plasma neutral amino acid levels, *Metabolism*, 31, 937, 1982.
30. Fernstrom, J.D., Carbohydrate ingestion and brain serotonin synthesis: relevance to a putative control loop for regulating carbohydrate ingestion and effects of aspartame consumption, *Appetite*, 11 (Suppl.), 35, 1988.
31. Fernstrom, J.D. et al., Twenty-four hour food intake in patients with anorexia nervosa and in healthy control subjects, *Biol. Psychiatry*, 36, 696, 1994.
32. Muno, H.N., Evolution of protein metabolism in mammals, in *Mammalian Protein Metabolism*, Vol. 3, Muro, H.N., Ed., Academic Press, New York, 1969, p. 133.
33. Young, S.N., Some effects of dietary components (amino acids, carbohydrate, folic acid) on brain serotonin synthesis, mood, and behavior, *Can. J. Physiol. Pharmacol.*, 69, 893, 1991.
34. Garattini, L., Nutrients affecting brain composition and behavior, *Integr. Psychiatry*, 6, 235, 1989.

35. Ree, J.M. et al., Endogenous opioids and reward, *Eur. J. Pharmacol.*, 405, 89, 2000.
36. Kelly, A.E. et al., APD modulation of taste hedonics within the ventral striatum, *Physiol. Behav.*, 76, 365, 2002.
37. Mercer, M.E. and Holder, M.D., Food cravings, endogenous aPD peptides, and food intake: a review, *Appetite*, 29, 325, 1997.
38. Berridge, K.C., Food reward: brain substrates of wanting and liking, *Neurosci. Biobehav. Rev.*, 20, 1, 1996.
39. Fantino, M., Hosotte, J., and Apfelbaum, M., An aPD antagonist, naltrexone, reduces preference for sucrose in humans, *Am. J. Physiol.*, 251, R91, 1986.
40. Drewnowski, A. et al., Taste responses and preferences for sweet high-fat foods: evidence for aPD involvement, *Physiol. Behav.*, 51, 371, 1992.
41. Carr, K.D., APD receptor subtypes and simulation-induced feeding, in *Drug Receptor Subtypes and Ingestive Behavior*, Cooper, S.J. and Clifton, P.G., Eds., Academic Press, New York, p. 167, 1996.
42. Drewnowski, A., Metabolic determinants of binge eating, *Addict. Behav.*, 20, 733, 1995.
43. Greeno, C.G. and Wing, R.R., Stress-induced eating, *Psychol. Bull.*, 115, 444, 1994.
44. Oliver, G. and Wardle, J., Perceived effects of stress on food choice, *Physiol. Behav.*, 66, 511, 1999.
45. Wardle, J. et al., Stress, dietary restraint and food intake, *J. Psychosom. Res.*, 48, 195, 2000.
46. Westover, A.N. and Marangell, L., A cross-national relationship between sugar consumption and major depression, *Depress. Anxiety*, 16, 118, 2002.
47. Christensen, L. and Somers, S., Comparison of nutrient intake among depressed and nondepressed individuals, *Int. J. Eat. Disord.*, 20, 105, 1996.
48. Fernstrom, M.H. and Kupfer, D.J., Imipramine treatment and preference for sweets, *Appetite*, 10, 149, 1988.
49. Weingarten, H.P. and Elston, D., The phenomenology of food cravings, *Appetite*, 15, 231, 1990.
50. Pelchat, M.L., Of human bondage: food craving, obsession, compulsion, and addiction, *Physiol. Behav.*, 76, 347, 352.
51. Christensen, L. and Pettijohn, L., Mood and carbohydrate cravings, *Appetite*, 36, 137, 2001.
52. Weingarten, H.P. and Elston, D., Food cravings in a college population, *Appetite*, 17, 167, 1991.
53. Gendall, K.A., Joyce, P.R., and Sullivan, P.F., Impact of definition on prevalence of food cravings in a random sample of young women, *Appetite*, 28, 63, 1997.
54. Pelchat, M.L., Food cravings in young and elderly adults, *Appetite*, 28, 103, 1997.
55. Pelchat, M.L. and Schaefer, S., Dietary monotony and food cravings in young and elderly adults, *Physiol. Behav.*, 68, 353, 2000.
56. Gendall, K.A. et al., Psychopathology and personality of young women who experienced food cravings, *Addict. Behav.*, 22, 545, 1997.
57. Paykel, E.S., Mueller, P.S., and De La Vergne, P.M., Amitiptyline, weight gain and carbohydrate craving: a side effec, *Br. J. Psychiatry*, 123, 501, 1973.
58. Fernstrom, J.D., Tryptophan, serotonin and carbohydrate appetite: will the real carbohydrate craver please stand up! *J. Nutr.*, 118, 1417, 1988.
59. Hill, A.J., Weaver, C.F.L., and Blundell, J.E., Food craving, dietary restraint and mood, *Appetite*, 17, 187, 1991.

60. Lafay, L. et al., Gender differences in the relation between food cravings and mood in an adult community: results from the Fleurbaix Laventie Ville Sante' study, *Eat. Disord.*, 29, 195, 2001.
61. Weinstein, A. et al., Integrating the cognitive and physiological aspects of craving, *J. Psychopharmacol.*, 12, 31, 1998.
62. Macdiarmid, J.I. and Hetherington, M.M., Mood modulation by food: an exploration of affect and cravings in '*chocolate* addicts,' *Br. J. Clin. Psychol.*, 34, 129, 1995.
63. Willner, P. et al., "Depression" increases "craving" for sweet rewards in animal and human models of depression and craving, *Psychopharmacology*, 36, 272, 1998.
64. Rosenthal, N.E. et al., Seasonal affective disorder: a description of the syndrome and preliminary findings with light therapy, *Arch. Gen. Psychiatry*, 41, 72, 1984.
65. Hippocrates, in *Hippocrates*, Jones, W.H.S., Trans., Harvard University Press, Cambridge, MA, 1931, p. 128.
66. Roccatagliata, G., *A History of Ancient Psychiatry*, Greenword Press, Westport, CT, p. 143, 1986.
67. Oren, D.A. and Rosenthal, N.E., Seasonal affective disorders, in *Handbook of Affective Disorders*, Paykel, E.S., Ed., Guilford Press, New York, 1992, p. 551.
68. Rosenthal, N.E. et al., Seasonal cycling in a bipolar patient, *Psychiatr. Res.*, 8, 25, 1983.
69. Garvey, M.J., Wesner, R., and Godes, M., Comparison of seasonal and nonseasonal affective disorders, *Am. J. Psychiatry*, 145, 100, 1988.
70. Thalen, B.F., Kjellman, B.F., and Morkrid, L., Seasonal and non-seasonal depression: a comparison of clinical characteristics in Swedish patients, *Eur. Arch. Psychiatry Clin. Neurosci.*, 245, 101, 1995.
71. Takahashi, K. et al., Multi-center study of seasonal affective disorders in Japan. A preliminary report, *J. Affect. Disord.*, 21, 57, 1991.
72. Boyce, P. and Parker, G., Seasonal affective disorder in the southern hemisphere, *Am. J. Psychiatry*, 145, 96, 1988.
73. Wehr, T.A. et al., Contrasts between symptoms of summer depression and winter depression, *J. Affect. Disord.*, 23, 173, 1991.
74. Rosenthal, N.E. et al., Disturbances of appetite and weight regulation in seasonal affective disorder, *Ann. N.Y. Acad. Sci.*, 575, 216, 1987.
75. Rosen, L.N. et al., Prevalence of seasonal affective disorder at four latitudes, *Psychiatr. Res.*, 31, 131, 1990.
76. Krauchi, K. and Wirz-Justice, A., The four seasons: food intake frequency in seasonal affective disorder in the course of a year, *Psychiatr. Res.*, 25, 323, 1988.
77. Krauchi, K., Wirz-Justice, A., and Graw, P., The relationship of affective state to dietary preference: winter depression and light therapy as a model, *J. Affect. Disord.*, 20, 43, 1990.
78. Leibenluft, E. et al., Depressive symptoms and the self-reported use of alcohol, caffeine, and carbohydrates in normal volunteers and four groups of psychiatric outpatients, *Am. J. Psychiatry*, 150, 294, 1993.
79. Rosenthal, N.E. et al., Psychobiological effects of carbohydrate- and protein-rich meals in patients with seasonal affective disorder and normal controls, *Biol. Psychiatry*, 25, 1029, 1989.
80. Mokdad, A.H. et al., Prevalence of obesity, diabetes, and obesity-related health risk factors, 2001, *JAMA*, 289, 76, 2003.
81. Simopoulos, A.P., Characteristics of obesity: an overview, in *Human Obesity*, Wurtman, R.J. and Wurtman, J.J., Eds., Ann. N.Y. Acad. Sci., New York, 1987, p. 4.

82. Drewnowski, A., Taste and food preferences in human obesity, in *Taste, Experience and Feeding*, Capaldi, E. and Powley, T.L., Ed., APA, Washington, DC, 1990, chap. 16.
83. Drewnowski, A. et al., Food preferences in human obesity: carbohydrates versus fats, *Appetite*, 18, 207, 1992.
84. Lieberman, H.R., Wurtman, J.J., and Chew, B., Changes in mood after carbohydrate consumption among obese individuals, *Am. J. Clin. Nutr.*, 44, 772, 1986.
85. Toornvliet, A.C. et al., Psychological and metabolic responses of carbohydrate craving obese patients to carbohydrate, fat and protein-rich meals, *Int. J. Obes.*, 21, 860, 1997.
86. Schlundt, D.G. et al., A sequential behavioral analysis of craving sweets in obese women, *Addict. Behav.*, 18, 67, 1993.
87. Wurtman, J. et al., Fenfluramine suppresses snack intake among carbohydrate cravers but not among noncarbohydrate cravers, *Int. J. Eat. Disord.*, 6, 687, 1987.
88. Hopkinson, G. and Bland, R.C., Depressive syndromes in grossly obese women, *Can. J. Psychiatry*, 27, 213, 1982.
89. Wurtman, R.J. and Wurtman, J.J., Carbohydrate craving, obesity and brain serotonin, *Appetite*, 7 (Suppl.), 99, 1986.
90. Ganley, R.M., Emotion and eating in obesity: a review of the literature, *Int. J. Eat. Disord.*, 8, 343, 1989.
91. Reid, M. and Hammersley, R., The effects of sugar on subsequent eating and mood in obese and non-obese women, *Psychol. Health Med.*, 3, 299, 1998.
92. Corsica, J.A., Carbohydrate self-administration, mood, and cognition in carbohydrate cravers, *Digital Dissertations*, publication AAT3032529, 2001.
93. Dennerstein, L. and Burrows, G.D., Affect and the menstrual cycle, *J. Affect. Disord.*, 1, 77, 1979.
94. Lauersen, N.H. and Stukane, E., *PMS: Premenstrual Syndrome and You*, Pennacle Books, New York, 1983.
95. American Psychiatric Association, *Diagnostic and Statistical Manual of Mental Disorders*, 4th ed., APA Press, Washington, DC, 1994.
96. Buffenstein, R. et al., Food intake and the menstrual cycle: a retrospective analysis with implications for appetite research, *Physiol. Behav.*, 58, 1067, 1995.
97. Vlitos, A.P. and Davies, G.J., Bowl function, food intake, and the menstrual cycle, *Nutr. Res. Rev.*, 9, 112, 1996.
98. Both-Orthman, B. et al., Menstrual cycle phase-related changes in appetite in patients with premenstrual syndrome and in control subjects, *Am. J. Psychiatry*, 145, 628, 1988.
99. Rubinow, D.R. et al., Premenstrual mood changes: characteristic patterns in women with and without premenstrual symptoms, *J. Affect. Disord.*, 10, 85, 1986.
100. Johnson, W.G. et al., Macronutrient intake, eating habits, and exercise as moderators of menstrual distress in healthy women, *Psychosom. Med.*, 57, 324, 1995.
101. Michener, W. et al., The role of low progesterone and tension as triggers of perimenstrual chocolate and sweets craving: some negative experimental evidence, *Physiol. Behav.*, 67, 417, 1999.
102. Dye, L., Warner, P., and Bancroft, J., Food craving during the menstrual cycle and its relationship to stress, happiness of relationship and depression; a preliminary enquiry, *J. Affect. Disord.*, 34, 157, 1995.
103. Cohen, I.T., Sherwin, B.B., and Fleming, A.S., Food cravings, mood, and the menstrual cycle, *Horm. Behav.*, 21, 457, 1987.
104. Choudhuri, S.M. and Tandon, B.N., A study of dietary intake in pre- and post menstrual period, *Hum. Nutr. Appl. Nutr.*, 40A, 213, 1986.

105. Tomelleri, R. and Grunewald, K.K., Menstrual cycle and food cravings in young college women, *J. Am. Diet. Assoc.*, 87, 311, 1987.
106. Hill, A.J. and Heaton-Brown, L., The experience of food craving: a prospective investigation in healthy women, *J. Psychosom. Res.*, 38, 801, 1994.
107. Wurtman, J.J. et al., Effect of nutrient intake on premenstrual depression, *Am. J. Obstet. Gynecol.*, 161, 1228, 1989.
108. Wurtman, J.J., Carbohydrate craving: relationship between carbohydrate intake and disorders of mood, *Drugs*, 39 (Suppl. 3), 49, 1990.
109. Smith, S.L. and Sauder, C., Food cravings, depression, and premenstrual problems, *Psychosom. Med.*, 31, 281, 1969.
110. Rossignol, A.M. and Bonnlander, H., Prevalence and severity of the premenstrual syndrome: effects of foods and beverages that are sweet or high in sugar content, *J. Reprod. Med.*, 36, 131, 1991.
111. Sayegh, R. et al., The effect of a carbohydrate-rich beverage on mood, appetite, and cognitive function in women with premenstrual syndrome, *Obstet. Gynecol.*, 86, 520, 1995.
112. Robins, L.N. et al., Lifetime prevalence of specific psychiatric disorders in three sites, *Arch. Gen. Psychiatry*, 41, 949, 1984.
113. Thalen, B.E. et al., Seasonal and non-seasonal depression: a comparison of clinical characteristics in Swedish patients, *Eur. Arch. Psychiatry Clin. Neurosci.*, 245, 101, 1995.
114. Moorhouse, M. et al., Carbohydrate craving by alcohol-dependent men during sobriety: relationship to nutrition and serotonergic function, *Alcohol Clin. Exp. Res.*, 24, 635, 2000.
115. Christensen, L. et al., Impact of a dietary change on emotional distress, *J. Abnorm. Psychol.*, 94, 565, 1985.
116. Christensen, L. and Burrows, R., Dietary treatment of depression, *Behav. Ther.*, 21, 183, 1990.
117. Krietsch, K., Christensen, L., and White, B., Prevalence, presenting symptoms, and psychological characteristics of individuals experiencing a diet-inducted mood-disturbance, *Behav. Ther.*, 19, 593, 1988.
118. Tuttle, W.W., Wilson, M., and Daum, K., Effect of altered breakfast habits on physiologic response, *J. Appl. Physiol.*, 1, 545, 1991.
119. Spring, B. et al., Effects of protein and carbohydrate meals on mood and performance: interactions with sex and age, *J. Psychiatr. Res.*, 17, 155, 1983.
120. Fischer, K. et al., Cognitive performance and its relationship with postprandial metabolic changes after ingestion of different macronutrients in the morning, *Br. J. Nutr.*, 85, 393, 2001.
121. Lloyd, H.M., Rogers, P.J., and Hedderley, D.I., Acute effects on mood and cognitive performance of breakfasts differing in fat and carbohydrate content, *Appetite*, 27, 151, 1996.
122. Reid, M. and Hammersley, R., Effects of carbohydrate intake on subsequent food intake and mood state, *Physiol. Behav.*, 58, 421, 1995.
123. Wells, A.S., Read, N.W., and MacDonald, I.A., Effects of carbohydrate and lipid on resting energy expenditure, heart rate, sleepiness, and mood, *Physiol. Behav.*, 63, 621, 1998.
124. Spring, B. et al., Effects of protein and carbohydrate meals on mood and performance: interactions with sex and age, *J. Psychiatr. Res.*, 17, 155, 1983.
125. Thayer, R.E., Energy, tiredness and tension: effects of a sugar snack versus moderate exercise, *J. Pers. Soc. Psychol.*, 52, 119, 1987.

126. Fischer, K. et al., Carbohydrate to protein ratio in food and cognitive performance in the morning, *Physiol. Behav.*, 75, 411, 2002.
127. Benton, D., Slater, O., and Donohoe, R.T., The influence of breakfast and a snack on psychological functioning, *Physiol. Behav.*, 74, 559, 2001.
128. Smith, A. et al., Effects of breakfast and caffeine on cognitive performance, mood and cardiovascular functioning, *Appetite*, 22, 39, 1994.
129. Reid, M. and Hammersley, R., The effects of sucrose and maize oil on subsequent food intake and mood, *Br. J. Nutr.*, 82, 447, 1999.
130. Pivonka, E.E.A. and Grunewald, K.K., Aspartame- or sugar-sweetened beverages: effects on mood in young women, *J. Am. Diet. Assoc.*, 90, 250, 1990.
131. Spring, B. et al., Effects of noon meals varying in nutrient composition on plasma amino acids, glucose, insulin, and behavior, *Psychopharmacol. Bull.*, 22, 1026, 1986.
132. Christensen, L. and Redig, C., Effect of meal composition on mood, *Behav. Neurosci.*, 107, 346, 1993.
133. Smith, A. et al., The influence of meal composition on post-lunch changes in performance efficiency and mood, *Appetite*, 10, 195, 1988.
134. Spring, B.J. et al., Effects of carbohydrates on mood and behavior, *Nutr. Rev.*, 44, (Suppl.), 51, 1986.
135. Deijen, J.B. et al., Dietary effects on mood and performance, *J. Psychiatr. Res.*, 23, 275, 1989.
136. Markus, R. et al., Effects of food on cortisol and mood in vulnerable subjects under controllable and uncontrollable stress, *Physiol. Behav.*, 70, 333, 2000.
137. Markus, C.R. et al., Does carbohydrate-rich, protein-poor food prevent a deterioration of mood and cognitive performance of stress-prone subjects when subjected to a stressful task? *Appetite*, 31, 49, 1988.
138. Lloyd, H.M., Green, M.W., and Rogers, P.J., Mood and cognitive performance effects of isocaloric lunches differing in fat and carbohydrate content, *Physiol. Behav.*, 56, 51, 1994.
139. Wells, A.S., Read, N.W., and Craig, A., Influences of dietary and intraduodenal lipid on alertness, mood, and sustained concentration, *Br. J. Nutr.*, 74, 115, 1995.
140. Kelly, T.H. et al., Effect of meal macronutrient and energy content on human performance, *Appetite*, 23, 97, 111.
141. Smith, A., Effects of meals on memory and attention, in *Practical Aspects of Memory: Current Research and Issues*, Gruneberg, M.M., Morris, P.E., and Sykes, R.N., Eds., John Wiley & Sons, New York, 1988, p. 477.
142. Garcia, J., Hankins, W.G., and Rusiniak, K.W., Behavioral regulation of the milieu interne in man and rat, *Science*, 185, 824, 1974.

14 Carbohydrates and Gastrointestinal Tract Function

Barbara O. Schneeman

CONTENTS

14.1 INTRODUCTION

The physiological activity of carbohydrates in the gastrointestinal tract (GIT) is determined by their physical and chemical properties and the extent and location of their digestion within the GIT. Carbohydrates include those digested by mammalian digestive enzymes, such as alpha-amylase and the disaccharidases, as well as the carbohydrates that are digested by enzymes of microbial origin. The primary carbohydrates found in the diet that are digested by mammalian enzymes are starch, sucrose, fructose, lactose, and maltose. These carbohydrates, referred to as digestible carbohydrates, are sources of energy; however, they can also influence GIT responses to diet and result in adaptive changes in the GIT. The carbohydrates that are digested by microbial enzymes include the polysaccharides associated with dietary fiber (cellulose, hemicelluloses, pectin, and gums) as well as oligosaccharides such as inulin. Because these compounds are not degraded until the large intestine, their physical and chemical properties have important physiologic effects throughout the GIT. The properties associated with nonstarch polysaccharides (NSPs) and oligosaccharides include their water-holding capacity, viscosity, bile acid binding, and bulk or nondigestibility. In addition, these compounds are the primary substrates for growth of the microflora in the large intestine. Many compounds in the NSP fraction are an integral part of the cell wall in plants, and hence a significant component of food structure. The Institute of Medicine (IOM) report on dietary fiber indicated that in addition to the effects of individual compounds in NSPs, this intact structure in foods is physiologically important.[1] Thus, one cannot ignore these structural aspects when considering the physiologic effects of carbohydrates. This chapter will outline the function of organs along the GIT and consider how the chemical and physical properties of carbohydrates will influence these functions.

14.2 GASTROINTESTINAL FUNCTION

14.2.1 Oral Cavity

In the oral cavity, foods are masticated and lubricated with saliva. This process initiates the breakdown of food in a manner that will allow penetration and action of digestive enzymes. Saliva contains alpha-amylase, which can initiate the hydrolysis of starch. Given the time that most food is in the oral cavity, it is unlikely that much starch hydrolysis occurs; however, the enzyme can remain active in the gastric contents until the pH is lowered significantly due to secretion of gastric acid. Foods rich in dietary fiber as a part of cell wall structures are likely to take longer to chew;[2,3] however, cooking plant foods will soften the cell wall structure, making mastication easier. It is possible that polysaccharides with a high water-holding capacity might stimulate more fluid secretion into the mouth during mastication, but such an effect has not been tested directly.

14.2.2 Stomach

After swallowing, the food bolus passes down the esophagus to the stomach. The stomach can hold several liters, allowing humans to consume food in meals. In the stomach, chyme is mixed with gastric juices, which contain acid as well as proteolytic and lipolytic enzymes. Salivary amylase can continue to hydrolyze starch until the pH is lowered, and the gastric secretions and motility will continue to solubilize and disperse food components and the polysaccharide matrix that is a part of foods containing NSPs and digestible carbohydrates. The rate of gastric filling is determined by the rate of food consumption, and, in turn, the rate of gastric emptying controls the rate of nutrient digestion and absorption from the small intestine. Properties such as water-holding capacity, nondigestible bulk, and viscosity can alter gastric distension and the rate of gastric emptying and contribute to feelings of fullness.[4] Carbohydrates that increase the viscosity of gastric contents have been shown to slow the rate of gastric emptying.[5,6] This effect of viscous polysaccharides on gastric emptying, leading to a slower small intestinal transit time, is associated with the ability of certain sources of NSPs to blunt glycemic and insulin responses, as well as contributes to lowering plasma cholesterol.[7–12]

When the food bolus or chyme leaves the stomach, the NSPs are relatively intact, some hydration of these compounds has occurred, starch hydrolysis has been initiated, and sugars have probably been solubilized. In addition, the pH of the bolus is relatively low due to mixing with gastric acid.

14.2.3 Small Intestine

In the small intestine, more fluid is mixed with the chyme to continue the process of digestion and facilitate mobility of compounds for absorption. The macromolecules in food are broken down by digestive enzymes to substrates that can be absorbed by the enterocytes. The presence of protein and fat as well as acid in the duodenum stimulates the secretion of pancreatic juice and bile into the small intestine. The presence of NSPs appears to facilitate the secretion of enymes into the small intestine; however, the effect may be indirect by slowing the digestion of lipids and proteins.[13,14] Pancreatic amylase hydrolyzes the alpha-1,4 linkages found in starch; in addition, saccharides are hydrolyzed by enzymes associated with the brush border of intestinal enterocytes. The net effect of the carbohydrate-digesting enzymes in the small intestine is to hydrolyze digestible carbohydrates to sugars that can be readily absorbed by the small intestine cells while leaving the nondigestible carbohydrates intact. During this process not all starch is digested; a certain amount remains in the small intestine and is referred to as resistant starch.[15] In the small intestine, carbohydrates are separated into those that are absorbed for use as energy substrates in the body and those that remain as bulk within the intestinal contents and pass as a part of the residue into the large intestine.

Thus, one of the important characteristics of carbohydrates that determine function within the small intestine is the nature of the chemical bonds and the ability of mammalian enzymes to hydrolyze these bonds. The digestible carbohydrates are absorbed for energy, but also promote adaptive changes in the digestive enzyme profile of the small intestine and pancreas.[13] Thus, a high-starch diet results in elevated specific activity of pancreatic amylase, and diets high in specific disaccharides can result in elevated activity of enzymes that hydrolyze these sugars. For those carbohydrates that are not digested by mammalian enzymes, their physical presence in the small intestine has physiologic importance. Digestive enzyme activity can be inhibited by interaction with certain polysaccharides or by specific enzyme inhibitors found in plant products that are also rich in dietary fiber.[16,17] In addition, the nondigestibility of NSPs contributes to an increase of material in the bulk phase and in the volume of the aqueous phase of the intestinal contents, which is likely to alter mixing and diffusion in the gut contents.[18,19] Certain polysaccharides, such as pectin and gums, can increase viscosity of the intestinal contents. Viscosity has been shown to delay absorption of sugar from the small intestine, in part due to alterations in mixing and diffusion.[20–23] In addition to interactions with digestive enzymes in the small intestine, certain NSPs can bind or adsorb bile acids and increase their excretion.[14,24,25] The effects of viscous polysaccharides on gastric emptying, transit and absorption from the small intestine, and on bile acids have been associated with the ability of these polysaccharides to blunt glucose and insulin responses and lower plasma cholesterol concentrations.

The ability of viscous polysaccharides to slow the rate of absorption from the small intestine is likely to alter the pattern of hormone release from the GIT.[4] Following consumption of a meal containing beta-glucan-enriched barley, cholecystokinin (CCK) concentrations remained elevated above baseline levels for a longer period of time than the concentrations from a meal without a fiber source.[26] CCK concentrations were twice as high after consuming a meal containing beans than a test meal lower in fiber.[27] Adding viscous polysaccharides to a low-fat test meal significantly increased the CCK response in women.[28] CCK plays a central role in orchestrating the GIT response to eating by delaying gastric emptying and stimulating pancreatic secretion and gall bladder contraction, and it is associated with satiety. The results suggest that slower lipid absorption from the intestine may be associated with prolonging the CCK response to a meal. Because ghrelin release may be suppressed by carbohydrates, it would be interesting to investigate the effect of NSPs on the release of this hormone from the stomach and the time course over which release is suppressed by the presence of NSPs, especially since elevated ghrelin levels are associated with feelings of hunger. In addition, hormones associated with regulation of food intake are released from the ileum, an effect that has been referred to as the ileal brake.[29,30] The presence of nondigested polysaccharides, as well as nutrients that are associated with this fraction of food in the ileum, can promote release of these peptides. Because of the unique chemical and physical properties of the NSPs, research is needed on their role in regulating the GIT response. In many cases, their effects on GIT function may be indirect due to viscosity and the ability to carry nutrients further down the small intestinal tract.

14.2.4 Large Intestine

From the small intestine, the residual diet material passes into the large intestine. A key function of the large intestine is reabsorption of water and electrolytes that have been secreted into the gut during digestion. Within the large intestine is a large and complex microflora that utilizes the residual diet, of which major components are NSPs and resistant starch, and digestive secretions for growth and metabolism. Cummings and coworkers estimated that approximately 15 to 60 g of carbohydrates that can be fermented enters the large intestine daily.[31,32] From this fermentation, approximately 200 to 700 mmol of short-chain fatty acids (SCFAs) can be produced. The primary SCFAs include acetate, propionate, and butyrate. Butyrate is utilized as an energy source by colonocytes, propionate is cleared from the portal blood by the liver, and acetate can be utilized by muscle cells and other peripheral tissues. In addition to the production of SCFAs, microbial activity in the large bowel results in decreased pH, modifications in microbial enzyme activity, and an increase in microbial cell mass.[33] SCFA production has been investigated as a source of energy as well as for its effects on mucosal cell proliferation.[33–36]

The significant component of large bowel function, in which nondigested carbohydrates have an important function, is the process of laxation and elimination of fecal material. The major components of fecal material are water, undigested diet residue, and microbial mass.[37–39] NSP is the only dietary component known to increase stool weight, which it can do either directly as undigested residue or by supporting the growth of microorganisms.[31] In both human and animal studies, stool weight is significantly correlated with consumption of NSPs.[40]

14.3 EMERGING AREAS

An emerging area of interest is the contribution of the large bowel to vitamin and mineral balance, and especially the impact of fermentation of carbohydrates on the bioavailability of these nutrients. Studies have indicated that calcium absorption can be enhanced by inclusion of fermentable carbohydrates in the diet.[41–43] In animal models, acidic fermentation in the cecum increases absorption of calcium and magnesium, contributing to a positive mineral balance.[42] Certain B vitamins as well as vitamin K are likely produced by the large intestine microflora. Although it is widely accepted that fermentation contributes to vitamin status, the exact role of fermentation is not well established.[44,45]

In summary, carbohydrates exert significant effects on GIT function and, in the process, contribute to energy balance, regulation of digestion and absorption, and the bioavailability of other nutrients. Whether carbohydrates are digested and absorbed in the small intestine or are resistant to digestion by mammalian enzymes, they impact the physiologic function of the GIT. For carbohydrates that are digested by mammalian enzymes, the focus is on the processes of digestion, absorption, and adaptation, whereas for carbohydrates that can only be digested by microbial enzymes, the focus is on their physical properties. The investigations on the NSP fraction of foods have led to a new understanding of the potential role of the GIT in determining metabolic response.

REFERENCES

1. Institute of Medicine (IOM), *Dietary Reference Intakes: Proposed Definition of Dietary Fiber*, National Academies Press, Washington, DC, 2001.
2. Duncan, K.H., Bacon, J.A., and Weinsier, R.L., The effects of high- and low-energy-density diets on satiety, energy intake, and eating time of obese and non-obese subjects, *Am. J. Clin. Nutr.*, 37, 763, 1983.
3. Heaton, K.W., Dietary fiber and control of energy intake, in *Handbook of Dietary Fiber in Human Nutrition*, Spiller, G.A., Ed., CRC Press, Boca Raton, FL, 165, 1986.
4. Burton-Freeman, B., Dietary fiber and energy regulation, *J. Nutr.*, 130, 272S, 2000.
5. Ebihara, K. et al., Correlation between viscosity and plasma glucose- and insulin-flattening activities of pectin from vegetables and fruits in rats, *Nutr. Rep. Int.*, 23, 985, 1981.
6. Schwartz, S.E. et al., Sustained pectin ingestion delays gastric emptying, *Gastroenterology*, 83, 812, 1982.
7. Carr, T. et al., Increased intestinal contents viscosity reduces cholesterol absorption efficiency in hamsters fed hydroxypropyl methylcellulose, *J. Nutr.*, 126, 1463, 1996.
8. Lairon, D., Dietary fibers: effects on lipid metabolism and mechanisms of action, *Eur. J. Clin. Nutr.*, 50, 125, 1996.
9. Leclére, C.J. et al., Role of viscous guar gums in lowering the glycemic response after a solid meal, *Am. J. Clin. Nutr.*, 59, 914, 1994.
10. Lin, H.C. et al., Sustained slowing effect of lentils on gastric emptying of solids in humans and dogs, *Gastroenterology*, 102, 787, 1992.
11. Marciani, L. et al., Gastric response to increased meal viscosity assessed by echo-planar magnetic resonance imaging in humans, *J. Nutr.*, 130, 122, 2000.

12. Tietyen, J.L. et al., Hypocholesterolemic potential of oat bran treated with endo-beta-D-glucanase from *Bacillus subtilis*, *J. Food Sci.*, 60, 560, 1995.
13. Schneeman, B.O., Carbohydrates: significance for energy balance and gastrointestinal function, *J. Nutr.*, 124, 1747S, 1994.
14. Lairon, D., Soluble fiber and dietary lipids, *Adv. Exp. Med. Biol.*, 427, 99, 1997.
15. Cummings, J.H. and Englyst, H.N., Gastrointestinal effects of food carbohydrate, *Am. J. Clin. Nutr.*, 61, 938S, 1995.
16. Schneeman, B.O. and Gallaher, D., Effects of dietary fiber on digestive enzymes, in *Handbook of Dietary Fiber*, Spiller, G., Ed., CRC Press, Boca Raton, FL, 305, 1986.
17. Schneeman, B.O. and Gallaher, D., Pancreatic response to dietary trypsin inhibitor: variations among species, in *Nutritional and Toxicological Significance of Enzyme Inhibitors in Foods*, Friedman, M., Ed., Plenum Press, New York, 1986, p. 185.
18. Edwards, C.A., Johnson, I.T., and Read, N.W., Do viscous polysaccharides reduce absorption by inhibiting diffusion or convection? *Eur. J. Clin. Nutr.*, 42, 307, 1988.
19. Phillips, D.R., The effect of guar gum in solution on diffusion of cholesterol and mixed micelles, *J. Sci. Food Agric.*, 37, 548, 1986.
20. Blackburn, N.A. and Johnson, I.T., The effect of guar gum on the viscosity of the gastrointestinal contents and on glucose uptake from the perfused jejunum in the rat, *Br. J. Nutr.*, 46, 239, 1981.
21. Ikegami, S.F. et al., Effect of viscous indigestible polysaccharides on pancreatic-biliary secretion and digestive organs in rats, *J. Nutr.*, 120, 353, 1990.
22. Johnson, L.R. and Gee, J.M., Effect of gel-forming gums on the intestinal unstirred layer and sugar transport *in vitro*, *Gut*, 22, 398, 1981.
23. Lund, E.K. et al., Effect of oat gum on the physical properties of the gastrointestinal contents and on the uptake of D-galactose and cholesterol by rat small intestine *in vitro*, *Br. J. Nutr.*, 62, 91, 1989.
24. Gallaher, D. and Schneeman, B.O., Intestinal interaction of bile acids, phospholipids, dietary fibers and cholestyramine, *Am. J. Physiol.*, 250, G420, 1986.
25. Story, J. and Furumoto, E.J., Dietary fiber and bile acid metabolism, in *Dietary Fiber*, Kritchevsky, D., Bonfield, C., and Anderson, J.W., Eds., Plenum Press, New York, 1990, p. 365.
26. Bourdon, I. et al., Postprandial lipid, glucose, insulin, and cholecystokinin responses in men fed barley pasta enriched with beta-glucan, *Am. J. Clin. Nutr.*, 69, 55, 2001.
27. Bourdon, I. et al., Beans, as a source of dietary fiber, increase cholecystokinin and apo B48 response to test meals in men, *J. Nutr.*, 131, 1485, 2001.
28. Burton-Freeman, B., Davis, P.A., and Schneeman, B.O., Plasma cholecystokinin is associated with subjective measures of satiety in women, *Am. J. Clin. Nutr.*, 76, 659, 2002.
29. Lin, H.C. et al., Fiber-supplemented enteral formula slows intestinal transit by intensifying inhibitory feedback from the distal gut, *Am. J. Clin. Nutr.*, 65, 1840, 1997.
30. Read, N.W. and Eastwood, M.A., Gastro-intestinal physiology and function, in *Dietary Fibre: A Component of Food*, Schweizer, T.F. and Edwards, C.A., Eds., Springer-Verlag, London, 1992, p. 103.
31. Cummings, J.H. et al., Fecal weight, colon cancer risk, and dietary intake of non-starch polysaccharides (dietary fiber), *Gastroenterology*, 103, 1783, 1992.
32. Cummings, J.H. and Englyst, H.N., Measurement of starch fermentation in the human large intestine, *Can. J. Physiol. Pharmacol.*, 69, 121, 1991.
33. Bergman, E.N., Energy contributions of volatile fatty acids from the gastrointestinal tract in various species, *Physiol. Rev.*, 70, 567, 1990.

34. Livesy, G., Energy values of unavailable carbohydrate and diets: an inquiry and analysis, *Am. J. Clin. Nutr.*, 51, 617, 1990.
35. Jacobs, L.R. and Lupton, J.R., Effect of dietary fibers on rat large bowel mucosal growth and cell proliferation, *Am. J. Physiol.*, 246, G378, 1984.
36. Velázquez, O.C., Lederer, H.M., and Rombeau, J.L., Butyrate and the colonocyte: production, absorption, metabolism and therapeutic implications, in *Dietary Fiber in Health and Disease*, Kritchevsky, D. and Bonfield, C., Eds., Plenum Press, New York, 1997, p. 123.
37. Stephen, A.M. and Cummings, J.H., Water holding by dietary *fibre in vitro* and its relationship to faecal output in man, *Gut*, 20, 722, 1979.
38. Chen, H.-L. et al., Mechanisms by which wheat bran and oat bran increase stool weight in humans, *Am. J. Clin. Nutr.*, 68, 711, 1998.
39. Cummings, J.H., The effect of dietary fiber on fecal weight and composition, in *CRC Handbook of Dietary Fiber in Human Nutrition*, 2nd ed., Spiller, G. and Arbor, A., Eds., CRC Press, Boca Raton, FL, 1993, p. 263.
40. Nyman, M. et al., Fermentation of dietary fibre in the intestinal tract: comparison between man and rat, *Br. J. Nutr.*, 55, 487, 1986.
41. Zafar, T.A. et al., Nondigestible oligosaccharides increase calcium absorption and suppress bone resorption in ovariectomized rats, *J. Nutr.*, 134, 399, 2004.
42. Younes, H., Demigne, C., and Remsey, C., Acidic fermentation in the caecum increases absorption of calcium and magnesium in the large intestine of the rat, *Br. J. Nutr.*, 75, 301, 1996.
43. van den Heuvel, E. et al., Oligofructose stimulates calcium absorption in adolescents, *Am. J. Clin. Nutr.*, 69, 544, 1999.
44. Institute of Medicine (IOM), *Dietary Reference Intakes for Thiamin, Riboflavin, Niacin, Vitamin B6, Folate, Vitamin B12, Pantothenic Acid, Biotin, and Choline*, National Academies Press, Washington, DC, 1998.
45. Institute of Medicine (IOM), *Dietary Reference Intakes for Vitamin A, Vitamin K, Arsenic, Boron, Chromium, Copper, Iodine, Iron, Manganese, Molybdenum, Nickel, Silicon, Vanadium,* and *Zinc*, National Academies Press, Washington, DC, 2001.

15 Probiotics, Prebiotics, and Synbiotics: Functional Ingredients for Microbial Management Strategies

G.C.M. Rouzaud

CONTENTS

15.1 OVERVIEW

The emergence of refined molecular tools has confirmed the complexity of the gut microflora and its involvement in gastrointestinal health and diseases. The increased occurrence of antibiotic-resistant pathogens and a rising incidence of diet-related gastrointestinal disorders have prompted interests in improving gut health by dietary means. The probiotic, prebiotic, and synbiotic approaches have been developed with the aim of increasing the activities and numbers of intestinal bacterial species exerting beneficial effects in the host. A growing number of appropriate epidemiological studies and randomized clinical trials strongly support the positive effect of such microbial management strategies in chronic and acute diseases. The potential for developing new compounds with probiotic and prebiotic effects is immense. It is expected that high-throughput techniques will provide a powerful means of exploration of the gut microflora, allowing an insight of the mechanism of action *in vivo* of pro-, pre-, and synbiotics. However, there is a need for establishing the framework within which efficacy of pro-, pre-, and synbiotics can be assessed and subsequent products commercialized. In this chapter, major issues concerning the safety and efficacy of pro-, pre-, and synbiotics will be examined. The technologies now available for measuring the biological effects of pro- and prebiotics will be reviewed. Current data available from clinical studies as well as future areas of utilization will be outlined.

15.2 INTRODUCTION

Public perceptions of the importance of gut microbiology in human health and nutrition have lead to a rapid expansion of the number of pro- and prebiotic products available on the consumer market. Management strategies geared toward the improvement of intestinal balance are applied not only to the enhancement of well-being, but also more and more as alternative or complementary treatments to allopathic therapies in acute and chronic diseases.

The gut microflora is extremely diverse and accounts for a whole array of functions.

Its fermentative activities allow utilization of substrates not digested by gastric and duodenal enzymes. The release of nutrients in the large bowel determines the maintenance of intestinal tract integrity and the intake of micronutrients such as vitamins B and K. Gut microflora plays an active role in the health of the host. Fermentation of undigested carbohydrates and proteins releases organic acids such as short-chain fatty acids (SCFAs). These are known to be beneficial for epithelial proliferation and exert bactericidal effects on potentially pathogenic organisms.

Until fairly recently, it was extremely difficult to measure diversity of microflora, but the refinement of molecular techniques has confirmed the presence of approximately 500 bacterial species in the intestinal community. In this community, major genera have been identified, but at the species level, the knowledge on colonization of the intestinal tract is still not complete. Bifidobacteria and lactobacilli are considered beneficial in the microbiota, and their prevalence is generally a good indicator of health promotion.

Prebiotics, probiotics, and synbiotics exert their purported effects in the large bowel by increasing bifidobacteria and lactobacilli populations and increasing immunoprotection of the host. They also act by preventing colonization by potential pathogens. Consequently, pro-, pre-, and synbiotic strategies have been implemented when gut microflora was immature or compromised, for instance, in infants and in the elderly, and also in patients undergoing abdominal surgery, chemotherapy, or antibiotherapy. Populations in developing countries with a prevalence of infectious agents may also benefit from the pre- and probiotic strategies. In healthy individuals, the effect may not be elicited as clearly. An example is the response of healthy individuals to ingestion of fructooligosaccharide (FOS) or lactulose. Bifidogenesis induced by the consumption of prebiotic was inversely correlated to the level of bifidobacteria present in the fecal microflora at the start of the study.[1,2] Consequently, only clear investigation in the target populations may clarify the biological significance of pro-, pre-, and synbiotic intake.

15.3 OVERVIEW OF PROBIOTIC CONCEPT

Current definition restricts probiotic classification to live microorganisms that, upon ingestion, elicit beneficial effects on the intestinal balance of the host.[3] Currently, most of the microorganisms corresponding to the definition of probiotics are from bifidobacteria and lactobacilli genera (Table 15.1).

Enteroccocus spp., *Escherichia coli*, and yeast such as *Saccharomyces boulardii* and *Saccharomyces cerevisiae* have also reported probiotic effects, but their positive effects are strongly counterbalanced by safety risks; therefore, there is much debate as to their compliance to the definition of probiotic *stricto sensu.*

Evidence supporting prophylactic action of characterized probiotics strains such as *Lactobacillus rhamnosus* and *Lactobacillus fermentum* in the vagina[18] may lead to extension of the probiotic definition to their use for topical application on tissues other than the digestive tract. It has been suggested that strains only be classified as probiotic once their beneficial effects have been established *in vivo*.[19] This accounts for the fact that probiotic traits are generally specific to a particular strain and may not be applicable to all probiotics.[20]

TABLE 15.1
Probiotic Bacteria with Documented Clinical Effects

Bacterial Group	Reference
Bifidobacteria	
Bifidobacterium bifidum	Marteau et al.[4]
Bifidobacterium breve Yakult strain	Shimakawa et al.[5]
Bifidobacterium lactis Bb-12	Isolauri et al.[6]
Bifidobacterium longum	Kiessling et al.[7]
Lactobacilli	
Lactobacillus acidophilus NCFM	Sui et al.[8]
Lactobacillus casei immunitass DN114001	Faure et al.[9]
Lactobacillus casei Shirota (YIT 0918)	Aso and Akazan[10]
Lactobacillus gasseri	Pedrosa et al.[11]
Lactobacillus johnsonii	Marteau et al.[12]
Lactobacillus LA-1	Bernet et al.[13]
Lactobacillus LB	Xiao et al.[14]
Lactobacillus plantarum 299v	Molin[15]
Lactobacillus reuteri	Casas and Dobrogosz[16]
Lactobacillus rhamnosus GG (ATCC 53103)	Goldin et al.[17]

15.3.1 Types of Probiotic Products

In the design of new product or dietary management strategy, the array of probiotic strains currently available is a notable advantage. Several strains can be combined in the same product. Food vehicles range from fermented milks, dairy products, and fruit juices to various forms of freeze-dried supplements (capsule, pills). Survival of live bacteria is, however, an issue. Probiotics are usually anaerobic bacteria and do not survive well during temperature changes. Fermented milks, for example, need refrigeration, which is a burden to their distribution in developing countries. Strain efficacy after freeze drying is debatable.[21]

15.3.2 Survival of Probiotics Strains

A point that is sometimes argued is the necessity for the probiotic to be alive not only at the time of ingestion, but also at the site of action, namely, the large bowel. This implies a resistance of the probiotic strains to chemical changes and secretions occurring during transit through the stomach and upper intestine. Some studies have shown that a stimulation of the immune system could be displayed by dead probiotic cells.[22,23] This passive mechanism, although desirable, is less potent than active health-promoting effects elicited by live microorganisms. For example, live probiotics strains may adhere more strongly to the intestinal mucosa[24] or produce bacteriocins against potential pathogens.[25] Determination of survivability *in vivo* still requires clarification. There is often confusion between the transient, persistent, and colonizing effects of studied strains. Gastric acid, bile salts, and pancreatic secretions are all barriers toward long-term persistence of probiotics. Most commercial probiotics have been tested *in vitro* for their resistance to gastric acidity and bile salts.[26,27] Fewer data are available relative to the survivability *in situ* (Table 15.2). Survival varies considerably between strains belonging to the same genus. Lactobacillus strains, for example, generally survive well *in vitro* in the presence of acidic pH and bile acid, confirming their strong potential as probiotics.[27] These specific traits may, however, not be sufficient to determine survivability *in situ*. Additional properties may be needed for effective competition of probiotic strains against the diversity of the indigenous microflora. Cell attachment and antimicrobial activities of the probiotic candidate may contribute greatly to the bacterial survival *in vivo*. The environment from which the probiotic strain originates may also affect chances of survival. In a study *in vitro* comparing the survival of 47 strains of *Lactobacillus* spp., strains isolated from the human gut were found to display a better survivability than probiotic strains isolated from food or dairy products.[20] Subsequently, five strains selected from the screening *in vitro* were studied in a feeding experiment. Findings confirmed that intestinal strains *Lactobacillus rhamnosus*, *Lactobacillus reuterii*, and *Lactobacillus GG* were persisting in the intestinal tract in higher numbers than dairy strains *Lactobacillus casei* subsp. *alactus* and *Lactobacillus delbrueckii* subsp. *lactis*.[20]

15.3.3 Probiotic Persistence: Dose Effect

The maximum amount of probiotics that can be ingested at each dose has not yet been Concentrated doses of 10^{10} colony-forming units (CFU) per day of *Lactobacillus rhamnosus GG* and *Lactobacillus johnsonii* La1 have been administrated to healthy volunteers with no adverse effect.[30,31] A dose–response study using *Lactobacillus johnsonii* La1 showed that a minimum amount of 10^{10} CFU per day was required to observe an immune response.[32] A highly concentrated probiotic preparation (VSL#3) containing eight strains of probiotics is currently gaining interest.[33] When a dose of $3*10^{12}$ per day of VSL#3 was administered to irritable bowel disease (IBD) patients in clinical trials, recorded side effects were deemed minor to nonexistent.[34–36] The administration of lower doses of probiotics in the range of 10^6 to 10^9 CFU per day may limit the viability of the probiotic in the colonic environment unless an appropriate delivery vehicle is administered.[37]

15.3.4 Transfer of Antimicrobial Resistance

There is some evidence that *Enterococcus* spp. could be applied as a probiotic therapy.[38] However, caution must be maintained because some strains of enterococci have been associated with nosocomial and antibiotic infections. One example is the emergence of a pathogenic vancomycin resistance enterococcus, which is known to cause severe infections in patients who have catheters, intravenous devices, or are undergoing dialysis.[39] The use of enterococci as a probiotic may raise health concerns. Such antibiotic resistance genes could be spread widely throughout the gut microbiota to susceptible recipient commensal bacteria.[40] It has been suggested that strains used as probiotics should be susceptible to at least two of the common molecules used in human antibiotherapies.[41]

TABLE 15.2
Studies of Probiotic Survival *In Situ*

Probiotic	Medium	Days of Feeding	Experimental Design	Survival in Feces and Effect on Gut Microflora	Reference
Lactobacillus casei Shirota	125 ml of fermented milk (10^8 CFU/ml)	3 days	Randomized Placebo controlled, n = 8	Recovery of 10^7 CFU/g feces	Yuki et al.[28]
Bifidobacterium breve strain Yakult	500 ml of fermented soy milk (10^9 CFU/ml)	14 days	Randomized, double-blind, placebo-controlled trial, n = 15	Recovery of 10^9 CFU/g feces; increase in bifidobacteria count	Shimakawa et al.[5]
Lactobacillus casei subsp. *rhamnosus* Lcr 35	10^8–10^{10} CFU/day	7 days	Randomized trial, 3 concentrations tested, n = 12	Recovery up to 3 weeks after trial; increase in Lactobacilli count; greatest variation in individuals with lower base level of Lactobacilli count	De Champs et al.[29]
Lactobacillus acidophilus NCFM	240 ml of skimmed milk (10^{10} CFU/day)	14 days	Randomized trial, n = 10	Recovery during feeding period; rapid decrease in *L. acidophilus* during washout period	Sui et al.[8]
Lactobacillus reuteri + *Lactobacillus rhamnosus* *Lactobacillus GG* + *Lactobacillus casei* subsp. *alactus* + *Lactobacillus delbrueckii*	2 freeze-dried granulates/day (10^{10} CFU/granulate)	35 days per mix	Double-blind crossover, n = 12	Recovery of 10^5–10^8 CFU/g for all strains during feeding; recovery during washout period only for *L. reuteri*, *L. rhamnosus*, and *L. GG*	Jacobsen et al.[20]
Lactobacillus GG (ATCC 53103)	Capsule (10^8–10^{10} CFU/day)	7 days	2 concentrations tested in 2 groups, n = 20	Recovery up to 3 days after trial; no change in Lactobacilli count	Saxelin et al.[30]

15.3.5 Safety of Probiotics

As live microorganisms, probiotics other than enterococci and yeasts have a proven safety record of use. Probiotic strains traditionally used in fermented milk have been granted GRAS (generally regarded as safe) status. With the emergence of less well known probiotic microorganisms, potential risks may need more careful investigations. The theoretical risk is bacteremia developing from overgrowth of probiotic organisms or indigenous colonic microbiota. Bifidobacteria and lactobacilli are not closely related to recognized human pathogenic bacteria.[40] Cases of infections by lactobacillus and bifidobacteria are extremely rare.[41] Seldom have cases of translocation of lactobacilli been reported as a result or complication of existing severe medical conditions.[41] Given their generally safe record of use throughout the world, a recent committee of experts has deemed the risk of death by probiotics of lactobacilli or bifidobacteria as negligible.[41]

15.4 OVERVIEW OF PREBIOTIC CONCEPT

In contrast to probiotics that have been the subject of research for a hundred years,[42] the development of prebiotics as a functional food is recent and rapidly expanding. The notion of prebiotics stemmed from the observation that resistant carbohydrates such as oligosaccharides are selectively fermented by bifidobacteria and can contribute to human health by inducing changes to the indigenous intestinal microflora without the need for ingestion of live microorganisms.[43] The latest definition accepts as prebiotic any "non-viable food component which evades digestion in the upper gut, reaches the colon intact and is selectively fermented by beneficial bacteria in the gastro-intestinal tract."[44] Currently, most of the food components classified as prebiotics are low to medium molecular weight carbohydrates. High molecular weight carbohydrates such as dietary fibers are not classified as prebiotics because they are not selectively fermented.[45] Recognized prebiotics are primarily built from glucose, galactose, xylose, and fructose (Table 15.3). Inulin, fructo-oligosaccharides (FOSs), and lactulose are prebiotics with the most documented effects *in vivo* due to their widespread commercial availability. Oligosaccharides containing other monosaccharides, such as arabinose, rhamnose, glucosamine, and galacturonic acid, are also under study (Table 15.4).

The mechanism underlying the selective fermentation process is still unclear. For example, the relationship between molecular weight and selectivity has not been clarified yet. An increase of selectivity is seen when polysaccharides decrease in size, for instance, with xylan to xylo-oligosaccharides,[46] dextran to isomaltose,[47] and pectins to pectic oligosaccharides.[48] A better understanding of the relation between structure of oligosaccharides and functional benefits in the gut may lead to the manufacture of prebiotics with enhanced selectivity that could be targeted toward particular strains in the commensal flora.

TABLE 15.3
Commercially Available Existing Prebiotics

Oligosaccharides	Structure	Natural Occurrence	Manufacturing
Inulin	Fruβ2 → $(1Fru)_n$, n = 20	Onion, chicory, banana, inulin from chicory	Hydrolysis of inulin by inulinase; synthesis from sucrose by β- fructosyl transferase
Fructo-oligosaccharides	Fruβ2 → $(1Fru)_n$, n = 2–5		
Lactulose	Galβ1 → 4Fru	None	Chemical isomerization of lactose
Lactosucrose	Galβ1 → 4Glcα1 ↔ 2βFru	None	Synthesis from lactose and sucrose by β-fructo-furanosidase
Trans-galacto-oligosaccharides	Tri- to pentasaccharides with:Galβ1 → 6GalGalβ1 → 3Gal linkages	Human and cow milk	Synthesis from lactose syrup by β-galactosidase
Soybean oligosaccharides	Galα1 → 6Glcα1 ↔ 2βFru	Soybean whey	Extraction from whey
Isomalto-oligosaccharides	Glcα1 → $(6Glc)_n$, n = 1–4	Cornstarch	Synthesis from starch using α-amilase, pullulanase, α-glucosidase
Gluco-oligosaccharides	Di- to heptasaccharides with: Glcα1 → 2Glc Glcα1 → 6Glc linkage	Oat β-glucans	Sucrose + maltose/glucosyl transferase

TABLE 15.4
Newly Developed Oligosaccharides for Use as Prebiotics

Oligosaccharides	Structure	Natural Occurrence	Manufacturing
Gentio-oligosaccharides	Gluβ1–6Gluβ1–6]$_n$, n = 1–5		
Chito-oligosaccharides	Glcβ1–4Glc	Mucopolysaccharides	
Xylo-oligosaccharides	Xylβ → 4Xyl	Corn cobs, oat spelt	Xylanase
Arabino-xylo-oligosaccharides		wheat	
Xylo-oligosaccharides	Xylβ → 4Xyl	Corn cobs, oat spelt	Xylanase
Arabino-xylo-oligosaccharides		wheat	
Oligodextrans	Glcα1 → (6Glc)$_n$, n = 1–4	Dextran	Endodextranase
Pectic-oligosaccharides		Pectins Soybeans	Endoglucanase Endoarabinanase
Arabino-galacto-oligosaccharides		Sugar beet	Rhamnogalacturonase
Arabino-oligosaccharide		Apple	Endogalacturonase
Rhamno-galacturo-oligosaccharides		Polygalacturonic acids	
Galacturonic-oligosaccharides			
Sialic acid oligosaccharides	N-acetyl neuraminic acid	Human milk, κ-casein, lactoferrin	

15.4.1 Ease of Use of Prebiotics

Unlike probiotics, prebiotics are nonviable and reach the colon intact. These properties confer to prebiotics considerable advantages. Their food manufacturing properties, such as thickening agents or sweeteners, make them more amenable to industrial processes. Prebiotics have a longer shelf-life than probiotics and can be incorporated into a large variety of food, such as infant formulae, weaning food, cereals, and confectionery, as well as beverages, dairy products, and dietary supplements. Their versatility has major potential, but their use is still underrepresented in countries other than Japan.

15.4.2 Side Effects

Possible side effects may be encountered when using prebiotics. Due to the resistance of prebiotics to digestion in the upper intestinal tract, the volume of material arising in the colon and the volume of fermentation end products are increased. An increase in stool frequency and stool weight is often reported in human feeding trials.[49–51] Absorption of a large dose (>20 g/day) of prebiotic such as inulin or lactulose may lead to a laxative effect.[52] It is critical that newly developed products are selective toward non-gas producer bacteria, as gas distension may discourage the intake of prebiotics.

15.4.3 Dose Effect

Optimum doses of prebiotics have been determined for common prebiotics such as FOSs and trans-galacto-oligosaccharides in various populations. Doses of FOSs administered in feeding and clinical trials range from 3 to 20 g per day in adults and 0.4 to 3.0 g per day in infants.[52–54] These doses were found in agreement with the amount of naturally occurring oligosaccharides ingested in a diet rich in vegetables.[55] A minimal intake of 4 to 10 g per day for induction of a bifidogenic effect is often suggested, but there is currently no recommended intake available.[44]

15.4.4 Purity and Safety of Prebiotics

Due to limitations in the manufacturing process, current prebiotic preparations are generally mixtures of polysaccharides of various chain lengths. The presence of mono- and disaccharides may hinder the specificity of the prebiotic. Chemical extraction of oligosaccharides from food may also result in undesirable color or flavor. To overcome these issues, new enzymatic technologies providing higher oligosaccharide selectivity and more palatable properties are developed.[56] The risk of bacteremia associated with prebiotics is probably negligible.

Bifidobacteria and lactobacilli are indigenous inhabitants of the gut microflora and also the major organisms targeted by prebiotic dietary management strategies; the proliferation of these populations has not lead to pathogenesis.

15.4.5 GRAS Status and Persistence of Effect

Existing prebiotics are granted GRAS status. The natural occurrence of prebiotic compounds in commonly consumed food products and the long history of use have so far justified the absence of risk assessment for this type of functional food.

The persistence of prebiotic effects when their intake is stopped has not been well established. In many feeding studies, colonic microbial changes are observed after treatment with prebiotics, but the effect generally ceased with the interruption of treatment.[2,34] Long-term daily intake seems to be necessary to achieve optimum efficiency; however, few studies have looked at long-term consequences of prebiotic ingestion.

15.5 OVERVIEW OF SYNBIOTIC CONCEPT

The concept of synbiotics is a combination of both probiotic and prebiotic approaches. A synbiotic aims at stimulating the growth or activity of indigenous bifidobacteria and lactobacilli by using an appropriate carbohydrate in conjunction with one or several probiotic strains. Prebiotics may provide additional protection during intestinal transit to ensure persistence of the probiotic strain to the lower intestinal tract. Prebiotics may enhance the growth of the probiotic strain and of the targeted commensal populations. Dual advantages of both approaches may thus be realized.

15.5.1 Type of Synbiotic

Few of the synbiotic products currently available have been designed specifically. Readily available oligosaccharides were used to complement a fermented milk. Technology input is required to achieve an optimal preparation. Table 15.5 and Table 15.6 summarize the synbiotic combinations currently in use or development.

15.5.2 Synbiotic Efficacy

The design of synbiotics should improve the survivability of probiotic strains, as the presence of prebiotic compounds positively influences the number of viable cells in the preparation. Resistant starch was found to enhance the survival of *Bifidobacterium lactis* in simulated gastric and intestinal contents.[45] A combination of fructo-oligosaccharide and *Lactobacillus acidophilus* increased the persistence of the probiotic *in vitro* in a model of the human gut.[57]

Consequently, probiotics, prebiotics, and synbiotics are dietary management tools that are recognized as harmless and have little contraindications or precautions of use.

15.6 DETERMINING EFFICACY

15.6.1 *In Vitro* Systems

Various *in vitro* model systems have been validated and allow a reproduction of the physicochemical events encountered in the different parts of the gastrointestinal tract. Available models are of various degrees of complexity, from single-stage

TABLE 15.5
Synbiotics Tested in Animals and *In Vitro*

Probiotic	Prebiotic	Experimental Model	Outcomes	Reference
Lactobacillus acidophilus 74-2	Fructo-oligosaccharides	*In vitro* batch culture using human fecal flora	Increased bifidobacteria populations; increased production of propionate; increased -galactosidase activity; decrease -glucuronidase activity	Gmeiner et al.[57]
Bifidobacterium longum	Inulin	Rats with carcinogen	Synergistic action of pro-/prebiotic	Rowland et al.[58]
Bifidobacterium	Fructo-oligosaccharides	Mice induced with colon cancer	Decrease in aberrant crypt foci	Gallaher and Khil[59]
Bifidobacterium breve	Trans-oligosaccharides	Mice infected with *Salmonella typhimerium*	Decrease in excretion of *Salmonella*	Asahara et al.[60]

TABLE 15.6
Synbiotic Tested in Clinical Trials

Probiotic	Prebiotic	Medium	Experimental Design	Primary Endpoint	Reference
Bifidobacterium breve + *Lactobacillus casei*	Galacto-oligosaccharides	Capsule, 3 g/day (1.10^9 CFU/g)	1 infant with laryngotracheo-esophageal cleft	Increase in short-chain fatty acid production; bowel movement resumed	Kanamori et al.[61]
Bifidobacterium sp.	Inulin	Fermented milk with 18 g/day prebiotic or placebo	Randomized, placebo controlled, n = 12	Increase in bifidobacteria with fermented milk; no additional effect of inulin	Bouhnik et al.[50]
Bifidobacterium lactis Bb12	Galacto-oligosaccharides	Yogurt	Randomized, n = 30	Decrease in *B. longum*; no persistence of probiotic after treatment was stopped	Malinen et al.[62]
Bifidobacterium longum 913 + *Lactobacillus acidophilus* 145	Fructo-oligosaccharides	300 g of yogurt with 1% prebiotic	Crossover placebo-controlled study, n = 29; 15 normocholesterolemic; 14 hypercholesterolemic	Increase in HDL concentration; decrease in LDL/HDL cholesterol	Kiessling et al.[7]
Lactobacillus paracasei			n = 12	Positive effect on microflora; persistence for a few days after treatment was stopped	Morelli[26]
Lactobacillus acidophilus + *Bifidobacterium infantis*	Fructo-oligosaccharides	Supplement	Parallel, double-blind, randomized study, n = 626	GI tolerance; decreased number of sick days; catch-up growth; decreased constipation	Fisberg[63]
Lactobacillus plantarum v299	Oat fiber		Parallel, randomized, double-blind study, n = 43	Decrease of pancreatic sepsis after surgical intervention	Oláh et al.[64]

fermenters to a cascade of bioreactors simulating the physiological differences between each part of the colon.[65–68] Generally, temperature, pH, redox potential, and transit time are controlled. Reproduction of peristaltic movements may vary from stirring to differential pulse movements. The gastrointestinal secretion and digestive absorption are also simulated with some models.[67] The latest advances in the area are models simulating the attachment of microbial cells to an artificial intestinal membrane (biofilm reactor).[69]

In vitro models allow the study of the human gut ecosystem in controlled laboratory conditions while preserving the diversity of gut microbiota. Fermentative activities of a large array of substrates can be tested; these models have thus been proven useful in the initial steps of development of new/emerging pre- and probiotics.

15.6.2 Animal Models

Animal models are a more realistic representation of the mammalian intestinal tract.[65] These models allow detailed studies of the systemic effects and of the host–response resulting from the manipulation of the gut microflora. Gastrointestinal disorders can be induced in some animals models, such as ulcerative colitis,[70] colorectal cancers,[71] and necrotizing enterocolitis.[72] Because immune response and microbial effects of probiotics, prebiotics, and synbiotics are often species specific, human flora-associated animals are of preferential use for assessing the effectiveness of organisms or carbohydrates under study.[73–75]

15.6.3 *Ex Vivo* Models

Biopsies of intact and pathologic tissues allow an investigation of the ecological niches present in the gut and a characterization of the microflora attached to the intestinal epithelium.[76] Most of the *in vitro* or animal models use fecal microflora as starting inoculum. Although fecal microflora is a good representation of the luminal microflora, bacteria adhering to the epithelium are likely to differ.[68] The development of molecular tools has greatly improved the possibilities of exploring microflora from biopsies' tissues. Characterization of the epithelium-adhering microflora may consequently advance greatly in the next few years.

Tissue culture is another system often employed to characterize attachment properties of emerging pro- and prebiotics. Tissue cultures, although validated systems, encounter limitations, as models are often derived from cancer cell lines. Limitations are also seen because tissue cultures need to be kept aerobic, whereas most of the probiotics are anaerobic strains. Tissue cultures may give an indication of the immune response after exposure to prebiotics or probiotics, but the models often need optimization to reflect the real conditions encountered at the intestinal site, particularly in a pathological phase (such as the inflammatory stage).

15.6.4 Human Trials

Definitive assessments of pro- and prebiotic effects are only achieved by results of well-designed human feeding studies. Ideally, trials should be double blind, randomized, and placebo controlled (similar to phase 2 trials in a drug development procedure). Comparative studies at multiple centers are advantageous. In the case of prebiotics, the choice of placebo is not always clarified; some studies use nondegradable polysaccharides (starch, maltodextrin) or readily digested saccharides (glucose/lactose).[2] Records of food intake and bowel habits during the trial period provide generally useful information.

If the product is designed to be used as replacement therapy or as a complement therapy in a particular disease state, sample size, exclusion criteria, and primary endpoints must be well defined. Criteria such as history of drug administration, genetic susceptibility, and family history must be taken into account. There may also be a need for comparative studies with standard therapy (phase 3 study). The effectiveness of probiotic studies may involve comparison between well-established strains and new strains or combinations of strains. Similarly, synbiotic trials may require a specific design measuring the potential synergistic effect between the synbiotic components. Follow-up studies (phase 4) are useful to determine the long-term effect of probiotic and prebiotic use.[19]

15.6.5 New Molecular Tools for Analyzing Gut Microflora (Biomarkers)

Modern molecular techniques have led to the possibility of characterizing the complete gut microflora *in situ*. They have enabled both the qualitative and quantitative monitoring of phylogenetically related bacterial groups without the need for traditional cultivation techniques that only select for those bacteria that are culturable in the laboratory (viable but nonculturable (VBNC)).

Analysis of 16S rDNA gene profiles obtained directly from feces has greatly expanded estimates of species diversity within the microflora. About 70% of clones correspond to novel bacterial lineages, whereby the majority fell into three dominant groupings: *Bacteroides* spp., *Clostridium coccoides*, and *Clostridium leptum*.[77] In feeding trials, a range of 16S rDNA gene probes designed to target the most important groups of bacteria present in the gut microflora are applied to monitor the changes in bacterial numbers. This technique, known as fluorescent *in situ* hybridization (FISH), allows a quantification of microbial populations. With the increase in the isolation and identification of novel bacterial species, however, the number of probes available is rapidly expanding, and so is the number of species monitored. With an increasing level of analysis, there is a need for high-throughput techniques allowing an integrated analysis of these changes while conserving the information on microbial diversity.

Bacterial community analysis using 16s rDNA gene fingerprinting techniques such as T/DGGE allows a qualitative whole community analysis of samples. T/DGGE separates polymerase chain reaction (PCR) amplicons according to their sequence variation. The profiles obtained can distinguish a change in a specific species or, indeed, monitor the overall changes in diversity in response to the application of a functional food or probiotic. Another approach that is being developed is real-time quantitative PCR, which, though expensive, is less time-consuming than FISH and ultimately will be more robust in quantifying bacterial numbers.[78] A combination of both quantitative and qualitative molecular approaches is extremely useful in evaluating the efficacy of pro-, pre-, and synbiotics in boosting human health. In addition to monitoring bacterial change following probiotic or prebiotic ingestion, there is a need to understand how these changes affect the expression of genes both in the microbial population and in the intestinal epithelium. Only this kind of information will lead us to understand the mechanisms whereby probiotics and prebiotics are effective. More and more bacterial genomes are being sequenced, and with the rapid development of DNA microarray technology, cross talk (transcriptomics, gene expression) between probiotics and human mucosa cells is very much on the horizon. Such data can be subsequently used to predict both the proteomics and the metabolomics of the effects of pre-, pro-, and synbiotics on gut health.

15.7 EVIDENCE FOR THE EFFECT OF PRO-, PRE-, AND SYNBIOTICS IN THE MAINTENANCE OF HEALTH

15.7.1 Infants and the Elderly

The gut microflora of breast-fed infants is primarily constituted of bifidobacteria.[79] The protective effect of bifidobacteria against enteropathogens has been demonstrated both *in vitro* and *in vivo*. Bifidobacteria are essential for the constitution of the infant gastrointestinal defense barrier and for stimulation of early immunological responses.[80] Differences exist between the intestinal microflora of breast-fed and formula-fed infants. Extensive comparative studies have repeatedly shown a more diverse bacterial community and lower numbers of bifidobacteria in feces of formula-fed infants.[81] These observations were correlated to a higher risk of pathogen colonization. Human breast milk contains oligosaccharides composed of sialic acid, N-acetyl glucosamine, L-fucose, D-glucose, and D-galactose.[82] These human oligosaccharides are naturally occurring prebiotics that enhance protection against pathogens by a treble mechanism: (1) they increase selectively the bifidobacteria and beneficial populations of the gut, (2) they act as decoy receptors to pathogenic bacteria by mimicking the oligosaccharide portion of epithelial glycoproteins, and (3) their colonic fermentation results in the production of SCFAs, such as acetate and lactate, which acidify the intestinal content and prevent the proliferation of pathogens. The composition and structure of human milk oligosaccharides cannot be reproduced industrially. Fructo-oligosaccharides, lactulose, and galacto-oligosaccharides are the best candidates for supplementing formula milk. Recent studies of prebiotic supplementation in term and preterm infants showed an increase in fecal bifidobacteria and an improvement of intestinal passage (Table 15.7). A mixture of fructo-oligosaccharides and galacto-oligosaccharides may reproduce more closely the composition of breast milk than the addition of one type of oligosaccharide only.[53] At weaning age, the gut microbiota acquires a diverse profile that remains relatively constant through adult life.

With aging, however, the intestinal homeostasis tends to be compromised. Investigations in the elderly showed a decrease in the proportion of bifidobacteria and a higher occurrence of enterobacteria species.[87,88] The use of oligosaccharides to maintain the prevalence of bifidobacteria in the gut microbiota is currently under

TABLE 15.7
Clinical Trials Investigating Change in the Microflora of the Infant

Prebiotic	Medium	Feeding Period	Dose	Study Design	Outcomes	Reference
Fructo-oligosaccharides + galacto-oligosaccharides	Formula	28 days	0.04 g/l and 0.08 g/l	Randomized, placebo-controlled, parallel trial, n = 60 term infants	Increase in bifidobacteria	Moro et al.[53]
Fructo-oligosaccharides + galacto-oligosaccharides	Formula	28 days	0.3 g/day	Randomized, placebo-controlled, parallel trial, n = 42 preterm infant	Increase in bifidobacteria	Boehm et al.[83]
Fructo-oligosaccharides	Formula	14 days	1–3 g/day	Randomized, placebo-controlled, parallel trial, n = 53 infants (1 month old)	No significant change of fecal microflora	Guesry et al.[84]
Oligosaccharides mix from fermentation of lactose by *Bifidobacterium breve* C50	Formula	120 days	N.D.	Randomized, double-blind, placebo-controlled, parallel trial, n = 35 newborn infants	Increase in bifidobacteria; higher production of anti-poliovirus IgA	Romond et al.[85]
Galacto-oligosaccharides	Formula	4 days	N.D.	Randomized, parallel trial, n = 42 newborn infants	No prevalence of bifidobacteria	Rubaltelli et al.[81]
Lactulose	Formula	2×21 days	5 and 10 g/l	Crossover trial, n = 6 infants (2–10 weeks old)	Increase in bifidobacteria; decrease in coliform species	Nagendra[86]

Note: N.D. = not determined.

TABLE 15.8
Clinical Trials Investigating Changes in the Elderly Gut Microflora with Consumption of Prebiotics

Prebiotic	Study Design	Outcomes	Reference
Inulin	n = 25 elderly	Increase in bifidobacteria	Klessen et al.[89]
Isomalto-oligosaccharides	n = 18 senile men + 6 healthy men	Increase in bifidobacteria	Kohmoto et al.[90]
Fructo-oligosaccharides	n = 19 elderly	Increase in bifidobacteria and bacteroides; decrease in inflammatory immune response	Guigoz et al.[91]

investigation. Few *in vivo* studies are available, but inulin, fructo-oligosaccharides, and isomalto-oligosaccharides showed a positive modulation on the fecal microflora of the elderly (Table 15.8). Development of specific synbiotic combinations geared toward protection against enterobacteria are currently under study.[92]

15.7.2 Immune Function

Probiotics induce a stimulation of the immune response at local and systemic levels. The effects of *L. GG*, *Bifidobacterium breve*, and *L. acidophilus* on the immune functions have been recently reviewed and demonstrated that probiotics interact with the intestinal tract to prevent intestinal permeability and exposure to allergen or pathogen translocation. The intestinal barrier is strengthened by an increase in intestinal immunoglobulin.[93] A stimulation of the systemic immune system is also observed with specific strains of probiotics. The stimulation of the nonspecific immune defense is particularly beneficial in subjects where the immune system needs consolidation, such as in infants (Table 15.9) and the elderly and in hospitalized patients. Prospective studies indicate that the development of atopic allergy is correlated to poor levels of lactic acid bacteria in the gut microflora. Higher counts of clostridia and lower counts of bifidobacteria were found in the feces of infants who later developed atopic dermatitis.[99] When infants at risk of allergy were exposed to *L. GG* during the perinatal period, a decrease in the occurrence of atopic allergy was seen.[100] Conversely, in the allergic patient where the immune response is exacerbated, probiotic intake led to a downregulation of the inflammatory response.[101]

Little is known about the immunostimulatory effects of prebiotics. Because they stimulate indigenous probiotic bacteria, they may have an indirect immunological effect. In the elderly, the feeding of fructo-oligosaccharides appeared to reduce nonspecific immunity, such as phagocytic activity of granulocytes and monocytes.[91]

15.7.3 Lipid and Energy Metabolism

Obesity, diabetes, and hypercholesterolemia are increasingly widespread and play a synergistic role in the onset of coronary heart disease. The potential modulating role of probiotics and prebiotics in lipid and energy metabolism has therefore raised

TABLE 15.9
Prebiotic, Probiotic, and Synbiotic in Children's Health

Prebiotic/Probiotic/Synbiotic	Medium	Duration	Study Design	Outcomes	Reference
Fructo-oligosaccharides (1.1 g/kg/day)	Infant cereal (3.32 g/kg/day)	N.D.	Prospective, double-blind, randomized, controlled study, n = 123 children (4–24 months old)	Decrease in constipation frequency; decrease in the occurrence of febrile illness; decrease in use of antibiotic; no change in incidence of diarrhea	Saavedra et al.[94]
Inulin (0.2 g/kg/day)	Cereal (5 g/kg/day)	10 weeks	Double-blind, placebo-controlled trial, n = 50 weaned infants (7–9 months)	Increase in antimeasles IgG	Firmansyah et al.[95]
L. acidophilus + *B. infantis* + fructo-oligosaccharides	Supplement	4 months	Multicenter, double-blind, randomized, placebo-controlled trial, n = 626 malnourished children (1–6 years old)	Decrease in number of sick days	Fisberg et al.[63]
L. rhamnosus + fructo-oligosaccharides	Formula	N.D.	Double-blind, placebo-controlled, randomized trial, n = 58 children with acute gastroenteritis	Decrease of diarrhea duration	Ahmad et al.[96]
B. bifidum + *Streptococcus thermophilus*	Formula	17 months	Double-blind, placebo-controlled, randomized trial, n = 55 hospitalized infants (5–24 months)	Decrease in risk of diarrhea and shedding of rotavirus	Saavedra et al.[94]
L. GG	Oral rehydration	N.D.	Multicenter, double-blind, placebo-controlled, randomized trial, n = 287 children with acute diarrhea	Decrease in duration of diarrhea	Guandalini et al.[97]
L. GG (2×10^{10} CFU)	Fruit puree	2 months	Double-blind, placebo-controlled study, n = 118 lower socioeconomic children	Decrease in diarrhea frequency	Costa-Ribeiro et al.[98]

interest for several decades. Probiotics are thought to have cholesterol binding properties that stimulate the excretion of cholesterol in feces, instead of reabsorption by the host.[102] Good correlation exists *in vitro* and in animals for a cholesterol-lowering effect following treatments with probiotics and prebiotics. Results of human studies, however, are conflicting. *Lactobacillus acidophilus*, *Lactobacillus bulgaricus*, *Bifidobacterium longum*, and *Enterococcus faecium* have been used in double-blind, randomized, placebo-controlled studies in humans, generating mixed results. A recent review summarizing these human studies has highlighted that a moderate decrease in serum total cholesterol may occur with specific strains of probiotics.[102] The effect was not systematically reflected by changes in the serum concentration of LDL cholesterol or triacylglycerols. A decrease in the LDL/HDL (low-density lipoprotein/high-density lipoprotein) ratio, an indicator of atherogenicity, may be observed in some studies, but this has been related to the fat content of the dairy product used as the vehicle.[7] Human studies investigating the cholesterol-lowering effect of prebiotics have essentially focused on fructo-oligosaccharides and inulin with inconsistent findings. The most significant change was a substantial decrease in circulating triacylglycerols. The proposed mechanism is an inhibition of *de novo* lipogenesis in the liver. Prebiotics may also have a low glycemic index and help in reducing blood glucose and preventing insulin resistance, but the findings in humans are still preliminary.[103] Both inhibition of the *de novo* lipogenesis and modulation of insulin release may be mediated by SCFAs produced in the gut following the fermentation of inulin and fructo-oligosaccharides.[104] The complexity and heterogeneity of lipid metabolism do not allow firm conclusions to be made on the protective effect of probiotics and prebiotics against coronary heart diseases.

15.7.4 Mineral and Vitamin Absorption

Oligosaccharides have been linked to an enhancement of mineral absorption in the large bowel, in particular calcium and magnesium. Feeding studies with fructo-oligosaccharides have used refined techniques such as stable-isotope markers and bone density measurements to demonstrate an improvement of bone metabolism with the ingestion of prebiotics during the growing adolescence and late menopause phases.[105,106] Recent reviews have highlighted the mechanism of actions relative to calcium and magnesium bioavailability and prebiotic intake.[107,108] No data suggest a direct involvement of probiotics in mineral metabolism, but the contribution of probiotic bacteria to the synthesis of exogenous vitamins such as biotin is perceived as an enhancer of the health effects.

15.7.5 Evidence for the Effect of Pro-, Pre-, and Synbiotics in Human Diseases

15.7.5.1 Acute Disorders

15.7.5.1.1 Traveller's Diarrhea

Attempts to prevent or reduce the occurrence of traveller's diarrhea have involved the intake of probiotic agents such *as L. acidophilus*, *B. bifidum*, *L. bulgaricus*, and

L. GG.[109,110] In a placebo-controlled study of two cohorts travelling at two different holiday destinations *L. GG* was proven efficacious in reducing the incidence of diarrhea in one cohort, but no significant change was seen in the second cohort. Etiological agents involved in this study were not identified, but were likely to be different at both sites. This may explain the unsuccessful prophylactic action. A specific probiotic may not elicit antipathogenic activities toward a wide range of pathogens. Multibiotherapy or synbiotic products may increase the efficacy of the dietary strategy. More structure–function studies are required to identify oligosaccharides that could mimic receptor sites and, therefore, prevent adhesion of the pathogen to the intestinal mucosa.

15.7.5.1.2 Antibiotic-Associated Diarrhea

Diarrhea often occurs as a side effect of antibiotherapy. The disruption of the intestinal microbial balance induces an attenuation of the natural defense barrier against pathogens. Opportunistic pathogens such as *Clostridium difficile* may then proliferate, and pseumembranous colitis may be seen as a complication.[111] *L. GG* and *B. longum* ingested solely have led to a significant decrease of diarrheal episodes in erythromycin-induced subjects.[112] A combination of *L. acidophilus* and *L. bulgaricus* has also reduced the incidence of diarrhea induced by ampicillin, neomycin, and amoxicillin-clavulanate.[113–115] An association of *B. longum* and *L. acidophilus* was also successfully used in the prevention of clyndamicin-induced diarrhea.[116]

15.7.5.1.3 Rotaviral Diarrhea in Children

Rotavirus is a common cause of gastroenteritis in infants. Probiotic strains *L. GG* and *B. bifidum* have been used effectively in clinical trials for prevention and shedding of rotavirus.[94,117] *L. GG* showed a drastic reduction of the duration of rotaviral diarrhea in several studies (Table 15.9), whereas the same probiotic seemed less effective in diarrhea unrelated to rotavirus.[97] The putative mechanism of action is a stimulation of the immune response specific to rotavirus and reinforcement of the mucosal integrity. A synbiotic combination of *Lactobacillus rhamnosus* and fructo-oligosaccharides has shown a significant reduction of duration of diarrhea in infants with acute gastroenteritis.[96]

15.7.5.1.4 Necrotizing Enterocolitis (NEC)

Enterocolitis is a gastrointestinal disorder of preterm neonates treated in intensive care units and receiving enteral feeds. The infectious agent is unclear, but bacteria from the commensal microflora, such as *Clostridium difficile* or *Clostridium butyricum*, may be involved.[72] The bifidobacterial population in the intestinal lumen of neonates fed enterally is abnormally low. This may result in poor immunity and favorable conditions for translocation of bacteria to the systemic environment.[118] Preventive administration of *L. acidophilus* and *Bifidobacterium infantis* to newborns admitted in intensive care unit significantly reduced the occurrence of NEC in comparison to the historical cohort.[119] Animal models of NEC have also shown an effect of fructo-oligosaccharide in the reduction of necrosis and ulceration. Additional clinical trials are needed to confirm observations in animals and underpin the mechanism of action.

15.7.5.2 Chronic Disorders

15.7.5.2.1 Enzyme Deficiencies

Probiotics have revealed a useful means of enzyme substitution therapy for digestive enzyme deficiency. Lactose intolerance is due to a deficiency in -galactosidase, an enzyme common to bacteria of *Lactobacillus*, *Bifidobacterium*, and *Streptococcus* groups often used as starter bacterial cultures in yogurts. Hydrolysis of lactose by bacterial -galactosidase justifies a better acceptance of fermented milk products than nonfermented milk by lactose-intolerant subjects. The ingestion of probiotic bacteria engineered to produce lipase or sucrase *in vivo* has been shown to improve lipid metabolism and sucrose digestion in enzyme-deficient subjects and animals.[120,121] To be efficient *in vivo*, the secretion of enzymes must take place in the upper intestinal tract, where the enzymatic digestion normally occurs. Appropriate combinations of synbiotics may be developed to ensure an optimal delivery of live probiotic bacteria capable of releasing the digestive enzymes in the intestinal chyme.

15.7.5.2.2 Autistic Children

Gastroenteritis is commonly associated with autistic children spectrum disorders. Studies of fecal samples from autistic children have highlighted an alteration of the intestinal microflora.[122] Microbial metabolites released in the gut may play a psychoactive role in autistic pathology.[123] A probiotic approach may contribute to the relief of gastrointestinal symptoms and help toward the normalization of the autistic intestinal flora. Clinical practices and carers of autistic individuals have circumstantially reported an improvement of autistic symptoms upon probiotic intake, but no appropriate epidemiological and feeding trials are currently available to confirm these observations.[123]

15.7.5.2.3 Ulcerative Colitis (UC)

Ulcerative colitis is characterized by an acute inflammation of the intestinal tract with no relation to infection. Individuals affected by the disease experience periods of relapse and remission throughout their lives. The location of main symptoms in the large intestine suggests a link between the local microflora and the disease. Animal models suggest that the normal microflora is needed for the disease to occur, but no specific etiological agent has been identified. One hypothesis is that UC is due to a partial breakdown of tolerance to the normal commensal colonic flora.[124] Several pathogens have been suggested as causative agents of the disease: *Escherichia coli*,[125] sulfate-reducing bacteria,[126] mycobacteria, and *Pseudomonas* and *Helicobacter* species.

The control of the disease involves heavy anti-inflammatory medication or surgery. The use of antibiotics is often not conclusive. Manipulation of the diet of colitis sufferers has shown an improvement at least in the maintenance of remission periods.

Randomized clinical trials have confirmed that pro- and prebiotic use may be effective in this pathology by prolonging the microbial composition of the remission state. The mode of actions of probiotics and prebiotics in UC are thought to be manifold and complementary. Probiotic studies in UC animal models have demonstrated a probiotic interaction with colonocytes leading to a possible downregulation of the inflammatory response.[127] Prebiotics such as germinated barley are a convenient and efficacious way of stimulating the production of butyrate at the colitis site. Butyrate produced by fermentation of germinated barley promotes the proliferation of colonocytes and restores the integrity of the intestinal mucosa in animal models of colitis.[128]

15.7.5.2.4 Colorectal Cancer (CRC)

The human microflora has an important role in the development of CRC. Microbial enzymes expressed by bacteria of the clostridium and bacteroides groups are able to convert dietary constituents to genotoxic or carcinogenic compounds. Probiotics are thought to act as a preventive agent of the carcinogenesis by inhibiting the activity of these enzymes. Most of the experimental evidence *in vivo* is based on animal models.[40] Rodents fed a diet supplemented with several probiotic strains of *Lactobacillus* spp. and *Bifidobacterium* spp. developed less colonic DNA damage and fewer tumors than placebo-fed counterparts.[129] An epidemiological study reported a negative correlation between the consumption of dairy probiotics and colonic adenomas.[130] The probiotic effect is likely to take place over a long time frame. Results of prospective studies investigating the long-term preventive effect of probiotics against CRC are lacking. Prebiotics such as inulin, fructo-oligosaccharides, lactulose, and galacto-oligosaccharides have been shown to be effective protection in CRC animal models. However, results of human feeding trials are less consistent. In a parallel, placebo-controlled study of 20 volunteers, the administration of lactulose at half the pharmacological dose (10 g/day) did not influence significantly the level of fecal genotoxicity.[2] The production of butyrate, a potent regulator of epithelial cells, is one of the mechanisms thought to underlie the preventive effect of prebiotics. Prebiotics may also maintain the metabolism of clostridia and bacteroides away from proteolysis toward carbohydrate hydrolysis, thereby releasing less harmful end products. A synbiotic combination containing *L. rhamnosus*, *B. bifidum* Bb12, and fructo-oligosaccharide has shown encouraging results *in vitro* and in animal models.[131] The synbiotic is currently being tested in a long-term prospective trial involving CRC patients and individuals at high risk of developing CRC (Syncan project).

15.7.5.2.5 Irritable Bowel Syndrome (IBS)

Irritable bowel syndrome defines a range of symptoms, including abdominal pain, flatulence, constipation, and diarrhea.[132] Unlike IBD, IBS is characterized by the absence of intestinal inflammation or pathology of the intestinal mucosa. The chronic stage occurs after gastroenteritis or a course of antibiotics. Women from Western countries are the major group of sufferers. The recurrence of yeast colonization by *Candida albicans* and a low prevalence of lactobacilli and bifidobacteria have been observed in IBS sufferers.[133] Probiotics may help the alleviation of the syndrome, but the effect is difficult to measure *in vivo*. *L. plantarum* 299v, *L. GG*, and VSL#3 were tested in separate randomized, placebo-controlled trials, and results showed only a limited effect on IBS symptoms.[134,135] Development of synbiotic products may be a more satisfactory strategy in IBS management, as it allows a dual action on the bowel motility and composition of intestinal microflora.[133]

15.8 CONCLUSIONS

Pre-, pro-, and synbiotics offer good opportunities for reducing the impact of food-related diseases. A strong body of *in vitro* and *in vivo* evidence now exists. A growing number of well-designed clinical trials have highlighted the positive outcomes against specific disease conditions. As the number of pro-, pre-, and synbiotic candidates increases, concerted international actions will be taken to assess their efficacy, to elevate manufacturing standards, and to regulate their use in the consumer market.

The application of newly developed molecular techniques in the field and the development of physiological biomarkers open up possibilities for screening new probiotic strains and for refining their target site in the gut. Multifunctional prebiotics designed with specific antiadhesive properties may inhibit the binding of specific pathogens to the intestinal mucosa, thereby allowing an increased resistance to infection. Probiotics engineered to carry specific immune-enhancing molecules may be used as vaccine vectors or as regulators of the immune response. However, in many disorders and diseases, the mode of action of pro- and prebiotics is still speculative and underexplored. Future studies will need to establish the interaction between host cells and pro- and prebiotic agents. Identification of bacterial metabolites, cell receptors sites, and genes activated by the pro-/prebiotic agent will provide a strong basis to ascertain the specificity of action and efficacy *in vivo*.

Finally, long-term and large-scale human studies examining clinical and molecular parameters should be undertaken to determine the relationship between maintenance of a beneficial gut microflora and human health.

REFERENCES

1. Tuohy, K.M. et al., The prebiotic effects of biscuits containing partially hydrolysed guar gum and fructo-oligosaccharides: a human volunteer study, *British Journal of Nutrition*, 86, 341, 2001.
2. Tuohy, K.M. et al., A human volunteer study to determine the prebiotic effects of lactulose powder on human colonic microbiota, *Microbial Ecology in Health and Disease*, 14, 165–173(9), 2002.
3. Fuller, R., Probiotics in man and animals, *Journal of Applied Bacteriology*, 66, 365, 1989.
4. Marteau, P. et al., Effect of chronic ingestion of a fermented dairy product containing *Lactobacillus acidophilus* and *Bifidobacterium bifidum* on metabolic activities of the colonic flora in humans, *American Journal of Clinical Nutrition*, 52, 685, 1990.
5. Shimakawa, Y. et al., Evaluation of *Bifidobacterium breve* strain Yakult: fermented soymilk as a probiotic food, *International Journal of Food Microbiology*, 81, 131, 2003.
6. Isolauri, E. et al., Probiotics in the management of atopic eczema, *Clinical and Experimental Allergy*, 30, 1604, 2000.
7. Kiessling, G., Schneider, J., and Jahreis, G., Long-term consumption of fermented dairy products over 6 months increases HDL cholesterol, *European Journal of Clinical Nutrition*, 56, 843, 2002.

8. Sui, J. et al., 16S ribosomal DNA analysis of the faecal lactobacilli composition of human subjects consuming a probiotic strain *Lactobacillus acidophilus* NCFM, *Journal of Applied Microbiology*, 93, 907, 2002.
9. Faure, G.C.M.M. et al., Specific IgA to lactic acid bacteria in feces of children consuming milk fermented by yoghurt symbiosis and *Lactobacillus casei* (Danone strain DN 114 001), *Advances in Experimental and Medical Biology*, 501, 385, 2002.
10. Aso, Y. and Akazan, H., Prophylactic effect of a *Lactobacillus casei* preparation on the recurrence of superficial bladder cancer, *Urology International*, 49, 125, 1992.
11. Pedrosa, M.C. et al., Survival of yogurt-containing organisms and *Lactobacillus gasseri* (ADH) and their effect on bacterial enzyme activity in the gastrointestinal tract of healthy and hypochlorhydric elderly subjects, *American Journal of Clinical Nutrition*, 61, 353, 1995.
12. Marteau, P. et al., Effects of intrajejunal perfusion and chronic ingestion of *Lactobacillus johnsonii* strain La1 on serum concentrations and jejunal secretions of immunoglobulins and serum proteins in healthy humans, *Gastroenterology and Clinical Biology*, 21, 293, 1997.
13. Bernet, M.F. et al., *Lactobacillus acidophilus* LA-1 binds to cultured human intestinal cell lines and inhibits cell attachment and cell invasion by enterovirulent bacteria, *Gut*, 35, 483, 1994.
14. Xiao, S.D. et al., Multicenter, randomized, controlled trial of heat-killed *Lactobacillus acidophilus* LB in patients with chronic diarrhea, *Advances in Therapeutics*, 20, 253, 2003.
15. Molin, G., Probiotics in foods not containing milk or milk constituents, with special reference to *Lactobacillus plantarum* 299v, *American Journal of Clinical Nutrition*, 73, 380S, 2003.
16. Casas, I.A. and Dobrogosz, W.J., Validation of the probiotic concept: *Lactobacillus reuteri* confers broad-spectrum protection against disease in humans and animals, *Microbial Ecology in Health and Disease*, 12, 247, 2000.
17. Goldin, B.R. et al., Survival of *Lactobacillus* species (strain GG) in human gastrointestinal tract, *Digestive Diseases and Sciences*, 37, 121, 1992.
18. Reid, G. et al., Oral probiotics can resolve urogenital infections, *FEMS Immunology Medicine Microbiology*, 30, 49, 2001.
19. Reid, G. et al., New scientific paradigms for probiotics and prebiotics, *Journal of Clinical Gastroenterology*, 37, 105, 2003.
20. Jacobsen, CN. et al., Screening of probiotic activities of forty-seven strains of *Lactobacillus* spp. by *in vitro* techniques and evaluation of the colonization ability of five selected strains in humans, *Applied and Environmental Microbiology*, 65, 4949, 1999.
21. Heyman, M., Effect of lactic acid bacteria on diarrheal diseases, *Journal of the American College of Nutrition*, 19, 137S, 2000.
22. Kaila, M. et al., Viable versus inactivated *Lactobacillus* strain GG in acute rotavirus diarrhoea, *Archives of Disease in Childhood*, 72, 51, 1995.
23. Ouwehand, A.C. and Salminen, S.J., The health effects of cultured milk products with viable and non-viable bacteria, *International Dairy Journal*, 8, 749, 1998.
24. Tuomola, E.M., Ouwehand, A.C., and Salminen, S.J., Chemical, physical and enzymatic pre-treatments of probiotic lactobacilli alter their adhesion to human intestinal mucus glycoprotein, *International Journal of Food Microbiology*, 60, 75, 2000.
25. Talarico, T.L. and Dobrogosz, W.J., Chemical characterization of an antimicrobial substance produced by *Lactobacillus reuteri*, *Antimicrobial Agents Chemotherapy*, 33, 674, 1989.

26. Morelli, L., *In vitro* selection of probiotic lactobacilli: a critical appraisal, *Current Issues in Intestinal Microbiology*, 1, 59, 2000.
27. Fernandez, M.F., Boris, S., and Barbes, C., Probiotic properties of human lactobacilli strains to be used in the gastrointestinal tract, *Journal of Applied Microbiology*, 94, 449, 2003.
28. Yuki, N. et al., Survival of a probiotic, *Lactobacillus casei* strain Shirota, in the gastrointestinal tract: selective isolation from faeces and identification using monoclonal antibodies, *International Journal of Food Microbiology*, 48, 51, 1999.
29. de Champs, C., Persistence of colonization of intestinal mucosa by a probiotic strain, *Lactobacillus casei* subsp. rhamnosus Lcr35, after oral consumption, *Journal of Clinical Microbiology*, 41, 1270, 2003.
30. Saxelin, M., Pessi, T., and Salminen, S., Fecal recovery following oral administration of *Lactobacillus GG* (AATC 53103) in gelatine capsules to healthy volunteers, *International Journal of Food Microbiology*, 25, 199, 1995.
31. Schiffrin, E., Immune system stimulation by probiotics, *Journal of Dairy Science*, 78, 1597, 1995.
32. Donnet-Hughes, A. et al., Modulation of nonspecific mechanisms of defense by lactic acid bacteria: effective dose, *Journal of Dairy Science*, 82, 863, 1999.
33. Gionchetti, P., Oral bacteriotherapy as maintenance treatment in patients with chronic pouchitis: a double-blind, placebo controlled trial, *Gastroenterology*, 119, 305, 2000.
34. Gionchetti, P. et al., Probiotics in infective diarrhoea and inflammatory bowel diseases, *Journal of Gastroenterology and Hepatology*, 15, 489, 2000.
35. Gionchetti, P., Prophylaxis of pouchitis onset with probiotic therapy: a double-blind, placebo controlled trial, *Gastroenterology*, 124, 1202, 2003.
36. Venturi, A. et al., Impact on the composition of the faecal flora by a new probiotic preparation: preliminary data on maintenance treatment of patients with ulcerative colitis, *Aliment Pharmacology and Therapeutics*, 13, 1103, 1999.
37. Lee, Y.K. and Salminen, S., The coming of age of probiotics, *Trends in Food Science and Technology* 6, 241, 1995.
38. Cremonini, F. et al., *In vitro* fermentation of cereal dietary fibre carbohydrates by probiotic and intestinal bacteria, *Journal of the Science of Food and Agriculture*, 82, 781, 2002.
39. Warren, D.K. et al., The epidemiology of vancomycin-resistant enterococcus colonization in a medical intensive care unit, *Infection Control and Hospital Epidemiology*, 24, 257, 2003.
40. Tuohy, K.M. et al., Using probiotics and prebiotics to improve gut health, *Drug Discovery Today*, 8, 692, 2003.
41. Borriello, S.P. et al., Safety of probiotics that contain lactobacilli or bifidobacteria, *Clinical Infectious Disease*, 36, 775, 2003.
42. Chen, T.S. and Chen, P.S., Intestinal autointoxication: a medical leitmotif, *Journal of Clinical Gastroenterology*, 11, 434, 1989.
43. Wang, X. and Gibson, G.R., Effects of the *in vitro* fermentation of oligofructose and inulin by bacteria growing in the human large intestine, *Journal of Applied Bacteriology*, 75, 373, 1993.
44. Roberfroid, M., van Loo, J.A.E., and Gibson, G.R., The bifidogenic nature of chicory inulin and its hydrolysis products, *Journal of Nutrition*, 128, 11, 1998.
45. Crittenden, R.G. et al., Selection of a *Bifidobacterium* strain to complement resistant starch in a synbiotic yoghurt, *Journal of Applied Bacteriology*, 90, 268, 2001.

46. Okazaki, K., Fujikawa, S., and Matsumoto, N., Effects of xylooligosaccharides on growth of bifidobacteria, *Journal of Japanese Society of Nutrition and Food Sciences*, 43, 395, 1990.
47. Olano-Martin, E. et al., *In vitro* fermentability of dextran, oligodextran and maltodextrin by human gut bacteria, *British Journal of Nutrition*, 83, 247, 2000.
48. Olano-Martin, E., Gibson, G.R., and Rastall, R.A., Comparison of the *in vitro* bifidogenic properties of pectins and pectic-oligosaccharides, *Journal of Applied Microbiology*, 93, 505, 2002.
49. Ito, M. et al., Effects of administration of galactooligosaccharides on the human faecal microflora, stool weight and abdominal sensation, *Microbial Ecology in Health and Disease*, 3, 285, 1995.
50. Bouhnik, Y. et al., Effects of fructo-oligosaccharides ingestion on fecal bifidobacteria and selected metabolic indexes of colon carcinogenesis in healthy humans, *Nutrition and Cancer*, 26, 21, 1996.
51. Chen, H.L. et al., Effects of isomalto-oligosaccharides on bowel functions and indicators of nutritional status in constipated elderly men, *Journal of American College of Nutrition*, 20, 44, 2001.
52. Bouhnik, Y. et al., Short-chain fructo-oligosaccharide administration dose-dependently increases fecal bifidobacteria in healthy humans, *Journal of Nutrition*, 129, 113, 1999.
53. Moro, G. et al., Dosage-related bifidogenic effects of galacto- and fructooligosaccharides in formula-fed term infants, *Journal of Pediatric Gastroenterology and Nutrition*, 34, 291, 2002.
54. Moore, N. et al., Effects of fructo-oligosaccharides-supplemented infant cereal: double-blind, randomized trial, *British Journal of Nutrition*, 90, 581, 2003.
55. Van Loo, J. et al., On the presence of insulin and oligofructose as natural ingredients in the Western diet, *CRC Critical Reviews in Food Science and Nutrition*, 35, 525, 1995.
56. Rastall, R.A. and Gibson, G.R., Prebiotic oligosaccharides: evaluation of biological activities and potential future developments, in *Probiotics and Prebiotics: Where Are We Going?* Tannock, W., Ed., Caister Academic Press, Wymondham, U.K., 2002, p. 107.
57. Gmeiner, M. et al., Influence of a synbiotic mixture consisting of *Lactobacillus acidophilus* 74-2 and a fructooligosaccharide preparation on the microbial ecology sustained in a simulation of the human intestinal microbiological system (SHIME), *Microbiology Biotechnology*, 53, 219, 2000.
58. Rowland, I.R. et al., Effect of *Bifidobacterium longum* and inulin on gut bacterial metabolism and carcinogen-induced aberrant crypt foci in rats, *Carcinogenesis*, 19, 281, 1998.
59. Gallaher, D.D. and Khil, J., The effects of synbiotics on colon carcinogenesis in rats, *Journal of Nutrition*, 129, 1483S, 1999.
60. Asahara, T. et al., Increased resistance of mice to *Salmonella enterica serovar Typhimirium* infection by synbiotic administration of bifidobacteria and transgalactosylated oligosaccharides, *Journal of Applied Microbiology*, 91, 985, 2001.
61. Kanamori, Y. et al., A novel synbiotic therapy dramatically improved the intestinal function of a pediatric patient with laryngotracheo-esophageal cleft (LTEC) in the intensive care unit, *Clinical Nutrition*, 21, 527, 2002.

62. Malinen, E.M.J. et al., PCR-ELISA II: analysis of *Bifidobacterium* populations in human faecal samples from a consumption trial with *Bifidobacterium lactis* Bb-12 and a galacto-oligosaccharide preparation, *Systematic Applied Microbiology*, 25, 249, 2002.
63. Fisberg, M. et al., Effect of oral supplementation with and without synbiotics on catch-up growth in preschool children, *Journal of Pediatric Gastroenterology and Nutrition*, 31, A987, 2000.
64. Oláh, A. et al., Randomized clinical trial of specific lactobacillus and fibre supplement to early enteral nutrition in patients with acute pancreatitis, *British Journal of Surgery*, 89, 1103, 2002.
65. Rumney, C.J. and Rowland, I.R., *In vivo* and *in vitro* models of the human colonic flora, *CRC Critical Reviews in Food Science and Nutrition*, 31, 299, 1992.
66. Molly, K. et al., Validation of the simulator of the human intestinal microbial ecosystem (SHIME) reactor using microorganism-associated activities, *Microbial Ecology in Health and Disease*, 7, 191, 1994.
67. Minekus, M. et al., A multicompartmental dynamic computer-controlled model stimulating the stomach and small intestine, *ATLA*, 23, 197, 1995.
68. Macfarlane, G.T., Macfarlane, S., and Gibson, G.R., Validation of a three stage compound continuous culture system for investigating the effect of retention time on the ecology and metabolism of bacteria in the human colon, *Microbial Ecology in Health and Disease*, 35, 180, 1998.
69. Probert, H.M. and Gibson, G.R., Development of a fermentation system to model sessile bacterial populations in the human colon, *Biofilms*, 1, 11, 2004.
70. Cummings, J.H., Macfarlane, G.T., and Macfarlane, S., Intestinal bacteria and ulcerative colitis, *Current Issues in Intestinal Microbiology*, 4, 9, 2003.
71. Hambly, R.J. et al., Effects of high- and low-risk diets on gut microflora-associated biomarkers of colon cancer in human flora-associated rats, *Nutrition and Cancer*, 27, 250, 1997.
72. Butel, M.J., Usefulness of an experimental model of the infant gut, *Journal of Pediatric Gastroenterology and Nutrition*, 37, 109, 2003.
73. Mallett, A.K. et al., The use of rats associated with a human faecal flora as a model for studying the effects of diet in the human gut microflora, *Journal of Applied Bacteriology*, 63, 39, 1987.
74. Djouzi, Z., Influence des probiotiques et des prébiotiques sur la composition et le métabolisme de la microflore humaine implantée chez le rat hétéroxénique, Ph.D. thesis, Université Paris-Sud., Orsay, 1995.
75. Edwards, C.A. et al., A human flora-associated rat model of the breast-fed infant gut, *Journal of Pediatric Gastroenterology and Nutrition*, 37, 168, 2003.
76. Schultsz, C. et al., The intestinal mucus layer from patients with inflammatory bowel disease harbors high numbers of bacteria compared with controls, *Gastroenterology*, 117, 1089, 1999.
77. Suau, A. et al., Direct analysis of genes encoding 16S rRNA from complex communities reveals many novel molecular species within the human gut, *Applied and Environmental Microbiology*, 65, 4799, 1999.
78. Matsuki, T. et al., Development of 16S rRNA-gene-targeted group-specific primers for the detection and identification of predominant bacteria in human feces, *Applied Environmental Microbiology*, 68, 5445, 2002.
79. Favier, C.F. et al., Molecular monitoring of succession of bacterial communities in human neonates, *Applied and Environmental Microbiology*, 68, 219, 2002.

80. Sudo, N. et al., The requirement of intestinal bacterial flora for the development of an IgE production system fully susceptible to oral tolerance induction, *Journal of Immunology*, 159, 1739, 1997.
81. Rubaltelli, F.F. et al., Intestinal flora in breast- and bottle-fed infants, *Journal of Perinatal Medicine*, 26, 186, 1998.
82. Sabbharwas, H., Sjoblad, S. and Lundblad, A., Affinity chromatographic identification and quantification of blood group A-active oligosaccharides in human milk and feces of breast-fed infants, *Journal of Pediatric Gastroenterology and Nutrition*, 12, 474, 1991.
83. Boehm, G. et al., Supplementation of a bovine milk formula with an oligosaccharide mixture increases counts of faecal bifidobacteria in preterm infants, *Archives of Diseases in Child, Fetal Neonatal Edition*, 86, F178, 2002.
84. Guesry, P.R. et al., Effect of 3 doses of fructo-oligosaccharides in infants, *Journal of Paediatric Gastroenterology and Nutrition*, 31, S252, 2000.
85. Romond, M.B. et al., Stimulation of endogenous bifidobacteria and enhancement of the intestinal immune response with a new fermented infant formula, FFC50, in infants aged 0 to 4 months: results of a double-blind randomised study, *Annals of Nutrition and Metabolism*, 45, 558, 2001.
86. Nagendra, R., Viswanatha, S., and Arun Kumar, S., Effect of feeding milk formula containing lactulose to infants on faecal bifidobacterial flora, *Nutrition Research*, 15, 15, 1995.
87. Mitsuoka, T. and Hayakawa, K., The faecal flora in man. I. Composition of the faecal flora of various age groups, *Zentralblatt fur Bakteriologie und Hygiene*, 223, 333, 1973.
88. Hopkins, M.J., Sharp, R., and Macfarlane, G.T., Age and disease related changes in intestinal bacterial populations assessed by cell culture, 16S rRNA abundance, and community cellular fatty acid profiles, *Gut*, 48, 198, 2001.
89. Klessen, B. et al., Effects of inulin and lactose on fecal microflora, microbial activity and bowel habit in the elderly constipated persons, *American Journal of Clinical Nutrition*, 65, 1397, 1997.
90. Kohmoto, T. et al., Effect of isomalto-oligosaccharides on human faecal flora, *Bifidobacteria Microflora*, 7, 61, 1988.
91. Guigoz, Y. et al., Effects of oligosaccharides on the faecal flora and non-specific immune system in elderly people, *Nutrition Research*, 22, 13, 2002.
92. Silvi, S. et al., EU project Crownalife: functional foods, gut microflora and healthy ageing. Isolation and identification of *Lactobacillus* and *Bifidobacterium* strains from faecal samples of elderly subjects for a possible probiotic use in functional foods, *Journal of Food Engineering*, 56(2–3), 195–200, 2003.
93. Isolauri, E. et al., Probiotics: effects on immunity, *American Journal of Clinical Nutrition*, 73, 444S, 2001.
94. Saavedra, J.M. et al., Feeding of *Bifidobacterium bifidum* and *Streptococcus thermophilus* to infants in hospital for prevention of diarrhoea and shedding of rotavirus, *Lancet*, 344, 8929, 1994.
95. Firmansyah, A. et al., Improved humoral immune response to measles vaccine in infants receiving infant cereal with fructooligosaccharides, *Journal of Paediatric Gastroenterology and Nutrition*, 31, A512, 2001.
96. Ahmad, A. et al., Effect of a combined probiotic, prebiotic and micronutrient supplementation in reducing duration of acute infantile diarrhea, *Journal of Pediatric Gastroenterology and Nutrition*, 31, A984, 2000.

97. Guandalini, S. et al., *Lactobacillus GG* administered in oral rehydration solution to children with acute diarrhea: a multicenter European trial, *Journal of Pediatric Gastroenterology and Nutrition*, 30, 54, 2000.
98. Costa-Ribeiro, H. et al., Prophylactic administration of Lactobacillus GG to children in a day care center, *Journal of Pediatric Gastroenterology and Nutrition*, 31, suppl. 2, S252, 2000.
99. Bjorksten, B. et al., Allergy development and the intestinal microflora during the first year of life, *Journal of Allergy and Clinical Immunology*, 108, 516, 2001.
100. Kalliomaki, M. et al., Probiotics and prevention of atopic disease: 4-year follow-up of a randomised placebo-controlled trial, *Lancet*, 31, 1869, 2003.
101. Pelto, L. et al., Probiotic bacteria down-regulate the milk-induced inflammatory response in milk-hypersensitive subjects but have an immunostimulatory effect in healthy subjects, *Clinical Experimental Allergy*, 28, 1474, 1998.
102. Pereira, D.I.A. and Gibson, G.R., Effects of consumption of probiotics and prebiotics on serum lipids levels in humans, *Critical Reviews in Biochemistry and Molecular Biology*, 37, 259, 2002.
103. Rumessen, J.J. et al., Fructans of the Jerusalem artichoke: intestinal transport, absorption, fermentation, and influence on blood glucose, insulin, and C-peptides responses in healthy subjects, *American Journal of Clinical Nutrition*, 52, 675, 1990.
104. Lovegrove, J.A. and Jackson, K.G., Coronary heart disease, in *Functional Foods: Concept to Product*, Williams, C.M. and Gibson, R., Eds., Woodhead Publishing Ltd., Cambridge, England, 2000.
105. Tahiri, M. et al., Effect of short chain fructooligosaccharides on intestinal calcium absorption and calcium status in postmenauposal women: a stable-isotope study, *American Journal of Clinical Nutrition*, 77, 449, 2003.
106. Griffin, I.J., Davila, P.M. and Abrams, S.A., Non-digestible oligosaccharides and calcium absorption in girls with adequate calcium intakes, *British Journal of Nutrition*, 87, S187, 2002.
107. Scholz-Ahrens, K.E. et al., Effects of prebiotics on mineral metabolism, *American Journal of Clinical Nutrition*, 73, 459S, 2001.
108. Cashman, K., Prebiotics and calcium bioavailability, *Current Issues in Intestinal Microbiology*, 4, 21, 2003.
109. Black, F.T., Prophylactic efficacy of lactobacilli on traveller's diarrhoea, *Travel Medicine*, 4, 3, 1989.
110. Hilton, E. et al., Efficacy of *Lactobacillus GG* as a diarrhoeal preventive in travellers, *Journal of Travel Medicine*, 4, 41, 1997.
111. Fooks, L.J. and Gibson, G.R., Probiotics as modulators of the gut flora, *British Journal of Nutrition*, 88, S39, 2002.
112. Marteau, P., Protection from gastro-intestinal diseases with the use of probiotics, *American Journal of Clinical Nutrition*, 73, 430S, 2001.
113. Witsell, D.L. et al., Effect of *Lactobacillus acidophilus* on antibiotic-associated gastrointestinal morbidity: a prospective randomized trial, *Journal of Otolaryngology*, 24, 230, 1995.
114. Clements, M.L. et al., Exogenous lactobacilli fed to man: their fate and ability to prevent diarrheal disease, *Progress in Food and Nutrition Science*, 7, 29, 1983.
115. Gotz, V. et al., Prophylaxis against ampicillin-associated diarrhea with a lactobacillus preparation, *American Journal of Hospital Pharmacology*, 36, 754, 1979.
116. Nord, C.E. et al., Oral supplementation with lactic acid-producing bacteria during intake of clindamycin, *Clinical Microbiology and Infection*, 3, 124, 1997.

117. Isolauri, E. et al., Oral bacteriotherapy for viral gastroenteritis, *Digestive Diseases Science*, 39, 2595, 1994.
118. Dai, D. and Walker, WA., Protective nutrients and bacterial colonization in the immature human gut, *Advances in Pediatrics*, 46, 353, 1999.
119. Hoyos, A.B., Reduced incidence of necrotizing enterocolitis associated with enteral administration of *Lactobacillus acidophilus* and *Bifidobacterium infantis* to neonates in an intensive care unit, *International Journal of Infectious Disease*, 3, 197, 1999.
120. Harms, H.K., Bertele-Harms, R.M., and Buer-Kleis, D., Enzyme substitution therapy with the yeast, *Saccharomyces cerevisiae* in congenital sucrase-isomaltase deficiency, *New England Journal of Medicine*, 316, 1306, 1987.
121. Drouault, S. et al., Survival, physiology and lysis of *Lactococcus lactis* in the digestive tract, *Applied and Environmental Microbiology*, 65, 4881, 1999.
122. Finegold, S.M., Gastrointestinal microbiota studies in late-onset autism, *Clinical Infectious Diseases*, 35, S6, 2002.
123. Bingham, M.O., Dietary Intervention in Autistic Spectrum Disorders, *Food Science and Technology Bulletin*, May 8, 2003.
124. Macpherson, A. et al., Mucosal antibodies in inflammatory bowel disease are directed against intestinal bacteria, *Gut*, 38, 365, 1996.
125. Darfeuille-Michaud, A. et al., Presence of adherent *Escherichia coli* strains in ileal mucosa of patients with Crohn's disease, *Gastroenterology*, 115, 1405, 1998.
126. Gibson, G.R., Cummings, J.H., and Macfarlane, G.T., Growth and activities of sulphate-reducing bacteria in gut contents from healthy subjects and patients with ulcerative colitis, *FEMS Microbiology Ecology*, 86, 103, 1991.
127. Steidler, L. et al., Treatment of murine colitis by *Lactococcus lactis* secreting interleukin-10, *Science*, 289, 1352, 2000.
128. Kanauchi, O. et al., Dietary fiber fraction of germinated barley foodstuff attenuated mucosal damage and diarrhea, and accelerated the repair of the colonic mucosa in an experimental colitis, *Journal of Gastroenterology and Hepatology*, 16, 160, 2001.
129. Pool-Zobel, B.L. et al., *Lactobacillus* and *Bifidobacterium* mediated antigenotoxicity in the colon of the rats, *Nutrition and Cancer*, 26, 365, 1996.
130. Burns, A.J. and Rowland, I.R., Anti-carcinogenicity of probiotics and prebiotics, *Current Issues in Intestinal Microbiology*, 1, 13, 2000.
131. Van Loo, J. and Jonkers, N., Evaluation in human volunteers of the potential anticarcinogenic activities of novel nutritional concepts: prebiotics, probiotics and synbiotics (the SYNCAN project QLK1-1999-00346), *Nutrition, Metabolism, and Cardiovascular Disease*, 11, 87, 2001.
132. Mercenier, A., Pavan, S., and Pot, B., Probiotics as biotherapeutic agents: present knowledge and future prospects, *Current Pharmaceutical Design*, 9, 175, 2003.
133. Smejkal, C.W. et al., Probiotics and prebiotics in female health, *Journal of British Menopause Society*, 9, 69, 2003.
134. Kim, H.J. et al., A randomized controlled trial of a probiotic, VSL#3, on gut transit and symptoms in diarrhoea-predominant irritable bowel syndrome, *Alimentary Pharmacology and Therapeutics*, 17, 895, 2003.
135. O'Sullivan, M.A. and O'Morain, C.A., Bacterial supplementation in the irritable bowel syndrome. A randomised double-blind placebo-controlled crossover study, *Digestive Liver Disease*, 32, 294, 2000.

16 Potential Use of Carbohydrates as Stabilizers and Delivery Vehicles of Bioactive Substances in Foods

Pirkko Forssell, Päivi Myllärinen, Piia Hakala, and Kaisa Poutanen

CONTENTS

16.1 INTRODUCTION

People are becoming increasingly health conscious and are looking for foods with special bioactive functions. Bioactive substances are often taken from their natural matrix by extraction or another concentration process, and need stabilization such as encapsulation to be successfully applied to food formulations. Encapsulation of a food ingredient means a process of forming a continuous, thin coating around the ingredient, or a process of entrapment of the ingredient within a matrix such as, e.g., a gel or crystal. Carbohydrates have been applied together with proteins and lipids as shell materials for encapsulation of food ingredients for decades. In fact, the most commonly used shell materials belong to the group of hydrophilic carbohydrates. Proteins are also used, while the hydrophobic lipids have so far found only special use, such as in fat coatings of larger particles.[1,2]

Initially, the main reasons for encapsulation of food ingredients were protection against the surrounding conditions, improvement of the overall food product quality, and facilitation of the production process. Today, specific protection systems, e.g., against oxidation, as well as targeted delivery matrices for human or animal nutrition have attracted much interest.

Health-promoting substances such as marine oils, vitamins, and microorganisms are often sensitive to oxygen, water, and light, which offer a challenge for long-time storage and formulation in food processes. If storage conditions are not controlled or if processing is performed under too harsh conditions, bioactive components may react and form degradation products that might even be carcinogenic. Especially when aiming at food fortifications, off-tastes or strong colors may also cause difficulties. It is obvious that to control the dissolution of the active substance during processing or when dealing with delivery systems related to human or animal nutrition, special encapsulation technologies are definitely needed.

In spite of the large number of carbohydrates available in nature, only a few in fact are used for encapsulation of food ingredients, mainly because of economic reasons. The choice of the shell material depends on the application: What is the component type to be protected (hydrophobic or hydrophilic, liquid or solid)? What are the conditions requested for potential release (dissolution-based, melting-based, or diffusion-based release)? Are there some special needs for textural characteristics of the product? Thus, the whole chain has to be evaluated, from the naked bioactive substance to the human digestive tract. Protection of the substance is needed against processing conditions and during storage, and finally in the digestive system.

16.2 MICROENCAPSULATION TECHNOLOGIES

Microencapsulation is the most common way to describe the processes used to prepare encapsulated food ingredients, because the particle size range varies from 1 to 2000 μm, depending on the technology applied to produce microcapsules. A wide variety of microencapsulation methods exist and, in addition, smaller, nanoscale particles can be produced using special technologies. Some selected encapsulation methods are listed in Table 16.1. The manufacturing technologies in the food area include spray drying, spray cooling or chilling, extrusion techniques, fluidized bed coating, coacervation, inclusion complexation, centrifugal coextrusion, and rotational suspension separation.[1]

Spray drying is the oldest and most commonly used method in the food industry. The first encapsulated flavors were produced by spray drying in the 1930s, while extrusion, the second most popular technique today, was not used in the flavor industry until the early 1960s.[4] The extrusion developed from the original method[5] involves mixing of the encapsulant into a molten carbohydrate, producing continuous filaments in which the encapsulant is entrapped within the shell matrix, similarly as in spray-dried particles (see Figure 16.1). The difference is that spray drying is a dehydration process, while extrusion is a melting process. The encapsulation methods that produce real coating membranes around the encapsulant are fluidized bed

TABLE 16.1
Encapsulation Technologies and Size Range of the Product Particles or the Particles That Can Be Coated[3]

Encapsulation Method	Size of the Final Particle or the Particle That Can Be Coated (μm)
Physical Methods	
Centrifugal coextrusion	150–2000
Rotational suspension separation	30–2000
Pan coating	>500
Fluidized bed	50–500
Spray drying	20–100
Chemical Methods	
Simple/complex coacervation	1–500
Interfacial polymerization	1–500
Liposome entrapment	0.1–1
Nanoencapsulation	<1

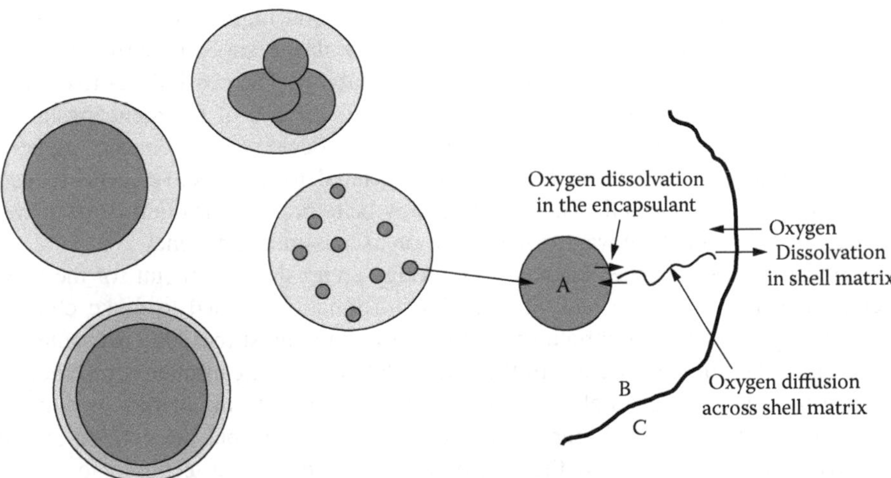

FIGURE 16.1 Two single-core, one multicore, and one matrix (entrapped) microcapsule. In the case of the matrix microcapsule, the possible mechanism for the oxidation of the encapsulant in a homogenous shell matrix is oxygen dissolvation from air (C) into the shell matrix (B), oxygen diffusion across the shell matrix, or oxygen dissolvation into the encapsulant droplet (A), which finally causes the oxidation reaction.

coating, centrifugal coextrusion, coacervation, and rotational suspension separation. The principal difference between spray cooling/chilling and spray drying is the shell material; lipids are used in the former, while carbohydrates and proteins are applied in the latter, which naturally means a difference in the process temperature.

In cosmetic applications, microcapsules have been traditionally prepared by coacervation or by precipitation or polymerization methods.[6] Today, liposomes are also used for improvement of performance of active substances in cosmetics. The pioneer in the field of microencapsulation applications is the pharmaceutical industry. In contrast to the food industries, the shell materials used in drug formulations are often synthetic polymers, as is normally the case also in cosmetic applications. Synthetic materials offer much wider functionality than biomaterials, but are not usually approved in food applications and may often be rather expensive for mass-distributed food ingredients.

The most common types of microcapsules are schematically presented in Figure 16.1, together with a schematic description of the challenge of protection of the encapsulant from oxidation. A possible mechanism for the oxidation of the encapsulant in a homogeneous single-phase shell matrix is also visualized in Figure 16.1. The three steps needed for oxidation are oxygen dissolution from air (C) into the shell matrix (B), oxygen diffusion across the shell matrix, and oxygen dissolution into the encapsulant droplet (A). When targeting both protection against oxidation and prevention of evaporation, control of very complicated transport mechanisms occurring in usually inhomogeneous matrices is required. Detailed research into these processes has only recently started to evolve.[7–9]

The general targets for microencapsulation of food ingredients have been (1) better technological performance, such as better flavor survival during baking, extruding, or deep frying, and (2) better storage stability. The possibilities to control the release of ingredients during processing with the aid of microencapsulation methods has also been of interest. The more recent developments concerning controlled release efforts are those dealing with targeted delivery of active agents related to human and animal health. The objective may be to prolong the duration of action of agents, to minimize unwanted reactions, or to maximize efficiency.[1]

Encapsulation has been claimed to be today an art that is difficult for the food scientist to master due to lack of available information needed to make choices concerning the most appropriate shell material and the most feasible encapsulation process.[10] The current progress in the food-related microencapsulation research is delivering new shell materials, such as β-cyclodextrins, into the market. A number of microencapsulated food ingredients exist in the market, but the volume is still relatively small compared with the huge potential that the various technologies offer.[3] The development of delivery vehicles for bioactive substances — whether focusing on improvement of overall storage stability or stability in processing or on more advanced desired release or targeted delivery systems — is inevitably a very challenging area of research, if aiming at feasible technology based on nature's own raw materials, such as carbohydrates.

16.3 CARBOHYDRATES AS ENCAPSULATING MATRICES

16.3.1 Functionality of Carbohydrates in Trigger Events

As compared with proteins and lipids, carbohydrates have been the most common choice of shell materials in food ingredient encapsulations, especially in flavor applications.[11] Some common examples from the three classes of natural materials are listed in Table 16.2. In practice, the best performance has most often been obtained by using special additives together with a single-carbohydrate material, or by applying many carbohydrates in combination with each other or with, e.g., proteins. Furthermore, several encapsulation processes can be applied to achieve better performance. Carbohydrates are, however, hydrophilic molecules, which means that they always interact with water. Changes in water activity can cause stickiness and caking of powders, and may also induce changes in the number of phases initially present in the system, which naturally can change the original functionality. The presence of water may also mean swelling, which might allow transport of water-soluble components across the matrix, or water can dissolve the shell matrix. In applications in which water interactions need to be avoided, instead of carbohydrates, lipid coatings could solve the problems.

Typically, glassy, highly viscous carbohydrates with low water contents form matrices, which have been used to prevent evaporation and minimize chemical reactions.[13,14] Amorphous glassy matrices are generally believed to be stable because of the low mobility of the matrix, which slows down, e.g., oxygen and flavor diffusion. In real coating-based encapsulations, film formation ability should be the inherent property of the shell material. In this case, the question is whether the membrane behaves according to the target functionality. In matrix systems, the shell material forms a continuous region in which ingredients are distributed, which is a much more complicated control system than a single-core particle. Various trigger events that may be exploited when developing targeted delivery systems for processes or special delivery vehicles for improving the biological performance are described in Table 16.3. In addition to its chemical and physical

TABLE 16.2
Approved Food-Grade Shell Materials for Microencapsulation according to Thies[12]

Polysaccharides	Fats and Waxes	Proteins
Gum arabic	Hydrogenated vegetable oil	Gelatins
Modified starches	Bee wax	Whey proteins
Hydrolyzed starches[a]		Soy proteins
Alginates		Sodium caseinate
Pectins		
Carrageenan		

[a]Maltodextrins.

properties and interactions between the shell material and the encapsulant, the final behavior of the shell material depends also on the technology applied to produce the encapsulated product.

The following discussion focuses on some selected carbohydrates that are already used in many food applications, or have been suggested to be exploited, or are potentials but so far not reported to be applied in the food area.

16.3.2 Starch

Starches have not been the usual choice of the encapsulation matrix for food ingredients in the most often applied spray-drying technology, because starches do not dissolve in cold water.[15] Instead of starch polymers, their water-soluble hydrolysis products, maltodextrins and slightly derivatized starch hydrolysates, have widely been used as encapsulating matrices in the food industry (see 16.3.5 and 16.3.6 in this chapter).

Starches are unique among polysaccharides because they occur in nature as discrete particles, called granules.[16] For dissolvation of granular starches under aqueous environments and at normal pH values, temperatures above 40°C are needed. Because amylases do not generally hydrolyze solids, starch granules are not easily digested in the human digestive tract. This, however, depends on starch origin. Native legume starches are more digestible than native potato or high-amylose cornstarch, but less digestible than native cereal or cassava starch. Native cornstarch granules have been shown to be hydrolyzed up to 75%, and legume starches 25 to 35% by porcine α-amylase.[17]

Not only starch granules differ in their digestibility, but also dissolved starch polymers after gelation vary in their accessibility to amylases. The linear starch polymer — amylose — typically forms networks that are much more resistant to digestive tract conditions than amylopectin, which is the main starch polymer in all starch granules and the only polymer in so-called waxy varieties. The resistant starch

TABLE 16.3
Possible Trigger Events That Can Release the Active Substance from the Microcapsule

Trigger Event	Release Mechanism
Mechanical treatment	Fragmentation and release
Solvent	Dissolvation of shell matrix and release, or swelling and diffusion across
Solvent + enzymes/microbes	Release only in the presence of specific enzymes or microorganisms due to degradation/dissolvation
Solvent + pH	E.g., stable at neutral pH, release at acid/alkaline pH, similar with solvent + temperature
Solvent + temperature	Release in solution at elevated temperature due to dissolvation or swelling
Temperature	Release from solids when heating or cooling due to melting

structures — granules, gels, and films — are fermented by the colonic microflora as part of dietary fiber.[18–20] On the other hand, films and gels prepared of amylopectin are mechanically weak and easily hydrolyzed by either acids or amylases.[21–23]

Investigations on the possible utilization of amylose for encapsulation of organic molecules have been conducted.[24,25] This behavior is based on the tendency of amylose to form helical conformations with an inner surface consisting predominantly of hydrogen atoms, making it hydrophobic, a structure that is closely related with cyclodextrins. About 6% unsaturated fatty acids were bound within amylose helix in both potato starch and high-amylose cornstarch and were very stable to oxygen.[25]

Native starch matrices have been developed to function as encapsulation carriers, e.g., for slow release.[26,27] Both methods were based on a similar process, which was total dissolvation of starch granules in water, followed by addition of the encapsulant. The final particles were produced by drying the rapidly cooled gel. Virtually any material may be encapsulated by the described process, and the release of the encapsulant could be controlled by the choice of the starch, high-amylose cornstarch giving prolonged release. The insolubility of amylose in gastric juices and its degradability by colonic bacteria were for the first time technologically exploited in a novel way when colon-specific drug delivery formulation — a macrocapsule — was developed about 10 years ago.[28,29] In another study based on starch polymers' ability to retrogradate from solutions, it was suggested that this phenomenon can be utilized when developing microencapsulation technology for lipophilic drug particles.[30]

Because of the exceptional granular nature of native starches, it is no wonder that specially treated starch granules have been suggested to be potential carriers for many active substances that could be utilized in the areas of food, cosmetics, agriculture, and medicine. Partially hydrolyzed and cross-linked starch granules were suggested to be suitable carriers for functional substances.[31] Hydrolysis was performed with the aid of amylases, and cornstarch was suggested to be the preferable starch used in the application. To improve the absorption capacity, the surface of the granules could be treated with proper agents or, in the case of absorbing lipophilic substances, chemically modified. Another investigation focused on various amylase-treated starch granules without any cross-link formation.[32] The produced granules were suggested to have broad application potential in food formulations due to low viscosity, and they also were claimed to be able to act as carriers for hydrophobic components.

In addition to exploiting partially hydrolyzed starch granules, starch granule aggregates were discovered to function as, e.g., a food ingredient carrier.[33,34] Water dispersion of starch granules, from small starch granules such as those of rice starch or the size-classified small starch granule fraction of, e.g., wheat starch, was particularly preferred to form spherical aggregates when dehydrated in a spray dryer in the presence of a proper binder. The active substance could be introduced into the porous aggregate either as a component of the spray-dried dispersion or as solution in an inert low-boiling solvent, which can be removed by evaporation following loading of the aggregate matrix. The aggregates filled with active components could further be coated with bio- or synthetic polymers to improve the performance of the overall product. Additionally, the surface of the granules could be pretreated before aggregate formation by proper agents to improve absorption of the encapsulant. Specially treated potato starch granules were observed to have the capacity to be filled with fragrant compounds up to 30%.[35]

A special technique for microencapsulation of living microorganisms in starch was developed using properties of both starch granules and starch polymers.[36] In this process, partially hydrolyzed granules offered the carrier matrix for the microbes, and after a coating process based on water-dissolved high-amylose cornstarch polymers, the final powdery microcapsules were produced. The presence of high-amylose cornstarch was also recently observed to increase survival of certain health-promoting bacteria at low pH, and during passage through the intestinal tract of mice.[37] Adhesion of the bacteria on the starch granule surface was considered to be a possible mechanism for increased bacterial survival.

A recently reported study demonstrated formation of small crystalline aggregates from, e.g., cornstarch water solution when the solution was slowly cooled after preparation in a jet cooker.[38] Crystalline aggregates were observed to be composed of amylose and suggested to be a result of crystallization of helical inclusion complexes formed from amylose and the native lipids present in cereal starch granules. The cavity of the amylose helices is chemically very similar to the hydrophobic cavity of β-cyclodextrin. Because it is known that β-cyclodextrin stabilizes various flavors (see section 16.3.6), jet cooking may offer a new process to exploit the binding ability of amylose.

Thus, starches offer porous carrier matrices, water-soluble and -insoluble film materials, and a special binding ability of the amylose helical structure, which all might be exploited when developing delivery vehicles for bioactive components. These materials may further be combined with proteins or polyelectrolytes to improve the overall performance.

16.3.3 Pectins

Pectins are anionic polyelectrolytes and resemble alginates. The key feature of all pectin molecules is a linear chain of (1-4)-linked α-D-galactopyranosyluronic acid units.[16] Thus, pectins are polygalactouronic acids, and the chain molecules are negatively charged at neutral pH. The pK values obtained have been in the range of 3.0 to 3.3.[39] Commercial pectins are mainly prepared from citrus peels and apple pomace.

The degree of esterification (DE) varies among pectins and controls gelation and film formation properties. Additionally, sugars have a large influence on pectin gelation because sugar competes with hydration water. Dry pectin coatings developed for drug formulations have been investigated.[40,41] High methyl ester pectins (methoxylation degree > 50%) are less water soluble and have been observed to have more potential in delivery formulations than pectins with lower degrees of esterification.[42] The gelation ability of high methyl esters makes it possible to reduce the penetration of water into the dosage form, and hence the dissolution of an active component incorporated into it. Gels prepared of high methyl ester pectins are also heat stable, which could be a beneficial property when exploiting encapsulated ingredients in processes.

The presence of calcium salts or other multivalent cations makes it possible for low methyl esters (methoxylation degree < 50%) to also form rigid gels.[43] The gelation is believed to occur in a manner similar to that of alginates, with formation of strong interchain binding resulting in the conformation known as the egg-box model. A very useful property of these gels is their stability in solutions with low pH and swellability under slightly alkaline conditions. The functionality of these matrices, however, depends much on the number of methyl ester groups and calcium ion content.

An important technological property related to development of delivery vehicles is that all pectins are resistant to digestion, but are fermented by the colonic microflora.[42] That is why development of pectin-based formulations aiming at colonic drug delivery has recently been reported. Studies on drug release have demonstrated that pectin salts (calcium or other metals) with different solubilities are able to form matrices that offer control of colon drug release.

pH-dependent swelling of pectin gels may be exploited in applications in which the target encapsulant is able to diffuse across the swollen matrix. Gels prepared of high methyl ester pectins may also offer delayed release properties for certain ingredients. The rate of release can be suggested to depend greatly on the molar mass of the encapsulant, as has been observed for the alginate gels.[44] This gives further possibilities to control the release characteristics. Furthermore, the observed colon drug delivery behavior — degradation of the gel and release of the encapsulant in the presence of bacteria — could perhaps be used in certain special food ingredient applications. Additionally, due to their anionic character, pectins could well be mixed with other biopolymers to achieve a wider variety of functionalities.

16.3.4 Other Polysaccharides

Gum arabic has been the standard of excellence as an encapsulating matrix for food ingredients, especially in applications based on spray-drying technology.[4] It is a very good emulsifier, bland in flavor, and provides good retention of volatiles during the drying process. Gum arabic is a complicated polymer and is composed not only of polysaccharides, but also of protein units, which probably explains its emulsifying potential.[16] The overall protein content is about 2%, but specific fractions may contain as much as 25% protein. Due to its high retention property, gum arabic has been much used for flavor carriers, especially for citrus and other flavor oils. An important characteristic is flavor load, which is closely related to retention because higher loadings generally tend to produce poor retention. The load ratio 4:1 (carrier:flavor) has most commonly been used in practice. Thus, gum arabic is well suited for the change of the physical form of an active substance, which normally is a change from a liquid to a solid to improve the processing. A beneficial property, as compared, for example, to protein carriers, is the stability of the emulsion formed by gum arabic under acidic pHs. Another important characteristic is its compatibility with a high concentration of sugars. Additionally, gum arabic is anionic and may react with other polymers.[44] Whether improvement of oxidation stability is achieved by binding food ingredients within gum arabic matrices cannot be answered based on reported literature,[4] in spite of the long history in using gum arabic as an encapsulating matrix. Gum arabic may or may not offer protection against oxidative deterioration, depending upon the gum. Some species are claimed to offer outstanding protection, while others offer little or no protection to the active substance.

Alginate is a linear polysaccharide and is composed of two monomeric units: β-D-mannuronic and α-L-guluronic acids.[45] Alginates prepared from various sources may differ in either molecular weight or proportion of monomers. The content of guluronic acid ranges from 10 to 80%, but the length of the guluronic sequence is the more important functional property, since it strongly influences gel formation. The viscosity of alginate solutions is mainly determined by the size of the molecule. The presence of calcium (or other divalent cations) increases drastically the viscosity due to gel formation, which is believed to occur via bridge formation between calcium and carboxyl, as well as hydroxyl groups of the parallel guluronic chains, resulting in an egg-box arrangement. The gel strength depends on the content of the guluronic units in the alginate polymer and the calcium concentration. Commercial alginate is usually sodium salt of the alginic acid and prepared from brown seaweeds.

The major development of encapsulation processes based on alginates has been with the aim to improve cell viability, but also vitamins, enzymes, and other active components have been of interest.[45] Many technologies exist to apply alginate in encapsulated form, but beads formed by external gelation, i.e., by dropping the alginate-encapsulant solution into a calcium chloride solution, is the most common technique. To achieve protection of the encapsulant against oxygen, denaturation, or light, alginate beads must be dried. Unfortunately, little information is available concerning either direct production of dry beads or drying of wet beads. In addition, the rehydration process of beads has not been dealt with in the literature. However, alginates are interesting, safe, and functional biopolymers offering matrices for food ingredient delivery vehicles, in which delivery may be controlled, e.g., by degree of swelling. Alginates could also be combined with other biopolymers to achieve better performance.

Water-soluble polysaccharides known as *β-glucans* are linear glucose chain polymers composed of 1→4 and 1→3 linkages.[16] Oat and barley brans are the source of commercial β-glucans. β-Glucan-rich fractions from oats were discovered to be able to function as a pH-depending encapsulation matrix for bioactive components such as living bacteria and enzymes.[46] The hydration of the matrix occurred under slightly alkaline conditions, where the encapsulant could also be added. No release under acidic conditions took place because no hydration occurred. Thus, β-glucans have similar functional properties as pectins in that their swelling depends on pH, offering pH-dependent release of the encapsulant from the matrices. It has also been reported that water-soluble polysaccharides from linseed (mucilage) could function as delivery systems for certain cosmetic, therapeutic, or nutritional substances.[47]

16.3.5 Maltodextrins, Syrups, and Sugars

Starch hydrolysis products with a dextrose equivalent (DE) below 20 are called maltodextrins. Maltodextrins have traditionally been used as encapsulation matrices, especially in spray drying. They are inexpensive, bland in flavor, very easily soluble in water (up to 75%), and exhibit low viscosity in solutions.[4,11] Generally in spray drying, maltodextrins have been used in combination with emulsifiers such as gum arabic for the preparation of stable emulsions of hydrophobic ingredients prior to the drying step to achieve best possible performance. Furthermore, it has been observed that the use of high-DE materials — syrups and sugars — together with maltodextrins often results in more stable formulations. There are also studies that have shown that higher-DE matrices give better protection against oil oxidation than lower-DE matrices.[4] This is not in agreement with the general thinking that glass transition temperature controls the stability; a possible explanation is the presence of trace minerals in the carrier matrix, or that the carrier may act as an antioxidant. Matrix porosity has also been suggested to affect oxidation. In addition, mixtures of sucrose and maltodextrins have been used in spray drying. They have also often been used in extrusion encapsulations.[48]

The major shortcomings of starch hydrolysates are a total lack of emulsifying capacity and low retention of volatiles. Caking of the microcapsule powders during storage is an additional problem associated with the higher-DE products. The lack of emulsification is not any problem if a water-soluble substance is the target encapsulant, or if a secondary emulsifier can be used in processing. As mentioned above, the efficiency to inhibit, e.g., oxidation cannot be clearly answered. Maltodextrins and sugars have been combined with gelatin in many commercially available bioactive substance preparations, especially for encapsulation of unsaturated lipids that are easily oxidized. These matrix formulations also contain various additives, such as antioxidants, for improvement of the performance of the powders.

Less commonly practiced encapsulation techniques include co-crystallization of flavors within sugars and adsorption of flavors into microporous carbohydrates such as sugars, of which the potential of the latter has only recently been recognized.[49] Although crystalline sucrose is a poor carrier for flavors, the co-crystallization process has been claimed to improve the stability. The binding of volatile flavors on highly porous carbohydrates is based on physical adsorption, which means reversible condensations of flavor gases onto the surface of solid carbohydrates due to weak attractive forces. Especially high porous sugar matrices using special drying technologies have been developed to function as adsorption carriers.

16.3.6 Starch Derivatives

Starch derivatives that are made more hydrophobic by replacing hydroxyl groups with more lipophilic groups were developed to function alone as microencapsulation matrices for lipophilic flavors. To perform the spray-drying process with enough solids, starch carrier can also be depolymerized. Starch octenyl succinate is one such derivative and, in fact, the only one allowed for emulsifying foods in Europe.[50] Starch octenyl succinates have excellent emulsifying and flavor retention properties, but unfortunately, they do not prevent much oxidation.[4,51,52] The protection efficiency against oxidation can be improved by combining glucose or glucose syrups with the starch derivative.[50] Starch derivatives have been reported to be used as carrier matrices alone or in combination with other carrier carbohydrates or proteins in producing many commercial powdered ingredients, such as fish oils, vitamins, and amino acids. Usually, the commercial products are recommended to be stored under dry and dark conditions and at temperatures below 15°C. The shelf-life reported varies from 18 to 24 months. The encapsulant load is in the range of 10 to 40%. After the package is opened, the product should be used within 1 month.

Cyclodextrins are a special group of carbohydrates that are produced enzymatically from starches and that are cyclic molecules made of glucose units.[48] These molecules have an inner hydrophobic cavity in which several hydrophobic substances can be solubilized. β-Cyclodextrin is approved to be used in food formulations. Garlic and onion oils can be complexed as odorless components by cyclodextrin, and stabilization of fat-soluble vitamins can also be performed with the aid of cyclodextrin.[1] Generally, cyclodextrin complexes can protect ingredients from oxidation, light-induced reactions, thermal decomposition, and evaporation losses.[48] Crystalline cyclodextrin complexes are stable and greatly improve processing performance, handling, and storage of food ingredients. The odorless complexes are claimed to be stable up to temperatures of 200°C. In mouth conditions, however, the dissociation of the complexes occurs. In contrast to starch hydrolysis products, cyclodextrin powders are nonhygroscopic and very heat stable, but the load is only in the range of 6 to 15%, which is one limitation of using β-cyclodextrin as a carrier. The other limitation is the size of the molecule, which has to fit exactly into the cavity.

16.4 FUTURE

Carbohydrates are already used for improvement of performance of many food ingredients, but they offer a much wider variety of functionality than exploited today. Amorphous glassy matrices, film formation capability, accessibility to enzymes or microbes, swelling or dissolution in water and pH dependency, specific binding ability, adsorption capacity, and temperature-dependent gel melting — all these functionalities could find use in specific high-value-added applications. Furthermore, to achieve better performance, carbohydrates may be combined with other biomaterials or modified with the aid of enzymes. One of the keys is to develop feasible technology designed to match the functional properties of carbohydrates. A successful application consists of a combination of proper technology with the tailored carbohydrate.

To make progress in this field, the encapsulation work needs to be performed in close interactive collaboration with experts from many fields, such as chemistry, biochemistry, chemical engineering, and material science. A better understanding of the basic phenomena related with dynamics of bioactive compounds, matrix rheology, and phase behavior would definitely facilitate the development work. If successfully applied, carbohydrates could serve not only as technological improvers of stability and delivery of sensitive food components, but also as nutritionally beneficial food ingredients.

REFERENCES

1. Gibbs, B.F. et al., Encapsulation in the food industry: a review, *Int. J. Food Sci. Nutr.*, 50, 213, 1999.

2. Brazel, C.S., Microencapsulation: offering solutions for the food industry, *Cereal Foods World*, 44, 388, 1999.
3. Vilstrup, P. Ed., *Microencapsulation of Food Ingredients*, Leatherhead Publ. LFRA, Leatherhead, England, 2001.
4. Reineccius, G.A., Multiple-core encapsulation. The spray drying of food ingredients, in *Microencapsulation of Food Ingredients*, Vilstrup, P., Ed., Leatherhead Publ. LFRA, Leatherhead, England, 2001, p. 151.
5. Schultz, T.H., Dimick, K.P., and Makower, B., Incorporation of natural fruit flavors into fruit juice powders. 1. Locking of citrus oil in sucrose and dextrose, *Food Technol.*, 10, 57, 1956.
6. Vinetsky, Y. and Magdassi, S., Microcapsules in cosmetics, in *Novel Cosmetic Delivery Systems*, Magdassi, S. and Touitou, E., Eds., Marcel Dekker, New York, 1999, p. 295.
7. Goubet, I., Le Quere, J.-L., and Voilley, A.J., Retention of aroma compounds by carbohydrates: influence of their physicochemical characteristics and of their physical state, a review, *J. Agric. Food Chem.*, 46, 1981, 1998.
8. Gunning, Y.M. et al., Phase behaviour and component partitioning in low water content amorphous carbohydrates and their potential impact on encapsulation of flavours, *J. Agric. Food Chem.*, 48, 395, 2000.
9. Andersen, A.B. et al., Oxygen permeation through an oil-encapsulating glassy food matrix studied by ESR line broadening using a nitroxyl spin probe, *Food Chem.*, 70, 499, 2000.
10. Vasishtha, N., Microencapsulation: Delivering a Market Advantage, *Prepared Foods*, July 28, 2002.
11. Runge, F.E., Multiple-core encapsulation, in *Microencapsulation of Food Ingredients*, Vilstrup, P., Ed., Leatherhead Publ. LFRA, Leatherhead, England, 2001, p. 133.
12. Thies, C., Microcapsule characterisation, in *Microencapsulation of Food Ingredients*, Vilstrup, P., Ed., Leatherhead Publ. LFRA, Leatherhead, England, 2001, p. 31.
13. Wallack, D.A. and King, C.J., Sticking and agglomeration of hygroscopic, amorphous carbohydrate food powders, *Biotechnol. Prog.*, 4, 31, 1988.
14. Slade, L. and Levine, H., Glass transition and water-food structure interactions, *Adv. Food Nutr. Res.*, 38, 103, 1995.
15. Kenyon, M.M., Modified starch, maltodextrin, and corn syrup solids as wall materials for food encapsulation, in *Encapsulation and Controlled Release of Food Ingredients*, ACS Symposium Series 590, Risch, S.J. and Reineccius, G.A., Eds., ACS, Washington, DC, 1995, p. 42.
16. Whistler, R.L. and BeMiller, J.N., *Carbohydrate Chemistry for Food Scientists*, American Association of Cereal Chemists, St. Paul, MN, 1997.
17. Hoover, R. and Zhou, Y., *In vitro* and *in vivo* hydrolysis of legume starches by α-amylase and resistant starch formation in legumes: a review, *Carbohydr. Polym.*, 54, 401, 2003.
18. Le Blay, G. et al., Enhancement of butyrate production in the rat caecocolonic tract by long-term ingestion of resistant potato starch, *Br. J. Nutr.*, 82, 419, 1999.
19. Ferguson, L.R. et al., Comparative effects of three resistant starch preparations on transit time and short-chain fatty acid production in rats, *Nutr. Cancer*, 36, 230, 2000.
20. Goni, I. et al., *In vitro* fermentation of different types of α-amylase resistant corn starches, *Eur. Food Res. Technol.*, 211, 316, 2000.
21. Ring, S.G. et al., Resistant starch: its chemical form in foodstuffs and effect on digestibility *in vitro*, *Food Chem.*, 28, 97, 1988.

22. Leloup, V.M., Colonna, P., and Marchis-Mouren, G., Mechanism of the adsorption of pancreatic alpha-amylase onto starch crystallites, *Carbohydr. Res.*, 232, 367, 1992.
23. Botham, R.L. et al., A comparison of the *in-vitro* and *in-vivo* digestibilities of retrograded starch, in *Gums and Stabilisers for the Food Industry 7*, Phillips, G.O., Williams, P.A., and Wedlock, D.J., Eds., Oxford University Press, Oxford, 1994, p. 187.
24. Kubik, S. et al., Molecular inclusion within polymeric carbohydrate matrices, in *Carbohydrates as Organic Raw Materials III*, Bekkum, H., Röper, H., and Voragen, F., Eds., VCH Publishers, New York, 1996, p. 169.
25. Höller, O., Molekulare Verkapselung von Peroxysäuren, PCB's und ungesättigten Lipiden mit Stärkederivaten, dissertation, Dusseldorf, 1997.
26. Eden, J., Trksak, R., and Williams, R., Starch Based Encapsulation Process, U.S. Patent 4,812,445, 1989.
27. Doane, W.M., Maiti, S., and Wing, R.E., Encapsulation by Entrapment within Matrix of Unmodified Starch, U.S. Patent 4,911,952, 1990.
28. Ring, S.G. et al., Delayed Release Formulations, WO91/07946, 1991.
29. Milojevic, S. et al., Amylose as a coating for drug delivery to the colon. Preparation and *in-vitro* evaluation using glucose pellets, *J. Controlled Release*, 38, 85, 1996.
30. Rein, H. and Steffens, K.-J., Surface modification of water-insoluble drug particles with starch, *Starch/Stärke*, 49, 364, 1997.
31. Whistler, R.L. and Lammert, S.R., Microporous Granular Starch Matrix Composition, U.S. Patent 4,985,082, 1989.
32. Kobayashi, S., Miwa, S., and Tsuzuki, W., Method of Preparing Modified Starch Granules, Patent publication EP 0539 910 A1, 1993.
33. Zhao, J. and Whistler, R.L., Spherical aggregates of starch granules as flavour carriers, *Food Technol.*, 48, 104, 1994.
34. Whistler, R.L., Porous Particle Aggregate and a Method Therefore, U.S. Patent 5,670,400,090, 1997.
35. Korus, J., Tomasik, P., and Lii, C.Y., Microcapsules from starch granules, *J. Microencapsulation*, 20, 47, 2003.
36. Myllärinen, P. et al., Starch Capsules Containing Micro-organisms and/or Polypeptides and/or Proteins and a Process of Producing Them, World patent WO9952511A1, 1999.
37. Wang, X. et al., The protective effects of high amylose maize starch granules on the survival of *Bifidobacterium* spp. in the mouse intestinal tract, *J. Appl. Microbiol.*, 87, 631, 1999.
38. Fanta, G.F., Felker, F.C., and Shogren, R.L., Formation of crystalline aggregates in slowly-cooled starch solutions prepared by steam jet cooking, *Carbohydr. Polym.*, 48, 161, 2002.
39. Voragen, A.G. et al., Pectins, in *Food Polysaccharides and Their Applications*, Stephen, A.M., Ed., Marcel Dekker, New York, 1995, p. 287.
40. Ashford, M. et al., An evaluation of pectin as carrier for drug targeting to the colon, *J. Controlled Rel.*, 26, 213, 1993.
41. Ashford, M. et al., Studies on pectin formulation for colonic drug delivery, *J. Controlled Rel.*, 30, 225, 1994.
42. Vandamme, T.F. et al., The use of polysaccharides to target drugs to the colon, *Carbohydr. Polym.*, 48, 219, 2002.
43. Edman, P., Kristensen, A., and Wideholt, B., Therapeutical Composition and Process for Its Preparation, WO 92/00732 Patent, 1992.

44. King, A.H., Encapsulation of food ingredients, in *Encapsulation and Controlled Release of Food Ingredients*, ACS Symposium Series 590, Risch, S.J. and Reineccius, G.A., Eds., ACS, Washington, DC, 1995, p. 26.
45. Poncelet, D. and Markvicheva, E., Multiple-core encapsulation, microencapsulation and alginate, in *Microencapsulation of Food Ingredients*, Vilstrup, P., Ed., Leatherhead Publ. LFRA, Leatherhead, England, 2001, p. 215.
46. Laakso, S.W. and Kangas, P.M., Menetelmä kauran käyttämiseksi kapselointiaineena, FI 914491 Patent, 1993.
47. O'Mullane, J.E. and Hayter, I.P., Linseed Mucilage, WO9316707 Patent, 1993.
48. Shahidi, F. and Han, X., Encapsulation of food ingredients, *Crit. Rev. Food Sci. Nutr.*, 33, 501, 1993
49. Zeller, B.L., Saleeb, F.Z., and Ludescher, R.D., Trends in development of porous carbohydrate food ingredients for use in flavour encapsulation, *Trends Food Sci. Technol.*, 9, 389, 1999.
50. Simon, J., Emulsifying Starches for Flavor Encapsulation, *The World of Ingredients*, November/December 19, 1998.
51. Partanen, R. et al., Microencapsulation of caraway extract in β-cyclodextrin and modified starches, *Eur. Food Res. Technol.*, 214, 242, 2002.
52. Partanen, R, e al., Effect of relative humidity on the oxidative stability of microencapsulated sea buckthorn seed oil, *J. Food Sci.*, 70, E37, 2005.

17 Food Regulations: Health Claims for Foods Fortified with Carbohydrates or Other Nutraceuticals

Jerzy Zawistowski

The real culprit of a disease is a poor diet.
—Hippocrates

In memory of Max

CONTENTS

17.1 INTRODUCTION

The past decade has seen an increase in consumer awareness to the benefits of healthy eating and, to a certain extent, self-medication through a specific diet, which includes the consumption of nutraceutical/functional foods. An offshoot of this activity has been the growth of numerous business opportunities for the food and pharmaceutical industry sectors. Industry and academia alike have been engaged in both product and process innovation research to yield new functional foods/nutraceuticals to satisfy customers' needs. The world market for nutraceutical products was estimated to gross an approximate $56 billion in 2001 alone, which represented 37% of the global nutrition industry. Three major markets: U.S. (33.2%), Europe (32.3%), and Japan (25.1%) accounted for over 90% of all functional food sales in 2001. The estimated growths of functional food sales in 2001 and 2002 were 7.3 and 9%, respectively, with growth as large as 35% for products such as fish oil and omega-3 fatty acids in 2002.[1]

The global regulatory framework governing foods is changing to include specific regulatory venues for functional foods and nutraceuticals. This is mainly in response to new trends and transformations of both the food industry and consumer eating patterns, which are difficult to accommodate under existing regulations. However, these changes are slow, and at this time, the regulatory framework is still very restrictive, which means the marketing of functional foods is severely impaired.

For centuries, an existing link between health, food, and diet has been well recognized. The functionality of carbohydrates was described ages ago by the philosopher Hippocrates, who advised men to eat bread made from bran containing whole grain meal due to its "salutary effects upon bowels."[2] In America, the consumption of bran (fiber) was encouraged as early as the 19th century. Physicians recommended this to persons who had diets rich in fat and simple sugar, and to those that complained about indigestion and constipation.[3]

In China, the functionality of foods was recognized as early as the seventh century. Tea drinking, which gained popularity during the Tang Dynasty (618 to 907), gave impetus for the Chinese to research and, consequently, introduce 10 main health functions associated with drinking tea. Some of these claims, such as "tea is beneficial to health and able to relieve fatigue" or "tea can be used to eliminate toxins from the body," are still recognized today by modern consumers.[4]

In spite of the health benefits that have been associated with food for many centuries, the concept of functional foods is a relatively new one. It was introduced in late 1980s in Japan and was followed in 1991 by the institution of an approval system for this category of foods. Japan is the only country that has legally defined functional foods, and the system has been established to help promote healthy foods with proper health claims. In 1995, Raiso Oy, the Finnish food company, introduced margarine containing phytosterols, marking the introduction of the first truly functional food introduced to the European market. This created havoc among global regulators who were not prepared for this new type of food. Since this date, many countries have tried to define functional foods as well as create a new regulatory venue to control the marketing and consumption of these kinds of products.

There is much controversy behind how functional foods should be discriminated from conventional foods. For the majority of consumers, orange juice is orange juice. The same orange juice with clinical support, however, is considered a functional food. In fact, some clinical studies show that flavonoids present in orange juice may favorably change a human lipid profile upon consumption.[5] Yet, juice enriched with calcium is a drug in countries like Canada, as evidenced by the regulatory requirement to place a Drug Identification Number (DIN) on the product. So, what are functional foods?

17.2 DEFINITIONS

Generally, functional foods can be defined as any food product that is marketed or perceived to deliver a health benefit in addition to nutritional value. A more specific definition, however, was proposed by Health Canada in 1998. It defines functional foods as food that is similar in appearance to conventional foods and is consumed as part of a usual diet, and has demonstrated physiological benefits or reduces the risk of chronic disease beyond basic nutritional functions. To distinguish between functional foods and nutraceuticals, Health Canada defines nutraceuticals as products produced from foods that are sold in pills, powders, potions, and other medicinal forms that are not generally associated with food, but that also demonstrate physiological benefits or protection against chronic diseases.[6] In late 2001, the latter definition was used to describe natural health products.

Stephen DeFelice of the Foundation for Innovation in Medicine introduced a definition of functional foods in the U.S. He defined a functional food as "any substance that is a food or part of a food that provides medical and/or health benefits, including the prevention and treatment of disease." This includes nutraceuticals, dietary supplements, and medical foods.[7]

Since unbalanced nutrition may lead to chronic diseases that take a long time to correct through a dietary approach, one of the important properties of functional foods is the ability to enhance or assist in the maintenance of human physiological conditions adjusted to a normal state in a relatively short period. These properties are the basis of the Japanese definition of Tokohu, known outside of Japan as Foods for Specified Health Use (FOSHU). As opposed to other definitions, the definition of FOSHU, Japanese functional foods, captures the essence of what makes foods functional in a more explicit way: "a food (not a capsule, tablet or powder) derived from naturally occurring ingredients that can and should be consumed as part of the daily diet." The definition[8] states that functional foods also have particular functions when ingested, serving to regulate a particular body process, such as:

1. Enhancement of the biological defense mechanisms
2. Prevention of a specific disease
3. Recovery from a specific disease
4. Control of physical and mental conditions
5. Slowing of the aging process

It is worthwhile to note that "prevention of diseases" is classified as a drug claim in the U.S and many other jurisdictions.

Functional foods can be roughly categorized into three groups:

1. Raw foods that contain endogenous constituents with health benefits, including tomatoes that contain lycopene, which may reduce the risk of cancer
2. Processed foods, such as oat bran cereal, which may reduce serum cholesterol
3. Processed foods with added ingredients, such as calcium-enriched orange juice, to reduce the risk of osteoporosis

17.3 JAPANESE FUNCTIONAL FOODS: TOKUHO SYSTEM

One of the very first functional foods in Japan was dietary fiber in the form of a soft drink called Fiber-Mini. The drink, which used water-soluble polydextrose as its functional ingredients and claimed to improve "gut's regulation," was launched in 1980 by Otsuka Pharmaceutical. The marketing success of this product led to a high demand for drinks with high levels of fiber and the start of the functional foods regulatory system in Japan.[9]

The concept of functional foods and food functions has been extensively studied under the sponsorship of Japanese government. Two years of studies (1984 to 1986) on "systematic analysis and development of functions of food" followed by studies (1988 to 1990) on "analysis of functions for adjusting physical conditions of the human body with food" led to the identification of the tertiary function of foods, which, unlike the conventional (primary, nutrition; secondary, taste) functions, is directly involved in the modulation of human physiological systems such as the immune and digestive.[10] It has become clear that food can be designed not only to satisfy primary functions, but also for adjusting conditions of the human body's function. The concept of physiologically functional foods (functional foods) was thus born and consequently led to the creation of the Japanese regulatory system of this category of foods. In 1993, the Ministry of Health and Welfare established a policy to regulate Foods for Specified Health Use (FOSHU), or Tokuho in its Japanese abbreviation. Under this system, the use of health claims for some selected functional foods is legally permitted.[11]

Unlike in Canada and some other countries, the Japanese definition for food — "foods are processed containing ingredients that aid specific bodily functions in addition to being nutrition" — allows the Japanese to use health claims, which in other countries would be allowed only for drugs.

Under the Japanese Nutrition Improvement Law, there are five categories of Foods for Specified Health Use:[11]

1. Foods for the ill
2. Milk powder for pregnant or lactating women
3. Formulated milk powder for infants
4. Foods for the aged
5. Foods for Specified Health Use (FOSHU)

FOSHU is the largest food group in Japan with health claims. As of October 2003, there were 398 foods approved under the FOSHU regulations.[12] Functional carbohydrates represent a significant portion of bioactive components of FOSHU (Table 17.1). Specifically, oligosaccharides and indigestible starch are functional ingredients of prebiotic and synbiotic foods to "help maintain good gastro-intestinal condition." Dietary fiber is used widely in FOSHU foods to "help improve bowel movements" and "maintain a healthy cholesterol level." It is also considered to be good for those who have "mildly higher blood glucose."

A group of fructooligosaccharides (FOSs) is an example of a dietary fiber that has been approved under the FOSHU regulations for use in a wide range of functional foods. Two different classes of FOS mixtures are produced commercially, based on inulin degradation or transfructosylation processes. In Japan, FOS is produced through the transfructosylating action of β-fructofuranosidases of *Aspergillus niger* on sucrose. This class of FOS is manufactured by Meiji Seika Kaisha Ltd. and marketed as Meioligo, Neosugar, Profeed, Actilight, or Nutraflora.[13]

Meioligo is used as a prebiotic, sweetening agent, flavor enhancer, and bulking agent. As a prebiotic, Meioligo is used in functional dairy drinks (prebiotics, synbiotics) to promote the growth of beneficial bacteria in the gastrointestinal tract. Because FOS molecules are not digested and thus not absorbed in the upper part of the gastrointestinal tract (or if absorbed, the quantity is negligible), they arrive unchanged to the colon, where the beneficial bacteria, most notably the bifidobacteria, utilize them as select food for growth and proliferation. Meioligo is also used as a low-calorie sucrose replacement in products for diabetics, such as cookies, cakes, breads, candies, table sugar, dairy products, and some beverages.[14] The approval of FOSs in Japan prompted the establishment of an acceptable daily intake of about 0.8 g/kg of body weight/day.

In addition to FOSs, there are a number of other oligosaccharides that are approved under the FOSHU for use as prebiotics. This list includes soy-oligosaccharides, galacto-, xylo-, and isomalto-oligosaccharides, as well as lactosucrose used in "one-a-night plain" soft drink. FOSHU approved dietary fiber includes polydextrose, indigestible dextrin, wheat bran, partially depolymerized guarana, and a psyllium seed coat. Chitosan is another example of functional carbohydrates bearing the FOSHU approval. This is a nondietary fiber that is extracted from the shell of crustaceans, such as crab and shrimp. Food products containing chitosan gain popularity among Japanese consumers for their ability to prevent fat absorption and

TABLE 17.1
Approved Foods for Specified Health Use as of October 2003[12]

Health Claims Permitted	Functional Ingredients	Number of FOSHU in a Specific Category	Percentage of Total FOSHU
Helps maintain good gastro-intestinal condition Helps improve bowel movement	Dietary fiber, oligosaccharides, indigestible starch, lactobacillus, bifidobacterium	196	49%
Helps prevent from accumulation of body fat Good for those who have higher serum cholesterol/triglycerides	Dietary fiber, soy protein, peptides, diacylglycerol, plant sterols/stanols esters, medium-chain triglycerides, tea catechins, EPA/DHA	66	17%
Good for those who have mildly higher blood glucose	Dietary fiber, albumin, polyphenols, L-arabinose	49	12%
Good for those who have mildly higher blood pressure	Peptides (Lacto-Tri, sardine, bonito, casein), tea glucosides, isoleucyltyrosine	38	10%
Helps maintain strong and healthy teeth May restrain mineralization and helps promote re-mineralization	Xylitol, polyols, tea polyphenols, CPP-ACP, Pos-Ca	26	7%
Helps improve absorption of calcium Good for those who have mild anemia	CPP, CCM, soy isoflavanoids, MBP, polyglutamin acid, V.K2, heme iron	23	6%
Total		398	100%

weight gain. Recently, Nissin Oil Mills Ltd. introduced an instant cup of noodle soup to the Japanese market that contains chitosan as the functional ingredient. This product bears the following health label: "Chitosan has the effect of inhibiting absorption of cholesterol and lowering blood cholesterol."[15]

The FOSHU approval process is quite complex and typically takes at least 1 year to complete. Although the Ministry of Health, Labour and Welfare (MHLW) is technically responsible for supervising the FOSHU process, the Japan Health Food and Nutritional Association (JHFNA), a private organization, actually administers the process. Subsequently, the applicant needs to consult JHNFA to review an application prior to submission to the local prefectural health authorities and MHLW. The application must contain scientific evidence (e.g., results of clinical studies) of the ingredient's efficacy and safety, physical and chemical characteristics, compositional analysis, and relevant analytical methods, as well as the basis for the recommended dose that will provide the efficacy from the ingredient. The application

TABLE 17.2
Information and Documentation Required in FOSHU Application[12]

1.	Sample of the entire package with labels and claims for which approval is sought
2.	Explanation of how the food contributes to the improvement, maintenance, or enhancement of human health
3.	Recommended daily intake of the product and its functional components
4.	Documentation that shows clinical and nutritional proof of the product's efficacy within recommended daily intake: 1) *In vitro* and *in vivo* animal studies explaining the effects, mechanisms of action, and metabolism and pharmacokinetics (absorption, distribution, metabolism, and excretion) on the constituents concerned, results of which should be statistically significant 2) Clinical studies using the food applied and confirming the specified health effects and the recommended daily intake, results of which should be statistically significant 3) Randomized, double-blind, controlled clinical study using local subjects may be required
5.	Documentation demonstrating the safety of the food and its constituents: 1) *In vitro* and *in vivo* animal studies confirming basic information on the limits of safe intake 2) Clinical studies confirming the safety in the case of excessive consumption 3) 3 to 5 times higher level of recommended daily intake of the food or its constituents concerned
6.	Documentation on the stability of the food and its constituents
7.	Documentation of the physicochemical properties of the food's constituents concerned, and description of the analytic methods used: 1) The identification of the food's constituents in question may be required
8.	Results of qualitative and quantitative determination of the food's constituents concerned, and description of the analytic methods used
9.	Results of tests determining the constituents concerned and nutrients, and energy value of the food
10.	Description of the quality control system explaining the facilities, equipment, and production methods used in manufacturing the food

must be accompanied by scientific publications in relevant peer-reviewed journals showing results to substantiate the health claim. After regulatory review, the approved foods are eligible to carry a label indicating the specified health benefit. The approval of the MHLW must also be displayed on the label of FOSHU food products. The documentation required for FOSHU approval is listed in Table 17.2.[12]

It is important to emphasize that the FOSHU approval process is voluntary, and approved products represent a small portion of the Japanese health food market. Nonapproved FOSHU functional foods, many of which contain approved functional components, are sold legally on the Japanese market without specific health claims. A lot of manufacturers sell functional foods as non-FOSHU products to avoid rather cumbersome, lengthy, and very expensive approval processes. In addition, many companies prefer to conduct market assessment for specific products before applying for the FOSHU approval. In 2003, FOSHU products accounted for about U.S.$4.1 billion of the total health food market, while other, non-health-claim functional foods represented the largest portion of the sector at an estimated U.S.$8 billion.[16]

17.4 U.S. REGULATORY FRAMEWORK FOR FUNCTIONAL FOODS

Although there is no legal definition for a functional food, nor are there specific regulations to govern functional foods in the U.S., the Food and Drug Administration (FDA) overlooks a number of regulatory policies that can be used as venues to approve functional foods and nutraceuticals for the American market. The FDA does not distinguish functional foods from conventional foods, since both categories, if qualified, may carry a health claim on their label. Any approval of functional foods and nutraceuticals has to be on a case-by-case basis under existing food regulations (Food, Drug, and Cosmetic Act) or the newly introduced regulation for dietary supplements (Dietary Supplement Health and Education Act (DSHEA)).

17.4.1 FOOD SAFETY: GENERALLY RECOGNIZED AS SAFE (GRAS)

Manufacturers of functional foods have to obtain premarket approval and demonstrate that all food additives, including functional ingredients, are generally recognized as safe (GRAS). If functional foods contain any non-GRAS ingredients, the potentially unsafe ingredient must undergo the FDA's approval process for new food ingredients. This requires manufacturers to conduct studies in food safety and submit the results of these experiments to the FDA for review before the ingredient can be used in marketed products. Based on a review of results, the FDA either authorizes or rejects the food's ingredients. Historically, to obtain a GRAS status for a food additive, the manufacturer must apply for approval under the Direct Food Additive Petition (21 CFR 171.1–171.8, 171.100–171.130, and 184.1).[17,18] This process is rather cumbersome, taking between 2 and 6 years for the GRAS status to be granted. In April 1997, however, the GRAS notification process was introduced to help speed up approval for food ingredients.[19] This is the most suitable system for approval of functional foods in the U.S. This approval process requires submission of a document (dossier) that contains safety and efficacy data, which is prepared by the GRAS panel. All pivotal data must be published prior to the application. The GRAS panel should be composed of at least two experts qualified by scientific training and experience to evaluate the safety of food and food ingredients using scientific procedures. The panel must be recognized by the FDA and should be independent of the company seeking approval. The FDA reserves 90 days to review submission and respond to the applying company by sending a letter of no objection or rejection. The functional ingredient in question has to be presented in context of the final food product. If needed, subsequent applications for other food products containing an ingredient under review may be prepared under the GRAS self-affirmation system. The latter approach is a policy of the FDA but not part of the Code of Federal Regulations (CFR). Health claims are not permitted unless a submission is made to the FDA under the Nutritional Labeling and Education Act (NLEA). Table 17.3 provides the information required for the GRAS notification approval process (21 CFR 170.35).[20]

TABLE 17.3
Requirements for the GRAS Notification Approval Process (21 CFR 170.35)[20]

	Requirements of the Proposed Rule
1.	Detailed information about the identity of the notified substance, composition, method of manufacture, characteristic properties, and specifications
2.	Information on any self-limiting levels of use
3.	Comprehensive discussion of, and citations to, generally available and accepted scientific data and information, including consideration of probable consumption
4.	The basis for concluding that there is a consensus of qualified experts that there is reasonable certainty that the substance is not harmful under the intended conditions of use

17.4.2 Health Claims

Prior to the introduction of functional foods in the American market, the U.S. was the first country to institute a law allowing health claims for constituents present in foods. The link between diet and health was a major topic of the White House Conference on Food, Nutrition, and Health in 1969.[21] Since then, consumers' interest in this area has grown. The issues of health, food safety, and nutrition labeling were the focus of attention for both the FDA and the U.S. Department of Agriculture (USDA) in the 1980s. Health messages on food labels were the subject of national debate, leading to mandatory nutrition labeling.

17.4.2.1 Nutrition Labeling and Education Act (NLEA)

In 1990, the American Congress passed the Nutrition Labeling and Education Act (NLEA) to ensure that food labels provide consumers accurate and reliable information regarding the nutritional and health values of foods.[22] The act states:

> The Secretary shall promulgate regulations authorizing claims characterizing the relationship of a nutrient to a disease or health-related condition which is diet-related only if the Secretary determines, based on the totality of publicly available scientific evidence (including evidence from well-designed studies conducted in a manner which is consistent with generally recognized scientific procedures and principles), that there is significant scientific agreement, among experts qualified by scientific training and experience to evaluate such claims, that the claim is supported by such evidence.

The NLEA clarified and strengthened the authority of the Food and Drug Administration to require nutrition labeling on foods, as well as the establishment of circumstances under which claims may be made about nutrients pertaining to foods. The legislation also provided a process for the orderly regulation of disease claims. This act describes the relationship between the food constituent and the disease or health-related condition. The health claim can be used only if it meets the requirements of the act (21 CFR 101.14):[23]

TABLE 17.4
Disqualifying Nutrient Levels[23]

Disqualifying Nutrients	Foods	Main Dishes	Meal Products
Fat	13 g	19.5 g	26 g
Saturated fat	4 g	6 g	8 g
Cholesterol	60 mg	90 mg	120 mg
Sodium	480 mg	720 mg	960 mg

Note: Food must contain less than the specified levels to make a health claim (21 CFR 101.14).

1. The health claim must be about the link between diet and certain serious chronic disease conditions.
2. The statement must be specific to describing the health benefit the compound, upon ingestion, has on the risk reduction and prevention of a particular chronic disease condition.
3. The health message must be made in the context of the total diet.
4. The health claim must be based on scientific evidence that is reviewed and accepted by the FDA.

In addition, food products that contain a specific health compound may carry a health claim on a label only if they do not contain excess amounts of undesirable components as specified by regulation (disclosure/disqualifier levels), i.e., they fulfill the "jelly bean rule" (21 CFR 101.14).[23] The undesirable components are total fat, saturated fat, cholesterol, and sodium. Food must contain less than the specified disqualifying levels of these four nutrients (Table 17.4). The disqualifying levels are the amount in the reference serving size, the label serving size, or (for foods with a reference size of 30 g or less or 2 tablespoons or less) the amount per 50 g. To qualify for a health claim, a food product must provide nutritional value to consumers. Therefore, it must contain, without fortification, at least 10% of the daily value for one of six nutrients, such as vitamin A, vitamin C, iron, calcium, protein, or fiber, per reference amount (Table 17.5). This rule is known as the jelly bean rule because jelly beans are free of disqualifying nutrients and, in theory, may carry a health claim. However, jelly beans are not nutritious and do not qualify as an appropriate food for the risk reduction of diseases.

There are currently 12 health claims approved under the NLEA rules (21 CFR 101.72–101.83).[24] It is worthwhile to note that 11 of these claims are in the form of final rulings, while the last claim, which links consumption of plant sterols to risk reduction of coronary heart disease (CHD), is still in purgatory and waits for the final version (Table 17.6). Four of the NLEA health claims are associated with the consumption of carbohydrates and the risk reduction of hearth disease and some types of cancers.

Fiber-containing foods have captured the interest of researchers for a long time. For almost 100 years, indigestible carbohydrates (equivalent to dietary fiber) were

TABLE 17.5
Jelly Bean Rule

Nutrient	DV	Nutrient	DV
Vitamin A	500 IU	Calcium	100 mg
Vitamin C	6 mg	Protein	5 g
Iron	1.8 mg	Fiber	2.5 g

Note: Food must contain, without fortification, 10% or more of the daily value (DV) for one of six nutrients, as specified in the table (dietary supplements excepted).

recognized to have positive effects on bodily functions. However, over the last 30 years, researchers proposed that this compound could be protective against many chronic diseases, such as cardiovascular disease, diabetes, and disorders of the gastrointestinal tract, including cancer of the colon. The FDA found this evidence nonconclusive, and petitions for approval of the dietary fiber and cancer claim, as well as the cardiovascular disease claim, have not been authorized and remain on a list of unauthorized claims (21 CFR 101.71).[24] However, the FDA recognized a health importance of fiber as a part of the diet, "fiber food," and approved two health claims that link fiber as a part of food products with the risk reduction of cancer and coronary heart disease:[24]

1. Fiber-containing grain products, fruits, and vegetables and cancer (21 CFR 101.76)
2. Fruits, vegetables, and grain products that contain fiber, particularly soluble fiber, and risk of coronary disease (21 CFR 101.77)

Both claims may be used only on grain, fruits, and vegetables that contain fiber without fortification. The type of fiber is not specified, but food must contain enough fiber to qualify as a good source of fiber (10 to 19% of the Daily Reference Values (DRVs) or 2.5 to 4.9 g of fiber per reference serving size) for the cancer claim to pertain, and at least 0.6 g of soluble fiber per reference amount (RA) for the coronary heart claim to pertain. The food itself must comply with the requirements discussed above, such as those regarding fat and cholesterol.

Dietary fiber is also the subject of the third claim, which links fruits and vegetables with the risk reduction of some types of cancer:[24]

3. Fruits and vegetables and cancer (21 CFR 101.78)

In this claim, fiber is one of the food constituents that, along with antioxidants such as vitamin A and vitamin C, may play an important role in the prevention of cancer.

In 1997, the FDA allowed for another health claim linking "soluble fiber from whole oats and risk of coronary health disease" (21 CFR 101.81).[24] In allowing this

TABLE 17.6
Generic Health Claims Approved under the NLEA Regulation in the U.S.[24]

Approved Claims	Model Claim	21 CFR
Calcium and osteoporosis	Regular exercise and a healthy diet with enough calcium help teens and young adult white and Asian women maintain good bone health and may reduce their high risk of osteoporosis later in life	§101.72
Dietary lipids and cancer	Development of cancer depends on many factors; a diet low in total fat may reduce the risk of some cancers	§101.73
Sodium and hypertension	Diets low in sodium may reduce the risk of high blood pressure, a disease associated with many factors	§101.74
Dietary saturated fat and cholesterol and risk of coronary heart disease	While many factors affect heart disease, diets low in saturated fat and cholesterol may reduce the risk of this disease	§101.75
Fiber-containing grain products, fruits, and vegetables and cancer	Low-fat diets rich in fiber-containing grain products, fruits, and vegetables may reduce the risk of some types of cancer, a disease associated with many factors	§101.76
Fruits, vegetables, and grain products that contain fiber, particularly soluble fiber, and risk of coronary heart disease	Diets low in saturated fat and cholesterol and rich in fruits, vegetables, and grain products that contain some types of dietary fiber, particularly soluble fiber, may reduce the risk of heart disease, a disease associated with many factors	§101.77
Fruits and vegetables and cancer	Low-fat diets rich in fruits and vegetables (foods that are low in fat and may contain dietary fiber, Vitamin A, or Vitamin C) may reduce the risk of some types of cancer, a disease associated with many factors; broccoli is high in vitamin A and C, and it is a good source of dietary fiber	§101.78
Folate and neural tube defects	Healthful diets with adequate folate may reduce a woman's risk of having a child with a brain or spinal cord defect	§101.79
Dietary sugar alcohols and dental caries	**Full claim**: Frequent between-meal consumption of foods high in sugars and starches promotes tooth decay; the sugar alcohols in [name of food] do not promote tooth decay **Shortened claim** (on small packages only): Does not promote tooth decay	§101.80

Soluble fiber from certain foods and risk of coronary heart disease	Soluble fiber from foods such as [*name of soluble fiber source and, if desired, name of food product*], as part of a diet low in saturated fat and cholesterol, may reduce the risk of heart disease; a serving of [*name of food product*] supplies __ grams of the [necessary daily dietary intake for the benefit] soluble fiber from [*name of soluble fiber source*] necessary per day to have this effect	§101.81
Soy protein and risk of coronary heart disease	(1) 25 grams of soy protein a day, as part of a diet low in saturated fat and cholesterol, may reduce the risk of heart disease; a serving of [*name of food*] supplies __ grams of soy protein (2) Diets low in saturated fat and cholesterol that include 25 grams of soy protein a day may reduce the risk of heart disease; one serving of [*name of food*] provides __ grams of soy protein	§101.82
Plant sterol/stanol esters and risk of coronary heart disease	(1) Foods containing at least 0.65 gram per serving of vegetable oil sterol esters, eaten twice a day with meals for a daily total intake of at least 1.3 grams, as part of a diet low in saturated fat and cholesterol, may reduce the risk of heart disease; a serving of [*name of food*] supplies __ grams of vegetable oil sterol esters (2) Diets low in saturated fat and cholesterol that include two servings of foods that provide a daily total of at least 3.4 grams of plant stanol esters in two meals may reduce the risk of heart disease; a serving of [*name of food*] supplies __ grams of plant stanol esters	§101.83

claim, the FDA concluded that there is enough scientific evidence to support the relationship between soluble fiber from whole oats and the risk reduction of CHD.

Oats are an excellent source of beta-glucan soluble fiber, which upon consumption forms a gel in the intestines. Subsequently, fiber gel interferes with the absorption of cholesterol in the intestine and helps to lower blood cholesterol levels. In allowing the whole oats–CHD health claim, the FDA assumed that soluble fiber from sources other than whole oats could positively modify blood lipids and consequently lower the risk of heart disease. However, since soluble dietary fibers are a family of heterogeneous substances that vary greatly in their effect on risk of CHD, a case-by-case approach was necessary to evaluate petitions for this health claim. Nevertheless, this health claim was amended in 1998 to "soluble fiber from certain foods and risk of coronary health disease." The authorization of the amendment was a response to a petition submitted by the Kellogg Company involving soluble fiber from psyllium seed husk. Psyllium, known also as blood or Indian psyllium, was introduced by Kellogg in a cereal product called Heartwise and in Kellogg's Bran Buds, as well as in a number of dietary supplements, such Metamucil and Fiberall. Once again, the FDA's decision was based on strong scientific evidence submitted by Kellogg with a petition. The FDA evaluated placebo-controlled clinical studies that tested an intake of over 10 g of psyllium (about 7 g of soluble fiber) per day as part of a diet low in saturated fat and cholesterol. These studies showed consistently significant blood total and LDL cholesterol-lowering effects. Foods carrying the health claim must provide at least 1.7 g of soluble fiber from psyllium per reference amount customarily consumed in the product. This single-serving size, multiplied by four eating occasions per day, totals the 7 g/day intake of the controlled studies.

On April 21, 2001, Pepsico's Quaker Oats Co. and the Cranbury, NJ-based Rhodia, Inc., petitioned the FDA to amend the 1998 modified health claim again, reporting that the subject of the fiber health claim was broader than what available evidence supported. They asked that this amendment be made with specific reference to Oatrim, known under the Quaker/Rhodia brand name Betatrim, which is processed either by alpha-amylase enzymes or by acid–base hydrolysis and has a beta-glucan soluble fiber content between 4 and 25%. Once again, the FDA amended the soluble fiber health claim linking fiber to a reduced risk of coronary heart disease (CHD). The amended claim included, in addition to the broad term *soluble fiber*, "soluble fraction of alpha-amylase hydrolyzed oat bran or whole oat flour with a beta-glucan content of up to 10 percent." The final rule of this health claim, "soluble dietary fiber from certain foods and coronary heart disease," was published on July 28, 2003, in the *Federal Register*.[25]

The above discussion about the soluble fiber claim indicates the interactive nature of the NLEA regulatory system. With proper scientific support, including clinical studies, some companies use this as a mechanism to gain approval for health claims in support of functional foods. The latest example of this interaction between the FDA and the food industry was an FDA-approved petition for a health claim linking consumption of sterols and stanols with the risk reduction of coronary heart disease. Although this claim is still an interim health claim, the ruling allows it to be used

TABLE 17.7
Health Claims Authorized Based on Authoritative Statements by Federal Scientific Bodies[26]

Approved Claims	Claim Requirements	Docket No.
Whole grain foods and risk of heart disease and certain cancers	Diets rich in whole grain foods and other plant foods and low in total fat, saturated fat, and cholesterol may reduce the risk of heart disease and some cancers	99P-2209
Potassium and the risk of high blood pressure and stroke	Diets containing foods that are a good source of potassium and that are low in sodium may reduce the risk of high blood pressure and stroke	00Q-1582

on products produced by McNeal/Raiso Oy. Unilever, Cargill, ADM, and Forbes Medi-Tech, Inc.

17.4.2.2 U.S. FDA Modernization Act

Further opportunity for food manufacturers to use health claims on functional foods came with the introduction of the FDA Modernization Act in 1997. Accordingly, health claims can be used based on authoritative statements published by a scientific institution of the U.S. government, such as the National Institutes of Health, the Centers for Disease Control and Prevention, and the National Academy of Sciences and its divisions.[26] These types of claims do not require approval by the FDA; however, the agency must be notified at least 120 days in advance of marketing a product that carries a potential health claim. During this period, the FDA may authorize a claim or object to it by issuing an interim final regulation prohibiting the claim. These claims must be solidly based on significant scientific evidence, similar to the NLEA claims. So far, the FDA has authorized two claims based on authoritative statements by federal scientific bodies (Table 17.7). One of these claims is associated with dietary fiber and links "whole grain foods and risk of heart disease and certain cancers." A food product may be illegible to carry a health claim on a label if it contains 51% or more of whole grain ingredients by weight per RA, and contains at least 3.0 g/RA of 55 g, 2.8 g/RA of 50 g, 2.5 g/RA of 45 g, or 1.7 g/RA of 35 g of dietary fiber. To qualify for such a claim, a food product must comply with the jelly bean and disclosure/disqualifier rules specified under NLEA regulations.

17.4.2.3 Dietary Supplement Health and Education Act (DSHEA)

In 1994, the U.S. Congress passed the Dietary Supplement Health and Education Act (DSHEA), which set up a new framework for FDA regulations concerning dietary supplements.[27] Traditionally, dietary supplements referred to a narrow group of nutrients such as vitamins and minerals. The act broadens the definition to include, with some exceptions, any product intended for ingestion as a supplement to the diet. This includes, in addition to vitamins and minerals, herbs, botanicals, plant-derived compounds, proteins, peptides, amino acids, fatty acids, carbohydrates, and other nutrients and constituents. Typically, dietary supplements are sold in medicinal forms, including tablets, capsules, soft gels, gel caps, powders, and liquids. However, this act allows a dietary supplement to be introduced into conventional food forms. A dietary supplement may be a product with physical attributes (e.g., product size, shape, taste, packaging) that are essentially the same as conventional food, so long as it is not represented for use as a conventional food. Thus, whether a product is a dietary supplement or a conventional food all depends on how it is labeled. For example, a product in the form of a bar that is labeled a dietary supplement, but also bears a label statement with reference to a snack food or cereal bar, would be subject to regulation as a conventional food. However, a cereal bar that bears a label referring to a dietary supplement and does not represent itself as a breakfast food or use the term *cereal* as a statement of identity would be classified as a dietary supplement.

17.4.2.3.1 Structure and Function Claims

DSHEA allows for the use of structure and function claims, which describe the role of a nutrient or dietary ingredient intended to affect the structure or function in humans.[28] Under this legislation, structure and function claims for dietary supplements require postmarket notification, with manufacturers obliged to substantiate their claims. The label must also bear an FDA disclaimer such as: "This statement has not been evaluated by the Food and Drug Administration. This product is not intended to diagnose, treat, cure, or prevent disease." Other information, such as directions for use and supplement facts (serving size, amount, list of ingredients including active ingredient), is also required on the label.[28]

Unlike the new food additives that have to undergo the FDA GRAS approval process, FDA review and approval of supplement ingredients are not required prior to marketing. To market a dietary supplement, manufacturers must notify the FDA at least 75 days before putting the product on the market, as well as provide information supporting the safety of the ingredient.[29] Safety means that a new ingredient does not present a significant risk of illness or injury under conditions of use recommended in the product's labeling. If the FDA feels a product may be harmful, it bears the onus of proving the danger.

The most important characteristic of DSHEA is that it allows the use of label structure–function claims for dietary supplements, including the use of publications or advertisements in connection with the sale of dietary supplements.[28] These claims link the relationship between nutrient and disease or any such related condition. They should not refer to the diagnosis, mitigation, cure, treatment, or prevention of disease. The following are examples of structure–function claims:

1. Fiber maintains bowel regularity
2. Maintains healthy cholesterol level
3. Use as a part of your diet to help maintain healthy blood sugar level
4. Calcium builds strong bones

Although the structure–function claims may be used without FDA authorization, manufacturers of dietary supplements that bear a label claim must inform the FDA no later than 30 days after the product is marketed.

17.4.2.3.2 Qualified Health Claims

Furthermore, dietary supplements are allowed to use qualified health claims. The use of this type of claim resulted from the U.S. Appeals Court of the D.C. Circuit 1999 decision in the case of *Pearson v. Shalala*.[30] The court decision ruled that the First Amendment of the U.S. Constitution (free speech amendment) does not permit the FDA to reject dietary supplement health claims that the agency determines fail to meet the significant scientific agreement validity standard unless the agency also reasonably determines that a disclaimer added to the claim would not eliminate the potentially misleading character of the claim. The qualified claims are based on the weight of the scientific evidence; i.e., there is more evidence for, than against, the relationship, but it falls short of the validity standard required for foods under the NLEA. In other words, this system applies to dietary supplement health claims that do not meet the "significant scientific agreement" standard of evidence under the NLEA approval system. Currently, there are four qualified health claims that are approved:

1. B vitamins and vascular disease
2. 0.8 mg of folate in dietary supplement form is more effective in reducing the risk of neural tube defects
3. Omega-3 fatty acids (from fish oils) and coronary heart disease
4. Olive oil and coronary heart disease

Dietary supplements of omega-3 fatty acids may bear the following label claim, approved by the FDA in 2002: "Consumption of omega-3 fatty acids may reduce the risk of coronary heart disease. The FDA evaluated the data and determined that, although there is scientific evidence supporting the claim, the evidence is not conclusive." By the end of 2004, the FDA approved the next qualified health claim that links consumption of olive oil to coronary disease. "Limited and not conclusive scientific evidence suggests that eating about 2 tablespoons (23 grams) of olive oil daily may reduce the risk of coronary heart disease due to the monounsaturated fat in olive oil. To achieve this possible benefit, olive oil is to replace a similar amount of saturated fat and not increase the total number of calories you eat in a day. One serving of this product contains [x] grams of olive oil."

17.5 CANADIAN APPROACH TO REGULATE FUNCTIONAL FOODS

In Canada, there are three components of regulatory framework that govern food products.

1. Regulations (laws) addressed in the Food and Drugs Act and Regulations
2. Guidelines
3. Premarket review requirements

The Food and Drugs Act and Regulations (FDAR) are the Canadian laws dealing with foods, drugs, cosmetics, and medical devices. This law was introduced to ensure food safety and integrity. The act addresses the sale and manufacture of food in a safe manner and also deals with deceptive and dishonest labeling, marketing, and advertisement. There are several guidelines that were introduced by Agriculture and Agri-Food Canada and Health Canada to interpret the provisions of this act. Although these guidelines and proposals are not enforced by law, a number of them are relevant to functional foods, and they may become law in the near future. Furthermore, there are mandatory requirements as a part of FDAR for premarket review of specific categories of products, such as all drugs, food additives, infant formulae, and irradiated foods.[31]

The concept of functional foods is well known in Canada, which was one of the first countries to recognize the importance of this food category and acknowledge it by issuing the official definition, which was discussed above. The definition does not have any legal meaning since there is no specific regulation dealing with functional foods. The current system exclusively offers manufacturers the option of licensing and selling functional foods as drugs if any health claims are attached. This is because the FDAR defines drugs as "any substance or mixture of substances manufactured, sold or represented for use in: the diagnosis, treatment, mitigation or prevention of a disease, disorder or abnormal physical state, or its symptoms, restoring, correcting or modifying organic function," while food is "any article manufactured, sold or represented for use as food or drink for human beings, chewing gum, and any ingredient that may be mixed with food for any purpose whatever." Section 3 (Part I) under the act specifically prohibits the sale or advertisement of any food to the general public represented as a treatment, preventive, or cure for any disease, disorder, or abnormal physical state as listed in Schedule A. Any health claims or descriptors such as *healthy*, *health*, etc., on food labels or in advertisements are not allowed. Any claims for a food product that involve prevention of a disease or modification of an organic function, for example, lowering serum cholesterol, are deemed to bring the product within the definition of a drug. The rationale behind this is that the public should not be self-medicating for diseases.

17.5.1 Health Claims for Functional Foods

In spite of the conservative nature of Canadian regulations, in 1996, Health Canada initiated a national debate concerning functional foods and health claims. In 1998, the Health Protection Branch of Health Canada established its final policy, proposing a policy framework to address the use of health claims on nutraceuticals/functional foods and related products that are currently regulated as either food or drugs.[32] This paper proposes that use of structure–function and risk reduction claims for foods should be permitted if supportive evidence of efficacy is produced. Other claims, regarding cure, treatment, mitigation and illness prevention, would continue to be regulated as drugs. The proposed regulatory framework was designed to exempt foods bearing certain "drug-like health claims" from the provisions of the act relating to drugs and to maintain them under the provisions of the act relating to foods to manage the health risk usually associated with foods.

17.5.1.1 Product-Specific Health Claims for Foods

Another initiative of Health Canada related to health claims for foods is a recent proposal for the product-specific authorization of health claims for foods, published in October 2001.[33] It has been proposed that the manufacturer of a food that is

> "... manufactured, sold or represented to have a direct, measurable effect on a body function or structure beyond normal growth and development or maintenance of good health, or reducing the risk of or facilitating the dietary management of disease or health-related conditions be required to submit detailed information to support such an effect being advertised or offered for sale."

A food meeting the criteria as supported by scientific evidence may bear a label health claim as identified by a Claim Identification Number (CIN). The CIN is granted on a product-by-product basis, and the health claim is authorized to a single food product. It cannot be used by similar foods unless acceptable supporting evidence is provided. To obtain a CIN for a specific health claim, a submission by the food's manufacturer to Health Canada is required. The conditions of submission are specified in the proposal and must contain a number of specifications, including the food product's composition, intended use, and target; the proposed claim and information required for assessing the product's safety; the claim of validity; and the quality assurance of the product. Once the application is made, a CIN may be granted within 90 days of receiving the submission. However, at this time it is unclear how a health claim may be authorized without amending Schedule A of the Canadian Food and Drug Act.

17.5.1.2 Foods for Special Dietary Use

Foods for special dietary use are regulated under Division 24 of Part B of the Food and Drug Regulations[31] and refer to foods

> "... that have been specially processed or formulated to meet the particular requirements of a person (a) in whom a physical or physiological condition exists as a result of a disease, disorder or injury, or (b) in whom a particular effect, including but not limited to weight loss, is to be obtained by a controlled intake of foods."

Health Canada proposed that different health claims may be used for this category of foods as part of the dietary management of a disease or health condition. For example, a product that lowers the blood cholesterol level may bear the following information on the label:

> This product is a food for special dietary use. It contains at least X grams per serving of ingredient Y (e.g., fiber). When eaten twice a day for a daily total of not less than Z grams of Y, this product has been shown to lower elevated blood cholesterol [in naming the target population where applicable]. Consult your doctor when using this product as part of dietary management of high blood cholesterol.

17.5.1.3 Generic Health Claims

In 2000, Health Canada announced approval for the use of five generic diet-related health claims. These five claims were adopted from the U.S. NLEA list after reviewing the scientific evidence supporting the U.S. generic health claims. The process involved Canadian experts who published a report for public review. Subsequently, in January 2003, regulations amending the FDAR to include five health claims became law:[34]

1. Sodium and hypertension
2. Calcium and osteoporosis
3. Saturated and trans fat and cholesterol and coronary heart disease
4. Fruits and vegetables and cancer
5. Sugar alcohols and dental caries

Generic claims apply to a food or a group of foods that have compositional characteristics that contribute to a dietary pattern associated with reducing the risk of disease or health condition. Once a claim is authorized, any food that meets the specified conditions for composition and labeling may carry the claim without further assessment. It is worthwhile to note that two carbohydrate-relating health claims — "fiber-containing grain products, fruits, and vegetables and cancer" and "fruits, vegetables, and grain products that contain fiber, particularly soluble fiber, and risk of coronary heart disease" — have not yet been authorized; however, they are being further reviewed and may be approved in the near future.

17.5.2 Novel Foods

The Novel Foods Regulations became law in Canada in 1999.[35,36] They have been primarily established to handle and regulate marketing of genetically modified foods or, in broader terms, foods derived from biotechnology. In fact, most of the applications for approval under Novel Foods in the past decade were related to genetically modified foods. However, these regulations are open to foods other than just the genetically modified organism (GMO) foods, as defined below (B.28.001):

1. A substance, including a microorganism, that does not have a history of safe use as a food
2. A food that has been manufactured, prepared, preserved, or packaged by a process that:
   i. Has not been previously applied to that food
   ii. Causes the food to undergo a major change

Most of all functional foods would correspond with this definition, and therefore, this set of regulations may be used for the premarket safety assessment of functional foods. The information requirements for the assessment of novel foods are contained in the *Guidelines for the Safety Assessment of Novel Foods* that were published in 1994 by Health Canada. Health Canada is currently reviewing these guidelines to reflect the advancement of methods and knowledge regarding product review. The process for novel food review is a notification process rather than a traditional premarket approval process. The review of all relevant technical information submitted in support of a novel food is much faster than that for food additives or ingredients. Health Canada is required to respond to the novel food application within 45 days. If additional information is required, Health Canada has a further 90 days in which to respond. It is worthwhile to note that the approval process does not require a public consultation or an amendment to the existing regulations, which usually takes between 1and 2 years. However, it is only used for the safety assessment of foods outside of the health claims that bare important characteristics of functional foods.

17.5.3 Natural Health Products Regulations

On January 2004, after nearly a decade of discussion, public consultation, and publication in *Canada Gazette* II, the Natural Health Products Directorate (NHPD) began its operation.[37] The NHPD is responsible for regulating dietary and nutritional supplements under the broad jurisdiction of Canadian drug legislation. Natural health products are now allowed to be manufactured, sold, or represented for use in the diagnosis, treatment, mitigation, or prevention of a disease, disorder, or abnormal physical state or its symptoms in humans — that is, claims that were only allowed in Canada for drug products. They may also be used in correcting organic functions in humans or maintaining or promoting health or otherwise modifying organic functions in humans. Unlike the regulations in many other countries, NHPD regulations allow for a full range of health claims, including structure–function, risk reduction, and therapeutic/treatment.

The category of natural health products includes:[37]

1. A plant or plant material, an alga, a bacterium, a fungus, or a nonhuman animal material
2. An extract or isolate of a substance described above, the primary molecular structure of which is identical to that which it had prior to its extraction or isolation
3. Any vitamins that are listed in the NHPD regulations
4. An amino acid, an essential fatty acid, a synthetic duplicate of a substance described above, a mineral, and a probiotic

Natural health products do not include:

1. A substance set out in Schedules C and D of the Food and Drug Act
2. A substance regulated under the Tobacco Act
3. A substance set out in any of Schedules I to V of the Controlled Drugs and Substances Act
4. A substance that is administered by puncturing the dermis
5. An antibiotic prepared from an alga, a bacterium, or a fungus, or a synthetic duplicate of that antibiotic

Although the regulations are not very clear regarding the delivery vehicle of a compound in question, they cover a range of acceptable dose forms, including capsules, tablets, and liquids. They also do not exclude the use of food matrices such as snack bars or beverages, as long as they are clearly defined with dosages. Traditional food forms are precluded from the use under the NHPD regulations. All products approved under the NHPD carry a unique identification number, in much the same way as pharmaceuticals. The product has either a NPN (Natural Product Number) or DIN-HM (Drug Identification Number — Homeopathic Medicine) on the label. These numbers let the consumer know that the product has undergone and passed a review of its formulation, labeling, and instruction for use. In addition, the numbers facilitate adverse event monitoring and allow easy recall of products for which safety concerns emerge. All manufacturers of natural health products will be required to notify the agency of their marketed products and will have 2 years to bring their operations into compliance with the regulations and obtain the appropriate product licenses.

17.6 EUROPEAN UNION REGULATORY INITIATIVE FOR FUNCTIONAL FOODS

There are currently 25 member states in the European Union (EU). At the present date, each country belonging to the EU has its own set of food laws and regulations. The majority of these member states do not have a regulatory and legislative definition of functional foods, nor do they allow the use of health claims for food products. A select few countries, however, such as Spain, the U.K., and the Netherlands, allow for specific health claims to be made as long as they do not pertain to the prevention or cure of disease. On the more progressive side of the scale, Sweden has already approved eight diet–disease-related claims, while France may approve health claims on a case-to-case basis as defined by the consumer code[38] (Table 17.8). This discrepancy in regulation is clearly a problem for the marketing and distribution of functional foods in the European market. Therefore, in the interest of efficient food product manufacturing and consumption, the regulations of all EU countries require unification under the Commission of the European Communities. A subsequent problem is the fact that there are no provisions allowing the labeling of nutrition and health claims on foods at the level of the European Commission. However, the publication of a Green Paper prompted the European Parliament to pass a resolution encouraging the European Commission to allow the use of health claims and to describe the circumstances for the use of them.[39] The newly introduced novel food regulations are currently used for the approval of functional foods. However, these regulations do not have the provisions to govern health claims.

17.6.1 Novel Foods

This set of regulations was specifically designed to govern genetically modified foods. However, the European authorities have introduced two clauses that allow manufacturers to use this system as a venue to introduce functional foods in the EU market. The Novel Food Regulations (258/57) were legislated in the EU in 1997[40]

TABLE 17.8
Functional Food Regulations in Selected European Countries[38]

Jurisdiction	Definition of Functional Food	Health Claims	Comments
Belgium	No definition	Royal decree of April 17 allows proven health claims to be used; however, there are no guidelines and no approvals	New code for the approval of health claims being drafted by an industry–government committee
Denmark	No definition	Claims on foods are prohibited whether documented or not	This country is very restrictive in its legislation and interpretations
Finland	No definition	It is illegal to imply a food influences health or has medicinal qualities	Leading country in marketing of functional foods with health claims; home of Benecol functional foods
France	Nutritional, functional nutrition, and health- and disease-related claims are defined in the consumer code	Claims are approved on a case-by-case basis; three government ministries cooperate to regulate health claims	
Germany	Functional foods are governed by the General Foodstuffs and Commodities Act, the Dietetic Foodstuffs Order, or legislation on novel foods	The General Foodstuffs and Commodities Act prohibits references to prevention, cure, or mitigation of disease, medical references, or opinions, unless for dietetic diets, or unless marketing is intended for physicians	
Italy	Functional foods may be governed by Legislative Decree 111 of January 27, 1992; this legislation is primarily for nutritional uses and enriched foods	Disease-related claims are not allowed	No great interest in functional foods since consumers believe the Mediterranean diet is very healthful
Netherlands	Health claims and medicinal claims are defined under the Warenwet, the Dutch Commodities Act	Medicinal claims pertaining to prevention or cure of disease are not permitted; codes relating to health claims have been written for industry self-regulation	

TABLE 17.8 (continued)
Functional Food Regulations in Selected European Countries[38]

Jurisdiction	Definition of Functional Food	Health Claims	Comments
Spain	General provisions for foods may be used for functional foods	Claims may be approved if they are not misleading; there is a list of diet–disease relationships for which claims cannot be approved	
Sweden	Health claims, nutritional physiological claims, and nutrition claims all defined in regulation	The Nutritional Labelling directive of the Food Act prohibits claims that a food can prevent, relieve, or cure diseases' otherwise, health claims that are true, not misleading, and not derogatory to other foods are allowed	Self-regulatory plan for marketing of foods with claims had been approved in Sweden before this country joined the EU; eight diet- and disease-related claims have been approved: obesity, cholesterol, blood pressure, atherosclerosis, constipation, osteoporosis, dental caries, and iron deficiency
U.K.	No legal definition of regulation for functional foods; safety is governed by the Food Safety Act of 1990	Health claims are allowed, if not misleading; the Food Labelling Regulations of 1996 prohibit disease-related claims; foods with health claims require a license under the Medicines Act	A code of practice allowing for the acceptable use of health claims has been determined and approved by joint industry–government committees

and are considered to be the first type of regulations introduced globally — well before the same regulations were introduced in Canada (1999) and Australia/New Zealand (2001).

Novel Food Regulations apply to novel foods and novel food ingredients[40] that have not been "previously used to a significant degree for human consumption" and that fall into one of the following categories:

1. Genetically modified organisms as defined by Directive 90/220/EEC;[41] i.e., "an organism in which genetic material has been altered in a way that does not occur naturally by mating or by natural recombination"
2. Food/ingredients produced from, but not containing, genetically modified organisms
3. Food/ingredients consisting of/or isolated from microorganisms, fungi, or algae
4. Food/ingredients consisting of/or isolated from plants and food ingredients isolated from animals, except for food ingredients obtained by traditional propagating or breeding practices and having a history of safe food use
5. Food/ingredients to which has been applied a production process not currently used, where that process gives rise to significant changes in the composition or structure of the food/ingredient that affect the nutritional value, metabolism, or level of undesirable substances

The Novel Food Regulations do not apply to food additives (89/107/EEC), flavorings (88/388/EEC), and extraction solvents (88/344/EEC).[42–44] They do, however, apply to any food or food ingredients that have properties that vary from their conventional counterparts, and consequently have no history of safe food use. New properties may be the result of the intentional alteration of the molecular structure of the food/food ingredients or of the novel production process. These regulations are therefore applicable to functional foods and functional food ingredients. It is worthwhile to emphasize that the overriding purpose of the Novel Food Regulations is to ensure consumer safety.

To determine whether functional foods/ingredients meet the novel food criteria, it must be decided whether the functional foods in question are foods that have not previously been used to a significant degree for human consumption, i.e., the history of the use of the product is examined. Although the regulations do not define precisely what is meant by "previously used to a significant degree," there are already functional ingredients approved and, therefore, there are some existing guidelines to follow. For example, in July 2000, the commission approved phytosterol esters as novel food ingredients and authorized their use in yellow fat spreads, as well as functional foods for lowering the cholesterol level in humans.[45] Phytosterols occur naturally in vegetable oils, corn, rice, and other cereals, and are consumed at approximately 100 to 300 mg/day as part of a typical Western diet.[46] Therefore, someone can argue that these ingredients have been "previously used for human consumption"; however, the EU Commission views phytosterols as novel food ingredients because they require a consumption of 2 to 3 g/day to show efficacy.

In addition, the functional food ingredients in question must fit into one of the five categories, listed above, that are defined by Novel Food Regulations. Depending on the manufacturing process, the 5th category may apply.

17.6.1.1 Novel Food Authorization in the EU

Manufacturers seeking to market a novel food must submit an application to the competent authority in the member state in which they wish to market the novel food. At the same time, the applicant must inform the commission of its request. Commission Recommendation 97/618/EEC sets out detailed guidelines on how the information should be presented and how the initial assessment reports should be prepared by member states.[47] In brief, the application should include:

1. Any material that shows compliance with the criteria set out in the Novel Food Regulations. This includes novel food specifications, production process, source organism, aspects of genetic modification, anticipated intake, information from previous exposure, and nutritional, microbiological, and toxicological information.
2. Any clinical and preclinical studies that have been carried out on food/ingredient in question to prove its functionality/efficacy.
3. Labeling and presentation proposals.

The initial assessment must be carried out within 3 months of receiving the application. The results of the initial application must be forwarded to the commission, which in turn forwards it to the other member states. Subsequently, the member states and the commission have 60 days to object to the marketing of the product or its presentation and labeling. If the decision is favorable and there are no objections, the product can be marketed. If there are objections or if an additional assessment is required, then the application is referred to the EU Standing Committee on Foodstuff involving the Scientific Committee for Food for its opinion, if necessary. The application process is time-consuming and requires at least 2 years, as can be judged based on the number of granted approvals for the non-GMO novel foods.

A total of 37 applications for novel food approval were made between 1997 and 2002.[48] So far, nine non-GMO novel foods have been approved for commercialization in the EU, while two products have been rejected. Two of the approved novel foods are carbohydrate ingredients: dextran produced by *Leuconostoc mesenteroides*, to be used up to 5% in bakery products,[49] and trehalose obtained from liquefied starch by a multistep enzymatic process, for particular uses in foods.[50]

17.6.1.2 Fast-Track Approval Procedure

The manufacturer may seek approval of the functional food/ingredient in question by using the fast-track procedure when there is evidence that a novel food product is considered to be "substantially equivalent" to an existing food or food ingredient. Although the legislation does not provide a definition of "substantially equivalent," this term must be justified by an opinion from the relevant competent authority.

Where a food/food ingredient can be shown to be substantially equivalent to a conventional food (or approved novel food), it is considered to be as safe as its counterpart and no further safety assessment will be required. However, if an objection is raised to the use of the fast-track procedure by any member state on safety grounds, marketing of the food may be suspended and the commission is required to refer the matter to the Standing Committee for Foodstuffs for a decision.

17.6.1.3 Labeling Requirements

In addition to the normal labeling requirements for foodstuffs (79/112/EEC), the novel foods have supplementary labeling requirements as follows:[51]

1. Labeling must inform the final consumer of any characteristic or food property that renders a novel food no longer equivalent to an existing food or food ingredient.
2. Labeling must indicate the modified characteristics or properties modified and the method by which these were obtained.
3. Labeling must inform the final consumer of the presence in the novel food/ingredient of material that is not present in an existing foodstuff that may have the implications for the health of certain sections of the population or give rise to ethical concerns.
4. Labeling must inform the final consumer of the presence of a genetically modified organism.

It is noteworthy that according to these requirements, a functional food label must provide information relating to the efficacy of the product (e.g., health claim); however, the health claims are not authorized under the Novel Foods Regulations. In addition, Directive 2000/13/EEC, which governs the labeling, advertising, and presentation of foodstuff, prohibits "attributing to foods any properties of prevention, treatment or cure of a human disease, or any reference to such properties."[52]

17.6.2 Health Claims Made on Foods: EU Proposal

The information appearing on food labels is an essential part of communication between the food manufacturer and its consumers. A label health claim that links food or one of its constituents with its effect on physiological functions of the body is indivisible with the functionality of functional foods. In 1996, the EU funded a concerted program entitled Functional Food Science in Europe (FUFOSE). FUFOSE acknowledged that claims are an important means of communicating the benefit of functional foods to the consumer.[53] The health claims, however, are not permitted at the EU level, although tolerance varies among member states for the use of health claims on labels and in advertising. Concerned about the inconsistency and the proliferation of the numbers and types of label health claims on foods, the EU Commission advocates the harmonization of rules and claims. In 2000, the commission proposed to consider "whether to introduce specific provisions in EU law to govern functional claims," claims related to beneficial effects of a nutrient on certain normal bodily functions.[54] This subject is still under discussion in the European Parliament, and the debate has been elevated to a higher level by the current proposal by the commission to regulate nutrition and health claims made on foods.[55]

In the current proposal on nutrition and health claims, the commission defines and sets conditions for nutrition and health claims to be used in the labeling, presentation, and advertising of foods in one single legislative proposal. One of the important aspects of this proposal is a new definition of nutrients. In addition to the traditional nutrients, such as protein, carbohydrate, fat, fiber, sodium, vitamins, and minerals, it is proposed that a new definition will also include "other substances with a nutritional or physiological effect." This will broaden the proposed legislation to include functional ingredients, which are already present in many products on the European market. Another important aspect of the proposal is to define a "desirable" nutritional profile of food to be eligible for bearing a health claim. Contrary to the U.S. FDA regulations, such as disclosure/disqualifier levels and the jelly bean rule, in most of the European countries there are no restrictions to prohibit the use of claims on certain foods on the basis of their nutritional profile. However, some restrictions on the use of health claims on food are under discussion. In particular, the amounts of total fat, saturated fat, trans fatty acids, sugars, and sodium may be used as criteria to qualify food products for their eligibility to bear a health claim. Furthermore, the presentation of the claim, including wording, logos, and images, and the actual communication, plays an important role in the way claims are perceived and understood by the consumer. A claim that is not understood is completely useless; a claim that is misunderstood could even be misleading, while a vague claim is often meaningless and also not verifiable. Health claims should therefore "only be approved for use on the labeling, presentation and advertising of foods on the Community market after a scientific evaluation of the highest possible standard." It is proposed that the European Food Safety Authority should carry out the scientific assessment of health claims.

To develop criteria for the assessment of scientific support for claims on foods, to make them available for food manufacturers and regulators, and to provide credibility for claims for consumers, the European Commission has funded the Process for the Assessment of Scientific Support for Claims on Foods (PASSCLAIM) project. This project involves academic experts in various aspects of the physiological functions relating to health claims, representatives of public interest groups, regulatory experts, and the food industry. The objectives of PASSCLAIM are:[56]

1. To produce a generic tool with principles for assessing the scientific support for health-related claims for foods and food components
2. To evaluate critically the existing schemes that assess the scientific substantiation of claims
3. To select common criteria for how markers should be identified, validated, and used in well-designed studies to explore the links between diet and health

As a result of phase 1 of the PASCLAIM project, principles for assessing the scientific support of health claims have been outlined. This includes a definition of health claims[57] and criteria for the scientific substantiation of claims.[58] Three categories of health claims are defined:

1. A nutrient function claim
2. An enhanced function claim
3. A reduction of disease risk claim

A nutrient function claim describes the role of a nutrient or another substance in growth, development, and the normal functions of the body, which are based on generally accepted scientific data. For example, calcium aids in the development of strong bones and teeth.

An enhanced function claim describes the specific health benefits of foods or food components on physiological and psychological, cognitive function, and biological activities. In addition, this claim links nutrients and other food components with health benefits beyond their generally accepted nutritional effect. The following examples may be considered as enhanced function claims:

1. An additional function of food nutrient, usually at a higher level of intake
2. A function of a food component (e.g., fiber prebiotic effect or phytosterols' cholesterol-lowering properties)
3. A specific physical or chemical property of food or food components (e.g., modified starch with low glycemic index)

A reduction of disease risk claim refers to the fact that the consumption of a food may help reduce the risk of a disease. The disease or disorder must be named and the risk reduction must be stated. For example, the consumption of phytosterol-containing food may reduce the risk of coronary heart disease (CHD). Risk reduction means significantly altering a major risk factor for a disease or health-related condition. In the case of CHD, this risk factor may be a high level of plasma cholesterol. However, diseases have multiple risk factors; therefore, it has been proposed that for reduction of disease risk claims, the label shall also "bear a statement indicating that diseases have multiple risk factors and that altering one of these risk factors may or may not have a beneficial effect."[58]

Regardless of a health claim, a food manufacturer making a claim should justify the use of the claim and substantiate the claim by generally accepted scientific data.

17.7 CONCLUSIONS

As a new category of foods, functional foods have attracted the attention of the consumer, food manufacturers, and regulatory authorities in various jurisdictions. Foods fortified with carbohydrates such as prebiotics and fiber-containing breakfast cereal with proven efficacy, as well as other functional foods that contain phytosterols, antioxidants, and other nutraceuticals, are gaining popularity on the global market. There are a number of reasons for this phenomenon. First, consumers look to functional foods to better their health, whereas manufacturers consider functional foods as the "new big thing" for market differentiation, and regulators consider functional foods as a new "headache" that needs to be regulated. The regulations set down to manage this new food category, along with rules that will govern health claims, are an important factor in marketing functional foods and communicating their health benefits to the consumer. Although it has been more than a decade since the introduction of this concept, regulators exercise caution in the introduction of new laws and regulations.

For many functional foods ingredients, such as fiber and phytosterols, sufficient scientific evidence links consumption of functional foods containing these ingredients with acceptable efficacy standards. This is in contrast to other potential functional ingredients that food manufacturers consider for use in fortifying functional foods, need more research and clinical studies. There is a general consensus among the scientific community that functional foods are sustainable on the global market; however, the extended life span of this kind of food is only possible with solid scientific support.

REFERENCES

1. Fourth Biennial Overview of the Global Nutrition Industry, Global Nutrition Market IV, *Nutrition Business Journal*, VIII (4), August/September 2001.
2. Vetter, J., Claims: health, nutrient content, and other claims, in *Food Labeling: Requirements for FDA Regulated Products*, American Institute of Baking, Manhattan, KS, 1993, chap. 9.
3. Asp, N.G., Dietary fiber, in *Healthy Lifestyles: Nutrition and Physical Activity*, International Life Sciences Institute Press, Washington, DC, 1998, chap. 6.
4. Ling, W., The sprouting and blooming of China's tea culture, in *Chinese Tea Culture*, Foreign Languages Press, Beijing, 2002, chap. 2.
5. Kurowska, E.M. et al., HDL-cholesterol-raising effect of orange juice in subjects with hypercholesterolemia, *Am. J. Clin. Nutr.*, 72, 1095, 2000.
6. Nutraceuticals/Functional Foods and Health Claims on Foods, Policy Paper, Therapeutic Products Programme and the Food Directorate from the Health Protection Branch, *Health Canada*, November 2, 1998.
7. DeFelice, S.L., Rationale and Proposed Guidelines for the Nutraceutical Research and Education Act, Foundation for Innovative Medicine, conference, New York, November 10, 2002.
8. Furukawa, T., The nutraceutical rules: health and medical claims: Food for Specified Health Use in Japan, *Regulatory Affairs*, 5, 189, 1993.
9. Heasman, M. and Mellentin, J., *The Functional Foods Revolution. Healthy People, Healthy Profits*, Earthscan Publication Ltd., London, 2001.
10. Arai, S., Studies of functional foods in Japan: state of the art, *Biosci. Biotechnol. Biochem.*, 60, 9, 1996.
11. The Nutrition Improvement Law Enforcement Regulations, Ministerial Ordinance 41, July 1991; Amendment to Ministerial Ordinance 33, May 25, 1996.
12. Hamano, H., Regulatory Framework for Health Claims in Japan, *Japan Health Food and Nutrition Food Association*, November 24, 2003.
13. Hidaka, H. et al., Effects of fructooligosaccharides on intestinal flora and human health, *Bifidobacteria Microflora*, 5, 37, 1986.

14. Oku, T., Tokunaga, T., and Hosoya, N., Nondigestibility of a new sweetener fructooligosaccharide (Neosugar) in rat, *J. Nutr.*, 114, 1574, 1984.
15. Bailey, R., Marine Products in Japan, *Nutraceuticals World*, 24, July/August 2003.
16. Yamaguchi, P., *Japan's Nutritional Supplements Market Report: Opportunities at Sunrise*, Paul Yamaguchi & Associates, 2003.
17. Code of Federal Regulations, Food and Drugs, 21, Vol. 3, Chap. I, Part 171, Food Additive Petitions, §171.1–171.130, pp. 21–26, revised as of April 1, 2002.
18. Code of Federal Regulations, Food and Drugs, 21, Vol. 3, Chap. I, Part 184, Direct Food Substances Affirmed as Generally Recognized as Safe, Subpart A, General Provisions, §184.1, pp. 464–466, revised as of April 1, 2002.
19. Foulke, J., F*DA Proposed Simplified GRAS Notification System*, Food Talk Paper T97-15, April 17, 1997.
20. Code of Federal Regulations, Food and Drugs, 21, Vol. 3, Chap. I, Part 170, Food Additives, §170.35, Affirmation of Generally Recognized as Safe (GRAS) Status, pp. 15–16, revised as of April 1, 2002.
21. *White House Conference on Food, Nutrition and Health*, Final Report, Superintendent of Documents, U.S. Government Printing Office, Washington, DC, 1969.
22. Pub. L. 101-535, 104 Stat. 2353, codified as amended at 21 U.S.C. ßß301, 321, 337, 343, 371, 1990.
23. Code of Federal Regulations, Food and Drugs, 21, Vol. 2, Chap. I, Part 101, Food Labeling, §101.14, Health Claims: General Requirements, pp. 62–65, revised as of April 1, 2002.
24. Code of Federal Regulations, Food and Drugs, 21, Vol. 2, Chap. I, Part 101, Food Labeling, §101.71–101.83, Health Claims, pp. 124–149, revised as of April 1, 2002.
25. Code of Federal Regulations, Food and Drugs, 21, Vol. 2, Chap. I, Part 101, Food Labeling, Docket 2001Q-0313, Health Claims: Soluble Dietary Fiber from Certain Foods and Coronary Heart Disease, pp. 44207–44209, *Federal Register*, Vol. 68, No. 144, July 28, 2003.
26. Food and Drug Modernization Act, Pub. L. 105-115, 111 Stat. 2296, codified at 21 U.S.C. §343(r)(3)(C) and (D).
27. One Hundred Third Congress of the United State of America, An Act to Amend the Federal Food, Drug, and Cosmetic Act to Establish Standards with Respect to Dietary Supplements, and for Other Purposes, Dietary Supplement Health and Education Act of 1994, S.784-1/11, January 25, 1994.
28. Code of Federal Regulations, Food and Drugs, 21, Vol. 2, Chap. I, Part 101, Food Labeling, Subpart F, Specific Requirements for Descriptive Claims That Are neither Nutrient Content Claims nor Health Claims, §101.93, Certain Types of Statements for Dietary Supplements, pp. 149–151, revised as of April 1, 2002.
29. Code of Federal Regulations, Food and Drugs, 21, Vol. 3, Chap. I, Part 190, Dietary Supplements, Subpart B, New Dietary Ingredient Notification, §190.6, Requirements for Premarket Notification, pp. 571–572, revised as of April 1, 2002.
30. United States District Court for the District of Columbia, Civil Action 00-2724 (GK), *Durk Pearson, and Sandy Shaw, the American Preventive Medical Association, Julian M. Whitaker, M.D., Pure Encapsulations, Inc., and XCEL Medical Pharmacy Ltd. – Plaintiffs v. Donna E. Shalala, Secretary, United States Department of Health and Human Services, Jane E. Henney, Commissioner of Food Drugs, Food and Drug Administration, and the United States of America – Defendants*, Order, February 2001.
31. Food and Drugs Act and Regulations, Department of Health, Minister of Public Works and Government Services, Canada, 2002.

32. *Nutraceuticals/Functional Foods and Health Claims on Foods*, Policy Paper, Therapeutic Products Programme and the Food Directorate from the Health Protection Branch, Health Canada, November 2, 1998.
33. *Product-Specific Authorization of Health Claims for Foods, A Proposed Regulatory Framework*, Bureau of Nutritional Sciences, Food Directorate, Health Products and Food Branch, Health Canada, October 2001.
34. Regulations Amending the Food and Drug Regulations (Nutrition Labeling, Nutrition Claims, and Health Claims), Registration SOR/2003-11, Food and Drugs Act, Department of Health, December 12, 2002, *Canada Gazette*, Part II, Vol. 137, Issue 1, pp. 2118–2123, January 1, 2003, last modified October 29, 2003.
35. Novel Foods Regulations, Schedule 948, *Canada Gazette*, Part II, Food and Drugs Regulations and Amendments, October 27, 1999.
36. Novel Foods, Part B, Section 28, pp. 387–389, Food and Drugs Act and Regulations, Department of Health, Minister of Public Works and Government Services, Canada, 2002.
37. Natural Health Products Regulations, *Canada Gazette*, Part II, Vol. 137, No. 13, Natural Health Products Directorate, Health Products and Food Branch, Health Canada, June 18, 2003.
38. Jarvis, D. et al., *Business and Market Impact of the Food and Drugs Act and Regulations on Functional Foods in Canada*, Inter/Sect Alliance, Inc., 2001.
39. Commission Green Paper, *The General Principles of Food Law in the European Union*, COM (97) 176, April 1997.
40. Novel Foods Regulations, European Parliament and Council Regulation 258/97, *Official Journal L 043*, January 27, 1997, pp. 0001–0006.
41. Council Directive 90/220/EEC, *Official Journal L 117*, May 8, 1990, p. 15; directive as last amended by Directive 94/15/EC, *Official Journal L 103*, April 23, 1990, p. 20.
42. Council Directive 89/107/EEC, *Official Journal L 40*, February 11, 1989, p. 27; directive as last amended by Directive 94/34/EC, *Official Journal L 237*, September 10, 1994, p. 1.
43. Council Directive 88/388/EEC, *Official Journal L 184*, July 15, 1988, p. 6140; directive as last amended by Directive 91/71/EEC, *Official Journal L 42*, February 15, 1991, p. 25.
44. Council Directive 88/344/EEC, *Official Journal L 157*, June 28, 1988, p. 28; directive as last amended by Directive 92/115/EEC, *Official Journal L 409*, December 31, 1992, p. 31.
45. Commission Decision 2000/500/EC, *Official Journal L 200/59*, August 8, 2000, p. 1.
46. Zawistowski, J. and Kitts, A., Functional soup, *Can. Chem. News*, 54, 17, 2002.
47. Council Directive 97/16/EEC, *Official Journal L 253*, September 16, 1997, pp. 1–33.
48. Applications under Regulation (EC) 258/97 of the European Parliament and of the Council as of December 7, 2002, Press Release, June 10, 2003.
49. Commission Decision 2001/122/EC, *Official Journal L 44/46*, February 15, 2001, p. 1.
50. Commission Decision 2001/721/EC, *Official Journal L 269/17*, October 10, 2001, p. 1.
51. Council Directive 79/112/EEC, *Official Journal L 33*, February 8, 1979, p. 1; directive as last amended by Directive 49/2000/EEC, *Official Journal L 6/13*, November 1, 2000, p. 1.
52. Council Directive 2000/13/EEC, *Official Journal L 109*, May 6, 2000, p. 29.
53. Bellisle, F. et al., Functional food science in Europe: consensus document, *Br. J. Nutr.*, 80, 1, 1998.

54. White Paper on Food Safety, *Commission of the European Communities*, COM (1999) 719, January 12, 2000.
55. *Proposal for a Regulation of the European Parliament and of the Council on Nutrition and Health Claims Made on Foods*, Commission of the European Communities, COM (2003) 424, 2003/0165 (COD), July 16, 2003.
56. Asp, N.G. and Contor, L., Process for the assessment of scientific support for claims on foods (PASSCLAIM): overall introduction, *Eur. J. Nutr.*, 42 (Suppl. 1), I/2, 2003.
57. Richardson, D.P. et al., PASSCLAIM: synthesis and review of existing processes, *Eur. J. Nutr.*, 42 (Suppl. 1), I/96, 2003.
58. Cummings, J., Pannemans, D., and Persin, Ch., PASSCLAIM: report on the first plenary meeting including a set of interim criteria to scientifically substantiate claims on foods, *Eur. J. Nutr.*, 42 (Suppl. 1), I/112, 2003.

Index

A

B

C

D

E

F

G

H

I

N

O

P

R

S

T

U

V

W

X

Y

Z